Encyclopedia of Molecular Cell Biology and Molecular Medicine

Edited by Robert A. Meyers

**Volume 11
Proteasomes to Receptor, Transporter and Ion Channel Diseases**

Encyclopedia of Molecular Cell Biology and Molecular Medicine

Editorial Board

*Werner Arber, Biozentrum, University of Basel, Switzerland

*David Baltimore, California Institute of Technology, Pasadena, USA

*Günter Blobel, The Rockefeller University, New York, USA

Martin Evans, Cardiff University, United Kingdom

*Paul Greengard, The Rockefeller University, New York, USA

*Avram Hershko, Technion, Israel Institute of Technology, Haifa, Israel

*Robert Huber, Max Planck Institute of Biochemistry, Martinsried, Germany

*Aaron Klug, MRC Laboratory of Molecular Biology Cambridge, United Kingdom

*Stanley B. Prusiner, University of California, San Francisco, USA

*Bengt Samuelsson, Karolinska Institutet, Stockholm, Sweden

*Phillip A. Sharp, Massachusetts Institute of Technology, Cambridge, USA

Alexander Varshavsky, California Institute of Technology, Pasadena, USA

Akiyoshi Wada, RIKEN, Yokohama, Japan

Shigeyuki Yokoyama, RIKEN, Yokohama, Japan

*Rolf M. Zinkernagel, University Hospital Zurich, Switzerland

*Nobel Laureate

Encyclopedia of Molecular Cell Biology and Molecular Medicine

Edited by Robert A. Meyers

Second Edition

Volume 11
Proteasomes to Receptor, Transporter and Ion Channel Diseases

WILEY-VCH Verlag GmbH & Co. KGaA

Editor:

Dr. Robert A. Meyers
President, Ramtech Limited
122 Escalle Lane
Larkspur, CA 94939
USA

■ All books published by Wiley-VCH are carefully produced. Nevertheless, authors, editors, and publisher do not warrant the information contained in these books, including this book, to be free of errors. Readers are advised to keep in mind that statements, data, illustrations, procedural details or other items may inadvertently be inaccurate.

Library of Congress Card No.: applied for

British Library Cataloguing-in-Publication Data: A catalogue record for this book is available from the British Library.

Bibliographic information published by Die Deutsche Bibliothek
Die Deutsche Bibliothek lists this publication in the Deutsche Nationalbibliografie; detailed bibliographic data is available in the internet at http://dnb.ddb.de.

©WILEY-VCH Verlag GmbH & Co. KGaA Weinheim, 2005

All rights reserved (including those of translation into other languages). No part of this book may be reproduced in any form – nor transmitted or translated into machine language without written permission from the publishers. Registered names, trademarks, etc. used in this book, even when not specifically marked as such are not to be considered unprotected by law.

Printed in the Federal Republic of Germany.
Printed on acid-free paper.

Composition: Laserwords Private Ltd, Chennai, India
Printing: Druckhaus Darmstadt GmbH, Darmstadt
Bookbinding: Litges & Dopf Buchbinderei GmbH, Heppenheim
ISBN-13: 978-3-527-30648-0
ISBN-10: 3-527-30648-X

Preface

The *Encyclopedia of Molecular Cell Biology and Molecular Medicine*, which is the successor and second edition of the *Encyclopedia of Molecular Biology and Molecular Medicine* (VCH Publishers, Weinheim), covers the molecular and cellular basis of life at a university and professional researcher level. The first edition, published in 1996–97, was very successful and is being used in libraries around the world. This second edition will almost double the first edition in length and will comprise the most detailed treatment of both molecular cell biology and molecular medicine available today. The Board Members and I believe that there is a serious need for this publication, even in view of the vast amount of information available on the World Wide Web and in text books and monographs. We feel that there is no substitute for our tightly organized and integrated approach to selection of articles and authors and implementation of peer review standards for providing an authoritative single-source reference for undergraduate and graduate students, faculty, librarians, and researchers in industry and government.

Our purpose is to provide a comprehensive foundation for the expanding number of molecular biologists, cell biologists, pharmacologists, biophysicists, biotechnologists, biochemists, and physicians, as well as for those entering molecular cell biology and molecular medicine from majors or careers in physics, chemistry, mathematics, computer science, and engineering. For example, there is an unprecedented demand for physicists, chemists, and computer scientists who will work with biologists to define the genome, proteome, and interactome through experimental and computational biology.

The Board Members and I first divided the entire study of molecular cell biology and molecular medicine into primary topical categories and further defined each of these into subtopics. The following is a summary of the topics and subtopics:

- *Nucleic Acids:* amplification, disease genetics overview, DNA structure, evolution, general genetics, nucleic acid processes, oligonucleotides, RNA structure, RNA replication and transcription.
- *Structure Determination Technologies Applicable to Biomolecules:* chromatography, labeling, large structures, mapping, mass spectrometry, microscopy, magnetic resonance, sequencing, spectroscopy, X-ray diffraction.
- *Biochemistry:* carbohydrates, chirality, energetics, enzymes, biochemical genetics, inorganics, lipids, mechanisms, metabolism, neurology, vitamins.

Encyclopedia of Molecular Cell Biology and Molecular Medicine, 2nd Edition. Volume 11
Edited by Robert A. Meyers.
Copyright © 2005 Wiley-VCH Verlag GmbH & Co. KGaA, Weinheim
ISBN: 3-527-30648-X

- *Proteins, Peptides, and Amino Acids:* analysis, enzymes, folding, mechanisms, modeling, peptides, structural genomics (proteomics), structure, types.
- *Biomolecular Interactions:* cell properties, charge transfer, immunology, recognition, senses.
- *Cell Biology:* developmental cell biology, diseases, dynamics, fertilization, immunology, organelles and structures, senses, structural biology, techniques.
- *Molecular Cell Biology of Specific Organisms:* algae, amoeba, birds, fish, insects, mammals, microbes, nematodes, parasites, plants, viruses, yeasts.
- *Molecular Cell Biology of Specific Organs or Systems:* excretory, lymphatic, muscular, nervous, reproductive, skin.
- *Molecular Cell Biology of Specific Diseases:* cancer, circulatory, endocrinal, environmental stress, immune, infectious, neurological, radiational.
- *Pharmacology:* chemistry, disease therapy, gene therapy, general molecular medicine, synthesis, toxicology.
- *Biotechnology:* applications, diagnostics, gene-altered animals, bacteria and fungi, laboratory techniques, legal, materials, process engineering, nanotechnology, production of classes or specific molecules, sensors, vaccine production.

We then selected some 400 article titles and author or author teams to cover the above topics. Each article is designed as a self-contained treatment which begins with a keyword section including definitions, to assist the scientist or student who is unfamiliar with the specific subject area. The Encyclopedia includes more than 3000 key words, each defined within the context of the particular scientific field covered by the article. In addition to these definitions, the glossary of basic terms found at the back of each volume, defines the most commonly used terms in molecular cell biology. These definitions, along with the reference materials (the genetic code, the common amino acids, and the structures of the deoxyribonucleotides) printed at the back of each volume, should allow most readers to understand articles in the Encyclopedia without referring to a dictionary, textbook, or other reference work. There is, of course, a detailed subject index in Volume 16 as well as a cumulative table of contents and list of authors, as well as a list of scientists who assisted in the development of this Encyclopedia.

Each article begins with a concise definition of the subject and its importance, followed by the body of the article and extensive references for further reading. The references are divided into secondary references (books and review articles) and primary research papers. Each subject is presented on a first-principle basis, including detailed figures, tables and drawings. Because of the self-contained nature of each article, some articles on related topics overlap. Extensive cross-referencing is provided to help the reader expand his or her range of inquiry.

The articles contained in the Encyclopedia include core articles, which summarize broad areas, directing the reader to satellite articles that present additional detail and depth for each subject. The core article Brain Development is a typical example. This 45-page article spans neural induction, early patterning, differentiation, and wiring at a molecular through to cellular and tissue level. It is directly supported, and cross-referenced, by a number of molecular neurobiology satellite articles, for example, Behavior Genes, and further supported by other core presentations, for example,

Developmental Cell Biology; Genetics, Molecular Basis of, and their satellite articles. Another example is the core article on Genetic Variation and Molecular Evolution by Werner Arber. It is supported by a number of satellite articles supporting the evolutionary relatedness of genetic information, for example, Genetic Analysis of Populations.

Approximately 250 article titles from the first edition are retained, but rewritten, half by new authors and half by returning authors. Approximately 80 articles on cell biology and 70 molecular biology articles have been added covering areas that have become prominent since preparation of the first edition. Thus, we have compiled a totally updated single source treatment of the molecular and cellular basis of life.

Finally, I wish to thank the following Wiley-VCH staff for their outstanding support of this project: Andreas Sendtko, who provided project and personnel supervision from the earliest phases, and Prisca-Maryla Henheik and Renate Dötzer, who served as the managing editors.

July 2005

Robert A. Meyers
Editor-in-Chief

Editor-in-Chief

Robert A. Meyers

Dr. Meyers earned his Ph.D. in organic chemistry from the University of California Los Angeles, was a post-doctoral fellow at California Institute of Technology and manager of chemical processes for TRW Inc. He has published in *Science*, written or edited 12 scientific books and his research has been reviewed in the *New York Times* and the *Wall Street Journal*. He is one of the most prolific science editors in the world having originated, organized and served as Editor-in-Chief of three editions of the *Encyclopedia of Physical Science and Technology*, the *Encyclopedia of Analytical Chemistry* and two editions of the present *Encyclopedia of Molecular Cell Biology and Molecular Medicine*.

Editorial Board

Werner Arber
Biozentrum, University of Basel, Switzerland
Nobel Prize in Physiology/Medicine for the discovery of restriction enzymes and their application to problems of molecular genetics

David Baltimore
California Institute of Technology, Pasadena, USA
Nobel Prize in Physiology/Medicine for the discoveries concerning the interaction between tumor viruses and the genetic material of the cell

Günter Blobel
The Rockefeller University, New York, USA
Nobel Prize in Physiology/Medicine for the discovery that proteins have intrinsic signals that govern their transport and localization in the cell

Martin Evans
Cardiff University, United Kingdom
Lasker Award for the development of a powerful technology for manipulating the mouse genome, which allows the creation of animal models of human disease

Paul Greengard
The Rockefeller University, New York, USA
Nobel Prize in Physiology/Medicine for the discoveries concerning signal transduction in the nervous system

Avram Hershko
Technion – Israel Institute of Technology, Haifa, Israel
Nobel Prize in Chemistry for the discovery of ubiquitin-mediated protein degration

Robert Huber
Max Planck Institute of Biochemistry, Martinsried, Germany
Nobel Prize in Chemistry for the determination of the three-dimensional structure of a photosynthetic reaction centre

Aaron Klug
MRC Laboratory of Molecular Biology Cambridge, United Kingdom
Nobel Prize in Chemistry for the development of crystallographic electron microscopy and his structural elucidation of biologically important nucleic acid-protein complexes

Stanley B. Prusiner
University of California, San Francisco, USA
Nobel Prize in Physiology/Medicine for the discovery of Prions – a new biological principle of infection

Bengt Samuelsson
Karolinska Institute, Stockholm, Sweden
Nobel Prize in Physiology/Medicine for the discoveries concerning prostaglandins and related biologically active substances

Editorial Board | xiii

Phillip A. Sharp
Massachusetts Institute of Technology, Cambridge, USA
Nobel Prize in Physiology/Medicine for the discoveries of split genes

Alexander Varshavsky
California Institute of Technology, Pasadena, USA
Lasker Award for the discovery and the recognition of the significance of the ubiquitin system of regulated protein degradation

Akiyoshi Wada
RIKEN Yokohama Institute, Japan
Director of the RIKEN Genomic Science Center

Shigeyuki Yokoyama
RIKEN Yokohama Institute, Japan
Head of the RIKEN Structural Genomics Initiative

Rolf M. Zinkernagel
University Hospital Zurich, Switzerland
Nobel Prize in Physiology/Medicine for the discoveries concerning the specificity of the cell mediated immune defence

Contents

Preface *v*

Editor-in-Chief *ix*

Editorial Board *xi*

List of Contributors *xix*

Color Plates *xxiii*

Proteasomes 1
Martin Rechsteiner

Protein and Nucleic Acid Enzymes 27
Daniel L. Purich

Protein Expression by Expansion of the Genetic Code 45
Jason W. Chin, Thomas J. Magliery

Protein Mediated Membrane Fusion 69
Reinhard Jahn

Protein Microarrays 101
Jens R. Sydor, David S. Wilson, Steffen Nock

Protein Modeling 133
Marian R. Zlomislic, D. Peter Tieleman

Protein NMR Spectroscopy 159
Thomas Szyperski

Encyclopedia of Molecular Cell Biology and Molecular Medicine, 2nd Edition. Volume 11
Edited by Robert A. Meyers.
Copyright © 2005 Wiley-VCH Verlag GmbH & Co. KGaA, Weinheim
ISBN: 3-527-30648-X

Protein Purification *181*
Richard R. Burgess

Protein Repertoire, Evolution of *199*
Christine Vogel, Rajkumar Sasidharan, Emma E. Hill

Protein Splicing *241*
Kenneth V. Mills

Protein Structure Analysis: High-throughput Approaches *267*
Andrew P. Turnbull, Udo Heinemann

Protein Translocation Across Membranes *287*
Carla M. Koehler, David K. Hwang

Proteomics *309*
Paul Cutler, Israel S. Gloger, Christine Debouck

Proton Translocating ATPases *333*
Masamitsu Futai, Ge-Hong Sun-Wada, Yoh Wada

Pufferfish Genomes: *Takifugu* and *Tetraodon* *361*
Melody S. Clark, Hugues Roest Crollius

Quantitative Analysis of Biochemical Data *381*
Albert Jeltsch, Jim Hoggett, Claus Urbanke

Radioisotopes in Molecular Biology *411*
Robert James Slater

Rat Genome (*Rattus norvegicus*) *433*
Kim C. Worley, Preethi Gunaratne

Real-Time Quantitative PCR: Theory and Practice *481*
Gregory L. Shipley

RecA Superfamily Proteins *525*
Dharia A. McGrew, Kendall L. Knight

Receptor Biochemistry *551*
Tatsuya Haga, Kimihiko Kameyama

Receptor Targets in Drug Discovery 593
Michael Williams, Christopher Mehlin, Rita Raddatz, David J. Triggle

Receptor, Transporter and Ion Channel Diseases 637
J. Jay Gargus

Glossary of Basic Terms 713

The Twenty Amino Acids that are Combined to Form Proteins in Living Things 721

The Twenty Amino Acids with Abbreviations and Messenger RNA Code Designations 725

Complementary Strands of DNA with Base Pairing 727

List of Contributors

Richard R. Burgess
University of Wisconsin,
Madison, WI,
USA

Jason W. Chin
MRC Laboratory of Molecular Biology,
Cambridge,
UK

Melody S. Clark
Biological Sciences Division,
British Antarctic Survey Cambridge,
UK

Hugues Roest Crollius
Laboratoire Dynamique et Organisation
des Genomes (LDOG),
Department of Biology,
Ecole Normale Superiere, Paris,
France

Paul Cutler
GlaxoSmithKline Pharmaceutical
Research & Development,
Stevenage, Hertfordshire,
UK

Christine Debouck
GlaxoSmithKline Pharmaceutical
Research & Development,
Collegeville, PA,
USA

Masamitsu Futai
Microbial Chemistry Research
Foundation and CREST,
Japan Science and Technology Agency,
Kamiosaki, Shinagawa, Tokyo,
Japan

J. Jay Gargus
University of California,
Irvine, CA,
USA

Israel S. Gloger
GlaxoSmithKline Pharmaceutical
Research & Development,
Harlow, Essex,
UK

Preethi Gunaratne
Human Genome Sequencing Center,
Baylor College of Medicine,
Houston, TX,
USA

Tatsuya Haga
Gakushuin University,
Tokyo,
Japan

Udo Heinemann
Max Delbrück Center for Molecular
Medicine,
Berlin,
Germany

Encyclopedia of Molecular Cell Biology and Molecular Medicine, 2nd Edition. Volume 11
Edited by Robert A. Meyers.
Copyright © 2005 Wiley-VCH Verlag GmbH & Co. KGaA, Weinheim
ISBN: 3-527-30648-X

List of Contributors

Emma E. Hill
University of California,
Berkeley, California,
USA

Jim Hoggett
University of York,
York,
UK

David K. Hwang
University of California Los Angeles,
Los Angeles, CA,
USA

Reinhard Jahn
Max-Planck-Institute for Biophysical Chemistry,
Göttingen,
Germany

Robert James Slater
University of Hertfordshire,
Hatfield,
UK

Albert Jeltsch
International University Bremen,
Bremen,
Germany

Kimihiko Kameyama
National Institute of Advanced Industrial Science and Technology,
Tsukuta,
Japan

Kendall L. Knight
University of Massachusetts Medical School,
Worcester, MA,
USA

Carla M. Koehler
University of California Los Angeles,
Los Angeles, CA,
USA

Thomas J. Magliery
Yale University,
New Haven, CT,
USA

Dharia A. McGrew
University of Massachusetts Medical School,
Worcester, MA,
USA

Christopher Mehlin
University Of Washington,
Seattle, WA,
USA

Kenneth V. Mills
College of the Holy Cross,
Worcester, MA,
USA

Steffen Nock
Absalus Inc.,
Mountain View, CA,
USA

D. Peter Tieleman
University of Calgary,
Albeta,
Canada

Daniel L. Purich
University of Florida College of Medicine,
Gainesville, FL,
USA

Rita Raddatz
Cephalon Incorporated,
West Chester, PA,
USA

Martin Rechsteiner
Department of Biochemistry University of Utah,
Salt Lake, UT,
USA

Rajkumar Sasidharan
MRC Laboratory of Molecular Biology,
Cambridge, MA,
USA

Gregory L. Shipley
The University of Texas Health
Science Center,
Houston, TX,
USA

Ge-Hong Sun-Wada
ISIR, Osaka University,
Ibaraki, Osaka,
Japan

Jens R. Sydor
Infinity Pharmaceuticals Inc.,
Cambridge, MA,
USA

Thomas Szyperski
State University of New York,
Buffalo, NY,
USA

David J. Triggle
State University of New York,
Buffalo, NY,
USA

Andrew P. Turnbull
Max Delbrück Center for Molecular
Medicine,
Berlin,
Germany

Claus Urbanke
Medizinische Hochschule Hannover,
Zentrale Einrichtung
Biophysikalisch-Biochemische Verfahren,
Hannover,
Germany

Christine Vogel
MRC Laboratory of Molecular Biology,
Cambridge, MA,
USA

Yoh Wada
ISIR,
Osaka University,
Ibaraki, Osaka,
Japan

Michael Williams
Cephalon Incorporated,
West Chester, PA,
USA

David S. Wilson
Absalus Inc.,
Mountain View, CA,
USA

Kim C. Worley
Human Genome Sequencing Center,
Baylor College of Medicine,
Houston, TX,
USA

Marian R. Zlomislic
University of Calgary,
Albeta,
Canada

Color Plates

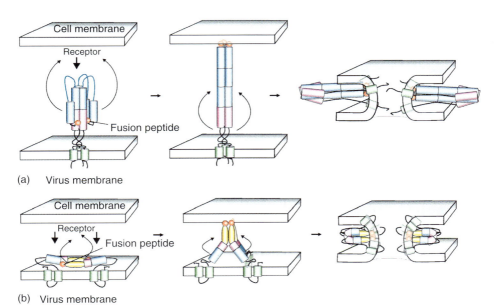

Fig. 3 (p. 77) Models for membrane fusion mediated by class I (a) and class II (b) viral fusion proteins involving a two-step mechanism. In the first step, the fusion peptides are exposed and are inserted in the target membrane. In the second step, the C-terminal part of the proteins undergoes another conformational change, resulting in bending/pulling that results in fusion. Red balls depict the fusion peptides, green cylinders the transmembrane domains. Note that the cellular receptors and the viral receptor–binding proteins are not shown. The colors indicate different domain of the proteins. Note that in class I proteins the colored boxes indicate α-helices, whereas in class II proteins they define folded domains that are not helical. Arrows indicate the presumed movements.

Encyclopedia of Molecular Cell Biology and Molecular Medicine, 2nd Edition. Volume 11
Edited by Robert A. Meyers.
Copyright © 2005 Wiley-VCH Verlag GmbH & Co. KGaA, Weinheim
ISBN: 3-527-30648-X

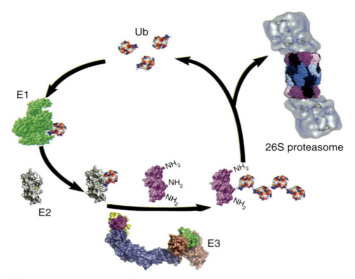

Fig. 3 (p. 8) Schematic representation of the ubiquitin-proteasome pathway. Ubiquitin molecules are activated by an E1 enzyme (shown in green at one-third scale) in an ATP-dependent reaction, transferred to a cysteine residue (yellow) on an E2 or Ub carrier protein and subsequently attached to amino groups (NH$_2$) on a substrate protein (lysozyme shown in purple) by an E3 or ubiquitin ligase, (the multicolored SCF complex). Note that chains of Ub are generated on the substrate, and these are recognized by the 26S proteasome depicted in the upper right at one-twentieth scale.

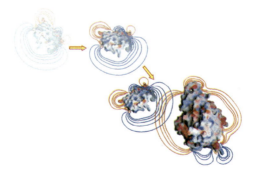

Fig. 4 (p. 150) Schematic illustration of the association of fasciculin (on the left) with acetylcholinesterase. Blue and red contour lines indicate regions of positive and negative electrostatic potential, respectively. Figure courtesy of D. Sept and A. Elcock. See Elcock, A.H., Gabdoulline, R.R., Wade, R.C., McCammon, J.A. (1999) Computer simulation of protein-protein association kinetics: acetylcholinesterase-fasciculin, *J. Mol. Biol.* **291**, 149–162 for more details.

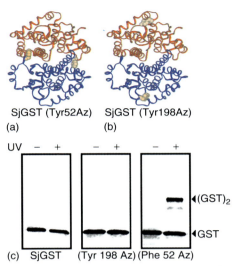

Fig. 6 (p. 63) An example of the utility of expanded genetic codes: *p*-azido-L-phenylalanine (1) is inserted into glutathionine-*S*-transferase at position 52 (a) or position 198 (b). On irradiation with UV light (c) the azido group photocross-links protein monomers in the dimeric GST complex. This cross-linking is dependent on the azido group being at the protein interface (position 52). Similar experiments have been used to map protein–protein interactions both *in vitro* and in cells.

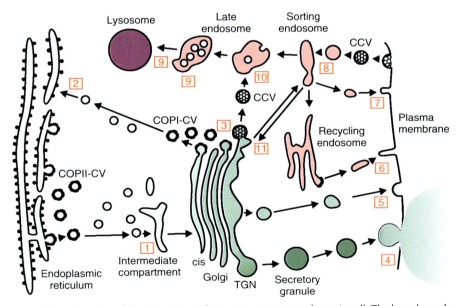

Fig. 4 (p. 80) Overview of the secretory pathway in a prototype eukaryotic cell. The boxed numbers indicate distinct fusion steps. CCV, clathrin-coated vesicle, COPI/II-CV, COPI/II-coated vesicle, TGN, *trans*-Golgi network. The biosynthetic route is depicted in green, the endocytotic pathway in pink.

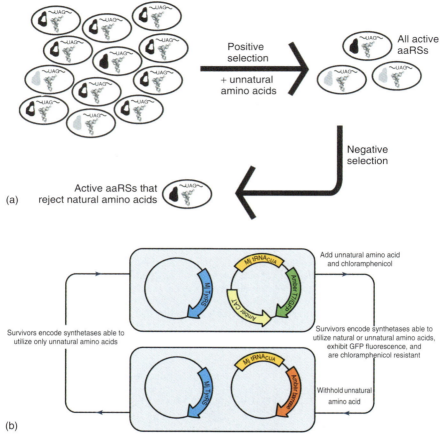

Fig. 2 (p. 53) "Double-sieve" selections for unnatural amino acid activity: (a) Schematic of a generic double-sieve selection. A library of orthogonal aminoacyl-tRNA synthetases is transformed into cells with an orthogonal tRNA. First, positive selection is done in the presence of unnatural amino acids, such that the selective phenotype requires acylation of the orthogonal tRNA and suppression of the unique codon. Cell with inactive synthetases (white) die, but those with synthetases active toward natural (grey) or unnatural (black) amino acids survive. Second, a negative selection is carried out in the absence of unnatural amino acid, where the selective phenotype requires that the tRNA *not* be acylated. Only cells with synthetases that lack activity toward natural amino acids survive. Thus, survival of both selections implies activity only toward unnatural amino acids. (b): A double-sieve selection in *E. coli*. (c): A double-sieve selection in yeast.

Color Plates | xxvii

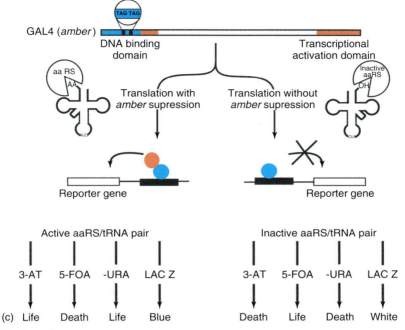

Fig. 2 (Continued)

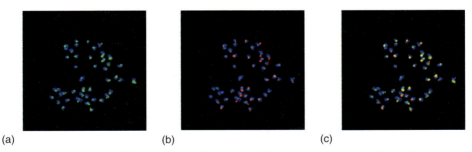

Fig. 2 (p. 367) *In situ* hybridization of two different non-LTR retrotransposons to *Tetraodon* chromosomes. (a) Green signals of Zebulon retrotransposon; (b) red signals of Barbar retrotransposon; (c) combined signal (yellow) of both retrotransposons. Both of these elements share the same sites in the genome, at the tip of some subtelocentric chromosomes. Babar is more abundant (in the shotgun data) and has more signals here, as seen by the additional red hybridizations in the combined data.

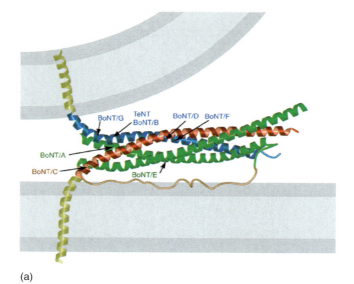

(a)

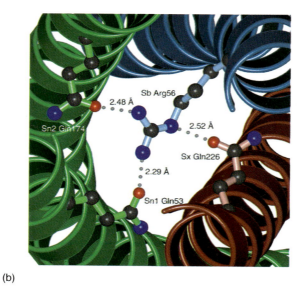

(b)

Fig. 6 (p. 84) (a) Crystal structure of the neuronal SNARE complex containing syntaxin 1 (red), SNAP-25 (green), and synaptobrevin/VAMP (blue). The structure is modeled between two membranes, the transmembrane domains and immediately adjacent residues are not part of the structure. (a) Structure of the ionic "0" layer that is conserved among all SNAREs and that is the basis for the classification of SNAREs into Q- and R-SNAREs, respectively. Modified from Sutton, B., Fasshauer, D., Jahn, R., Brünger, A.T. (1998) Crystal structure of a core synaptic fusion complex at 2.4 Å resolution, *Nature* **395**, 347–353.

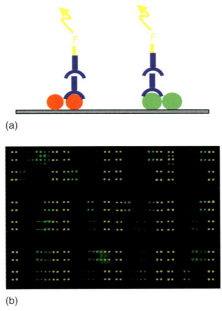

Fig. 4 (p. 111) Scheme and example of an antigen array. (a) Antigens are arrayed on a two-dimensional surface. Autoantibodies in serum are captured and detected with a fluorescently labeled antihuman antibody-antibody. (b) "Arthritis Array" characterization of autoantibodies in systemic lupus erythematosus. Arthritis arrays were produced by printing common lupus antigens including DNA, histone proteins, and additional nuclear proteins on microscope slides. Arrays were incubated with patient serum, and binding of autoimmune antibodies was detected with green fluorescent markers (green spots). The yellow spots are used as "marker features" to orient the arrays. (Figure courtesy Dr. William Robinson, Stanford University).

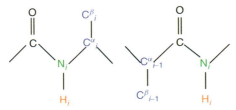

Fig. 3 (p. 164) Example of chemical shifts of nuclei (in color) correlated in triple resonance NMR experiments for obtaining sequence-specific backbone resonance assignments of proteins. A first experiment correlates the shifts of amide proton, nitrogen, α-carbon, and β-carbon within a given residue i (on the left). A second experiment correlates the shifts of amide proton and nitrogen of residue i with those of the α- and β-carbons of the preceding residue $i-1$. Combining the two experiments allows one to obtain sequence-specific resonance assignments for the backbone (and $^{13}C^{\beta}$ shifts) of the entire polypeptide.

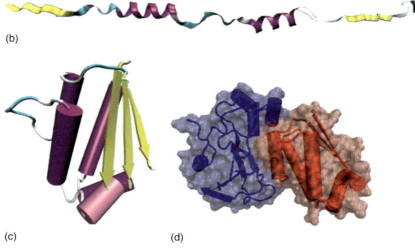

Fig. 1 (p. 137) (a) The primary sequence of the protein barstar in FASTA format. (b) The secondary structure of the first 60 residues of barstar, colored purple for helices, and yellow for β-strands. All are shown in ribbon format. (c) The complete tertiary structure of barstar in cartoon format. The helices are now bundled together, and the β-strands have formed a three-stranded parallel β-sheet. (d) Example of quaternary structure: the barnase/barstar complex. Barstar (red) and barnase (blue) area shown in ribbon and spacefilling formats. Images created with VMD using the PDB entry 1B2U.

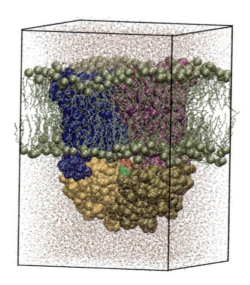

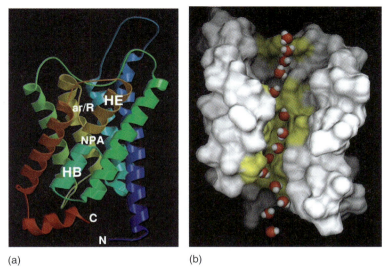

(a)　　　　　　　　　(b)

Fig. 5 (p. 151) (a) Ribbon diagram of the aquaporin monomer and (b) spacefilling representation illustrating a water file through the channel pore. (a) Starting at the N-terminus of the monomer, there are two transmembrane helices, followed by the coil–NPA–helix motif. A third transmembrane helix completes the first half of the protein. The second half of the monomer has two transmembrane helices, followed by the coil–NPA–helix motif, followed by another transmembrane helix. (b) The path of the water channel is through the core of the protein, following the path of the coil motifs, which meet at the NPA signature. This is more clearly illustrated in 5(b). The protein is rendered as a molecular surface in white, with the front surface of the protein cut away so that we can clearly see the pore in the middle of the protein. Surfaces colored yellow are those that interact most strongly with passing water molecules. The waterfile displayed is an overlay of a number of snapshots from the 10 ns simulation. The dipole inversion of water at the NPA motif is clearly illustrated here. Figure courtesy of B. de Groot. See De Groot, B.L., Grubmuller, H. (2001) Water permeation across biological membranes: mechanism and dynamics of aquaporin-1 and GlpF, *Science* **294**, 2353–2357. De Groot, B.L., Frigato, T., Helms, V., Grubmuller, H. (2003) The mechanism of proton exclusion in the aquaporin-1 water channel, *J. Mol. Biol.* **333**, 279–293 for more details.

Fig. 3 (p. 147) An orthographic view of the periodic box for the ATP-bound BtuCD simulation, showing water (red and white), lipid with phosphorus atoms enlarged, and the protein. The transporter consists of two transmembrane domains (blue and purple) and two nucleotide binding domains (orange and ochre). The two docked MgATP molecules are partially visible (green and red). Figure courtesy of E. Oloo. See Oloo, E.O., Tieleman, D.P. (2004) Conformational transitions induced by the binding of MgATP to the vitamin B_{12} ABC-transporter BtuCD, *J. Biol. Chem.* **279**, 45013–45019 for more details. Rendered with VMD.

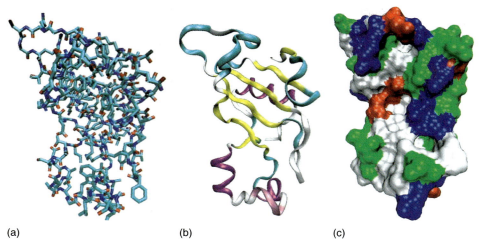

(a) (b) (c)

Fig. 2 (p. 143) Barnase rendered in three formats: (a) colored by atom-type (carbon – blue, nitrogen – navy blue, oxygen – red); (b) colored by secondary structure features in ribbon format, where β-sheets are colored yellow, helices are colored purple, and turns and random coil features are colored blue and white; (c) rendered in surface format, colored by residue type, where charged residues are colored red or blue, polar residues are green, and nonpolar residues are white. All figures are rendered with the same view of the protein, looking into the barstar binding pocket. All figures were rendered with VMD.

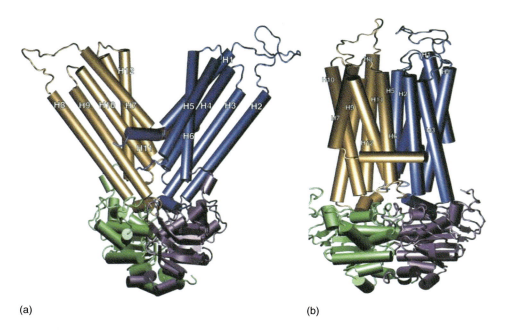

(a) (b)

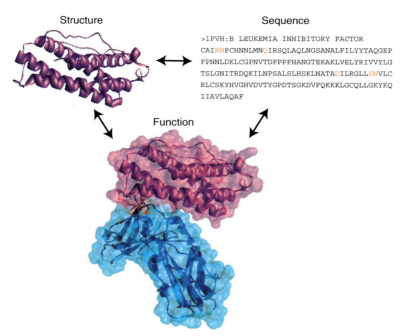

Fig. 1 (p. 204) Proteins are characterized by their sequence, structure, and function. The sequence, individual domain structure (colored purple), and interacting structure (with Gp130 colored blue) of leukemia inhibitory factor are shown (Protein Data Bank (PDB) ID: 1pvh). Within the sequence and interacting domains residues that are colored red form hydrogen bonds with the receptor.

Fig. 6 (p. 154) (a) Construction of a molecular model for P-glycoprotein (P-gp). Each half of P-gp was modeled by homology to the crystal structure of MsbA (PDB code 1JSQ), which had been extended to a full atom representation. The two halves of P-gp were assembled such that the NBDs adopt the ATP-dependent orientation observed in MJ0796 (PDB code 1L2T). The constituent domains are individually colored. (b) Reconciliation of cross-linking data with a P-gp model. P-gp-Model-B was generated by rotation of each NBD with respect to its cognate TMD. The final model contains a parallel TMD:TMD interface (blue and gold subunits) and a consensus NBD:NBD interface (green and purple subunits). Adapted from Stenham, D.R., Campbell, J.D., Sansom, M.S.P., Higgins, C.F., Kerr, I.D., Linton, K.J. (2003) An atomic detail model for the human ATP binding cassette transporter P-glycoprotein derived from disulphide cross-linking and homology modeling, *FASEB J.* **17**, 2287–2289.

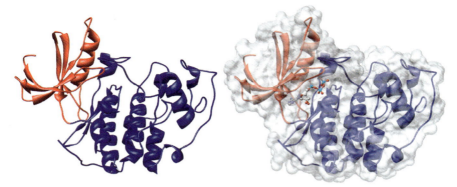

Fig. 4 (p. 210) Protein kinases, as defined in SCOP and CATH. On the left is the kinase domain (PDB ID: 1hck) colored according to the CATH definition of two distinct structural domains (first CATH domain residues 1–84 red, the second CATH domain residues 85–297 blue). On the right, the same coloring is implemented; however, we also show the surface of the entire kinase domain and the location of the bound ATP. This highlights the fact that in terms of surface the entire domain appears as one globular structure, and the location of the ATP binds at the interface of the two CATH defined domains. Functionally, both domains are necessary for ATP to bind and therefore the single-domain SCOP definition seems more accurate in terms of functional, evolutionary domain definition.

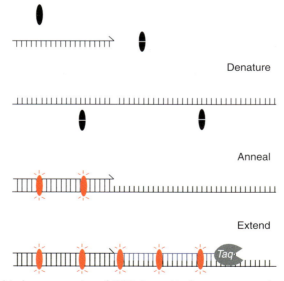

Fig. 3 (p. 492) Graphical representation of SYBR Green I in fluorescent signal generation during a PCR. Free dye has very low fluorescence and will not bind to single-stranded, denatured DNA. During primer annealing, a double-stranded structure is formed and SYBR green dye begins to bind and emit a fluorescent signal. During primer extension by *Taq* DNA polymerase, the number of SYBR green dye molecules bound per double-stranded molecule and the fluorescent signal increases proportionally. The process is repeated in each cycle with increasing total fluorescence.

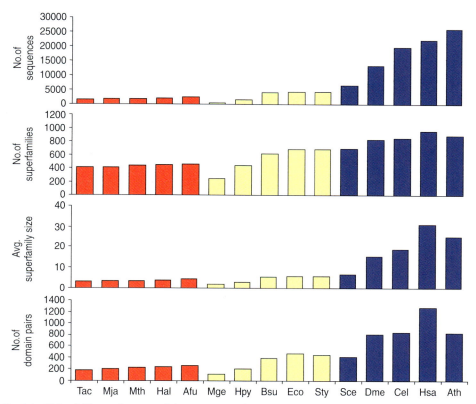

Fig. 6 (p. 214) Different features of the genome and protein repertoire. From top: The total number of sequences (genes) predicted for each organism. The total number of different SCOP domain superfamilies predicted with SUPERFAMILY version 1.65 (http://supfam.mrc-lmb.cam.ac.uk/SUPERFAMILY/). The average number of proteins per domain superfamily that is, number of duplicates per domain superfamily. Finally, the number of different two-domain combinations. Red – Archaea: Tac, *Thermoplasma acidophilum*; Mja, *Methanocaldococcus jannaschii*; Mth, *Methanothermobacter thermautotrophicus* Delta H; Hal, *Halobacterium* sp. NRC-1; Afu, *Archaeoglobus fulgidus* DSM 4304. Yellow – Bacteria: Mge, *Mycoplasma genitalium*; Hpy, *Helicobacter pylori* 26695; Bsu, *Bacillus subtilis* ssp. subtilis 168; Eco, *Escherichia coli* K12; Sty, *Salmonella typhimurium* LT2. Blue – Eukarya: Sce, *Saccharomyces cerevisiae*; Dme, *Drosophila melanogaster* 3.2; Cel, *Caenorhabditis elegans* WS123; Hsa, *Homo sapiens* 22.34d; Ath, *Arabidopsis thaliana* 5.

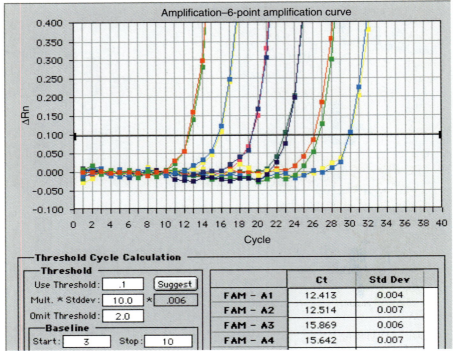

Fig. 10 (p. 515) Magnified view of the fluorescent signal in the baseline phase in relation to the threshold setting. A demonstration of setting the threshold high enough to be above the fluorescent signals (noise) in the baseline, but low enough to be within the geometric phase for each amplification curve. The threshold is the black horizontal line at 0.1 ΔRn.

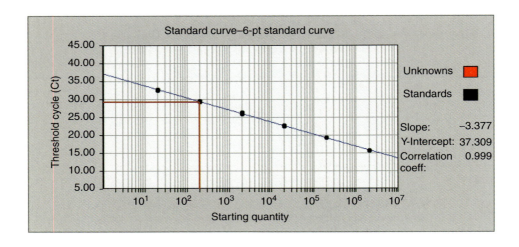

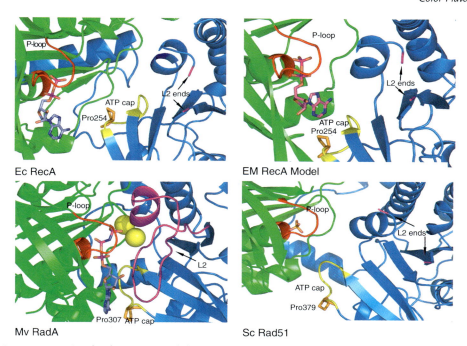

Fig. 3 (p. 533) ATP binding region and the "ATP cap." This view of subunit interface is seen from the inside of the filament. The ribbon diagram of the α-carbon backbone is blue for one subunit and green for the neighbor. In the green subunit, the Walker A motif, or P-loop, is in red, and the following ligands are seen bound to this motif: ADP in EcRecA and EcRecA model; AMP-PNP in MvRadA; a sulfate ion in ScRad51. In the blue subunit, the L2 region (MvRadA) or residues that flank L2 (other 3 panels) are in pink, the "ATP cap" is in yellow, and its universally conserved Proline is in orange. The yellow spheres in the MvRadA structure correspond to two K^+ atoms that are bound to the AMP-PNP. Images are made from PDB files: 1REA, 1N03, 1XU4, and 1SZP.

Fig. 11 (p. 519) A 6-log standard curve with a graphic demonstration of how the software determines molecules from the Ct values measured by real-time instruments. A Ct (threshold cycle) value of approximately 29.5 (*y*-axis) would result in an interpolated value of 200 molecules (*x*-axis) using this standard curve. A change in the slope of the curve or the measured Ct values of the standards would change the number of molecules obtained from the same Ct reading.

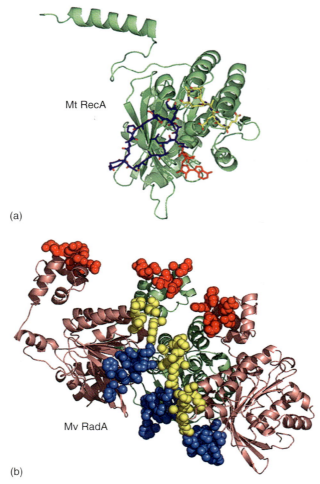

Fig. 4 (p. 535) DNA binding domains of RecA and RadA. (a) Monomer of MtRecA, with the L1 and L2 DNA binding regions shown in yellow and blue respectively. (b) A trimer of MvRadA as seen from the inside of the filament, showing amino acid side chains of L1 (yellow), L2 (blue) and the N-terminal Helix-hairpin-Helix domain (red) as space-filling spheres. Images are made from PDB file: 1G18/1G19 and 1XU4. The green MtRecA subunit (a) is in approximately the same orientation as the green MvRadA subunit (b).

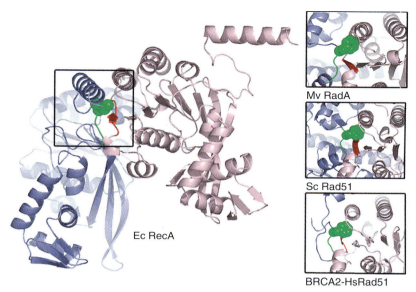

Fig. 5 (p. 537) N-terminal Polymerization motif. The left-hand image is two subunits of the EcRecA filaments (viewed from the outside surface of the filament), showing the N-terminal polymerization motif (green) and the Core Polymerization Motif (red). Isoleucine 26 is shown in space-filling spheres. The boxes on the right show the equivalent interface region in MvRadA (Phe64), ScRad51 (Phe144), and the BRC4-HsRad51 fusion protein (Phe1524) with indicated phenylalanine residues interacting with the neighboring subunit. Images were made from PDB files: 1REA, 1XU4, 1SZP and 1NOW.

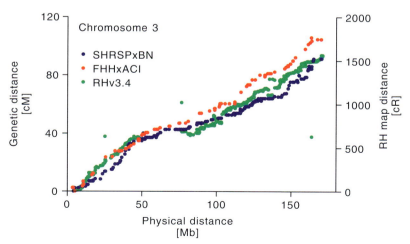

Fig. 2 (p. 441) Map correspondence. Correspondence between positions of markers on two genetic maps of the rat (SHRSPxBN intercross and FHHxACI intercross), on the rat radiation hybrid map, and their position on the rat genome assembly (Rnor3.1). Used with permission from *Nature* **428**, 493–521 (2004).

xl | Color Plates

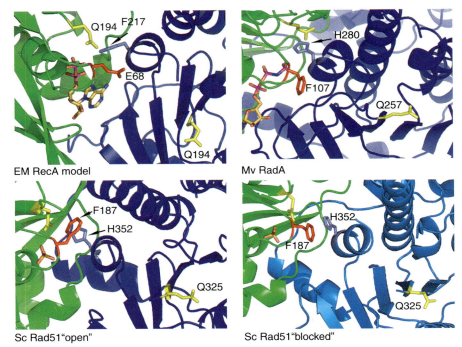

Fig. 6 (p. 539) Allosteric mechanism. Four panels showing different conformations of residues important in sensing ATP binding and conformational change of the filament structure. Glutamine 194 (EcRecA) and its homologous residues in MvRadA (Gln257) and ScRad51 (Gln325) are shown in yellow, while Phe217 (EcRecA) and the homologous histidine residues (MvRadA, His280; ScRad51, His352) are shown in purple. Glutamate 68 (EcRecA) and equivalent phenylalanine residues are in red. The region between Q194 and Phe217 (loop2) is unstructured in all but MvRadA. Images are made from PDB files: 1N03, 1XU4, and 1SZP.

Fig. 1 (p. 558) Structural model of the nicotinic acetylcholine receptor (nAChR: an ion channel-coupled receptor, ICCR). (A) An atomic model with the closed pore was determined from electron images by electron cryo-microscopy of crystalline postsynaptic membranes and their reconstruction by optical diffraction and computation. The side (a) and cross-sectional (c) views show an extracellular ligand-binding domain (green), an inner, and the transmembrane segments composed of pore-facing M2 helices (blue) and lipid-facing M1, M3, M4 helices (red). (b) Stereo view of the pore, as seen from the extracellular face. (c) Cross-sectional view through the pentamer at the middle of the membrane. The pore is shaped by an inner ring of five α helices (M2), which curve radially to create a tapering path for ions, and an outer ring of 15 α-helices (M1, M3, M4), which coil around each other and shield the inner ring from the lipid layer. (B) A model for the gating mechanism. With binding of acetylcholine (ACh), two M2 α-helices of α-subunits rotate to opposite directions. This movement triggers the concerted movement of M2 α-helices of the other subunits resulting in the opening of the gate that consists of hydrophobic residues in the central part of five M2 segments. (From Miyazawa, A. et al. (2003) Structure and gating mechanism of the acetylcholine receptor pore, *Nature* **424**, 949–955).

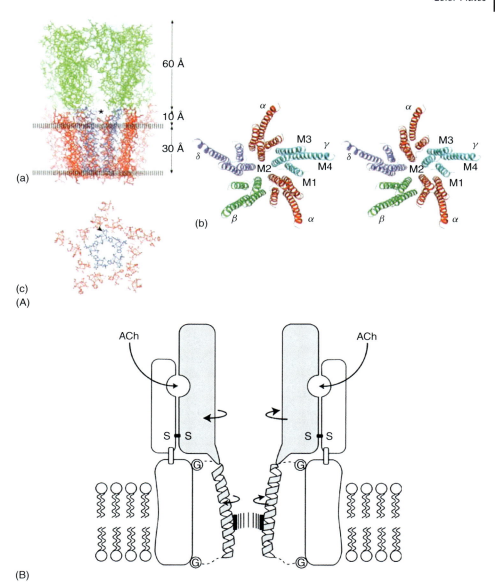

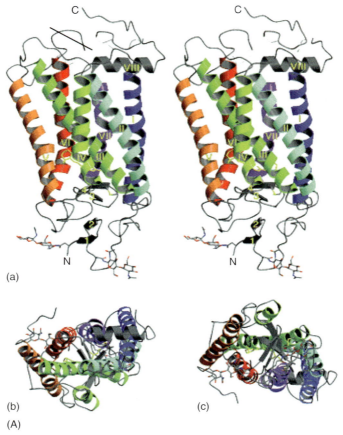

Fig. 2 (p. 567) Structures of rhodopsin and a heterotrimeric G-protein. (A) An atomic model for rhodopsin was obtained by X-ray analysis of the rhodopsin crystal. (a) A stereoview in parallel to the plane of the membrane. The upper and lower parts of this model correspond to the intra- and extracellular domains respectively. There are seven transmembrane α-helixes (I-VII), another α-helix in the C-terminal region (VIII), 11-cis-retinal (yellow) in the transmembrane region, and two β-sheets (1–2 and 3–4) and sugars in the N-terminal region. View of the receptor from the cytoplasmic (b) and intradiscal (extracellular) side (c) of the membrane. (From Palczewski, K. et al. (2000) Crystal structure of rhodopsin: A G protein-coupled receptor, *Science* **289**, 739–745). (B) Receptor-G protein interaction. (a) A structural model for the cytoplasmic surface of the M_1 muscarinic receptor. Residues important for interaction with G protein are shown in red, magenta (i3 loop), and yellow. The arrows indicates the outward movement of TMb and increased mobility of TM7 and helix 8, which are thought to accompany receptor activation. (b) A structural model for the interaction between G protein-coupled receptor (GPCR) and G-protein $\alpha\beta\gamma$ trimer. Structures are taken from crystal structures of rhodopsin and a chimera of transducin (Gt) and Gi. The N- and C-termini of the transducin α-subunit and the $\alpha 4$-$\beta 6$ loop sequences, which are thought to be involved in the interaction with GPCR, are shown in red. The switch-II helix in the G-protein α-subunit is bound to the β-subunit when the α-subunit is bound to GDP, and is subject to conformational change and is released from the β-subunit when the α-subunit is bound with GTP. (From Lu, Z.L. et al. (2002) Seven-transmembrane receptors: crystals clarify, *Trends Pharmacol. Sci.* **23**, 140–146).

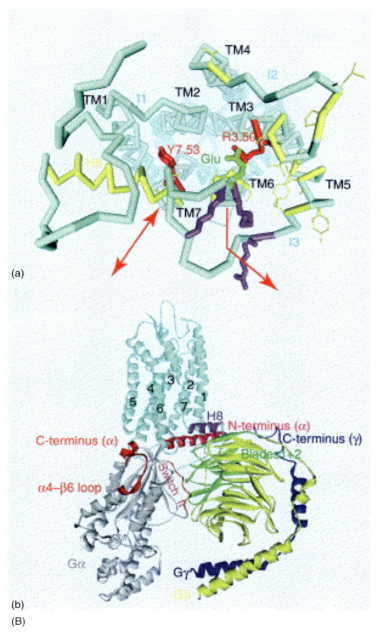

Fig. 2 (Continued)

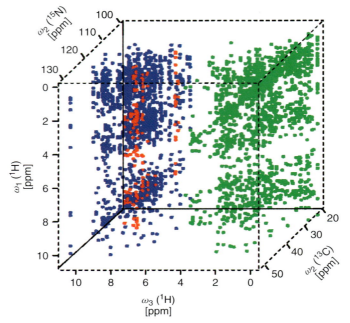

Fig. 8 (p. 172) Cube with dots representing NOEs detected in 3D heteronuclear resolved [^{1}H,^{1}H]-NOESY, which were used for calculating the NMR solution structure of YqfB shown in Fig. 9. Blue: signals detected along ω_3 on backbone amide protons; Red: detected on aromatic protons; Green: detected on aliphatic protons. Courtesy of Hanudatta Atreya.

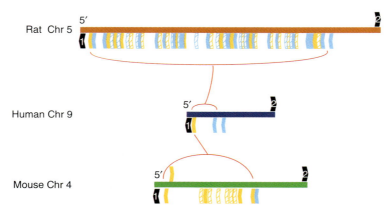

Fig. 13 (p. 468) Adaptive remodeling of genomes and genes. Orthologous regions of rat, human and mouse genomes encoding pheromone-carrier proteins of the lipocalin family (α_{2u} globulins in rat and major urinary proteins in mouse) shown in brown. Zfp37-like zinc finger genes are shown in blue. Filled arrows represent likely genes, whereas striped arrows represent likely pseudogenes. Gene expansions are bracketed. Arrow head orientation represents transcriptional direction. Flanking genes 1 and 2 are TSCOT and CTR1 respectively.

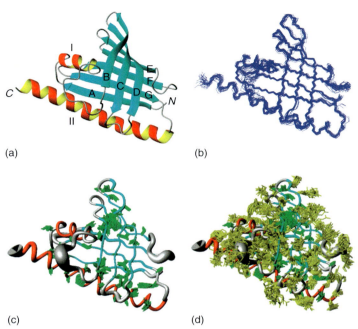

Fig. 9 (p. 173) Commonly used representations of NMR solution structures, exemplified for the 18-kDa Northeast Structural Genomics consortium target protein CC1736. (a) Ribbon drawing of the conformer exhibiting the smallest residual constraints violations. α-Helices and β-strands are depicted, respectively, in red/yellow, and cyan. (b) Superposition of the polypeptide backbone of the 20 conformers chosen to represent the NMR solution structure. The higher precision of the atomic coordinates of the regular secondary structure elements is apparent when compared with the loops. (c) The polypeptide backbone is represented by a spline function through the mean C^α coordinates, where the thickness represents the rmsd values for the C^α-coordinates after superposition of the regular secondary structure elements for minimal rmsd. The superpositions of the best defined side chains of the molecular core are shown in green to indicate the precision of their structural description. (d) Same as in (c), also showing the superposition of the more flexibly disordered side chains on the protein surface. The protein sample was provided by Drs. Acton and Montelione, Rutgers University. Courtesy of Yang Shen.

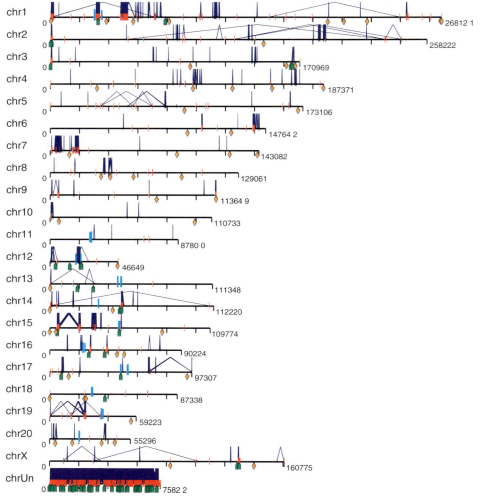

Fig. 3 (p. 442) Distribution of segmental duplications in the rat genome. Interchromosomal duplications (red) and intrachromosomal duplications (blue) are depicted for all duplications with ≥90% sequence identity and ≥20 kb length. The intrachromosomal duplications are drawn with connecting blue line segments; those with no apparent connectors are local duplications very closely spaced on the chromosome (below the resolution limit for the figure). P arms on the left and the q arms on the right. Chromosomes 2, 4–10, and X are telocentric; the assemblies begin with pericentric sequences of the q arms, and no centromeres are indicated. For the remaining chromosomes, the approximate centromere positions were estimated from the most proximal STS/gene marker to the p and q arm as determined by fluorescent *in situ* hybridization (FISH) (cyan vertical lines; no chromosome 3 data). The chrUn sequence is contigs not incorporated into any chromosomes. Green arrows indicate 1-Mb intervals with more than tenfold enrichment of classic rat satellite repeats within the assembly. Orange diamonds indicate 1-Mb intervals with more than tenfold enrichment of internal (TTAGGG)*n*-like sequences. For more details, see http://ratparalogy.cwru.edu. Used with permission from *Nature* **428**, 493–521 (2004).

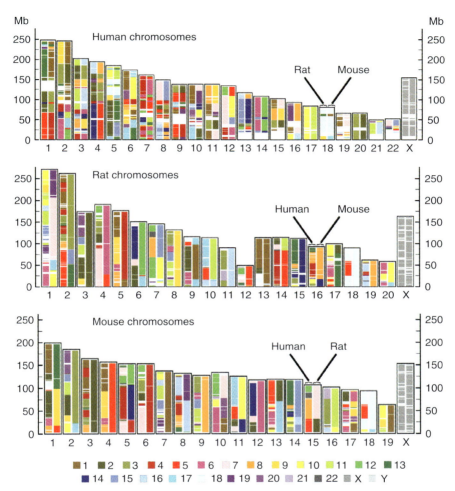

Fig. 4 (p. 444) Map of conserved synteny between the human, mouse, and rat genomes: For each species, each chromosome (x-axis) is a two column boxed pane (p-arm at the bottom) colored according to conserved synteny to chromosomes of the other two species. The same chromosome color code is used for all species (indicated below). For example, the first 30 Mb of mouse chromosome 15 is shown to be similar to part of human chromosome 5 (by the red in left column) and part of rat chromosome 2 (by the olive in right column). An interactive version is accessible (http://www.genboree.org). Used with permission from *Nature* **428**, 493–521 (2004).

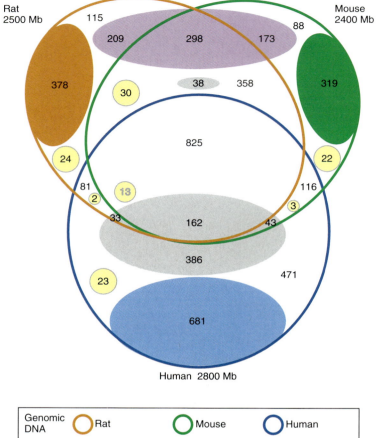

Fig. 7 (p. 447) Aligning portions and origins of sequences in rat, mouse, and human genomes. Each outlined ellipse is a genome, and the overlapping areas indicate the amount of sequence that aligns in all three species (rat, mouse, and human) or in only two species. Nonoverlapping regions represent sequences that do not align. Types of repeats classified by ancestry: those that predate the human–rodent divergence (gray), those that arose on the rodent lineage before the rat–mouse divergence (lavender), species-specific (orange for rat, green for mouse, blue for human), and simple (yellow), placed to illustrate the approximate amount of each type in each alignment category. Uncolored areas are nonrepetitive DNA – the bulk is assumed to be ancestral to the human–rodent divergence. Numbers of nucleotides (in Mb) are given for each sector (type of sequence and alignment category). Used with permission from *Nature* **428**, 493–521 (2004).

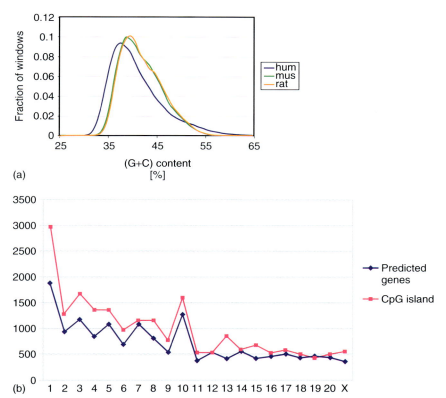

Fig. 8 (p. 450) Base composition distribution analysis. (a) The fraction of 20 kb nonoverlapping windows with a given G + C content is shown for human, mouse, and rat. (b) The number of Ensembl predicted genes per chromosome and the number of CpG islands per chromosome. The density of CpG islands averages 5.9 islands per Mb across chromosomes and 5.7 islands per Mb across the genome. Chromosome 1 has more CpG islands than other chromosomes, yet neither the island density nor ratio to predicted genes exceeds the normal distribution. The number of CpG islands per chromosome and the number of predicted genes are correlated ($R^2 = 0.96$). Used with permission from *Nature* **428**, 493–521 (2004).

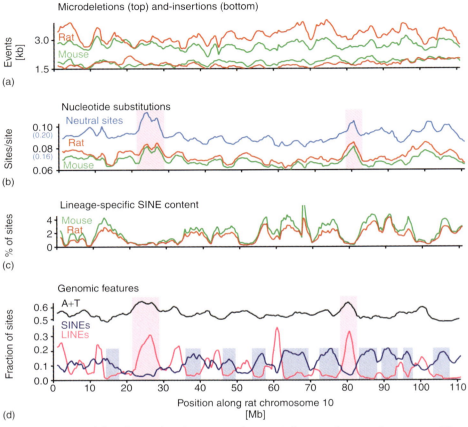

Fig. 9 (p. 451) Variability of several evolutionary and genomic features along rat chromosome 10. (a) Rates of microdeletion and microinsertion events (less than 11 bp) in the mouse and rat lineages since their last common ancestor, revealing regional correlations. (b) Rates of point substitution in the mouse and rat lineages. Red and green lines represent rates of substitution within each lineage estimated from sites common to human, mouse, and rat. Blue represents the neutral distance separating the rodents, as estimated from rodent-specific sites. Note the regional correlation among all three plots, despite being estimated in different lineages (mouse and rat) and from different sites (mammalian vs rodent-specific). (c) Density of SINEs inserted independently into the rat or mouse genomes after their last common ancestor. (d) A + T content of the rat, and density in the rat genome of LINEs and SINEs that originated since the last common ancestor of human, mouse, and rat. Pink boxes highlight regions of the chromosome in which substitution rates, AT content, and LINE density are correlated. Blue boxes highlight regions in which SINE density is high but LINE density is low. Used with permission from *Nature* **428**, 493–521 (2004).

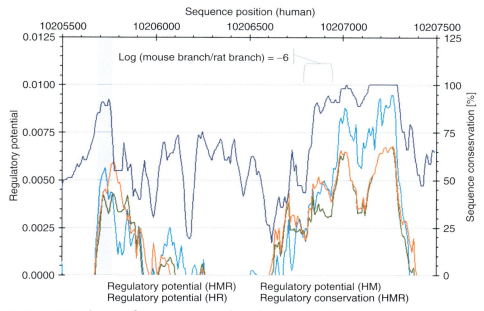

Fig. 11 (p. 460) Close-up of PEX14 (peroxisomal membrane protein) locus on human chromosome 1 (with homologous mouse chromosome 4 and rat chromosome 5). Conservation score computed on 3-way human−mouse−rat alignments presents a clear coding exon peak (gray bar) and very high values in a 504 bp noncoding, intronic segment (right; last 100 bp of alignment are identical in all three organisms). The latter segment showed a striking difference between the inferred mouse and rat branch lengths: the gray bracket corresponds to a phylogenetic tree where the logarithm of mouse to rat branch length ratio is −6. Regulatory potential (RP) scores that discriminate between conserved regulatory elements and neutrally evolving DNA are calculated from 3-way (human−mouse−rat) and 2-way (human-rodent) alignments. Here the 3-way regulatory potential scores are enhanced over the 2-way scores. Used with permission from *Nature* **428**, 493−521 (2004).

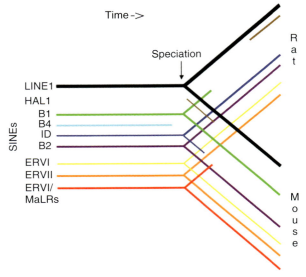

Fig. 12 (p. 465) Historical view of rodent repeated sequences. Relationships of the major families of interspersed repeats (Table 7) are shown for the rat and mouse genomes, indicating losses and gains of repeat families after speciation. The lines indicate activity as a function of time. Note that HAL1-like elements appear to have arisen both in the mouse and rat lineage. Used with permission from *Nature* **428**, 493–521 (2004).

Fig. 14 (p. 470) Evolution of cytochrome P450 (CYP) protein families in rat, mouse, and human. (a) Dendrogram of topology from 234 full-length sequences. The 279 sequences of 300 amino acids; subfamily names and chromosome numbers are shown. Black branches have >70% bootstrap support. Incomplete sequences (they contain Ns) are included in counts of functional genes (84 rat, 87 mouse, and 57 human) and pseudogenes (including fragments not shown; 77 rat, 121 mouse, and 52 human). Thus, 64 rat genes and 12 pseudogenes were in predicted gene sets. Human CYP4F is a null allele due to an in-frame STOP codon in the genome, although a full-length translation exists (SwissProt P98187). Rat CYP27B, missing in the genome, is "incomplete" since there is a RefSeq entry (NP_446215). Grouped subfamilies CYP2A, 2B, 2F, 2G, 2T, and CYP4A, 4B, 4X, 4Z, occur in gene clusters; thus nine loci contain multiple functional genes in a species. One (CYP1A) has fewer rat genes than human, seven have more rodent than human, and all nine have different copy numbers. CYP2AC is a rat-specific subfamily (orthologs are pseudogenes). CYP27C has no rodent counterpart. Rodent-specific expansion, rat CYP2J is illustrated below. (b) The neighbor-joining tree, with the single human gene, contains clear mouse (Mm) and rat (Rn) orthologous pairs (bootstrap values >700/1000 trials shown). Bar indicates 0.1 substitutions per site. (c) All rat genes have a single mouse counterpart except for CYP2J 3, which has further expanded in mouse (mouse CYP2J 3a, 3b, and 3c) by two consecutive single duplications. The genes flanking the CYP2J orthologous regions (rat chromosome 5, 126.9–127.3 Mb; mouse chromosome 4, 94.0–94.6 Mb; human chromosome 1, 54.7–54.8 Mb) are hook1 (HOOK1; pink) and nuclear factor I/A (NFIA; cyan). Genes (solid) and gene fragments (dashed boxes) are shown above (forward strand) and below (reverse strand) the horizontal line. No orthology relation could be concluded for most of these cases. Used with permission from *Nature* **428**, 493–521 (2004).

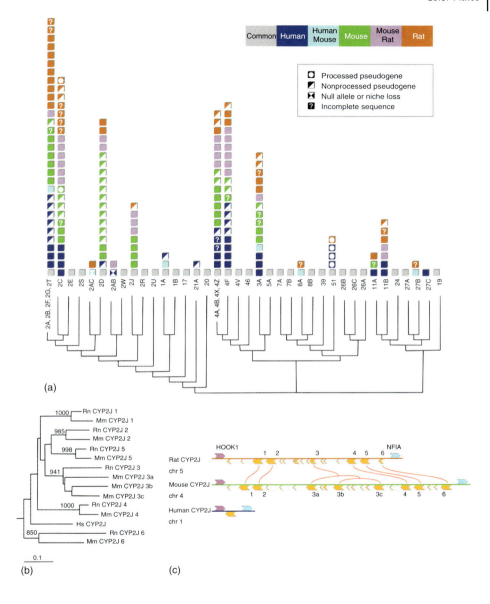

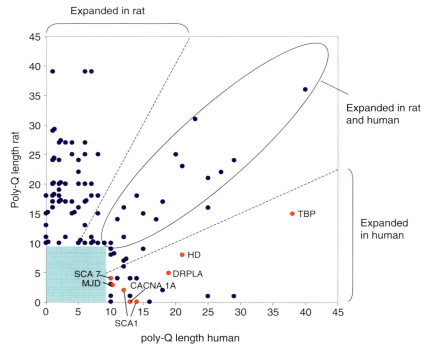

Fig. 16 (p. 473) Polyglutamine repeat length comparison between human and rat. Points represent protein poly-Q length for rat and human. Red points correspond to repeats in genes associated with human disease: SCA1, spinocerebellar ataxia 1 protein, or ataxin1; SCA7, spinocerebellar ataxia 7 protein; MJD, Machado–Joseph disease protein; CACNA1A, spinocerebellar ataxia 6 protein, or calcium channel α-1A subunit isoform 1; DRPLA, dentatorubro–pallidoluysian atrophy protein; HD, Huntington's disease protein, or huntingtin; TBP, TATA binding protein or spinocerebellar ataxia 17 protein. Repeat lengths over 10 were examined; green shading delineates the range not included in the analysis. Also noted are a set that are expanded in rat and human (black circle) and a set where repeats are expanded in the rat. Used with permission from *Nature* **428**, 493–521 (2004).

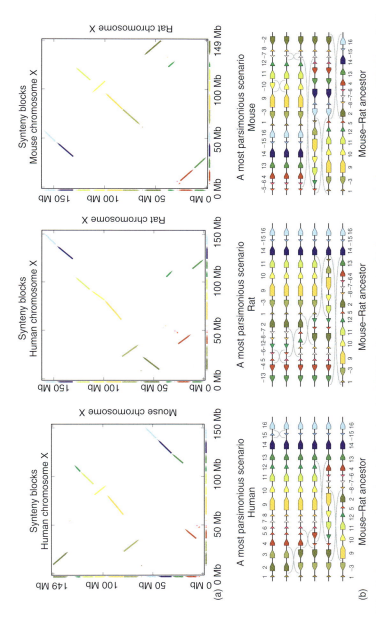

Fig. 6 (p. 446) X chromosome in each pair of species: (a) GRIMM–Synteny computes 16 three-way orthologous segments (≥300 kb) on the X chromosome of human, mouse, and rat, shown for each pair of species, using consistent colors. (b) The arrangement (order and orientation) of the 16 blocks implies that at least 15 rearrangement events occurred during X chromosome evolution of these species. The program MGR determined that evolutionary scenarios with 15 events are achievable and all have the same median ancestor (located at the last common mouse–rat ancestor). Shown is a possible (not unique) most parsimonious inversion scenario from each species to that ancestor. Note the last common ancestor of human, mouse, and rat should be on the evolutionary path between this median ancestor and human. Used with permission from *Nature* **428**, 493–521 (2004).

Encyclopedia of Molecular Cell Biology and Molecular Medicine

Second Edition

Proteasomes

Martin Rechsteiner
Department of Biochemistry University of Utah, Salt Lake, UT, USA

1	**Proteasomes – A Quick Summary** 5	
2	**The 20S Proteasome** 5	
2.1	Structure 5	
2.2	Enzyme Mechanism and Proteasome Inhibitors 6	
2.3	Immunoproteasomes 6	
3	**The 26S Proteasome** 7	
3.1	The Ubiquitin–proteasome System 7	
3.2	Ultrastructure of the 26S Proteasome and Regulatory Complex 8	
3.3	The 19S Regulatory Complex (RC) 9	
3.4	ATPases of the RC 9	
3.5	The non-ATPase Subunits 11	
3.6	Biochemical Properties of the Regulatory Complex 12	
3.6.1	Nucleotide Hydrolysis 12	
3.6.2	Chaperone-like Activity 12	
3.6.3	Proteasome Activation 12	
3.6.4	Ubiquitin Isopeptide Hydrolysis 12	
3.6.5	Substrate Recognition 13	
4	**Proteasome Biogenesis** 13	
4.1	Subunit Synthesis 13	
4.2	Biogenesis of the 20S Proteasome 13	
4.3	Biogenesis of the RC 13	
4.4	Posttranslational Modification of Proteasome Subunits 13	
4.5	Biogenesis of the 26S Proteasome 14	

5	**Substrate Recognition by Proteasomes**	**14**
5.1	Degradation Signals (Degrons)	14
5.2	Ubiquitin-dependent Recognition of Substrates	14
5.3	Substrate Selection Independent of Ubiquitin	15
6	**Proteolysis by the 26S Proteasome – Mechanism**	**16**
6.1	Contribution of Chaperones to Proteasome-mediated Degradation	16
6.2	Presumed Mechanism	16
6.3	Processing by the 26S Proteasome	18
7	**Proteasome Activators**	**18**
7.1	REGs or PA28s	18
7.2	PA200	19
7.3	Hybrid Proteasomes	19
7.4	ECM29	19
8	**Protein Inhibitors of the Proteasome**	**19**
9	**Physiological Aspects of Proteasomes**	**20**
9.1	Tissue and Subcellular Distribution of Proteasomes	20
9.2	Physiological Importance	20
9.3	The Ubiquitin–proteasome System and Human Disease	20
	Bibliography	21
	Books and Reviews	21
	Primary Literature	22

Keywords

AAA ATPases
A large subfamily of ATPases involved in separating and unfolding proteins. The six ATPases of the 19S regulatory complex are members of the AAA subfamily.

Aggresome
"Abnormal proteins" – pericentrosomal accumulations of abnormal proteins that recruit chaperones and proteasomes.

Base
A nine-protein subcomponent of the 19S regulatory complex, containing six ATPases and three proteins that recognize ubiquitin or ubiquitin-like proteins.

Chaperone
Chaperones are proteins that assist in the folding of nascent proteins or the refolding of denatured proteins. The heat shock proteins such as hsap90 or hsp70 are prime examples.

COP9 Signalosome (Csn)
A protein complex containing a metalloisopeptidase that removes the ubiquitin-like protein NEDD8 from ubiquitin ligases. Each of the eight Csn subunits is evolutionarily related to the eight lid subunits in the 19S regulatory complex.

Degron
Short peptide motifs (or even single N-terminal amino acids) that confer rapid proteolysis on the polypeptides bearing them.

Ecm29
A proteasome-associated protein proposed to function either as a clamp holding the 19S RC to the 20S proteasome or as an adaptor that localizes the 26S proteasome to specific membrane compartments within cells.

Epoxomicin
A bacterial compound that is a specific inhibitor of the proteasome.

Hybrid Proteasome
A 20S proteasome with a 19S regulatory complex at one end and either PA28 or PA200 at the other.

Immunoproteasome
A 20S proteasome where each of the three constitutive catalytic subunits is replaced by an active subunit with altered substrate specificity. Immunoproteasomes are induced by immune cytokines and play a role in specifying epitopes presented by the Class-I immune pathway.

Lactacystin
A fungal metabolite that is a reasonably specific inhibitor of the proteasome's active sites.

Lid
An eight-protein subcomponent of the 19S regulatory complex, containing a metalloisopeptidase that removes ubiquitin chains from the substrate protein.

Metabolic Regulation
Metabolic regulation refers to the control of biological processes by changes in the concentration of metabolites or proteins or by changes in the activities of enzymes. Phosphorylation and ubiquitilation are widespread mechanisms for controlling enzyme activity.

PA28s (aka 11S REGs)
Donut shaped, heptameric protein complexes that bind the ends of the 20S proteasome and promote peptide entry to or efflux from its central proteolytic chamber.

PA200
A large nuclear protein that binds the ends of the 20S proteasome and is thought to play a role in DNA repair.

Proteasome Activators
Single polypeptide chains or small protein complexes that bind 20S proteasomes and stimulate peptide hydrolysis. To date, three have been discovered.

20S Proteasome
A cylindrical proteolytic particle composed of 28 subunits arranged as a stack of four heptameric rings. The enzyme's active sites face an internal chamber.

26S Proteasome
The only ATP-dependent protease discovered so far in the nuclear and cytosolic compartments of eukaryotic cells. The enzyme consists of one or two 19S regulatory complexes attached to the ends of the 20S proteasome.

19S Regulatory Complex (aka PA700)
A multisubunit particle containing six ATPases and eleven additional proteins that functions to bind, unfold, and translocate substrate proteins into the central proteolytic chamber of the 20S proteasome.

Ubiquitin
An exceptionally conserved 76-amino acid protein that is covalently attached to a wide variety of other eukaryotic proteins. Chains of ubiquitin attached to substrate proteins target them for destruction by the 26S proteasome.

VCP
A large hexameric ATPase that transfers some polyubiquitylated substrates to the 26S proteasome.

Velcade
A peptide boronic acid inhibitor of the proteasome used clinically to treat multiple myeloma.

The 20S proteasome was discovered in 1980 and the 26S proteasome six years later. Research over the past two decades has made it abundantly clear that the Ub-proteasome system is of central importance in eukaryotic cell physiology and medicine. At the cellular and biochemical levels, there are a number of unresolved problems. We need a crystal structure of the 19S RC, or better yet, the 26S proteasome, for they would surely provide an insight into the mechanism by which the 26S proteasome degrades its substrates. How the 26S proteasome itself is regulated and the extent to which proteasomal components vary among tissues in higher eukaryotes are other important unresolved problems. As for the medical perspective, we need to know how many diseases will be found to arise because of defects in the ubiquitin–proteasome system. Hopefully, these problems will generate research on the UPS by some readers of this article.

1
Proteasomes – A Quick Summary

Proteasomes are multisubunit, cylindrical proteases found in eukaryotes, eubacteria, and archaebacteria. The proteasome's active sites face a central chamber buried within the cylindrical particle. Thus, the proteasome is an ideal intracellular protease because cellular proteins can only be degraded if they are actively transferred to the enzyme's central chamber. Eukaryotic proteasomes come in two sizes, the 20S proteasome and the considerably larger ATP-dependent 26S proteasome. The latter is formed when the 20S proteasome binds one or two multisubunit ATPase-containing particles known as *19S regulatory complexes*. The 26S proteasome is responsible for degrading ubiquitylated proteins and is therefore essential for a vast array of cellular processes including cell-cycle traverse, control of transcription, regulation of enzyme levels, and apoptosis. Being the key protease of the ubiquitin system, the 26S proteasome also impacts a number of human diseases, especially cancer, cachexia, and neurodegenerative diseases. Both 20S and 26S proteasomes can associate with other protein complexes. As their name implies, proteasome activators stimulate peptide hydrolysis and may serve to localize 20S and 26S proteasomes within cells. Hybrid proteasomes consist of the 20S proteasome with a 19S regulatory complex bound at one end and a proteasome activator at the other. Immunoproteasomes are formed when the catalytic subunits found in constitutive proteasomes are replaced by interferon-inducible subunits with different substrate specificities. Immunoproteasomes are found mainly in immune tissues where they play a role in Class-I antigen presentation.

2
The 20S Proteasome

2.1
Structure

We know the molecular anatomy of archaebacterial, yeast, and bovine proteasomes in great detail since high-resolution crystal structures have been determined for all three enzymes. The archaebacterial proteasome is composed of two kinds of subunits, called α and β. Each subunit forms heptameric rings that assemble into the 20S proteasome by stacking four deep on top of one another to form a "hollow" cylinder. Catalytically inactive α-rings form the ends of the cylinder, while proteolytic β-subunits occupy the two central rings. The quaternary structure of the 20S proteasome can therefore be described as $\alpha7\beta7\beta7\alpha7$. The active sites of the β-subunits face a large central chamber about the size of serum albumin (see Fig. 1). The α-rings seal off the central proteolytic chamber and two smaller antechambers from the external solvent. Archaebacterial proteasomes, with

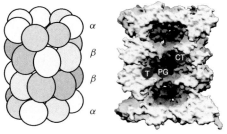

Fig. 1 20S Proteasomes. At the left is a schematic representation of a eukaryotic 20S proteasome. Different colors serve to emphasize the diversity of α and β subunits. At the right is a cutaway view of the yeast proteasome, showing the three internal cavities and the three protease activities: T (trypsin-like), PG (post glutamyl cleaving) and CT (chymotrypsin-like).

their fourteen identical β-subunits, preferentially hydrolyze the peptide bonds following hydrophobic amino acids and are therefore said to have chymotrypsin-like activity. Eukaryotic proteasomes maintain the overall structure of the archaebacterial enzyme, but they exhibit a more complicated subunit composition. There are seven different α-subunits and at least seven distinct β-subunits arranged in a precise order within their respective rings (see Fig. 1). Although current evidence indicates that only three of its seven β-subunits are catalytically active, the eukaryotic proteasome cleaves a wider range of peptide bonds containing, as it does, two copies each of trypsin-like, chymotrypsin-like, and post-glutamyl-hydrolyzing subunits. For this reason, it is capable of cleaving almost any peptide bond, having difficulty only with proline-X, glycine-X, and, to a lesser extent, with glutamine-X bonds.

2.2
Enzyme Mechanism and Proteasome Inhibitors

Whereas standard proteases use serine, cysteine, aspartate, or metals to cleave peptide bonds, the proteasome employs an unusual catalytic mechanism. N-terminal threonine residues are generated by self-removal of short peptide extensions from the active β-subunits, and they act as nucleophiles during peptide-bond hydrolysis. Given its unusual catalytic mechanism, it is not surprising that there are highly specific inhibitors of the proteasome. The fungal metabolite lactacystin and the bacterial product epoxomicin covalently modify the active-site threonines and inhibit the enzyme. Both compounds are commercially available; other inhibitors include vinylsulfones and various peptide aldehydes that are generally less specific. Recently, Velcade, a peptide boronate inhibitor of the proteasome, has been approved for the treatment of multiple myeloma.

2.3
Immunoproteasomes

Interferonγ is an immune cytokine that increases expression of a number of cellular components involved in Class-I antigen presentation. Among the IFNγ inducible components are three catalytically active β-subunits of the proteasome, called LMP2, LMP7, and MECL1. Each replaces its corresponding constitutive subunit, resulting in altered peptide-bond cleavage preferences of 20S immunoproteasomes. For example, immunoproteasomes exhibit much reduced cleavage after acidic residues and enhanced hydrolysis of peptide bonds following branch-chain amino acids such as isoleucine or valine. Class-I molecules preferentially bind peptides with hydrophobic or positive C-termini, and proteasomes generate the vast majority of Class-I peptides. Hence, the observed β-subunit exchanges are well suited for producing peptides able to bind Class-I molecules.

3
The 26S Proteasome

3.1
The Ubiquitin–proteasome System

Bacteria can express as many as five ATP-dependent proteases. By contrast, the 26S proteasome is the only ATP-dependent protease discovered so far in the nuclear and cytosolic compartments of eukaryotic cells. Because the 20S proteasome's internal cavities are inaccessible to intact proteins, openings must be generated in the enzyme's outer surface for proteolysis to occur. A number of protein complexes have been found to bind the proteasome and stimulate peptide hydrolysis. (see Fig. 2). The most important of the proteasome-associated components is the 19S regulatory complex (RC), for it is a major part of the 26S ATP-dependent enzyme that degrades ubiquitin-tagged proteins in eukaryotic cells. Ubiquitin (Ub) is treated in a separate chapter of this Encyclopedia. Still, this important protein must be briefly covered since it plays a central role in substrate recognition by the 26S proteasome.

Ubiquitin is a small, evolutionarily conserved eukaryotic protein that can be attached to a wide variety of intracellular proteins, including itself. Although Ub serves nonproteolytic roles, such as histone modification or viral budding, the protein's major function is targeting proteins for destruction. To do so,

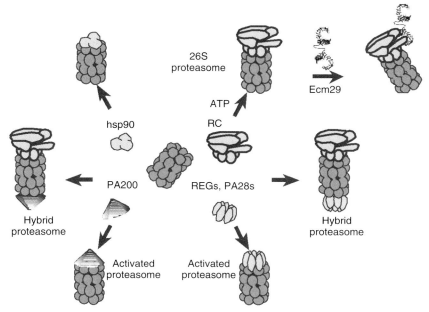

Fig. 2 Schematic representation of the 20S proteasome assembling with various activator proteins (RC, REGs, PA200) or with the chaperone hsp90, a protein that inhibits the enzyme.

the carboxyl terminus of ubiquitin is activated by an ATP-consuming enzyme (E1) and is transferred to one of several small carrier proteins (E2s) in the form of a reactive thiolester. The carboxyl terminus of an activated Ub then forms an isopeptide bond with lysine amino groups on proteolytic substrates (S) that have been selected by members of several large families of Ub ligases or E3s. Chains of Ub are formed, and the Ub-conjugated substrate is recognized by the 26S proteasome and degraded; Ub is recycled for use in further rounds of proteolysis (see Fig 3). It is important to note that ubiquitin contains seven lysine residues and polyUb chains formed via Lys6, Lys27, Lys29, Lys48, and Lys63 exist in nature. Lys29 and Lys48 chains form directly on proteolytic substrates and target them for destruction. Lys27 chains have been found on the cochaperone BAG1, and they target degradation of misfolded proteins bound by the Hsp70 chaperone to the 26S proteasome. Ub monomers linked to each other through Lys63 are involved in endocytosis and DNA repair, but not, apparently, in targeting proteins to the 26S proteasome.

3.2
Ultrastructure of the 26S Proteasome and Regulatory Complex

Electron micrographs (EMs) of purified 26S proteasomes reveal a dumbbell-shaped particle, approximately 40 nm in

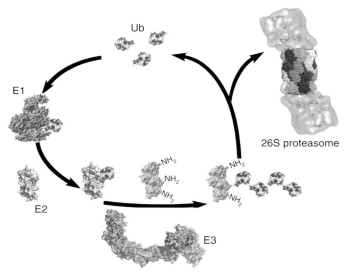

Fig. 3 Schematic representation of the ubiquitin-proteasome pathway. Ubiquitin molecules are activated by an E1 enzyme (shown in green at one-third scale) in an ATP-dependent reaction, transferred to a cysteine residue (yellow) on an E2 or Ub carrier protein and subsequently attached to amino groups (NH_2) on a substrate protein (lysozyme shown in purple) by an E3 or ubiquitin ligase, (the multicolored SCF complex). Note that chains of Ub are generated on the substrate, and these are recognized by the 26S proteasome depicted in the upper right at one-twentieth scale. (See color plate p. xxiv).

length, in which the central 20S proteasome cylinder is capped at one or both ends by asymmetric regulatory complexes looking much like Chinese dragonheads (see Fig 4). In doubly capped 26S proteasomes, the regulatory complexes face in opposite directions, indicating that contact between the proteasome's α-rings and the RC is highly specific. However, the contacts may not be especially tight since image analysis of Drosophila 26S proteasomes suggests movement of the RCs relative to the 20S proteasome. EM images of the 26S proteasome appear the same in all organisms, indicating that the overall architecture of the enzyme has been conserved in evolution. This conclusion is also supported by sequence conservation among RC subunits (see Sect. 3.4). A yeast mutant lacking the RC subunit Rpn10 contains a salt-labile RC that dissociates into two subcomplexes called the *lid* and the *base* (see Fig 4). The base contains 9 RC subunits, which include six ATPases described below in Sect. 3.4, the two largest RC subunits S1 and S2, as well as S5a; the lid contains the remaining RC subunits. Thus, the RC is composed of two subcomplexes separated on one side by a cavity, that is, the dragon's mouth. Ultrastructural studies have also been performed on the lid and on a related protein complex called the *COP9 signalosome*. Both particles lack obvious symmetry. Some particles exhibit a negative-stain-filled central groove; other classes of particles exhibit seven or eight lobes in a disc-like arrangement. Since both particles are composed of eight subunits, the lobes may represent individual subunits.

3.3
The 19S Regulatory Complex (RC)

The regulatory complex is also called the 19S cap, PA700, and the μ-particle. As its most common name suggests, the 19S regulatory complex is roughly the same size as the 20S proteasome. In fact, it is a more complicated protein assembly containing 17 or 18 different subunits, ranging in size from 25 to about 110 kDa. In animal cells, the subunits are designated S1 through S15. Homologs for each of these subunits are present in budding yeast where an alternate nomenclature has been adopted (see Table 1). Sequences for the 18 RC subunits permit their classification into a group of 6 ATPases and another group containing the 12 nonATPases.

3.4
ATPases of the RC

The six ATPases belong to the rather large family of AAA ATPases (for ATPases

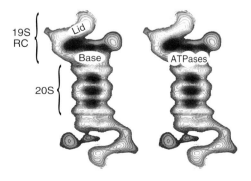

Fig. 4 Electron microscopic reconstructions of the 26S proteasome. Two images of negatively stained, doubly capped 26S proteasomes are presented to illustrate the positions of the 19S RC lid and base subcomplexes and to identify the most probable location of the RC ATPases.

Tab. 1 Subunits of the 19S regulatory complex.

Mammalian nomenclature	Yeast nomenclature	Function	Motifs
S1	Rpn2	Ub/Ubl binding	Leucine-rich repeats, KEKE
S2	Rpn1	Ub/Ubl binding	Leucine-rich repeats, KEKE
S3	Rpn3	?	PCI
p55	Rpn5	?	PCI
S4	Rpt2	ATPase	AAA nucleotidase
S5a	Rpn10	polyUb binding	UIM, KEKE
S5b	none	?	
S6	Rpt3	ATPase	AAA nucleotidase
S6'	Rpt5	ATPase	AAA nucleotidase
S7	Rpt1	ATPase	AAA nucleotidase
S8	Rpt6	ATPase	AAA nucleotidase
S9	Rpn6	?	PCI
S10a	Rpn7	?	PCI, KEKE
S10b	Rpt4	ATPase	AAA nucleotidase
S11	Rpn9	?	PCI
S12	Rpn8	?	MPN, KEKE
S13	Rpn11	Isopeptidase	MPN
S14	Rpn12	?	PCI

Associated with a variety of cellular Activities) whose members include the motor protein dynein, the membrane fusion factor NSF, and the chaperone VCP/Cdc48. The six ATPases, denoted S4, S6, S6', S7, S8, and S10b in mammals, are about 400 amino acids in length and homologous to one another. On the basis of their sequences, one can distinguish three major regions: (1) a central nucleotide binding domain of about 200 amino acids, which is roughly 60% identical among members of the RC subfamily; (2) the C-terminal region, approximately 100 amino acids in length and with a lesser, though significant, degree of conservation ($\sim$40%); and (3) a highly divergent N-terminal region (<20% identity) around 120 amino acids in length; this region contains heptad repeats characteristic of coiled-coil proteins. Despite sequence differences among RC ATPases within an organism, each ATPase has been conserved during evolution with specific subunits being almost 75% identical between yeast and humans. The high degree of conservation encompasses the entire sequence, making it likely that even the divergent N-terminal regions play an important role in RC function. Conceivably, they are used to select substrates for degradation by the 26S proteasome. The "helix-shuffle" hypothesis proposes that the N-terminal coiled-coils of S4 subfamily ATPases bind unassembled substrate proteins such as fos or jun through the latter's unpaired leucine zippers. Alternatively, the variable N-terminal regions in the RC ATPases may be involved in the assembly of the RC by promoting the specific placement of the ATPase subunits within the complex. In this regard, the six ATPases associate with one another in highly specific pairs: S4 binds S7, S6 binds S8, and S6' binds to S10b. Moreover, the N-terminal regions of RC ATPases are required for partner-specific binding. Staining patterns of two-dimensional gels show the six RC ATPases to be present at comparable levels and the affinity capture of yeast 26S proteasomes indicate the presence one copy of each in the regulatory complex. Mutational analysis in yeast demonstrates that the ATPases are not functionally redundant since mutation of yeast S4 has a particularly profound effect on peptide hydrolysis. It is probable that the ATPases form a hexameric ring like other members of the AAA family of ATPases such as NSF or VCP/Cdc48. But this assumption has not been experimentally verified. Finally, it is quite likely that the ATPases directly bind the α-ring of the 20S proteasome. Evidence favoring this arrangement comes from chemical cross-linking experiments and the presence of the ATPases in the base subcomplex of the yeast 26S proteasome.

3.5
The non-ATPase Subunits

Whereas the six RC ATPases are homologous and relatively uniform in size, the nonATPases are heterogeneous in size and sequence. Nonetheless, they can be grouped on the basis of their location, for example, lid versus base, on the presence or absence of certain sequence motifs, for example, PCI and MPN domains, and on their affinity for Ub chains or Ub-like proteins. Eight RC subunits are found in the lid subcomplex. One of these subunits (S13) is a metalloisopeptidase that removes Ub chains from the tagged substrate prior to its translocation into the proteasome for degradation. Each of the eight lid subunits is homologous to a subunit in a separate protein complex called the COP9 signalosome. Six of the eight lid subunits contain PCI domains, stretches of about 200 residues so named from their occurrence in Proteasome, COP9 signalosome, and the eukaryotic Initiation factor 3 subunits. The PCI domains are thought to mediate subunit–subunit interactions. Two lid subunits contain 140 amino acid long MPN domains (Mpr1p and Pad1p N-terminal regions), with one of these subunits being the S13 isopeptidase. Although several models have been proposed from 2-hybrid screens of lid and COP9-signalosome subunits, the arrangement of the eight subunits within the lid subcomplex is not known with certainty. S13 stands out because it exhibits isopeptidase activity. Functions of the remaining seven lid subunits have not been discovered, although two lid subunits, S3 and S9, are critical for the degradation of specific substrates. The presence of the S13 isopeptidase in the lid explains why the lid is necessary for degradation of ubiquitylated proteins, even though the RC–base complex supports the ATP-dependent degradation of some small nonubiquitylated proteins. Interestingly, the COP9-signalosome also exhibits isopeptidase activity that removes the ubiquitin-like protein, NEDD8, from certain ubiquitin ligases.

In addition to the six ATPases, the base subcomplex contains the two largest RC subunits (S1, S2) and a smaller subunit called S5a. Besides their common location, these three subunits share the property of binding polyUb chains or Ub-like domains. S5a binds polyubiquitin chains even after it has been transferred from SDS-PAGE gels, and it displays many features that match polyubiquitin recognition by the 26S proteasome. However, S5a cannot be the only ubiquitin recognition component in the 26S proteasome because deletion of the gene-encoding yeast S5a has only a modest impact on proteolysis. This strongly suggests that there are other ubiquitin recognition components in the RC, with S1 and S2 being prime candidates. S1 and S2 display significant homology to each other, and both can be modeled as α-helical toroids. They have been shown to bind the ubiquitin-like domains of RAD23 and Dsk2, adapter proteins that target ubiquitylated proteins to the 26S proteasome. It has also been found that the S6′ ATPase can be cross-linked to Ub. Currently, then it appears that the RC contains three, or possibly four, subunits able to recognize Ub or Ub-like proteins. As discussed below, there are other ways in which the RC can select proteins for destruction.

3.6 Biochemical Properties of the Regulatory Complex

3.6.1 Nucleotide Hydrolysis

Both the 26S proteasome and the regulatory complex hydrolyze all four nucleotide triphosphates, with ATP and CTP preferred over GTP and UTP. K_ms for hydrolysis by the 26S proteasome are two- to fivefold lower for each nucleotide and virtually identical to the Kms required for Ub-conjugate degradation. Although ATP hydrolysis is required for conjugate degradation, the two processes are not strictly coupled. Complete inhibition of the peptidase activity of the 26S proteasome by calpain inhibitor I has little effect on the ATPase activity of the enzyme. The nucleotidase activities of the RC and the 26S proteasome closely resemble those of *Escherichia coli* Lon protease, which is composed of identical subunits that possess both proteolytic and nucleotidase activities in the same polypeptide chain. Like the regulatory complex and 26S proteasome, Lon hydrolyzes all four ribonucleotide triphosphates, but not ADP or AMP.

3.6.2 Chaperone-like Activity

AAA nucleotidases share the common property of altering the conformation or association state of proteins. So, it is not surprising that the RC has been shown to prevent aggregation of several denatured proteins including citrate synthase and ribonuclease A. The chaperone activity of the RC may explain why the RC plays a role in DNA repair even in the absence of an attached 20S proteasome.

3.6.3 Proteasome Activation

The 20S proteasome is a latent protease because of the barrier imposed by the α-subunit rings on peptide entry. Consequently, a readily measured activity of the RC is activation of fluorogenic peptide hydrolysis by the 20S proteasome. The extent of activation is generally found to be in the range 3- to 20-fold. Activation is relatively uniform for all three proteasome catalytic subunits and presumably reflects opening by the attached RC of a channel, leading to the proteasome's central chamber.

3.6.4 Ubiquitin Isopeptide Hydrolysis

The channel through the proteasome α-ring into the central chamber measures 1.3 nm in diameter, a size too small to permit passage of a folded protein even as small as ubiquitin. This consideration coupled with the fact that Ub is recycled intact upon substrate degradation requires an enzyme to remove the polyUb chain prior to or concomitant with proteolysis. Several isopeptidases that remove Ub from substrates has been found associated with the 26S proteasome. Of these, S13 is an integral component of the enzyme. S13 is a metalloisopeptidase stimulated by nucleotides and is active only in the fully assembled 26S proteasome. Thus, it is not strictly correct to list isopeptidase activity as a property of the RC.

3.6.5 Substrate Recognition

It is clear that the RC plays a predominant role in selecting proteins for degradation. This important topic is covered in Sect. 5 in the context of substrate recognition by both 20S and 26S proteasomes.

4 Proteasome Biogenesis

4.1 Subunit Synthesis

The synthesis of proteasome subunits is markedly affected by proteasome function. For example, inhibition of proteasome activity by lactacystin induces coordinate expression of both RC and 20S proteasome subunits. Similarly, impaired synthesis of a given RC subunit results in overexpression of all RC subunits. Proteasome subunit synthesis in yeast is controlled by Rpn4p, a short-lived positive transcription factor that binds PACE elements upstream of proteasome genes. Rpn4p is a substrate of the 26S proteasome, suggesting that the transcription factor functions in a feedback loop in which proteasome activity limits its concentration, thereby regulating proteasome levels. To date, an Rpn4-like factor has not been identified in higher eukaryotes; however, the presence of such a factor could explain why proteasome inhibition results in higher expression of proteasome subunits in mammalian and *Drosophila* cells.

4.2 Biogenesis of the 20S Proteasome

Proteasome β-subunits are synthesized with N-terminal extensions, and they are inactive because a free N-terminal threonine is required for peptide-bond hydrolysis. The precursor β-subunits assemble with α-subunits to form half proteasomes composed of one α-and one β-ring. These two ring intermediates dimerize to form the 20S particle, and the N-terminal extensions are removed, thereby generating a new unblocked N-terminal threonine in the catalytically active β-subunits. A small accessory protein called Ump1 in yeast and POMP or proteassemblin in mammalian cells assists in the final assembly of the 20S proteasome. Interestingly, Ump1/POMP is apparently trapped in the proteasome's central chamber and degraded upon maturation of the enzyme.

4.3 Biogenesis of the RC

Assembly pathways for the RC are virtually unknown. As mentioned above, the ATPases interact with one another and complexes containing all six S4 subfamily members have been observed following *in vitro* synthesis. Impaired synthesis of a lid subunit can result in the absence of the entire lid, so presumably, lid and base subcomplexes assemble independently and associate in the final stages of RC formation cells. *In vivo*, 26S proteasomes assemble from preformed regulatory complexes and 20S proteasomes.

4.4 Posttranslational Modification of Proteasome Subunits

Proteasome and RC subunits are subjected to a variety of posttranslational modifications including phosphorylation, acetylation, and even myristoylation in the case of the RC ATPase S4. In yeast, all seven α-subunits as well as two β-subunits are acetylated. Since acetylation of the N-terminal threonine in an active β-subunit would poison catalysis, it has been suggested that the propeptide extensions function to prevent acetylation. Three members of the S4 ATPase subfamily (S4, S6, and S10b) and two 20S α-subunits (C8 and C9) are known to be phosphorylated. Phosphorylation appears to be particularly important for 26S proteasome assembly since the kinase inhibitor staurosporine reduces 26S proteasome levels in mouse lymphoma cells.

4.5
Biogenesis of the 26S Proteasome

The RC and 20S proteasome associate to form the 26S proteasome in the presence of ATP. Comparison of the cross-linking patterns of RC and assembled 26S proteasomes indicates that this association is accompanied by subunit rearrangement. In yeast, two proteins play a special role in 26S proteasome assembly or stability. Nob1p is a nuclear protein required for biogenesis of the 26S proteasome. It is degraded following assembly of the 26S enzyme, suffering the same fate as Ump1 does following 20S maturation. The molecular chaperone Hsp90 also plays a role in the assembly and maintenance of yeast 26S proteasomes since functional loss of Hsp90 results in 26S proteasome dissociation.

5
Substrate Recognition by Proteasomes

5.1
Degradation Signals (Degrons)

One of the major insights of twentieth century cell biology was the recognition that proteins possess built-in signals, targeting them to specific locations within cells. Selective proteolysis can be considered to be targeting out of existence, and a number of short peptide motifs have been discovered to confer rapid proteolysis on their bearers. These include PEST sequences, destruction boxes, KEN boxes, and even the N-terminal amino acid. These motifs are recognized by one or more ubiquitin ligases that mark the substrate protein by addition of a polyUb chain. However, some proteins are degraded by the 26S proteasome without prior marking by Ub. Denatured proteins are also selectively degraded by both 26S and 20S proteasomes. It is not clear what features of denatured proteins are recognized by proteasomes or by components of the Ub proteolytic system.

5.2
Ubiquitin-dependent Recognition of Substrates

Most well-characterized substrates of the 26S proteasome are ubiquitylated proteins, so our discussion starts with them. Efficient proteolysis of ubiquitylated proteins by the 26S proteasome requires a chain containing at least four Ub monomers. This matches well with the Ub-binding characteristics of the RC subunit S5a. It also selectively binds ubiquitin polymers composed of four or more ubiquitin moieties and exhibits increased affinity for longer chains. In addition, binding to S5a is impaired by mutations in ubiquitin that allow chain formation but reduce the targeting competence of the chains. These properties and the fact that denatured S5a can readily regain affinity for polyUb suggest that multiple, short sequences within S5a form "loops" that are able to bind "grooves" in the polyubiquitin chain. S5a molecules from a number of higher eukaryotes contain two repeated motifs that are independent polyubiquitin binding sites. Each motif is approximately 30-residues long and is characterized by five hydrophobic residues that consist of alternating large and small hydrophobic residues, for example, leu-ala-leu-ala-leu. Similar motifs have been found in other proteins of the ubiquitin system and are now called UIMs, for Ubiquitin Interacting Motifs. Because a hydrophobic patch on the surface of ubiquitin is critical for substrate targeting, current models envision direct hydrophobic interaction between the UIMs in S5a and the hydrophobic patches on ubiquitin molecules in polyUb chains.

Whereas S5a provides for direct recognition of polyubiquitylated substrates, a second mechanism involves adaptor proteins possessing both a UbL (a Ubiquitin-like domain) and one or more UbA domains (Ubiquitin-associated domains are polyUb binding domains found in several proteins of the Ub system). The adaptor proteins include RAD23 and Dsk2 in yeast and recruit substrates to the 26S proteasome through UbL binding to 26S proteasome subunits, while the UbA domain binds substrate-tethered Ub chains. In yeast, the RC subunits S1 and S2 serve as UbL binding components; in mammals, S5a also serves this purpose.

The cochaperone BAG1 illustrates a third way in which polyUb can target substrate proteins to the 26S proteasome. In this case, the substrate is not polyubiquitylated; rather, it is bound to the chaperone Hsp70. A polyUb chain, linked through Lys27, is attached to the Hsp70-associated cochaperone BAG1. Apparently, the Lys27 chain promotes association of the chaperone–substrate complex with the 26S proteasome, after which the substrate is degraded while BAG1, Hsp70, and ubiquitin are recycled. Direct interaction between E3 ubiquitin ligases and RC subunits can also deliver ubiquitylated substrates to the protease. The yeast E3 UFD4 binds RC ATPases, and UFD4-mediated delivery of substrates bypasses the requirement for S5a. In what appears to be a similar delivery system, the mammalian E3 Parkin uses a UbL to bind the 26S proteasome. Mutational analyses in yeast have shown that whereas deletion of either S5a or Rad23 has mild impact on proteolysis, loss of both proteins produces a severe phenotype. Yeast lacking S5a, RAD23, and Dsk2 are not viable, indicating that direct delivery by E3s cannot compensate for the absence of all three proteins.

5.3
Substrate Selection Independent of Ubiquitin

The 26S proteasome also degrades nonubiquitylated proteins. The short-lived enzymeornithine decarboxylase (ODC) and the cell-cycle regulator p21Cip provide well-documented examples of Ub-independent proteolysis by the 26S enzyme. ODC degradation is stimulated by a protein called *antizyme* that binds to both ODC and the 26S proteasome. Antizyme functions as an adapter much like RAD23 and Dsk2 except that polyUb chains are not involved, although interestingly, free Ub chains compete with antizyme–ODC for degradation. p21Cip is also degraded in a nonUb-dependent reaction. This was clearly demonstrated by substitution of arginines for all the lysine residues in p21Cip, thereby preventing ubiquitylation of p21Cip. The lysine-less protein was still degraded in human fibroblasts by the proteasome. The C-terminal region of p21Cip binds to the proteasome α-subunit C8, and in vitro p21Cip is degraded by the 20S proteasome alone. Direct binding of p21Cip to the 20S proteasome may open a channel through the α-ring, allowing the loosely structured protein to enter the central chamber. c-Jun, "aged" calmodulin, troponin C, and p53 are other proteins that can be degraded by the 26S proteasome absent marking by Ub. The destruction of p53 and p21Cip can proceed by both Ub-dependent and Ub-independent pathways. Other 20S proteasome substrates, *in vitro* at least, include oxidized proteins, small denatured proteins, and loosely folded proteins such as casein. Whether the 20S proteasome degrades proteins within cells is an unresolved problem.

6
Proteolysis by the 26S Proteasome – Mechanism

6.1
Contribution of Chaperones to Proteasome-mediated Degradation

Chaperones are connected to proteasomes in at least four ways. First, chaperones can deliver substrates to the proteasome as described above for the cochaperone BAG1. In a similar fashion, the chaperone VCP/Cdc48 is required for the degradation of several Ub pathway substrates. VCP is a member of the AAA family of ATPases. The large hexameric ATPase appears to function as a protein separase able to remove ubiquitylated monomers from multisubunit complexes. In some cases, the liberated proteins are degraded by the 26S proteasome; in other cases, the separated proteins may change their intracellular location. The proteasome also degrades endoplasmic-reticulum membrane proteins. If these ER membrane proteins possess a large cytoplasmic domain, their proteasomal degradation can require Hsps 40, 70, and 90 as well as VCP. Hsp90 is required to assemble and stabilize the yeast 26S proteasome, providing a third connection between chaperones and proteasomes. Hsp90 is also able to bind and suppress peptide hydrolysis by the 20S proteasome. Finally, both chaperones and proteasomes are induced by the accumulation of denatured proteins within eukaryotic cells.

6.2
Presumed Mechanism

Proteolysis of ubiquitylated proteins by the 26S proteasome can be thought to be consisting of seven steps: (1) chaperone-mediated substrate presentation; (2) substrate association with RC subunits; (3) substrate unfolding; (4) detachment of polyUb from the substrate; (5) translocation of the substrate into the 20S proteasome central chamber, (6) peptide-bond cleavage; and (7) release of peptide products as well as polyUb (see Fig 5). Step 1 is optional, depending on the substrate, and in principle, steps 3 and 4 could occur in either order. Step 4 is unnecessary for substrates like ODC that are not ubiquitylated. The other steps almost have to occur as presented. Although it is easy to conceptualize the reaction sequence, few experimental findings bear directly on any of the proposed subreactions, and virtually nothing is known about molecular movements within the 26S proteasome. It has been shown that steps 6 and 7 are not required for sequestration of ODC by the 26S proteasome. But which RC subunits actually recognize ODC–antizyme complexes has not been discovered. For ubiquitylated substrates, it is very likely

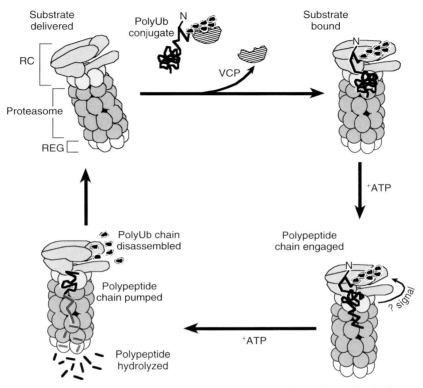

Fig. 5 Hypothetical reaction cycle for the 26S proteasome. A polyubiquitylated substrate is delivered to the 26S proteasome possibly by chaperones such as VCP (step 1). Substrate is bound by polyubiquitin-recognition components of the RC until the polypeptide chain is engaged by the ATPases (step 2). As the polypeptide chain is unfolded and pumped down the central pore of the proteasome, a signal is conveyed to the metalloisopeptidase to remove the polyUb chain (step 3). The unfolded polypeptide is eventually degraded within the inner chamber of the proteasome (step 4) and peptide fragments exit the enzyme.

that S1, S2, and S5a are recognition components, but this is not certain. Even if they are, we do not know whether they recognize just polyUb or polyUb and portions of the substrate. Translocation is thought to proceed by the six ATPases threading the polypeptide through a channel in the 20S proteasome's α-ring, but the possibility that convulsive movements transfer substrate has not been ruled out. It is also thought that the RC ATPases processively unravel substrates from degradation signals within the polypeptide chain and are able to "pump" the polypeptide chain in either N-terminal to C-terminal or the opposite direction. The ATPases may even be capable of transferring loops into the 20S enzyme. One thing that is certain is that the peptide fragments generated in the central chamber are generally 5 to 10 residues in length, but fragments as long as 35 amino acids can be present. How these fragments exit the central chamber is not known. Clearly, there is much to be discovered about the internal workings of the 26S proteasome.

6.3 Processing by the 26S Proteasome

In some cases, the 26S proteasome partially degrades the substrate protein releasing processed functional domains. The best-studied example of processing involves the transcriptional activator NFκB. The C-terminal half of a 105-kDa precursor is degraded by the 26S proteasome to yield a 50-kDa N-terminal domain that is the active transcription component. A glycine-rich stretch of amino acids at the C-terminal boundary of p50 is an important factor in limiting proteolysis. It is possible that polypeptide translocation by the RC starts at the glycine-rich region and proceeds in only one direction because of the presence of the tightly folded N-terminal domain. Or the RC may start translocation at the C-terminus and stop when the ATPases encounter the glycine-rich region. Another example of partial processing involves SPT23, a yeast protein embedded in the endoplasmic-reticulum membrane. SPT23 controls unsaturated fatty acid levels and membrane fluidity regulates 26S proteasomal generation of a freely diffusible transcription factor from the SPT23 precursor. Partial processing may be a more widespread regulatory mechanism than is currently thought.

7 Proteasome Activators

In addition to the RC, there are two protein complexes, REG$\alpha\beta$ and REGγ, and a single polypeptide chain, PA200, that bind the 20S proteasome and stimulate peptide, but not protein, degradation (see Fig. 2). Like the RC, proteasome activators bind the ends of the 20S proteasome, and importantly, they can form mixed or hybrid 26S proteasomes in which one end of the 20S proteasome is associated with a 19S RC and the other is bound to a proteasome activator. This latter property raises the possibility that proteasome activators serve to localize the 26S proteasome within eukaryotic cells.

7.1 REGs or PA28s

There are three distinct REG subunits called $\alpha\beta\gamma$. REG$\alpha\beta$ form donut-shaped hetero-heptamers found principally in the cytoplasm, whereas, REGγ forms a homo-heptamer located in the nucleus. REG$\alpha\beta$ are abundantly expressed in immune tissues while REGγ is highest in brain. The REGs also differ in their activation properties. REG$\alpha\beta$ activate all three proteasome active sites; REGγ only activates the trypsin-like subunit. There is reasonably solid evidence that REG$\alpha\beta$ play a role in Class-I antigen presentation, but we have no idea what REGγ does, especially since REGγ knockout mice have almost no phenotype. The crystal structure of REGα reveals that the seven subunits form a donut-shaped structure with a central aqueous channel, and the structure of a REG–proteasome complex provides important insight into the mechanism by which REGα activates the proteasome. The carboxyl tail on each REG subunit fits into a corresponding cavity on the α-ring of the proteasome and loops on the REG subunits cause N-terminal strands on several proteasome α-subunits to reorient upward into the aqueous channel of the REG heptamer. These movements open a continuous channel from the exterior solvent to the proteasome central chamber.

7.2
PA200

A new proteasome activator, called PA200, was recently purified from bovine testis. Human PA200 is a nuclear protein of 1843 amino acids that activates all three catalytic subunits, with preference for the PGPH active site. Homologs of PA200 are present in budding yeast, worms, and plants. A single chain of PA200 can bind each end of the proteasome, and when bound, PA200 molecules look like volcanoes in negatively stained EM images. Evidence from both yeast and mammals suggests that PA200 is involved in DNA repair.

7.3
Hybrid Proteasomes

As the α-rings at each end of the 20S proteasome are equivalent, the 20S proteasome is capable of binding two RCs, two PA28s, two PA200s, or combinations of these components. In fact, 20S proteasomes simultaneously bound to RC and PA28 or PA200 have been observed, and they are called *hybrid proteasomes*. In HeLa cells, hybrid proteasomes containing PA28 at one end and an RC at the other have been estimated to be twice as abundant as 26S proteasomes capped at both ends by 19S RCs. Hybrid 26S proteasomes containing PA200 appear to be much less abundant in HeLa cells.

7.4
ECM29

Another proteasome-associated protein, called ecm29, has been identified in several recent proteomic screens. Ecm29p is reported to stabilize the yeast 26S proteasome by clamping the RC to the 20S cylinder. However, in human culture cells, ecm29p is found predominately associated with the endoplasmic-reticulum Golgi intermediate compartment, a location suggesting a role in secretion rather than stability of 26S proteasome. Ecm29p clearly associates with 26S proteasomes; whether it activates proteasomal peptide hydrolysis is unknown.

8
Protein Inhibitors of the Proteasome

A number of proteins have been found to suppress proteolysis by the proteasome. One of these is PI31, another is the abundant cytosolic chaperone, Hsp90, and a third is a proline/arginine-rich 39-residue peptide called PR39. Both PI31 and Hsp90 may affect how the proteasome functions in Class-I antigen presentation. PI31 is a 30-kDa proline-rich protein that inhibits peptide hydrolysis by the 20S proteasome and can block activation by both RC and REG$\alpha\beta$. Although surveys of various cell lines show PI31 to be considerably less abundant than RC or REG$\alpha\beta$, when overexpressed, PI31 is reported to inhibit Class-I antigen presentation by interfering with the assembly of immunoproteasomes. A number of studies have shown that Hsp90 can bind the 20S proteasome and inhibit its chymotrypsin-like and PGPH activities. Interestingly, inhibition by Hsp90 is observed with constitutive but not with immunoproteasomes, a finding consistent with proposals that Hsp90 shuttles immunoproteasome-generated peptides to the endoplasmic reticulum for Class-I presentation. PR39 was originally isolated from bone marrow as a factor able to induce angiogenesis and inhibit inflammation. Two hybrid screens showed that PR39 binds the 20S proteasome. Apparently, PR39 affects angiogenesis and inflammation by inhibiting respectively the degradation of HIF1 or IkBα, the latter being an inhibitor of NFkB. Finally, HIV's Tat protein inhibits the 20S proteasome's peptidase activity. Tat also competes with REGαβ for proteasome binding, and by doing so, Tat can inhibit Class-I presentation of certain epitopes.

9
Physiological Aspects of Proteasomes

9.1
Tissue and Subcellular Distribution of Proteasomes

Proteasomes are found in all organs of higher eukaryotes, but the degree to which the composition of proteasomes and its activators varies among tissues is largely unexplored territory. Proteasomes are very abundant in testis since the organ contains almost fivefold more 20S subunits than skeletal muscle. At the cellular level, there are about 800 000 proteasomes in a HeLa cell and roughly 20 000 proteasomes in a yeast cell. At the subcellular level, 26S proteasomes are present in cytosol and nucleus where they appear to be freely diffusible. They are not found in the nucleolus or within membrane-bound organelles other than the nucleus. When large amounts of misfolded proteins are synthesized by a cell, the aberrant polypeptides often accumulate around the centrosome in what are called "aggresomes." Under these conditions, 26S proteasomes, chaperones, and proteasome activators also redistribute to the aggresomes presumably to refold and/or degrade the misfolded polypeptides.

9.2
Physiological Importance

Deletion of yeast genes encoding 20S proteasome and 19S RC subunits is usually lethal, indicating that the 26S proteasome is required for eukaryotic cell viability. Known substrates of the 26S proteasome include transcription factors, cell-cycle regulators, protein kinases, and so on – essentially, most of the cell's important regulatory proteins. Surprisingly, even proteins secreted into the endoplasmic reticulum are returned to the cytosol for degradation by the 26S proteasome. Given the scope of its substrates, it is hardly surprising that the ubiquitin–proteasome system contributes to the regulation of a vast array of physiological processes, ranging in higher eukaryotes. Discussion of these fascinating regulatory mechanisms is beyond the scope of this article. A brief review of proteasomes and disease does, however, seem appropriate.

9.3
The Ubiquitin–proteasome System and Human Disease

As the central protease in the ubiquitin system, the 26S proteasome impacts a number of diseases, especially cancer, neurological diseases, and muscle wasting. Ubiquitin-mediated destruction of proteins driving cell division imparts directionality to the cell cycle. So it is not surprising that several cancers arise because of an impaired or overactive ubiquitin–proteasome system. VonHippel-Lindau (VHL) disease is a hereditary cancer syndrome, characterized by a wide range of malignant tumors. The protein defective in VHL is the substrate-recognition component of a Ub–ligase complex that targets the transcription-component hypoxia-inducible factor 1 (HIF1) for destruction. HIF1 induces synthesis of VEGF, a growth factor important for angiogenesis, and many of the tumors in VHL involve the vasculature. The human papilloma virus oncoprotein E6 provides an example of an overactive Ub ligase causing cancer. E6 is a positive regulator of the Ub ligase that targets the tumor-suppressor protein p53 for degradation. Downregulation of p53 prevents apoptosis of HPV-transformed cells, thereby promoting malignancy. Angelman and Liddle's syndromes and juvenile-onset Parkinson's disease are three other human diseases involving E3 Ub ligases.

Impairment of the proteasome occurs in a number of neurodegenerative diseases, especially those characterized by formation of protein inclusions, such as Parkinson's and Huntington's diseases. In these diseases, proteasomes and ubiquitylated proteins accumulate at the inclusions. It is thought that sequestration and/or inhibition of proteasomes may impair normal protein turnover, eventually resulting in neuronal cell death.

The ubiquitin–proteasome system also plays a key role in cellular immunity. There is considerable evidence that proteasomes generate the majority of peptides presented on Class-I molecules to cytotoxic lymphocytes. Class-I antigen presentation is a major defense against viral infection, and viruses have devised a number of ways to prevent Class-I molecules from reaching the cell surface. In several cases, viruses actually hijack the Ub-proteasome system to degrade newly synthesized MHC Class-I molecules, thereby preventing surface presentation of viral peptides. Given its pervasive role in metabolic regulation, there is no doubt that many more human diseases will be shown to result from malfunction of the Ub-proteasome system.

See also Electron Microscopy in Cell Biology; Electron Microscopy of Biomolecules; Membrane Traffic: Vesicle Budding and Fusion; Motor Proteins; Ubiquitin-Proteasome System for Controlling Cellular Protein Levels.

Bibliography

Books and Reviews

Attaix, D., Briand, Y. (2003) The proteasome in the regulation of cell function, *Int. J. Biochem.* **35**, 545–755.

Bashir, T., Pagano, M. (2003) Aberrant ubiquitin-mediated proteolysis of cell cycle regulatory proteins and oncogenesis, *Adv. Cancer Res.* **88**, 101–144.

Blobel, G. (2000) Protein targeting (Nobel lecture), *Chembiochem* **1**, 86–102.

Cope, G.A., Deshaies, R.J. (2003) COP9 signalosome: a multifunctional regulator of SCF and other cullin-based ubiquitin ligases, *Cell* **114**, 663–671.

Ferrell, K., Wilkinson, C.R., Dubiel, W., Gordon, C. (2000) Regulatory subunit interactions of the 26S proteasome, a complex problem, *Trends Biochem. Sci.* **25**, 83–88.

Goldberg, A.L., Cascio, P., Saric, T., Rock, K.L. (2002) The importance of the proteasome and subsequent proteolytic steps in the generation of antigenic peptides, *Mol. Immunol.* **39**, 147–164.

Hampton, R.Y. (2002) ER-associated degradation in protein quality control and cellular regulation, *Curr. Opin. Cell Biol.* **14**, 476–482.

Hartmann-Petersen, R., Seeger, M., Gordon, C. (2003) Transferring substrates to the 26S proteasome, *Trends Biochem. Sci.* **28**, 26–31.

Hegde, A.N., DiAntonio, A. (2002) Ubiquitin and the synapse, *Nat. Rev. Neurosci.* **3**, 854–861.

Hershko, A., Ciechanover, A., Varshavsky, A. (2000) Basic Medical Research Award. The ubiquitin system, *Nat. Med.* **6**, 1073–1081.

Hill, C.P., Masters, E.I., Whitby, F.G. (2002) The 11S regulators of 20S proteasome activity, *Curr. Top. Microbiol. Immunol.* **268**, 73–89.

Hilt, W., Wolf, D.H. (2004) The ubiquitin-proteasome system: past, present and future, *Cell Mol. Life Sci.* **61**, 1545–1632.

Kopito, R.R. (2000) Aggresomes, inclusion bodies and protein aggregation, *Trends Cell Biol.* **10**, 524–530.

Lipford, J.R., Deshaies, R.J. (2003) Diverse roles for ubiquitin-dependent proteolysis in transcriptional activation, *Nat. Cell Biol.* **5**, 845–850.

McCracken, A.A., Brodsky, J.L. (2003) Evolving questions and paradigm shifts in endoplasmic-reticulum-associated degradation (ERAD), *BioEssays* **25**, 868–877.

Muratani, M., Tansey, W.P. (2003) How the ubiquitin-proteasome system controls transcription, *Nat. Rev. Mol. Cell Biol.* **4**, 192–201.

Orlowski, M., Wilk, S. (2003) Ubiquitin-independent proteolytic functions of the proteasome, *Arch. Biochem. Biophys.* **415**, 1–5.

Peters, J.M., Harris, J.R. Finley, D. (1998) *Ubiquitin and the Biology of the Cell*, Plenum Press, New York, pp. 472.

Rechsteiner, M., Realini, C., Ustrell, V. (2000) The proteasome activator 11S REG (PA28) and class I antigen presentation, *Biochem. J.* **345**, 1–15.

Rechsteiner, M., Rogers, S.W. (1996) PEST sequences and regulation by proteolysis, *Trends Biochem. Sci.* **21**, 267–271.

Reed, S.I. (2003) Ratchets and clocks: the cell cycle, ubiquitylation and protein turnover, *Nat. Rev. Mol. Cell Biol.* **4**, 855–864.

Varshavsky, A. (1996) The N-end rule: Functions, mysteries, uses, *Proc. Natl. Acad. Sci. USA* **93**, 12142–12149.

Vierstra, R.D. (2003) The ubiquitin/26S proteasome pathway, the complex last chapter in the life of many plant proteins, *Trends Plant Sci.* **8**, 135–142.

Yewdell, J.W., Hill, A.B. (2002) Viral interference with antigen presentation, *Nat. Immunol.* **3**, 1019–1025.

Zwickl, P., Baumeister, W. (2002) The Proteasome-Ubiquitin Protein Degradation Pathway, *Curr. Top. Microbiol. Immunol.* **268**, 1–213.

Primary Literature

Bogyo, M., McMaster, J.S., Gaczynska, M., Tortorella, D., Goldberg, A.L., Ploegh, H.L. (1997) Covalent modification of the active site threonine of proteasomal β subunits and the *Escherichia coli* homolog HsIV by a new class of inhibitors, *Proc. Natl. Acad. Sci. USA* **94**, 6629–6634.

Braun, B.C., Glickman, M., Kraft, R., Dahlmann, B., Kloetzel, P.M., Finley, D., Schmidt, M. (1999) The base of the proteasome regulatory particle exhibits chaperone-like activity, *Nat. Cell Biol.* **1**, 221–226.

Brooks, P., Fuertes, G., Bose, S., Murray, R.Z., Knecht, E., Rechsteiner, M.C., Hendil, K.B. (2000) Subcellular localization of proteasomes and their regulatory complexes in mammalian cells, *Biochem. J.* **346**, 155–161.

Chau, V., Tobias, J.W., Bachmair, A., Marriott, D., Ecker, D.J., Gonda, D.K., Varshavsky, A. (1989) A multiubiquitin chain is confined to specific lysine in a targeted short-lived protein, *Science* **243**, 1576–1583.

Cope, G.A., Suh, G.S., Aravind, L., Schwarz, S.E., Zipursky, S.L., Koonin, E.V., Deshaies, R.J. (2002) Role of predicted metalloprotease motif of Jab1/Csn5 in cleavage of Nedd8 from Cul1, *Science* **298**, 608–611.

Deveraux, Q., Ustrell, V., Pickart, C., Rechsteiner, M. (1994) A 26 S protease subunit that binds ubiquitin conjugates, *J. Biol. Chem.* **269**, 7059–7061.

Elsasser, S., Gali, R.R., Schwickart, M., Larsen, C.N., Leggett, D.S., Muller, B., Feng, M.T., Tubing, F., Dittmar, G.A., Finley, D. (2002) Proteasome subunit Rpn1 binds ubiquitin-like protein domains, *Nat. Cell Biol.* **4**, 725–730.

Fenteany, G., Standaert, R.F., Lane, W.S., Choi, S., Corey, E.J., Schreiber, S.L. (1995) Inhibition of proteasome activities and subunit-specific amino-terminal threonine modification by lactacystin, *Science* **268**, 726–731.

Flynn, J.M., Neher, S.B., Kim, Y.I., Sauer, R.T., Baker, T.A. (2003) Proteomic discovery of cellular substrates of the ClpXP protease reveals five classes of ClpX-recognition signals, *Mol. Cell* **11**, 671–683.

Gaczynska, M., Osmulski, P.A., Gao, Y., Post, M.J., Simons, M. (2003) Proline- and arginine-rich peptides constitute a novel class of allosteric inhibitors of proteasome activity, *Biochemistry* **42**, 8663–8670.

Glickman, M.H., Rubin, D.M., Coux, O., Wefes, I., Pfeifer, G., Cjeka, Z., Baumeister, W., Fried, V.A., Finley, D. (1998) A subcomplex of the proteasome regulatory particle required for ubiquitin-conjugate degradation and related to the COP9-signalosome and elF3, *Cell* **94**, 615–623.

Groll, M., Ditzel, L., Lowe, J., Stock, D., Bochtler, M., Bartunik, H.D., Huber, R. (1997) Structure of 20S proteasome from yeast at 2.4 A resolution, *Nature* **386**, 463–471.

Guterman, A., Glickman, M.H. (2004) Complementary roles for rpn11 and ubp6 in deubiquitination and proteolysis by the proteasome, *J. Biol. Chem.* **279**, 1729–1738.

Harris, J.L., Alper, P.B., Li, J., Rechsteiner, M., Backes, B.J. (2001) Substrate specificity of the human proteasome, *Chem. Biol.* **8**, 1131–1141.

Hendil, K.B., Khan, S., Tanaka, K. (1998) Simultaneous binding of PA28 and PA700 activators to 20 S proteasomes, *Biochem. J.* **332**, 749–754.

Hofmann, K., Falquet, L. (2001) A ubiquitin-interacting motif conserved in components of the proteasomal and lysosomal protein degradation systems, *Trends Biochem. Sci.* **26**, 347–350.

Hoppe, T., Matuschewski, K., Rape, M., Schlenker, S., Ulrich, H.D., Jentsch, S. (2000) Activation of a membrane-bound transcription factor by regulated ubiquitin/proteasome-dependent processing, *Cell* **102**, 577–586.

Johnson, E.S., Ma, P.C., Ota, I.M., Varshavsky, A. (1995) A proteolytic pathway that recognizes ubiquitin as a degradation signal, *J. Biol. Chem.* **270**, 17442–17456.

Kane, R.C., Bross, P.F., Farrell, A.T., Pazdur, R. (2003) Velcade: U.S. FDA approval for the treatment of multiple myeloma progressing on prior therapy, *Oncologist* **8**, 508–513.

Kenniston, J.A., Baker, T.A., Fernandez, J.M., Sauer, R.T. (2003) Linkage between ATP consumption and mechanical unfolding during the protein processing reactions of an AAA+ degradation machine, *Cell* **114**, 511–520.

Kimura, Y., Saeki, Y., Yokosawa, H., Polevoda, B., Sherman, F., Hirano, H. (2003) N-Terminal modifications of the 19S regulatory particle subunits of the yeast proteasome, *Arch. Biochem. Biophys.* **409**, 341–348.

Kleijnen, M.F., Alarcon, R.M., Howley, P.M. (2003) The ubiquitin-associated domain of hPLIC-2 interacts with the proteasome, *Mol. Biol. Cell* **14**, 3868–3875.

Lam, Y.A., Lawson, T.G., Velayutham, M., Zweier, J.L., Pickart, C.M. (2002) A proteasomal ATPase subunit recognizes the polyubiquitin degradation signal, *Nature* **416**, 763–767.

Lee, C., Schwartz, M.P., Prakash, S., Iwakura, M., Matouschek, A. (2001) ATP-dependent proteases degrade their substrates by processively unraveling them from the degradation signal, *Mol. Cell* **7**, 627–637.

Leggett, D.S., Hanna, J., Borodovsky, A., Crosas, B., Schmidt, M., Baker, R.T., Walz, T., Ploegh, H., Finley, D. (2002) Multiple associated proteins regulate proteasome structure and function, *Mol. Cell* **10**, 495–507.

Liu, C.W., Millen, L., Roman, T.B., Xiong, H., Gilbert, H.F., Noiva, R., DeMartino, G.N., Thomas, P.J. (2002) Conformational remodeling of proteasomal substrates by PA700, the 19 S regulatory complex of the 26 S proteasome, *J. Biol. Chem.* **277**, 26815–26820.

Lowe, J., Stock, D., Jap, B., Zwickl, P., Baumeister, W., Huber, R. (1995) Crystal structure of the 20S proteasome from the archaeon T. acidophilum at 3.4 A resolution, *Science* **268**, 533–539.

Meng, L., Mohan, R., Kwok, B.H., Elofsson, M., Sin, N., Crews, C.M. (1999) Epoxomicin, a potent and selective proteasome inhibitor, exhibits *in vivo* anti-inflammatory activity, *Proc. Natl. Acad. Sci. USA* **96**, 10403–10408.

Mueller, T.D., Feigon, J. (2003) Structural determinants for the binding of ubiquitin-like domains to the proteasome, *EMBO. J.* **22**, 4634–4645.

Murakami, Y., Matsufuji, S., Kameji, T., Hayashi, S., Igarashi, K., Tamura, T., Tanaka, K., Ichihara, A. (1992) Ornithine decarboxylase is degraded by the 26S proteasome without ubiquitination, *Nature* **360**, 597–599.

Neher, S.B., Sauer, R.T., Baker, T.A. (2003) Distinct peptide signals in the UmuD and UmuD' subunits of UmuD/D' mediate tethering and substrate processing by the ClpXP protease, *Proc. Natl. Acad. Sci. U S A* **100**, 13219–13224.

Peng, J., Schwartz, D., Elias, J.E., Thoreen, C.C., Cheng, D., Marsischky, G., Roelofs, J., Finley, D., Gygi, S.P. (2003) A proteomics approach to understanding protein ubiquitination, *Nat. Biotechnol.* **21**, 921–926.

Petroski, M.D., Deshaies, R.J. (2003) Context of multiubiquitin chain attachment influences the rate of Sic1 degradation, *Mol. Cell* **11**, 1435–1444.

Reits, E.A., Benham, A.M., Plougaste, B., Neefjes, J., Trowsdale, J. (1997) Dynamics of proteasome distribution in living cells, *EMBO J.* **16**, 6087–6094.

Rubin, D.M., Glickman, M.H., Larsen, C.N., Dhruvakumar, S., Finley, D. (1998) Active site mutants in the six regulatory particle ATPases reveal multiple roles for ATP in the proteasome, *EMBO J.* **17**, 4909–4919.

Sakata, E., Yamaguchi, Y., Kurimoto, E., Kikuchi, J., Yokoyama, S., Yamada, S., Kawahara, H., Yokosawa, H., Hattori, N., Mizuno, Y., Tanaka, K., Kato, K. (2003) Parkin binds the Rpn10 subunit of 26S proteasomes through its ubiquitin-like domain, *EMBO Rep.* **4**, 301–306.

Seemuller, E., Lupas, A., Stock, D., Lowe, J., Huber, R., Baumeister, W. (1995) Proteasome from Thermoplasma acidophilum: a threonine protease, *Science* **268**, 579–582.

Sheaff, R.J., Singer, J.D., Swanger, J., Smitherman, M., Roberts, J.M., Clurman, B.E. (2000) Proteasome turnover of p21^{Cip1} does not require p21^{Cip1} ubiquitination, *Mol. Cell* **5**, 403–410.

Studemann, A., Noirclerc-Savoye, M., Klauck, E., Becker, G., Schneider, D., Hengge, R. (2003) Sequential recognition of two distinct sites in sigma(S) by the proteolytic targeting factor RssB and ClpX, *EMBO J.* **22**, 4111–4120.

Taxis, C., Hitt, R., Park, S.H., Deak, P.M., Kostova, Z., Wolf, D.H. (2003) Use of modular substrates demonstrates mechanistic diversity and reveals differences in chaperone requirement of ERAD, *J. Biol. Chem.* **278**, 35903–35913.

Thrower, J.S., Hoffman, L., Rechsteiner, M., Pickart, C.M. (2000) Recognition of the polyubiquitin proteolytic signal, *EMBO J.* **19**, 94–102.

Unno, M., Mizushima, T., Morimoto, Y., Tomisugi, Y., Tanaka, K., Yasuoka, N., Tsukihara, T. (2002) The structure of the mammalian 20S proteasome at 2.75 A resolution, *Structure (Camb.)* **10**, 609–618.

Ustrell, V., Hoffman, L., Pratt, G., Rechsteiner, M. (2002) PA200, a nuclear proteasome activator involved in DNA repair, *EMBO J.* **21**, 3403–3412.

van Nocker, S., Sadis, S., Rubin, D.M., Glickman, M., Fu, H., Coux, O., Wefes, I., Finley, D., Vierstra, R.D. (1996) The Multiubiquitin-Chain-Binding Protein Mcb1 is a Component of the 26S Proteasome in *Saccharomyces cerevisiae* and Plays a Nonessential, Substrate-Specific Role in Protein Turnover, *Mol. Cell Biol.* **16**, 6020–6028.

Verma, R., Chen, S., Feldman, R., Schieltz, D., Yates, J., Dohmen, J., Deshaies, R.J. (2000) Proteasomal proteomics: identification of nucleotide-sensitive proteasome-interacting proteins by mass spectrometric analysis of affinity-purified proteasomes, *Mol. Biol. Cell* **11**, 3425–3439.

Walz, J., Erdmann, A., Kania, M., Typke, D., Koster, A.J., Baumeister, W. (1998) 26S proteasome structure revealed by three-dimensional electron microscopy, *J. Struct. Biol.* **121**, 19–29.

Whitby, F.G., Masters, E.I., Kramer, L., Knowlton, J.R., Yao, Y., Wang, C.C., Hill, C.P. (2000) Structural basis for the activation of 20 S proteasomes by 11 S regulators, *Nature* **408**, 115–120.

Wojcik, C., Yano, M., DeMartino, G.N. (2004) RNA interference of valosin-containing protein (VCP/p97) reveals multiple cellular roles linked to ubiquitin/proteasome-dependent proteolysis, *J. Cell Sci.* **117**, 281–292.

Wu-Baer, F., Lagrazon, K., Yuan, W., Baer, R. (2003) The BRCA1/BARD1 heterodimer assembles polyubiquitin chains through an unconventional linkage involving lysine residue K6 of ubiquitin, *J. Biol. Chem.* **278**, 34743–34746.

Xie, Y., Varshavsky, A. (2001) RPN4 is a ligand, substrate, and transcriptional regulator of the 26S proteasome: a negative feedback circuit, *Proc. Natl. Acad. Sci. U S A* **98**, 3056–3061.

Xie, Y., Varshavsky, A. (2002) UFD4 lacking the proteasome-binding region catalyses ubiquitination but is impaired in proteolysis, *Nat. Cell Biol.* **4**, 1003–1007.

Yamano, T., Murata, S., Shimbara, N., Tanaka, N., Chiba, T., Tanaka, K., Yui, K., Udono, H. (2002) Two distinct pathways mediated by PA28 and hsp90 in major histocompatibility complex class I antigen processing, *J. Exp. Med.* **196**, 185–196.

Young, P., Deveraux, Q., Beal, R., Pickart, C., Rechsteiner, M. (1998) Characterization of two polyubiquitin binding sites in the 26S protease subunit 5a, *J. Biol. Chem.* **273**, 5461–5467.

Zaiss, D.M., Standera, S., Kloetzel, P.M., Sijts, A.J. (2002) PI31 is a modulator of proteasome formation and antigen processing, *Proc. Natl. Acad. Sci. U S A* **99**, 14344–14349.

Zhang, M., Pickart, C.M., Coffino, P. (2003) Determinants of proteasome recognition of ornithine decarboxylase, a ubiquitin-independent substrate, *EMBO J.* **22**, 1488–1496.

Protein Aggregation: *see* Aggregation, Protein

Protein and Nucleic Acid Enzymes

Daniel L. Purich
University of Florida College of Medicine, Gainesville, FL, USA

1	Enzymes as Life's Actuators	29
2	Chemical Nature of Enzymes	30
3	A Brief History of Enzyme Science	30
4	Systematic Classification of Enzyme-catalyzed Reactions	32
5	Mechanochemical Enzyme Reactions	33
6	The Catalytic Reaction Cycle	35
7	Basic Enzyme Kinetics	41
8	Enzyme Regulation	41
9	Inborn Errors in Enzyme Function	43
	Bibliography 43	
	Books and Reviews 43	

Keywords

Catalytic Antibody
An antibody that catalyzes chemical reactions.

Encyclopedia of Molecular Cell Biology and Molecular Medicine, 2nd Edition. Volume 11
Edited by Robert A. Meyers.
Copyright © 2005 Wiley-VCH Verlag GmbH & Co. KGaA, Weinheim
ISBN: 3-527-30648-X

Catalytic Proficiency
A measure of the rate enhancement achieved in the presence of an enzyme; the dimensionless parameter v_{enz}/v_{non}, where v_{enz} and v_{non} are the reaction velocities in the presence and absence of an enzyme respectively.

Energase
A mechanochemical enzyme that catalyzes energy-dependent translocation, energy-dependent transport, and/or energy-dependent force production.

Enzyme
A biological catalyst that speeds up the making and breaking of chemical bonds without altering the reaction's equilibrium.

Protein Enzyme
A biological catalyst composed of one or more polypeptide chains.

RNA Enzyme
A biological catalyst composed of ribonucleic acid.

■ As is true of any catalyst, the defining quality of every enzyme is the ability to enhance chemical reactivity without altering the thermodynamic equilibrium of the reaction catalyzed. An enzyme's catalytic proficiency is measured by its rate enhancement factor, a parameter that equals v_{enz}/v_{non}, where v_{enz} and v_{non} are the reaction velocities in the presence and absence of an enzyme respectively. Although the rates of uncatalyzed reactions are often too slow to be determined under physiologic conditions, enhancement factors of enzymes are thought to range typically from 10^9 to 10^{17}, with a few approaching 10^{22} and others falling well below 100. Each catalytic cycle shares common features: (1) the enzyme must combine with its substrate, which is then converted into the transition state; (2) and this least stable reactant configuration then undergoes equipartition (i.e. half of the time, reforming EnzymeSubstrate complex, and just as often proceeding to form EnzymeProduct complex, the latter followed by product release). After each catalytic reaction cycle, an enzyme returns to its original form, such that an enzyme's action is without any effect on the reaction's equilibrium constant. Because product release is so often the slowest step in catalysis, discrete bond-making and bond-breaking steps, and even the formation of covalent intermediates, rarely hinder catalysis. The rates of highly perfected enzymes are limited only by the diffusion-limited availability of substrate; other enzymes are so sluggish that each complete catalytic cycle operates on the seconds to minutes timescale. In terms of natural selection, however, an organism accrues no competitive advantage should one of its enzymes evolve catalytic proficiency beyond that needed for efficient metabolism.

1
Enzymes as Life's Actuators

Because normal physiologic conditions (e.g. neutral pH, aqueous environment, moderate ionic strength, pressures near one atmosphere, and a narrow 0–50 °C temperature range) conspire to deactivate most chemical reactions, their uncatalyzed rates are frequently extraordinarily low. Consider, for example, the hexokinase reaction: $Mg·ATP^{2-}$ + D-Glucose = D-Glucose 6-Phosphate + $Mg·ADP^{1-}$. At cellular glucose and ATP concentrations, detectable glucose-6-P would require hundreds or thousands of years in the absence of hexokinase, but occurs in seconds at cellular enzyme concentrations. Even when nonenzymatic reactions are fairly fast, as is true for the reversible hydration of carbon dioxide to form bicarbonate anion, the presence of an enzyme (carbonic anhydrase, in this case) assures the reaction rate is always sufficient for effective metabolism.

The speed and specificity of enzyme-catalyzed reactions provides cells with the capacity to organize a series of these reactions into pathways responsible for the synthesis (*anabolism*) and breakdown (*catabolism*) of essential life-sustaining intermediates. With hexokinase as an example once again, exclusive phosphorylation at position-6 of the β-D-glucopyranose form would be impossible in the absence of enzyme. Even when nonenzymatic reactions are reasonably rapid, as in the case of reversible hydration of carbon dioxide to form bicarbonate anion, an enzyme is required to assure the reaction rate is sufficiently fast for efficient metabolism. In this respect, enzymes play roles analogous to on–off switches and amplifiers within electronic circuits. Substrate specificity also reduces the formation of toxic substances that would otherwise accumulate in side reactions. By coupling energy-requiring and energy-yielding reactions, enzymes also assist in managing cellular bioenergetics, thus permitting cells to store energy in many forms (e.g. as low molecular–weight metabolites, as macromolecules, as ion and solute concentration gradients, etc.). Enzymes catalyzing the first committed step in a metabolic pathway often possess regulatory sites that bind activators and inhibitors, allowing them to respond to regulatory cues as substances are produced or consumed within their own and other metabolic pathways. Moreover, because some enzymes are oligomeric (i.e. they are composed of multiple subunits), regulatory information can be transferred across their subunit–subunit interfaces in a manner that alters their catalytic efficiency in response to small fluctuations in the concentration of regulatory (or effector) molecules. This phenomenon, known as *cooperativity*, is attended by enzyme shape changes (i.e. allosteric transitions) that can greatly alter catalytic activity.

Certain highly specialized mechanochemical enzymes interact with their binding partners to adopt low- and high-affinity interaction states, the interconversion of which is driven by the Gibb free energy liberated upon the hydrolysis of certain metabolites, most often ATP or GTP. These affinity-modulated systems can also derive free energy during the dissipation of transmembrane solute concentration gradients. Such energy-transducing enzymes can also generate the forces needed for cell division and motility, solute transport, as well as intracellular trafficking, and sorting of molecules and/or organelles. By generating the forces and mechanical work needed to power sensory processes, transport and translocation, and the ability to divide, these mechanochemical enzymes bestow upon organisms the essentially animating qualities of irritability, motility, and proliferation.

2
Chemical Nature of Enzymes

From earliest times, enzymes were invariably found to be associated with proteins. On the basis of his Nobel prize–winning work establishing the structures of heme and chlorophyll molecules, the German chemist Willstätter incorrectly asserted what he believed to be a universal principle, namely, that the catalytic entity of every enzyme would prove to be some small molecule that is bound to a protein that serves only passively as a stabilizing colloidal support. Sumner's crystallization of urease proved that the active enzyme is itself a protein, and this accomplishment earned him a Nobel Prize. However, while most known enzymes are proteins, others are partly or entirely composed of nucleic acid. Nobel laureates Cech and Altman demonstrated that certain RNA molecules are efficient catalysts of RNA self-splicing, phosphotransfer, and even peptide bond-forming reactions. These catalytic RNA molecules, also known as *ribozymes*, often achieve rate enhancements approaching 10^{11}. The hammerhead-shaped ribozyme was the first RNA motif observed to catalyze sequence-specific self-cleavage. With only around 30 nucleotides in their catalytic cores, these enzymes are among the smallest known catalytic RNAs. They display Michaelis–Menten kinetics, with Michaelis constants values ranging from 20–200 nM and turnover numbers in the range of $0.03\ \text{s}^{-1}$. Product release is generally fast, suggesting that the rate-determining step is phosphodiester bond-scission. Ribozyme-catalyzed phosphoryl transfer appears to involve destabilization of the substrate's ground state. This transesterification reaction is analogous to the catalytic mechanisms used by the mRNA spliceosome as well as DNA topoisomerases. Some evolution experts have argued that the likely early appearance of nucleic acids necessitated reliance on their catalytic properties in what might have been a "preprotein" biotic world.

One cannot dismiss the possibility that other biological substances (e.g. polysaccharides, complex lipids, etc.) may prove to be biological catalysts. The phenomenon of micellar catalysis, for example, is already firmly rooted in modern organic chemistry. In fact, some micellar catalysts even exhibit chiral recognition (i.e. the capacity to combine with and transform substrate molecules in a stereoselective manner). Whether the same holds for cellular membranes remains to be determined.

3
A Brief History of Enzyme Science

On the basis of his studies of diastase, a crude preparation of the enzyme α-amylase, the Swedish chemist Berzelius advanced the germinal concept that, to hasten product formation, a catalyst must first combine with its reactant. In keeping with his idea that a catalyst must weaken the bonds within a reactant, he coined the term *catalysis* by combining the Greek words *kata* ("break") and *lyein* ("release") to describe its ability to break apart chemical reactants. The birth of the field of enzymology can be traced to 1810, when the French chemist and physicist Gay-Lussac demonstrated that anaerobic growth of yeast on sugars produces mainly ethanol and carbon dioxide. In 1878, Kühne coined the name enzyme, taken from the Greek words *en* ("within") and *zyme* ("yeast"), to designate the fermentative catalyst within yeast. In 1896, Büchner successfully opened yeast cells in a mortar and pestle through the abrasive action of fine grains of sand on the microorganism's cell wall. Upon addition of sugars to this cell-free extract, he noted the evolution of CO_2 bubbles, a property that he took as a telltale sign of active metabolism. Büchner's sensational conclusion that biochemical reactions can occur outside a living cell gave birth to investigative enzyme chemistry and earned him the Nobel Prize. The discipline of enzyme kinetics was firmly established in 1913, when Michaelis and Menten published the famous rate law that now bears their names. A decade later, Briggs and Haldane introduced the concept of a steady state to enzyme kinetics and metabolism. By improving rapid mixing kinetic methods, Chance detected EnzymeSubstrate complexes spectrally and characterized their subsequent conversion to product. About the same time, Racker also discovered an enzyme (glyceraldehydes 3-phosphate dehydrogenase) that formed covalent enzyme–substrate intermediates during catalysis, thus dispelling notions that formation of covalently bound intermediates would be an unproductive detour (or *umweg*, as the German Nobelist Warburg put it), serving only to retard catalysis. Surging interest in shock waves and sonar during World War II, along with Einstein's extension of Maxwell's theory of chemical relaxation to include pressure-induced perturbations, provided an impetus for investigating the individual steps (or elementary reactions) within multistep mechanisms. For their pioneering work in developing and exploiting novel techniques that detect and quantify changes in rapid chemical and biological processes, Eigen, Norrish, and Porter shared the Nobel Prize in chemistry.

Enzymology is now a highly sophisticated and innovative discipline that continues to provide powerful insights into enzyme catalysis and metabolic regulation. Among its most notable contributors are: Abeles – for mechanism-based (or suicide) inhibitor design; Albery and Knowles – for enzyme energetics and catalytic perfection; Altman and Cech – for the discovery of catalytic RNA; Benkovic and Lerner – for catalytic antibodies; Boyer – for isotope tracer kinetics, equilibrium exchange theory, and ATP synthase binding change mechanism; Chance – for stopped-flow rapid mixing techniques, heme–protein kinetics; Cleland – for nomenclature, exchange-inert metal-nucleotides, exchange kinetics, and kinetic isotope effects; Cohn – for oxygen-18 probes of phosphotransfer reactions and NMR distance measurements; Eigen, De Maeyer, and Hammes – for temperature-jump reaction techniques to enzyme systems; Fersht – for site-directed mutagenesis as mechanistic probes as well as kinetic proofreading mechanism of protein folding; Frieden – for enzyme hysteresis; Fromm – for reversible inhibitors as tools for discriminating multisubstrate systems; Hartley – for burst kinetic methods for detecting reaction intermediates; Jencks – for catalytic strategies in chemistry and enzymology, as well as the conceptual basis for catalytic antibodies; Koshland – for the induced-fit hypothesis and cooperative ligand-binding mechanisms; Mitchell – for the chemiosmotic principle; Monod, Wyman and Changeux – for cooperativity and allosteric models; Rose – for dynamic stereochemical probes and isotope trapping methods; Westheimer – for enzyme stereochemistry, stereochemistry of phosphoryl transfer, and photoaffinity labeling; and Wolfenden – for transition-state inhibitor theory and the conceptual basis of catalytic proficiency.

4
Systematic Classification of Enzyme-catalyzed Reactions

The number of classified protein enzymes now approaches 7000; fewer than several hundred reactions are known to require RNA enzymes. Given the diversity of living organisms, there must be thousands more that catalyze reactions remaining to be defined. The Enzyme Commission and the Joint Commission on Biochemical Nomenclature have distinguished enzymes into six chemically distinct classes, based entirely on the ways that enzymes break, rearrange, and form covalent bonds (See Table 1).

For more than a half century, enzymes have been named and classified only on the basis of the making and breaking of covalent bonds. There are, however, numerous examples of biological catalysis involving the interconversion of noncovalent substrate-like and product-like states. Consider the hundreds of mechanochemical enzyme reactions (see following text) that catalyze reactions that accomplish mechanical work or that generate force. Another prominent example is the entire group of so-called nucleotide exchange factors. These proteins play important physiologic roles, as illustrated by the facilitated exchange of solution-phase GTP with tightly bound GDP on GTP-regulatory proteins (or G-Proteins), resulting in the formation of G-ProteinGTP complex and solution-phase GDP. Another example is the adenine nucleotide exchange factor known as profilin (Reaction: Actin·ADP + $\mathbf{ATP}_{\text{solution-phase}}$ = Actin·$\mathbf{ATP}$ + $ATP_{\text{solution-phase}}$, where boldface type is used to indicate that exchange is the purely physical release of bound ADP, followed by binding of $\mathbf{ATP}$). This uncatalyzed exchange reaction is too slow to sustain the high filament elongation rates (400–500 monomer per filament per second) during actin-based motility. To overcome this kinetic barrier, the protein profilin accelerates the ADP ↔ ATP exchange reaction by a factor of 150, without any effect on the equilibrium constant for the exchange reaction. These features indicate that G-protein exchange factors and profilin must be regarded as biological catalysts, and hence as enzymes.

In the face of the growing number of instances where biological catalysis is not attended only by changes in covalent bonding, Purich offered a definition in 2001: "An enzyme as a biological catalyst that facilitates the making and/or breaking of chemical bonds." While appearing to be no more encompassing than existing definitions of enzyme catalysis, the crucial difference lies in the use of *chemical* in place of *covalent* to describe the bonding changes. In *The Nature of the Chemical Bond*, Linus Pauling offered the following definition of chemical bonds: "We shall say that there is a chemical bond between two atoms or groups of atoms in case that the forces acting between them are such as to lead to the formation of an aggregate with sufficient stability to make it convenient

Tab. 1 Systematic classification of enzyme-catalyzed reactions

Class 1: Oxidoreductases catalyze redox reactions.	
Dehydrogenase:	Ethanol + NAD$^+$ = Acetaldehyde + NADH + H$^+$
Class 2: Transferases catalyze the transfer of functional groups.	
Sugar kinase:	Glucose + MgATP^{2-} = Glucose = 6-P^{1-} + MgADP^{1-}
Aminotransferase:	Oxaloacetate + Glutamate = Aspartate + 2-Oxoglutarate
Class 3: Hydrolases catalyze hydrolysis reactions.	
Esterase:	Ester + H$_2$O = Acid + Alcohol
Amidase:	Glutamiine + H$_2$O = Glutamic Acid + Ammonia
Phosphatase:	Glucose-6-P^{2-} + H$_2$O = Glucose + HPO$_4^{2-}$
Class 4: Lyases catalyze addition/elimination reactions.	
Fumarase:	Malate = Fumarate + H$_2$O.
Aspartase:	Aspartate = Oxaloacetate + NH$_3$
Class 5: Isomerases catalyze intramolecular rearrangements.	
Isomerase:	Glucose-6-P^{2-} = Fructose-6-P^{2-}
Class 6: Ligases catalyze joining reactions.	
Synthetase:	Glutamate + NH$_3$ + MgATP^{2-} = Glutamine + MgADP^{1-}
Polymerase:	DNA$_n$ + dNTP = DNA$_{n+1}$ + PP$_i$

for the chemist to consider it as an independent molecular species." Note that Pauling made no comment about covalent bonding. In this respect, because many protein conformational states and numerous ProteinLigand complexes are sufficiently long lived to exhibit chemically definable properties, their formation and/or transformation must be considered as chemical reactions. The same logic applies to the persistent, chemically definable position of a solute relative to the faces of a membrane, representing substrate-like and product-like states. For this reason, transporters should be regarded as enzymes. Finally, Purich in 2001 suggested that each mechanochemical ATPase can be termed an *energase*, a name he offered for those specialized enzymes catalyzing energy-dependent translocation, energy-dependent transport, energy-dependent force production, and so on. He also suggested that these mechanochemical enzymes should be considered as an entirely separate class, as these enzymes cannot be rationally accommodated in the current Enzyme Commission classification.

5
Mechanochemical Enzyme Reactions

There are literally hundreds of catalyzed force-generating and/or work-producing reactions that culminate in a change of some molecular condition (i.e. transport against a concentration gradient, locomotion against an opposing load, or stabilization of an otherwise unstable conformation/configuration). These reactions are typically driven by the Gibbs free energy liberated upon the hydrolysis of such molecules as ATP, GTP, P-enolpyruvate, and acetyl-phosphate. With the input of sufficient Gibbs free energy, these mechanochemical enzymes facilitate their reactions by interconverting between two or more noncovalent substrate- and product-like interactions states:

$$\text{Condition}_A + \cdot ENZ \cdot \text{ATP} = \text{Condition}_A \cdot ENZ \cdot \textbf{ATP} \quad (1)$$

$$\text{Condition}_A \cdot ENZ \cdot \text{ATP} = \text{Condition}_A \cdot ENZ \cdot \text{ADP} \cdot \text{Pi} \quad (2)$$

$$\text{Condition}_A \cdot ENZ \cdot \text{ADP} \cdot \text{Pi} = \text{Condition}_B \cdot ENZ \cdot \text{ADP} + \text{Pi} \quad (3)$$

$$\text{Condition}_B \cdot ENZ \cdot \text{ADP} + \text{ATP} = \text{Condition}_B + \cdot ENZ \cdot \text{ATP} \quad (4)$$

$$\text{Condition}_A + \text{ATP} = \text{Condition}_B + \text{ADP} + \text{Pi} \quad \sum(1-4)$$

Note that the above reaction forms a catalytic reaction cycle, culminating in recycling of the enzyme to its initial form. In some cases (e.g. microtubule assembly/disassembly), the entire process may also require a considerable period of time to complete its cycle.

These reactions utilize Gibbs free energy liberated upon the hydrolysis of molecules such as ATP, GTP, P-enolpyruvate, and acetyl-phosphate. While certain bonds within these substances are thermodynamically highly unstable, they are highly stable kinetically. Mechanochemical enzymes share the capacity to catalyze the hydrolysis of these P—O—P or C—O—P anhydride bonds, thereby triggering a conformational transition from one interaction state to another.

Active solute transporters promote the transport of ionic and neutral solutes by forming a high-affinity TransporterSolute complex on the side of the membrane where the solute concentration is low. Hydrolysis of transporter-bound ATP subsequently drives the conversion of the high-affinity TransporterSolute complex into its low-affinity TransporterSolute complex, such that solute release is facilitated on the membrane's high-solute concentration side (see Table 2). In a similar manner, molecular motors power contractile processes, intracellular organelle trafficking, and cell crawling.

Likewise, the building up and tearing down of supramolecular structures depends on ATP/GTP-driven affinity-modulated interactions (e.g. regulatory G-proteins, chaperonins, and proteasomes, and various other latching and switching molecules). The field of cell biology can be regarded as structural metabolism, where the supramolecular components are continually formed, remodeled, and degraded. Endocytosis and organelle traffic, cell crawling, signal transduction, and mitosis/meiosis are processes that are taking on the appearance of the pathways of intermediary metabolism. Even long-term potentiation, a neuronal process lying at the root of our memory and consciousness, is now known to depend on actin polymerization motors that remodel and maintain structures known as dendritic spines into functional neuronal synapses.

The essence of these energase-type enzymes is a molecular discrimination (or affinity-modulating) mechanism comprised of a succession of conformational states that transit between high- and low-affinity ligand interaction states. In a similar fashion, electron transport systems convert the free energy of NADH oxidation into the proton-motive force (pmf) driving ATP synthesis in oxidative phosphorylation. In its well-choreographed succession of electron transfer steps, a

Tab. 2 Mechanochemical enzymes: Affinity-modulated clamps, latches and switches

Reaction	Energy source	High-affinity interaction partner
Transporters	ATP	Ligand on low-concentration side of membrane
ATP synthase	ATP	Itself ($\alpha\beta-\gamma$ subunit interface)
Myosin	ATP	Actin filament
Electron transport chain	emf	Proton on high pH side of gradient
Dynein and kinesin	GTP	Microtubule
GTP-regulatory proteins	GTP	Target protein
DNA topoisomerase (type II)	ATP	Itself (at its dimer interface)
ATP sulfurylase	GTP	Active site
Actoclamplin	ATP	Clampin (Ena/VASP, N-WASP and formin protein families)

membrane-bound electron transport protein complex undergoes conformational change(s) altering the acidity of one or more acid-base groups, thus permitting the transmembrane protein to extrude one or more protons across a membrane bilayer. After NADH is formed upon the oxidation of foodstuffs, it undergoes oxidation (a net loss of electrons) and initiates the chemical reduction (a net gain of electrons) of electron transport proteins (ETP) situated in the inner mitochondrial membrane. Subsequent chemical reduction of molecular oxygen is mediated by various ETP complexes, resulting in the accumulation of protons within a submitochondrial compartment. Precisely how the electromotive force (emf) of electrochemical gradient generates the proton-motive force, remains an area of intense investigation. The current view is that electron transport drives a conformational change that lowers the affinity of carboxyl groups within the ETP complex for their protons. When viewed in this manner, the action of these multiprotein ETP complexes represents a highly specialized example of molecular discrimination.

6
The Catalytic Reaction Cycle

The long-standing problem in biological catalysis concerns the origin of enzyme rate enhancement and the functional coupling of enzyme and reactant in transformations of EnzymeSubstrate, EnzymeIntermediate, and EnzymeProduct complexes. The following table (Table 3) lists some of the ways in which an enzyme is thought to increase chemical reactivity. There is presently no single enzyme-catalyzed reaction for which the precise contribution of each factor has been unambiguously determined. The alert reader will also recognize that many of these factors are surely acting simultaneously and synergistically, and in various degrees, at defined stages in the reaction cycle.

Of all the explanations advanced to account for the catalytic proficiency of enzymes, the most appealing is that enzymes promote catalysis by stabilizing reaction transition states. This concept was assumed implicitly in writings of Haldane in 1930 and was explicitly stated by Pauling in 1940: "In order to catalyze a reaction an enzyme must recognize the transition state in a selective way (i.e. it must

Tab. 3 Physicochemical factors thought to increase catalytic proficiency

Approximation	Enzymes increase the local effective reactant concentration by bringing reactants (as well as essential coenzymes and metal ion cofactors) into the vicinity of catalytic groups within the enzyme's active site.
Orientation	Enzymes adjust the mutual orientation of reactive groups on the substrate(s) and enzyme.
Push–pull catalysis	Enzymes have suitably positioned acid and base groups that can provide ("push") or withdraw ("pull") protons to and from the substrate, product or reaction intermediate at appropriate times in the reaction cycle.
Stereoselection	Enzymes possess the capacity to serve as three-dimensional templates that discriminate between various reactants on the basis of the stereochemical arrangement of their atoms.
Chelation	Enzymes and substrates often employ transition metal ions to form "bridge" complexes (e.g. enzyme–metal–ion substrate; enzyme–substrate–metal ion; or cyclic complexes as well).
Lewis acid effects	Enzymes use metal ions as electron-pair acceptors to polarize otherwise unreactive groups on the substrate or enzyme and/or to alter their acid/base properties.
Redox effects	Enzymes alter the redox properties of certain metal ions to transiently achieve oxidation states that would be otherwise unavailable in aqueous solutions.
Desolvation	Enzymes remove water molecules from the active site in order to increase the nucleophilic reactivity of functional groups on the substrate or enzyme.
Hydrophobic effects	Enzymes can alter the polar and apolar microenvironment within the active site to increase/decrease reactivity.
Proton shuttling	Enzymes create extended chains of hydrogen bonded water molecules that serve as "proton wires" for providing or removing protons from the active site on an as-needed basis.
Transition-state stabilization	Enzymes have active sites that are structurally complementary to one or more activated complexes formed along the reaction trajectory.
Ground-state stabilization	Enzymes can promote catalysis by destabilizing an otherwise unreactive ground-state enzyme–substrate complex.
Entropic effects	Enzymes can temporarily alter the thermodynamic properties of bound reactants in a manner that favors attainment of a highly reactive configuration.
Covalent catalysis	Enzymes can transiently form covalent intermediates that facilitate catalysis by managing the group transfer potential of reactants or by masking/unmasking functional groups to promote catalysis.
Entatic effects	Enzymes can speed up reactions by exerting substantial push–pull forces on reacting species, especially over short distances.

stabilize it better than the substrate)." The following diagram (Fig. 1) illustrates this fundamental concept by comparing an uncatalyzed reaction proceeding through a highly unstable transition-state ($X^{\ddagger}$) and a hypothetical enzyme operating by a three-stage mechanism entailing the formation of an Enzyme·Substrate complex (E·S), an Enzyme·Transition State (E·$X^{\ddagger}$), and an Enzyme·Product complex (E·P). Note that the stabilization free energy may be written as: $\Delta\Delta G^{\ddagger} = \Delta G^{\ddagger}_{non} - \Delta G^{\ddagger}_{enz}$, where $\Delta G^{\ddagger}_{non}$ is the energy change for forming the nonenzymatic transition state, and $\Delta G^{\ddagger}_{enz}$ is the energy change for forming the enzymatic transition state. On the basis of absolute rate theory, the ν_{non}/ν_{enz} may be estimated from the relationship: $\nu_{non} = \nu_{enz} e^{-\Delta\Delta G^{\ddagger}/RT}$, where R and T are the universal gas constant and the absolute temperature.

Hammes in 2002 offered the view that "when a substrate binds to an enzyme, it becomes an integral part of the macromolecule. The subsequent dynamics of the macromolecular conformational changes are then the catalytic process itself. This view of catalysis means that the making and breaking of noncovalent bonds within the structure are part of the catalytic process, and that these events can occur close to and far from the active site. The advantage of having hundreds of intramolecular interactions dynamically involved in catalysis is that the energetics of the reaction can be easily manipulated to produce catalysis, and extremely fine tuning is provided by hundreds of intramolecular interactions. This mechanism could be viewed as a "gear shift" mechanism: the conformational transitions are analogous to shifting gears, and the interactions between the enzyme and substrate correspond to the gear coupling mechanism. Asking what "drives" the reaction is not terribly meaningful, as the essence of cooperative processes is that many events are occurring essentially simultaneously."

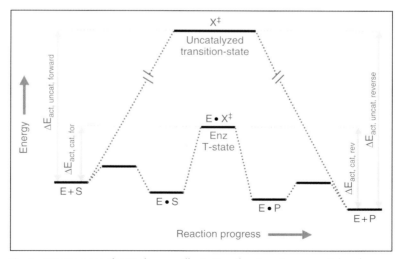

Fig. 1 Reaction coordinate diagram illustrating the steps in an uncatalyzed reaction (upper pathway) and those in an enzyme-catalyzed process (lower pathway). Note that the energy of the initial and final states are the same for the enzymatic and nonenzymatic processes.

These fundamentally important ideas are tantamount to saying that an enzyme creates a catalytic environment, one that dynamically promotes the reaction trajectory from substrate to product. Although E·X‡ has been widely employed to represent the transition state, the latter should be represented as (E·X)‡ to indicate that enzyme and substrate are mutually altered as the catalytic reaction cycle proceeds. That enzyme and substrate jointly reach the transition state also complies with the principle of thermodynamic reciprocity, which requires that substance a cannot affect substance without the latter likewise affecting the former. Traditional use of E·X‡ leaves a mistaken impression that the enzyme is an immutable template, such that only the substrate reaches the activated transition-state configuration.

The chief ambition of enzyme chemists is to understand the catalytic process used by enzymes by obtaining a complete description of all of the physical and chemical events occurring during an enzyme-catalyzed reaction. Such a daunting undertaking inevitably leads to the formulation of a reaction mechanism, a chemical model or framework consisting of all of the elementary reactions comprising the overall catalytic process. The mechanism must be consistent with the reaction's overall stoichiometry and stereochemistry, its kinetic and thermodynamic properties, the structures of detected intermediates and inferred transition state(s), the energetics of intermediates involved in rate-determining step(s), as well as effects of temperature, ph, ionic strength, and solvent. modern enzymology has benefited enormously from the atomic level molecular structures, as provided by X-ray crystallography and high-resolution, multidimensional nmr spectroscopy. even so, while structural biologists have glimpsed various stages of catalysis, there is no such thing as a "tell-all motion picture" of even the simplest catalytic process. Enzyme chemists must also rely on stereochemistry to learn whether certain covalent intermediates are formed during catalysis. The optimal approach is to construct hypotheses offering testable predictions that can be recursively tested in kinetic experiments designed to discriminate among rival explanations or at least to suggest a more definitive experiment.

Intensive investigation of proteolytic enzymes over the past half century has culminated in a self-consistent mechanism for chymotrypsin catalysis (Fig. 2). This enzyme cleaves peptide bonds where the carboxyl group–donating residue contains a hydrophobic side chain. The reaction is facilitated by push–pull acid/base interactions involving three residues common to hundreds of other mechanistically related "serine" proteases. An acyl-serine intermediate permits one product (designated by the R-group) to dissociate such that water can replace the departing amino group in a manner that leads to hydrolysis of the peptidyl-acyl-enzyme and subsequent release of the second peptide fragment (designated by R′).

Enzyme chemists are reasonably confident of the general outline of this reaction mechanism, especially in the light of the wealth of structural, chemical, and kinetic information gleaned from persistent and systematic investigation. That said, there remain obvious voids in our understanding of chymotrypsin catalysis. A major limitation stems from an almost exclusive reliance on data obtained from the use of synthetic chromogenic substrates (i.e. molecules that generate spectral change

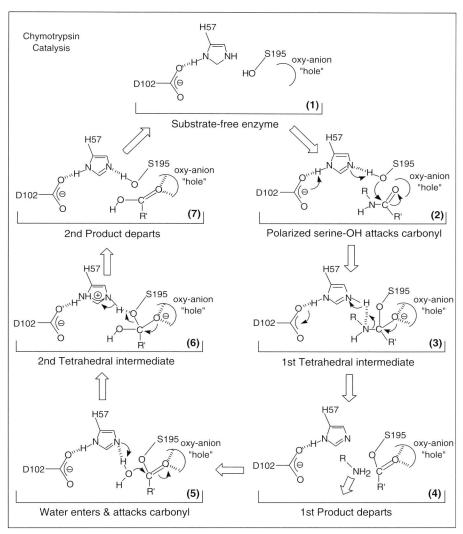

Fig. 2 Catalytic mechanism of the proteolytic enzyme chymotrypsin. The substrate-free enzyme (**1**) with its triad of aspartate, histidine, and serine residues, as well as a diagrammatic representation of the positively charged residues making up the "oxy-anion hole". The triad is thought to facilitate nucleophilic attack of the serine hydroxyl group, as shown in (**2**), thereby forming a negatively charged tetrahedral intermediate (**3**) that rearrange to the *O*-acyl-enzyme (**4**). The reaction cycle proceeds after the amine-containing product departs and is replaced by a water molecule (**5**). Nucleophilic attack by water then forms the second tetrahedral intermediate (**6**), which rearranges to form the second product-bound (**7**). The same mechanism applies to "serine" protease/esterase superfamily.

in the visible and/or ultraviolet spectrum upon peptide bond cleavage). Virtually nothing is known about chymotrypsin's action on natural protein substrates.

Whenever it becomes feasible to explain the nature of biological catalysis, biochemists can also begin to understand the structural, dynamic, and catalytic features responsible for effective regulation. Activators and inhibitors of enzymes have the effect of respectively lowering and raising reaction barrier(s), as do the activating and inhibitory effects of posttranslational covalent modifications of enzymes. Another benefit of developing detailed, well-reasoned models for enzyme catalysis is the improved prospect for designing highly effective inhibitors that may be employed as therapeutic agents. Not surprisingly, pharmaceutical companies continually make huge investments of time and money in their investigative efforts to gain such valuable information about key enzyme reactions. An excellent example is angiotensinogen-converting enzyme (ACE), for which well over a million compounds have been synthesized and evaluated in terms of their potential to become reliable blood pressure–regulating drugs.

Finally, a long-standing aspiration in chemistry is the creation of synthetic molecules that attain the stereochemical selectiveness, if not the catalytic proficiency, of naturally occurring enzymes. The rationale for this approach is to circumvent the limited stability of proteins, especially in nonaqueous solutions. One approach for obtaining novel catalysts for specific organochemical reactions involves the creation of catalytic antibodies (or "abzymes"). The idea that enzymes promote catalysis by stabilizing reaction transition states led Jencks in 1969 to offer the following prediction that appropriately selected antibodies would prove to be effective catalysts. "If complementarity between the active site and the transition state contributes significantly to enzymatic catalysis, it should be possible to synthesize an enzyme by constructing such an active site. One way to do this is to prepare an antibody to a antigenic group which resembles the transition state of a given reaction. The combining sites of such antibodies should be complementary to the transition state and should cause an acceleration by forcing bound substrates to resemble the transition state." Because transition states are intrinsically unstable, catalytic antibodies are selected by using chemically stable transition-state analogs as immunigens. For example, antibodies generated against a bent porphyrin ring were found to catalyze the metal ion insertion into heme groups, presumably by straining the planar substrate toward a bent transition-state conformation. Extensive investigations, chiefly in the laboratories of Lerner and Benkovic, demonstrated that the observed rate enhancements correlated with an antibody's affinity for analog versus that for the reactant. While many details of antibody catalysis remain to be elucidated, it is safe to assert that antibodies are unlikely to approach the efficiency of enzymes, at least not in the foreseeable future. Beyond stabilizing one or more transition states in a catalytic reaction cycle, synthetic enzymes must also undergo an intricate series of energetically and temporally controlled conformational changes, which permit naturally occurring enzymes the capacity to control their acid/base, nucleophilic, electrophilic, and/or redox chemistry. Achieving the skill and insight to "teach" synthetic catalysts the finely honed art of catalysis remains a stubbornly elusive goal in modern enzyme science.

7
Basic Enzyme Kinetics

The Michaelis–Menten treatment of enzyme-catalyzed reactions is based on the following key assumptions: (1) the enzyme combines rapidly and reversibly with the substrate to form a single E·X complex, such that the Michaelis constant K_S equals $\{([E_{total}] - [E \cdot X]) \times ([S_{total}] - [E \cdot X])\}/[E \cdot X]$; (2) the total enzyme concentration is negligible relative to the total substrate concentration (i.e. $[E_{total}] \ll [S_{total}]$), thereby simplifying the earlier expression to: $K_S = \{([E_{total}] - [E \cdot X]) \times [S_{total}]\}/[E \cdot X]$; (3) the reaction rate (i.e. $v = k_{cat}[E \cdot X]$) is constant during the initial reaction phase, when substrate consumption and product accumulation is minimal; and (4) the concentration of active enzyme remains unchanged. The resulting Michaelis–Menten rate equation is:

$$v = V_{max}[S]/\{K_S + [S]\}$$

where V_{max} is the maximal rate (units = Molarity per second). [Note: $V_{max}/[E_{total}]$ equals the catalytic constant k_{cat} (units = second^{-1}), a measure of the frequency of catalytic cycling.] In more rigorous steady state treatments, particularly those accounting for isomerizations among many enzyme-bound species (e.g. $E + S = E \cdot S_1 = E \cdot S_2 = \cdots E \cdot S_n \cdots = (E \cdot X)^{\ddagger} = E \cdot P_n \cdots = E \cdot P_2 = E \cdot P_1 = E + P$), the Michaelis constant K_M (units = Molarity) is more complicated than the thermodynamic parameter K_S. For this reason, the Michaelis constant should not be regarded as an accurate measure of the dissociation constant for the E·S complex.

Most enzyme-catalyzed reactions actually utilize more than one substrate, especially when ATP, NAD$^+$, Coenzyme A, and similar cosubstrates are required. The kinetic behavior of multisubstrate enzyme systems depends on the order in which substrates bind and products are released. In hydrolase-type reactions, water is the second substrate; however, its concentration (55.5 M) is so high that, for kinetic purposes, water cannot be treated as a variable.

8
Enzyme Regulation

Homeostasis arises from the ability of enzymes, membrane-bound and mobile receptors, as well as key regulatory proteins to sense changes in the concentration of substrates, coenzymes, activators, as well as inhibitors, to integrate these input signals and to make appropriate responses that modify metabolic activity. Of particular significance are the enzymes catalyzing the first committed reactions within a pathway, and often play a pace-making role in controlling metabolic flux (or throughput). Other enzymes within a pathway may simply adjust their individual catalytic properties in response to changes in the availability of pathway intermediates. Because most pathways are interconnected at branch points (or nodes), enzymes catalyzing first steps of pathways must be regulated in a hierarchical manner. Fast signaling occurs on the millisecond-to-second timescale and requires rapid binding and release of regulatory molecules to and from specific binding sites on target enzymes. On the second-to-minute timescale, cellular regulation is achieved by hormone and/or effector controlled enzymes catalyzing posttranslational modifications (e.g. phosphorylation, methylation, nucleotidylation, ADPRibosylation, acetylation, proteolysis, etc.) of target enzymes, thereby altering the latter's catalytic performance. Subsequent removal of the transferred group by certain phosphatases, phosphodiesterases, glycosylases, and hydrolases restores the target enzyme's original activity state. Regulatory processes taking place on the minutes-to-days timescale often reflect changes in the rates of biosynthesis or degradation of regulated enzymes. Some enzymes have short biological half-lives (e.g. 12 min for ornithine decarboxylase), whereas others, like lactate dehydrogenase, are highly stable and persist for weeks or longer.

Feedback inhibition allows enzymes catalyzing committed steps to be inhibited as pathway end products eventually accumulate. Feedback inhibitors most often bind at discrete regulatory sites that are topologically remote from the active site, and they possess the ability to modify catalytic activity by altering the properties of active-site residues. Inhibitors have a net effect of raising the height of energy barriers limiting the rate of substrate binding and/or catalysis; activators have the opposite effect. It should be stressed that feedback inhibition at pathway branch points presents the opportunity for highly coordinated metabolic regulation of highly branched metabolic networks.

The most common mode of enzyme regulation results from spontaneous or ligand-induced conformational changes. In the case of spontaneous changes, a protein may exist in two states X and Y that are in thermodynamic equilibrium with each other. The equilibrium concentrations of X and Y are determined by the relative stabilities of each conformation. In other cases, a ligand molecule selectively alters the stability of one conformational state of the enzyme. In either case, structural alterations may arise from short-range effects (i.e. at distances much less than a nanometer) or they can occur from long-range changes at the quaternary structural level (i.e. through rearrangements in subunit–subunit interactions). The term *allostery* is derived from the Greek prefix *allo* (meaning "another") and the suffix "steros" (meaning "shape" or "form"). Allosteric interactions are attended by changes in substrate binding affinity and/or catalytic efficiency, by raising or lowering thermodynamic barriers for substrate binding and/or catalysis. Allosteric activators have the net effect of increasing an enzyme's affinity for its substrate(s) and/or its catalytic performance; allosteric inhibitors have the opposite effect.

Unlike those enzymes lacking regulatory mechanism, almost every allosteric enzyme possesses multiple, mutually interactive subunits, each with its own active site and one or more sites for activators and/or inhibitors. Cooperativity is the phenomenon wherein effector molecule binding on one subunit of an oligomeric protein results in local structural changes that are communicated to and from sites on other subunits. In the Monod–Wyman–Changeux concerted transition model, an oligomeric protein undergoes an all-or-none transition in response to changes in the concentration of effector molecules; alternatively, in the Adair–Koshland model, the conformational changes occur sequentially, again in response to changes in regulatory molecule concentration. Such binding interactions alter catalytic activity, often over narrow ranges of substrate and/or effector concentrations. The need for such responsiveness becomes clear where one realizes that most enzymes are not particularly responsive to changes in the concentration of input signals. Enzymes obeying the Michaelis–Menten kinetics (see preceding text) require large changes in substrate concentration, whereas the activity of highly cooperative enzymes responds over a more narrow concentration range.

9
Inborn Errors in Enzyme Function

Genetic mechanisms minimize the occurrence of mutant enzymes that improperly catalyze a reaction or inappropriately respond to regulatory cues. Such renegade enzymes can wreak havoc on cells, tissues, and organs. Biomedical research has uncovered countless instances wherein improper catalysis/control of even a single enzyme can greatly distress a living organism. In fact, many animal and plant diseases arise from point mutations altering enzyme efficiency through an incorrect substitution of a single amino acid. Other injurious mutations result in over/underproduction of enzymes, defective regulatory interactions, impaired stability, incorrect posttranslational modification, improper subcellular targeting, and so on. Diseases are linked to the failure of specialized enzymes designed to remediate any DNA damage arising from photolysis, oxidation, hydrolysis, racemization, and other molecular rearrangements. A notable example of an enzyme-associated pathophysiology is amyotrophic lateral sclerosis (ALS) (also known as *Lou Gehrig's disease*), a terrible neurodegenerative disorder linked to impaired action of superoxide dismutase. This enzyme plays a powerful role in mitigating oxidative stress by catalyzing the conversion of two superoxide ($O_2^{\cdot -}$) ions to produce dioxygen and the less toxic substance, hydrogen peroxide. Excess superoxide destroys essential cellular components, especially in neurons where oxidative stress has profound pathological sequelae. Another example, taken from research on Alzheimer's disease, is the very finding that enzyme-catalyzed prolyl isomerization can regulate folding of the cytoskeletal protein Tau in axons, such that enzyme knockouts in mice bring about progressive age-dependent neuropathy characterized by motor and behavioral deficits, tau hyperphosphorylation, paired helical filament formation, and neurodegeneration. Finally, the devastating effects observed in primary oxalemia have been traced to the mistargeting of an oxalate-degrading enzyme from peroxisomes to mitochondria.

See also Heme Enzymes; Metalloenzymes; Organic Cofactors as Coenzymes; Ribozymes; Telomerase.

Bibliography

Books and Reviews

Abeles, R.H., Frey, P.A., Jencks, W.P. (1992) *Biochemistry*, Jones & Bartlett, Boston, MA, p. 838.

Fersht, A.R. (1998) *Structure and Mechanism in Protein Science: A Guide to Enzyme Catalysis and Protein Folding*, W. H. Freeman, New York, p. 650.

Haldane, J.B.S. (1930) *Enzymes*, Longmans-Green, London.

Hammes, G.G. (2002) Multiple conformational changes in enzyme catalysis, *Biochemistry* **41**, 8221.

Jencks, W.P. (1969) *Catalysis in Chemistry and Enzymology*, McGraw-Hill, San Francisco, CA.

Lerner, R.A., Benkovic, S.J. (1988) Principles of antibody catalysis, *Bioessays* **9**, 107–112.

Purich, D.L. (2001) Enzyme catalysis: a new definition accounting for noncovalent substrate- and product-like states, *Trends Biochem. Sci.* **26**, 417.

Scriver, C.R., Sly, W.S., Childs, B., Beaudet, A.L., Valle, D., Kinzler, K.W., Vogelstein, B. (2000) *The Metabolic & Molecular Basis of Inherited Disease*, 7th edition, McGraw-Hill, New York.

Voet, D., Voet, J.G. (2003) *Biochemistry Volume 1 Biomolecules, Mechanisms of Enzyme Action, and Metabolism* 3rd edition, J. Wiley, New York, p. 1178.

Wolfenden, R. (2001) The depth of chemical time and the power of enzymes as catalysts, *Accts Chem. Res.* **34**, 938–945.

Protein Circular Dichroism: *see* Circular Dichroism in Protein Analysis

Protein Complexes, Supramolecular: *see* Supermolecular Protein Complexes

Protein-DNA Interactions: *see* DNA–Protein Interactions

Protein Expression by Expansion of the Genetic Code

Jason W. Chin[1] and Thomas J. Magliery[2]
[1] *MRC Laboratory of Molecular Biology, Cambridge, UK*
[2] *Yale University, New Haven, CT, USA*

1	**Expanding the Genetic Code** 47	
1.1	The Chemical and Biosynthetic Incorporation of Unnatural Amino Acids into Proteins 47	
1.2	A General Strategy for Expanding the Genetic Code of Organisms 48	
1.3	Unique Codons 49	
2	**Expanding the Genetic Code of Bacteria** 50	
2.1	Orthogonal Aminoacyl-tRNA-Synthetase/tRNA Pairs for Bacteria 50	
2.2	Methods for Evolving Synthetases with Unnatural Amino Acid Specificity in *E. coli* 52	
3	**Expanding the Genetic Code of Eukaryotic Cells** 58	
3.1	Stoichiometric Aminoacylation Methods 58	
3.2	Catalytic Aminoacylation Methods 58	
3.3	Orthogonal Aminoacyl-tRNA Synthetase/tRNA Pairs in Eukaryotic Cells 58	
3.4	Methods for Evolving Synthetases with Unnatural Amino Acid Specificity in Eukaryotic Cells 60	
4	**Using Expanded Genetic Codes** 62	
4.1	The Utility of Expanded Genetic Codes 62	
4.2	Biophysical Probes 63	
4.3	Unique Reactive Handles 64	
4.4	Posttranslational Modifications 64	
4.5	Organism Evolution 64	
5	**Summary** 65	

Encyclopedia of Molecular Cell Biology and Molecular Medicine, 2nd Edition. Volume 11
Edited by Robert A. Meyers.
Copyright © 2005 Wiley-VCH Verlag GmbH & Co. KGaA, Weinheim
ISBN: 3-527-30648-X

Acknowledgment   65

Bibliography   66
Books and Reviews   66
Primary Literature   66

Keywords

Amber
The name given to the UAG stop codon, which is the least used of the three stop codons in *E. coli* and yeast.

Amber-suppressor tRNA
A tRNA that is capable of decoding the *amber* stop codon, UAG, is said to "suppress" *amber* nonsense mutations.

Aminoacyl-tRNA Synthetase (aaRS)
An enzyme capable of esterifying an amino acid onto a tRNA. These enzymes activate free amino acids to adenylates using ATP, and some actively "edit" misacylated tRNAs by hydrolysis of the errant amino acid.

Codon
The unit of mRNA that codes for an amino acid. Natural codons are three consecutive ribonucleotides (G, A, U and C); there are therefore 64 natural codons of which three (UAG, UGA, and UAA) code for termination of protein synthesis (i.e. "stop"). Codons are read by complementarity to anticodons on tRNAs at the ribosome.

Identity Element
The portions of tRNAs responsible for conferring specificity for a cognate aminoacyl-tRNA synthetase.

Library
A collection of related molecules that vary in a specified way, typically synthesized in a highly parallel fashion. For example, a library might consist of a large collection of mutants of a specific protein, such as an aminoacyl-tRNA synthetase.

Orthogonal Pairs
Aminoacyl-tRNA synthetases and their tRNA substrates that do not interact with the tRNAs and synthetases of the host cell in which they are expressed.

Selection
A procedure for sorting cells on the basis of their ability to survive under specified conditions. Selections can be used to sort libraries of enzymes expressed in cells,

if the survival of the organism requires the activity of the enzyme. In contrast, a screen is a procedure for sorting cells on the basis of phenotype, such as fluorescence.

tRNA
Small RNAs, typically about 75 nt ending with the sequence CCA-3', which deliver amino acids for protein synthesis at the ribosome in response to the mRNA codons to which they correspond. tRNAs are acylated with amino acid by aminoacyl-tRNA synthetases.

Unnatural Amino Acids
Protein monomer building blocks not inserted by the biosynthetic machinery of living organisms. There are 20 natural amino acids normally incorporated into proteins.

> The Genetic Code consists of 64 three-base codons that code for 20 amino acids and three stop signals. Recent engineering of the cell's protein synthesis machinery has allowed unnatural amino acids to be encoded in the genetic code of prokaryotic and eukaryotic organisms. Importantly, the approaches discussed add unnatural amino acids to the genetic code, without detracting from the cell's ability to incorporate the common 20 amino acids into proteins. Here we discuss the intellectual and practical strategies that have been employed in the challenging, but ultimately successful expansion of the genetic code. We also discuss the emerging applications of expanded genetic codes in solving basic biological and biophysical problems and in creating a new synthetic biology.

1
Expanding the Genetic Code

1.1
The Chemical and Biosynthetic Incorporation of Unnatural Amino Acids into Proteins

One of the remarkable chemical mysteries of life is that molecules composed of just 20 amino acid monomers carry out the entire repertoire of protein function – which spans structural roles, metabolism, signal transduction, immunity, replication and degradation, to name just a few. The chemical repertoire of proteins is naturally enlarged by the occasional insertion of selenocysteine or pyrrolysine, and by the common addition of posttranslational modifications (including phosphorylation, sulfation, lipidation, biotinylation, glycosylation, hydroxylation, methylation and acetylation), particularly in higher organisms. Those taking a chemical approach to biology have long recognized that expanding and precisely controlling the functionalities that can be incorporated into proteins would allow the creation of powerful tools for enzymatic, biophysical, and cellular studies. Despite its obvious utility, achieving the facile incorporation of "unnatural" amino acids has been and continues to be a difficult problem.

A number of methods have been developed for the insertion of unnatural amino acids into proteins, from synthetic to biosynthetic. Solid-phase peptide synthesis combined with enzymic or chemoselective ligation now permits the *in vitro* synthesis of small proteins with total control over the amino acid composition. Larger proteins are accessible *in vitro* using semisynthetic methods such as expressed protein ligation, wherein a synthetic peptide is chemoselectively ligated to a recombinant protein fragment. Biosynthetic approaches give access to essentially any size protein, but have their own limitations. For example, addition of near-homologs of natural amino acids (such as selenomethionine) to the growth medium sometimes results in widespread replacement of a natural amino acid, but, while useful, this does not actually expand the genetic code. Site-specific biosynthetic expansion of the genetic code has been achieved by chemical esterification of a tRNA with an unnatural amino acid, both in cell-free translation systems and with microinjection or electroporation of cells. Here, the rarest stop codon is used as a "blank" in the genetic code, read by the esterified tRNA. While this is a powerful approach, its use has been limited by the heroic synthetic effort required. Recent ribozyme-based approaches for acylation of tRNAs with unnatural amino acids may expand the scope of this method.

A quantum leap forward has occurred in recent years in this field – the first organisms have been engineered that site specifically utilize a twenty-first amino acid monomer. One advantage to this approach is that the researcher need only transform an expression vector for the protein of interest into a strain with the appropriate biosynthetic machinery, and addition of unnatural amino acid to the growth medium results in protein containing unnatural amino acid. The ease of use finally makes available proteins with unique reactive handles, homogeneous posttranslational modifications, and intrinsic biophysical probes, to the experimentalists such as enzymologists and cell biologists who can make the best use of them. Here we describe the engineering of prokaryotic and, more recently, eukaryotic organisms with expanded genetic codes, highlighting uses, challenges and future directions.

1.2
A General Strategy for Expanding the Genetic Code of Organisms

Expanding the genetic code of an organism involves the addition of an unnatural amino acid to the organism's genetic repertoire without loss of the ability to insert any of the 20 natural amino acids into proteins. An organism with an expanded genetic code will incorporate unnatural amino acids at specific sites in a protein in response to a unique codon, through a unique enzyme (an aminoacyl-tRNA synthetase) capable of acylating a tRNA for delivery of the unnatural amino acid in response to that codon (see Fig. 1). Ideally, the amino acid will be incorporated with a fidelity rivaling that for natural amino acid–incorporation. To achieve an organism with an expanded genetic code, the following are required:

1. A host organism: Owing to the extensive engineering required to expand the genetic code, well-understood systems like *Escherichia coli* and yeast have proven good starting points.
2. An insertion signal (codon): The *amber* stop codon is a good first choice, since it is rarely used in *E. coli* and yeast, and

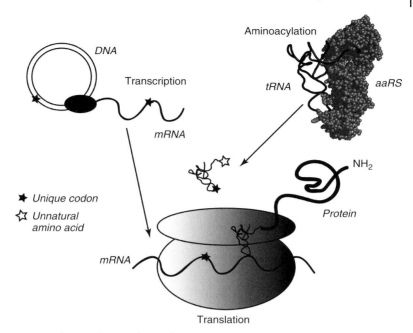

Fig. 1 Scheme of protein biosynthesis: Expansion of the genetic code first requires a unique codon (e.g. *amber* stop codon or 4-base codon), which is specified in the DNA and transcribed into the mRNA. A unique aminoacyl-tRNA synthetase (aaRS) must acylate a unique tRNA with the unnatural amino acid. This tRNA then delivers the unnatural amino acid to the ribosome in response to the unique codon, incorporating it into protein site specifically.

since robust "suppressors" of *amber* mutations are known. Expansion of the genetic code with multiple distinct unnatural amino acids is possible using extended insertion signals such as four-base codons.

3. An "orthogonal" tRNA corresponding to the insertion signal that is neither acylated nor deacylated ("edited") by any endogenous aminoacyl-tRNA synthetase (aaRS).
4. An orthogonal aaRS, that acylates only the orthogonal tRNA, and only with an unnatural amino acid. For simplicity, an aaRS without hydrolytic editing is a good choice.
5. An unnatural amino acid that is neither toxic nor incorporated into proteins by endogenous aaRSs, but that is readily taken up by the cell (or produced there).

1.3 Unique Codons

The genetic code consists of 64 three-base codons that code for 20 amino acids and three stop signals. Since only one stop codon is needed for translational termination, the other two can in principle be used to encode unnatural amino acids. The *amber* stop codon (UAG) has been extensively used in the *in vitro* and *in vivo* biosynthesis of unnatural amino acid–containing proteins. This codon is also the least used stop codon in both *E. coli* and in yeast, and several laboratory *E. coli* strains contain suppressors that insert a natural amino acid in response to UAG without significantly affecting growth rates. Suppressors have also been discovered for *amber* codons in mammalian cells and in higher organisms, such as yeast, worms and mice. It should therefore be possible to incorporate unnatural amino acids in response to an *amber* codon in a range of host organisms from prokaryotes to eukaryotes without significant perturbation of the host. In fact, orthogonal pairs that decode *amber* codons and insert unnatural amino acids have been developed for *E. coli*, yeast, and mammalian cells.

An alternative and potentially complementary set of codons to the *amber* codon are four-base and other extended codons. Naturally occurring frameshift suppressors in yeast and *Salmonella* are known, typically functioning by decoding a 4-base codon using a tRNA with an extended anticodon loop (8 nt instead of 7 nt). Suppression of 4-base and 5-base codons has been demonstrated in *E. coli* and in *E. coli* cell-free translation systems. Variants of tRNA^{Ser} with randomized, extended anticodon loops (8 or 9 nt instead of 7 nt) have been selected for the ability to suppress randomized four-base codons in the gene for β-lactamase. The selected variants included suppressors for the codons AGGA, UAGA, CCCU, and CUAG, with efficiencies comparable to *amber* suppression. A similar approach afforded fairly efficient 5-base codon suppressors of CUAGU, CUACU, and AGGAU.

In practice, it has been demonstrated that a four-base codon AGGA and the *amber* codon UAG can be used to direct the incorporation of two distinct unnatural amino acids into a single site in protein in *E. coli* with high fidelity and specificity. Moreover, simultaneous insertion of two unnatural amino acids has been accomplished in cell-free transcription/translation systems with two 4-base codons. The number of 4-base codons with reasonable suppression efficiency probably places a small, finite upper limit (perhaps 4) on the number of amino acids that can be inserted this way. An alternative approach is to develop "unnatural codons" that utilize a nonnatural base pair, and this is actively being pursued by several groups. The ability to generate a stable base pair that can faithfully be enzymatically incorporated into both DNA and RNA in living cells (which must make or uptake the monomers) is a difficult problem that is still far from a practical solution.

2
Expanding the Genetic Code of Bacteria

2.1
Orthogonal Aminoacyl-tRNA-Synthetase/tRNA Pairs for Bacteria

The first stage in expanding the genetic code of an organism is to generate an orthogonal tRNA/aaRS pair, and second, to change the specificity of the aaRS with respect to amino acid. (Hereafter, the notation *Oo*tRNA^{Aaa}(NNN) for tRNAs and *Oo*AaaRS for aminoacyl-tRNA synthetases shall be used, where *Oo* denotes the organism, Aaa denotes the amino acid, and NNN denotes the anticodon.) All initial work was carried out in *E. coli* using *amber*-suppressing tRNAs. The Schultz laboratory first employed an engineering approach to generate an orthogonal tRNA/synthetase pair based on the exceedingly well-characterized *E. coli* glutaminyl-tRNA synthetase/tRNA^{Gln}(CUA) pair. (GlnRS naturally acylates an efficient *amber* suppressor in *E. coli*.) Despite a remarkable change in specificity toward the orthogonal tRNA, the engineered mutant GlnRS was still capable of efficiently acylating the wild-type tRNA^{Gln} as well. If the amino acid specificity of this enzyme had been subsequently altered, the unnatural amino acid would have been inserted at Gln codons throughout the proteome, in addition to *amber* codons.

Selection against aminoacylation of endogenous wild-type tRNAs might lead to aaRSs with specificity for the orthogonal tRNA. RajBhandary and coworkers pursued such a strategy with a mutant *E. coli* initiator tRNA that is not aminoacylated by endogenous synthetases and the tyrosyl-tRNA synthetase from *Saccharomyces cerevisiae*. The wild-type ScTyrRS also aminoacylates *E. coli* tRNA^{Pro}, and is therefore toxic in *E. coli*. However, when a library of ScTyrRS mutants was screened simultaneously for a lack of toxicity and the ability to suppress *amber* mutations in the gene for chloramphenicol acetyltransferase (which confers resistance to the antibiotic chloramphenicol), nontoxic variants of TyrRS were isolated, which discriminated 15-fold against *E. coli* tRNA^{Pro}.

A second strategy for generating new orthogonal pairs involves importing a heterologous *amber*-suppressor aaRS/tRNA pair from another organism. This approach was, in part, driven by the observation that the identity elements in tRNAs for eukaryotic aaRSs often differ significantly from their prokaryotic counterparts. Many eukaryotic aminoacyl-tRNA synthetases therefore do not charge *E. coli* tRNAs and should be orthogonal in *E. coli*. In particular, several yeast synthetases (notably glutamine, aspartic acid, arginine, and tyrosine) do not aminoacylate total *E. coli* tRNA, and might be useful for the construction of orthogonal tRNA-synthetase pairs.

On the basis of the observation that *S. cerevisiae* tRNA^{Gln} (SctRNA^{Gln}) is not acylated by *E. coli* GlnRS (EcGlnRS), Liu and Schultz demonstrated that a modified SctRNA^{Gln}(CUA) and ScGlnRS constitute a functional orthogonal pair in *E. coli*. Despite nearly 10 man-years of work engineering this pair, it was not possible to alter the amino acid specificity of ScGlnRS, even though much of the selection and library construction technology was worked out on this system.

Archaeal tRNA synthetases are more similar to their eukaryotic than prokaryotic counterparts in terms of homology and tRNA identity elements. However, unlike synthetases from eukaryotic cells, synthetases from archaea can be expressed in good yields in *E. coli* and can be readily purified in active form. Numerous archaeal genome sequences are currently available, which, together with the absence of introns in archeal genomes, facilitate the cloning of the archaeal synthetase genes. Moreover, early work on the halophile *Halobacterium cutirebrum* indicated that almost all its tRNAs (notably those for leucine, arginine, tyrosine, serine, histidine, and proline) cannot be charged by *E. coli* aminoacyl-tRNA synthetases, suggesting that these tRNAs are potentially orthogonal in *E. coli*. Consequently, attention has recently focused on the archaea, rather than on eukaryotes, as a source of orthogonal tRNA-synthetase pairs for *E. coli*.

It was shown that a tRNA^{Tyr}(CUA) and TyrRS from the archea *Methanococcus jannaschii* constitute an orthogonal pair in *E. coli*. However, while the MjTyrRS is considerably more active than ScGlnRS, the MjtRNA^{Tyr}(CUA) was not "as orthogonal" as SctRNA^{Gln}(CUA); that is: the

MjtRNA^{Tyr}(CUA) is detectably acylated by E. coli synthetases. A semirational design strategy was used to increase the orthogonality of this tRNA using a variant of the "double-sieve" selection for unnatural amino acids (see the following section). The resulting orthogonal tRNA/synthetase pair has been the most useful one so far in E. coli, due to the high activity of the synthetase and good orthogonality of the tRNA. The relatively large, hydrophobic amino acid binding pocket of TyrRS has allowed the incorporation of many useful amino acids. Recently, solution of the MjTyrRS crystal structure permitted design of a mutant synthetase with better recognition of the MjtRNA^{Tyr}(CUA), which may be useful in further work altering the amino acid specificity of the synthetase.

A number of other pairs have also been introduced into E. coli including yeast tRNA^{Asp}(CUA) and a point-mutant of yeast AspRS; derivatives of Halobacterium NRC-1 tRNA^{Leu} and Methanobacterium thermoautotrophicum LeuRS that result in suppression of amber, opal (UGA), and 4-base (AGGA) codons; and a consensus-derived archaeal tRNA^{Glu}(CUA) and Pyrococcus horikoshii GluRS. However, there has been no success in altering the amino acid specificity of any of these synthetases. Recently a mutant Pyrococcus horikoshii LysRS derivative has been selected for the incorporation of homoglutamine in E. coli in response to the four-base codon AGGA using a consensus-derived archaeal tRNA^{Lys}(UCCU). This is only the second orthogonal pair shown to be useful for unnatural amino acid incorporation in E. coli, and the fact that it allows four-base suppression means that it is now possible to simultaneously, site specifically introduce two unnatural amino acids into proteins.

2.2
Methods for Evolving Synthetases with Unnatural Amino Acid Specificity in E. coli

Once an orthogonal tRNA-synthetase pair has been developed for use in E. coli, the next step is the modification of the synthetase specificity such that it aminoacylates the orthogonal tRNA with a desired unnatural amino acid and no other endogenous amino acid.

This problem is complicated considerably by at least five factors. (1) The small amino acid substrates are contacted by a large number of residues concentrated in three-dimensional space but often spread over the primary structure of the synthetase. This makes it technically difficult to construct reasonable mutagenic libraries. (2) In spite of this, mutations far from the active site of aaRSs are known to affect aminoacylation kinetics. (3) Mutations to the synthetase can also affect recognition of tRNA, thereby altering the orthogonality of the enzyme. (4) It is extremely difficult to alter the specificity of an aaRS without reducing its aminoacylation activity, but a certain level of activity is required to support high-level protein synthesis. Thus, the engineering of synthetases with weak aminoacylation activity toward native substrates has proven to be exceedingly difficult. (5) Selecting for unnatural amino acid specificity requires that the survival of the organism be tied to the insertion of an unnatural amino acid, and no direct way to do this has been developed.

A small number of attempts have been made to alter the specificity of aaRSs by inspection or semiempirically, with varying results. For example, from genetic screens, it was known that the EcPheRS Ala294 → Ser mutant resisted incorporation of p-F-Phe. Ibba and Hennecke showed that the Ala294 → Gly mutant of this enzyme, with an expanded binding pocket, is capable of inserting *p*-Cl-Phe and *p*-Br-Phe, and the Tirrell group has shown that *para*-iodo, azido-, cyano-, and ethynyl-Phe are also accepted. Furter used the associated *p*-F-Phe resistant *E. coli* strain to engineer the first bacterium able to site-selectively insert an unnatural amino acid by introducing a *Sc*PheRS/*Sc*tRNA^{Phe}(CUA) pair that accepts *p*-F-Phe. However, the *Ec*PheRS(A294S) accepts *p*-F-Phe weakly,

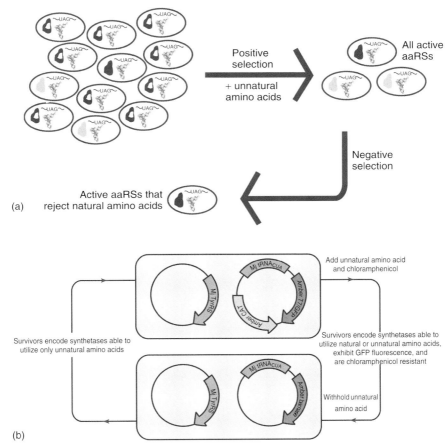

Fig. 2 "Double-sieve" selections for unnatural amino acid activity: (a) Schematic of a generic double-sieve selection. A library of orthogonal aminoacyl-tRNA synthetases is transformed into cells with an orthogonal tRNA. First, positive selection is done in the presence of unnatural amino acids, such that the selective phenotype requires acylation of the orthogonal tRNA and suppression of the unique codon. Cell with inactive synthetases (white) die, but those with synthetases active toward natural (grey) or unnatural (black) amino acids survive. Second, a negative selection is carried out in the absence of unnatural amino acid, where the selective phenotype requires that the tRNA *not* be acylated. Only cells with synthetases that lack activity toward natural amino acids survive. Thus, survival of both selections implies activity only toward unnatural amino acids. (b): A double-sieve selection in *E. coli*. (c): A double-sieve selection in yeast. (See color plate p. xxvi.)

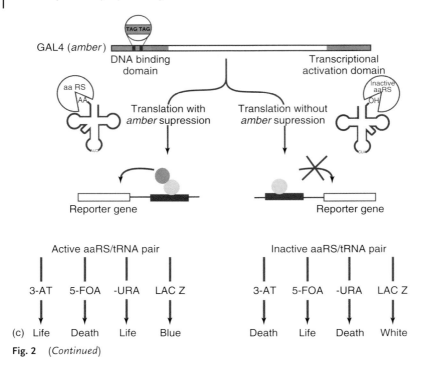

Fig. 2 (Continued)

and the SctRNA^{Phe}(CUA) is also acylated with Phe and Lys (presumably by EcLyrsRS, in the latter case). Kiga et al. converted EcTyrRS into an enzyme capable of accepting 3-iodotyrosine (3-I-Tyr) as a substrate by selecting three sites for mutagenesis based on structure, and examining the aminoacylation properties of 50 mutants *in vitro*. Interestingly, while >95% of *amber*-encoded sites were occupied by 3-I-Tyr in a wheat-germ extract translation reaction, Tyr-containing protein was produced if 3-I-Tyr was left out of the reaction mixture. This is potentially a problem when moving this technology into cells, since the concentration of unnatural amino acid cannot be controlled arbitrarily. A number of attempts have been made to change GlnRS into GluRS, but the resulting enzymes still prefer Gln (although by a considerably smaller amount than wild-type GlnRS). Additionally, some attempts to computationally predict mutations to alter amino acid specificity have been reported.

By far the most successful and general method for altering the specificity of aminoacyl-tRNA synthetases has been directed combinatorial mutagenesis of the amino acid binding pocket of synthetase enzymes coupled to a general "double-sieve" selection (see Fig. 2). Several general methods have been developed to alter the specificity of orthogonal aminoacyl-tRNA synthetase/tRNA(CUA) pairs that function in *E. coli* so that they use only a desired unnatural amino acid.

The selection of a library of aaRS variants was first carried out by a two-step procedure. In the first step, the library of synthetases is selected in bacteria bearing tRNA(CUA) and an *amber* mutant of β-lactamase, with ampicillin and unnatural amino acid supplementation of the medium. Use of a permissive site in β-lactamase ensures that survivors of the selection contain aaRSs that are capable of acylating the tRNA(CUA) – thus producing β-lactamase and conferring resistance to ampicillin – but which amino acid was esterified onto the tRNA is unknown. The synthetases from the survivors of this positive selection are then expressed in bacteria bearing the tRNA(CUA) and a barnase gene containing multiple amber codons (wild-type barnase is a lethal ribonuclease in *E. coli*). No unnatural amino acids are added to the medium, so survival at this stage ensures that no cellular amino acids (proteinogenic or otherwise) are a substrate for the mutant synthetases. Clones that survive *both* steps of the selection should therefore be active *only* toward the unnatural amino acid. The initial implementation of this strategy (with the *Sc*GlnRS/*Sc*tRNA^{Gln}(CUA) orthogonal pair, a large library of mostly commercially available amino acids, and synthetase libraries created by random mutation) resulted in no useful synthetases.

Two critical technical improvements to the methodology were (1) the replacement of the β-lactamase positive selection with a chloramphenicol acetyltransferase-based selection, and (2) use of a directed, semirational library-construction method. Chloramphenicol is bacteriostatic and acts cytosolically, allowing a broad range of chloramphenicol concentration to be suitable for selection. Random mutagenesis of wild-type GlnRS had two negative consequences. First, most of the library members were very active toward Gln, requiring exceptional performance from the negative selection. Second, the probability of mutation in the proximity of the amino acid substrate was low. Instead, guided by the ternary X-ray crystal structure of *Ec*GlnRS, tRNA^{Gln}, and a Gln-AMP analog, 5–10 residues proximal to the substrate were first mutated to Ala and then randomized to all 20 amino acids. The resulting libraries contained few synthetases with activity toward Gln, allowing one to forego the negative selection in early rounds, and were almost certain to have an altered amino acid binding pocket.

Unfortunately, libraries designed for use with carboxamide N-alkylated Gln analogs, analogs elaborated from the γ-carbon, and the α-hydroxy acid analog of Gln were not found to contain any synthetases capable of activating the unnatural amino acids at useful levels. However, using the very same library construction and selection methodology, it has been possible to alter the specificity of *Mj*TyrRS and *Ph*KRS, allowing the delivery of several useful unnatural amino acids (see section 4 and Fig. 4). Santoro et al. extended the selection technology by developing a fluorescence-based screen for *amber* suppression. An *amber* mutant of T7 RNA polymerase was used to drive the transcription of green fluorescent protein (GFP). In combination with fluorescence-activated cell sorting (FACS), this system can be used as both a positive screen for *amber* suppression (where one sorts for fluorescent cells in the presence of unnatural amino acids) and a negative screen against *amber* suppression (sorting for dim cells in the absence of unnatural amino acids). A multivalent system, where chloramphenicol-based selection is possible in addition to screening, has been especially effective.

The fidelity of incorporation of each amino acid was established by several methods. First the clones were characterized by their ability to confer chloramphenicol resistance on cells transformed with the CAT(*amber*) reporter in the presence and absence of a particular unnatural amino acid. In all successful cases, the IC_{50} values with respect to chloramphenicol were large (100 µg mL^{-1}) in the presence of unnatural amino acid, and were generally less than 10 µg mL^{-1} in the absence of the unnatural amino acid. Similarly, cellular fluorescence can be measured with the T7/GFP system in the presence and absence of unnatural amino acid.

Expression of *amber* mutants of DHFR, myoglobin, and the Z-domain of protein A was also directly assessed. In the presence of unnatural amino acid, synthetase, and tRNA, milligrams of each protein could be purified from a liter of liquid culture. Yields were typically 25 to 30% of wild-type protein. In the absence of any one component synthetase, tRNA, or amino acid, no protein could be observed by Western blot, silver stain, or Coomassie stain. Proteins were subjected to mass analysis of intact protein, and tryptic digests confirmed that the unnatural amino acid and only this amino acid is incorporated in response to the UAG codon and at no other site in the protein.

It is likely that the high activity of the *Mj*TyrRS and the excellent orthogonality of the synthetase and tRNA account for the malleability of this enzyme's substrate specificity. Though all of the libraries of *Mj*TyrRS were designed using the *Bacillus stearothermophilus* structure, the structure of the *M. jannaschii* protein has recently been solved. There are a number of differences in the tyrosine binding pocket that will improve further library design with the enzyme, and may even suggest ways to improve some of the existing mutant synthetases (see Fig. 3).

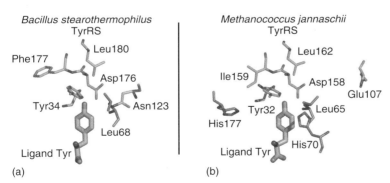

Fig. 3 The active site of TyrRS: Amino acids near the substrate tyrosine of (a) *Bacillus stearothermophilus* and (b) *Manococcus jannaschii* TyrRS. The Bacillus structure is a good model for the *Ec*TyrRS used to expand the genetic code of yeast. All the *Ec*TyrRS variants for the insertion of unnatural amino acids were derived from variants of Tyr37 (Tyr34), Asn126 (Asn123), Asp182 (Asp176), Phe183 (Phe177) and Leu186 (Leu180), where the corresponding amino acids in *Bs*TyrRS are noted parenthetically. Nearly all of the *Mj*TyrRS derivatives used for insertion of unnatural amino acids in *Escherichia coli* were generated by randomization of the Tyr32 (Tyr34), Glu107 (Asn123), Asp158 (Asp176), Ile159 (Phe177), and Leu162 (Leu180). These amino acids were selected from the *Bs*TyrRS structure, which was available at the time, but the recent *Mj*TyrRS structure suggests some possible improvements. Rendered using PyMOL (Warren Delano, 2002, http://www.pymol.org) from PDB entries 4TS1 and 1J1U.

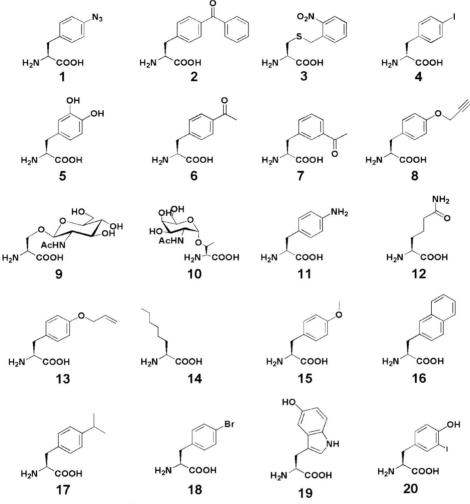

Fig. 4 Unnatural amino acids incorporated site specifically in living cells with modified aminoacyl-tRNA synthetases:
(1) *p*-azido-L-phenyalanine,
(2) *p*-benzoyl-L-phenylalanine,
(3) *o*-nitrobenzyl-L-cysteine,
(4) *p*-iodo-L-phenylalanine,
(5) *m*-hydroxy-L-tyrosine,
(6) *p*-acetyl-L-phenylalanine,
(7) *m*-acetyl-L-phenylalanine,
(8) *p*-propargyloxy-L-phenylalanine,
(9) β-GlcNAc-L-serine,
(10) β-GlcNAc-L-threonine,
(11) *p*-amino-L-phenylalanine,
(12) homoglutamine, (13) *O*-allyl-L-tyrosine,
(14) α-aminocaprylic acid,
(15) *O*-methyl-L-tyrosine,
(16) 3-(2-naphthyl)-L-alanine,
(17) *p*-isopropyl-L-phenylalanine,
(18) *p*-bromo-L-phenylalanine,
(19) 5-hydroxy-L-tryptophan,
(20) *m*-iodo-L-tyrosine.

3
Expanding the Genetic Code of Eukaryotic Cells

3.1
Stoichiometric Aminoacylation Methods

The first expansion of a cell's genetic code was performed in *Xenopus* oocytes by microinjection of a chemically misacylated *Tetrahymena thermophila* tRNA(CUA), and a mutant mRNA containing the *amber* nonsense codon. An advantage of this approach, shared with the *in vitro* incorporation methods from which it is derived, is that the chemical groups that can be incorporated are constrained only by the translational machinery of the cell (which is highly permissive to a variety of amino acids with altered side chains) and by the chemistry of aminoacylation. This methodology allowed the detailed biophysical study of the nicotinic acetylcholine receptor in oocytes by the introduction of amino acids containing side chains with unique physical or chemical properties and the study their effect using patch clamp methods. A disadvantage of this approach for studying other cellular proteins is that it is not applicable to producing mutant versions of many proteins at their normal *in vivo* levels or for producing enough protein to assay using standard cell biological approaches.

Recent extensions of this approach have extended the type of eukaryotic cell that it can be used to study, by altering the method of delivery of chemically acylated tRNAs and mRNA. It has been shown that chemically aminoacylated tRNAs can be electroporated or transfected into mammalian cells (including neurons) in culture. Again, since the tRNA cannot be re-acylated *in vivo*, the yield of protein produced is very low and this potentially limits the range of cellular processes that these methods can be used to study.

3.2
Catalytic Aminoacylation Methods

There has been considerable interest in extending the principles for expanding the genetic code of bacteria to eukaryotic cells. This would have significant technical and practical advantages over stoichiometric incorporation methods, since tRNAs would be acylated by their cognate synthetases, leading to large amounts of mutant protein. Moreover, genetically encoded aminoacyl-tRNA synthetases and tRNAs are, in principle, heritable, allowing the unnatural amino acid to be incorporated into proteins through many cell divisions without exponential dilution.

The steps necessary to add new amino acids to the genetic code of *E. coli* have been described; similar principles are likely to be useful for expanding the genetic code of eukaryotes. However, because the cellular tRNAs and aminoacyl-tRNA synthetases are not well conserved between prokaryotes and eukaryotes, the translational components from *M. janaschii* added to *E. coli* are not orthogonal in eukaryotic cells. The challenges in extending this approach are therefore to discover new orthogonal pairs in which the tRNA is produced in a functional form *in vivo*, to devise selection methods for altering the specificity of the orthogonal synthetase, and to create large libraries of mutant synthetases in a eukaryotic cell.

3.3
Orthogonal Aminoacyl-tRNA Synthetase/tRNA Pairs in Eukaryotic Cells

The orthogonal pairs reported to date for eukaryotes are derived from bacterial pairs. This reflects the fact that the divergence of identity elements of tRNAs from prokaryotes to eukaryotes is greater than that from eukaryotes and archea and between one eukaryote and another. However, it is clear that in some instances bacterial tRNA cannot simply be cut and pasted into eukaryotic cells. In eukaryotes, the synthesis of tRNAs is dependent on RNA Polymerase III and its associated factors, which recognize A and B box sequences that are internal to the tRNA structural gene ("internal promoters"). The A and B box consensus sequences place limitations on the tRNA sequences that can be used in eukaryotic cells, and some bacterial tRNAs that diverge from the eukaryotic A and B box sequence are simply not transcribed in eukaryotic cells. Moreover, many eukaryotic tRNAs have the 3′ CCA sequence added enzymatically by CCA transferase, while all *E. coli* tRNAs encode this sequence in the DNA sequence, and it may be necessary to delete the 3′ CCA in the *tRNA* gene imported from *E. coli*. The processing or transcription of a tRNA appears, in some cases, to require organism-specific flanking sequences 5′ and 3′ to the tRNA gene. Posttranscriptional nucleotide modifications, that may be absent in the host eukaryote, may also be necessary to produce a functional tRNA molecule. In addition, since the tRNA is transcribed in the nucleus, it must be exported to the cytoplasm to function in translation via an exportin-dependent process. Nonetheless, several potential orthogonal pairs have been described in eukaryotes.

E. coli tyrosyl-tRNA(CUA) can be expressed and correctly processed, as judged by Northern blot, in *S. cerevisiae* when the flanking sequences 5′ and 3′ to the structural gene are included and the 3′ flanking sequence is altered to a poly T stretch to allow Pol III transcriptional termination. To demonstrate the functionality and orthogonality of *E. coli* tyrosyl-tRNA(CUA) in yeast, a reporter strain was constructed that contained three auxotrophic *amber*-suppressible alleles, allowing qualitative comparison of *amber* suppression on media lacking methionine, tryptophan, or histidine. When this strain was transformed with a plasmid containing the *E. coli* tRNA(CUA) at a variety of copy numbers, all transformants were MET$^-$ HIS$^-$ TRP$^-$, indicating that any aminoacylation of this tRNA by yeast cytoplasmic synthetases was at a low level. When the same yeast strain was cotransformed with plasmids encoding the *E. coli* tRNA(CUA) and its cognate synthetase, suppression of all three alleles was observed, and shown to be correlated with the expression level of the tRNA. These experiments demonstrated that *E. coli* tyrosyl-tRNA is orthogonal in *S. cerevisiae*, and can be aminoacylated by its cognate synthetase. Numerous *in vitro* experiments have demonstrated that *E. coli* tyrosyl-tRNA synthetase does not aminoacylate total *S. cerevisiae* tRNA. Thus, the *E. coli* tyrosyl-tRNA synthetase/tRNA pair is a candidate orthogonal pair. Two other orthogonal pairs have subsequently been reported for *S. cerevisiae*. One is derived from *E. coli* glutamine-tRNA synthetase and an amber suppressor derived from human initiator tRNA while the other is derived from an *E. coli* leucine pair. Thus far the tyrosine and leucine orthogonal pairs form *E. coli* have been used to expand the genetic code of *S. cerevisiae*.

There have also been efforts to create orthogonal pairs in higher eukaryotic cells. A derivative of the *E. coli* glutamine pair is orthogonal and functional in mammalian cells. The tRNA of this pair, a variant of *E. coli* tRNA^{Gln}(CUA), was created by replacing C9 in the bacterial tRNA with the mammalian A box consensus base A9, and the native flanking sequences were replaced with those of the human initiator tRNA. These changes were shown to drastically increase the transcription of the tRNA in a HeLa cell-free system. When COS-1 cells were transformed with a CAT reporter containing an amber codon at a permissive site, a tRNA^{Gln}(CUA) gene driven to high copy on an SV40-based vector and an *Ec*GlnRS gene, *amber* suppression was observed in a CAT assay. CAT activity was dependent on the presence of all three components. Since *Ec*GlnRS does not charge total mammalian tRNA, the *Ec*GlnRS/tRNA(CUA) constitute an orthogonal pair in mammalian cells. However, the amount of suppression observed with this pair was much lower than observed for a human serine amber suppressor, suggesting that this synthetase/tRNA pair is far from optimal. Yokoyama and coworkers have subsequently investigated bacterial tyrosyl-tRNA synthetase tRNA(CUA) pairs for their potential as orthogonal pairs in mammalian cells. These experiments showed that *E. coli* tyrosyl synthetase/*Bs*tRNA(CUA) form a potential orthogonal pair in mammalian cells. The efficiency of this pair appears to be good, with Ras yields from an amber mutant of the *ras* gene being about 25% of those from the wild-type *ras* gene in the same system. The synthetase of this pair has subsequently been altered to allow the incorporation of numerous unnatural amino acids in mammalian cells (see the following). In addition, it has subsequently been shown that the tryptophanyl synthetase and cognate opal suppressor derived from *B. subtilis* are orthogonal in mammalian cells. The efficiency of the pair appears to be comparable to the *E. coli* tyrosine pair and it was used to incorporate an unnatural amino acid in mammalian cells.

3.4
Methods for Evolving Synthetases with Unnatural Amino Acid Specificity in Eukaryotic Cells

With an orthogonal aminoacyl-tRNA synthetase that aminoacylates the orthogonal suppressor tRNA in hand, a method is required to alter the specificity of the aminoacyl-tRNA synthetase so that it aminoacylates its cognate tRNA, and only its cognate tRNA with desired unnatural amino acid and no endogenous amino acid present in a eukaryotic cell.

The first approach to this problem was to screen a small number of active site variants. Using this method, Yokoyama and coworkers evaluated a small collection of *Ec*TyrRS mutants in an *in vitro* wheat-germ translation system and discovered an *Ec*TyrRS variant that utilizes 3-iodotyrosine more effectively than tyrosine. However, this enzyme still incorporates tyrosine in the absence of the unnatural amino acid. More recently, Yokoyama and coworkers have also demonstrated that this *Ec*TyrRS mutant functions with a tRNA(CUA) from *B. stearothermophilus* (see preceding text for a discussion of this orthogonal pair) to suppress *amber* codons in mammalian cells. The fidelity of incorporation was shown by LC-MS to be greater than 95%. A similar approach was subsequently used to discover mutants of the orthogonal *B. subtilis* TrpRS/tRNA(UCA) that incorporate 5-hydroxy-L-tryptophan (5-HTPP). On the basis of the crystal structure of the homologous enzyme from *B. stearothermus*, Val144 that points toward C5 of bound tryptophan adenylate was mutated to all other 19 amino acids. One mutant Val144 → Pro was able to suppress a TGA codon placed in the *foldon* gene in the presence of 5-HTPP. Moreover, in the absence of the unnatural amino acid, no full-length Foldon protein was synthesized, suggesting the mutant synthetase is specific for 5-HTPP. Mass spectrometry confirmed the incorporation with a fidelity >97%.

To add numerous unnatural amino acids to the genetic code of eukaryotic cells, Chin et al. developed a general, *in vivo* selection method for the discovery of variants of *Ec*TyrRS/tRNA(CUA) or other orthogonal pairs that function in yeast to incorporate unnatural amino acids, but no endogenous amino acids, in response to the *amber* codon UAG. A major advantage of a selection is that enzymes that incorporate unnatural amino acids can be rapidly selected and enriched from libraries of 10^8 *Ec*TyrRS active site variants, 6 to 7 orders of magnitude more diverse than can be easily screened *in vitro*. This increase in diversity vastly increases the likelihood of isolating *Ec*TyrRS variants for the incorporation of a diverse range of useful functionality with very high fidelity.

A key to the selection was a strain of *S. cerevisiae*, MaV203:pGADGAL4 (2TAG) (see Fig. 2), that contains a plasmid encoded gene for the transcriptional activator protein GAL4 in which the codons for Thr44 and Arg110 (for which substitution with numerous other amino acids has minimal effects on GAL4 activity) were converted to amber nonsense codons. Suppression of these amber codons leads to the production of full-length GAL4, which in turn leads to transcriptional activation of genomic GAL4 responsive *HIS3*, *URA3*, and *LACZ* reporter genes. Expression of *HIS3* and *URA3* complement the histidine and uracil auxotrophy in this strain, allowing clones expressing active aaRS/tRNA pairs to be selected. Upon addition of 5-fluoroorotic acid (5-FOA), which is converted to a toxic product by the URA3 protein, clones expressing inactive aaRS/tRNA pairs can be selected. The *lacZ* reporter allows active and inactive synthetase/tRNA pairs to be discriminated colorimetrically. Thus, a single strain of *S. cerevisiae* can be used for positive and negative selections, minimizing the number of manipulations necessary to identify aaRS/tRNA$_{CUA}$ pairs that use unnatural amino acids.

To use this selection method to expand the eukaryotic genetic code, a library of *Ec*TyrRS mutants was generated. On the basis of the crystal structure of the homologous *Bs*TyRS in complex with tyrosine, five residues within 7 Å of the *para* position of the bound tyrosine were randomly mutated to all amino acids, and the library was transformed into the selection strain yielding 10^8 transformants. To select for cells containing mutant synthetases that selectively aminoacylate the orthogonal *E. coli* tRNA(CUA), a two-step selection was carried out. In the first step, a positive selection for cells able to grow on media lacking uracil or containing 20 mM 3-aminotriazole (3-AT, a competitive inhibitor of the HIS3 protein) in the absence of histidine was performed. If a mutant TyrRS charges tRNA(CUA) with an amino acid, then the cell biosynthesizes histidine and uracil and survives. In a second step aimed at removing clones that incorporate endogenous amino acids in response to an amber codon, a negative selection was applied. Cells were grown on media containing 0.1% 5-FOA, but lacking the unnatural amino acid. Those cells expressing URA3, as a result of suppression of the GAL4 amber mutations with natural amino acids, convert 5-FOA to a toxic product and cause cell death. Surviving clones were amplified in the presence of unnatural amino acid and reapplied to the positive selection. Synthetases for the incorporation of numerous unnatural amino acids have been developed through this method.

The specificity and activity of each mutant synthetase/tRNA pair was tested in several ways. First, individual clones were tested for the ability to confer survival on MaV203:pGADGAL4 on − Ura or − His +20 mM 3-AT in the presence and absence of an unnatural amino acid, and the blue/white phenotype was tested on X-Gal in the presence and absence of the amino acid. Clones that grow on selective media and turn blue on X-Gal only in the presence of the unnatural amino acid were selected for further characterization. Second, an amber mutant of hexahistidine tagged human superoxide dismutase (SOD) was overexpressed in the presence of each synthetase/tRNA pair with and without cognate unnatural amino acid (see Fig. 5). In the presence of the unnatural amino acid, SOD-His$_6$ was produced, but in the absence of added amino acid no SOD could be purified. Moreover, the SOD purified from yeast cell growth in the presence of each unnatural amino acid was shown, by mass spectrometry, to contain the desired unnatural amino acid at the desired position with fidelity in excess of 99.9%. The total yield of purified SOD was about 20% of that produced from the analogous wild-type SOD gene.

Synthetases evolved in yeast can be readily transplanted to higher eukaryotes when paired with a functional cognate orthogonal tRNA for the higher organism. Using this strategy, the unnatural amino acid dependent synthetases evolved in yeast have been transplanted into mammalian cells, along with the cognate amber-suppressor tRNA derived from *B. stearothermophilus*, and incorporation of each unnatural amino acid demonstrated.

4
Using Expanded Genetic Codes

4.1
The Utility of Expanded Genetic Codes

Several applications of expanded genetic codes have been demonstrated. Numerous future applications can be predicted from applications that have been achieved with other methods including the *in vitro* incorporation of amino acids, multisite amino acid–incorporation, and expressed protein ligation. Perhaps even more exciting are the entirely novel applications that are beginning to emerge, in which *in vivo* incorporation of unnatural amino

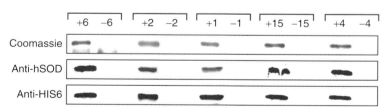

Fig. 5 Unnatural amino acid dependent protein expression in yeast: An *hSOD* gene containing an amber codon at position 33 was used to produce protein in yeast strains with aminoacyl-tRNA synthetase/tRNA pairs for specific amino acids (see Fig. 4). HSOD was purified from yeast in the presence (+) and absence (−) of the unnatural amino acid indicated and SOD expression was visualized by Coomassie stain or Western blot using antibodies against hSOD and the C-terminal hexahistidine tag.

acids are used to interrogate processes inside living cells with molecular precision. Here we focus on the utility of the expanded genetic code in E. coli using pairs derived from *Mj*TyrRS/tRNA(CUA) or in yeast using the *Ec*TyrRS/tRNA(CUA) pair or EcLeuRS/tRNA(CUA).

4.2
Biophysical Probes

Aryl iodides have been site specifically incorporated into proteins in both E. coli and yeast, and have subsequently been used for X-ray structure determination via single wavelength anomalous diffraction on a home X-ray source. The method appears to require significantly less data than selenomethionine incorporation. Both p-benzoyl-phenylalanine and p-azido-phenylalanine have been incorporated into proteins in bacteria and yeast (see Fig. 6). These amino acids allow photoaffinity labeling of interacting proteins upon irradiation at 360 nm. The covalently linked proteins can then be isolated or visualized electrophoretically after in vitro or in vivo cross-linking. This utility of this method was demonstrated in a model system, and has subsequently been used to determine a binding site of a client protein on the chaperone ClpB. Isotopic labeling (e.g. ^{19}F) of unnatural amino acids should aid NMR structure determination. Other biophysical probes such as spin labels for EPR studies are likely future targets.

The genetic code of yeast has been expanded to include a nitrobenzyl cysteine. This photocaged cysteine can be deprotected, allowing the temporal control of protein function. Several unnatural amino acids have been useful for engineering chemical scission sites into proteins, though these have not yet been incorporated into proteins in vivo. Hydroxyacids inserted through chemical acylation of tRNA result in an ester linkage that can be cleaved in base. Similarly, specific cleavage at aminooxyacetic acid can be achieved with zinc and acetic acid, and cleavage of the protein backbone at allylglycine

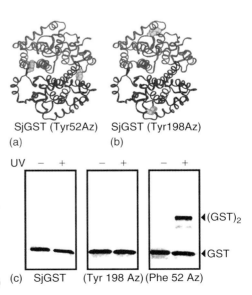

Fig. 6 An example of the utility of expanded genetic codes: p-azido-L-phenylalanine (1) is inserted into glutathionine-S-transferase at position 52 (a) or position 198 (b). On irradiation with UV light (c) the azido group photocross-links protein monomers in the dimeric GST complex. This cross-linking is dependent on the azido group being at the protein interface (position 52). Similar experiments have been used to map protein–protein interactions both in vitro and in cells. (See color plate p. xxv.)

is possible upon treatment with iodine. Photochemical proteolysis occurs upon irradiation at 2-nitrophenylglycines.

4.3
Unique Reactive Handles

Expanding the reactivity of proteins, particularly in a site-specific way, is a potentially useful alteration to proteins. Chemical modification of natural proteins is basically limited to use of the nucleophilic groups: cysteine's thiol, amines on lysine, and at the N terminus, histidine's imidazole, and the alcohols, depending upon pH. However, it is often difficult to prevent cross-reaction with other nucleophilic groups in the protein, and extensive mutation (for example, to yield a single-Cys version of a protein) is often required. It is possible to introduce an electrophile into a protein by mild oxidation of N-terminal serine or threonine, but more general approaches to unique reactive handles have many potential applications.

The genetic codes of *E. coli* and yeast have been expanded to include a ketone that allows the chemoselective formation of hydrazones and oximes using hydrazides and hydroxylamines respectively. Hydrazone formation is possible in living cells to label ketone-containing protein with fluorophore hydrazides. Both cytosolic and outer-membrane proteins have been labeled in this fashion. Ketones also allow for the possibility of protein stabilization by formation of an imine with lysine. The addition of an azides and alkynes to the genetic code of both *E. coli* and yeast allows site specific–protein derivatization via a Cu(I) catalyzed 3 + 2 cycloaddition. This reaction has been used to specifically PEGylate proteins or to add biotin or fluoroscein groups. An alternative derivatization strategy for azides would involve a Staudinger-like ligation, which has been used to selectively modify sugar moieties on cell surfaces, in mice, and for the labeling of azidohomoalanine introduced into proteins using *in vivo* amino acid replacement.

4.4
Posttranslational Modifications

Understanding the roles of proteins in cellular function necessarily means understanding the extent and function of posttranslational modification, particularly in eukaryotes. However, it is difficult to obtain specifically, homogeneously modified protein. The genetic code of *E. coli* has been expanded for the incorporation of β-GlcNAc-serine, where proteins are not normally glycosylated, permitting homogeneous preparation of large amounts of glycoprotein. Expressed protein ligation has also been used to study posttranslational modifications via the synthesis of phosphorylated or nonhydrolyzable phosphorylation mimic amino acids, or via geranylgeranyl amino acids. It is likely that the expansion of the genetic code will allow the direct incorporation of these and other posttranslational modifications into proteins *in vivo*.

4.5
Organism Evolution

In all the above studies, the amino acid must be added to the growth medium and taken up by the cell in order to be incorporated into a protein in response to the *amber* codon. An alternative strategy is to engineer a pathway for the biosynthesis of the unnatural amino acid in the host organism. In this way, a completely autonomous bacterium has been generated that contains genes for the biosynthesis of *p*-amino-L-phenylalanine (pAF) from simple carbon sources, an aminoacyl-tRNA synthetase that uses this new amino acid and no other endogenous amino acids, and a tRNA that delivers into proteins in response to the *amber* codon.

To biosynthesize pAF, the *papA*, *papB*, *papC* genes from *S. venezuelae* were imported into *E. coli*. The PapA protein, 4-amino-4-deoxychorismate, was expected to convert chorismate to 4-amino-4-deoxychorismic acid using ammonia from glutamine in a simple addition-elimination reaction. PapB should convert 4-amino-4-deoxychorismic acid to 4-amino-4-deoxyprephenic acid, and PapC should then convert 4-amino-4-deoxyprephenic acid to *p*-aminophenylpyruvic acid. The endogenous aminotransferases (*tyrB*, *asps*, *ilvE*) were expected to convert this to pAF. *E. coli* containing these genes were found to produce *p*-amino-L-phenylalanine at levels comparable to those of other aromatic amino acids and had normal growth rates. Cells producing *p*-amino-L-phenylalanine were able to incorporate it into myoglobin in response to an *amber* codon with high fidelity, as judged by mass spectrometry of the mutant protein and in high yield of 2 mg L^{-1} of culture. These yields are comparable to those observed when the unnatural amino acid is added to the growth media directly. In addition to *p*-amino-L-phenylalanine, it should be possible to biosynthesize and genetically encode other amino acids *in vivo* as well including methylated, acetylated, and glycosylated amino acids.

The creation of a 21 or 22 amino acid code removes a billion year constraint on the evolution of life, allowing the experimental assessment of the fitness of organisms with expanded genetic codes. By placing *amber*, *opal*, or four-base codons throughout the genome, we can compare the survival of various 20, 21, and 22 amino acid organisms in response to selective pressures. Some unnatural amino acids, such as ketones, expand the reactivity of proteins, permitting new mechanisms of stabilization (imines with lysine, for example) and new chemistries (ketones and aldehydes are found in cofactors but not among the natural amino acids). It will be interesting to see how nature can make use of these new reactivities in response to stresses like extremes of temperature of pH.

5
Summary

Expanding the natural repertoire of amino acids is creating a powerful set of tools for biochemical, biophysical, and cellular studies far beyond the possibilities afforded by site-directed mutagenesis. Engineering the biosynthetic machinery also tests our understanding of how organisms achieve fidelity in protein synthesis, and has resulted in some of the most profound changes in enzyme function attained by any method of protein engineering. However, it is still not possible to easily generate a synthetase with substrate specificity for any desired amino acid, and there is a strict limit on code expansion imposed by the small number of blanks in the genetic code from suppressible stop codons and 4-base codons. The development of further highly active and orthogonal tRNA/synthetase pairs and the development of unnatural codons will be major areas of effort in the upcoming decade.

Acknowledgment

Thomas J. Magliery is an N.I.H. Postdoctoral Fellow (GM065750), and he thanks Lynne Regan (Yale University) for additional support while writing this manuscript.

See also Bacterial Cell Culture Methods; Synthetic Peptides: Chemistry, Biology, and Drug Design.

Bibliography

Books and Reviews

Cornish, V.W., Mendel, D., Schultz, P.G. (1995) Probing protein-structure and function with an expanded genetic-code, *Angew. Chem. Int. Ed. Engl.* **34**, 621–633.

Cropp, T.A., Schultz, P.G. (2004) An expanding genetic code, *Trends Genet.* **20**, 625–630.

Dawson, P.E., Kent, S.B. (2000) Synthesis of native proteins by chemical ligation, *Annu. Rev. Biochem.* **69**, 923–960.

Hohsaka, T., Sisido, M. (2002) Incorporation of non-natural amino acids into proteins, *Curr. Opin. Chem. Biol.* **6**, 809–815.

Magliery, T.J. (2005) Unnatural protein engineering: producing proteins with unnatural amino acids, *Med. Chem. Rev. Online* (expected August, 2005).

Magliery, T.J., Pasternak, M., Anderson, J.C., Santoro, S.W., Herberich, B., Meggers, E., Wang, L., Schultz, P.G. (2003) In vitro Tools and in vivo Engineering: Incorporation of Unnatural Amino Acids into Proteins, in: Lapointe, J., Brakier-Gingras, L., (Eds.) *Translation Mechanisms*, Landes Bioscience/Eurekah.com, Georgetown, TX and Kluwer Academic/Plenum, New York, pp. 95–114.

Muir, T.W. (2003) Semisynthesis of proteins by expressed protein ligation, *Annu. Rev. Biochem.* **72**, 249–289.

Soll, D., RajBhandary, U.L. (Eds.) (1995) *tRNA: Structure, Biosynthesis and Function*, ASM Press, Washington, DC.

Primary Literature

Anderson, J.C., Schultz, P.G. (2003) Adaptation of an orthogonal archaeal leucyl-tRNA and synthetase pair for four-base, amber, and opal suppression, *Biochemistry* **42**, 9598–9608.

Anderson, J.C., Magliery, T.J., Schultz, P.G. (2002) Exploring the limits of codon and anticodon size, *Chem. Biol.* **9**, 237–244.

Anderson, J.C., Wu, N., Santoro, S.W., Lakshman, V., King, D.S., Schultz, P.G. (2004) An expanded genetic code with a functional quadruplet codon, *Proc. Natl. Acad. Sci. U.S.A.* **101**, 7566–7571.

Chin, J.W., Schultz, P.G. (2002) In vivo photocrosslinking with unnatural amino acid mutagenesis, *ChemBioChem* **11**, 1135–1137.

Chin, J.W., Martin, A.B., King, D.S., Wang, L., Schultz, P.G. (2002) Addition of a photocrosslinking amino acid to the genetic code of *Escherichia coli*, *Proc. Natl. Acad. Sci. U.S.A.* **99**, 11020–11024.

Chin, J.W., Cropp, T.A., Chu, S., Meggers, E., Schultz, P.G. (2003) Progress toward an expanded eukaryotic genetic code, *Chem. Biol.* **10**, 511–519.

Chin, J.W., Santoro, S.W., Martin, A.B., King, D.S., Wang, L., Schultz, P.G. (2002) Addition of p-Azido-L-phenylalanine to the genetic code of *Escherichia coli*, *J. Am. Chem. Soc.* **124**, 9026–9027.

Chin, J.W., Cropp, T.A., Anderson, J.C., Mukherji, M., Zhang, Z., Schultz, P.G. (2003) An expanded eukaryotic genetic code, *Science* **301**, 964–967.

Deiters, A., Cropp, T.A., Summerer, D., Mukherji, M., Schultz, P.G. (2004) Site-specific PEGylation of proteins containing unnatural amino acids, *Bioorg. Med. Chem. Lett.* **14**, 5743–5745.

Deiters, A., Cropp, T.A., Mukherji, M., Chin, J.W., Anderson, J.C., Schultz, P.G. (2003) Adding amino acids with novel reactivity to the genetic code of *Saccharomyces cerevisiae*, *J. Am. Chem. Soc.* **125**, 11782–11783.

Drabkin, H.J., Park, H.J., RajBhandary, U.L. (1996) Amber suppression in mammalian cells dependent upon expression of an *Escherichia coli* aminoacyl-tRNA synthetase gene, *Mol. Cell. Biol.* **16**, 907–913.

Edwards, H., Schimmel, P. (1990) A bacterial amber suppressor in *Saccharomyces cerevisiae* is selectively recognized by a bacterial aminoacyl-tRNA synthetase, *Mol. Cell. Biol.* **10**, 1633–1641.

Furter, R. (1998) Expansion of the genetic code: site-directed p-fluoro-phenylalanine incorporation in *Escherichia coli*, *Protein Sci.* **7**, 419–426.

Henry, A.A., Romesberg, F.E. (2003) Beyond A, C, G and T: augmenting nature's alphabet, *Curr. Opin. Chem. Biol.* **7**, 727–733.

Hohsaka, T., Ashizuka, Y., Sisido, M. (1999) Incorporation of two nonnatural amino acids into proteins through extension of the genetic code, *Nucleic Acids Symp. Ser.* **42**, 79–80.

Ibba, M., Hennecke, H. (1995) Relaxing the substrate specificity of an aminoacyl-tRNA synthetase allows in vitro and in vivo synthesis of proteins containing unnatural amino acids, *FEBS Lett.* **364**, 272–275.

Kiga, D., Sakamoto, K., Kodama, K., Kigawa, T., Matsuda, T., Yabuki, T., Shirouzu, M., Harada, Y., Nakayama, H., Takio, K., Hasegawa, Y., Endo, Y., Hirao, I., Yokoyama, S. (2002) An engineered *Escherichia coli* tyrosyl-tRNA synthetase for site-specific incorporation of an unnatural amino acid into proteins in eukaryotic translation and its application in a wheat germ cell-free system, *Proc. Natl. Acad. Sci. U.S.A.* **99**, 9715–9720.

Kobayashi, T., Nureki, O., Ishitani, R., Yaremchuk, A., Tukalo, M., Cusack, S., Sakamoto, K., Yokoyama, S. (2003) Structural basis for orthogonal tRNA specificities of tyrosyl-tRNA synthetases for genetic code expansion, *Nat. Struct. Biol.* **10**, 425–432.

Kohrer, C., Xie, L., Kellerer, S., Varshney, U., RajBhandary, U.L. (2001) Import of amber and ochre suppressor tRNAs into mammalian cells: a general approach to site-specific insertion of amino acid analogues into proteins, *Proc. Natl. Acad. Sci. U.S.A.* **98**, 14310–14315.

Kowal, A.K., Kohrer, C., RajBhandary, U.L. (2001) Twenty-first aminoacyl-tRNA synthetase-suppressor tRNA pairs for possible use in site-specific incorporation of amino acid analogues into proteins in eukaryotes and in eubacteria, *Proc. Natl. Acad. Sci. U.S.A.* **98**, 2268–2273.

Link, A.J., Mock, M.L., Tirrell, D.A. (2003) Non-canonical amino acids in protein engineering, *Curr. Opin. Biotechnol.* **14**, 603–609.

Liu, D.R., Schultz, P.G. (1999) Progress toward the evolution of an organism with an expanded genetic code, *Proc. Natl. Acad. Sci. U.S.A.* **96**, 4780–4785.

Liu, D.R., Magliery, T.J., Pasternak, M., Schultz, P.G. (1997) Engineering a tRNA and aminoacyl-tRNA synthetase for the site-specific incorporation of unnatural amino acids into proteins in vivo, *Proc. Natl. Acad. Sci. U.S.A.* **94**, 10092–10097.

Magliery, T.J., Anderson, J.C., Schultz, P.G. (2001) Expanding the genetic code: selection of efficient suppressors of four-base codons and identification of "shifty" four-base codons with a library approach in *Escherichia coli*, *J. Mol. Biol.* **307**, 755–769.

Mehl, R.A., Anderson, J.C., Santoro, S.W., Wang, L., Martin, A.B., King, D.S., Horn, D.M., Schultz, P.G. (2002) Generation of a bacterium with a 21 amino acid genetic code, *J. Am. Chem. Soc.* **125**, 935–939.

Monahan, S.L., Lester, H.A., Dougherty, D.A. (2003) Site-specific incorporation of unnatural amino acids into receptors expressed in Mammalian cells, *Chem. Biol.* **10**, 573–580.

Noren, C.J., Anthonycahill, S.J., Griffith, M.C., Schultz, P.G. (1989) A general-method for site-specific incorporation of unnatural amino-acids into proteins, *Science* **244**, 182–188.

Nowak, M.W., Kearney, P.C., Sampson, J.R., Saks, M.E., Labarca, C.G., Silverman, S.K., Zhong, W., Thorson, J., Abelson, J.N., Davidson, N., Schultz, P.G., Dougherty, D.A., Lester, H.A. (1995) Nicotinic receptor-binding site probed with unnatural amino-acid-incorporation in intact-cells, *Science* **268**, 439–442.

Pasternak, M., Magliery, T.J., Schultz, P.G. (2000) A new orthogonal suppressor tRNA/aminoacyl-tRNA synthetase pair for evolving an organism with an expanded genetic code, *Helv Chim Acta* **83**, 2277.

Sakamoto, K., Hayashi, A., Sakamoto, A., Kiga, D., Nakayama, H., Soma, A., Kobayashi, T., Kitabatake, M., Takio, K., Saito, K., Shirouzu, M., Hirao, I., Yokoyama, S. (2002) Site-specific incorporation of an unnatural amino acid into proteins in mammalian cells, *Nucleic Acids Res.* **30**, 4692–4699.

Santoro, S.W., Wang, L., Herberich, B., King, D.S., Schultz, P.G. (2002) An efficient system for the evolution of aminoacyl-tRNA synthetase specificity, *Nat. Biotechnol.* **20**, 1044–1048.

Schlieker, C., Weibezahn, J., Patzelt, H., Tessarz, P., Strub, C., Zeth, K., Erbse, A., Schneider-Mergener, J., Chin, J.W., Schultz, P.G., Bukau, B., Mogk, A. (2004) Substrate recognition by the AAA+ chaperone ClpB, *Nat. Struct. Mol. Biol.* **11**, 607–615.

Trezeguet, V., Edwards, H., Schimmel, P. (1991) A single base pair dominates over the novel identity of an *Escherichia coli* tyrosine tRNA in *Saccharomyces cerevisiae*, Mol. Cell. Biol. **11**, 2744–2751.

Wang, L., Schultz, P.G. (2001) A general approach for the generation of orthogonal tRNAs, Chem. Biol. **8**, 883–890.

Wang, L., Magliery, T.J., Liu, D.R., Schultz, P.G. (2000) A new functional suppressor tRNA/aminoacyl-tRNA synthetase pair for the in vivo incorporation of unnatural amino acids into proteins, J. Am. Chem. Soc. **122**, 5010–5011.

Wang, L., Brock, A., Herberich, B., Schultz, P.G. (2001) Expanding the genetic code of *Escherichia coli*, Science **292**, 498–500.

Wolin, S.L., Matera, A.G. (1999) The trials and travels of tRNA, Genes Dev. **13**, 1–10.

Wu, N., Deiters, A., Cropp, T.A., King, D., Schultz, P.G. (2004) A genetically encoded photocaged amino acid, J. Am. Chem. Soc. **126**, 14306–14307.

Xie, J., Wang, L., Wu, N., Brock, A., Spraggon, G., Schultz, P.G. (2004) The site-specific incorporation of p-iodo-L-phenylalanine into proteins for structure determination, Nat. Biotechnol. **22**, 1297–1301.

Zhang, Z., Alfonta, L., Tian, F., Bursulaya, B., Uryu, S., King, D.S., Schultz, P.G. (2004) Selective incorporation of 5-hydroxytryptophan into proteins in mammalian cells, Proc. Natl. Acad. Sci. U.S.A. **101**, 8882–8887.

Zhang, Z., Gildersleeve, J., Yang, Y.Y., Xu, R., Loo, J.A., Uryu, S., Wong, C.H., Schultz, P.G. (2004) A new strategy for the synthesis of glycoproteins, Science **303**, 371–373.

Protein Mediated Membrane Fusion

Reinhard Jahn
Max-Planck-Institute for Biophysical Chemistry, Göttingen, Germany

1	**Introduction and Overview** 71	
2	**Physical Principles of Bilayer Fusion** 72	
3	**Transition States of Protein-mediated Fusion and Fusion Pores** 74	
4	**Viral Fusion Proteins** 75	
4.1	Structure of Class I Fusion Proteins 75	
4.2	Fusion Mechanism of Class I Proteins 76	
4.3	Class II Viral Fusion Proteins 78	
5	**Mitochondrial Fusion** 79	
6	**Fusion Reactions in the Secretory Pathway** 79	
6.1	Overview 79	
6.2	Tethering and Docking – the Role of Rab Proteins and Their Effectors 81	
6.3	Fusion – SNARE Proteins as Fusion Catalysts 83	
6.3.1	The SNARE Assembly– Disassembly Cycle 84	
6.3.2	Mechanism of SNARE-mediated Membrane Fusion 86	
6.3.3	Conformational Intermediates of SNAREs 87	
6.3.4	Topology and Specificity of SNAREs in Intracellular Fusion Reactions 89	
6.4	Regulators of SNAREs 91	
6.4.1	SM Proteins 91	
6.4.2	Synaptotagmin and Complexin – Regulators of Ca^{2+}-dependent Exocytosis 92	
6.4.3	''Pseudo'' SNAREs 94	
7	**Conclusion and Outlook** 94	

Encyclopedia of Molecular Cell Biology and Molecular Medicine, 2nd Edition. Volume 11
Edited by Robert A. Meyers.
Copyright © 2005 Wiley-VCH Verlag GmbH & Co. KGaA, Weinheim
ISBN: 3-527-30648-X

Bibliography   95
Books and Reviews   95
Primary Literature   95

Keywords

Exocytosis
Fusion of an intracellular organelle (usually a trafficking vesicle or a specialized secretory vesicle) with the plasma membrane. Exocytosis can be further subdivided into regulated and constitutive exocytosis. Examples for the former are neurotransmitter release by exocytosis of synaptic vesicles or hormone and enzyme release by secretion from exocrine and endocrine glands. Examples for the latter include the continuous incorporation of trafficking vesicles as required for receptor recycling and cell growth. For many years, neuronal exocytosis has served as a paradigm for studying the mechanisms of membrane fusion.

Fusion Peptides
Stretches of usually 20–60 amino acids that are part of viral fusion proteins and are thought to be inserted into the target membrane upon activation. Fusion peptides are amphiphilic, that is, when forming an α-helix one face contains hydrophobic/large-side chains whereas the rest of the surface is hydrophilic. Isolated fusion peptides insert into membranes, and several of them were shown to be potent fusogens, that is, they are capable of fusing membranes even without being attached to a protein.

Fusion Pore
The opening of an aqueous connection between the distal sides of the two fusing membranes. In a protein-mediated fusion, fusion pores are tight so that no leakage occurs between the distal and proximal sides of the fusing membranes. While fusion pores can be characterized with electrophysiological methods, the molecular structure of these intermediates in the fusion pathway is unknown. Both proteinaceous and lipidic intermediates are discussed.

Hemifusion
A hypothetical intermediate in bilayer fusion in which there is already a free exchange of membrane lipids, although an aqueous connection (fusion pore) is not yet formed.

SNAREs
A superfamily of small proteins (molecular weight between 10 and 35 kDa) that catalyze membrane fusion in the secretory pathway of eukaryotic cells. Each fusion step involves a specific set of SNAREs that spontaneously assemble into ternary or quaternary complexes, bridging the membrane and initiating fusion. After fusion, SNARE complexes are disassembled by the ATPase NSF (NEM-sensitive factor) in conjunction with cofactors termed SNAPs (soluble NSF attachment proteins).

Viral Fusion Proteins
Membrane proteins on the surface of enveloped viruses, that is, viruses that are surrounded by a phospholipid bilayer. Viral fusion proteins are metastable. Upon activation during cell contact or after endocytosis they undergo a spontaneous conformational change that mediates the fusion of the viral with the cell membrane, resulting in the injection of the nucleocapsid into the host cell cytoplasm. Viral fusion proteins are divided into class I and class II fusion proteins, which differ with respect to overall structure and fusion kinetics.

> The fusion of biological membranes is one of the most fundamental reactions of life. It involves the initial recognition of the membranes, the close apposition of the membranes followed by the generation of nonbilayer intermediates, and finally the opening of an aqueous fusion pore. While the intermediates in various fusion reactions share common features, diverse classes of proteins evolved independently to carry out the task. Single proteins may be responsible for the entire reaction sequence as is the case for fusion proteins of enveloped viruses. In contrast, intracellular fusion reactions are mediated by cascades of reactions carried out by multiprotein complexes allowing for tight spatial and temporal regulation. Although the precise mechanism of fusion proteins is not yet fully understood, it is becoming clear that fusion proteins are nanomachines that exert mechanical force upon the membranes destined to fuse. Intriguingly, all characterized fusion proteins undergo major refolding during the reaction. Refolding proceeds downhill an energy gradient and yields highly stable oligomeric complexes with a stiff rodlike structure suggesting similarities in the fusion pathway.

1
Introduction and Overview

Membranes are fundamental to all forms of life. They form barriers that cannot be penetrated by hydrophilic molecules and that allow the enclosing of compartments containing biological macromolecules. Every cell is surrounded by a membrane that usually is represented by a bilayer of amphiphiles (phospholipids and related compounds). Furthermore, eukaryotic cells contain membrane-surrounded organelles.

Since the barrier function of membranes must be maintained at all times, special mechanisms that enable membranous compartments to merge or to split have been evolved. These mechanisms are referred to as *membrane fusion* and *membrane fission*. Both fusion and fission of biological membranes are characterized by specific features:

- no leakage occurs between the inside and the outside of the membrane-delineated spaces
- the asymmetry of the membrane is maintained, that is, amphiphiles do not randomize between the two monolayers
- they are mediated by specific proteins.

Biological membrane fusion events can be classified in

- extracellular fusions between cells, for example, sperm–egg fusion and fusions yielding polynucleated cells such as muscle cells or osteoclasts
- fusions of enveloped viruses with the plasma membrane of host cells
- intracellular fusions of and between organelles with the surrounding plasma membrane.

In recent years, significant progress has been made in identifying the proteins responsible for fusing biological membranes. Although there is still considerable debate about the precise role of many of these proteins, it has become clear that proteins capable of fusing membranes originated several times independently during evolution. Best understood are the fusion proteins of enveloped viruses. They fall into at least two major classes that are structurally unrelated to each other, and the diversity in either one of the classes is so great that even for these a common evolutionary origin is doubtful. Eukaryotic cells contain at least three classes of fusion proteins: those involved in extracellular fusions whose identities are largely unknown, those involved in the homotypic fusion of mitochondria, and finally those involved in fusion events of the so-called secretory pathway.

2
Physical Principles of Bilayer Fusion

Membranes are typically organized in bilayers composed of two leaflets of amphiphilic molecules such as phospholipids. These molecules have an elongated shape, with a hydrophilic head group and a hydrophobic tail that is often composed of hydrocarbon chains. The hydrophilic groups are exposed to the aqueous environment, whereas the hydrophobic tails face the hydrophobic interior of the bilayer.

Membrane fusion first requires that the two membranes destined to fuse be brought into close contact, requiring that electrostatic and hydration repulsion is overcome. Next, transition states that locally disrupt the bilayer structure must be generated, involving both hydrophilic and hydrophobic contacts between the membrane constituents. These transition states are governed by the hydrophobic effect, that is, by forces that minimize the exposure of hydrophobic surfaces to water. Finally, the bilayer structure is reinstated, which is associated with the local opening of an aqueous fusion pore.

Fusion sites are "dirty nanostructures" in that they have a variable composition involving hundreds of amphiphiles but with a size too small for resolving them with a light microscope. Furthermore, they dynamically change during the reaction, and it is not easy to arrest the fusion reaction at a defined intermediate state. Thus, the structure of the nonbilayer transition states cannot be directly determined. However, both theoretical and experimental approaches have led to well-founded models that describe the steps involved in membrane fusion in accordance with physical principles. The most widely accepted model is referred to as *stalk hypothesis* (Fig. 1). In its original version, the stalk hypothesis is based on macroscopic physics (continuum models) and describes the energy landscape of the fusion pathway on the basis of parameters such as bending energy, intrinsic curvature of the lipids, tilt of hydrocarbon chains, and it takes into account energy penalties due to the presence of hydrophobic voids in the

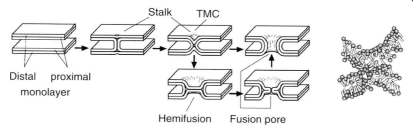

Fig. 1 Transition states in membrane fusion according to the stalk hypothesis. The monolayers are shown as smooth and bendable sheets. TMC, transmembrane contact. On the right, a snapshot of a nonbilayer intermediate is shown that is derived from a simulation of membrane fusion using simplified (coarse-grain) models for membrane lipids. Reprinted from Jahn, R., Lang, T., Südhof, T.C. (2003) Membrane fusion, *Cell* **112**, 519–533.

nonbilayer transition states. According to this hypothesis, the first steps involve the merger of proximal monolayers, a process that may be facilitated by thermal motions of individual amphiphiles perpendicular to the membrane plane. As a result, a fusion stalk is formed, followed by a symmetrical structure in which the first contact between the distal monolayers is established, which is also referred to as transmembrane contact (TMC). From this intermediate, a fusion pore may directly form or else, a hemifusion diaphragm is formed in which the contact zone of the distal monolayers expands, with a fusion pore forming at a later state (Fig. 1).

A large body of experimental evidence supports the overall description by the stalk hypothesis of the fusion pathway. For instance, structures similar to the intermediate involving the first TMC (Fig. 1) have been observed by X-ray diffraction in stacks of dehydrated bilayers. Similarly, such structures are part of the so-called rhombohedral/hexagonal phase of certain membrane lipids. Furthermore, addition of amphiphiles, which promotes either positive or negative curvature promotes or inhibits progression through the transition states as predicted by the model, and in many fusion events lipid mixing is observable before an aqueous connection is formed. However, the structure and precise orientation of the membrane lipids during the transition states is not known, and it has not been possible to show directly that stalks are intermediates of biological fusion reactions. Recent efforts to simulate membrane fusion using coarse-grain models also resulted in stalk-like intermediates, but the transition states appeared to be more disordered than originally assumed by the stalk hypothesis. While these models presently suffer from an arbitrary choice of fitting parameters, it is hoped that full atomistic simulations using refined force fields will yield more realistic and less biased structural models of fusion intermediates.

Unless strained or otherwise destabilized, membranes do not spontaneously fuse with each other under normal conditions (ambient temperature and pressure). However, amphiphilic molecules such as polyethylene glycol or certain amphipatic peptides can mediate spontaneous fusion of membranes. Similarly, certain amphipatic proteins (e.g. annexins) induce fusion of artificial membranes although there is no evidence that they act as fusion proteins in their native environment. Apparently, the activation energy barrier for membrane fusion is not very high, requiring less specificity for the catalyst than, for example, an enzyme-catalyzed metabolite conversion.

3
Transition States of Protein-mediated Fusion and Fusion Pores

In the 1980s, two models were proposed explaining how proteins may mediate fusion. They describe two possible extreme situations that are instructive and that are still widely used as reference (Fig. 2). The first model describes the initial steps of fusion analogous to the formation and activation of gap junction channels. Each membrane contains oligomeric fusion proteins that assemble into a closed, ringlike oligomeric structure. Upon membrane contact, these rings connect and form an aqueous channel that represents the first opening of the fusion pore. Capacitance patch-clamping experiments revealed that fusion pores with an initial diameter of about 2 to 5 nm may reversibly open and close at a millisecond timescale ("flickering"), supporting the view that fusion pores, at least in their initial state, resemble channel proteins. Subsequently, both the oligomeric assembly and the TMC need to be broken up in order for the fusion pore to expand and for fusion to be completed. According to this model, the initial states of fusion including the transition states are primarily governed by protein–protein interactions and subsequently by protein–lipid interactions.

The second model assumes that the role of proteins in fusion is confined to providing a scaffold that lowers the activation energy barrier, while the transition states and the initial fusion pore is exclusively lipidic and thus must be in accordance with lipid biophysics. Accordingly, there is no need for the proteins to span the membrane. Rather, the role of fusion proteins may be

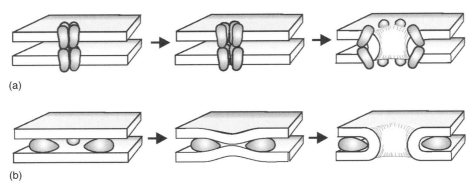

Fig. 2 Classical models for protein-mediated membrane fusion. (a) Channel-model. Here, the fusion proteins form a closed channel that connects in "trans," resulting in a tight connection between the two membranes that is reminiscent of gap junctions. Formation of a fusion pore is due to the opening of a proteinaceous channel. Enlargement is mediated by lipid invasion between the dissociating subunits. (b) Scaffold model. Here, the role of the proteins is confined to bending of the membrane (in the original version it was assumed that a dimple is created in one of the membranes), with all transition states governed by lipid biophysics.

confined to forming a tight connection between the membranes, or else, they may destabilize the hydrophilic–hydrophobic boundary of the membrane to facilitate nonbilayer transition states. Fusion pore flickering, originally considered to be a hallmark of protein-mediated fusion, was later also observed in fusion events of protein-free artificial membranes, thus depriving the "channel" model of one of its strongest supportive arguments.

As discussed in more detail below, it is still unknown which of the two models is more accurate. Furthermore, the hypothesis that different classes of fusion proteins operate by different mechanisms cannot be excluded. While most investigators agree that lipids play a major role in the early transition states as suggested by the scaffold model, modifications of fusion proteins were reported to directly affect fusion pore characteristics indicating that fusion proteins do more than just clamping membranes together.

4
Viral Fusion Proteins

Enveloped viruses are enclosed by a host cell–derived bilayer that protects the nucleocapsid containing the viral genome. In order to infect cells, the viral membrane needs to fuse with the host cell membrane, resulting in the delivery of the nucleocapsid into the cytoplasm of the host cell. For this task, enveloped viruses contain specific membrane proteins whose purpose is to execute the fusion of the viral with the host cell membrane. Viral fusion proteins can be divided into class I and class II proteins, with class I proteins being better understood.

4.1
Structure of Class I Fusion Proteins

Class I fusion proteins typically are homooligomeric glycoproteins with a single transmembrane domain. They are oriented perpendicular to the viral membrane and often form characteristic spikes. At or near the N-terminal end, these proteins contain an amphiphilic region of 15 to 35 amino acids, referred to as *fusion peptide*, which is essential for function. All class I fusion proteins have one remarkable feature in common: before the viruses bud from the host cell, the otherwise stably folded fusion proteins are posttranslationally processed by a single proteolytic cut yielding a stable globular N-terminal and a metastable C-terminal fragment containing the transmembrane domain. The metastable C-terminal fragments are the fusion catalysts. When these proteins are triggered by binding to a surface receptor or by low pH, as encountered after endocytosis of the virus particle, they spontaneously undergo major conformational changes. As a result, elongated helical bundles (usually consisting of trimeric coiled coils) are formed and the previously buried fusion peptides are exposed at the tip of the bundle. After completion of the conformational transition, the resulting protein is very stable and inactive. Thus, viral fusion proteins are "single-shot" devices resembling a loaded spring: they can be triggered only once. Biologically, this makes sense since they are needed only once in an infectious fusion reaction.

The fusion peptides play an essential role in virus-mediated fusion. These domains insert into the membrane of the host cell upon activation with the large hydrophobic side chains buried in the interior of the membrane. Perturbation of the amphiphilic nature of the fusion peptide, for example by site-directed mutagenesis, either cripples or completely inactivates fusion activity. Interestingly, many isolated fusion peptides derived from viral fusion proteins are efficient fusogens: when added to a suspension of liposomes, spontaneous fusion occurs. However, it is likely that their prime role is to provide a firm anchor in the target membrane.

Much of our understanding of viral membrane proteins is derived from the influenza hemagglutinin (HA) fusion protein that for many years served as a paradigm for class I fusion proteins. It is the only fusion protein for which fragments were crystallized both for the metastable prefusion state and the relaxed postfusion state. In the prefusion state, the HA trimer contains a central trimeric coiled coil that is connected to the fusion peptide by a long loop that folds back to the base of the molecule. Triggering of the protein by low pH results in two major rearrangements. First, the connecting loop is converted into a helix that now extends the coiled-coil region, effectively propelling the fusion peptide from the base to the top of the elongated molecule. Second, a reverse turn is generated in the middle of the original helix, rotating the C-terminal part by 180°, with these helical parts and some nonhelical extensions now fitting into the grooves of the central trimer (Fig. 3). While the purpose of the first molecular motion is obvious (exposure of the fusion peptide and its "shooting" into the target membrane), the significance of the second motion was only fully appreciated when structures of other viral fusion proteins (e.g. gp41 of the human immunodeficiency virus) were solved in which such back folding (resulting in an inner coiled coil and an outer ring of inverted helices) is more obvious than in HA.

It is thought that the second motion repositions the transmembrane domain from the base of the molecule to its tip, that is, adjacent to the fusion peptide. It should be borne in mind, however, that the region adjacent to the transmembrane domain was not part of the structure, and thus the precise position of the transmembrane domain in the relaxed form remains to be determined. In addition to HA, the crystal structures of several additional class I proteins have been determined, but for these only the relaxed structures are available. They all contain a central trimeric coiled coil and three surrounding outer helices although there is no sequence similarity between most of these proteins.

4.2
Fusion Mechanism of Class I Proteins

Despite a wealth of structural information and numerous studies on fusions mediated by wild-type and mutated class I fusion proteins, there is still no consensus about the mechanism by which these proteins fuse membranes. It is undisputed that any interference with the conformational change (or part of it) either inhibits fusion completely or leads to an arrested intermediate state, suggesting that these fusion proteins are nanomachines that exert mechanical force on the membranes destined to fuse. The most popular model ("jackknife model") envisions fusion as a two-step process progressing via defined intermediates (Fig. 3). In the first step, the fusion peptide is propelled at the tip of the protein, resulting in its insertion and thus anchoring in the host cell membrane. In this state, the fusion protein is in an extended conformation, being connected to the viral membrane by its transmembrane domain and to the host cell membrane by its fusion peptide. In the second step, the

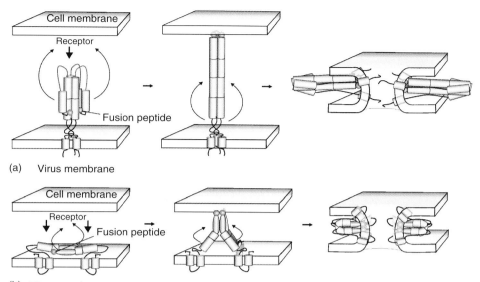

Fig. 3 Models for membrane fusion mediated by class I (a) and class II (b) viral fusion proteins involving a two-step mechanism. In the first step, the fusion peptides are exposed and are inserted in the target membrane. In the second step, the C-terminal part of the proteins undergoes another conformational change, resulting in bending/pulling that results in fusion. Red balls depict the fusion peptides, green cylinders the transmembrane domains. Note that the cellular receptors and the viral receptor–binding proteins are not shown. The colors indicate different domain of the proteins. Note that in class I proteins the colored boxes indicate α-helices, whereas in class II proteins they define folded domains that are not helical. Arrows indicate the presumed movements (see color plate p. xxiii).

outer helices revert and fold back on the rigid coiled coil, thus dragging the transmembrane domains from the base to the top of the molecule adjacent to the fusion peptide. As a result, the two membranes are forced into close apposition, with the fusion peptides probably destabilizing the monolayer surface by displacing water at the contact site, facilitating the formation of nonbilayer transition states.

Although widely accepted, the jackknife model suffers from some shortcomings. First, the rotational symmetry of the trimeric fusion proteins (see cartoon in Fig. 3) precludes bending to one side (as mandatory for the jackknife model) unless one assumes that the second folding step occurs asynchronously, that is, that back folding in only one of the subunits allows for lateral bending of the molecule. Second, it has been known for some time that activation of HA in the absence of a target membrane results in insertion of the fusion peptide into the viral membrane. Again, it is difficult to explain how such "reverse" insertion is prevented upon contact with a host membrane. To overcome these problems, alternative mechanisms were proposed that differ from the jackknife model in that the fusion peptide is either exclusively or partially inserted into the viral membrane. According to the first scenario, the formation of the coiled coil would then exert a strong bending force on the viral membrane, while the protein remains attached to the cell membrane by binding to the receptor. Fusion proteins grouped in a ring around the prospective fusion site would create a membrane protrusion that extends in the direction of the target membrane, resulting in its destabilization and initiation of fusion. A second alternative is that the fusion peptides of a given trimer are inserted randomly into the viral and the target membrane. In this scenario, the extension of the central coiled coil rather than the backfolding would exert the primary pulling force that brings the membranes in close apposition. While both of the latter models account for the "symmetry problem," experimental support for either of them presently is lacking.

4.3
Class II Viral Fusion Proteins

Class II fusion proteins including the E-glycoproteins of flaviviruses and alphaviruses are also activated by low pH to undergo conformational changes, but it appears that the mechanisms are very different from those of class I proteins. Recently, the structure of whole virus particles containing class II fusion proteins has been solved by a combination of cryoelectron microscopy and crystallography. These fusion proteins form an interconnected lattice on the viral surface. They are oriented horizontally rather than perpendicular to the viral membrane. Proteolytic activation usually occurs in an accessory protein rather than in the fusion protein itself. α-helices and coiled coils are not part of the protein structure, rather, they are primarily composed of β-strands. The E-glycoproteins have three domains: an N-terminal domain, a middle domain that carries the fusion peptide (rather than being positioned near the N-terminus), and a C-terminal domain containing the transmembrane domain. The fusion peptides form short loops, which, in the nonactivated form are located close to the viral membrane and are covered either by an accessory protein involved in receptor binding or by a dimerization interface. Upon activation, the fusion peptides are exposed by what appears to be a coordinated conformational change of neighboring molecules. In the case of the Semliki forest virus, activation reorients the E1-fusion protein from a dimer with its accessory protein E2 to a homotrimer, associated with a movement away from the viral membrane surface (Fig. 3). This motion probably causes the insertion of at least part of the fusion peptides into the target membrane, probably in a β-barrel conformation. *In vitro* fusion of viruses carrying class II proteins with artificial membranes is considerably faster than that of viruses with class I proteins.

Despite the differences in structure, recently solved crystal structures of class II fusion proteins have revealed striking similarities between class I and class II fusion proteins. In both cases, activation leads to conformational rearrangements that result in a juxtapositioning of the transmembrane domain with the fusion peptide, with the final conformation being very stable. In the relaxed low pH trimeric form, domains 1 and 2 form a thick, rodlike structure with the fusion peptides at their tips, again reminiscent of the class I protein structure. Thus, it is conceivable that despite a very different structural basis the fusion mechanisms of class I and class II proteins are similar (Fig. 3).

5
Mitochondrial Fusion

Mitochondria continuously undergo dynamic shape changes including frequent fusion and fission events. Since mitochondria are surrounded by an inner and an outer membrane with very different protein compositions, homotypic fusion of mitochondria must involve the consecutive fusion of the outer and inner membrane, respectively.

Until recently, the protein machinery involved in mitochondrial fusion has remained enigmatic although it is clearly distinct from the protein complexes mediating fusions in the secretory pathway. A few years ago, a first putative protein participating in mitochondrial fusion has been discovered. Originally identified as a gene involved in the formation of a special mitochondrial aggregate in the sperm of Drosophila, a conserved integral membrane protein, termed fzo (for "fuzzy onion"), was discovered to be essential for fusion. Fzo represents a small protein family of evolutionary highly conserved proteins (termed *mitofusins* in mammals) that possess two transmembrane domains and an N-terminal GTPase domain, with both the C- and N-terminus facing the cytoplasm. Adjacent to the membrane-anchor domain are regions that are predicted to have a high propensity for the formation of coiled coils, thus bearing superficial similarities with bona fide fusion proteins such as the class I viral fusion proteins and the SNAREs. While yeast possesses a single family member (Fzo1p), higher organisms and mammals contain two mitofusins.

Recently it has been discovered that mitofusins can form dimers in which the C-terminal regions of mitofusins form antiparallel coiled coils, thus tethering the outer membranes of two adjacent mitochondria. Furthermore, an active GTPase domain is required for mitochondria to fuse. However, it is not clear at present whether the role of mitofusins is confined to tethering or whether they are also involved in later states of the fusion reaction. Additional proteins are likely to be involved including, for example, Ugo1p, an outer membrane protein with several transmembrane domains, and Mgm1p, a dynamin-like GTPase on the outer surface of the inner membrane. Furthermore, recent screens in yeast have uncovered more genes that appear to be involved in mitochondrial fusion and whose molecular roles are yet to be elucidated.

6
Fusion Reactions in the Secretory Pathway

6.1
Overview

The term *secretory pathway* refers to all organelles that originate from the endoplasmic reticulum and that are connected by defined vesicular trafficking routes. They include both the biosynthetic route, leading from the endoplasmic reticulum to the Golgi complex, and from there lead either to the plasma membrane or to endosomal intermediates, and the endocytotic/degradative route leading from the plasma membrane to the lysosomes via early and late endosomal intermediates. Additional connections and shortcuts are provided by further specialized trafficking routes, some of which may be tissue and cell-type specific such as regulated exocytosis (Fig. 4).

Each trafficking step in the secretory pathway can be broken down into three

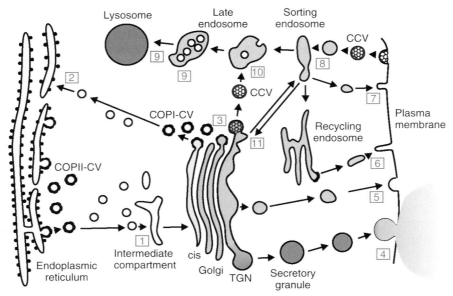

Fig. 4 Overview of the secretory pathway in a prototype eukaryotic cell. The boxed numbers indicate distinct fusion steps. CCV, clathrin-coated vesicle, COPI/II-CV, COPI/II-coated vesicle, TGN, *trans*-Golgi network. The biosynthetic route is depicted in green, the endocytotic pathway in pink (see color plate p. xxv).

elementary steps: the generation of a transport vesicle by budding from a precursor membrane, the movement of the vesicle to its target membrane mediated by the cytoskeletal machinery, and the consumption of the transport vesicle by fusion with its target compartment. In addition, there is "homotypic" fusion between organelles of the same class such as endosomes and cisternae of the endoplasmic reticulum. Intracellular trafficking pathways are highly organized, and they exhibit a high (but not absolute) degree of specificity. Thus, a vesicle derived from the endoplasmic reticulum will normally not fuse with the plasma membrane. Furthermore, vesicles originating from common sorting platforms such as the *trans*-Golgi network or the early endosome may have different destinations. This specificity is a feature of the organelles and trafficking vesicles, and as discussed in more detail below it is encoded in specific proteins that provide layers of regulation.

Unlike viral fusion proteins that single handedly carry out the jobs of docking and fusion, complex supramolecular assemblies mediate fusion in the secretory pathway, which act in concert with each other and carry out cascades of sequential reactions. These supramolecular machines are not represented by stable entities with fixed stoichiometries. Rather, they are assembled on demand and dissociate from the fusion site when the task is completed.

Despite the diversity of intracellular fusion reactions with respect to structure and kinetics, it appears that the fusion machines of the secretory pathway operate by principally similar mechanisms. This is suggested by the fact that several of the proteins required for docking and fusion represent evolutionarily conserved protein families that have diversified to adjust to the specific demands of individual fusion reactions. Such conserved proteins include the small GTPases of the Ypt/Rab family, the SM proteins, and the SNARE proteins.

Before intracellular trafficking organelles fuse, they need to be transported into close proximity. Furthermore, signaling between the membranes is required in order for them to "know" that another potential fusion partner is nearby. This also involves "proofreading," that is, mechanisms that ensure that the fusion partner is appropriate. These steps may or may not be associated with the establishment of physical contact that proceeds via first loose ("tethering") and then firm ("docking") contact. Next, the fusion reaction is executed which may require prior activation of the fusion machinery. After fusion, the fusion machinery is disassembled and regenerated to be available for another round of fusion.

Intriguingly, it appears that with exception of the presumed fusion catalysts (the SNARE proteins) and some regulatory proteins involved in specialized fusion reactions (such as the synaptotagmins), all other proteins are recruited on demand from the cytoplasm and thus are not permanently associated with the membrane. The advantage of such a mechanism is probably to minimize the "efforts" a cell needs to spend on protein recycling. After fusion of a trafficking vesicle, membrane-resident fusion proteins need to be returned by membrane trafficking to the precursor compartment for the proteins to be reused. This necessity also explains why membrane-bound fusion proteins (SNAREs) are not only present in the organelle for whose fusion they are responsible but also in all other organelles belonging to the recycling pathway required for returning these proteins to their original compartment.

6.2
Tethering and Docking – the Role of Rab Proteins and Their Effectors

What is the first step in the reaction cascade leading to fusion of two proximal organelles? Very little is known about the initial signaling events. However, it is well established that Rab GTPases play a central role in defining a site for fusion and in orchestrating the recruitment of protein complexes.

Rab proteins belong to the ras-superfamily of small monomeric GTPases. They operate as molecular switches that exist in an active GTP-bound and an inactive GDP-bound conformation. The switch between these conformations is made by regulatory proteins, the GTPase-activating proteins (GAPs) that stimulate the intrinsic GTPase activity of the rabs, and the guanine nucleotide exchange factors (GEFs) that replace GDP with GTP. Superimposed on the GTP-GDP cycle is a membrane association–dissociation cycle. Rab proteins carry posttranslationally attached polyprenyl chains that anchor the proteins in the membrane. GDP-Rab proteins are removed from the membrane by the action of a universal protein termed GDP-dissociation inhibitor (GDI) that forms soluble complexes with all GDP-Rabs by hiding the prenyl anchor in a hydrophobic pocket. Rebinding is associated with GTP-reloading by GEFs and may require assistance of additional proteins such as the members of the Yip/Pra1 family. These proteins displace GDI and thus are termed GDFs (for GDI displacement factors).

Rab proteins display the most organelle-specific localization of all known trafficking proteins, and they have therefore been assigned a major role in defining the identity of organelles. More than 60 Rab proteins are known in mammalian species. Individual organelles may contain multiple Rabs that probably define functionally distinct subdomains on the organellar surface. However, it is still not understood by which mechanisms this exquisite specificity is achieved.

The main job of active Rab proteins is to recruit specific proteins (termed Rab effectors) to defined sites at the membrane. While many effectors are exquisitely specific for a given Rab protein, others are shared between multiple rabs. Examples for the latter are rabaptin, the TRAPP complex, and mss4. The number of known Rab effectors is increasing steadily, with individual Rab proteins known to bind to dozens of different proteins. Presently, there are no common denominators between the many prospective Rab effectors. However, in some well worked out cases such as the homotypic fusion of early endosomes it is becoming clear that Rab effectors orchestrate the initial steps in docking and fusion. Effectors of the early endosomal Rab5 include large proteins such as EEA1, which is capable of forming homooligomeric structures, tethering prospective fusion partners. Other effectors include a phosphatidylinositol 3-kinase and proteins containing FYVE and PX domains that in turn bind to specific polyphosphoinositides. Phosphoinositides are known to be centrally involved in intracellular membrane traffic, and it is possible that together with an active Rab they provide a local binding site for further proteins that is highly specific for a given fusion step. For instance, EEA1 is both an effector for Rab5 and it contains a FYVE domain that binds to phosphatidylinositol-3-phosphate, suggesting that only the combination of both – active Rab5 and a sufficiently high concentration of phosphatidylinositol-3-phosphate – is effective in recruiting EEA1 to the membrane.

While it is generally agreed upon that active Rabs recruit docking and tethering complexes, the underlying molecular mechanisms are not known. Many Rab effectors are multiprotein complexes including the exocyst (presumably involved in constitutive exocytosis), the GARP, TRAPPI, TRAPPII, and COG complexes that are involved in trafficking steps at the Golgi apparatus, and the HOPS/VpsC-complex that functions in the homotypic fusion of yeast vacuoles. For the homotypic fusion of yeast vacuoles, sequentially occurring assembly and disassembly reactions were worked out involving components of the VpsC-complex and additional proteins such as LMA1. An example from the other end of the spectrum is represented by the RIM proteins, effectors of the Rab3 that function in neurotransmitter release at the synapse. RIM is part of the so-called active zone, an electron-dense structure at the presynaptic plasma membrane to which synaptic vesicles dock. Active zones consist of an insoluble supramolecular complex containing a set of large and insoluble proteins including Piccolo/aczonin, bassoon, ERCs (acronym for ELKS/Rab3-interacting molecule/CAST), RIM-binding proteins 1 and 2, and α-liprins. Taken together, there is little similarity between the components of these multiprotein complexes, and it is thus conceivable that the diversity of Rab effectors is responsible for imprinting specific features to each fusion reaction in the secretory pathway and to specialized fusions in certain cell types.

6.3
Fusion – SNARE Proteins as Fusion Catalysts

SNARE proteins (acronym for soluble NSF acceptor protein receptors) represent a superfamily of small and mostly membrane-bound proteins that are conserved throughout the eukaryotic kingdom, with 24 members in yeast, at least 35 members in mammals, and more than 50 in plants. SNARE proteins typically contain a C-terminal transmembrane domain, an adjacent stretch of 60–70 amino acids arranged in heptad repeats that is characteristic for the entire SNARE superfamily and that is referred to as *SNARE motif*, and an N-terminal region that either consists of only few amino acids or else, is composed of a separately folded domain (Fig. 5). There is considerable heterogeneity among SNARE family members both with respect to primary structure and to domain structure. In addition to the prototypical SNARE described above, some family members contain two instead of one SNARE motifs. Examples include the neuronal SNARE SNAP-25, its relatives SNAP-29 and SNAP-23, and the yeast SNAREs Sec9p and Spo20. In these proteins, the SNARE motifs are connected with a linker region. These SNAREs lack a transmembrane domain; instead, some of them carry hydrophobic palmitoyl side chains. Other exceptions include SNAREs with single SNARE motifs that carry either a posttranslationally added polyprenyl anchor (such as Ykt6) and the yeast SNARE Vam7 that lacks a membrane anchor altogether.

The importance of SNARE proteins was originally recognized because of three independent lines of evidence: first, the SNAREs involved in neuronal exocytosis (including the proteins synaptobrevin/VAMP, SNAP-25, and syntaxin 1) were identified as the targets of botulinum and tetanus toxins, second, the proteins were recognized in genetic screens in yeast as being essential for intracellular fusion reactions, and third, the SNAREs were identified as the membrane receptor

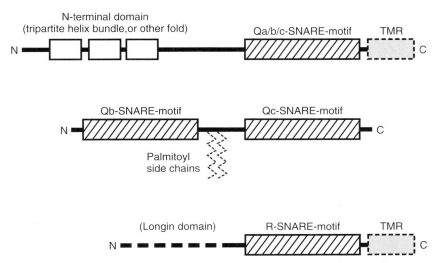

Fig. 5 Domain structures of SNARE proteins. Dotted lines delineate domains that may be absent in some of the family members. TMR, transmembrane region.

of the ATPase NSF (hence the name SNARE) that was previously recognized as an important component of intracellular fusion reactions.

6.3.1 The SNARE Assembly–Disassembly Cycle

SNARE proteins undergo an assembly–disassembly cycle that involves the SNARE motifs and that is essential for membrane fusion. Monomeric SNARE motifs are unstructured in solution. However, when appropriate sets of SNAREs are combined, they spontaneously form complexes in which four different SNARE motifs form a stable helical bundle. The crystal structures of two only distantly related SNARE complexes revealed an extraordinary degree of structural conservation despite limited sequence homology. The structures are represented by four elongated α-helices that form a coiled coil containing 16 layers of interacting amino acid side chains stacked on top of each other. These amino acids are mostly hydrophobic with the exception of an unusual layer in the middle of the complex, termed "0" layer, to which three SNARE motifs each contribute a glutamine, and one an arginine (Fig. 6). On the basis of this highly conserved layer, SNAREs are now classified into subfamilies termed

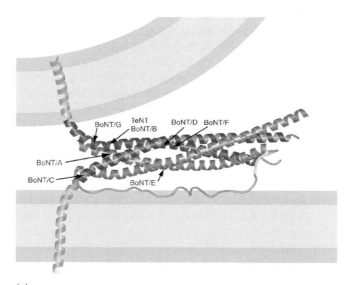

(a)

Fig. 6 (a) Crystal structure of the neuronal SNARE complex containing syntaxin 1 (red), SNAP-25 (green), and synaptobrevin/VAMP (blue). The structure is modeled between two membranes, the transmembrane domains and immediately adjacent residues are not part of the structure. (a) Structure of the ionic "0" layer that is conserved among all SNAREs and that is the basis for the classification of SNAREs into Q- and R-SNAREs, respectively. Modified from Sutton, B., Fasshauer, D., Jahn, R., Brünger, A.T. (1998) Crystal structure of a core synaptic fusion complex at 2.4 Å resolution, *Nature* **395**, 347–353 (see color plate p. xxviii).

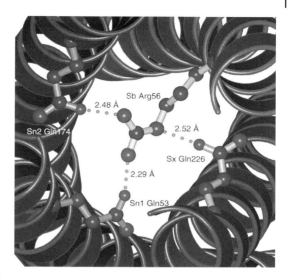

(b)

Fig. 6 (*Continued*)

Q-SNAREs and R-SNAREs, with the Q-SNAREs being further subdivided into Qa-SNAREs (representing the "true" syntaxins), Qb-SNAREs, Qc-SNAREs (homologs of the two SNARE motifs of SNAP-25, respectively), and R-SNAREs (relatives of VAMP/synaptobrevin). The division of SNAREs into these four subfamilies is strongly supported by additional similarities that were derived from sequence comparisons. As a general rule, SNARE complexes can only form if one SNARE motif of each subfamily is contributed. It replaces the previous classification of SNAREs to v-SNAREs (vesicular SNAREs) and t-SNAREs (SNAREs residing in target membranes) since the intracellular topology that was used for this classification does not agree with the subclasses as defined by activity, crystal structures, and sequence alignment. In fact, *in vitro* SNARE motifs of the appropriate subclasses can – within certain limits – substitute for each other.

Although due to an unusual hysteresis, it is not possible to determine the free energy of SNARE assembly, assembly proceeds downhill an energy gradient. The stability of assembled SNARE complexes is remarkable, requiring temperatures above 80 °C or very strong denaturants such as 5 M guanidinium chloride for dissociation. Both the elongated structure of the helical bundle and the stability toward heat, denaturants, and proteases is strikingly parallel to the relaxed conformation of class I viral fusion proteins discussed above. However, unlike the viral fusion proteins, SNAREs are reactivated by disassembly, involving the ATPase NSF (acronym for NEM-sensitive factor). On its own, however, NSF cannot interact with SNAREs; it requires cofactors termed SNAPs (acronym for soluble NSF attachment proteins). Both in yeast and in mammals only one NSF gene has been found. In contrast, mammalian species contain three variants of SNAP termed α-, β-, and γ-SNAP respectively, with α-SNAP being the ubiquitous isoform. Three SNAPs bind to one SNARE complex, resulting in the recruitment of NSF. Remarkably, the SNAP-NSF system interacts with all known SNARE complexes in this manner despite the fact that the surface amino acid composition is highly variable with respect to charge and size. The crystal structure of the yeast SNAP-homolog Sec17p revealed a rigidly packed protein with an N-terminal region in which nine α-helices are packed against each other in an antiparallel fashion, resulting in a bent sheet and a C-terminal globular region. The N-terminal region is responsible for SNARE binding, whereas the C-terminal region recruits NSF. Several SNAP mutants are known, which are defective in binding and/or activation of NSF, they are strong inhibitors of all fusion reactions of the secretory pathway.

NSF is a hexameric ATPase that belongs to the AAA superfamily (ATPases associated with other activities). AAAs frequently operate as unfoldases, that is, they dissociate tightly packed or aggregated proteins. NSF monomers consist of three domains: an N-terminal region that binds to the substrate and that undergoes large conformational changes upon ATP hydrolysis, and two homologous ATP binding regions termed D1 and D2, with only D1 being catalytically active and D2 being responsible for hexamer formation. Single-particle and cryoelectron microscopy as well as crystallographic studies on NSF domains and on VCP (valosin-containing protein), a relative of NSF, have revealed that the protein is represented by a double barrel structure with a conspicuous hole in the middle. Presently it is unknown how AAAs manage to disentangle proteins as tightly packed and stable as SNARE complexes. It has been proposed that unfolding involves threading of the substrate through the central channel (resembling some chaperones). Alternatively, it is possible that the barrel opens sideways ("clamp loader") thus getting a "grip" on the substrate and exposing it to hydrophobic side chains in the interior of the oligomeric molecule. The ATP-requirement has not yet been precisely determined, and thus it is not known whether a full catalytic cycle of all six subunits is required for disassembly. Mutations of NSF are known both in yeast and in Drosophila, which loose activity at mildly elevated temperature, resulting in a general impairment of membrane traffic. In Drosophila, the dominant effect is in synapses where exocytotic release of neurotransmitters is affected first, resulting in paralysis ("comatose") and an accumulation of SNARE complexes.

6.3.2 Mechanism of SNARE-mediated Membrane Fusion

The most widely accepted model describing how SNAREs mediate membrane fusion implies that it is the spontaneous assembly of SNAREs into SNARE complexes that drives the membrane merger. According to this model, each of the membranes destined to fuse contains complementary sets of SNAREs that upon close contact spontaneously interact with each other in "trans" and form a SNARE complex by zippering up from the N- toward the C-terminal ends of the respective SNARE motifs. Such zippering would force the transmembrane domains in close apposition. The energy being released during assembly is used to overcome the repulsive energy barrier separating the membranes, thus exerting strain on the membranes. During fusion, the SNAREs change from a strained "trans" into a relaxed "cis" conformation, in a remarkable parallelism to the class I viral fusion proteins. After fusion is complete, the NSF-SNAP disassembly system then dissociates the complex and makes available the SNAREs for another round of fusion.

This model, although not undisputed, is supported by a large body of evidence such as

- experiments using yeast vacuolar fusion *in vitro* have shown that NSF is not required for the fusion reaction but rather is needed for a preceding activation step;
- many SNARE mutants from yeast, flies, and worms causing functional impairments map to amino acids of the central interacting layers, supporting the view that the formation of the helical bundle is critical for fusion;
- cells can be enticed to fuse when SNAREs are "flipped," that is, expressed by facing the exterior rather than the cytoplasm;
- any interference with SNAREs in intact cells or in cell-free assays using antibodies, clostridial neurotoxins, or recombinant SNARE motifs as competitors invariably inhibits fusion;
- proteoliposomes reconstituted with appropriate sets of SNAREs spontaneously fuse with each other, and this fusion requires the formation of SNARE complexes.

6.3.3 Conformational Intermediates of SNAREs

Recent studies have shown that the zippering model described above, while providing a useful overall description of the SNARE mechanism, is an oversimplification that needs to be refined. For instance, while SNAREs spontaneously form complexes and fuse liposomes *in vitro*, these reactions are many orders of magnitude slower than their biological counterparts, although high protein concentrations were used in the assays, and artificial docking of the liposomes also does not accelerate the reaction. Similar to viral fusion proteins, the attention has therefore focused on conformational intermediates of the SNARE cycle.

A model outlining potential intermediate steps is shown in Fig. 7. This model is primarily based on studies of the SNARE proteins involved in neuronal exocytosis that served as paradigms. It is possible that certain depicted states do not exist in other SNARE complexes. The following features are noteworthy:

- *Hotspots of concentrated clusters of SNAREs.* In plasma membranes, SNAREs are not uniformly distributed across the membrane but rather concentrated in microdomains. These microdomains are dependent on the presence of cholesterol in the membrane but are not resistant to detergent, that is, they are different from rafts. The local concentration of SNAREs in these hotspots appears to be very high. This is particularly noteworthy in neurons where the two Q-SNAREs syntaxin 1 and SNAP-25 each contribute at least 1% of total brain protein, making them by far the most abundant membrane proteins in the nervous system. In PC12 cells, secretory granules exclusively undergo exocytosis on such hotspots suggesting that for fusion to be efficient many SNARE complexes need to cooperate. In addition, the SNAREs in these hotspots readily form complexes with both exogenous and endogenous SNAREs suggesting that activation of SNAREs prior to fusion is not rate limiting.
- *SNARE acceptor complexes.* Considering that the SNARE motifs are unstructured in solution, it is not surprising that SNARE complex formation *in vitro*

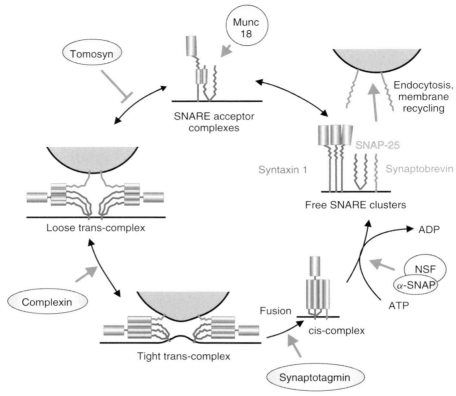

Fig. 7 Model of the cycle of the neuronal SNAREs and its regulation by accessory proteins. See text for details.

is exceedingly slow as it requires the simultaneous association of four different molecules. *In vitro*, the neuronal Q-SNAREs syntaxin 1 and SNAP-25 form a binary complex consisting of two syntaxin and one SNAP-25 molecule. This complex resembles the SNARE core complex but is less stable. For synaptobrevin to bind, one of the syntaxins needs to be displaced which appears to be rate limiting. Recent work revealed that an unstable intermediate is formed between SNAP-25 and a single syntaxin molecule. Binding of synaptobrevin to this complex appears to be considerably faster. This intermediate may represent the "true" acceptor complex that can either recruit a second syntaxin, resulting in a nonconstructive "dead-end" reaction, or bind synaptobrevin in a pathway leading to fusion. Intriguingly, independent lines of evidence suggest that such acceptor complexes are stabilized by SM proteins (see below).

- *SNARE trans-complexes.* Regulated exocytosis implies that the pathway leading to fusion is arrested at some step, requiring activation by a trigger that is usually an elevation of intracellular Ca^{2+}. In neuronal exocytosis, the delay time between Ca^{2+} triggering and membrane fusion is extremely short (<1 ms). The question arises whether SNARE assembly can be arrested at an intermediate state, with the N-terminal parts of the SNARE motifs being already assembled, whereas the C-terminal parts are still apart. Evidence for such intermediates primarily stems from studies of exocytosis of chromaffin cells, which can be monitored with high time resolution using electrophysiological techniques. Following a jump in intracellular Ca^{2+}, distinct pools of granules are observed that undergo exocytosis with a characteristic time constant: a cytoplasmic depot pool, a reserve pool representing membrane-attached vesicles that still need to be activated ("primed"), and two pools of release-ready vesicles, referred to as slowly releasable pool (SRP) and rapidly releasable pool (RRP). These pools are in dynamic equilibria with each other. Recent studies suggested that in the RRP, SNAREs are associated with each other in a trans-complex prior to fusion, and similar conclusions have also been reached from studies at the crayfish neuromuscular junction. While all these studies are indirect (i.e. conclusions about molecular states are drawn from kinetic data), they suggest that metastable folding intermediates may exist in the SNARE assembly pathway, again paralleling similar observations on class I viral fusion proteins.

- *Contribution of SNARE transmembrane domains.* For the final step in membrane fusion, the model predicts that a pulling force is exerted on the membranes, suggesting that the linkers between the transmembrane domains and the SNARE motifs are stiff. Both theoretical calculations and experimental evidence suggest that this may be the case. Furthermore, it is possible that the transmembrane domains do more than just providing a membrane anchor. For instance, peptides corresponding to SNARE transmembrane domains have recently been shown to be capable of fusing liposomes, suggesting that they may also be involved in destabilizing the bilayer. Furthermore, recent studies indicate that the transmembrane domains of syntaxin may contribute to the fusion pore structure. Also, transmembrane domains of syntaxin and synaptobrevin have the capacity to form homooligomers and heterooligomers, again supporting the view that in addition to providing membrane attachment they may participate in transition states in a specific manner.

6.3.4 Topology and Specificity of SNAREs in Intracellular Fusion Reactions

The model outlined above for SNARE function implies that of the fusing membranes each one carries at least one of a complementary set of SNAREs that contains a transmembrane domain. In neuronal exocytosis this requirement is fulfilled, with synaptobrevin (R-SNARE) being localized primarily to the synaptic vesicle and syntaxin 1 (Qa-SNARE) to the plasma membrane. Intracellular fusion reactions, however, often involve SNARE complexes in which three or all four SNARE motifs are contributed by SNAREs possessing a transmembrane domain, allowing, at least in theory, variable distributions between the membranes. However, it is conceivable that the need for preformation of acceptor complexes limits the choices, for example, if, as in the case of neuronal exocytosis, an acceptor complex consists of Qa/Qb/Qc-SNARE motifs. Indeed, liposome fusion experiments suggested that only one type of topology of a given SNARE complex results in fusion. However, in yeast there are examples of SNAREs that may function in different complexes with different topology. For instance, the yeast R-SNARE Sec22p functions both in anterograde as well as in retrograde traffic between the endoplasmic reticulum and the Golgi apparatus. In anterograde traffic, Sec22p appears to be colocalized with the Qb- and Qc-SNAREs Bos1p and Bet1p respectively, on the transport vesicle, with the Qa-SNARE Sed5p residing in the acceptor membrane. In contrast, all three Q-SNAREs for retrograde traffic including Qa (Ufe1p), Qb (Sec20p), and Qc (Use1p) are localized to the endoplasmic reticulum, with Sec22p being the sole functional SNARE on the retrograde transport vesicle.

The example of yeast anterograde and retrograde traffic between the endoplasmic reticulum and the Golgi apparatus also raises the question to which extent SNARE pairing is specific. Although this issue continues to be controversially discussed, the following points can be made:

- *In vitro*, SNAREs can be assembled into SNARE complexes rather promiscuously as long as the QabcR-SNARE rule for SNARE complexes is obeyed. However, while in some experiments no differences were observed between different R- or Q-SNAREs, respectively, it is likely that promiscuity is not absolute, that is, that not every SNARE of a given subfamily is capable of substituting for a particular family member in a given complex.
- Certain SNAREs are known to be involved in different trafficking steps involving different SNARE partners. In addition to Sec22p discussed above, other yeast examples include Sed5p (a Qa-SNARE operating in ER to Golgi, and intra Golgi trafficking steps), Vti1p (a Qb-SNARE operating in traffic to the *cis*-Golgi, and from the Golgi to the prevacuole (late endosome) and the vacuole compartment), and Ykt6p (an R-SNARE operating in retrograde traffic to the Golgic, in traffic from the Golgi to the prevacuole (late endosome) and the vacuole compartment, and in anterograde traffic from the ER to the Golgi).
- Conversely, in a given SNARE complex certain SNAREs can functionally substitute for each other, albeit the efficiency appears to be lower than with the "cognate" SNARE. For instance, in yeast Ykt6p can substitute for Sec22p in anterograde but not retrograde ER to Golgi traffic, and SNAP-23 can substitute for SNAP-25 in exocytosis of chromaffin granules. Such functional substitutions of one SNARE for another one may explain why genetic deletion of individual SNAREs in some cases is known to yield surprisingly mild phenotypes.

The question then arises to which extent SNARE complexes are specific for a given fusion step. Although a definitive answer cannot be given at this time, it appears that the SNAREs active in a specific intracellular fusion event are precisely regulated, that is, a given SNARE complex functions only in one defined step even if some flexibility is allowed in the identity of some of its participating SNARE proteins. It should be noted, however, that presently it cannot be excluded that more than one SNARE complex operates in parallel in certain fusion events.

It is not known how nonspecific pairing (as observed *in vitro*) is prevented. Furthermore, it is not known by which mechanism the correct SNAREs are selected for a given fusion step. As integral membrane proteins, SNAREs need to recycle by membrane traffic to their resident membrane after fusion, explaining the widespread intracellular distribution of many SNAREs. Probably every trafficking vesicle carries "passenger" SNAREs in addition to the ones needed for the forthcoming fusion step. For instance, in neurons and neuroendocrine cells the recycling pathway of synaptic vesicles includes at least two fusion steps – exocytosis and fusion with an early endosomal intermediate that is reached after endocytosis via clathrin-coated vesicles (CCV). The SNAREs involved in these two fusion steps are distinct (exocytosis: syntaxin 1 (Qa), SNAP-25 (Qbc), synaptobrevin (Qc); endosome fusion: syntaxin 13 or 16 (Qa), Vti1a (Qb), syntaxin 6 (Qc), and VAMP4 (R)). Functionally, there appears to be no cross talk at least between the R-SNAREs as only exocytosis but not endosome fusion is inhibited by clostridial neurotoxins cleaving the R-SNARE synaptobrevin.

6.4
Regulators of SNAREs

An almost overwhelming panoply of proteins is reported to interact with SNAREs in a functionally relevant manner, either regulating fusion or else, being regulated by the SNAREs. In particular, the neuronal SNARE proteins were the target of intense investigations, with syntaxin 1 alone being thought to bind to more than 40 different proteins including many ion channels and receptors. However, in most of these cases the evidence is limited to qualitative binding studies, complemented with perturbation studies in intact cells, making it difficult to assess the specificity of the reported interactions. As discussed above, SNAREs can assume different conformations, and for proteins thought to specifically interact with SNAREs it is essential to determine where exactly these proteins act in the conformational cycle of SNAREs. While the discussion will be restricted here to those proteins for which experimentally supported molecular models are available, it needs to be borne in mind that this is an actively ongoing field of research, with additional proteins likely to be involved.

6.4.1 SM Proteins

SM proteins comprise a small group of conserved cytoplasmic proteins that, as far known, are as essential as SNAREs for all fusion steps of the secretory pathway. In fact, impairment or knock out of SM proteins leads to dramatic phenotypes. For instance, genetic deletion of Munc-18, the SM protein involved in neuronal exocytosis, leads to mice whose nervous system by and large develops normally but in which synapses are completely silent, that is, no exocytotic release of neurotransmitter occurs. There are fewer SM proteins than SNAREs, with four members in yeast and seven in mammals. Crystallographic studies on several SM proteins revealed a high degree of structural conservation, being represented by globular, arch-shaped molecules with a conspicuous cleft.

Numerous genetic and biochemical studies suggest that SM proteins regulate SNAREs. However, major confusion has arisen from the fact that despite a high degree of structural conservation among both SNAREs and SM proteins, the interaction between these proteins, at least *in vitro*, occurs by different mechanisms:

- Munc-18, the SM protein involved in neuronal exocytosis binds to the N-terminal region of the Qa-SNARE syntaxin 1 with high affinity. Like many other SNAREs, syntaxin 1 possesses an antiparallel three-helix bundle at its N-terminus that is connected to the

SNARE motif by a linker region. In a reversible intramolecular transition between an "open" and a "closed" conformation, this N-terminal domain can fold back onto the SNARE motif, preventing the latter from entering SNARE complexes. *In vitro*, Munc-18 exclusively binds to the closed conformation, effectively preventing it from interacting with its SNARE partners. Indeed, the crystal structure of the Munc-18/syntaxin complex reveals that both the N-terminal helix bundle and part of the SNARE motif are bound to the large cleft in the middle of Munc-18, with numerous crystal contacts between Munc-18 and both regions of syntaxin.

- The yeast SM proteins Sly1p and Vps45 (and probably also their mammalian counterparts) bind to the respective Qa-SNAREs Sed5p and Tlg2p in a completely different manner. Here, a short sequence at the N-terminal end of the Qa-SNAREs (i.e. beyond the folded part) binds to the outside of the SM protein. This interaction does not prevent formation of SNARE complexes; rather, it appears to facilitate them.
- In trafficking steps of the endocytotic pathway, the SM proteins do not appear to interact directly with SNAREs. Rather they form complexes with other proteins. Interestingly, this includes not only the SM protein Vps33p that is part of the VpsC or HOPS complex involved in the fusion of vacuoles but also Vps45p that forms a complex with Vac8p. Thus, one and the same SM protein may either bind directly to a SNARE, or else, operate via multiprotein complexes without displaying direct SNARE binding.

Recent evidence suggests that, at least in the case of Sly1p, direct binding to the SNARE is not required for function. Thus, at least some of the bindings described above may be recruiting mechanisms to ensure a sufficiently high concentration of SM proteins at the site of action rather than reflecting part of the mechanism these proteins are involved in, explaining the surprising structural diversity of the SM-SNARE complexes. How then are SM proteins regulating SNAREs? Several lines of evidence suggest that some SM proteins facilitate SNARE assembly, raising the possibility that the main job of SM proteins is to form and maintain SNARE acceptor complexes. Such acceptor complexes are probably unstable and thus hard to measure – the more stable they are, the less energy is left for the ensuing fusion reaction. Furthermore, it has recently been suggested that SM proteins may be part of a proofreading system that ensures pairing of appropriate SNAREs.

6.4.2 Synaptotagmin and Complexin – Regulators of Ca^{2+}-dependent Exocytosis

Exocytosis of synaptic vesicles represents a membrane fusion event that is tightly regulated by the intracellular Ca^{2+} concentration. Delay times between increases in intracellular Ca^{2+} and fusion are extremely short (below 1 ms), suggesting that the system is arrested at a very late step in the pathway of protein–protein interactions leading to fusion. The Ca^{2+} receptor responsible for triggering exocytosis was identified as synaptotagmin, an integral membrane protein of synaptic vesicles. Genetic deletion of synaptotagmin 1 leads to a massive loss of Ca^{2+}-stimulated exocytosis, whereas spontaneous transmitter release persists. Intriguingly, stimulation of exocytosis by means of bypassing the Ca^{2+} trigger (e.g. using the active ingredient of black widow spider venom) persisted but was sensitive to clostridial neurotoxins, documenting that SNARE function in fusion was not affected in the deletion mutant.

Synaptotagmin is represented by a protein family of hitherto 15 members in mammals. Synaptotagmin 1 and 2 that function in neuronal exocytosis are type I transmembrane glycoproteins, that is, with the C-terminus facing the cytoplasm and the glycosylated N-terminus being exposed to the vesicle lumen. Characteristic for synaptotagmins are two tandem C2-domains (C2A and C2B) that represent Ca^{2+} binding modules binding 3 and 2 Ca^{2+} ions, respectively. Whereas the intrinsic Ca^{2+} affinity of both C2-domains is very low (close to 1 mM), it is increased by almost three orders of magnitude in the presence of acidic phospholipids. Structural studies revealed that the C2-domains provide only a partial coordinations sphere for Ca^{2+} ions that requires acidic phospholipids for completion. C2-domains are present in many additional proteins, and in some instances (e.g. protein kinase C) they are known to convey Ca^{2+}-dependent translocation of the protein to the membrane. Furthermore, Ca^{2+} binding is very rapid and it is not associated with a conformational change in the C2-domain.

While it is undisputed that Ca^{2+} binding to the C2-domains of synaptotagmin triggers exocytosis, the mechanism by which this occurs is still not clear. One possibility is that Ca^{2+}-dependent binding to phospholipids at a site where SNAREs are already forming trans-complexes suffices to further destabilize the membrane surface, leading to the formation of a fusion pore. Indeed, both mutant analysis and ion dependence suggest that phospholipid binding is critical for synaptotagmin function. On the other hand, both Ca^{2+}-dependent and Ca^{2+}-independent binding of synaptotagmin to SNAREs has been observed, raising the possibility that synaptotagmin directly acts upon the SNAREs, for example, by driving the assembly of a partially zippered trans-complex to completion. Intriguingly, Sr^{2+} ions, known to substitute for Ca^{2+} in stimulating exocytosis, promote only phospholipid but not SNARE binding of synaptotagmin. While this finding shows that Ca^{2+}-dependent phospholipid interaction is essential for function, it is conceivable that binding to SNAREs increases efficiency, for example, by positioning the C2-domains precisely before the arrival of the Ca^{2+} signal.

Intriguingly, a phenotype similar to synaptotagmin deletion albeit weaker was observed upon the deletion of complexins, two small and highly homologous isoforms that bind to the surface of assembled neuronal SNARE complexes. Structural studies revealed that part of the complexin forms a helix positioned in the groove formed by syntaxin I and synaptobrevin, which bends away from the SNARE complex toward the C-terminal transmembrane domains. Thus, it is possible that complexins bind and stabilize trans-SNARE complexes, maintaining them in a metastable state and allowing for synaptotagmin to act efficiently in promoting the reaction. Lack of complexins may thus result in less stable SNARE complexes (i.e. complexes that reversibly loosen and tighten), requiring more Ca^{2+} binding to synaptotagmin for efficient fusion.

It remains to be established whether similar regulatory mechanisms also apply to other SNARE-mediated fusion reactions. Many intracellular fusion reactions are known to require Ca^{2+} albeit at a concentration well below 1 µM, that is, the normal cytoplasmic concentration suffices for fusion to proceed. It is conceivable that other ubiquitously expressed synaptotagmin isoforms participate in such intracellular fusion reactions. However, some synaptotagmins do not bind Ca^{2+} at all, raising the possibility that synaptotagmins may have other functions. Conversely, several additional synaptic proteins possess tandem C2-domains including the active zone components RIM (an effector of Rab3) and Piccolo/Aczonin.

6.4.3 "Pseudo" SNAREs

Two soluble proteins, termed *tomosyn* and *amisyn* respectively are known, which possess C-terminal R-SNARE motifs but otherwise have no similarities to SNAREs. Tomosyn is a large protein of approximately 110 kDa (SNAREs rarely exceed 35 kDa) whose large N-terminal region bears similarity to the lethal giant larvae proteins, whereas amisyn is represented by a smaller protein. Both proteins, when introduced into secretory cells, inhibit exocytosis. Detailed studies on the R-SNARE motif of tomosyn revealed that it directly competes with synaptobrevin in the formation of SNARE complexes, thus reducing the number of active acceptor complexes that are available for fusion. Presently, it cannot be decided whether competition by soluble SNARE motifs (resulting in the formation of nonconstructive cis-complexes) is a general physiological means to downregulate SNARE complexes or whether these proteins are specialized for neuronal exocytosis.

7 Conclusion and Outlook

Despite the diversity of fusion proteins, the examples discussed above show that there appear to be some common denominators between viral fusion proteins and SNAREs. In both cases, fusion is associated with a spontaneous conformational change that involves movement of membrane-anchor domains (transmembrane domains or fusion peptides) over large distances. At the end of the conformational change, an elongated and stiff rodlike structure is generated in which all membrane-anchor domains are localized at one end. Furthermore, both viral fusion proteins and SNAREs are oligomers that exhibit at least some rotational symmetry (trimers in the case of viral fusion proteins, trimers and tetramers in the case of SNAREs). These intriguing features suggest that the basic mechanism of operation is principally similar in all cases. However, there are still many unresolved issues. For instance, it is unknown whether all fusion reactions transit through hemifusion intermediates. Furthermore, it remains to be established whether the fusion proteins are either part of the initial fusion pores or at least directly influence their diameter and enlargement kinetics.

There is still only scant knowledge about the proteins involved in cell–cell fusion. Several candidate proteins were identified during the last years but in most cases it is not clear whether these proteins are primarily functioning in attachment rather than fusion. These include the tetraspanins CD 9 and CD 81, the single-pass membrane proteins of the ADAM family that may be involved in sperm-egg fusion, and the *Caneorhabditis elegans* protein EEF-1. Of these, EEF-1 is presently the strongest candidate for a bona fide fusion protein. It is both necessary and sufficient for cell–cell fusion, and it is structurally related to class II viral fusion proteins, containing an internal putative fusion peptide. Indeed, the search for cellular fusion proteins was driven for many years by the expectation that cellular fusion proteins must be similar to viral fusion proteins. After all, viral fusion proteins must have originated from primordial eukaryotes.

See also Membrane Traffic: Vesicle Budding and Fusion; Membrane Transport; Protein Translocation Across Membranes.

Bibliography

Books and Reviews

Bonifacino, J.S., Glick, B.S. (2004) The mechanisms of vesicle budding and fusion, *Cell* **116**, 153–166.

Chernomordik, L.V., Kozlov, M.M. (2003) Protein-lipid interplay in fusion and fission of biological membranes, *Annu. Rev. Biochem.* **72**, 175–207.

Earp, L.J., Delos, S.E., Park, H.E., White, J.M. (2005) The many mechanisms of viral membrane fusion proteins, *Curr. Top. Microbiol. Immunol.* **285**, 25–66.

Eckert, D.M., Kim, P.S. (2001) Mechanisms of viral membrane fusion and its inhibition, *Annu. Rev. Biochem.* **70**, 777–810.

Fasshauer, D. (2003) Structural insights into the SNARE mechanism, *Biochim. Biophys. Acta* **1641**, 87–97.

Jahn, R., Lang, T., Südhof, T.C. (2003) Membrane fusion, *Cell* **112**, 519–533.

Maxfield, F.R., McGraw, T.E. (2004) Endocytic recycling, *Nat. Rev. Mol. Cell Biol.* **5**, 121–132.

Smith, A.E., Helenius, A. (2004) How viruses enter animal cells, *Science* **304**, 237–242.

Südhof, T.C. (2004) The synaptic vesicle cycle, *Annu. Rev. Neurosci.* **27**, 509–547.

Tamm, L.K., Crane, J., Kiessling, V. (2003) Membrane fusion: a structural perspective on the interplay of lipids and proteins, *Curr. Opin. Struct. Biol.* **13**, 453–466.

Primary Literature

Almers, W., Tse, F.W. (1990) Transmitter release from synapses: does a preassembled fusion pore initiate exocytosis? *Neuron* **4**, 813–818.

Antonin, W., Fasshauer, D., Becker, S., Jahn, R., Schneider, T.R. (2002) Crystal structure of the endosomal SNARE complex reveals common structural principles of all SNAREs, *Nat. Struct. Biol.* **9**, 107–111.

Atlashkin, V., Kreykenbohm, V., Eskelinen, E.L., Wenzel, D., Fayyazi, A., Fischer von Mollard, G. (2003) Deletion of the SNARE vti1b in mice results in the loss of a single SNARE partner, syntaxin 8, *Mol. Cell. Biol.* **23**, 5198–5207.

Bai, J., Chapman, E.R. (2004) The C2 domains of synaptotagmin – partners in exocytosis, *Trends Biochem. Sci.* **29**, 143–151.

Bock, J.B., Matern, H.T., Peden, A.A., Scheller, R.H. (2001) A genomic perspective on membrane compartment organization, *Nature* **409**, 839–841.

Bracher, A., Weissenhorn, W. (2002) Structural basis for the Golgi membrane recruitment of Sly1p by Sed5p, *EMBO J.* **21**, 6114–6124.

Breckenridge, L.J., Almers, W. (1987) Currents through the fusion pore that forms during exocytosis of a secretory vesicle, *Nature* **328**, 814–817.

Bressanelli, S., Stiasny, K., Allison, S.L., Stura, E.A., Duquerroy, S., Lescar, J., Heinz, F.X., Rey, F.A. (2004) Structure of a flavivirus envelope glycoprotein in its low-pH-induced membrane fusion conformation, *EMBO J.* **23**, 728–738.

Brose, N., Petrenko, A.G., Südhof, T.C., Jahn, R. (1992) Synaptotagmin: a Ca^{2+} sensor on the synaptic vesicle surface, *Science* **256**, 1021–1025.

Brunger, A.T. (2001) Structural insights into the molecular mechanism of calcium-dependent vesicle-membrane fusion, *Curr. Opin. Struct. Biol.* **11**, 163–173.

Brunger, A.T., DeLaBarre, B. (2003) NSF and p97/VCP: similar at first, different at last, *FEBS Lett.* **555**, 126–133.

Bullough, P.A., Hughson, F.M., Skehel, J.J., Wiley, D.C. (1994) Structure of influenza hemagglutinin at the pH of membrane fusion, *Nature* **371**, 37–43.

Chanturiya, A., Chernomordik, L.V., Zimmerberg, J. (1997) Flickering fusion pores comparable with initial exocytotic pores occur in protein-free phospholipid bilayers, *Proc. Natl. Acad. Sci. U.S.A.* **94**, 14423–14428.

Chen, X., Tomchick, D.R., Kovrigin, E., Arac, D., Machius, M., Südhof, T.C., Rizo, J. (2002) Three-dimensional structure of the complexin/SNARE complex, *Neuron* **33**, 397–409.

Christoforidis, S., McBride, H.M., Burgoyne, R.D., Zerial, M. (1999) The Rab5 effector EEA1 is a core component of endosome docking, *Nature* **397**, 621–625.

Dilcher, M., Köhler, B., von Mollard, G.F. (2001) Genetic interactions with the yeast Q-SNARE VTI1 reveal novel functions for the R-SNARE YKT6, *J. Biol. Chem.* **276**, 34537–34544.

Dimmer, K.S., Fritz, S., Fuchs, F., Messerschmitt, M., Weinbach, N., Neupert, W., Westermann, B. (2002) Genetic basis of mitochondrial function and morphology in Saccharomyces cerevisiae, *Mol. Biol. Cell* **13**, 847–853.

Fasshauer, D., Margittai, M. (2004) A transient N-terminal interaction of SNAP-25 and syntaxin nucleates SNARE assembly, *J. Biol. Chem.* **279**, 7613–7621.

Fasshauer, D., Sutton, B., Brünger, A.T., Jahn, R. (1998) Conserved structural features of the synaptic fusion complex: SNARE proteins reclassified as Q- and R-SNAREs, *Proc. Natl. Acad. Sci. U.S.A.* **95**, 15781–15786.

Fasshauer, D., Antonin, W., Subramaniam, V., Jahn, R. (2002) SNARE assembly and disassembly exhibit a pronounced hysteresis, *Nat. Struct. Biol.* **9**, 144–151.

Fasshauer, D., Otto, H., Eliason, W.K., Jahn, R., Brunger, A.T. (1997) Structural changes are associated with soluble N-ethylmaleimide-sensitive fusion protein attachment protein receptor complex formation, *J. Biol. Chem.* **272**, 28036–28041.

Fasshauer, D., Antonin, W., Margittai, M., Pabst, S., Jahn, R. (1999) Mixed and non-cognate SNARE complexes. Characterization of assembly and biophysical properties, *J. Biol. Chem.* **274**, 15440–15446.

Fernandez, I., Ubach, J., Dulubova, I., Zhang, X., Südhof, T.C., Rizo, J. (1998) Three-dimensional structure of an evolutionarily conserved N-terminal domain of syntaxin 1A, *Cell* **94**, 841–849.

Fujita, Y., Shirataki, H., Sakisaka, T., Asakura, T., Ohya, T., Kotani, H., Yokoyama, S., Nishioka, H., Matsuura, Y., Mizoguchi, A., Scheller, R.H., Takai, Y. (1998) Tomosyn: a syntaxin-1-binding protein that forms a novel complex in the neurotransmitter release process, *Neuron* **20**, 905–915.

Gallwitz, D., Jahn, R. (2003) The riddle of the Sec1/Munc-18 proteins – new twists added to their interactions with SNAREs, *Trends Biochem. Sci.* **28**, 113–116.

Garner, C.C., Kindler, S., Gundelfinger, E.D. (2000) Molecular determinants of presynaptic active zones, *Curr. Opin. Neurobiol.* **10**, 321–327.

Geppert, M., Goda, Y., Hammer, R.E., Li, C., Rosahl, T.W., Stevens, C.F., Südhof, T.C. (1994) Synaptotagmin I: a major Ca^{2+} sensor for transmitter release at a central synapse, *Cell* **79**, 717–727.

Gibbons, D.L., Erk, I., Reilly, B., Navaza, J., Kielian, M., Rey, F.A., Lepault, J. (2003) Visualization of the target-membrane-inserted fusion protein of Semliki forest virus by combined electron microscopy and crystallography, *Cell* **114**, 573–583.

Gibbons, D.L., Vaney, M.C., Roussel, A., Vigouroux, A., Reilly, B., Lepault, J., Kielian, M., Rey, F.A. (2004) Conformational change and protein-protein interactions of the fusion protein of Semliki Forest virus, *Nature* **427**, 320–325.

Han, X., Wang, C.T., Bai, J., Chapman, E.R., Jackson, M.B. (2004) Transmembrane segments of syntaxin line the fusion pore of Ca^{2+}-triggered exocytosis, *Science* **304**, 289–292.

Hanson, P.I., Roth, R., Morisaki, H., Jahn, R., Heuser, J.E. (1997) Structure and conformational changes in NSF and its membrane receptor complexes visualized by quick-freeze/deep-etch electron microscopy, *Cell* **90**, 523–535.

Hardwick, K.G., Pelham, H.R.B. (1992) SED5 encodes a 39-kD integral membrane protein required for vesicular transport between the ER and the Golgi complex, *J.Cell Biol.* **119**, 513–521.

Hatsuzawa, K., Lang, T., Fasshauer, D., Bruns, D., Jahn, R. (2003) The R-SNARE motif of tomosyn forms SNARE core complexes with syntaxin 1 and SNAP-25 and down-regulates exocytosis, *J. Biol. Chem.* **278**, 31159–31166.

Hsu, S.C., TerBush, D., Abraham, M., Guo, W. (2004) The exocyst complex in polarized exocytosis, *Int. Rev. Cytol.* **233**, 243–265.

Hu, C., Ahmed, M., Melia, T.J., Söllner, T.H., Mayer, T., Rothman, J.E. (2003) Fusion of cells by flipped SNAREs, *Science* **300**, 1745–1749.

Jahn, R., Grubmuller, H. (2002) Membrane fusion, *Curr. Opin. Cell Biol.* **14**, 488–495.

Jardetzky, T.S., Lamb, R.A. (2004) Virology: a class act, *Nature* **427**, 307–308.

Koshiba, T., Detmer, S.A., Kaiser, J.T., Chen, H., McCaffery, J.M., Chan, D.C. (2004) Structural basis of mitochondrial tethering by mitofusin complexes, *Science* **305**, 858–862.

Kozlov, M.M., Chernomordik, L.V. (1998) A mechanism of protein-mediated fusion – coupling between refolding of the influenza hemagglutinin and lipid rearrangements, *Biophys. J.* **75**, 1384–1396.

Laage, R., Rohde, J., Brosig, B., Langosch, D. (2000) A conserved membrane-spanning amino acid motif drives homomeric and supports heteromeric assembly of presynaptic SNARE proteins, *J. Biol. Chem.* **275**, 17481–17487.

Lang, T., Margittai, M., Holzler, H., Jahn, R. (2002) SNAREs in native plasma membranes are active and readily form core complexes with endogenous and exogenous SNAREs, *J. Cell Biol.* **158**, 751–760.

Lang, T., Bruns, D., Wenzel, D., Riedel, D., Holroyd, P., Thiele, C., Jahn, R. (2001) SNAREs are concentrated in cholesterol-dependent clusters that define docking and fusion sites for exocytosis, *EMBO J.* **20**, 2202–2213.

Langosch, D., Crane, J.M., Brosig, B., Hellwig, A., Tamm, L.K., Reed, J. (2001) Peptide mimics of SNARE transmembrane segments drive membrane fusion depending on their conformational plasticity, *J. Mol. Biol.* **311**, 709–721.

Lehman, K., Rossi, G., Adamo, J.E., Brennwald, P. (1999) Yeast homologues of tomosyn and lethal giant larvae function in exocytosis and are associated with the plasma membrane SNARE, Sec9, *J. Cell Biol.* **146**, 125–140.

Lin, R.C., Scheller, R.H. (1997) Structural organization of the synaptic exocytosis core complex, *Neuron* **19**, 1087–1094.

Littleton, J.T., Barnard, R.J., Titus, S.A., Slind, J., Chapman, E.R., Ganetzky, B. (2001) SNARE-complex disassembly by NSF follows synaptic-vesicle fusion, *Proc. Natl. Acad. Sci. U.S.A.* **98**, 12233–12238.

Liu, Y., Barlowe, C. (2002) Analysis of Sec22p in endoplasmic reticulum/Golgi transport reveals cellular redundancy in SNARE protein function, *Mol. Biol. Cell* **13**, 3314–3324.

Markin, V.S., Albanesi, J.P. (2002) Membrane fusion: stalk model revisited, *Biophys. J.* **82**, 693–712.

Masuda, E.S., Huang, B.C.B., Fisher, J.M., Luo, Y., Scheller, R.H. (1998) Tomosyn binds t-SNARE proteins via a VAMP-like coiled coil, *Neuron* **21**, 479–480.

McNew, J.A., Parlati, F., Fukuda, R., Johnston, R.J., Paz, K., Paumet, F., Söllner, T.H., Rothman, J.E. (2000) Compartmental specificity of cellular membrane fusion encoded in SNARE proteins, *Nature* **407**, 153–159.

Misura, K.M., Scheller, R.H., Weis, W.I. (2000) Three-dimensional structure of the neuronal-Sec1-syntaxin 1a complex, *Nature* **404**, 355–362.

Modis, Y., Ogata, S., Clements, D., Harrison, S.C. (2004) Structure of the dengue virus envelope protein after membrane fusion, *Nature* **427**, 313–319.

Monck, J.R.F. (1995) The exocytotic fusion pore and neurotransmitter release, *Neuron* **12**, 707–716.

Novick, P., Guo, W. (2002) Ras family therapy: Rab, Rho and Ral talk to the exocyst, *Trends Cell Biol.* **12**, 247–249.

Pelham, H.R. (2001) SNAREs and the specificity of membrane fusion, *Trends Cell Biol.* **11**, 99–101.

Peng, R., Gallwitz, D. (2004) Multiple SNARE interactions of an SM protein: Sed5p/Sly1p binding is dispensable for transport, *EMBO J.* **23**, 3939–3949.

Pfeffer, S. (2003) Membrane domains in the secretory and endocytic pathways, *Cell* **112**, 507–517.

Pobbati, A.V., Razeto, A., Boddener, M., Becker, S., Fasshauer, D. (2004) Structural basis for the inhibitory role of tomosyn in exocytosis, *J.Biol.Chem.* **279**, 47192–47200.

Reim, K., Mansour, M., Varoqueaux, F., McMahon, H.T., Südhof, T.C., Brose, N., Rosenmund, C. (2001) Complexins regulate a late step in Ca^{2+}-dependent neurotransmitter release, *Cell* **104**, 71–81.

Rizo, J. (2003) SNARE function revisited, *Nat. Struct. Biol.* **10**, 417–419.

Rizo, J., Südhof, T.C. (2002) Snares and Munc18 in synaptic vesicle fusion, *Nat. Rev. Neurosci.* **3**, 641–653.

Roy, R., Laage, R., Langosch, D. (2004) Synaptobrevin transmembrane domain dimerization-revisited, *Biochemistry* **43**, 4964–4970.

Scales, S.J., Hesser, B.A., Masuda, E.S., Scheller, R.H. (2002) Amisyn, a novel syntaxin-binding protein that may regulate SNARE complex assembly, *J. Biol. Chem.* **277**, 28271–28279.

Schiavo, G., Matteoli, M., Montecucco, C. (2000) Neurotoxins affecting neuroexocytosis, *Physiol. Rev.* **80**, 717–766.

Schoch, S., Deak, F., Königstorfer, A., Mozhayeva, M., Sara, Y., Südhof, T.C., Kavalali, E.T. (2001) SNARE function analyzed in synaptobrevin/VAMP knockout mice, *Science* **294**, 1117–1122.

Seabra, M.C., Wasmeier, C. (2004) Controlling the location and activation of Rab GTPases, *Curr. Opin. Cell Biol.* **16**, 451–457.

Sesaki, H., Jensen, R.E. (2004) Ugo1p links the Fzo1p and Mgm1p GTPases for mitochondrial fusion, *J. Biol. Chem.* **279**, 28298–28303.

Skehel, J.J., Wiley, D.C. (2000) Receptor binding and membrane fusion in virus entry: the influenza hemagglutinin, *Annu. Rev. Biochem.* **69**, 531–569.

Söllner, T., Bennet, M.K., Whiteheart, S.W., Scheller, R.H., Rothman, J.E. (1993) A protein assembly-disassembly pathway in vitro that may correspond to sequential steps of synaptic vesicle docking, activation, and fusion, *Cell* **75**, 409–418.

Söllner, T., Whiteheart, S.W., Brunner, M., Erdjument-Bromage, H., Geromanos, S., Tempst, P., Rothman, J.E. (1993) SNAP receptors implicated in vesicle targeting and fusion, *Nature* **362**, 318–324.

Sorensen, J.B. (2004) Formation, stabilization and fusion of the readily releasable pool of secretory vesicles, *Pflugers Arch.* **448**, 347–362.

Sørensen, J.B., Nagy, G., Varoqueaux, F., Nehring, R.B., Brose, N., Wilson, M.C., Neher, E. (2003) Differential control of the releasable vesicle pools by SNAP-25 splice variants and SNAP-23, *Cell* **114**, 75–86.

Stein, K.K., Primakoff, P., Myles, D. (2004) Sperm-egg fusion: events at the plasma membrane, *J. Cell Sci.* **117**, 6269–6274.

Südhof, T.C. (2002) Synaptotagmins: why so many? *J. Biol. Chem.* **277**, 7629–7632.

Sutton, B., Fasshauer, D., Jahn, R., Brünger, A.T. (1998) Crystal structure of a core synaptic fusion complex at 2.4 Å resolution, *Nature* **395**, 347–353.

Toonen, R.F., Verhage, M. (2003) Vesicle trafficking: pleasure and pain from SM genes, *Trends Cell Biol.* **13**, 177–186.

Verhage, M., Maia, A.S., Plomp, J.J., Brussaard, A.B., Heeroma, J.H., Vermeer, H., Toonen, R.F., Hammer, R.E., van den Berg, T.K., Missler, M., Geuze, H.J., Südhof, T.C. (2000) Synaptic assembly of the brain in the absence of neurotransmitter secretion, *Science* **287**, 864–869.

Wang, Y., Okamoto, M., Schmitz, F., Hofmann, K., Südhof, T.C. (1997) Rim is a putative Rab3 effector in regulating synaptic-vesicle fusion, *Nature* **388**, 593–598.

Washbourne, P., Thompson, P.M., Carta, M., Costa, E.T., Mathews, J.R., Lopez-Bendito, G., Molnar, Z., Becher, M.W., Valenzuela, C.F., Partridge, L.D., Wilson, M.C. (2002) Genetic ablation of the t-SNARE SNAP-25 distinguishes mechanisms of neuroexocytosis, *Nat. Neurosci.* **5**, 19–26.

Weber, T., Paesold, G., Galli, C., Mischler, R., Semenza, G., Brunner, J. (1994) Evidence for H^+-induced insertion of influenza hemagglutinin HA2 N-terminal segment into viral membrane, *J. Biol. Chem.* **269**, 18353–18358.

Weber, T., Zemelman, B.V., McNew, J.A., Westermann, B., Gmachl, M., Parlati, F., Söllner, T., Rothman, J.E. (1998) SNAREpins: minimal machinery for membrane fusion, *Cell* **92**, 759–772.

Weisman, L.S. (2003) Yeast vacuole inheritance and dynamics, *Annu. Rev. Genet.* **37**, 435–460.

Weissenhorn, W., Dessen, A., Harrison, S.C., Skehel, J.J., Wiley, D.C. (1997) Atomic structure of the ectodomain from HIV-1 gp41, *Nature* **387**, 426–430.

Westermann, B. (2003) Mitochondrial membrane fusion, *Biochim. Biophys. Acta* **1641**, 195–202.

Whiteheart, S.W., Matveeva, E.A. (2004) Multiple binding proteins suggest diverse functions for the N-ethylmaleimide sensitive factor, *J. Struct. Biol.* **146**, 32–43.

Whyte, J.R., Munro, S. (2002) Vesicle tethering complexes in membrane traffic, *J. Cell Sci.* **115**, 2627–2637..

Wickner, W. (2002) Yeast vacuoles and membrane fusion pathways, *EMBO J.* **21**, 1241–1247.

Wickner, W., Haas, A. (2000) Yeast homotypic vacuole fusion: a window on organelle trafficking mechanisms, *Annu. Rev. Biochem.* **69**, 247–275.

Xu, T., Rammner, B., Margittai, M., Artalejo, A.R., Neher, E., Jahn, R. (1999) Inhibition of SNARE complex assembly differentially affects kinetic components of exocytosis, *Cell* **99**, 713–722.

Yang, B., Gonzalez, L. Jr., Prekeris, R., Steegmaier, M., Advani, R.J., Scheller, R.H. (1999) SNARE interactions are not selective. Implications for membrane fusion specificity, *J. Biol. Chem.* **274**, 5649–5653.

Zerial, M., McBride, H. (2001) Rab proteins as membrane organizers, *Nat. Rev. Mol. Cell Biol.* **2**, 107–117.

Protein Microarrays

Jens R. Sydor[1], *David S. Wilson*[2], *and Steffen Nock*[2]
[1]*Infinity Pharmaceuticals Inc., Cambridge, MA*
[2]*Absalus Inc., Mountain View, CA*

1 **Introduction** 103

2 **Protein Microarrays for Protein Expression Profiling** 104
2.1 Principle 104
2.2 Content for Protein Expression Profiling Chips 104
2.2.1 Source of Antigen 105
2.2.2 Types of Protein-binding Agents 105
2.3 Protein-immobilization Methods 107
2.4 Detection Techniques 109
2.5 Applications 110
2.5.1 Antibody Arrays 110
2.5.2 Antigen Arrays 111
2.6 Alternative Technologies 112

3 **Protein Microarrays for Enzyme Activity Profiling** 113
3.1 Principle 113
3.2 Immobilization of Enzymes or their Substrates 114
3.3 Detection of Enzymatic Activity on Microarrays 117
3.4 Applications 118
3.5 Alternative Technologies 119

4 **Protein Microarrays for Protein Interaction Profiling** 120
4.1 Principle 120
4.2 Applications 120
4.3 Alternative Technologies 124

5 **Concluding Remarks** 124

Encyclopedia of Molecular Cell Biology and Molecular Medicine, 2nd Edition. Volume 11
Edited by Robert A. Meyers.
Copyright © 2005 Wiley-VCH Verlag GmbH & Co. KGaA, Weinheim
ISBN: 3-527-30648-X

Bibliography 125
Books and Reviews 125
Primary Literature 126

Keywords

Capture Agents
An antibody or antibody mimetic that is immobilized on a surface for the purpose of binding an analyte of interest from a test sample.

Enzyme Activity Profiling
Determination of the activities of multiple enzymes in parallel, or of a single enzyme with respect to multiple substrates and/or potential inhibitors.

Multiplexed Protein Assay
An assay in which a single sample aliquot is simultaneously tested with respect to the abundance or activity of multiple proteins.

Protein Expression Profiling
Determination of the expression levels of multiple proteins in parallel and changes of expression levels.

Protein Interaction Profiling
Identification of the interactions between proteins and other molecules such as DNA, RNA, metabolites, or other proteins.

Protein Microarray
A two-dimensional surface with different antibodies or proteins attached at different defined locations, used for the study of the abundance or activity of proteins.

■ Protein microarrays allow for the parallel analysis of a large number of proteins with respect to their abundance or activity and, therefore, are playing an emerging role in proteome research.

Protein expression profiling, detection of enzymatic activity and protein interaction mapping are the most important applications of protein microarrays. The biological content, as well as the detection strategies, vary significantly between these applications, making the protein microarray technologies much more varied than those used with DNA microarrays.

In most cases, the biological content on the microarray, as well as the analytes, are proteins, which, due to their delicate nature and the heterogeneity of their properties, complicate the manufacturing and assay design. The potential advantages of protein microarrays over nonparallel, macroscopic technologies cannot, however, be overestimated.

1
Introduction

Microarrays are miniaturized surface-based assay devices that allow for the multiplexed analysis of biological samples. In the case of DNA microarrays, which were developed in the 1990s, sequences of DNA are immobilized on a glass surface in a two-dimensional (2D) array format. These DNA microarrays are used for a number of applications in genome research.

In gene expression profiling, for example, differences in the expression levels of genes can be measured. DNA probes, each of which is composed of a sequence that is complementary to an mRNA sequence of interest, are immobilized on a glass surface in an array format. In order to detect different expression levels, mRNA from different biological samples is then extracted, amplified into cDNA and labeled with different fluorophores representing the different biological samples. Differences in gene expression can then be detected by an increase or decrease in fluorescence.

Today, DNA microarrays are commercially available and have been proven to be a very important tool in genomic research for the multiplexed comparative analysis of gene expression and gene polymorphism. Many important insights into gene expression patterns associated with disease states have been derived from such technology.

It is the proteins, however, that are the functional product of almost all genes, and mRNA levels correlate only weakly with protein levels. The measurement of mRNA also does not provide information about the activity or posttranslational modification of proteins. To gain a realistic understanding of biological systems, it is necessary to understand the expression and activity of the full complement of proteins in different biological systems.

Inspired by the power of DNA arrays to measure mRNA expression, several academic and industrial groups have created protein microarrays, with the aim of multiplexing and miniaturizing protein analysis. Protein microarrays allow researchers to quantify the abundance or activity of many proteins in parallel while expending minimal amounts of precious sample.

The main types of protein arrays are (1) expression profiling arrays, consisting of immobilized protein-capture molecules such as antibodies, which provide information about protein abundance; (2) functional arrays, consisting of arrays of correctly folded proteins, peptides or other molecules, which can be used to characterize enzymatic activities; and (3) protein interaction arrays, which can be used to characterize protein–protein, protein–DNA, protein–carbohydrate and protein–small-molecule interactions. Protein array technology has been developed to such an extent that it now has the requisite robustness, breadth, reproducibility, and flexibility to play a leading role in proteomics research.

2
Protein Microarrays for Protein Expression Profiling

2.1
Principle

The rationale for using protein arrays for expression profiling has been explained in the previous section, and can be justified by several studies showing that more information can be culled from the analysis of multiple molecular markers than could be obtained from analyzing individual ones.

The principle behind the construction of arrays for measuring multiple proteins in parallel is similar to that used in the construction of DNA arrays for multiplexed mRNA analysis. To construct DNA arrays, each feature is derivatized with a probe that hybridizes to a single mRNA species. The production of these DNA-based "capture agents" is relatively straightforward: one simply needs to design and synthesize a DNA sequence that is complementary to the mRNA that is to be measured. For protein arrays, however, there is no comparable method for creating capture agents using bioinformatics tools alone. Protein-capture agents – usually antibodies – must be developed experimentally, usually by immunizing an animal with the protein target and then isolating the resulting antibodies for immobilization.

The other major difference between arrays for measuring mRNA versus those for measuring proteins relates to how the captured analytes are detected. Captured nucleic acids can be labeled with fluorescent probes and directly detected using a confocal fluorescence scanner. Proteins in a sample can also be directly labeled, but the process is much less reproducible than is the labeling of nucleic acids. Alternatively, proteins captured on an array can be indirectly detected by the use of labeled antibodies that recognize a different epitope on the captured target proteins. These and other issues regarding the development of the biological content for expression profiling applications are discussed in the following section.

2.2
Content for Protein Expression Profiling Chips

Most protein-profiling platforms rely on "sandwich pair" reagents – two different antibodies that can simultaneously bind to a target protein. For each target protein, one of these antibodies is immobilized onto the array, and the other one, which contains a label, is used to detect the target protein that has been captured on the array (Fig. 1). The principle behind the sandwich assay, and its advantages, are discussed in Sect. 2.4.

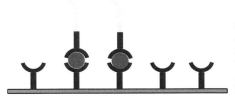

Fig. 1 Principle of a sandwich assay. A capture antibody is immobilized on a solid support. The captured analyte is detected using a second, fluorescently labeled antibody.

2.2.1 Source of Antigen

The first step in developing such sandwich pairs is usually to express and purify significant amounts of the target protein in a form as close as possible to its native state. This task can require as much, if not more, work than creating the capture agents themselves. Most proteins of interest need to be expressed in systems such as mammalian or insect cells, which require the investment of considerable time and cost. Most mammalian proteins are misfolded and insoluble when expressed in bacteria, and refolding protocols need to be individually tailored to each protein. The possibility of forgoing protein production altogether by using peptide antigens is rarely successful since most of the resulting antibodies have insufficient affinity against native proteins.

There are several sources of cDNAs for cloning into expression vectors, such as the NIH Mammalian Gene Collection, the Harvard Institute of Proteomics, the IMAGE Consortium, and the American Tissue Culture Collection (ATCC), and private companies such as Open Biosystems (Huntsville, AL) and Invitrogen (Carlsbad, CA). Modern cloning methods that avoid the use of restriction enzymes and contain various characteristics that favor correct insert orientation can make the cloning as high throughput and generic as possible. Multihost vectors and recombination-based transfer systems allow for the production of target proteins, including purification and detection tags for expression in multiple hosts. The first step is to transfer the gene into a "transfer" or "donor" vector and confirm its sequence; in the second step, the insert is shuttled into vectors for expressing the proteins in mammalian or insect cells. Both Invitrogen (Carlsbad, CA) and BD Biosciences Clontech (Palo Alto, CA) have commercialized such systems. The inclusion of N- or C-terminal purification tags in the encoded constructs allow for "generic" purification systems. The use of two affinity tags, such as the His tag plus the Glutathione-S-transferase (GST), or FLAG tag, can result in >95% pure protein using generic operating procedures.

For proteins that are large or contain important posttranslational modifications, it is often necessary to use expression systems based on higher eukaryotes. Recent improvements in technologies for introducing DNA constructs into cells, either virally or through transient transfection, have allowed for yields of up to 50 mg L^{-1} in mammalian cells or even higher in baculovirus-infected insect cells.

2.2.2 Types of Protein-binding Agents

Monoclonal antibodies (mAbs) are the most commonly used protein-binding agents for protein expression profiling platforms since they are monospecific and can be produced in unlimited quantities at low cost once a hybridoma cell line is developed. In some cases, proteolytic fragments of antibodies (Fabs) are used, since these can give improved performance due to the fact that an oriented presentation of the binding agent is possible. For sandwich-based approaches, the capture agent is almost always a mAb but the detection agents are occasionally polyclonal antibodies.

Monoclonal antibodies used in protein biochips should have dissociation constants (K_D) in the single-digit nanomolar range. Lower affinity antibodies generally have rapid ($>10^{-3}$ s^{-1}) dissociation rates (k_{off}), which result in the loss of the vast majority of captured antigen during the assay steps of washing and (in the case of sandwich assays) incubation with detection antibodies. Capture antibodies should preferably have $k_{off} < 10^{-4}\ s^{-1}$ and $K_D < 10^{-10}$ M. It usually requires repeated attempts at immunization to obtain such reagents.

Polyclonal antibodies are generally poorly suited as capture agents for protein biochips because only 0.1 to 10% of the antibodies in a preparation are specific to the immunogen used to create them. The only way to improve on this is to affinity-purify the antigen-specific antibodies away from the bulk immunoglobulin, but this requires large amounts of antigen, and such affinity-purified polyclonal antibodies cannot compare to mAbs in terms of homogeneity and lot-to-lot consistency. For these reasons, polyclonal antibodies instead find a more suitable role as the detection agents in multiplexed platforms using arrays of mAbs. In the detection step, the captured antigen will be selectively targeted by the highest affinity antibodies in the polyclonal preparation, and the fact that >95% of the antibodies in this preparation have limited or no affinity to the antigen is relatively unimportant.

The high expense and slow nature of mAb development has inspired the invention of alternative methods for making antibodies and antibody mimetics. Display technologies in which recombinant antibody fragments are linked to the genetic information encoding them are increasingly used. Such systems include phage display, ribosome display, mRNA display, yeast display, and bacterial display. The displayed libraries are usually in a Fab or single-chain variable region (scFv) configuration, the latter of which is less suitable due to its instability and conformational heterogeneity, which leads to lower affinities toward antigens. Artificial libraries of "antibody mimetics" have been created in which certain surface loops of nonantibody proteins are replaced by random sequences to make artificial complementarity-determining regions (CDRs). Although there have been several demonstrations of the creation of high-affinity protein-binding agents from naïve libraries using display methods, typical dissociation constants are in the 10^{-6} to 10^{-8} M range and therefore not suitable for multiplexed protein analysis platforms. Another drawback of these methods is that antibody fragments are significantly more expensive to manufacture than traditional mAbs.

Perhaps the most radical approach for creating the protein-binding agents is to use nucleic acid–based molecules. Libraries of random nucleic acid sequences can be subjected to systematic evolution of ligands by exponential enrichment (SELEX) to identify "aptamers" – rare single-stranded molecules that fold into a structure capable of binding to protein targets. Arrays of aptamers can be used to create protein expression profiling arrays. "Photoaptamers" are bromodeoxyuridine-substituted DNA molecules that can photo-crosslink to their target proteins. The photo-crosslinking event provides a second dimension of specificity since it requires close juxtaposition of reactive groups. The covalent crosslink allows for stringent washing of the arrays to remove non-specifically bound proteins. Nonspecific protein-reactive dyes can be used to detect the captured proteins on the arrays. Photoaptamers can crosslink low picomolar concentrations of target proteins with very high specificity, even in the presence of high concentrations of noncognate proteins from serum. A fluorescent image of a photoaptamer array is shown in Fig. 2.

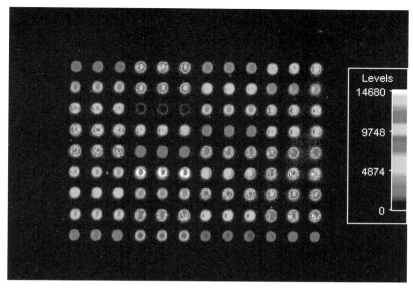

Fig. 2 Example of a photoaptamer array. Universal protein stain (UPS) photoaptamer array for the simultaneous assay of 26 analytes. The array was exposed to 10% human serum. After photo-crosslinking, the array was washed and stained with Alexa 647 dye and read on a TECAN LS300 scanner. (Figure courtesy SomaLogic Inc).

2.3
Protein-immobilization Methods

Proteins are very fragile molecules that are easily denatured at liquid–solid and liquid–air interfaces, and therefore the nature of the surfaces to which proteins are attached, as well as the dispensing method, can have pronounced effects on the surface protein density and the fraction of immobilized protein that is active. This section discusses dispensing methods and surface chemistries used for two-dimensional protein arrays.

Three approaches to protein immobilization are in common use.

Physical adsorption. The easiest way to immobilize proteins is by physical adsorption to surfaces with inherent nonspecific protein-binding activity. The surface-binding energy is hydrophobic in nature when using materials such as plastic, or electrostatic in nature when using charged surfaces such as polylysine-coated glass or nitrocellulose. A protein film is spontaneously formed on such surfaces when brought into contact with a protein solution, making this the simplest procedure for creating arrays. In such films, each protein is attached via multiple weak noncovalent interactions, but since the overall interaction surface on each protein is large, the sum of these weak interactions results in binding energies that are orders of magnitude stronger than typical specific protein–protein interactions. The main disadvantage of these approaches is that the interaction energies are so great that they frequently denature or otherwise inactivate the immobilized proteins. Generally, 75 to 90% of proteins attached to surfaces in this way are inactivated.

Random chemical attachment. The second most facile protein-immobilization method is based on random covalent attachment of proteins to substrates. Most methods take advantage of the fact that proteins typically contain several surface-exposed lysine residues. The amino group on the lysine side chains can react with surfaces containing, for example, *N*-hydroxysuccinimide moieties. The random chemical coupling results in heterogeneity with respect to the number of crosslinking sites per protein, the position of the crosslinks, and, as a result, the orientation of the proteins with respect to the surface. Many of the proteins are therefore inactivated by modification, by inaccessibility of the binding site, or by denaturation arising from strain caused by multiple attachment sites. A variation on this theme relies on random chemical biotinylation of lysine residues followed by exposure to streptavidin-coated surfaces.

Site-specific attachment. Optimal protein activity-retention is achieved by using single point oriented surface attachment of the binding agents. The most facile way of achieving this is to conjugate a binding agent through a unique cysteine residue that is either naturally present or has been introduced by genetic engineering. Surfaces can be derivatized by a number of thiol-reactive functionalities such as maleimide. Coupling can be made to occur almost exclusively to the thiol group on the unique cysteine, giving homogeneously oriented binding agents. Several studies have demonstrated the superior performance of protein arrays made using this method. Alternatively, the glycosylated functionalities on the Fc region of antibodies, which is distal from the antigen-binding site, can be used as an immobilization handle. Recombinant proteins can also be attached to surfaces in an oriented fashion by incorporating affinity tags such as the hexahistidine tag or the biotin-accepting peptide, which can be enzymatically biotinylated. These tags allow for the oriented immobilization on surfaces coated with Ni-complexes or streptavidin.

The choice of methods for immobilization depends highly on the sensitivity required in the assay and the number of binding agents that will be immobilized. For example, the amount of work involved in creating protein arrays with site-specifically attached proteins may not be justified when the number of proteins is high. Alternatively, arrays of 10 to 20 proteins in which a 2 to 5-fold increased sensitivity would be highly valued would justify the added effort required to create arrays with specifically oriented binding agents.

Flat surfaces have an inherent limitation on the number of protein molecules that can be immobilized per unit area. To extend beyond this limit, one must build a brush or gel structure onto the surface and attach the proteins to it. One theoretical drawback to this approach may be poor penetration into and out of the structure, leading to slow equilibration times and high nonspecific binding. This approach led to the commercialization of a "hydrogel" slide based on polyacrylamide, now sold by Perkin-Elmer Life Sciences (Foster City, CA). Slides such as these have been shown capable of immobilizing active proteins at a density at least fivefold higher than is possible using optimized flat surfaces. Another advantage is that the hydrogel can protect proteins from dehydration and therefore provide for a higher specific activity.

2.4 Detection Techniques

Nearly all reports of protein arrays rely on detection by fluorescence or chemiluminescence. Charge-coupled device-based cameras or laser scanners with confocal detection optics are commercially available. The proteins in a biological sample can be directly labeled with fluorescent dyes or with biotin, captured by the binding agents on the features of the array, and then detected using such fluorescent scanners (Fig. 3). To compare expression patterns between two different samples, each of the two samples is labeled with a different fluorescent dye. After labeling, the two samples are mixed together and then exposed to the array. A confocal fluorescence scanner then quantifies the signal from the two dyes (typically Cy3 and Cy5). However, such methods generally have very poor sensitivity for the following reasons: (1) fewer labels can be conjugated per protein without causing loss of solubility than for nucleic acids; (2) many proteins in samples are prone to nonspecific binding to surfaces, resulting in high background; and (3) different proteins are labeled with different efficiencies, usually relating to the number of accessible lysine residues. Owing to these reasons, chemical labeling of proteins has very limited applications for protein arrays.

Orders of magnitude higher sensitivity and reproducibility can be obtained with the use of "sandwich pair" antibodies (Fig. 1). In this case, two different antibodies, reacting to different regions (epitopes) of the target protein, are exploited. One antibody – the "capture agent" – is immobilized onto the array feature and captures the target protein; the other one – the "detection agent" – carries a label and binds to the captured target protein, forming a "sandwich" around it. The use of two antibodies greatly increases both the absolute amount of signal per captured target protein (since the detection antibody can be derivatized with very bright labels such as phycoerythrin), as well as the specificity of the signal (since two antibody-target protein-binding events must occur to produce signal). Such an assay format has the additional advantage of lending itself to high sample throughput since the biological sample need not be chemically labeled. The disadvantage of the use of sandwich reagent pairs is that two antibodies, both of which must recognize the target protein in its native form, must be developed and validated.

The sensitivity of sandwich assays can be pushed to new limits by amplifying the signal from the detection antibody using rolling circle amplification (RCA). To accomplish this, the detection antibody is labeled with a DNA primer annealed to a circular DNA template. Upon localization of the detection antibody on the array, the DNA primer on the antibody can be extended by a DNA polymerase on the circular template. Since the template has no end, the polymerase creates very

Fig. 3 Fluorescently labeled proteins are captured on an antibody array.

long DNA products in a "rolling circle amplification" reaction. Hybridization of fluorescently labeled oligonucleotides to this DNA product leads to a high density of fluorophores on the feature. Using this approach, it is possible to detect dozens of cytokines in parallel in the low picomolar range.

Another strategy for the fluorescent detection of proteins on array surfaces is to use planar waveguide technology in conjunction with fluorescently labeled detection antibodies. In this case, only the surface-bound fluorescently labeled molecules are detected, so no washing step is required. This allows one to take measurements under equilibrium conditions, thus avoiding the time-consuming array-washing steps.

In contrast with fluorescence and other optical readouts, mass spectrometry (MS) provides information as to the molecular weights of the captured proteins. The most widely used system relies on matrix-assisted laser desorption–ionization (MALDI) or surface-enhanced laser desorption–ionization (SELDI). The target proteins are captured by surface immobilized antibodies. One limitation of this system is the relatively low sensitivity in comparison to sandwich assay–based methods and the need to treat the surface-captured proteins with proteases prior to MALDI analysis. It is also difficult to obtain quantitative data on the basis of a purely MS-based approach.

2.5
Applications

2.5.1 **Antibody Arrays**

Antibody arrays have been used to monitor relatively small numbers of proteins in parallel, mainly due to the lack of developed antibody content for these platforms. Most arrays investigating >100 specificities have relied upon direct sample labeling since there are few proteins for which high quality sandwich pair antibodies are available. Generally, the two-dye labeling method mentioned above is used to compare two different samples. Several groups have analyzed changes in protein expression among cells as they become cancerous, or progress through different cytological stages, or are exposed to radiation.

Protein expression changes have been identified in squamous cell carcinoma, for example, with the aid of laser capture microdissection to separate out the different cell types prior to sample labeling. Samples were applied to an array of 368 antibodies against known signaling proteins. Eleven different proteins were identified that changed expression levels, some of which were in the cancer cells and others were in the surrounding stroma. A similar approach of combining laser capture microdissection with antibody arrays to find cancer biomarkers was used to study hepatocellular carcinoma. Antibodies to 83 proteins likely to be involved in proliferation made up the array, and differences in expression were observed for 32 of them. The data was confirmed by the use of western blots and tissue microarrays.

Most published reports of protein expression profiling arrays focus on a smaller number (10–75) of specificities, generally all within a restricted class of proteins, and use sandwich-based antibody pairs for detection. One high-density array that allowed for the measurement of expression levels of 75 cytokines was constructed. Similar products have been reported, but with a smaller number of specificities. Such cytokine-detection arrays have made it possible to uncover the complex changes in cytokine abundance that result from exposure to proinflammatory factors such as lipopolysaccharide or tumor necrosis factor-α. Similar arrays have also delineated differences in cytokine production between TH1 and TH2 helper cells, as well as uncovered an unexpected relationship between vitamin E levels and the cytokine monocyte chemoattractant protein-1.

In another study, the relationship between 78 inflammatory mediators and cerebral palsy was studied. Cord blood from 19 cerebral palsy children and 19 gestation matched controls were analyzed and significant differences were observed in the expression of 11 of the 78 proteins studied. In addition, it was found that cord blood from preterm infants with cerebral palsy showed even more elevated levels of certain inflammatory mediators than the gestation matched controls.

2.5.2 Antigen Arrays

Antigen arrays, sometimes referred to as *reverse arrays*, are used to measure the levels of antibodies with defined specificities in serum. Proteins of interest, allergens, or proteins known to elicit an autoimmune response are arrayed on a substrate. Serum samples of patients are applied, allowing the antibodies from the patient to bind to the arrayed proteins, if antibodies with such binding specificity are present. Then, a fluorescently labeled antihuman antibody–antibody is added to reveal which types of antibodies are present in the serum (Fig. 4).

Antigen arrays have been successfully applied to the fields of allergy, autoimmune disease, and inflammation. An array of 430 distinct proteins and overlapping peptides that represent the simian-human immunodeficiency virus (SHIV) proteome, for example, was used to analyze serum from vaccinated and nonvaccinated primates after challenge with SHIV. It was observed that the vaccinated versus nonvaccinated animals could be distinguished and survival predicted. A

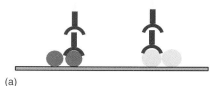

Fig. 4 Scheme and example of an antigen array. (a) Antigens are arrayed on a two-dimensional surface. Autoantibodies in serum are captured and detected with a fluorescently labeled antihuman antibody-antibody. (b) "Arthritis Array" characterization of autoantibodies in systemic lupus erythematosus. Arthritis arrays were produced by printing common lupus antigens including DNA, histone proteins, and additional nuclear proteins on microscope slides. Arrays were incubated with patient serum, and binding of autoimmune antibodies was detected with green fluorescent markers (green spots). The yellow spots are used as "marker features" to orient the arrays. (Figure courtesy Dr. William Robinson, Stanford University) (see color plate p. xxix).

second study used an array of proteins and peptide fragments representing all the known proteins in myelin to monitor autoantibody development in experimental autoimmune encephalomyelitis, a model for multiple sclerosis.

Antigen arrays were used to profile antibodies from patients diagnosed with various rheumatoid diseases, including rheumatoid arthritis, systemic lupis, systemic sclerosis, and Sjogrens syndrome. A characteristic autoantibody pattern for each of the diseases was observed, giving hope that such an array could be used as a diagnostic and monitoring tool.

It is well known that cancer patients often develop antibodies against proteins expressed in their tumors. In an effort to use such immunological information in the diagnosis of cancer, Qiu and colleagues fractionated protein lysates from a human lung cancer cell line and then arrayed them onto nitrocellulose-coated slides. The serum from individuals with or without lung cancer were then applied to the arrays to visualize the autoreactivity to cancer antigens. The signals from duplicate spots (within-slide) and duplicate slides were highly reproducible, exhibiting correlation values >0.9. Of the 1840 arrayed fractions, 63 demonstrated increased reactivity in cancer patients relative to normal, as measured by a rank-based statistic ($p < 0.008$). This type of approach, or one based on arraying purified cancer antigens, may be useful in the future in the identification and classification of patients.

2.6
Alternative Technologies

To increase the sample throughput and still maintain at least a modest degree of multiplexing, several groups have constructed antibody arrays within the wells of microtiter plates, generally containing 96 wells/plate. In such cases, the readout is typically based on a sandwich assay in which the detection antibodies are labeled with an enzyme that catalyzes a reaction to form locally precipitating fluorescent or chemiluminescent products. A commercial system for cytokine analysis has been commercialized by Pierce (Rockford, IL).

A more radical departure from the array configurations considered in the previous chapters is the so-called "liquid array" – suspended, encoded particles, each coated with a single antibody specificity. The most common way to encode beads relies on the incorporation of different ratios of two or more fluorophores with distinguishable spectral properties. A flow cytometer is exploited to read particles one at a time, thus decoding the identity of each particle (and thereby the identity of the immobilized antibody), while simultaneously reading a third fluorescent signal that represents binding of the target protein. This third signal is generally produced by the binding of a detection antibody labeled with the strongly fluorescent protein phycoerythrin, in a sandwich setup. The most popular suspension array platform is marketed by Luminex (Austin, TX). Several companies sell kits for multiplexed detection of cytokines and other proteins. Instruments and reagents are also available commercially to allow researchers to construct their own liquid arrays. Some of the key similarities of the liquid versus the more standard two-dimensional array platforms are shown in Table 1. The key advantages of the liquid array system are as follows: (1) manufacturing is easier since antibodies can be incubated in batch mode with the beads, thus overcoming difficulties in reproducible spotting due to denaturation, evaporation, and so on; (2) commercial systems are already

Tab. 1 Comparison of multiplexed protein analysis platforms.

	Two-dimensional arrays	*Suspended particle arrays*
Assay type	Direct capture, sandwich assay, functional assays	Direct capture, sandwich assay
Array construction method	Robotic spotting	Batch incubation of encoded particles with antibodies followed by particle mixing
Detection	Fluorescence, radioactivity, SPR, mass spectrometry	Fluorescence
Multiplexing	High	Medium
Sample throughput	Low	High
Storage	Dry	In buffer

available for fully automating the assay procedure; (3) ability to "mix and match" beads with a variety of specificities, as needed; and (4) compatibility with microtiter plates, since samples are loaded from 96-well plates and instruments can analyze an entire plate of samples in less than one day. Because of these advantages, the liquid array systems are becoming more popular and useful in life sciences than the two-dimensional arrays and will continue to be so for multiplexing applications in which the number of specificities is <100 and when sample throughput is important.

There are numerous publications using liquid arrays for multiplexed protein expression profiling, mostly involving the measurement of a limited number of cytokines or other secreted signaling molecules. One example is the measurement of cytokines in the serum of children infected with rotavirus. Three cytokines showed significantly higher expression levels in infected children. In another study, Soldan and colleagues profiled seven cytokines from stimulated peripheral blood mononuclear cells from females with relapsing-remitting multiple sclerosis (RRMS), and secondary progressive multiple sclerosis (SPMS). Two cytokines were significantly different between the RRMS and SPMS patients, which may indicate therapeutic directions for treating these two ailments.

3
Protein Microarrays for Enzyme Activity Profiling

3.1
Principle

Although it is of great value to determine the abundance of proteins in complex biological systems, it is also important to measure the enzymatic activity of proteins. Kinases, for example, play an essential role in almost all signal transduction processes in cells. The multiplexing capability of microarrays is especially suited for these screening efforts.

Two microarray formats are most commonly used for enzyme activity screening. In one case, the substrates of enzymes such as kinases or proteases are immobilized (Fig. 5a). The reaction can be monitored by a variety of different detection strategies that depend on the modification. If used for the determination of enzyme abundance, this format can also lead to

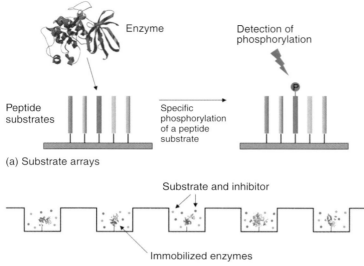

Fig. 5 The two most common formats for enzyme activity screening on microarrays. (a) Substrates, in this case of kinases, are arrayed on a surface, and their modification by enzymes detected. (b) Enzymes are immobilized, typically in nanowells, and their activity is measured by incubation with substrates in the absence or presence of inhibitors.

improved sensitivity, as the enzyme abundance is measured by its activity, which only requires the presence of a specific substrate and not through its binding to an immobilized capture agent, in which case the sensitivity depends on the affinity.

In the other case, an array of enzymes is prepared and their activity detected at their location, usually in a nanowell format (Fig. 5b). This format can also be used for inhibitor screening.

3.2 Immobilization of Enzymes or their Substrates

As in the case of protein-profiling microarrays, the immobilization strategy is also a key feature for the sensitivity and reproducibility of enzyme assays in microarray formats. The advantage in the manufacturing of peptide or small-molecule arrays, however, is the lack of complex and environmentally dependent three-dimensional structures of these molecules as compared with proteins. To ensure reproducibility and allow for a quantitative comparison of results across microarrays, however, it is still crucial to immobilize peptide substrates in an oriented and uniform fashion.

Peptide substrates for enzyme activity screening can be readily produced in a high-throughput fashion by solid-phase synthesis, which also allows for the incorporation of a variety of functional groups for covalent surface attachment. Several elegant approaches for covalent immobilization have been developed using different reaction schemes (Table 2).

Tab. 2 Selected reactions for the covalent immobilization of peptides on surfaces.

Functional group in substrate	Functional group on surface	Reaction product for immobilization of substrate on surface
Alkoxyamino group	Aldehyde group	Oxime group
1,2-Aminothiol group	Aldehyde group	Thiazolidine group
Cysteine group	Thioester group	Amide group

(continued overleaf)

Tab. 2 (Continued)

Functional group in substrate	Functional group on surface	Reaction product for immobilization of substrate on surface
Biotin group	Avidin – surface	High-affinity interaction between avidin on the surface and biotin in the substrate
Cyclopentadiene group	Benzoquinone group	Bicyclic group
NHS-activated ester group	Amino group (H₂N—Surface)	Amide group

An advantage of these approaches is the oriented immobilization, thereby decreasing the fraction of peptides that are immobilized in an orientation that inhibits their participation in the enzymatic reaction.

Another approach towards manufacturing peptide arrays is the direct synthesis of the peptides on surfaces. In the so-called SPOT synthesis, reaction reagents, together with the amino acid monomers, are spotted onto a supporting membrane, such as cellulose, allowing for the construction of arrays with about 100 sequences/cm^2. An additional on-chip peptide synthesis method has been described by Gao and coworkers, which uses t-Boc-based chemistry to generate acids on the surface by photolithography for spatial deprotection of the amino acids during the synthesis process.

Nitrocellulose has been used to create arrays, not only of proteins, but also of peptides. This surface, like many others, has a very high nonspecific protein-binding tendency, which can however, interfere with assays. Bovine serum albumin (BSA) is a very common blocking agent in protein microarray assays to minimize nonspecific protein binding to microarrays surfaces. In peptide microarrays, however, it can mask the much smaller peptides and prevent enzymes from gaining access to them. The use of oligo(ethylene glycol) groups on the surface has proven to be very useful in the prevention of nonspecific adsorption. These groups can also act as a linker between the surface and the immobilized peptides. Another approach exploits the nonspecific adsorption of BSA onto surfaces such as glass; the film of BSA can then be modified by attaching N-hydroxysuccinimide groups that can react with amino groups on the peptide substrates.

In cases where the enzyme, rather than the substrate, is arrayed, it is often necessary to create a diffusion boundary between the different features of the array; as a result, any modified substrate remains associated with the feature containing the arrayed enzyme. Fabricated chips with microwells containing covalently attached kinases or proteases have been used to characterize the specificity of enzymes and to identify inhibitors. In this case, the assay strategy is very similar to plate-based enzyme assays, in which the enzyme is incubated with its substrate in the presence of different inhibitors and the extent of substrate modification is an indication of the inhibitory potential of the inhibitors.

3.3
Detection of Enzymatic Activity on Microarrays

The strategy for detecting enzymatic activity on a microarray depends on the nature of the catalyzed reaction. To profile proteolytic or otherwise hydrolytic reactions, one can take advantage of the reaction releasing an immobilized fluorescent dye from the substrate. Coumarin-based derivatives are the most commonly exploited dyes for such studies and have been used to monitor proteolysis of peptide substrates (Fig. 6).

Kinase activity results in a phosphorylation of a specific target sequence in a peptide or protein substrate on the microarray. Such a reaction requires the presence of a cosubstrate such as ATP, from which the phosphate group is transferred to the substrate on the surface. A very sensitive approach for the detection of phosphorylation events is to use radioactive [γ^{32}P]-ATP and quantify the amount of radioactive phosphate transferred to different arrayed

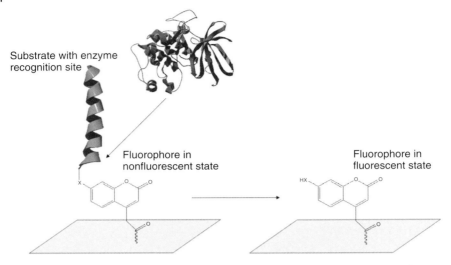

Fig. 6 Detection of protease activity on microarray surfaces with fluorescently labeled peptide substrates. The peptide substrates are immobilized on the array surface with fluorophores adjacent to the enzyme recognition site. Cleavage releases the peptide substrate and the immobilized fluorophores become fluorescent.

substrates using phosphorimaging analysis or autoradiography. Because these detection methods suffer from low spatial resolution, an alternative is to apply a photographic emulsion directly to the arrays (after the enzymatic phosphotransfer reactions were complete), and visualize the results by light microscopy.

Owing to the inconvenience and potential health hazard of working with radioactivity, nonradioactive alternatives to the detection of phosphotransfer reactions on arrays can also be applied. Phosphorylated peptide substrates can be detected using fluorescently labeled antiphosphopeptide antibodies. The universal application of this technique compared with radioactive detection, however, is problematic, as the affinity of antibodies varies depending on the phosphoamino acid sequence context. Another possible detection strategy for kinase microarray assays exploits the use of phosphospecific fluorescent dyes; such dyes are used in gel electrophoresis for the detection of phosphorylation.

Figure 7 summarizes the different detection technologies for kinase activity arrays.

3.4
Applications

Microarrays of substrates or enzymes are becoming increasingly important for the characterization, abundance determination and inhibitor screening of enzymes. Perhaps because it is far easier to array the substrates than the enzymes themselves, most published applications to date describe peptide microarrays used to detect protease or kinase activity. Mrksich and coworkers, for example, developed a peptide chip used to characterize inhibitors of the nonreceptor tyrosine kinase c-Src. The chip was prepared by immobilizing the kinase substrate onto a self-assembled

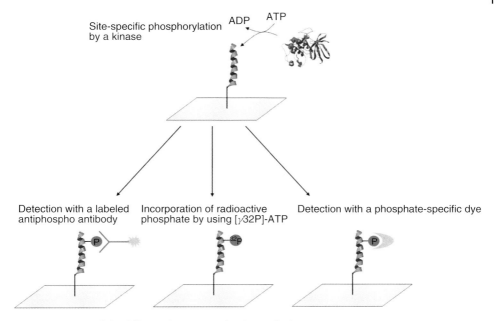

Fig. 7 Overview of the different detection technologies for kinase activity arrays.

monolayer of alkanethiolates on gold. Phosphorylation of the immobilized peptides was characterized by surface plasmon resonance, fluorescence, and phosphorimaging. Three inhibitors of the enzyme were quantitatively evaluated in an array format on a single, homogeneous substrate.

Most applications that have been described so far, however, are proof-of-concept studies evaluating either the immobilization or the detection strategies that have been described above. Examples of these applications to characterize kinases or proteases with substrate microarrays can be found in the list of the primary articles listed at the end of this chapter.

Novel microwell-based protein chip technology for high-throughput analysis of biochemical activities has also been described for the analysis of nearly all protein kinases from the yeast *Saccharomyces cerevisiae*. Of the 122 known and predicted yeast protein kinases, 119 were overexpressed and analyzed using 17 different substrates. Studies of this nature have shown that many novel activities can be identified and that a large number of protein kinases, not known to be capable of phosphorylating tyrosine peptide sequences, do possess this biochemical activity.

3.5
Alternative Technologies

The advent of microarrays for enzyme activity profiling has made tremendous progress in the last few years, as reflected by the number of recent publications, especially concerning peptide microarrays for kinase activity profiling and protease fingerprinting. The use of this technology in industrial research, however, is still limited owing to the fact that most laboratories today perform assays in 384- or 1536-well plates, and existing automation technology facilitates high-throughput screening. Such plate-based assays do not require as much technology and assay development as surface-based setups.

One interesting alternative to microarrays for multiplexing the analysis of enzymatic activity was reported by Kozarich, who applied the principles of enzyme chemistry, mechanism of action and inhibitor design to create a set of activity-based probes. The use of suicide inhibitors allows for interrogation of enzyme family members, both known and unknown, in cells and protein extract fractions without the need for individual assay development and isolation. The serine hydrolases and cysteine proteases have provided the proofs of concept for this type of activity-based proteomics platform, and other studies are rapidly following.

4
Protein Microarrays for Protein Interaction Profiling

4.1
Principle

The interactions of proteins with each other and with other molecules constitute an essential component of all biological processes. In addition to the profiling of enzymatic activity, functional proteomics is therefore also focused on the elucidation of protein interaction networks, also called protein interaction maps. Different research groups, using different methods, however, often report conflicting results, emphasizing the importance of using a variety of methods in order to verify results.

Because of the complexity of these networks and the large number of interactions, high-throughput methods are essential for generating comprehensive data sets on any of the critical pathways controlling biological events. Microarrays are ideally suited for the parallel analysis of thousands of interactions of proteins with each other and with other key metabolic and regulatory molecules such as DNA, RNA, lipids, carbohydrates, metabolites, and allosteric effectors. Most of the reports on the use of protein microarrays for interaction profiling have, to date, focused on the interaction of proteins with non-protein molecules. The main reasons for this are (1) high-throughput methods already exist for studying protein–protein interactions, most notably the yeast two-hybrid system and (2) the difficulty in creating and purifying large collections of active proteins (as discussed in Sect. 2.2.1) in comparison to the relative ease of producing small molecules, peptides, RNA, or DNA.

The on-chip detection of substances, whether proteins or other molecules, that bind to the proteins on the arrays poses an additional challenge. If the analyte contains a complex mixture of molecules, it is not trivial to identify which ones are binding to the arrayed proteins. In the simplest case, for example, the analyte contains only one substance that is labeled by a fluorescent group–in which case the readout is straightforward (Fig. 8).

The used label is usually very similar to the ones that were described in Chapter 2 for protein expression profiling microarrays.

4.2
Applications

To date, at least one whole proteome microarray has been developed, consisting of

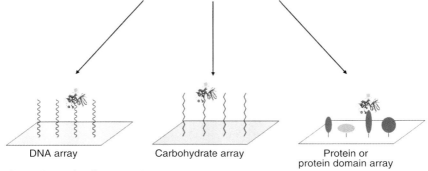

Fig. 8 Principle of a protein interaction microarray.

almost the entire yeast proteome on a chip. High-throughput protein expression and purification methods are critical for the success of such endeavors. To establish such a proteome microarray, open reading frames encoding the proteins must be cloned into an expression vector that allows for the production of fusion proteins with affinity tags that can be used in high-throughput purification (e.g. GST tag for purification on glutathione columns) and immobilization (e.g. His tag for immobilization on Ni^{2+}-nitrilotriacetic acid (Ni-NTA), coated glass slides). These proteome microarrays can then be probed, for example, with biotin-labeled proteins or other molecules and the interactions detected by adding fluorescently labeled streptavidin.

More commonly, a subset of proteins from an organism, defined by a protein or disease class are immobilized and then probed for interactions.

An elegant approach for alleviating the difficulties with fabrication of protein microarrays, as opposed to the relatively straightforward methods for arraying DNA, is the following: DNA sequences encoding proteins of interest, genetically fused to an immobilization tag, are incubated in a cell-free expression lysate on a tag-binding surface. Proteins are synthesized *in situ* and directly immobilized by the interaction of the tag with the surface (Fig. 9).

Despite these efforts, the high-throughput production of correctly folded proteins still poses an obstacle in the fabrication of protein microarrays (Sect. 2.2.1). One way to alleviate this is to synthesize and immobilize smaller, functional domains of proteins that are responsible for a specific activity or interaction. The so-called SPOT synthesis, for example (see Sect. 3.2), has been used for the preparation of microarrays of protein domains that can then be probed for interaction with target molecules.

Protein–DNA interactions are crucial for the regulation of gene transcription and DNA replication. The fact that the methods for the construction of DNA microarrays are well established makes

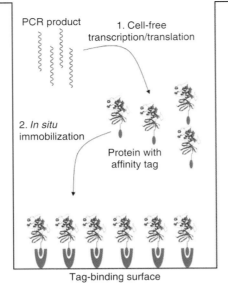

Fig. 9 Example of *in situ* immobilization of proteins on surfaces by the DiscernArray™ technology described by He and colleagues.

them an ideal tool for the screening of DNA–protein interactions. This approach can, for example, be used to screen the interaction of transcription factors or DNA-modifying enzymes.

Carbohydrates constitute another important class of molecules that interact with proteins. Advancements in the synthesis of carbohydrates, as well as carbohydrate-compatible immobilization strategies, led to the first manufacturing of arrays in 2002 and established carbohydrate microarrays as an important tool in "glycomics." Carbohydrates can be immobilized on microarray surfaces and remain accessible to interacting with proteins. The application of carbohydrate arrays to answer important biological questions has been demonstrated by Seeberger and coworkers, who used them to characterize the binding specificity of different gp120-binding proteins, important for HIV entry into host cells.

Another potentially important application in drug discovery will be the use of small-molecule microarrays for the screening of protein inhibitors. Small molecules can be routinely synthesized in large numbers by parallel synthesis technologies and modified with functional groups for specific immobilization onto surfaces. Small-molecule microarrays are therefore of growing interest to the research community. Most of the groundbreaking work in this field has been performed by Schreiber and coworkers, who developed a variety of immobilization schemes for small molecules and demonstrated the use of small-molecule microarrays in elucidating pathways in glucose signaling and in discovering transcription factor inhibitors.

Additional recent innovations with respect to the immobilization of small molecules have been reported. The so-called Staudinger ligation has been successfully used selectively to immobilize azide-functionalized molecules on phosphane-derivatized glass slides by forming an amide bond. Another strategy, which allows for the immobilization of non-functionalized molecules, including natural products, can also be applied for the production of small-molecule microarrays by using photoaffinity labeling. Although the small molecules are not immobilized in an oriented fashion, this technology is widely applicable and the random orientation enables one to display the molecules such that most, if not all, of the functional groups will be available on some proportion of the immobilized compounds.

A different but elegant approach has been to use a DNA array to derive information about protein–small-molecule interactions. To accomplish this, each of the small molecules is coupled to a unique PNA (peptide nucleic acid hybrid molecule) sequence and a fluorescent dye. A library of fluorescent small molecule-PNA hybrids are then mixed together and incubated with a target enzyme to allow binding to occur. Next, size-exclusion chromatography is used to separate the bound from the unbound library members. The bound members are then hybridized to an array of DNA sequences that hybridize to the different PNA tags, and a fluorescence scan of the array reveals which molecules have affinity for the target protein (Fig. 10).

The possibility of screening large libraries of small molecules with a variety of target proteins, under conditions of minimal sample consumption, could potentially have an important impact on the efficiency of the early steps in the drug discovery process.

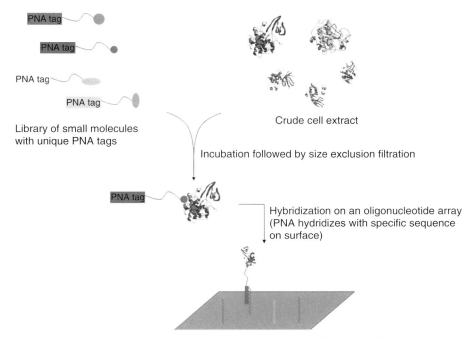

Fig. 10 Screening of protein interactions with PNA-encoded small molecule libraries as described by Winssinger and colleagues.

4.3
Alternative Technologies

The most important technology to date for the elucidation of protein–protein interactions are the two-hybrid systems. These are genetic-based *in vivo* screening systems that have been used extensively for the discovery of new interactions between proteins. The most widely used such system is in yeast and is based on the functional reconstitution of the transcriptional activator Gal4p. The interaction of two proteins is detected by fusing one protein to the DNA-binding domain of Gal4p and the other to the transcriptional activation domain. The interaction of the two proteins reconstitutes the activity of Gal4p, allowing for the transcription of reporter genes. Using the yeast two-hybrid system in large-scale screens, a detailed analysis of the whole "interactome" of yeast *S. cerevisiae* has already been performed. A disadvantage of the yeast-based system, however, is the fact that all interactions take place in the yeast nucleus, which is not an optimal environment for the proper folding of certain proteins, especially with respect to post-translational modifications. To solve this problem, other two-hybrid systems have been developed that allow for the interaction of two proteins in the cytoplasm of yeast or in mammalian cells.

Another *in vivo* screening system is based on reconstituting an enzyme from two polypeptide fragments by genetically fusing one protein of interest to one fragment and another protein to the other fragment. If the two proteins of interest interact, they bring together and reconstitute the genetically fragmented enzyme, making it active. The activity of the enzyme can then be followed by using a substrate that is converted into a fluorescent product, for example.

Many protein–protein interactions are not binary in nature but rather require the formation of multiprotein complexes. Different approaches have been developed to tackle this problem in which a large number of proteins are cloned into expression vectors and expressed with fusion tags, such that they can form protein complexes under physiological conditions. The tag can then be used to extract the protein complexes from cell lysates and then identify the proteins in these complexes by gel electrophoresis and mass spectrometry.

Each of these techniques has its advantages and disadvantages. The two-hybrid system, for example, can only elucidate binary interactions but is very efficient at detecting transient or weak interactions, in contrast to methods requiring the extraction of protein complexes. Only a combination of techniques can result in a more complete picture of protein interaction maps for complex biological systems. Protein microarrays are especially useful for the determination of protein interactions with other molecules such as DNA, RNA or carbohydrates. Even for the detection of protein–protein interactions, protein arrays can be *in vitro* alternatives to the cell-based two-hybrid system and may be useful where the latter is deficient, such as for identifying interactions involving secreted proteins.

5
Concluding Remarks

The emergence of proteomics has generated the need for new technologies that introduce parallel measurements and miniaturization into protein analysis. Protein expression profiling is particularly well suited to miniaturization, especially for the analysis of soluble proteins such as growth factors and cytokines. Dozens of platforms that rely on different types of configurations and detection methods have been established. Both planar and suspension-based antibody arrays, using fluorescent sandwich pair antibodies for detection, have demonstrated that dozens of proteins can be analyzed quantitatively at picomolar concentrations in parallel. Further broadening the applications of this powerful technology will require the development of hundreds to thousands of new high quality antibodies. Multiplexed analysis of enzyme activity and protein interaction profiling is also technically well established and already has enabled important advances. For these technologies also, the rate-limiting step in developing further applications is the development of biological content, that is, collections of functional recombinant proteins. As such content becomes available, there is little doubt that protein microarrays will become a mainstream tool for biological research.

See also Microarray-Based Technology: Basic Principles, Advantages and Limitations; Peptide and Non-Peptide Combinatorial Libraries; Total Analysis Systems, Micro.

Bibliography

Books and Reviews

Amaratunga, D., Cabrera, J. (2003) *Exploration and Analysis of DNA Microarray and Protein Array Data*, Wiley, New York.

Cahill, D.J., Nordhoff, E. (2002) Protein arrays and their role in proteomics, *Adv. Biochem. Eng. Biotechnol.* **83**, 177–187.

Drickamer, K., Taylor, M.E. (2002) Glycan arrays for functional glycomics, *Genome Biol.* **3**, pp. 1034.1–1034.4.

Feizi, T., Fazio, F., Chai, W., Wong, C.H. (2003) Carbohydrate microarrays – a new set of technologies at the frontiers of glycomics, *Curr. Opin. Struct. Biol.* **13**, 637–645.

Fodor, S.P., Rava, R.P., Huang, X.C., Pease, A.C., Holmes, C.P., Adams, C.L. (1993) Multiplexed biochemical assays with biological chips, *Nature* **364**, 555–556.

Fung, E. (2004) *Protein Arrays: Methods and Protocols (Methods in Molecular Biology)*, Humana Press, Totowa, NJ.

Goddard, J.-P., Reymond, J.-L. (2004) Enzyme assays for high-throughput screening, *Curr. Opin. Biotechnol.* **15**, 314–322.

Haab, B.B. (2003) Methods and applications of antibody microarrays in cancer research, *Proteomics* **3**, 2116–2122.

Houseman, B.T., Mrksich, M. (2002) Towards quantitative assays with peptide chips: a surface engineering approach, *Trends Biotechnol.* **20**, 279–281.

Huang, R.P. (2003) Cytokine antibody arrays: a promising tool to identify molecular targets for drug discovery, *Comb. Chem. High Throughput Screen.* **6**, 769–775.

Jain, K.K. (2004) Applications of biochips: from diagnostics to personalized medicine, *Curr. Opin. Drug Discov. Dev.* **7**, 285–289.

Jenkins, R.E., Pennington, S.R. (2001) Arrays for protein expression profiling: towards a viable alternative to two-dimensional gel electrophoresis? *Electrophoresis* **1**, 13–29.

Jessani, N., Cravatt, B.F. (2004) The development and application of methods for activity-based protein profiling, *Curr. Opin. Chem. Biol.* **8**, 54–59.

Jona, G., Snyder, M. (2003) Recent developments in analytical and functional protein microarrays, *Curr. Opin. Mol. Ther.* **5**, 271–277.

Kambhampati, D. (2004) *Protein Microarray Technology*, Wiley, New York.

Koch, J., Mahler, M. (2002) *Peptide Arrays on Membrane Supports: Synthesis and Applications*, Springer-Verlag, Heidelberg, Germany.

Kozarich, J.W. (2003) Activity-based proteomics: enzyme chemistry redux., *Curr. Opin. Chem. Biol.* **7**, 78–83.

Kumble, K.D. (2003) Protein microarrays: new tools for pharmaceutical development, *Anal. Bioanal. Chem.* **377**, 812–819.

Mantripragada, K.K., Buckley, P.G., de Stahl, T.D., Dumanski, J.P. (2004) Genomic microarrays in the spotlight, *Trends Genet.* **20**, 87–94.

Michnik, S.W. (2003) Protein fragment complementation strategies for biochemical network mapping, *Curr. Opin. Biotechnol.* **14**, 610–617.

Miller, J., Stagljar, I. (2004) Using the yeast two-hybrid system to identify interacting proteins, *Methods Mol. Biol.* **261**, 247–262.

Mirzabekov, A., Kolchinsky, A. (2002) Emerging array-based technologies in proteomics, *Curr. Opin. Chem.* **6**, 70–75.

Ng, J.H., Ilag, L.L. (2003) Biochips beyond DNA: technologies and applications, *Biotechnol. Annu. Rev.* **9**, 1–149.

Nielsen, U.B., Geierstanger, B.H. (2004) Multiplexed sandwich assays in microarray format, *J. Immunol. Methods* **90**, 107–120.

Panicker, R.C., Huang, X., Yao, S.Q. (2004) Recent advances in peptide-based microarray technologies, *Comb. Chem. High Throughput Screen.* **7**, 547–556.

Petach, H., Gold, L. (2002) Dimensionality is the issue: use of photoaptamers in protein microarrays, *Curr. Opin. Biotechnol.* **13**, 309–314.

Phelan, M.L., Nock, S. (2003) Generation of bioreagents for protein chips, *Proteomics* **3**, 2123–2134.

Predki, P.F. (2004) Functional protein microarrays: ripe for discovery, *Curr. Opin. Chem. Biol.* **8**, 8–13.

Robinson, W.H., Steinman, L., Utz, P.J. (2003) Protein arrays for autoantibody profiling and fine-specificity mapping, *Proteomics* **3**, 2077–2084.

Schena, M. (2002) *Microarray Analysis*, Wiley, New York.

Schena, M. (2004) *Protein Microarrays*, Jones & Bartlett Publishers, Boston, MA.

Skerra, A. (2000) Engineered protein scaffolds for molecular recognition, *J. Mol. Recognit.* **13**, 167–187.

Sydor, J.R., Nock, S. (2003) Protein expression profiling arrays: tools for the multiplexed high-throughput analysis of proteins, *Proteome Sci.* **1**, 3.

Templin, M.F., Stoll, D., Bachmann, J., Joss, T.O. (2004) Protein microarrays and multiplexed sandwich immunoassays: what beads the bead? *Comb. Chem. High Throughput Screen.* **7**, 223–229.

Templin, M.F., Stoll, D., Schwenk, J.M., Potz, O., Kramer, S., Joos, T.O. (2003) Protein microarrays: promising tools for proteomic research, *Proteomics* **3**, 2155–2166.

Tucker, C.L., Gera, J.F., Uetz, P. (2001) Towards an understanding of complex protein networks, *Trends Cell Biol.* **11**, 102–106.

Uttamchandani, M., Chen, G.Y.J., Lesaicherre, M.-L., Yao, S.Q. (2004) Site-specific peptide immobilization strategies for the rapid detection of kinase activity on microarrays, *Methods Mol. Biol.* **264**, 191–204.

Walhout, A.J.M., Vidal, M. (2001) High-throughput yeast two-hybrid assays for large-scale protein interaction mapping, *Methods* **24**, 297–306.

Walsh, D.P., Chang, Y.T. (2004) Recent advances in small molecule microarrays: applications and technology, *Comb. Chem. High Throughput Screen.* **7**, 557–564.

Wiesner, A. (2004) Detection of tumor markers with ProteinChip technology, *Curr. Pharm. Biotechnol.* **5**, 45–67.

Wilson, D.S., Nock, S. (2003) Recent developments in protein microarray technology, *Angew. Chem., Int. Ed.* **42**, 494–500.

Winter, G., Griffiths, A.D., Hawkins, R.E., Hoogenboom, H.R. (1994) Making antibodies by phage display technology, *Annu. Rev. Immunol.* **12**, 433–455.

Yeo, D.S., Panicker, R.C., Tan, L.P., Yao, S.Q. (2004) Strategies for immobilization of biomolecules in a microarray, *Comb. Chem. High Throughput Screen.* **7**, 213–221.

Zhou, F.X., Bonin, J., Predki, P.F. (2004) Development of functional protein microarrays for drug discovery: progress and challenges, *Comb. Chem. High Throughput Screen.* **7**, 539–546.

Zhu, H., Snyder, M. (2003) Protein chip technology, *Curr. Opin. Chem. Biol.* **7**, 55–63.

Primary Literature

Adams, E.W., Ratner, D.M., Bokesch, H.R., McMahon, J.B., O'Keefe, B.R., Seeberger, P.H. (2004) Oligosaccharide and glycoprotein microarrays as tools in HIV glycobiology: glycan-dependent gp120/protein interactions, *Chem. Biol.* **11**, 875–881.

Alizadeh, A.A., Eisen, M.B., Davis, R.E., Ma, C., Lossos, I., Rosenwald, S.A., Boldrick, J.C., Sabet, H., Tran, T., Yu, X., Powell, J.I.,

Yang, L., Marti, G.E., Moore, T. Jr., Hudson, J., Lu, L., Lewis, D.B., Tibshirani, R., Sherlock, G., Chan, W.C., Greiner, T.C., Weisenburger, D.D., Armitage, J.O., Warnke, R., Staudt, L.M. (2000) Distinct types of diffuse large B-cell lymphoma identified by gene expression profiling, *Nature* **403**, 503–511.

Anderson, L., Seilhamer, J. (1997) A comparison of selected mRNA and protein abundances in human liver, *Electrophoresis* **18**, 533–537.

Barnes-Seeman, D., Park, S.B., Koehler, A.N., Schreiber, S.L. (2003) Expanding the functional group compatibility of small-molecule microarrays: discovery of novel calmodulin ligands, *Angew. Chem., Int. Ed.* **42**, 2376–2379.

Bernard, A., Fitzli, D., Sonderegger, P., Delamarche, E., Michel, B., Bosshard, H.R., Biebuyck, H. (2001) Affinity capture of proteins from solution and their dissociation by contact printing, *Nat. Biotechnol.* **19**, 866–869.

Bertozzi, C.R., Kiessling, L.L. (2001) Chemical glycobiology, *Science* **291**, 2357–2364.

Bhaduri, A., Sowdhamini, R. (2003) A genome-wide survey of human tyrosine phosphatases, *Protein Eng.* **16**, 881–888.

Boisguerin, P., Leben, R., Ay, B., Radziwill, G., Moelling, K., Dong, L., Volkmer-Engert, R. (2004) An improved method for the synthesis of cellulose membrane-bound peptides with free C termini is useful for PDZ domain binding studies, *Chem. Biol.* **11**, 449–459.

Borrebaeck, C.A., Ekstrom, S., Hager, A.C., Nilsson, J., Laurell, T., Marko-Varga, G. (2001) Protein chips based on recombinant antibody fragments: a highly sensitive approach as detected by mass spectrometry, *BioTechniques* **30**, 1126–1130, 1132.

Boutell, J.M., Hart, D.J., Godber, B.L.J., Kozlowski, R.Z., Blackburn, J.M. (2004) Functional protein microarrays for parallel characterization of p53 mutants, *Proteomics* **4**, 1950–1958.

Bryan, M.C., Lee, L.V., Wong, C.-H. (2004) High-throughput identification of fucosyltransferase inhibitors using carbohydrate microarrays, *Bioorg. Med. Chem. Lett.* **14**, 3185–3188.

Bulyk, M.L., Gentalen, E., Lockhart, D.J., Church, G.M. (1999) Quantifying DNA-protein interactions by double-stranded DNA arrays, *Nat. Biotechnol.* **17**, 573–577.

Bulyk, M.L., Huang, X., Choo, Y., Church, G.M. (2001) Exploring the DNA-binding specificities of zinc fingers with DNA microarrays, *Proc. Natl. Acad. Sci. U.S.A.* **98**, 7158–7163.

Butler, J.E., Ni, L., Brown, W.R., Joshi, K.S., Chang, J., Rosenberg, B., Voss, E.W. (1993) The immunochemistry of sandwich ELISAs-VI. Greater than 90% of monoclonal and 75% of polyclonal anti-fluoresceyl capture antibodies (CAbs) are denatured by passive adsorption, *Mol. Immunol.* **30**, 1165–1175.

Carson, R.T., Vignali, D.A. (1999) Simultaneous quantitation of 15 cytokines using a multiplexed flow cytometric assay, *J. Immunol. Methods* **227**, 41–52.

Daugherty, P.S., Olsen, M.J., Iverson, B.L., Georgiou, G. (1999) Development of an optimized expression system for the screening of antibody libraries displayed on the Escherichia coli surface, *Protein Eng.* **12**, 613–621.

de Haard, H.J., van Neer, N., Reurs, A., Hufton, S.E., Roovers, R.C., Henderikx, P., de Bruine, A.P., Arends, J.W., Hoogenboom, H.R. (1999) A large non-immunized human Fab fragment phage library that permits rapid isolation and kinetic analysis of high affinity antibodies, *J. Biol. Chem.* **274**, 18218–18230.

Dhanasekaran, S.M., Barrette T.R., Ghosh, D., Shah, R., Varambally, S., Kurachi, K., Pienta, K.J., Rubin, M.A., Chinnaiyan, A.M. (2001) Delineation of prognostic biomarkers in prostate cancer, *Nature* **412**, 822–826.

Durocher, Y., Perret, S., Kamen, A. High-level and high-throughput recombinant protein production by transient transfection of suspension-growing human 293-EBNA1 cells, *Nucleic Acids Res.* **30**, E9.

Duveneck, G.L., Bopp, M.A., Ehrat, M., Balet, L.P., Haiml, M., Keller, U., Marowsky, G., Soria, S. (2003) Two-photon fluorescence excitation of macroscopic areas on planar waveguides, *Biosens. Bioelectron.* **18**, 503–510.

Eppinger, J., Funeriu, D.P., Miyake, M., Denizot, L., Miyake, J. (2004) Enzyme microarrays: On-chip determination of inhibition constants based on affinity-label detection of enzymatic activity, *Angew. Chem., Int. Ed.* **43**, 3806–3810.

Falsey, J.R., Renil, M., Park, S., Li, S., Lam, K.S. (2001) Peptide and small molecule microarray for high-throughput cell adhesion and functional assays, *Bioconjug. Chem.* **12**, 346–353.

Feldhaus, M.J., Siegel, R.W., Opresko, L.K., Coleman, J.R., Feldhaus, J.M., Yeung, Y.A., Cochran, J.R., Heinzelman, P., Colby, D., Swers, J., Graff, C., Wiley, H.S., Wittrup, K.D.

(2003) Flow-cytometric isolation of human antibodies from a nonimmune Saccharomyces cerevisiae surface display library, *Nat. Biotechnol.* **21**, 163–170.

Feng, Y. Ke, X., Ma, R., Chen, Y., Hu, G., Liu, F. (2004) Parallel detection of autoantibodies with microarrays in rheumatoid diseases, *Clin. Chem.* **50**, 416–422.

Fields, S., Song, O. (1989) A novel genetic system to detect protein-protein interactions, *Nature* **340**, 245–246.

Figeys, D. (2004) Combining different 'omics' technologies to map and validate protein-protein interactions in humans, *Brief. Funct. Genomic. Proteomic.* **2**, 357–365.

Fotin-Mleczek, M., Rottmann, M., Rehg, G., Rupp, S., Johannes, F.-J. (2000) Detection of protein-protein interactions using a green fluorescent protein-based mammalian two-hybrid system, *BioTechniques* **29**, 22–26.

Frank, R. (1992) Spot synthesis an easy technique for positionally addressable, parallel chemical synthesis on a membrane, *Tetrahedron* **48**, 9217–9232.

Frank, R. (2002) The SPOT-synthesis technique. Synthetic peptide arrays on membrane supports-principles and applications, *J. Immunol. Methods* **267**, 13–26.

Fukui, S., Feizi, T., Galustian, C., Lawson, A.M., Chai, W. (2002) Oligosaccharide microarrays for high-throughput detection and specificity assignments of carbohydrate-protein interactions, *Nat. Biotechnol.* **20**, 1011–1017.

Gao, X., Zhou, X., Gulari, E. (2003) Light directed massively parallel on-chip synthesis of peptide arrays with t-Boc chemistry, *Proteomics* **3**, 2135–2141.

Gavin, A.C., Bosche, M., Krause, R., et al. (2002) Functional organization of the yeast proteome by systematic analysis of protein complexes, *Nature* **415**, 141–147.

Golden, M.C., Collins, B.D., Willis, M.C., Koch, T.H. (2000) Diagnostic potential of PhotoSELEX-evolved ssDNA aptamers, *J. Biotechnol.* **81**, 167–178.

Golub, T.R., Slonim, D.K., Tamayo, P., Huard, C., Gaasenbeek, M., Mesirov, J.P., Coller, H., Loh, M.L., Downing, J.R., Caligiuri, M.A., Bloomfield, C.D., Lander, E.S. (1999) Molecular classification of cancer: class discovery and class prediction by gene expression monitoring, *Science* **286**, 531–537.

Gygi, S.P., Rochon, Y., Franza, B.R., Aebersold, R. (1999) Correlation between protein and mRNA abundance in yeast, *Mol. Cell. Biol.* **19**, 1720–1730.

He, M. (2004) Generation of protein in situ arrays by DiscernArray technology, *Methods Mol. Biol.* **264**, 25–31.

Hergenrother, P.J., Depew, K.M., Schreiber, S.L. (2003) Small-molecule microarrays: Covalent attachment and screening of alcohol-containing small molecules on glass slides, *J. Am. Chem. Soc.* **122**, 7849–7850.

Ho, Y., Gruhler, A., Heilbut, A., Bader, G.D., Moore, L., Adams, S.L., Millar, A., Taylor, P., Bennett, K., Boutilier, K., Yang, L., Wolting, C., et al. (2002) Systematic identification of protein complexes in Saccharomyces cerevisiae by mass spectrometry, *Nature* **415**, 180–183.

Houseman, B.T., Mrksich, M. (2002) Carbohydrate arrays for the evaluation of protein binding and enzymatic modification, *Chem. Biol.* **9**, 443–454.

Houseman, B.T., Huh, J.H., Kron, S.J., Mrksich, M. (2002) Peptide chips for the quantitative evaluation of protein kinase activity, *Nat. Biotechnol.* **20**, 270–274.

Huang, R. (2001) Detection of multiple proteins in an antibody-based protein microarray system, *J. Immunol. Methods* **255**, 1–13.

Huang, R.P., Huang, R., Fan, Y., Lin, Y. (2001) Simultaneous detection of multiple cytokines from conditioned media and patient's sera by an antibody-based protein array system, *Anal. Biochem.* **294**, 55–62.

Hubsman, M., Yudkovsky, G., Aronheim, A. (2001) A novel approach for the identification of protein-protein interaction with integral membrane proteins, *Nucleic Acids Res.* **29**, E18.

Ito, T., Chiba, T., Ozawa, R., Yoshida, M., Hattori, M., Sakaki, Y. (2001) A comprehensive two-hybrid analysis to explore the yeast protein interactome, *Proc. Natl. Acad. Sci. U.S.A.* **98**, 4569–4574.

Jiang, B., Snipes-Magaldi, L., Dennehy, P., Keyserling, H., Holman, R.C., Bresee, J., Gentsch, J., Glass, R.I. (2003) Cytokines as mediators for or effectors against rotavirus disease in children, *Clin. Diagn. Lab Immunol.* **10**, 995–1001.

Kanoh, N., Kumashiro, S., Simizu, S., Kondoh, Y., Hatakeyama, S., Tashiro, H., Osada, H. (2003) Immobilization of natural products on glass slides by using a photoaffinity reaction and the detection of protein-small-molecule interactions, *Angew. Chem., Int. Ed.* **42**, 5584–5587.

Kaukola, T., Satyaraj, E., Patel, D.D., Tchernev, V.T., Grimwade, B.G., Kingsmore, S.F., Koskela, P., Tammela, O., Vainionpaa, L., Pihko, H., Aarimaa, T., Hallman, M. (2004) Cerebral palsy is characterized by protein mediators in cord serum, *Ann. Neurol.* **55**, 186–194.

Kettman, J.R., Davies, T., Chandler, D., Oliver, K.G., Fulton, R.J. (1998) Classification and properties of 64 multiplexed microsphere sets, *Cytometry* **33**, 234–243.

Knappik, A., Ge, L., Honegger, A., Pack, P., Fischer, M., Wellnhofer, G., Hoess, A., Wolle, J., Pluckthun, A., Virnekas, B. (2000) Fully synthetic human combinatorial antibody libraries (HuCAL) based on modular consensus frameworks and CDRs randomized with trinucleotides, *J. Mol. Biol.* **296**, 57–86.

Knezevic, V., Leethanakul, C., Bichsel, V.E., Worth, J.M., Prabhu, V.V., Gutkind, J.S., Liotta, L.A., Munson, P.J. III, Petricoin, E.F., Krizman, D.B. (2001) Proteomic profiling of the cancer microenvironment by antibody arrays, *Proteomics* **1**, 1271–1278.

Koehler, A.N., Shamji, A.F., Schreiber, S.L. (2003) Discovery of an inhibitor of a transcription factor using small molecule microarrays and diversity-oriented synthesis, *J. Am. Chem. Soc.* **125**, 8420–8421.

Koehn, M., Wacker, R., Peters, C., Schroeder, H., Soulere, L., Breinbauer, R., Niemeyer, C.M., Waldmann, H. (2003) Staudinger ligation: A new immobilization strategy for the preparation of small-molecule arrays, *Angew. Chem., Int. Ed.* **42**, 5830–5834.

Kohl, A., Binz, H.K., Forrer, P., Stumpp, M.T., Pluckthun, A., Grutter, M.G. (2003) Designed to be stable: crystal structure of a consensus ankyrin repeat protein, *Proc. Natl. Acad. Sci. U.S.A.* **100**, 1700–5.

Koide, A., Bailey, C.W., Huang, X., Koide, S. (1998) The fibronectin type III domain as a scaffold for novel binding proteins, *J. Mol. Biol.* **284**, 1141–1151.

Kramer, A., Reineke, U., Dong, L., Hoffmann, B., Hoffmuller, U., Winkler, D., Volkmer-Engert, R., Schneider-Mergener, J. (1999) Spot synthesis: observations and optimizations, *J. Pept. Res.* **54**, 319–327.

Kuruvilla, F.G., Shamji, A.F., Sternson, S.M., Hergenrother, P.J., Schreiber, S.L. (2002) Dissecting glucose signalling with diversity-oriented synthesis and small-molecule microarrays, *Nature* **416**, 653–657.

Landgraf, C., Panni, S., Montecchi-Palazzi, L., Castagnoli, L., Schneider-Mergener, J., Volkmer-Engert, R., Cesareni, G. (2004) Protein interaction networks by proteome peptide scanning, *PLoS Biol.* **2**, 94–103.

Lesaicherre, M.-L., Uttamchandani, M., Chen, G.Y.J., Yao, S.Q. (2002a) Developing site-specific immobilization strategies of peptides in a microarray, *Bioorg. Med. Chem. Lett.* **12**, 2079–2083.

Lesaicherre, M.-L., Uttamchandani, M., Chen, G.Y.J., Yao, S.Q. (2002b) Antibody-based fluorescence detection of kinase activity on a peptide array, *Bioorg. Med. Chem. Lett.* **12**, 2085–2088.

Lin, Y., Huang, R., Santanam, N., Liu, Y.G., Parthasarathy, S., Huang, R.P. (2002) Profiling of human cytokines in healthy individuals with vitamin E supplementation by antibody array, *Cancer Lett.* **187**, 17–24.

MacBeath, G., Koehler, A.N., Schreiber, S.L. (1999) Printing small molecules as microarrays and detecting protein-ligand interactions en masse, *J. Am. Chem. Soc.* **121**, 7967–7968.

MacBeath, G., Schreiber, S.L. (2000) Printing proteins as microarrays for high-throughput function determination, *Science* **289**, 1760–1763.

Manning, G., Whyte, D.B., Martinez, R., Hunter, T., Sudarsanam, S. (2002) The protein kinase complement of the human genome, *Science* **298**, 1912–1934.

Martin, K., Steinberg, T.H., Cooley, L.A., Gee, K.R., Beechem, J.M., Patton, W.F. (2003) Quantitative analysis of protein phosphorylation status and protein kinase activity on microarrays using a novel fluorescent phosphorylation sensor dye, *Proteomics* **3**, 1244–1255.

Mendoza, L.G., McQuary, P., Mongan, A., Gangadharan, R., Brignac, S., Eggers, M. (1999) High-throughput microarray-based enzyme-linked immunosorbent assay (ELISA), *BioTechniques* **27**, 778–788.

Miller, J.C., Zhou, H., Kwekel, J., Cavallo, R., Burke, J., Butler, E.B., The, B.S., Haab, B.B. (2003) Antibody microarray profiling of human prostate cancer sera: Antibody screening and identification of potential biomarkers, *Proteomics* **3**, 56–63.

Moody, M.D., van Ardel, S.W., Orencole, S.F., Burns, C. (2001) Array-based ELISAs for high-throughput analysis of human cytokines, *BioTechniques* **31**, 186–194.

Neuman de Vegvar, H.E., Amara, R.R., Steinman, L., Utz, P.J., Robinson, H.L., Robinson, W.H. (2003) Microarray profiling of responses against simian-human immunodeficiency virus: postchallenge convergence of reactivities independent of host histocompatibility type and vaccine regime, *J. Virol.* **77**, 11125–11138.

Nielsen, T.O., West, R.B., Linn, S.C., Alter, O., Knowling, M.A., O'Connell, J.X., Zhu, S., Fero, M., Sherlock, G., Pollack, J.R., Brown, P.O., Botstein, D., van de Rijn, M. (2002) Molecular characterisation of soft tissue tumours: a gene expression study, *Lancet* **359**, 1301–1307.

Nord, K., Gunneriusson, E., Ringdahl, J., Stahl, S., Uhlen, M., Nygren, P.A. (1997) Binding proteins selected from combinatorial libraries of an alpha-helical bacterial receptor domain, *Nat. Biotechnol.* **15**, 772–777.

Pawlak, M., Schick, E., Bopp, M.A., Schneider, M.J., Oroszlan, P., Ehrat, M. (2002) Zeptosens' protein microarrays: a novel high performance microarray platform for low abundance protein analysis, *Proteomics* **2**, 383–393.

Pellois, J.P., Zhou, X., Srivannavit, O., Zhou, T., Gulari, E., Gao, X. (2002) Individually addressable parallel peptide synthesis on microchips, *Nat. Biotechnol.* **20**, 922–926.

Peluso, P., Wilson, D.S., Do, D., Tran, H., Venkatasubbaiah, M., Quincy, D., Heidecker, B., Poindexter, K., Tolani, N., Phelan, M., Witte, K., Jung, L.S., Wagner, P., Nock, S. (2003) Optimizing antibody immobilization strategies for the construction of protein microarrays, *Anal. Biochem.* **312**, 113–124.

Perou, C.M., Sorlie, T., Eisen, M.B., van de Rijn, M., Jeffrey, S.S., Rees, C.A., Pollack, J.R., Ross, D.T., Johnsen, H., Akslen, L.A., Fluge, O., Pergamenschikov, A., Williams, C., Zhu, S.X., Lonning, P.E., Borresen-Dale, A.L., Brown, P.O., Botstein, D. (2000) Molecular portraits of human breast tumours, *Nature* **406**, 747–752.

Petricoin, E.F., Ardekani, A.M., Hitt, B.A., Levine, P.J., Fusaro, V.A., Steinberg, S.M., Mills, G.B., Simone, C., Fishman, D.A., Kohn, E.C., Liotta, L.A. (2002) Use of proteomic patterns in serum to identify ovarian cancer, *Lancet* **359**, 572–577.

Pomeroy, S.L., Tamayo, P., Gaasenbeek, M., Sturla, L.M., Angelo, M., McLaughlin, M.E., Kim, J.Y., Goumnerova, L.C., Black, P.M., Lau, C., Allen, J.C., Zagzag, D., Olson, J.M., Curran, T., Wetmore, C., Biegel, J.A., Poggio, T., Mukherjee, S., Rifkin, R., Califano, A., Stolovitzky, G., Louis, D.N., Mesirov, J.P., Lander, E.S., Golub, T.R. (2002) Prediction of central nervous system embryonal tumour outcome based on gene expression, *Nature* **415**, 436–442.

Qiu, J., Madoz-Gurpide, J., Misek, D.E., Kuick, R., Brenner, D.E., Michailidis, G., Haab, B.B., Omenn, G.S., Hanash, S. (2004). Development of natural protein microarrays for diagnosing cancer based on an antibody response to tumor antigens, *J. Proteome Res.* **3**, 261–267.

Ramachandran, N., Hainsworth, E., Bhullar, B., Eisenstein, S., Rosen, B., Lau, A.Y., Walter, J.C., LaBaer, J. (2004) Self-assembling protein microarrays, *Science* **305**, 86–90.

Robertson, M.P., Ellington, A.D. (2001) In vitro selection of nucleoprotein enzymes, *Nat. Biotechnol.* **19**, 650–655.

Rowe, C.A., Scruggs, S.B., Feldstein, M.J., Golden, J.P., Ligler, F.S. (1999) An array immunosensor for simultaneous detection of clinical analytes, *Anal. Chem.* **71**, 433–439.

Rowe, C.A., Tender, L.M., Feldstein, M.J., Golden, J.P., Scruggs, S.B., MacCraith, B.D., Cras, J.J., Ligler, F.S. (1999) Array biosensor for simultaneous identification of bacterial, viral, and protein analytes, *Anal. Chem.* **71**, 3846–3852.

Salisbury, C.M., Maly, D.J., Ellman, J.A. (2002) Peptide microarrays for the determination of protease substrate specificity, *J. Am. Chem. Soc.* **124**, 14868–14870.

Schaffitzel, C., Hanes, J., Jermutus, L., Pluckthun, A. (1999) Ribosome display: an in vitro method for selection and evolution of antibodies from libraries, *J. Immunol. Methods* **231**, 119–135.

Schena, M., Shalon, D., Davis, R.W., Brown, P.O. (1995) Quantitative monitoring of gene expression patterns with a complementary DNA microarray, *Science* **270**, 467–470.

Schweitzer, B., Wiltshire, S., Lambert, J., O'Malley, S., Kukanskis, K., Zhu, Z., Kingsmore, S.F., Lizardi, P.M., Ward, D.C. (2000) Immunoassays with rolling circle DNA amplification: a versatile platform for ultrasensitive antigen detection, *Proc. Natl. Acad. Sci. U.S.A.* **97**, 10113–10119.

Schweitzer, B., Roberts, S., Grimwade, B., Shao, W., Wang, M., Fu, Q., Shu, Q., Laroche, I., Zhou, Z., Tchernev, V.T., Christiansen, J., Velleca, M., Kingsmore, S.F. (2002) Multiplexed protein profiling on microarrays by rolling-circle amplification, *Nat. Biotechnol.* **20**, 359–365.

Sheets, M.D., Amersdorfer, P., Finnern, R., Sargent, P., Lindquist, E., Schier, R., Hemingsen, G., Wong, C., Gerhart, J.C., Marks, J.D., Lindqvist, E. (1998) Efficient construction of a large nonimmune phage antibody library: the production of high-affinity human single-chain antibodies to protein antigens, *Proc. Natl. Acad. Sci. U.S.A.* **95**, 6157–6162.

Soldan, S.S., Alvarez-Retuerto, A.I., Sicotte, N.L., Voskuhl, R.R. (2004) Dysregulation of IL-10 and IL-12p40 in secondary progressive multiple sclerosis, *J. Neuroimmunol.* **146**, 209–215.

Sorlie, T., Perou, C.M., Tibshirani, R., Aas, T., Geisler, S., Johnsen, H., Hastie, T., Eisen, M.B., van de Rijn, M., Jeffrey, S.S., Thorsen, T., Quist, H., Matese, J.C., Brown, P.O., Botstein, D., Eystein Lonning, P., Borresen-Dale, A.L. (2001) Gene expression patterns of breast carcinomas distinguish tumor subclasses with clinical implications, *Proc. Natl. Acad. Sci. U.S.A.* **98**, 10869–10874.

Sreekumar, A., Nyati, M.K., Varambally, S., Barrette, T.R., Ghosh, D., Lawrence, T.S., Chinnaiyan, A.M. (2001) Profiling of cancer cells using protein microarrays: discovery of novel radiation-regulated proteins, *Cancer Res.* **61**, 7585–7593.

Stillman, B.A., Tonkinson, J.L. (2000) FAST slides: a novel surface for microarrays, *BioTechniques* **29**, 630–635.

Strausberg, R.L., Feingold, E.A., Grouse, L.H., Derge, J.G., Klausner, R.D., Collins, F.S., Wagner, L., Shenmen, C.M., Schuler, G.D., et al. (2002) Generation and initial analysis of more than 15 000 full-length human and mouse cDNA sequences, *Proc. Natl. Acad. Sci. U.S.A.* **99**, 16899–16903.

Sydor, J.R., Mariano, M., Sideris, S., Nock, S. (2002) Establishment of intein-mediated protein ligation under denaturing conditions: C-terminal labeling of a single-chain antibody for biochip screening, *Bioconjug. Chem.* **13**, 707–712.

Sydor, J.R., Scalf, M., Sideris, S., Mao, G.D., Pandey, Y., Tan, M., Mariano, M., Moran, M.F., Nock, S., Wagner, P. (2003) Chip-based analysis of protein-protein interactions by fluorescence detection and on-chip immunoprecipitation combined with microLC-MS/MS analysis, *Anal. Chem.* **75**, 6163–6170.

Tam, S.W., Wiese, R., Lee, S., Gilmore, J., Kumble, K.D. (2002) Simultaneous analysis of eight human Th1/Th2 cytokines using microarrays, *J. Immunol. Methods* **261**, 157–165.

Tannapfel, A., Anhalt, K., Hausermann, P., Sommerer, F., Benicke, M., Uhlmann, D., Witzigmann, H., Hauss, J., Wittekind, C. (2003) Identification of novel proteins associated with hepatocellular carcinomas using protein microarrays, *J. Pathol.* **201**, 238–249.

Tegge, W.J., Frank, R. (1998) Analysis of protein kinase substrate specificity by the use of peptide libraries on cellulose paper (SPOT-method), *Methods Mol. Biol.* **87**, 99–106.

Toepert, F., Knaute, T., Guffler, S., Pires, J.R., Matzdorf, T., Oschkinat, H., Schneider-Mergener, J. (2003) Combining SPOT synthesis and native peptide ligation to create large arrays of WW protein domains, *Angew. Chem., Int. Ed.* **42**, 1136–1140.

Uetz, P., Giot, L., Cagney, G., Mansfield, T.A., Judson, R.S., Knight, J.R., Lockshon, D., Narayan, V., Srinivasan, M., Pochart, P., Qureshi-Emili, A., Li, Y., Godwin, B., Conover, D., Kalbfleisch, T., Vijayadamodar, G., Yang, M., Johnston, M., Fields, S., Rothberg, J.M. (2000) A comprehensive analysis of protein-protein interactions in Saccharomyces cerevisiae, *Nature* **403**, 623–627.

Uttamchandani, M., Chan, E.W.S., Chen, G.Y.J., Yao, S.Q. (2003) Combinatorial peptide microarrays for the rapid determination of kinase specificity, *Bioorg. Med. Chem. Lett.* **13**, 2997–3000.

Vaughan, T.J., Williams, A.J., Pritchard, K., Osbourn, J.K., Pope, A.R., Earnshaw, J.C., McCafferty, J., Hodits, R.A., Wilton, J., Johnson, K.S. (1996) Human antibodies with subnanomolar affinities isolated from a large non-immunized phage display library, *Nat. Biotechnol.* **14**, 309–314.

von Mering, C., Krause, R., Snel, B., Cornell, M., Oliver, S.G., Fields, S., Bork, P. (2002) Comparative assessment of large-scale data sets of protein-protein interactions, *Nature* **417**, 399–403.

Wahler, D., Badalassi, F., Crotti, P., Reymond, J.-L. (2001) Enzyme fingerprints by fluorogenic and chromogenic substrate arrays, *Angew. Chem. Int. Ed.* **40**, 4457–4460.

Wang, D., Liu, S., Trummer, B.J., Deng, C., Wang, A. (2002) Carbohydrate microarrays for the recognition of cross-reactive molecular markers of microbes and host cells, *Nat. Biotechnol.* **20**, 275–281.

Wiese, R., Belosludtsev, Y., Powdrill, T., Thompson, P., Hogan, M. (2001) Simultaneous multianalyte ELISA performed on a microarray platform, *Clin. Chem.* **47**, 1451–1457.

Winssinger, N., Ficarro, S., Schultz, P.G., Harris, J.L. (2002) Profiling protein function with small molecule microarrays, *Proc. Natl. Acad. Sci. U.S.A.* **99**, 11139–11144.

Xu, L., Aha, P., Gu, K., Kuimelis, R., Kurz, M., Lam, T., Lim, A., Liu, H., Lohse, P., Sun, L., Weng, S., Wagner, R., Lipovsek, D. (2002) Directed evolution of high-affinity antibody mimics using mRNA display, *Chem. Biol.* **9**, 933–942.

Zhu, H., Bilgin, M., Bangham, R., Hall, D., Casamayor, A., Bertone, P., Lan, N., Jansen, R., Bidlingmaier, S., Houfek, T., Mitchell, T., Miller, P., Dean, R.A., Gerstein, M., Snyder, M. (2001) Global analysis of protein activities using proteome chips, *Science* **293**, 2101–2105.

Zhu, Q., Uttamchandani, M., Li, D., Lesaicherre, M.L., Yao, S.Q. (2003) Enzymatic profiling in a small-molecule microarray, *Org. Lett.* **5**, 1257–1260.

Zhu, H., Klemic, J.F., Chang, S., Bertone, P., Casamayor, A., Klemic, K.G., Smith, D., Gerstein, M., Reed, M.A., Snyder, M. (2000) Analysis of yeast protein kinases using protein chips, *Nat. Genet.* **26**, 283–289.

Protein Modeling

Marian R. Zlomislic and D. Peter Tieleman
University of Calgary, Albeta, Canada

1	**Introduction**	136
1.1	Primary, Secondary, Tertiary, and Quaternary Structure of Proteins	136
1.2	Relationship between Structure and Function	136
1.3	Experimental Structure Determination Methods	138
1.4	Role of Modeling	139
2	**Structure Prediction Methods**	139
2.1	The Protein-folding Problem	139
2.2	*De Novo/Ab Initio* Methods	140
2.3	Fold Recognition Methods	141
2.4	Homology Modeling	142
3	**Structure-based Modeling**	143
3.1	Molecular Graphics-based Methods	143
3.2	Poisson–Boltzmann/Electrostatics Calculations	144
3.3	Ligand Docking	145
3.4	Protein–Protein Interactions	146
4	**Simulations of Protein Dynamics**	146
4.1	Molecular Dynamics Simulation	146
4.2	Free-energy Calculations	148
4.3	Brownian Dynamics Calculations	148
5	**Example Applications**	149
5.1	Acetylcholinesterase	149
5.2	Water Transport Through Aquaporins	150
5.3	ABC Transporters and Multidrug Resistance Proteins	152
6	**Perspectives**	154

Acknowledgment 155

Bibliography 156
Books and Reviews 156
Primary Literature 156

Keywords

Ab Initio Modeling
The goal of *ab initio* modeling is to predict the structure of a protein from only its amino acid sequence.

Brownian Dynamics Simulation
Brownian dynamics simulations are related to molecular dynamics, but do not include the same amount of detail. Solvent in particular is normally represented as a continuous medium, which significantly decreases the complexity of the models and allows much longer simulations.

CASP, CAFASP, CAPRI
Critical Assessment of Structure Prediction (CASP), Critical Assessment of Fully Automated Structure Prediction (CAFASP), and Critical Assessment of PRedicted Interactions (CAPRI) are large-scale assessments in which hundreds of research groups submit their best structure predictions on the same set of targets, allowing a critical comparison between methods and an assessment of strengths and weaknesses of different methods and the field as a whole.

Docking
The term docking incorporates a group of methods for studying the interactions of small molecules and proteins. The process typically involves extensive searching of possible orientations of the small molecule in a binding site of the protein, and a scoring function that ranks different orientations by their likelihood of occurrence.

Fold Recognition
The goal of fold recognition is to predict which fold, out of ca. 1000 observed protein folds, a given sequence will adopt.

Homology Modeling
Predicting the three-dimensional structure of a protein on the basis of known structure of a related protein.

Molecular Dynamics Simulation
A computer simulation method in which the motions of all atoms in a molecular model, including water and ions are calculated over a period of up to a microsecond.

This computationally expensive method is one of the most detailed ways of studying the dynamics of macromolecules.

Molecular Graphics
Modern computer graphics are used to visualize molecular structures and models. The ability to highlight different features or properties in a three-dimensional model is a powerful way to explore molecules. Stereo hardware and immersive environments further enhance the usefulness of molecular graphics.

Poisson–Boltzmann Equation
The Poisson–Boltzmann equation describes electrostatic interactions between charges in a continuous medium (e.g. water) in the presence of salt. It can be solved numerically by several popular software packages to give insight into electrostatic interactions in proteins, between proteins and their substrates, between proteins and membranes, and in other systems. A common application is the calculation of pK_a-shifts for acidic and basic residues in proteins due to their environment.

Primary, Secondary, Tertiary, and Quaternary Protein Structure
The amino acid sequence of a protein is its primary structure. The α-helices, β-strands, and random coils are typical secondary structures that form between consecutive residues. They are defined by the hydrogen-bonding pattern and dihedral angles observed in the protein backbone. The tertiary structure of a protein describes the relative position of all the protein atoms in three-dimensional space. Elements of secondary structure that were far apart sequentially may now form higher order structures such as parallel or antiparallel β-sheets, or helix bundles. The quaternary structure describes how multiple proteins are arranged together in larger complexes.

QM/MM Simulation
A computer simulation method in which an important part of a system, such as the active site of an enzyme, is described by quantum mechanics, with the rest of the system described by molecular dynamics methods. The part of the system treated by QM includes explicit treatment of electrons.

Structural Genomics
A concerted effort to determine experimentally the high-resolution structure of all proteins in a genome, often with a focus on proteins that are predicted to be structurally different from all structures that are already in the database. This greatly increases the likelihood that a given sequence can be modeled using homology modeling.

> Protein modeling consists of a broad range of computational techniques to understand the properties of proteins and has become an integral part of structural biology and drug design. Modeling can be used to predict the secondary structure or fold of a protein on the basis of its sequence alone, to predict the three-dimensional structure of a protein on the basis of knowledge of the structure of a related protein,

to design new proteins, and to predict properties that depend on the experimentally determined three-dimensional structure of a protein. Examples of such properties include drug binding, protein–protein interactions, and interactions with elements in a protein's environment, including ions, lipids, carbohydrates, and nucleic acids. Conformational changes in proteins can be investigated by molecular dynamics simulations to provide detailed insight into the dynamics of proteins, a crucial aspect of protein function. In recent developments, quantum mechanical calculations are used more and more frequently to study reactions in proteins. With the ever-increasing power of computers, increasingly detailed aspects of protein function can now be investigated by modeling methods, at a scale and level of detail that is often very difficult or impossible to achieve by experimental methods.

In this chapter, the main principles and techniques involved in protein modeling are introduced. A few examples from the literature will highlight how protein modeling can be used in complement with other methods.

1
Introduction

1.1
Primary, Secondary, Tertiary, and Quaternary Structure of Proteins

Proteins are structured at different levels. The primary structure of a protein is its amino acid sequence, encoded by DNA. All proteins are constructed from about 20 common amino acids as well as some amino acids that are formed through chemical modification. The secondary structure of a protein is a sequence of common structural, three-dimensional elements or building blocks. These secondary structure elements include α-helices and β-strands as best-known elements, but there are many other building blocks. The tertiary structure of a protein is its three-dimensional structure in space, which can be thought of as placing the secondary structure elements together in space. The quaternary structure of a protein describes how multiple proteins interact through noncovalent forces to form larger protein complexes. An example of the different levels of structure is shown in Fig. 1, based on the protein barstar. It is shown in a protein complex with barnase to illustrate quaternary structure. The minimal requirement for modeling proteins is knowledge of the primary structure. Since the primary sequence is one of the main results of genome sequencing, this is a very modest requirement. As more experimental structural information becomes available, modeling typically becomes more accurate by using that experimental information to identify plausible models. Any experimental information that puts limitations on possible models is useful.

1.2
Relationship between Structure and Function

A basic assumption in protein science is that the function of a protein follows from its structure. Yet, proteins with very similar three-dimensional structure can have very different functions, so

MKKAVINGEQIRSISDLHQTLKKELALPEYYGENLAALWDCLTGWVEYPLVLEWRQ
FEQSKQLTENGAESVLQVFREAKAEGCDITIILS
(a)

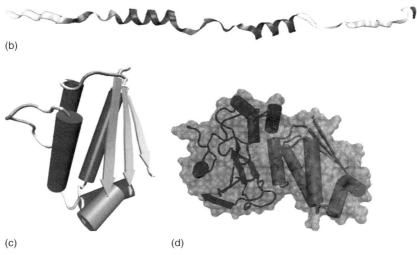

Fig. 1 (a) The primary sequence of the protein barstar in FASTA format. (b) The secondary structure of the first 60 residues of barstar, colored purple for helices, and yellow for β-strands. All are shown in ribbon format. (c) The complete tertiary structure of barstar in cartoon format. The helices are now bundled together, and the β-strands have formed a three-stranded parallel β-sheet. (d) Example of quaternary structure: the barnase/barstar complex. Barstar (red) and barnase (blue) area shown in ribbon and spacefilling formats. Images created with VMD using the PDB entry 1B2U. (See color plate p. xxx.)

precise details of the three-dimensional (tertiary/quaternary) structure are very important. Nonetheless, there is a large amount of information contained in the primary and secondary structure of a protein. Indeed, nature in most cases only seems to need the primary sequence, but computational methods are not yet able to reliably predict the three-dimensional structure of proteins based on primary sequence alone. At the current state of the art, a primary sequence is useful to compare proteins of known function with a similar sequence and sometimes allows assignment of a tentative function. Knowing the secondary structure of a protein significantly narrows down the range of possible three-dimensional structures. It allows a comparison with databases of the secondary sequence of known proteins, which often makes it possible to recognize the three-dimensional fold of a protein on the basis of the secondary structure alone. Interestingly, although in principle the number of possible three-dimensional structures is practically infinite, only about 1000 different "folds" seem to occur in nature. This is particularly useful for whole-genome studies: if we can use the primary sequence and predictions or knowledge of the secondary structure, chances are reasonable that the fold of the protein can be recognized. This is a constructive step toward attaining a three-dimensional structure.

The three-dimensional structure of a protein is the most useful level for molecular modeling. At this level, the interactions between small molecules (drugs, substrate) and an enzyme, permeation properties of ion channels and transporters, activation of receptors by ligand binding, and chemical reactions in enzymes and other processes that require a very detailed knowledge of atomic structure to understand them can be studied. This level is also required to study interactions between protein subunits or for protein–protein or protein–peptide interactions. In practice, the best way to obtain knowledge of the three-dimensional structure is from experimental structure determination, but prediction methods primarily based on homology modeling (see Sect. 2) are increasingly becoming accurate enough to provide useful starting points for further computational studies.

1.3
Experimental Structure Determination Methods

There are several experimental methods to determine the structure of proteins. The primary structure is determined by sequencing, either of the protein itself or of the DNA that encodes the protein. The secondary structure can be measured by spectroscopic methods, which usually determine a percentage of various secondary structure elements. Common techniques to do this are CD (circular dichroism) spectroscopy, IR (infrared) spectroscopy, and NMR (nuclear magnetic resonance) spectroscopy.

As a basis for detailed modeling problems, high-resolution 3-dimensional structures are typically required. The two main methods to experimentally determine 3D structures are X-ray crystallography and nuclear magnetic resonance. Electron microscopy is a third method that has been used to solve the structure of several membrane proteins. Crystallography is a very powerful method that can be used on proteins and protein complexes of any size. The main limitation of the method is that a protein must form regular crystals. Not all proteins do this (especially membrane proteins), and sometimes crystallization forces proteins in structures that are probably not physiologically relevant. NMR experiments can determine the structure of a protein in solution, which is usually a more realistic environment. The main limitations of solution NMR are that it is difficult to apply the method in practice to proteins that are larger than ca. 40 kDa, and that it is usually considerably more labor intensive to determine a structure by NMR compared to crystallography. These limitations have spurred the development of new methods that are catered to work with larger proteins, and to make the process of structure calculation from a measured spectrum more automated. Solid-state protein NMR is a relatively new field that can be used to determine the structure of proteins in ordered systems, such as membranes. Electron microscopy as a method to determine 3-D structures has been mainly used for membrane proteins, which are especially difficult to crystallize for X-ray crystallography. The method is very labor intensive. It does have a significant strength for very large complexes, where the overall arrangement of large subunits can be seen at low resolution. The details of the structure can then be added from crystal structures of individual proteins that make up the complex.

1.4
Role of Modeling

Modeling has many potential uses, at different levels of protein structure. Protein modeling is such a broad field it may be beneficial to distinguish two separate general goals.

One set of goals consists of predicting aspects of the structure of a protein based on less detailed information. If only the primary structure (the sequence) of a protein is known, we can try to predict the secondary structure with reasonable accuracy, often in the 75 to 85% range. On the basis of the degree of hydrophobicity of predicted helices, it is usually also possible to predict transmembrane segments, identifying a protein as a membrane protein, with reasonable accuracy. The secondary structure can be useful to predict the general fold of a protein, which may make it possible to assign a tentative function to the protein. Predicting the tertiary structure directly from the sequence is a very difficult problem with a low success rate. If a high-resolution structure of a related protein is known, then homology modeling can be used to provide a good model of the target protein. In principle, such modeling efforts make it possible to circumvent experimental structure determination. One major goal of protein structure research at the moment is to improve all steps in the process that leads to an accurate three-dimensional model, both by improving *ab initio* prediction methods and by experimentally determining enough protein structure that homology modeling becomes feasible for most proteins in the genome – ideally, one could then obtain structural models for all proteins in an organism directly from its genome sequence. As the number of structures solved experimentally increases, the algorithms to detect homology between remotely related proteins and built homology models will improve, making this goal increasingly feasible.

A second set of goals starts from an experimentally determined high-resolution structure (or a very high-quality homology model) and uses physics-based methods to model aspects of a protein that cannot be easily determined experimentally. This type of approach could include electrostatics calculations to understand how proteins interact with substrates or other proteins, docking calculations to understand differences in binding constants and design new inhibitors for enzymes, or detailed molecular dynamics simulations to investigate the dynamics of proteins. This type of goal usually involves much more biochemical knowledge of the protein and is often centered on very specific questions: why does mutating residue x to alanine change the binding affinity of a particular drug by a factor of 1000? Why is this ion channel selective for potassium over sodium, even though the only difference between potassium and sodium superficially is a small difference in radius?

In the following sections we consider each set of goals in more detail and give specific examples.

2
Structure Prediction Methods

2.1
The Protein-folding Problem

One of the most challenging problems in biophysics is to understand how it is possible that proteins fold rapidly (microseconds to seconds) into a well-defined structure when based on their primary sequence alone, a practically infinite number of structures is possible. Levinthal showed that folding on a realistic timescale could therefore not occur through a systematic search of all possible conformations (Levinthal's paradox). Major research efforts are devoted to understanding protein folding, with a significant focus on the physics of simplified models. Clearly, if it is possible to fold a protein *in vivo* from the sequence only, then with a proper understanding of the laws governing protein folding, we should be able to predict any protein structure computationally. This is still a lofty goal, but in the past years major progress has been made in improving prediction methods at all levels of structure. This progress has been documented strikingly in the proceedings of the semiannual CASP competition (Critical Assessment of Structure Prediction).

CASP is an interesting and exciting venture involving a large scientific community. Its main goal is to obtain an in-depth and objective assessment of our current abilities and inabilities in the area of protein structure prediction (http://predictioncenter.llnl.gov). It involves a neutral organizing committee that collects unpublished, but already solved protein structures from experimental structure determination groups. Specific information, such as the protein's primary sequence, is published on a Web site, and modeling groups can attempt to predict the structures. A panel of assessors judges the predictions against the real structures, and at the CASP scientific meeting the results are discussed, areas of progress and areas of problems are identified, and the progress and directions of the field as a whole are examined. This procedure makes CASP a very fair way of establishing the merits of particular methods and of identifying where future efforts can be most productively focused. CASP6 was held in December 2004 in Italy. In the last few years, CASP has begun to introduce automated prediction software as a separate exercise (CAFASP – Critical Assessment of Fully Automated Structure Prediction). The process has also been adapted to predicting protein–protein interactions (CAPRI – Critical Assessment of PRediction of Interactions), which saw its first edition in 2001. More details about CAPRI can be found at http://capri.ebi.ac.uk. These competitions give a novice in the field of structure prediction an excellent starting point to compare the many protein modeling programs and help a user to decide which program would work best with their requirements. Despite differences in the performance of these programs, many of them draw from the same fundamental approaches to protein structure prediction. The three main approaches are discussed in the following.

2.2
De Novo/Ab Initio Methods

De novo methods are a general designation for methods that predict the structure of a protein on the basis of the primary sequence alone. The ideal method to do this would be based on simple physical laws only. Although this would be a true *de novo* method, it is exceedingly difficult and has only been modestly successful for peptides and small proteins. The difficulty is the extremely large number of possible conformations for a polypeptide chain; predicting the structure of a protein from the sequence only would indeed be equivalent to solving the protein-folding problem. In practice, *de novo* methods incorporate a variety of information derived from a database of existing structures. This can be very direct information, such as taking the conformation of a loop with the same sequence directly from the database, or less direct, such as through heuristical rules derived from statistical analysis of all known protein structures.

2.3
Fold Recognition Methods

Fold recognition methods attempt to identify structures within a new protein (the query) by comparing its sequence to proteins with known structure (the template). This can be called the *inverse folding problem*, since the goal is to find probable folds that might fit the sequence instead of trying to determine how the sequence will fold.

There are two broad approaches to fold recognition methods. The first approach is sequence-based. It relies on finding homologous sequences and then assumes that strong sequence similarity equates to strong structure and function similarity. While this may seem straightforward, current algorithms may not be sensitive enough to recognize distant homologs, and therefore the "correct" template might be missed; as well, mutations between these sequences, including gaps and insertions, sometimes make it difficult to identify the correct sequence alignment. It can be misleading to rely on the sequence alignment alone for fold recognition. It is possible to identify proteins with some sequence identity that do not have any structural or functional similarity to the query, and so, blind sequence alignments may not yield a successful prediction. Classifying protein sequences into families, and then identifying the patterns within the family such as strongly conserved residues (which might be important to function) and the pattern of mutations (For example, is a hydrophobic residue always replaced by another hydrophobic residue? That is an acceptable mutation.) can improve the confidence of your alignment. PSI-BLAST is one of the most popular homology recognition algorithms, although there are a number of others that also perform well.

The second approach to fold recognition is structure-based. "Threading" is a popular term for this approach, which evokes the image of the thread-like primary sequence being "threaded" through the three-dimensional structure of template proteins. Finding the optimal fit of the sequence with each framework is a complex process, particularly if there are gaps and deletions among the sequences of the two proteins. It is very common to use sequence-based methods in conjunction with threading to identify the best matches. Each iteration of the threading process is scored and this score is used to determine the best matches. Following Anfinsen's hypothesis that the native state of the protein is the lowest energy structure, the evaluated energy of the query's fit to the template is a strong predictor of which models are "good" within an experiment. There are a number of threading algorithms available, varying in their choice of protein model (backbone atoms vs just α-carbon atoms, side chains vs interaction sites), their method of alignment, as well as their method of evaluating the energy of the proposed structure, just to name a few differences.

When embarking on a fold recognition study, the state of the art of current fold recognition methods can be judged from results of the previously discussed CASP competition, and also of the Live Bench Project. Unlike CASP, the targets in this exercise are recently deposited structures from the Protein Data Bank. All of the methods assessed are fully automated. An expanding number of fold recognition servers (as well as metapredictors and sequence comparison servers) participate in this weekly assessment exercise, and the most current results can be found on the Web at http://bioinfo.pl/livebench/. It is important to recognize that fold recognition methods are mostly knowledge-based. They are making predictions based only on what is already known. This means that it is impossible to identify any novel folds. Novel fold prediction would fall into the category of an *ab initio* method, where the predictions are based primarily on physical principles.

2.4
Homology Modeling

Homology modeling is the only generally useful method to predict the total three-dimensional structure of a protein. It is based on the observation that the structure of a protein during evolution is much more conserved than the sequence, so that sequences that differ somewhat are still likely to have the same structure. "Somewhat" can be quantified on the basis of the database of known structures: if two proteins with more than ca. 50 residues have a sequence identity of ca. 20 to 30%, they will likely have the same structure. Thus, homology modeling can be used if there is a protein with a known structure (the template) that is sufficiently homologous to the protein whose structure we want to model (the target). In practice, homology modeling follows a series of steps, each of which involve choices by the modeler, or an automated computer server. Following Krieger et al., homology modeling can be thought of as involving seven steps:

1. Template recognition and initial alignment
2. Alignment correction
3. Backbone generation
4. Loop modeling
5. Side-chain modeling
6. Model optimization
7. Model validation

The overall strategy is to begin with aligning the sequence of the target with that of the template, to build the overall chain structure of the target, and then to consider "details." The accuracy of this alignment is the most important determinant of the accuracy of the model. Step 2 is a manual correction of the alignment if there is a compelling reason to do so. For example, in potassium channels there is a nearly universally conserved motif TVGYG that is likely to be aligned even if multiple-sequence alignment algorithms offer alternative alignments. The backbone of the target is typically initially copied from the template, followed by algorithms that allow some flexible adjustments where necessary. The major problems in homology modeling typically involve insertions that are not present in the template. In potassium channels, the loops between transmembrane helices can be very different between channels, even if the transmembrane segments are very similar. Similarly, the loops connecting the transmembrane helices in G-protein coupled receptors convey much of the specificity of different receptors, but this most interesting part generally cannot be modeled by homology modeling. Thus, being able to model loops for the targets that are missing in the template is an important step, and a significant research focus. Once the backbone has been optimized as much as possible, side chains can be modeled and the model tested against experimental information. In many cases, steps have to be repeated.

It is now usually fairly easy to generate homology models, provided a suitable template is available. Commonly used programs include Modeller and WhatIf, as well as automated Web servers such as SwissModel. The key measure of success for homology models is their experimental validation, but less than perfect models can be very useful in interpreting experiments and guiding the design of new experiments.

3 Structure-based Modeling

3.1 Molecular Graphics-based Methods

Over the past 5 to 10 years, the personal computer has begun to replace the large supercomputers and dedicated workstations once needed to use molecular visualization tools. Today, a number of molecular graphics programs are available for use on personal computers, often as freeware. Any program with basic rendering capabilities can be a valuable tool to study a particular protein. A variety of representations, illustrated in Fig. 2 for the protein barnase, can help the modeler understand the link between the protein's structure and its function. By manipulating colors and representations, one can interrogate the structure to see where conserved residues/sequences are in the protein's structure, look at the overall surface of the protein as a guide to potential protein–protein interactions, and look for cavities that might be ligand-docking sites, just to name a few examples. In this section, we aim to point out a few commonly used programs for protein

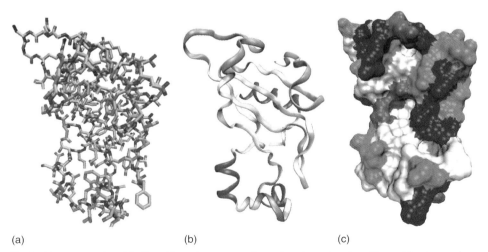

Fig. 2 Barnase rendered in three formats: (a) colored by atom-type (carbon – blue, nitrogen – navy blue, oxygen – red); (b) colored by secondary structure features in ribbon format, where β-sheets are colored yellow, helices are colored purple, and turns and random coil features are colored blue and white; (c) rendered in surface format, colored by residue type, where charged residues are colored red or blue, polar residues are green, and nonpolar residues are white. All figures are rendered with the same view of the protein, looking into the barstar binding pocket. All figures were rendered with VMD. (See color plate p. xxxii.)

modeling. Besides those mentioned here, a few more are described by Tate, and even more are listed on the PDB Web site (http://www.rcsb.org/pdb/software-list.html).

Rasmol is one of the most widely used programs for molecular visualization. Besides the basic options in which the display of the protein can be altered from sticks to spacefilling to ribbon format, and so on, there are a variety of tools that can be used through the command line interface. Examples include selection of specific residues of the protein, measuring distances, and highlighting parts of the protein within a certain radius. Rasmol can be used to probe all of the information contained in a ".pdb file," but it has few tools to alter that information. With Swiss-PDB Viewer, protein structures can be manipulated in a number of ways. Residues can be "mutated," and dihedral angles can be modified. This program provides a graphical interface within which several protein structures can be analyzed and manipulated at the same time. Moreover, Swiss-PdbViewer is tightly linked to SwissModel, an automated homology-modeling server. Working with these two programs greatly reduces the amount of work necessary to generate models, as it is possible to thread a protein primary sequence onto a three-dimensional template and get an immediate feedback of how well the threaded protein will be accepted by the reference structure before submitting a request to build missing loops and refine side-chain packing.

VMD is another very powerful visualization program, which can also be used to view trajectories from molecular dynamics simulations. It has a broad range of representations for proteins, including molecular surfaces, electrostatic potential maps, and crystal information. Other useful programs are PyMol and MolMol.

3.2
Poisson–Boltzmann/Electrostatics Calculations

Poisson–Boltzmann theory (PB) has become an important tool for studying biomolecular systems. This type of calculation provides a view of a protein in terms of regions of positive or negative potential. PB calculations are useful for identifying interaction sites on proteins with charged ligands or other proteins. They can also be used for calculating amino acid pK_as in different environments, such as in enzyme active sites where catalytic residues often have large pK_a shifts due to their local environment. PB calculations also provide a basis for Brownian dynamics simulations (see Sect. 4).

PB is based on the Poisson equation, which describes electrostatic interactions in general, but includes implicitly the presence of ions in solution. It is assumed that at equilibrium, the distribution of mobile ions in the system can be approximated by a continuous charge density $\rho_{eq}(\mathbf{r})$, determined by the Boltzmann factor:

$$\rho_{eq}(\mathbf{r}) = \sum_i z_i e n_{0i} \exp\left[-\frac{z_i e \phi(\mathbf{r})}{kT}\right]$$

where i is an ionic species, z_i the charge of ion species i, n_{0i} a reference number density for species i, e is the unit charge, and $\rho(\mathbf{r})$ and $\phi(\mathbf{r})$ are the local equilibrium charge density and the average electrostatic potential, respectively. The electrostatic potential $\phi(\mathbf{r})$ is the solution of Poisson's equation:

$$\varepsilon_0 \nabla \cdot [\varepsilon(\mathbf{r}) \nabla \phi(\mathbf{r})] = -\rho_{eq}(\mathbf{r}) - \rho_{ex}(\mathbf{r})$$

Assuming a 1:1 electrolyte, these two equations can be combined to give:

$$\varepsilon_0 \nabla \cdot [\varepsilon(\mathbf{r}) \nabla \phi(\mathbf{r})] = 2en_0 \times \sinh\left[\frac{e\phi(\mathbf{r})}{kT}\right] - \rho_{ex}(\mathbf{r})$$

Here the mobile charges represented by ρ_{eq} are in equilibrium, and the fixed charges on the protein or membrane are represented by ρ_{ex}. This is the full nonlinear Poisson–Boltzmann equation, which can be solved numerically. It can be linearized by expanding sinh and retaining only the leading term. When there are no fixed charges and assuming a spatially homogeneous dielectric, this simplifies to:

$$\nabla^2 \phi = \kappa^2 \phi$$

with κ^{-1} is Debye screening length, given by:

$$\kappa^{-1} = \sqrt{\frac{\varepsilon_0 \varepsilon kT}{2e^2 n_0}}$$

The Debye screening length $\kappa^{-1} = 8$ Å for 150 mM salt at 298 K in water with a dielectric constant of 80. The screening of a central charge in bulk solution is about 80% at $r = 3\kappa^{-1}$, or ca. 25 Å. This distance decreases to about 9 Å for 1 M salt solution.

The Poisson–Boltzmann equation, either in its linear or nonlinear form, can be solved for macromolecules by a number of commonly used programs, including DelPhi, UHBD, and APBS.

3.3
Ligand Docking

The binding of small molecules to proteins is a key determinant of many biological processes. The ligand-docking problem is the challenge to predict where, in what orientation, and with what affinity a small molecule binds to a binding site in a protein. In the ultimate state of the art, one would like to predict all three with high accuracy, for a high computational cost if necessary. In addition, we also need methods that allow rapid (computationally inexpensive) screening of a library of millions of compounds to search for drug leads that may interact favorably with potential drug targets. Broadly speaking, ligand docking can be approached from two angles. In the first approach, compounds of known structure are docked into the active site of a protein of interest, and some form of scoring function is a measure of how likely the compound will bind tightly to the protein. In the second approach, starting from the known structure of the protein, a molecule can be designed by fitting functional groups in the active site to optimize the interactions with the protein target. The resulting "computational" molecule can then be synthesized and tested. Several classes of scoring functions exist. One class is based on a combination of interaction energies, with terms that are similar to the potential function used in molecular dynamics simulations. Simpler functions based on, for example, the number of hydrogen bonds or penalties for steric clashes are also useful and can be computationally faster. A second class tries to directly calculate the free energy of binding using a parameterized function or an approximate description of the free energy of binding derived from a potential function. A third class uses statistical information about similar molecules and general features of the binding site to rank ligand–protein complexes. At the moment a combination of methods seems to give the best results, but to our knowledge there is no consensus on which method is the best in any given case.

3.4
Protein–Protein Interactions

Protein–protein interactions are emerging as one of the most important themes in biochemistry. Some interactions are sufficiently stable that a complete complex can be crystallized, but this is relatively rare. More often, proteins interact transiently with other proteins at some point in their lives, often in critical processes such as signaling. The protein database contains mainly single proteins, because to crystallize them they are normally purified to a high degree, thus disrupting all but the strongest noncovalent interactions.

The general goal of protein–protein docking could be described as predicting correctly the structure of a protein–protein complex, given only the structures of the two independent proteins. This assumes that these structures are already known. Whether the proteins are allowed to change their structure somewhat upon binding, which seems realistic, is an additional complexity. There are many approaches in the literature, with two common themes. First, there is a remarkable degree of surface complementarity in most stable complexes. This suggests that it should be possible to draw a molecular surface on both proteins (e.g. see Fig. 1d) and search for all possible relative orientations of the two proteins for the most favorable orientation. This type of search is expensive, and the challenge is to identify the most favorable orientation among many possible orientations. Usually, some form of energy function is used, which gives more favorable values for orientations that optimize interactions such as hydrogen bonding, burying of hydrophobic exposed area, and matching up complementary electrostatic potentials on the surfaces of the proteins that form the complex. A second approach combines experimental data with energy functions to guide the two proteins to a solution that is compatible with experimental data. A recent successful example of this approach is HADDOCK, which combines a molecular mechanics energy function, a simplified form of dynamics to enable flexibility of parts of the interface, and incorporation of "ambiguous interaction restraints." These are "soft" restraints that, by themselves, are insufficient to guide docking but combine to yield more reliable solutions than energy functions alone, provided there is experimental information that can be used as restraints. As mentioned above, progress in the area of protein–protein docking is monitored in the CAPRI challenge (http://capri.ebi.ac.uk).

4
Simulations of Protein Dynamics

Three common simulation methods used to study protein dynamics are explained here.

4.1
Molecular Dynamics Simulation

The most common atomistic simulation technique is molecular dynamics (MD). In MD simulations, the interactions between all atoms in the system are described by empirical potentials. An example of a frequently used potential function is:

$$V(\mathbf{r}^N) = \sum_{\text{bonds}} \frac{k_i}{2}(l_i - l_{i,0})^2$$

$$+ \sum_{\text{angles}} \frac{k_i}{2}(\theta_i - \theta_{i,0})^2$$

$$+ \sum_{\text{torsions}} \frac{V_n}{2}(1 + \cos(n\omega - \gamma))$$

$$+ \sum_{i=1}^{N} \sum_{j=i+1}^{N} \left(4\varepsilon_{ij} \left[\left(\frac{\sigma_{ij}}{r_{ij}}\right)^{12} - \left(\frac{\sigma_{ij}}{r_{ij}}\right)^{6} \right] \right.$$
$$\left. + \frac{q_i q_j}{4\pi \varepsilon_0 r_{ij}} \right)$$

This potential function contains harmonic terms for bonds and angles, a cosine expansion for torsion angles, and Lennard–Jones and Coulomb interactions for nonbonded interactions. The constants k_i are harmonic force constants, l_i, is the current bond length, $l_{i,0}$ the reference bond length, θ_i the current angle, $\theta_{i,0}$ the reference angle, V_n, n and γ are the barrier height, multiplicity, and off-set from the origin for the cosine function used to describe dihedral angles (rotations around a central bond), ε and σ are Lennard–Jones parameters (a different pair for each possible combination of two different atom types), q_i and q_j are (partial) atomic charges, and r_{ij} is the distance between two atoms, and ε_0 (different than ε in the Lennard–Jones potential) is the dielectric constant of the medium.

Using this potential function, the forces (the derivative of the potential with respect to position) on all atoms in the system of interest are calculated and used to solve classical equations of motions to generate a trajectory of all atoms in time. An example of a system studied with MD is shown in Fig. 3: a simulation snapshot of the ABC transporter BtuCD in a realistic environment consisting of lipids, water, and ions.

The primary result of the simulation is a trajectory of all atoms in time, from which specific details of the system can be analyzed. This is an exciting idea, because atoms can be followed as they move in real time on a timescale of up to ca. 100 ns, although longer simulations have also been reported. In principle, any properties that depend on coordinates, velocities, or forces can be calculated, given sufficient simulation time. No assumptions are required about the nature of the solvent, there is no need to choose dielectric boundaries because all atoms are explicitly present, and in principle all interactions (water–ions,

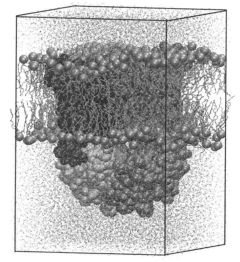

Fig. 3 An orthographic view of the periodic box for the ATP-bound BtuCD simulation, showing water (red and white), lipid with phosphorus atoms enlarged, and the protein. The transporter consists of two transmembrane domains (blue and purple) and two nucleotide binding domains (orange and ochre). The two docked MgATP molecules are partially visible (green and red). Figure courtesy of E. Oloo. See Oloo, E.O., Tieleman, D.P. (2004) Conformational transitions induced by the binding of MgATP to the vitamin B$_{12}$ ABC-transporter BtuCD, *J. Biol. Chem.* **279**, 45013–45019 for more details. Rendered with VMD. (See color plate p. xxx.)

water–protein, water–lipid, lipid–protein etc.) are incorporated.

The main limitations of MD are its computational cost, the limited time and length scale that can be treated in a simulation, and technical limitations such as the accuracy of current empirical force fields.

4.2
Free-energy Calculations

Several useful extensions of the basic simulation method make it possible to calculate free-energy differences from simulations. If a reaction coordinate can be identified, processes that are orders of magnitude slower can be studied than would be possible by direct simulation. Such a reaction coordinate could be a concerted conformational change, or a pathway for ion permeation in ion channels. In umbrella sampling, a biasing potential is used to restrict a simulated system to sample phase space within a specified region (called a *window*). By placing windows along the reaction coordinate, one can generate a free-energy profile (also called *potential of mean force*), which will quantitatively describe why one region of space is more favorable than another.

Relative free energies of different side-chain mutations in a protein can be studied using free-energy perturbation by slowly changing one side chain into another within the computer (also known as computational alchemy). This makes it possible to investigate in the computer the effect of mutations on the stability of a protein or on, for example, the binding affinity of a ligand. The theory behind this kind of calculations is well developed and the algorithms have been implemented in many molecular dynamics software packages.

4.3
Brownian Dynamics Calculations

In Brownian dynamics (BD) simulations, the trajectories of individual particles (ions, molecules) are calculated using the Langevin equation:

$$m_i \frac{d\mathbf{v}_i}{dt} = -\gamma_i \mathbf{v}_i + \mathbf{F}_R + \mathbf{F}_i$$

where m_i, and $\mathbf{v}_i$ are the mass and velocity of atom i. Water molecules are not included explicitly, but are present implicitly in the form of a friction coefficient $\gamma = kT/D_i m_i$ (where k is the Boltzmann constant, T is temperature, and D is the diffusion coefficient) and a stochastic force $\mathbf{F}_R$ arising from random collisions of water molecules with ions, obeying the fluctuation-dissipation theorem. $\mathbf{F}_i$ is the force due to other particles in the system as well as external sources, such as an applied electric field. When the friction is large and the motions are overdamped, the inertial term $m_i d\mathbf{v}/dt$ may be neglected, and the simplified form

$$\mathbf{v}_i = \frac{D_i}{kT}\mathbf{F}_i + \mathbf{F}_R$$

may be used. This is the approximation made in Brownian dynamics.

BD simulations require only a few input parameters: in its simplest form, only the diffusion coefficients of the different species of ions and the charge on the ions are needed. However, the model can be refined. To study a protein such as an ion channel, the residues of the protein can be modeled as a set of partial charges, and some form of interaction potential between the mobile ions and the protein must be specified (as seen earlier).

The result of BD simulations is a large set of trajectories for diffusion particles such as ions, proteins, or ligands. By averaging over these trajectories macroscopic properties such as the conductance of ion channels or association rate of protein–protein or protein–ligand encounters can be calculated. In addition, the simulations yield molecular details of the paths through space for the diffusing particles.

Although Brownian dynamics simulations are conceptually simple, replacing solvent by a continuum description and ignoring internal conformational changes in proteins are significant assumptions that require careful consideration. When BD simulations are valid, they are a very powerful method to study biological processes on a timescale that is much longer than can be reached by MD.

5
Example Applications

The literature is now very extensive, and hundreds of papers are published that are based on each of the methods described above. For illustration purposes, we give a few examples of biochemical problems that have been addressed by modeling and simulation methods.

5.1
Acetylcholinesterase

Acetylcholinesterase (AChE) is an extremely fast enzyme that hydrolyzes the neurotransmitter acetylcholine to terminate signaling in cholinergic synapses. It has been studied in great detail by a number of computational methods, including molecular dynamics simulations, continuum electrostatics calculations and Brownian dynamics simulations.

One of the first puzzles posed by the crystal structure of AChE is the question of how the substrate acetylcholine gains access to the active site, since in the crystal structure there is no unobstructed pathway to the active site. This original observation for AChE from the fish *Torpedo californica* has subsequently been reiterated in mouse and human AChE. Simulations have shown how breathing motions in the enzyme facilitate the displacement of substrate from the surface of the enzyme to the buried active site. These motions appear quite complex and spatially extensive, which suggests possible modes of regulation of the activity of the enzyme. Such a mechanism has been observed in other proteins, including hemoglobin and myoglobin, but the fast reaction rates of AChE are hard to reconcile with major structural rearrangements. MD simulations suggest that the primary point of access opens and closes on a timescale that is fast enough not to slow down the entrance of the substrate substantially. Interestingly, the protein appears to have several secondary channels that allow water to enter and leave the active site as the substrate enters.

The reaction rate of AChE is limited by the diffusion of the neurotransmitter acetylcholine. Brownian dynamics are a suitable method to study encounters of acetylcholine with AChE. The acetylcholine diffuses under the influence of intermolecular interactions between acetylcholine and the protein and a random force. The net effect is that the protein "guides" the acetylcholine to its active site through an electrostatic mechanism that enhances the rate of finding acetylcholine beyond the pure diffusional limit. A similar mechanism has since been identified for several fast-reacting proteins.

AChE is the target of the neurotoxin fasciculin, a peptide. Brownian dynamics of the peptide and AChE give insight into

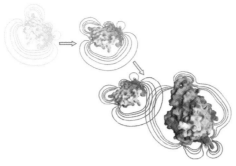

Fig. 4 Schematic illustration of the association of fasciculin (on the left) with acetylcholinesterase. Blue and red contour lines indicate regions of positive and negative electrostatic potential, respectively. Figure courtesy of D. Sept and A. Elcock. See Elcock, A.H., Gabdoulline, R.R., Wade, R.C., McCammon, J.A. (1999) Computer simulation of protein-protein association kinetics: acetylcholinesterase-fasciculin, *J. Mol. Biol.* **291**, 149–162 for more details. (See color plate p. xxiv.)

the binding kinetics of fasciculin and the structure of the resulting complex. In a series of Brownian dynamics simulations, Elcock et al. investigated the effect of mutations in the protein on association rate constants. The electrostatic interaction between AChE and fasciculin that promotes association is illustrated in Fig. 4. In its simplest form, the Brownian dynamics simulations reproduced the correct order of rate constants for different mutants, although the absolute values were too large by about a factor of 30. In a more accurate treatment of the continuum electrostatics problem, this discrepancy can be reduced greatly. Interestingly, these calculations make it possible to distinguish kinetic and thermodynamic effects, and suggest that for the case of fasciculin-AChE-binding mutations can separately affect the rate of association and the binding constant.

5.2
Water Transport Through Aquaporins

Aquaporins are ubiquitous membrane proteins that aid in the maintenance of crucial osmotic balance in cells, on whose discovery the Nobel Prize in chemistry was awarded in 2003. The road to high-resolution structures of these proteins, and to understanding with atomic detail how they function, is a nice example of the link between experiment, modeling, and simulations.

In early 2000, De Groot and coworkers used 4.5-Å resolution data, insufficient to identify the fold of the protein directly, to predict the fold of aquaporin based on the constraints of helix packing, atomic force microscopy data, and primary sequence. From 1440 possible folds, they were able to identify a maximum of 8 possible folds. Later that year, a 3.8-Å resolution structure of aquaporin from cryoelectron microscopy became available, as well as a 2.2-Å X-ray crystal structure for a homolog of aquaporin, GlpF, the glycerol transport facilitator. The fold of aquaporin is illustrated in Fig. 5(a). It has six transmembrane helices in an "hourglass" arrangement. The protein pore is shaped by two reentrant loops on either side of the membrane that have a conserved coil–NPA–helix motif, in which the Asn-Pro-Ala (NPA) residues of these loops meet in the middle of the channel. The partial helices of this motif are labeled HB and HE. The aromatic rich constriction region (ar/R), labeled in Fig. 5(a), is suspected of contributing to aquaporin's highly specific behavior.

Direct simulation of the AQP structure obtained from cryoelectron microscopy

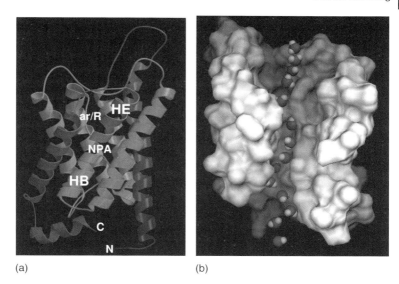

Fig. 5 (a) Ribbon diagram of the aquaporin monomer and (b) spacefilling representation illustrating a water file through the channel pore. (a) Starting at the N-terminus of the monomer, there are two transmembrane helices, followed by the coil–NPA–helix motif. A third transmembrane helix completes the first half of the protein. The second half of the monomer has two transmembrane helices, followed by the coil–NPA–helix motif, followed by another transmembrane helix. (b) The path of the water channel is through the core of the protein, following the path of the coil motifs, which meet at the NPA signature. This is more clearly illustrated in 5(b). The protein is rendered as a molecular surface in white, with the front surface of the protein cut away so that we can clearly see the pore in the middle of the protein. Surfaces colored yellow are those that interact most strongly with passing water molecules. The waterfile displayed is an overlay of a number of snapshots from the 10 ns simulation. The dipole inversion of water at the NPA motif is clearly illustrated here. Figure courtesy of B. de Groot. See De Groot, B.L., Grubmuller, H. (2001) Water permeation across biological membranes: mechanism and dynamics of aquaporin-1 and GlpF, *Science* **294**, 2353–2357. De Groot, B.L., Frigato, T., Helms, V., Grubmuller, H. (2003) The mechanism of proton exclusion in the aquaporin-1 water channel, *J. Mol. Biol.* **333**, 279–293 for more details. (See color plate p. xxxi.)

showed a discontinuous water file through the channel and obvious defects in the structure, particularly at the NPA motifs, because of instability observed through the MD simulations. Further homology modeling between the available structures yielded improved stability in the core of the channel. Simulations of the high-resolution GlpF structure, and the improved homology model of AQP1 both observed continuous water files diffusing through the water channel, illustrated in Fig. 5(b), at rates comparable to those observed experimentally.

In late 2001, a 2.2-Å resolution X-ray crystal structure was obtained of human aquaporin-1. With confidence in the three-dimensional structure, the simulation studies that followed were able to focus on measuring physical properties of the channel using classical simulation methods, and also tried to pinpoint the origin of aquaporin's ability to transport water at high diffusion rates, while restricting proton transport. The basis of proton transport is a topic better addressed by quantum mechanical simulations, and so combinations of classical dynamics methods with semiempirical methods have been employed by a number of groups to determine the basis for proton exclusion by these channels.

What is particularly exciting about aquaporin as a case study is that the biological function of this protein, to act as a water channel across the cell membrane, occurs at a rate which is accessible to study by the current state of the art in molecular dynamics simulations. Through simulations, researchers were able to point out deficiencies in the low-resolution crystal structures on the basis of structural instability, and because the observed behavior *in silico* did not agree with that measured experimentally. This observation helped to point out the significance of certain structural motifs in the protein (i.e. the reentrant loops with the NPA motif). The evolution of our understanding of the aquaporin structure through experiments, homology modeling, and simulations has lead to similar methods being applied to other aquaporins with yet unsolved structures.

5.3
ABC Transporters and Multidrug Resistance Proteins

Many human proteins of major medical interest are difficult to obtain in large enough quantities for structural studies. The only exceptions have been proteins that occur in large amounts, such as bovine lens protein rhodopsin or Aqp1 from blood (as seen earlier). An emerging theme in membrane protein structural biology over the past years has been the use of bacterial homologs of human proteins, with, as striking examples, several potassium channel structures and several structures of ABC transporters. ATP-binding cassette (ABC) transporters are modular mechanical machines that couple the hydrolysis of ATP with the transport of molecules across membranes. They consist of two nucleotide binding domains (NBDs), two transmembrane domains, and optional additional domains, organized in a varying number of polypeptide chains. Mutations in the genes encoding many of the 48 ABC transporters of human cells are associated with several diseases, including cystic fibrosis. Increased expression of certain ABC transporters is a major cause of resistance to peptide antibiotics, antifungals, herbicides, anticancer drugs, and other cytotoxic agents. Interestingly, ABC transporters probably form the largest group of homologous proteins and exist in all species. In an exciting recent development, three crystal structures of ABC transporters have been determined: MsbA from *Escherichia coli* at 4.5-Å resolution, MsbA from *Vibrio cholera* at 3.8-Å resolution, and BtuCD from *E. coli* at 3.2-Å resolution. In addition, about 10 structures of nucleotide binding domains (NBDs) have been solved as well as ATP binding domains from other proteins. Despite this impressive progress in structural studies, several of the key questions about the basic mechanism of action of ABC transporters are unresolved. Simulations and homology modeling can be used to make some progress toward understanding the dynamics of ABC transporters and to translating the bacterial structures to human homologs like P-glycoprotein.

Molecular dynamics simulation studies have been used to investigate the "real-time" dynamics of the vitamin B_{12} importer BtuCD from *E. coli*, based on the single snapshot captured in the crystal structure. An example of a simulation model that incorporates BtuCD is shown in Fig. 3. In this model, the crystal structure is incorporated in a realistic environment of phospholipids, water, and ions, and simulated for 20 ns, both in the absence and presence of MgATP in the two binding sites. The results demonstrate that the docking of ATP to the catalytic pockets progressively draws the two cytoplasmic nucleotide-binding cassettes toward each other. Movement of the cassettes into closer opposition in turn induces conformational rearrangement of α-helices in the transmembrane domain. The shape of the translocation pathway consequently changes in a manner that could aid the vectorial movement of vitamin B_{12}. These results suggest that ATP binding may indeed represent the power stroke in the catalytic mechanism. Moreover, occlusion of ATP at one catalytic site is mechanically coupled to opening of the nucleotide-binding pocket at the second site. This may indicate that asymmetric behavior at the two catalytic pockets forms the structural basis by which the transporter is able to alternate ATP hydrolysis from one site to the other. While this remains to be tested, this study is an example of the use of dynamics simulations that build on a single crystal structure to obtain more information about the full process the protein is involved in.

The multidrug resistance P-glycoprotein mediates the extrusion of chemotherapeutic drugs from cancer cells. Characterization of the drug binding and ATPase activities of the protein have made it the paradigm ATP binding cassette (ABC) transporter. Although no high-resolution structure is known, P-glycoprotein has been imaged at low resolution by electron cryomicroscopy and extensively analyzed by disulfide cross-linking. Stenham et al. used an interesting combination of approaches to create a molecular model of P-glycoprotein (shown in Fig. 6) that fits with most experimental data from cross-linking and imaging, helps interpret these experiments, and provides insight into possible mechanisms of drug transport. As described earlier, the only ABC transporter structures whose high-resolution structures are known are MsbA and BtuCD. There is no homology between the transmembrane domains of BtuCD and P-glycoprotein, but the homology between MsbA and P-glycoprotein is 23% sequence identity for the transmembrane domains and 51% sequence identity for the nucleotide binding domains. MsbA is the closest known bacterial relative of P-glycoprotein. Unfortunately, the nucleotide binding domains in MsbA are poorly resolved and the structure has a low resolution. In contrast, the nucleotide binding domains in BtuCD are well resolved. The situation is further complicated by the tertiary and quaternary organization of the domains of MsbA, which is not consistent with either the BtuCD crystal or the extensive cross-linking data already published for P-gp. Nevertheless, Stenham et al. have shown that it is possible to generate atomic scale models of P-gp by combining experimental and theoretical methods. Their homology models are based on an MsbA template but with a tertiary organization that reflects the increasingly accepted consensus NBD dimer interface. This model will be useful in designing and interpreting experimental work on P-gp. In turn, results

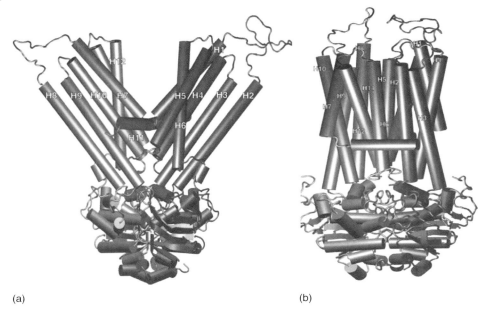

(a)   (b)

Fig. 6 (a) Construction of a molecular model for P-glycoprotein (P-gp). Each half of P-gp was modeled by homology to the crystal structure of MsbA (PDB code 1JSQ), which had been extended to a full atom representation. The two halves of P-gp were assembled such that the NBDs adopt the ATP-dependent orientation observed in MJ0796 (PDB code 1L2T). The constituent domains are individually colored. (b) Reconciliation of cross-linking data with a P-gp model. P-gp-Model-B was generated by rotation of each NBD with respect to its cognate TMD. The final model contains a parallel TMD:TMD interface (blue and gold subunits) and a consensus NBD:NBD interface (green and purple subunits). Adapted from Stenham, D.R., Campbell, J.D., Sansom, M.S.P., Higgins, C.F., Kerr, I.D., Linton, K.J. (2003) An atomic detail model for the human ATP binding cassette transporter P-glycoprotein derived from disulphide cross-linking and homology modeling, *FASEB J.* **17**, 2287–2289. (See color plate p. xxxii.)

from experimental work can be used directly to improve the model, driving an iterative process that may advance our understanding of medically important ABC transporters.

6
Perspectives

Molecular modeling methods have become standard techniques to address biochemical problems at the level of biomolecular structure. A number of factors have contributed to this success. Computers themselves have become vastly more powerful in the past decades. Initially, simulations were restricted to very simple systems on timescales and levels of detail that were typically not adequate to answer specific questions about proteins. With a 10-fold increase in speed every 5 years a wide range of problems is now accessible directly by computer simulation, such as water transport in aquaporins. The fastest folding proteins can now be studied directly by simulations on the same timescale as the fastest experiments, with unprecedented progress in our understanding of protein folding. Computer graphics hardware has enjoyed a similar increase in speed, enabling complex surface representations of large biomolecular complexes and interactive 3D graphics. A second factor is the improvement in simulation and modeling software, which has led to the development of software that is vastly more user friendly than even a few years ago and runs on commodity hardware instead of exotic (expensive) workstations. A third factor is the improvement in algorithms and parameters that describe interactions between atoms or predict the structure of proteins. With increasing accuracy or more efficient algorithms, molecular modeling is able to answer increasingly more detailed and complicated questions. Finally, the massive advances in experimental studies at the level of biomolecular structure feed back into computer modeling. The expanding database of high-resolution protein structures is a key resource for further development of modeling methods. Advanced single-molecule experiments provide a direct link with simulations that was very rare in the past, and an increasing number of experimental techniques can now be linked to simulation results directly. An example of this is data from nuclear magnetic resonance spectroscopy, which provides both structural and dynamical data on timescales that are comparable to those of simulations.

Because molecular modeling is such a broad area, it is difficult to indicate exactly where most progress can be expected. Technical advances in hardware and software engineering, algorithm and parameter development will continue to increase the number of problems that can be addressed by molecular modeling. These advances will also facilitate incorporation of computationally intense calculations such as QM and QM/MM methods into accessible programs that are user-friendlier to the molecular modeler. The use of Grid technology, distributed computing, and massively parallel commodity-based Beowulf clusters will bring CPU-intensive calculations within reach of an increasing number of researchers, allowing us to study bigger and more complex systems. We expect that important areas of growth will be a stronger link between modeling and experiments, the routine incorporation of modeling and simulation in otherwise mostly experimental studies, and improved methods to study protein–protein interactions and chemical reactions.

Acknowledgment

Tieleman is a Scholar of the Alberta Heritage Foundation for Medical Research and a Sloan Foundation Fellow. Work in his laboratory is supported by the National Sciences and Engineering Research Council and the Canadian Institutes of Health Research.

See also Mass Spectrometry of Proteins (Proteomics); Protein NMR Spectroscopy; Protein Structure Analysis: High-throughput Approaches; Proteomics; RNA Three-Dimensional Structures, Computer Modeling of.

Bibliography

Books and Reviews

Bourne P.E., Weissig, H. (Eds.) (2003) *Structural Bioinformatics*, John Wiley & Sons Inc., New Jersey, MI.

Day, R., Daggett, V. (2003) All-atom Simulations of Protein Folding and Unfolding, in: Eisenberg, D., Kim, P. (Eds.) *Advances in Protein Chemistry*, Vol. 66, Elsevier Academic Press, New York, pp. 373–403.

Frenkel, D. Smit, B. (2001) *Understanding Molecular Simulation. From Algorithms to Applications*, 2nd edition, Academic Press.

Honig, B., Nicholls, A. (1995) Classical electrostatics in biology and chemistry, *Science* **268**, 1144–1149.

Jorgensen, W.L. (2004) The many roles of computation in drug discovery, *Science* **303**, 1813–1818.

Karplus, M., McCammon, J.A. (2002) Molecular dynamics simulations of biomolecules, *Nat. Struct. Biol.* **9**(9), 646–652.

Leach, A.R. (2001) *Molecular Modeling, Principles and Applications*, 2nd edition, Pearson Education Limited, England.

Petsko, G.A., Ringe, D. (2004) *Primers in Biology: Protein Structure and Function*, New Science Press Ltd.

Schlick, T. (2002) *Molecular Modeling and Simulation – An Interdisciplinary Guide*, Springer, New York.

Primary Literature

Agre, P. (2004). Aquaporin Water Channels (Nobel Lecture). *Angew. Chem. Int. Ed.* **43**, 4278–4290.

Altschul, S.F., Madden, T.L., Schaeffer, A.A., Zhang, J., Zhang, Z., Miller, W., Lipman, D.J. (1997) Gapped BLAST and PSI-BLAST: a new generation of protein database search programs, *Nucleic Acid Res.* **25**, 3389–3402.

Ash, W.L., Zlomislic, M.R., Oloo, E.O., Tieleman, D.P. (2004) Computer simulations of membrane proteins, *Biochim. Biophys. Acta-Biomembranes* **1666**, 158–189.

Baker, N.A., McCammon, J.A. (2003) Electrostatic Interactions, in: Bourner, P.E., Weissig, H. (Eds.) *Structural Bioinformatics*, Wiley & Sons, Inc., New Jersey, MI.

Baker, D., Sali, A. (2001) Protein structure prediction and structural genomics, *Science* **294**, 93–96.

Baker, N.A., Sept, D., Joseph, S., Holst, M.J., McCammon, J.A. (2001) Electrostatics of nanosystems: application to microtubules and the ribosome, *Proc. Natl. Acad. Sci. U.S.A.* **98**, 10037–10041.

Berendsen, H.J.C. (2001) Bioinformatics – Reality simulation – Observe while it happens, *Science* **294**, 2304–2305.

Berman, H.M., Westbrook, J., Feng, Z., Gilliland, G., Bhat, T.N., Weissig, H., Shindyalov, I.N., Bourne, P.E. (2000) The protein data bank, *Nucleic Acids Res.* **28**, 235–242.

Bradley, P., Chivian, D., Meiler, J., Misura, K.M.S., Rohl, C.A., Schief, W.R., Wedemeyer, W.J., Schueler-Furman, O., Murphy, P., Schonbrun, J., Strauss, C.E.M., Baker, D. (2003) Rosetta predictions in CASP5: Successes, failures, and prospects for complete automation, *Proteins: Struct., Funct., Genet.* **53**, 457–468.

Burykin, A., Warshel, A. (2003) What really prevents proton transport through aquaporin? Charge self-energy versus proton wire proposals, *Biophys. J.* **85**, 3696–3706.

Campbell, J.D., Biggin, P.C., Baaden, M., Sansom, M.S.P. (2003) Extending the structure of an ABC transporter to atomic resolution: modeling and simulation studies of MsbA, *Biochemistry* **42**, 3666–3673.

Chakrabarti, N., Tajkhorshid, E., Roux, B., Pomes, R. (2004) Molecular basis of proton blockage in aquaporins, *Structure* **12**, 65–74.

Chance, M.R., Fiser, A., Sali, A., Pieper, U., Eswar, N., Xu, G.P., Fajardo, J.E., Radhakannan, T., Marinkovic, N. (2004) High-throughput computational and experimental techniques in structural genomics, *Genome Res.* **14**, 2145–2154.

Chang, G. (2003) Structure of MsbA from Vibrio cholera: a multidrug resistance ABC transporter homolog in a closed conformation, *J. Mol. Biol.* **330**, 419–430.

Davis, M.E., McCammon, J.A. (1990) Electrostatics in Biomolecular Structure and Dynamics, *Chem. Rev.* **90**, 509–521.

De Groot, B.L., Grubmuller, H. (2001) Water permeation across biological membranes: mechanism and dynamics of aquaporin-1 and GlpF, *Science* **294**, 2353–2357.

De Groot, B.L., Engel, A., Grubmuller, H. (2003) The structure of the aquaporin-1 water channel: a comparison between cryo-electron microscopy and x-ray crystallography, *J. Mol. Biol.* **325**, 485–493.

De Groot, B.L., Frigato, T., Helms, V., Grubmuller, H. (2003) The mechanism of proton exclusion in the aquaporin-1 water channel, *J. Mol. Biol.* **333**, 279–293.

De Groot, B.L., Heymann, J.B., Engel, A., Mitsuoka, K., Fujiyoshi, Y., Grubmuller, H. (2000) The fold of human aquaporin-1, *J. Mol. Biol.* **300**, 987–994.

Dill, K.A., Chan, H.S. (1997) From Levinthal to pathways to funnels, *Nat. Struct. Biol.* **4**, 10–19.

Dominguez, C., Boelens, R., Bonvin, A.M. (2003) HADDOCK: a protein-protein docking approach based on biochemical or biophysical information, *J. Am. Chem. Soc.* **125**, 1731–1737.

Elcock, A.H. (2004) Molecular simulations of diffusion and association in multimacromolecular systems, *Numerical Comput. Methods, Pt D* **383**, 166–198.

Elcock, A.H., Sept, D., McCammon, J.A. (2001) Computer simulation of protein-protein interactions, *J. Phys. Chem. B* **105**, 1504–1518.

Elcock, A.H., Gabdoulline, R.R., Wade, R.C., McCammon, J.A. (1999) Computer simulation of protein-protein association kinetics: acetylcholinesterase-fasciculin, *J. Mol. Biol.* **291**, 149–162.

Ermak, D.L., McCammon, J.A. (1978) Brownian dynamics with hydrodynamic interactions, *J. Chem. Phys.* **69**, 1352–1360.

Fiser, A.S., Sali, A. (2003) MODELLER: generation and refinement of homology-based protein structure models, *Macromol. Crystallogr., Pt D* **374**, 461–46.

Gabdoulline, R.R., Wade, R.C. (2001) Protein-protein association: Investigation of factors influencing association rates by brownian dynamics simulations, *J. Mol. Biol.* **306**, 1139–1155.

Godzik, A. (2003) Fold Recognition Methods, in: Bourner, P.E., Weissig, H. (Eds.) *Structural Bioinformatics*, John Wiley & Sons, Inc., New Jersey, MI, pp. 525–546.

Gues, N., Peitsch, M.C. (1997) Swiss-model and the Swiss-PDB viewer: an environment for comparative protein modeling, *Electrophoresis* **18**, 2714–2723.

Hartley, R.W. (1989) Barnase and barstar – 2 small proteins to fold and fit together, *Trends Biochem. Sci.* **14**, 450–454.

Holland, I.B., Cole, S.P.C., Kuchler, K., (Eds.) (2002). *ABC Proteins: From Bacteria to Man.* Academic Press.

Honig, B., Nicholls, A. (1995) Classical electrostatics in biology and chemistry, *Science* **268**, 1144–1149.

Humphrey, W., Dalke, A., Schulten, K. (1996) VMD – visual molecular dynamics, *J. Mol. Graph.* **14**, 33–38.

Ilan, B., Tajkhorshid, E., Schulten, K., Voth, G.A. (2004) The mechanism of proton exclusion in aquaporin channels, *Proteins* **55**, 223–228.

Janin, J., Henrick, K., Moult, J., Eyck, L.T., Sternberg, M.J., Vajda, S., Vakser, I., Wodak, S.J. (2003) CAPRI: a critical assessment of predicted interactions, *Proteins* **52**, 2–9.

Jensen, M.O., Tajkhorshid, E., Schulten, K. (2003) Electrostatic tuning of permeation and selectivity in aquaporin water channels, *Biophys. J.* **85**, 2884–2899.

Kollman, P. (1993) Free-energy calculations – applications to chemical and biochemical phenomena, *Chem. Rev.* **93**, 2395–2417.

Krieger, E., Nabuurs, S.B., Vriend, G. (2003) Homology Modeling, in: Bourner, P.E., Weissig, H. (Eds.) *Structural Bioinformatics*, Wiley & Sons, Inc, New Jersey, MI.

Kuhlman, B., Dantas, G., Ireton, G.C., Varani, G., Stoddard, B.L., Baker, D. (2003) Design of a novel globular protein fold with atomic-level accuracy, *Science* **302**, 1364–1368.

Lee, L.P., Tidor, B. (2001) Barstar is electrostatically optimized for tight binding to barnase, *Nat. Struct. Biol.* **8**, 73–76.

Levinthal, C. (1969). How to Fold Graciously. *Mossbauer Spectroscopy in Biological Systems*, University of Illinois Press, Urbana, Illinois, MN.

Locher, K.P., Lee, A.T., Rees, D.C. (2002) The E-coli BtuCD structure: a framework for ABC transporter architecture and mechanism, *Science* **296**, 1091–1098.

MacKinnon, R. (2004) Potassium channels and the atomic basis of selective ion conductance (Nobel lecture), *Angew. Chem.-Int. Ed.* **43**, 4265–4277.

Marti-Renom, M.A., Stuart, A.C., Fiser, A., Sanchez, R., Melo, F., Sali, A. (2000) Comparative protein structure modeling of genes and genomes, *Annu. Rev. Biophys. Biomol. Struct.* **29**, 291–325.

McCammon, J.A., Gelin, B.R., Karplus, M. (1977) Dynamics of folded proteins, *Nature* **267**, 585–590.

McCammon, J.A., Gelin, B.R., Karplus, M., Wolynes, P.G. (1976) Hinge-bending mode in lysozyme, *Nature* **262**, 325–326.

Oloo, E.O., Tieleman, D.P. (2004) Conformational transitions induced by the binding of MgATP to the vitamin B12 ATP-binding cassette (ABC) transporter BtuCD, *J. Biol. Chem.* **279**, 45013–45019.

Rost, B. (2003) Prediction in 1D: Secondary Structure, Membrane Helices, and Accessibility, in: Bourner, P.E., Weissig, H. (Eds.) *Structural Bioinformatics*, John Wiley & Sons, Inc, New Jersey, MI.

Rychelewski, L., Fischer, D., Elofsson, A. (2003) LiveBench-6: large-scale automated evaluation of protein structure prediction servers, *Proteins: Struct., Funct., Genet.* **53**, 542–547.

Sali, A., Blundell, T.L. (1993) Comparative protein modeling by satisfaction of spatial restraints, *J. Mol. Biol.* **234**, 779–815.

Sayle, R., Milner-White, E.J. (1995) RasMol: biomolecular graphics for all, *Trends Biochem. Sci.* **20**, 374.

Schreiber, G., Fersht, A.R. (1995) Energetics of protein-protein interactions – analysis of the barnase-barstar interface by single mutations and double mutant cycles, *J. Mol. Biol.* **248**, 478–486.

Sharp, K.A., Honig, B. (1990) Electrostatic interactions in macromolecules – theory and applications, *Annu. Rev. Biophys. Biophys. Chem.* **19**, 301–332.

Shen, T.Y., Tai, K.H., Henchman, R.H., McCammon, J.A. (2002) Molecular dynamics of acetylcholinesterase, *Acc. Chem. Res.* **35**, 332–340.

Smith, G.R., Sternberg, M.J. (2002) Prediction of protein–protein interactions by docking method, *Curr. Opin. Struct. Biol.* **12**, 28–35.

Smith, G.R., Sternberg, M.J. (2003) Evaluation of the 3D-dock protein docking suite in rounds 1 and 2 of the CAPRI blind trial, *Proteins* **52**, 74–79.

Stenham, D.R., Campbell, J.D., Sansom, M.S.P., Higgins, C.F., Kerr, I.D., Linton, K.J. (2003) An atomic detail model for the human ATP binding cassette transporter P-glycoprotein derived from disulphide cross-linking and homology modeling, *FASEB J.* **17**, 2287–2289.

Tajkhorshid, E., Nollert, P., Jensen, M.O., Miercke, L.J.W., O'Connell, J., Stroud, R.M., Schulten, K. (2002) Control of the selectivity of the aquaporin water channel family by global orientational tuning, *Science* **296**, 525–530.

Tajkhorshid, E., Aksimentiev, A., Balabin, I., Gao, M., Israelwitz, B., Phillips, J.C., Zhu, F., Schulten, K. (2003) Large Scale Simulation of Protein Mechanics and Function, in: Eisenberg, D., Kim, P. (Eds.) *Advances in Protein Chemistry*, Vol. 66, Elsevier Academic Press, New York, pp. 195–247.

Tate, J. (2003) Molecular Visualization, in: Bourner, P.E., Weissig, H. (Eds.) *Structural Bioinformatics*, Wiley & Sons, Inc, New Jersey, MI.

Taylor, R.D., Jewsbury, P.J., Essex, J.W. (2002) A review of protein-small molecule docking methods, *J. Comput. Aided Mol. Des.* **16**, 151–166.

Tieleman, D.P., Marrink, S.J., Berendsen, H.J.C. (1997) A computer perspective of membranes: molecular dynamics studies of lipid bilayer systems, *Biochim. Biophys. Acta-Rev. Biomembranes* **1331**, 235–270.

Van Gunsteren, W.F., Berendsen, H.J.C. (1990) Computer-simulation of molecular-dynamics - methodology, applications, and perspectives in chemistry, *Angew. Chem.-Int. Ed. Engl.* **29**, 992–1023.

Vaughan, C.K., Buckle, A.M., Fersht, A.R. (1999) Structural response to mutation at a protein-protein interface, *J. Mol. Biol.* **286**, 1487–1506.

Vriend, G. (1990) What if – a molecular modeling and drug design program, *J. Mol. Graph.* **8**, 52–55.

Warshel, A., Aqvist, J. (1991) Microscopic Simulations of Chemical Processes in Proteins and the Role of Electrostatic Free Energy, in: Beveridge, D.L., Lavery, R. (Eds.) *Theoretical Biochemistry and Molecular Biophysics*, Vol. 2, Adenine Press, New York, pp. 257.

Protein NMR Spectroscopy

Thomas Szyperski
State University of New York, Buffalo, NY

1	Aspects of Multidimensional Protein NMR Spectroscopy	161
2	NMR Instrumentation	165
3	Experimental Protein NMR Parameters	165
4	Recently Developed Techniques	167
5	Structure Determination	172
6	Dynamics	175
7	Hydration	176
8	Folding	177
9	Structural Genomics	177
	Bibliography	178
	Books and Reviews	178
	Primary Literature	178

Keywords

Correlation Spectroscopy
"COSY": NMR experiment devised to correlate chemical shifts via through-bond scalar nuclear spin–spin couplings.

Encyclopedia of Molecular Cell Biology and Molecular Medicine, 2nd Edition. Volume 11
Edited by Robert A. Meyers.
Copyright © 2005 Wiley-VCH Verlag GmbH & Co. KGaA, Weinheim
ISBN: 3-527-30648-X

Fourier Transformation
Mathematical transformation that provides an analysis of the angular frequencies encoded in a given function.

Nuclear Magnetic Moment
Since atomic nuclei are positively charged, the nuclear spin is proportional to a nuclear magnetic moment.

Nuclear Magnetic Resonance
Owing to their magnetic moment, atomic nuclei interact with external magnetic fields. This enables one to pursue nuclear magnetic resonance (NMR) spectroscopy.

Nuclear Overhauser Enhancement Spectroscopy
"NOESY": NMR experiment to measure through-space dipolar interactions between the magnetic moments of protons. This allows one to estimate distances between protons in proteins.

Nuclear Spin
The nuclear spin is a nonclassical angular momentum associated with atomic nuclei possessing a spin quantum number larger than zero.

Triple Resonance NMR Experiment
Experiment devised to correlate the chemical shifts of three types of nuclei, that is, ^{1}H, ^{13}C, and ^{15}N.

▪ Nuclear magnetic resonance (NMR) spectroscopy provides unique information about protein structure, dynamics, hydration, and folding in aqueous solution, and has become a pivotal biophysical technique to investigate biological macromolecules. Although the first application of NMR spectroscopy to study proteins date back to the 1960s, its widespread use for proteins was fostered only in the late 1970s and early 1980s by introduction of two-dimensional NMR spectroscopy conducted on spectrometers equipped with superconducting high-field magnets. Advances made in the subsequent 25 years until today have established NMR as an indispensable tool to study even large proteins with molecular masses above 100 kDa. These advances were due to new approaches for efficient production of stable isotope labeled proteins, novel spin relaxation, optimized and rapid sampling of NMR data collection strategies, and the advent of highly sensitive cryogenic probes used for signal detection at field strengths corresponding to 500–900 MHz proton resonance frequency. Recently, the outstanding value of NMR was demonstrated for high-throughput structure determination in the newly emerging field of structural genomics, which makes protein NMR spectroscopy also a key tool for systems biology.

1
Aspects of Multidimensional Protein NMR Spectroscopy

Nuclear magnetic resonance (NMR) spectroscopy is based on the existence of a nonclassical angular momentum, the *nuclear spin*. All atomic nuclei with a nonzero nuclear spin quantum number possess such a spin. Since nuclei are positively charged, the spin generates a colinearly oriented magnetic moment, which interacts with magnetic fields as compass needles interact with the earth's magnetic field. "Spin-$\frac{1}{2}$ nuclei," with a spin quantum number of $\frac{1}{2}$, exhibit neither electric dipole nor electric quadrupole moments. Hence, spin-$\frac{1}{2}$ nuclei are quite weakly coupled to the environment (often called the *lattice*) via their magnetic dipole moments only: these interact with random fluctuations of magnetic fields that arise from thermal motions. In proteins, the weak coupling enables the observation of coherent spatial reorientations of spins (briefly named *coherences*) about the axis of a strong external magnetic field, B_0, for time periods up to several hundred milliseconds. For a typical study, this allows accurate measurement of hundreds or even thousands of NMR parameters. This wealth of information is unrivaled among solution spectroscopic techniques and is a central reason for the success of modern protein NMR spectroscopy.

Protein NMR relies on the three spin-$\frac{1}{2}$ nuclei ^{1}H, ^{13}C, and ^{15}N. Since these nuclei possess the same spin quantum number of $\frac{1}{2}$, the absolute value of their spin angular momenta is equal. However, owing to varying charges (and charge distributions within the nuclei), the resulting magnetic moments differ greatly. ^{1}H possesses the largest magnetic moment, while the magnetic moments of ^{13}C and ^{15}N are fourfold and tenfold smaller, respectively. Since resonance frequencies scale accordingly, this has far-reaching consequences for the design of protein NMR experiments: highest sensitivity is achieved for experiments starting with proton polarization and ending with proton signal detection. Furthermore, only ^{1}H is 100% abundant in nature, while ^{13}C (natural abundance: ~1%; mainly naturally occurring is ^{12}C, a spin-0 nucleus which is "invisible" in NMR) and ^{15}N (~0.5%; mainly naturally occurring is ^{14}N, a spin-1 nucleus with an electric quadrupole moment that results in broad NMR lines) need to be enriched if a protein has to be studied with more demanding heteronuclear NMR spectroscopy based on ^{13}C and/or ^{15}N.

Radio frequency (rf) pulses created with an rf coil can induce the coherent precession of spin magnetic moments about the axis of an external magnetic field. This precession and its decay can be detected with an rf coil, and the resulting signal ("free induction decay") yields, after Fourier Transformation (FT), a one-dimensional (1D) NMR spectrum. For proteins, the resonance lines provide chemical shifts and, in favorable cases, estimates for scalar spin–spin couplings from resonance fine structures, and nuclear spin relaxation times from line widths. However, owing to the large number of nuclei, 1D NMR spectra of proteins and other biological macromolecules are very crowded. Hence, protein NMR primarily relies on two-, three-, and four-dimensional (2D, 3D, and 4D) spectral information. The high dimensional NMR spectra afford increased spectral resolution and correlate several chemical shifts in a single data set.

A 2D data set is obtained if the signals of a 1D spectrum are modulated by the evolution of an NMR parameter (usually a chemical shift) in an "indirect dimension" prior to acquisition (Fig. 1). Fourier transformation along the indirect dimension yields the 2D frequency domain spectrum. In the same fashion, 3D (4D) NMR data sets are obtained by recording many 2D (3D) data sets, which are modulated by an additional NMR parameter. As a result, we have the minimal measurement time, T_m, increasing steeply with dimensionality: acquiring 32 points in each indirect dimension (with one scan per 1D spectrum each second) yields T_m(3D) ~0.5 h, T_m(4D) ~9 h, T_m(5D) ~12 days and T_m(6D) ~1 year. This obstacle for acquiring the highest dimensional NMR spectral information has been named the *NMR sampling problem*.

Figure 2(a) shows a 2D [^{15}N,^{1}H]-correlation spectrum ("COSY"), which was recorded for a ^{15}N-labeled 21-kDa globular protein on a spectrometer operating at 900 MHz ^{1}H resonance

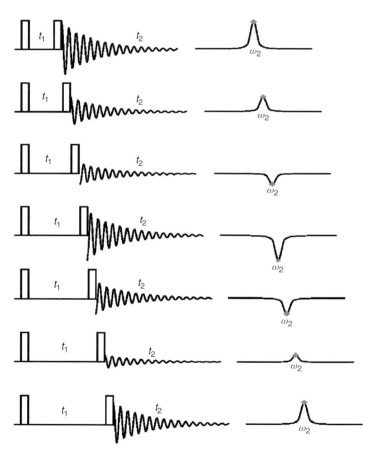

Fig. 1 Principle of 2D NMR. 1D spectra (on the right), obtained after Fourier transformation (FT) of the free induction decays along t_2, are modulated by the evolution of an NMR parameter (e.g. a chemical shift) in an indirect dimension ("t_1"). The frequency domain 2D spectrum is obtained after a second FT along t_1. Courtesy of Jeff Mills.

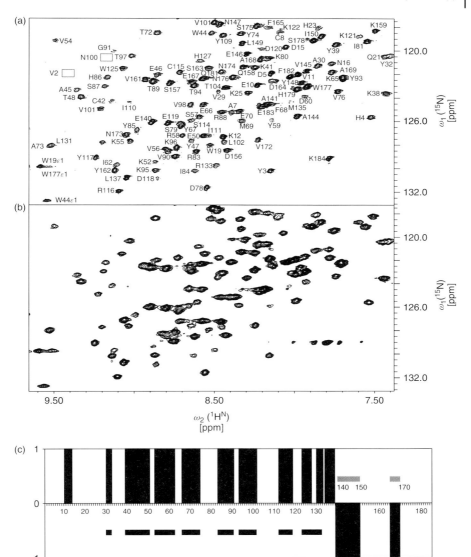

Fig. 2 (a) 2D [^{15}N,^{1}H]-COSY spectrum recorded in the transverse relaxation optimized ("TROSY") acquisition mode (see Sect. 4) on a 900-MHz Varian INOVA spectrometer (25 °C) for the uniformly ^{15}N-labeled 21-kDa protein FluA(R95K). The peaks are labeled using the one-letter code for amino acids. (b) For comparison, a 2D [^{15}N,^{1}H]-COSY spectrum recorded without "TROSY" at 600 MHz is shown. (c) Chemical shift index (CSI) consensus plot for identification of regular secondary structure elements. The eight β-strands forming a β-barrel (black bars) as well as the α-helices (gray bars) are indicated. (a and b are reproduced with permission from Liu, G., Mills, J.L., Hess, T.A., Kim, S., Skalicky, J.J., Sukumaran, D.K., Kupce, E., Skerra, A., Szyperski, T. (2003) Resonance assignments for the 21 kDa engineered fluorescein-binding lipocalin FluA, *J. Biomol. NMR* **27**, 187–188).

frequency. To generate this spectrum, 1D ^{1}H NMR spectra of amide protons are modulated with the chemical shift of the covalently attached ^{15}N nucleus, and one signal is detected for each amino acid residue (except for prolinyl residues and the N-terminal residue). Owing to high intrinsic sensitivity, one can rapidly obtain site-specific information along the entire polypeptide backbone. This feature makes 2D [^{15}N,^{1}H]-COSY (correlation spectroscopy) one of the most important experiments in modern NMR-based structural biology. First, the signal dispersion allows one to assess the degree to which the protein is folded: unfolded polypeptide segments tend to exhibit signals in a comparably narrow range attributed to "random coils." Secondly, the comparison of expected and detected number of signals enables one to investigate if parts of the protein, for example, loops connecting regular secondary structure elements, are affected by slow motional modes in the microsecond to millisecond time range. Such motions quite often lead to broadening of signals beyond detection.

Prior to using NMR spectroscopy for deriving site-specific information on structure or dynamics of proteins, the nuclear magnetic resonances must be "assigned," that is, one has to identify the chemical shifts (see Sect. 3) of spins. Since proteins contain each type of amino acid several times at various positions along the polypeptide chain, it is not sufficient to identify the type of proton (e.g. ^{1}H$^\alpha$ of glutamate) belonging to a given resonance. Instead, one has to obtain "sequence-specific" resonance assignments (e.g. ^{1}H$^\alpha$ of the glutamate *in position 79*). Obtaining (nearly complete) sequence-specific resonance assignments is generally considered a prerequisite for determining protein structures (see Sect. 5). This so-called "assignment problem" is comparable to the crystallographic "phase problem" in the sense that a structural interpretation of experimental data depends on solving these problems. Nowadays, sequential resonance assignment of ^{15}N/^{13}C-labeled proteins relies largely on recording ^{15}N/^{13}C/^{1}H triple resonance NMR experiments (Fig. 3). In these experiments, several proton, carbon, and nitrogen chemical shifts are correlated. These originate either from the same, or from two sequentially neighboring residues. Combining the information from the two experiments then allows one to effectively "walk" along the polypeptide chain to obtain sequence-specific resonance assignments for backbone (and often ^{13}C$^\beta$ shifts). Importantly, these chemical shifts can provide the location of regular secondary structure elements (Fig. 2c) without reference to nuclear Overhauser effects (NOEs). The

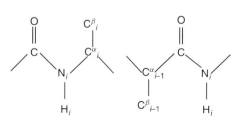

Fig. 3 Example of chemical shifts of nuclei (in color) correlated in triple resonance NMR experiments for obtaining sequence-specific backbone resonance assignments of proteins. A first experiment correlates the shifts of amide proton, nitrogen, α-carbon, and β-carbon within a given residue i (on the left). A second experiment correlates the shifts of amide proton and nitrogen of residue i with those of the α- and β-carbons of the preceding residue $i - 1$. Combining the two experiments allows one to obtain sequence-specific resonance assignments for the backbone (and ^{13}C$^\beta$ shifts) of the entire polypeptide. (See color plate p. xxix).

backbone NMR experiments are then combined with other experiments to assign resonances of side chains, thus yielding a *complete* protein resonance assignment.

2
NMR Instrumentation

Transitions between nuclear spin states are associated with radio frequency energy quanta that are, at ambient temperature, several orders of magnitude smaller than kT (k and T represent Boltzman constant and temperature). Hence, NMR transitions are *lowest energy* transitions. This results in both, the greatest strength and the greatest weakness of NMR spectroscopy. The strength is due to the fact that the low energy required for observing the system (using the nuclear spins as "spies") hardly affects even the subtlest conformational equilibria in their structural and dynamic manifestation. The weakness associated with the low interaction energy is due to the low sensitivity of NMR spectroscopy when compared with, for example, optical spectroscopy. This problem is further aggravated in protein NMR, since slowly tumbling macromolecules exhibit short transverse spin relaxation times and thus exhibit broad resonance lines. Hence, to alleviate this drawback, modern protein NMR depends on using large superconducting magnets (Fig. 4a). In 2005, the largest commercially available magnets have a price tag of about four million US dollars and can induce a ^{1}H resonance frequency of 900 MHz. Apart from increased sensitivity, which scales with $B_0^{3/2}$, high magnetic fields yield increased signal dispersion scaling with B_0^N, where N is the dimensionality of the NMR experiment. A second major breakthrough in hardware development was the construction of cryogenic NMR probes (Fig. 4b). In such probes, the rf coil for signal detection is cooled to about 25 K, which leads to a large reduction in thermal noise in the rf circuitry. In spite of the fact that the insulation between rf coil and NMR sample (which needs to be kept at ambient temperature) reduces the filling factor of the coil, 2 to 3 fold gains in sensitivity can be routinely achieved in protein NMR. This corresponds to about 5 to 10 fold reduced NMR data collection times, so that rapid NMR data collection techniques (see Sect. 4) are best suited to take advantage of the high sensitivity of such probes.

3
Experimental Protein NMR Parameters

Protein NMR spectroscopy aims at measurement of parameters, which provide the desired structural and/or dynamic information of a system under consideration. *Chemical shifts* are due to the site-specific shielding of the external magnetic field at the location of the nucleus. Hence, chemical shifts depend on the covalent and conformational environment in which a spin is embedded and are affected by long-range electrostatic and ring current effects. The dispersion of chemical shifts arising from the folding of a polypeptide into a tertiary structure is pivotal for using NMR: without such conformation-dependent dispersion, one could not resolve the site-specific information. Chemical shifts are routinely measured in multidimensional NMR spectra (see Sect. 1), and the resonance assignment is quite generally considered a prerequisite for the site-specific interpretation of other NMR parameters described below. Importantly, backbone and $^{13}C^\beta$ chemical shifts provide the location of the regular secondary structure elements

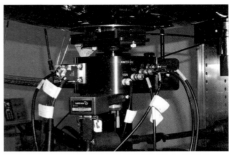

Fig. 4 Modern high-field NMR equipment used for protein NMR spectroscopy. (a) Spectrometer at the New York Structural Biology Center operating at 900 MHz ^{1}H resonance frequency (Courtesy of David Cowburn). (b) Cryogenic probe installed at the University at Buffalo high-field NMR facility (Courtesy of Dinesh Sukumaran). The probe is inserted from below in the magnet, and its rf coil is cooled to 25 K, as can be read off the monitor shown in (c).

(Fig. 2c). Moreover, chemical shifts are not isotropic, that is, the shielding of the external magnetic field depends on the orientation of the molecule relative to the field. Chemical shifts are thus accurately described by using a "chemical shift tensor."

Scalar nuclear spin–spin couplings, J, arise from the through-bond coupling of the nuclear magnetic moments. The J-coupling is the most important interaction used to devise multidimensional experiments for measurement and correlation of chemical shifts (Fig. 2a). Moreover, three-bond J-couplings are related through the "Karplus-relations" to dihedral angles and thus encode valuable structural information.

^{1}H–^{1}H NOEs arise from through-space coupling of the magnetic dipole moments of protons. Owing to the r^{-6} dependence of the dipolar coupling (with r being the distance between two protons), the measurement of NOEs in [^{1}H,^{1}H]-NOESY (nuclear Overhauser enhancement spectroscopy) allows one to estimate distances between protons: NOEs effectively constitute a "molecular ruler." Thus, having a large number of such ^{1}H–^{1}H distance constraints, one can calculate three-dimensional molecular structures (see Sect. 5).

Residual dipolar couplings (RDCs) arise if a protein is partially aligned in a dilute liquid crystalline medium. Notably, for paramagnetic metalloproteins, such alignment takes place at high magnetic fields without the presence of a liquid crystalline medium. Owing to the partial alignment, the dipolar through-space coupling between dipole moments is not completely averaged out, as in the case of molecules tumbling in an isotropic medium. RDCs encode orientational constraints, that is, they provide information regarding the orientation of the internuclear axis of the dipolarly coupled nuclei relative to the principle axes of the alignment tensor. RDC-derived orientational constraints are thus quite distinct from – and complementary to – NOE-derived distance constraints for NMR structure calculation and validation. Since RDCs depend on motional averaging, they also provide valuable information about internal motional modes.

Nuclear spin relaxation parameters, such as longitudinal/transverse spin relaxation times and heteronuclear Overhauser effects reflect dynamic features of the protein under investigation. Hence, the majority of our insights into protein dynamics, be it overall rotational reorientation or internal motions, have thus far been obtained from measurement of these parameters.

4
Recently Developed Techniques

The introduction of (1) multidimensional triple resonance NMR and heteronuclear resolved [^{1}H,^{1}H]-NOESY, and (2) efficient ^{13}C/^{15}N labeling and protein deuteration protocols until the mid 1990s, made NMR a key tool to study structure and dynamics of proteins up to about 15 to 20 kDa. During the last decade, new NMR techniques were introduced to study much larger systems and to dramatically increase NMR data collection speed.

Transverse relaxation optimized NMR spectroscopy (TROSY) takes advantage of the mutual cancellation of nuclear spin relaxation pathways in slowly tumbling macromolecules studied at high magnetic fields (Fig. 5). 2D [^{15}N,^{1}H]-TROSY (Fig. 2a), and triple resonance NMR variants thereof, are most prominent and shall be discussed. Both, the ^{15}N and the ^{1}H nucleus of a polypeptide backbone H–N moiety exhibit substantial chemical shift anisotropy (CSA). As a result, two major pathways for transverse ^{15}N and ^{1}H spin relaxation are encountered: one owing to the dipolar ^{1}H–^{15}N interaction and the other owing to the CSA. While spin relaxation due to random fluctuation of the shielding is independent of spin states, the sign of the fluctuating local B-field arising from the spatial modulation of the dipolar interaction does depend on spin states. Hence, the transverse relaxation rate of a given ^{15}N or ^{1}H spin is accelerated when both the CSA and dipolar interaction random B-field fluctuations add up, whereas the transverse rate is reduced in cases where the two random fluctuations (partially) cancel. Since short transverse relaxation times lead to broad NMR lines, this phenomenon can be directly monitored in a 2D [^{15}N,^{1}H]-COSY experiment in which the one-bond ^{1}H–^{15}N *J*-coupling ($\sim$95 Hz) places the various transitions into separate spectral regions (no "^{15}N–^{1}H" decoupling; Fig. 5). Out of the four transitions registered for a given ^{15}N–^{1}H moiety, one is broadened in both dimensions, two are broadened in one dimension but narrow in the other,

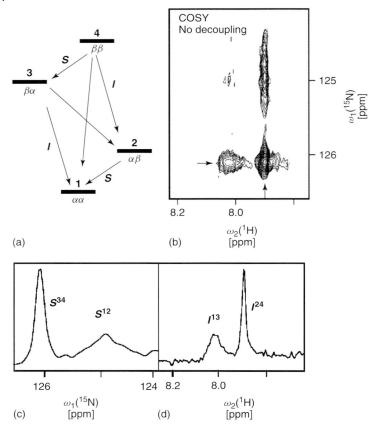

Fig. 5 (a) Energy levels of a scalarly coupled two spin-$\frac{1}{2}$ system IS (for example, $I = {}^1$H, $S = {}^{15}$N). The nuclear spin states are indicated as "α" and "β" depending on the orientation of the magnetic moments relative to the external magnetic field. (b) Cross-peak registered for a backbone N—H group in a [^{15}N,^{1}H]-COSY spectrum recorded (without decoupling of the ^{15}N—^{1}H scalar coupling interaction) for the 110-kDa protein aldolase at 20 °C. In this protein preparation, aliphatic and aromatic ^{1}H were replaced by ^{2}H ("deuterated") in order to eliminate dipolar interactions between backbone amide and carbon-bound protons. (c) and (d) 1D cross sections taken along ω_1 and ω_2 at the positions indicated in (b) by arrows. Comparison of line widths of the four components reflects the different transverse relaxation rates between the transitions shown in (a). A [^{15}N,^{1}H]-TROSY experiment relies on selecting the two transitions S^{34} and I^{24}. Reproduced with permission from Pervushin, K. (2000) Impact of transverse relaxation optimized spectroscopy (TROSY) on NMR as a technique in structural biology, *Q. Rev. Biophys.* **33**, 161–197.

and one is narrow in both dimensions. TROSY relies on selecting the latter narrow component, while discarding the others.

Since CSA-based transverse relaxation scales with B_0^2, TROSY is best performed at a magnetic field strength where the two relaxation pathways cancel (nearly) entirely. Fortunately, ^{1}H and ^{15}N CSA are of comparable magnitude (in Hz), so that about the same optimal TROSY field strength is predicted for both ^{1}H and ^{15}N transitions (corresponding to 900–1100 MHz ^{1}H resonance frequency). This makes triple resonance [^{15}N,^{1}H]-TROSY feasible. In conjunction with protein deuteration, this allows one to study systems with molecular masses well above 100 kDa, that is, up to about an order of magnitude larger than what could be approached without TROSY.

Methodology for rapid acquisition of NMR data focuses on resolving the "NMR sampling problem" (see Sect. 1), that is, on making high dimensional spectral information available at short measurement times. Currently, two broader classes of rapid data sampling approaches can be distinguished.

First, *G-matrix Fourier Transform* (GFT) NMR is a projection technique, which can also serve to reconstruct the higher-dimensionality parent NMR spectra one would obtain if no projection is applied. GFT NMR is based on the joint sampling of indirect chemical shift evolution periods that leads to signals with a multiplet fine structure, encoding several chemical shifts. A "G-matrix transformation" is applied in order to sort the components of the shift multiplets into subspectra, so that the peaks in each subspectrum encode a distinct linear combination of the jointly measured chemical shifts (Fig. 6). Hence, monitoring of chemical shifts in FT NMR

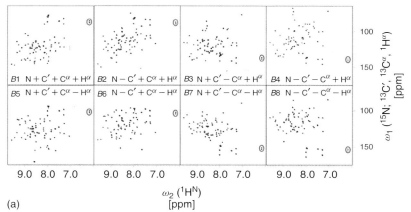

Fig. 6 15 2D planes constituting a (5,2)D HACACONHN GFT NMR experiment, which can provide the information of a 5D FT NMR experiment within less than an hour of measurement time. The type of linear combination of chemical shifts detected in a given plane is indicated. (a) The basic spectra. (b), (c) and (d) First, second, and third order central peak spectra in which linear combinations with fewer shifts are monitored to resolve assignment ambiguities. (e) Cross sections showing that signals do not broaden with an increasing number of shifts being jointly sampled. Reproduced with permission from Kim, S., Szyperski, T. (2003) GFT NMR, a new approach to rapidly obtain precise high dimensional NMR spectral information, *J. Am. Chem. Soc.* **125**, 1385–1393.

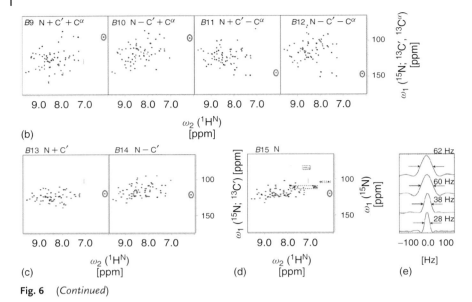

Fig. 6 (Continued)

spectroscopy is replaced in GFT NMR spectroscopy by measuring permutations of linear combinations of chemical shifts. This leads to approximately an order of magnitude reduction of measurement times *per* indirect dimension that is included in the joint sampling scheme.

Second, *single-scan acquisition* of 2D NMR spectra, named *ultrafast NMR*, was introduced. This approach constitutes an interface between high-resolution NMR and magnetic resonance imaging. The indirect chemical shift evolution is spatially encoded by use of pulse-field gradients applied along the axis of the external magnetic field B_0, and then "read out" together with the chemical shift evolution in the direct dimension in a single scan. In principle, this single-scan acquisition scheme can be extended to an arbitrary number of dimensions, that is, to the implementation of 3D and 4D single-scan acquisition. For proteins, however, reduced sensitivity currently limits this concept to the use of 2D single-scan acquisition. Importantly, ultrafast NMR is the only technique providing multidimensional NMR spectral information without sampling of indirect chemical shift evolution periods (Fig. 7). This allows one to acquire a multidimensional spectrum within the fraction of a second, and enables high-throughput acquisition of such spectra at unprecedented speed.

5
Structure Determination

NMR solution structures are usually solved in several major steps as outlined in the following. At the outset, a suitable sample (usually about 500 µL of a 1-mM protein solution) is prepared. Nowadays, both prokaryotic and eukaryotic high-yield overexpression systems serve for that purpose. If the molecular weight of a protein exceeds about 10 kDa, labeling with the NMR active spin-$\frac{1}{2}$ nuclei ^{13}C and ^{15}N is required to resolve spectral

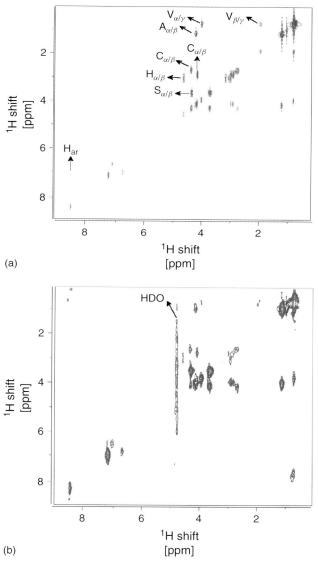

Fig. 7 Comparison of a conventionally acquired 2D [^{1}H,^{1}H] total correlation spectrum (a) acquired in 90 min for the hexapeptide CSHAVC (in the one-letter amino acid code) with the corresponding ultrafast congener recorded in 150 ms (b). In the upper panel, resonance assignments are indicated. Courtesy of Lucio Frydman.

overlap and to measure heteronuclear chemical shift correlations. For systems above about 25 kDa, one would usually consider deuterating the protein.

The sample thus obtained is used to record a set of multidimensional NMR experiments at temperatures around 30 °C. These allow (nearly) complete sequential NMR assignments to be obtained, and the conformation-dependent dispersion of the shifts enables one to derive experimental constraints for the NMR structure calculation. In most cases, structures are calculated on the basis of ^{1}H–^{1}H upper distance limit constraints derived from [^{1}H,^{1}H]-NOESY (Fig. 8).

The assignment of NOESY cross peaks and the calculation of the NMR structure is quite generally pursued in an iterative fashion. An initial set of distance constraints, which can be unambiguously obtained from the chemical shift data alone, is used to determine a low-resolution description of the protein. In turn, this initial structural model allows one to resolve the remaining NOE assignment ambiguities. This improves the precision of the structural model and allows the identification of yet additional constraints in the next round. Iterations involving structure calculations and identification of new constraints are usually pursued until (nearly) all experimentally derived constraints are in agreement with a bundle of protein conformations representing the NMR solution structure, in which conformational

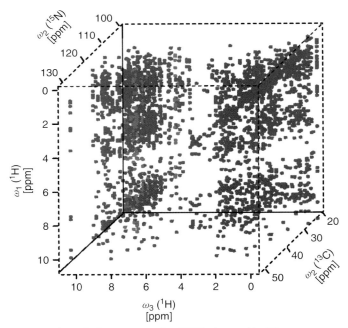

Fig. 8 Cube with dots representing NOEs detected in 3D heteronuclear resolved [^{1}H,^{1}H]-NOESY, which were used for calculating the NMR solution structure of YqfB shown in Fig. 9. Blue: signals detected along ω_3 on backbone amide protons; Red: detected on aromatic protons; Green: detected on aliphatic protons. Courtesy of Hanudatta Atreya. (See color plate p. xliv).

variations reflect the precision of the NMR structure determination. Finally, the NMR structure can be refined using molecular force fields, which in essence reflect our knowledge about conformational preferences in proteins.

In order to sample the "conformational space" allowed by experimental constraints, NMR structures are usually represented by an ensemble of conformers (typically ~10–30). Various representations (Fig. 9) are used to (1) display the structural information and (2) indicate the local and global precision of the structure determination, which is reflected by rmsd values calculated for sets of atom

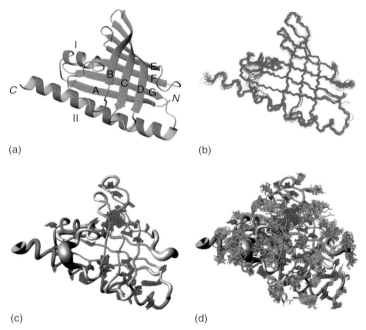

Fig. 9 Commonly used representations of NMR solution structures, exemplified for the 18-kDa Northeast Structural Genomics consortium target protein CC1736. (a) Ribbon drawing of the conformer exhibiting the smallest residual constraints violations. α-Helices and β-strands are depicted, respectively, in red/yellow, and cyan. (b) Superposition of the polypeptide backbone of the 20 conformers chosen to represent the NMR solution structure. The higher precision of the atomic coordinates of the regular secondary structure elements is apparent when compared with the loops. (c) The polypeptide backbone is represented by a spline function through the mean C^α coordinates, where the thickness represents the rmsd values for the C^α-coordinates after superposition of the regular secondary structure elements for minimal rmsd. The superpositions of the best defined side chains of the molecular core are shown in green to indicate the precision of their structural description. (d) Same as in (c), also showing the superposition of the more flexibly disordered side chains on the protein surface. The protein sample was provided by Drs. Acton and Montelione, Rutgers University. Courtesy of Yang Shen. (See color plate p. xlv).

positions for the ensemble of conformers. Additional criteria, such as the number and size of residual constraint violations or the completeness of stereo-specific resonance assignments, can be used to assess the quality of an NMR structure.

6 Dynamics

Proteins are not rigid but quite dynamic biological macromolecules. In fact, the internal mobility of proteins is of key importance for their function. Nuclear spin relaxation experiments provide unique insights into protein dynamics on the sub-nanosecond to millisecond timescale. This is because the KHz to MHz random magnetic field fluctuations associated with the motions can be monitored by measuring relaxation parameters. In particular, for ^{15}N, ^{13}C, and ^{2}H-labeled proteins, the measurement of corresponding non-proton relaxation parameters can afford a rather complete coverage of the entire protein.

In a typical protein dynamics study, longitudinal (T_1) and transverse relaxation times (T_2) and heteronuclear Overhauser effects are measured. These measurements can be complemented by use of multidimensional exchange spectroscopy, which provides information regarding very slow motional modes on the ~100 ms timescale. The first key challenge that is encountered for proper interpretation of relaxation parameters is the dissection of the impact arising from the overall rotational reorientation and the internal motional modes.

While the overall tumbling of the protein is important for "finding" an interaction partner, the internal motional modes are likely the key to linking function with dynamics. Proteins are usually not of spherical shape, so that their overall tumbling can be highly anisotropic. Any failure to accurately consider this anisotropy leads to a biased picture of the internal motions. Once the overall tumbling properties of a protein are known, the proper selection of an internal motional model remains as the second key challenge. In the so-called "model-free" approach, the internal motional mode is described by a minimal set of two parameters: an order parameter reflecting spatial motional restriction and a correlation time. Importantly, the model-free approach is limited to situations in which the internal motions occur on a faster timescale than the overall tumbling. In contrast, detailed assumptions on the nature of the internal motions are made within the framework of analytical models. These are inferred from either intuition or molecular dynamics simulations.

Comparable to reaction mechanisms in chemical kinetics, motional models can never be truly "proven" by measurement of spin relaxation parameters. Instead, one can only attest consistency between model and experimental parameters. This is because the complexity of possible spatial motional modes far exceeds the complexity that can be encoded in a rather small number of scalar relaxation parameters. In the same spirit, a more extensive body of data, for example, relaxation parameters measured for various types of nuclei at different magnetic field strengths, allows one to develop more refined multiparameter motional models. Very recently, the measurement of RDCs (see Sect. 3) has been established as a valuable complement to characterize protein dynamics.

Among the currently better characterized motional modes are low-amplitude segmental motions of loops, large-amplitude flipping of aromatic rings and disulfide bonds, as well as the relative motion of entire domains in multidomain proteins. Moreover, important progress has been made in recent years to link the investigation of protein motion with functional aspects.

7
Hydration

Proteins have evolved in aqueous solution, and their surface interacts with water molecules. Hence, protein hydration plays a key role for understanding structure, dynamics, and function of proteins. In particular, macromolecular recognition involves protein surfaces: the hydration water molecules need to be "stripped off" prior to the formation of direct contacts between biological molecules. Hence, thermodynamics and kinetics of protein hydration deserve proper consideration in structural biology. Two NMR techniques provide complementary insights into protein hydration: measurement of $^1H-^1H$ NOEs between water molecules and protein, and measurement of ^{17}O and ^2H nuclear spin relaxation times at different magnetic field strengths.

For a few selected systems, such studies have shown that more than 95% of the water molecules being in contact with the protein surface are *less than* twofold motionally retarded compared to bulk water molecules. Hence, it appears that stripping off the hydration layer can hardly constitute a rate-limiting step for biomolecular interactions. This finding has evidently far-reaching consequences for understanding the kinetic properties of biological systems. In contrast to the very mobile hydration water molecules with "lifetimes" on the picosecond timescale, water molecules with longer lifetimes on the millisecond timescale have been identified and characterized. These may well play an integral structural role for proteins.

8
Folding

The rules governing the folding of a protein from an ensemble of "random coil" state into its native, functional conformation remain a major enigma of structural biology. NMR spectroscopy has been extensively employed to investigate thermodynamics and kinetics of protein folding. Measurement of chemical shifts (see Sect. 3), spin relaxation parameters (see Sect. 7), exchange spectroscopy (see Sect. 6), as well as determination of backbone amide proton–deuteron exchange rates can provide valuable insights. The majority of NMR studies focused on a few small model systems. These studies fostered the development of intriguing theoretical concepts shaping our view of how protein folding may take place, but have as yet not lead to algorithms enabling one to predict protein structure from sequence. In fact, even the existence of a concise algorithm to predict protein structure, for example, from first physical principles, could be called in question. John von Neumann noted in the framework of his automata theory, that systems may exhibit a degree of complexity, which is so high that their (empiric) description represents the easiest approach to assess them. Such complex systems have subsequently been named *von Neumann systems*, and one may ask the question whether proteins are possibly such systems. If so, it appears that the solution of the protein folding problem through structural genomics (see Sect. 9) might be the most viable way to proceed. Notably, all methods currently available for "*ab initio*" prediction of protein structures, that is, without direct reference to an experimental template, rely on our empiric knowledge of protein structures. Recently, NMR studies have also proven the common existence of intrinsically unstructured but functional proteins, and showed that folding of proteins may take place upon binding to their physiological target molecules.

9
Structural Genomics

Since the genetic code is known, genomic DNA sequences can provide the amino acid sequence of all proteins encoded in a genome. However, since the protein folding problem (see Sect. 8) is not solved, the three-dimensional structure of the proteins cannot be inferred from sequence alone. One central aim of the new discipline of "structural genomics" is to make available at least one experimentally determined protein structure for each naturally occurring family of sequence homologs. This shall allow one to (1) homologically model the remaining members of the family using an experimental X-ray or NMR structure as a template, and (2) support the functional annotation of the gene encoding the protein. One may thus argue that structural genomics corresponds to seeking a "semi-empirical" solution of the protein folding problem.

Structural genomics relies on high-throughput structure determination

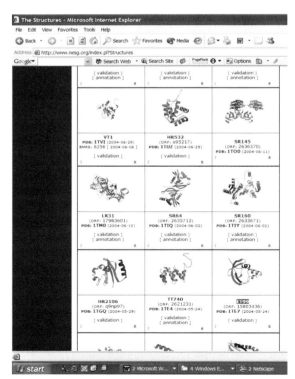

Fig. 10 "Structure gallery" accessible at the Web site of the Northeast Structural Genomics consortium in the United State (http://www.nesg.org), documenting successful implementation of a high-throughput NMR structure production pipeline.

(Fig. 10). Protein NMR spectroscopy has only very recently been shown to be a valuable technique for structural genomics (and thus also for systems biology). The ability to achieve protein structure determination (see Sect. 5) in high throughput depends primarily on: (1) sensitive high-field NMR spectrometers equipped with cryogenic probes (see Sect. 2); (2) methodology for rapid NMR data acquisition (see Sect. 4); (3) software for efficient analysis of NMR spectra and fast structure calculation; and (4) methods for automated structure quality assessment, validation, and data bank deposition. The setup of an efficient NMR-based structural genomics pipeline thus requires development of methodology, which strengthens the scientific infrastructure available for NMR-based structural biology in general.

See also Protein Modeling; Protein Structure Analysis: High-throughput Approaches; Proteomics; Structure-based Drug Design and NMR-based Screening.

Bibliography

Books and Reviews

Atreya, H., Szyperski, T. (2005) Rapid NMR data collection, *Methods Enzymol.* **394**, 78–108.

Bruschweiler, R. (2003) New approaches to the dynamic interpretation of NMR relaxation data from proteins, *Curr. Opin. Struct. Biol.* **13**, 175–183.

Cavanagh, J., Fairbrother, W.J., Palmer, A.G., Skelton, N.J. (1996) *Protein NMR Spectroscopy*, Academic Press, San Diego, CA.

Dyson, H.J., Wright, P.E. (2004) Unfolded proteins and protein folding studied by NMR, *Chem. Rev.* **104**, 3607–3622.

Ernst, R.R., Bodenhausen, G., Wokaun, A. (1987) *Principles of Nuclear Magnetic Resonance in One and Two Dimensions*, Clarendon Press, Oxford.

Montelione, G.T., Zheng, D., Huang, Y., Gunsalus, K.C., Szyperski, T. (2000) Protein NMR spectroscopy for structural genomics, *Nat. Struct. Biol.* **7**, 982–984.

Otting, G. (1997) NMR studies of water bound to biological molecules, *Prog. Nucl. Magn. Reson. Spectrom.* **31**, 259–285.

Pervushin, K. (2000) Impact of transverse relaxation optimized spectroscopy (TROSY) on NMR as a technique in structural biology, *Q. Rev. Biophys.* **33**, 161–197.

Sandström, J. (1982) *Dynamic NMR Spectroscopy*, Academic Press, London, UK.

Wüthrich, K. (1986) *NMR of Proteins and Nucleic Acids*, Wiley, New York.

Primary Literature

Akke, M., Palmer, A. (1996) Monitoring macromolecular motions on microsecond to millisecond time scales by R(1)rho-R(1) constant relaxation time NMR spectroscopy, *J. Am. Chem. Soc.* **118**, 911–912.

Arora, A., Abildgaard, F., Bushweller, J.H., Tamm, L.K. (2001) Structure of outer membrane protein A transmembrane domain by NMR spectroscopy, *Nat. Struct. Biol.* **8**, 334–338.

Balbach, J., Forge, V., Lau, W.S., van Nuland, N.A., Brew, K., Dobson, C.M. (1996) Protein folding monitored at individual residues during a two-dimensional NMR experiment, *Science* **274**, 1161–1163.

Bax, A. (2003) Weak alignment offers new NMR opportunities to study protein structure and dynamics, *Protein Sci.* **12**, 1–16.

Bruschweiler, R., Liao, X., Wright, P.E. (1995) Long-range motional restrictions in a multidomain zinc-finger protein from anisotropic tumbling, *Science* **268**, 886–889.

Cano, K.E., Thrippleton, M.J., Keeler, J., Shaka, A.J. (2004) Cascaded z-filters for efficient single-scan suppression of zero-quantum coherence, *J. Magn. Reson.* **167**, 291–297.

Castellani, F., van Rossum, B., Diehl, A., Schubert, M., Rehbein, K., Oschkinat, H. (2002) Structure of a protein determined by solid-state magic-angle-spinning NMR spectroscopy, *Nature* **420**, 98–102.

Chou, J.J., Li, H., Salvesen, G.S., Yuan, J., Wagner, G. (1999) Solution structure of BID, an intracellular amplifier of apoptotic signaling, *Cell* **96**, 615–624.

Fiaux, J., Bertelsen, E.B., Horwich, A.L., Wuthrich, K. (2002) NMR analysis of a 900K GroEL GroES complex, *Nature* **418**, 207–211.

Frydman, L., Scherf, T., Lupulescu, A. (2002) The acquisition of multidimensional NMR spectra within a single scan, *Proc. Natl. Acad. Sci. U.S.A.* **99**, 15858–15862.

Gardner, K.H., Kay, L.E. (1998) The use of ^{2}H, ^{13}C, ^{15}N multidimensional NMR to study the structure and dynamics of proteins, *Annu. Rev. Biophys. Biomol. Struct.* **27**, 357–406.

Gutmanas, A., Billeter, M. (2004) Specific DNA recognition by the Antp homeodomain: MD simulations of specific and nonspecific complexes, *Proteins* **57**, 772–782.

Hare, B.J., Wyss, D.F., Osburne, M.S., Kern, P.S., Reinherz, E.L., Wagner, G. (1999) Structure, specificity and CDR mobility of a class II restricted single-chain T-cell receptor, *Nat. Struct. Biol.* **6**, 574–581.

Kalodimos, C.G., Biris, N., Bonvin, A.M., Levandoski, M.M., Guennuegues, M., Boelens, R., Kaptein, R. (2004) Structure and flexibility adaptation in nonspecific and specific protein-DNA complexes, *Science* **305**, 386–389.

Kim, S., Szyperski, T. (2003) GFT NMR, a new approach to rapidly obtain precise high dimensional NMR spectral information, *J. Am. Chem. Soc.* **125**, 1385–1393.

Kitahara, R., Yokoyama, S., Akasaka, K. (2005) NMR snapshots of a fluctuating protein structure: ubiquitin at 30bar-3kbar, *J. Mol. Biol.* **347**, 277–285.

Korzhev, D.M., Salvatella, X., Vendruscolo, M., DiNardo, A.A., Davidson, A.R., Dobson, C.M., Kay, L.E. (2004) Low-populated folding intermediates of Fyn SH3 characterized by relaxation dispersion NMR, *Nature* **29**, 586–590.

Kupše, E., Freeman, R. (2003) Projection-reconstruction of three-dimensional NMR spectra, *J. Am. Chem. Soc.* **125**, 13958–13959.

Lipari, G., Szabo, A. (1982) Model-free approach to the interpretation of nuclear magnetic resonance relaxation in macromolecules. 1. Theory and range of validity, *J. Am. Chem. Soc.* **104**, 4546–4559.

Liu, G., Mills, J.L., Hess, T.A., Kim, S., Skalicky, J.J., Sukumaran, D.K., Kupce, E., Skerra, A., Szyperski, T. (2003) Resonance assignments for the 21 kDa engineered fluorescein-binding lipocalin FluA, *J. Biomol. NMR* **27**, 187–188.

Luginbühl, P., Pervushin, K.V., Iwai, H., Wuthrich, K. (1997) Anisotropic molecular rotational diffusion in ^{15}N spin relaxation studies of protein mobility, *Biochemistry* **36**, 7305–7312.

Markley, J.L., Bax, A., Arata, Y., Hilbers, C.W., Kaptein, R., Sykes, B.D., Wright, P.E., Wüthrich, K. (1998) Recommendations for the presentation of NMR structures of proteins and nucleic acids, *J. Mol. Biol.* **280**, 933–952.

Modig, K., Liepinsh, E., Otting, G., Halle, B. (2003) Dynamics of protein and peptide hydration, *J. Am. Chem. Soc.* **126**, 102–114.

Monleon, D., Colson, K., Moseley, H.N.B., Anklin, C., Oswald, R., Szyperski, T., Montelione, G.T. (2002) Rapid data collection and analysis of protein resonance assignments using AutoProc, AutoPeak, and AutoAssign Software, *J. Struct. Funct. Genom.* **2**, 93–101.

Montelione, G.T. (2001) Structural genomics: an approach to the protein folding problem, *Proc. Natl. Acad. Sci. U.S.A.* **98**, 13488–13489.

Mueller, T.D., Feigon, J. (2003) Structural determinants for the binding of ubiquitin-like domains to the proteasome, *EMBO J.* **22**, 4634–4645.

Orekhov, V., Pervushin, K.V., Arseniev, A.S. (1994) Backbone dynamics of (1–71)bacterioopsin studied by two-dimensional 1H-15N NMR spectroscopy, *Eur. J. Biochem.* **219**, 887–896.

Otting, G., Wüthrich, K. (1991) Protein hydration in aqueous solution, *Science* **254**, 974–980.

Perez Canadillas, J.M., Varani, G. (2003) Recognition of GU-rich polyadenylation regulatory elements by human CstF-64 protein, *EMBO J.* **22**, 2821–2830.

Pervushin, K., Riek, R., Wider, G., Wüthrich, K. (1997) Attenuated T-2 relaxation by mutual cancellation of dipole-dipole coupling and chemical shift anisotropy indicates an avenue to NMR structures of very large biological macromolecules in solution, *Proc. Natl. Acad. Sci. U.S.A.* **94**, 12366–12371.

Peti, W., Meiler, J., Bruschweiler, R., Griesinger, C. (2002) Model-free analysis of protein backbone motion from residual dipolarcouplings, *J. Am. Chem. Soc.* **124**, 5822–5833.

Powers, R., Garrett, D.S., March, C.J., Frieden, E.A., Gronenborn, A.M., Clore, G.M. (1992) Three-dimensional solution structure of human interleukin-4 by multidimensional heteronuclear magnetic resonance spectroscopy, *Science* **256**, 1673–1677.

Reif, B., Hennig, M., Griesinger, C. (1997) Direct measurement of angles between bond vectors in high-resolution NMR, *Science* **276**, 1230–1233.

Serber, Z., Dötsch, V. (2001) In-cell NMR spectroscopy, *Biochemistry* **40**, 14317–14323.

Skalicky, J.J., Sukumaran, D.K., Mills, J.L., Szyperski, T. (2000) Toward structural biology in supercooled water, *J. Am. Chem. Soc.* **122**, 3230–3231.

Szyperski, T., Luginbühl, P., Otting, G., Güntert, P., Wüthrich, K. (1993) Protein dynamics studied by rotating frame ^{15}N spin relaxation times, *J. Biomol. NMR* **3**, 151–164.

Tjandra, N., Bax, A. (1997) Direct measurement of distances and angles in biomolecules by NMR in a dilute liquid crystalline medium, *Science* **278**, 1111–1114.

Tolman, J.R., Flanagan, J.M., Kennedy, M.A., Prestegard, J.H. (1995) Nuclear magnetic dipole interactions in field-oriented proteins: information for structure determination in solution, *Proc. Natl. Acad. Sci. U.S.A.* **92**, 9279–9283.

Volkman, B.F., Lipson, D., Wemmer, D.E., Kern, D. (2001) Two-state allosteric behavior in a single-domain signalling protein, *Science* **291**, 2429–2433.

Von Neumann, J. (1966) *Theory of Self-Reproducing Automata*, University of Illinois Press, Urbana, IL.

Wüthrich, K. (2003) NMR studies of structure and function of biological macromolecules (Nobel lecture), *Angew. Chem., Int. Ed. Engl.* **42**, 3340–3363.

Yabuki, T., Kigawa, T., Dohmae, N., Takio, K., Terada, T., Ito, Y., Laue, E.D., Cooper, J.A., Kainosho, M., Yokoyama, S. (1998) Dual amino acid-selective and site-directed stable-isotope labeling of the human c-Ha-Ras protein by cell-free synthesis, *J. Biomol. NMR* **11**, 295–306.

Yee, A., Chang, X., Pineda-Lucena, A., Wu, B., Semesi, A., Le, B., Ramelot, T., Lee, G.M., Bhattacharyya, S., Gutierrez, P., Denisov, A., Lee, C.H., Cort, J.R., Kozlov, G., Liao, J., Finak, G., Chen, L., Wishart, D., Lee, W., McIntosh, L.P., Gehring, K., Kennedy, M.A., Edwards, A.M., Arrowsmith, C.H. (2002) An NMR approach to structural proteomics, *Proc. Natl. Acad. Sci. U.S.A.* **99**, 1825–1830.

Zheng, D., Aramini, J.M., Montelione, G.T. (2004) Validation of helical tilt angles in the solution NMR structure of the Z domain of Staphylococcal protein A by combined analysis of residual dipolar coupling and NOE data, *Protein Sci.* **13**, 549–554.

Protein Phosphorylation: *see* Biological Regulation by Protein Phosphorylation

Protein Purification

Richard R. Burgess
University of Wisconsin, Madison, WI, USA

1	Introduction	183
2	Types of Molecular Interactions and Variables that Affect Them	183
2.1	Hydrogen Bonds 184	
2.2	Hydrophobic Interactions 184	
2.3	Ionic Interactions 184	
2.4	Variables that Affect Molecular Forces 185	
3	Protein Properties that Can Be Used as Handles for Purification	185
3.1	Size 186	
3.2	Shape 186	
3.3	Charge 186	
3.4	Isoelectric Point 187	
3.5	Charge Distribution 187	
3.6	Hydrophobicity 187	
3.7	Solubility 188	
3.8	Density 188	
3.9	Ligand Binding 189	
3.10	Metal Binding 189	
3.11	Reversible Association 189	
3.12	Posttranslational Modifications 189	
3.13	Specific Sequence or Structure 189	
3.14	Unusual Properties 189	
3.15	Genetically Engineered Purification Handles 190	
3.16	What Can Be Learnt from the Amino Acid Sequence of a Protein that is Useful in Purification? 190	

Encyclopedia of Molecular Cell Biology and Molecular Medicine, 2nd Edition. Volume 11
Edited by Robert A. Meyers.
Copyright © 2005 Wiley-VCH Verlag GmbH & Co. KGaA, Weinheim
ISBN: 3-527-30648-X

4	Types of Separation Methods 191
5	Protein Inactivation and How to Prevent It 191
6	Protein Purification Strategy 194
7	Overproducing Recombinant Proteins 194
8	**Refolding Proteins Solubilized from Inclusion Bodies** 196
8.1	Increasing Production of Soluble Protein 196
8.2	Refolding Inclusion Bodies 196
8.3	A General Procedure for Refolding Proteins from Inclusion Bodies 197

Bibliography 197
Books and Reviews 197
Primary Literature 198

Keywords

Fractionation
The process of separating different proteins based on differences in their properties, such as size, charge, shape, solubility, and binding properties.

Inclusion Body
An insoluble form of a recombinant protein often observed when the protein is overproduced in a bacterial expression host.

Protein
A macromolecule formed by the covalent joining of amino acids to make a polypeptide. The properties of a protein are determined by the number and sequence of its amino acids, and how the polypeptide chain folds into a native structure. Proteins can function as enzymes, as structural components, or as regulators.

Protein Refolding
Protein in an inclusion body is solubilized with a denaturant and then the denaturant is removed to allow the protein to reform its native structure, determined by its primary sequence, and its biological activity.

■ Protein purification is the process of separating a given protein from all the other proteins, nucleic acids, polysaccharides, lipids, metabolites, and other small molecules in a cell extract. This purification is done ideally with reasonable speed, yield, final purity, and economy, while retaining the biological activity and chemical and structural integrity. A purification scheme involves a series of fractionation steps that exploit the many differences in the properties of the protein of interest and the other components in the mixture. Design of such fractionation steps must take into account the molecular forces that determine molecular interactions between proteins and various materials used in purification. Care must be taken to prevent inactivation, degradation or absorptive losses during the purification. In recent years, most researchers purify recombinant proteins whose genes have been cloned into an expression vector and expressed in a suitable expression host organism, often the bacterium *Escherichia coli*. The use of purification tags to aid in affinity purification is common. Sometimes, the overproduced protein is found as an insoluble inclusion body that must be isolated, solubilized by denaturants, and then refolded.

1
Introduction

The magnitude of the challenge of protein purification becomes clearer when one considers the mixture of macromolecules present in a cell extract. In addition to the protein of interest, several thousand other proteins with different properties are present in the extract, along with nucleic acids (DNA and RNA), polysaccharides, lipids, and small molecules. The proteins present in the bacterium *Escherichia coli* may be dramatically visualized after resolution by two-dimensional gel electrophoresis as shown in Fig. 1. A given protein may be present at more than 10% or at less than 0.001% of the total protein in the cell. Enzymes are found in different states and locations: soluble, insoluble, membrane bound, DNA bound, in organelles, cytoplasmic, periplasmic, and nuclear. The challenge, therefore, is to separate the protein of interest from all of the other components in the cell, especially the unwanted contaminating proteins, with reasonable efficiency, speed, yield, and purity, while retaining the biological activity and chemical integrity of the polypeptide.

Sections 1 to 6 provide background on classical protein fractionation and purification; that is, the isolation of a protein from its natural source. Sections 7 and 8 give a brief introduction to overproduction and purification of recombinant proteins cloned and overexpressed in a bacterial host expression system.

2
Types of Molecular Interactions and Variables that Affect Them

With regard to protein structure and stability and the interaction between an individual protein and other proteins, DNA, or materials used in protein purification, one must understand the molecular forces involved and how the strength of these forces varies as one varies conditions such as temperature, pH, and ionic strength of a solution. The atomic interactions that seem to be the most important with regard to protein interactions are hydrogen bonds, hydrophobic interactions, and ionic interactions. These are described briefly below.

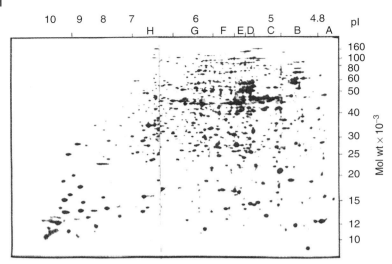

Fig. 1 *E. coli* proteins resolved on a two-dimensional gel. The approximate isoelectric point and molecular weight scales are indicated. *E. coli* K12 strain W3110 was labeled with $^{35}SO_4$ during growth in glucose minimal medium at 37 °C. A composite autoradiogram was made from nonequilibrium (*left side*) and pH 5–7 (*right side*) isoelectric focusing gels. (Adapted, with permission, from Neidhardt, F.C., Phillips, T.A. (1985) The Protein Catalog of *E. coli*, in: Celis, J.E., Bravo, R. (Eds.) *Two-Dimensional Gel Electrophoresis of Proteins*, Academic Press, New York, pp. 417–444.)

2.1
Hydrogen Bonds

Hydrogen bonds (Fig. 2) occur when a proton is shared between a proton donor (−NH and −OH) and a proton acceptor (O=C− and :N−). Optimal hydrogen bonds have a linear geometry and a distance between the donor and acceptor atoms between 2.6 and 3.1 Å. Hydrogen bonds are stronger at low temperature and are weakened as the temperature is raised.

2.2
Hydrophobic Interactions

Nonpolar residues (isoleucine, leucine, valine, phenylalanine, and tryptophan) cannot make favorable hydrogen bonds with water. In order to avoid water, they tend to come together in a so-called hydrophobic interaction (see Fig. 2) usually resulting in their being buried in the interior of a protein. Hydrophobic interactions are strengthened at high salt and high temperature.

2.3
Ionic Interactions

Ionic interactions (see Fig. 2) occur between charged molecules, with like charges repelling and opposite charges attracting. The force of the electrostatic interaction is given by an approximation of Coulomb's law, $E = Z_A Z_B e^2 / D r_{AB}$, where r_{AB} is the distance between two charges, A and B, Z_A, and Z_B are their respective number of unit charges, e is one

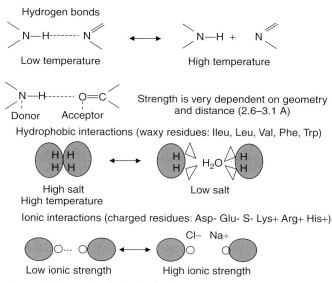

Fig. 2 Hydrogen bonds, hydrophobic interactions, ionic interactions.

unit of electronic charge, and D is the dielectric constant of the solvent. The strength of ionic interactions is therefore inversely proportional to the distance between the charges and the dielectric constant of the solvent, which varies from 2 in nonpolar solvents like hexane to 80 in highly polar solvents such as water. Ionic interactions are weakened as the ionic strength of the solvent increases and the charge is shielded by counterions. Ionic interactions are affected by the pH of the solution, since pH determines the number of charged residues.

2.4 Variables that Affect Molecular Forces

One can vary conditions to affect the relative strength of the above molecular forces. One can easily vary temperature, ionic strength, ion type, dielectric constant, and pH. In a few cases, researchers have also varied pressure.

3 Protein Properties that Can Be Used as Handles for Purification

The reason one is able to purify one protein from a mixture of thousands of proteins is that proteins vary tremendously in a number of their physical and chemical properties. These properties are the result of proteins having different numbers and sequences of amino acids. The amino acid residues attached to the polypeptide backbone may be positively or negatively charged, neutral and polar, or neutral and hydrophobic. In addition, the polypeptide is folded in a very definite secondary structure (α-helices, β-sheets, and various turns) and tertiary structure to create a unique size, shape, and distribution of residues on the surface of the protein. By exploiting the differences in properties between the protein of interest and other proteins in the mixture, one can design a rational series of fractionation steps. These properties include the following.

3.1 Size

Proteins may vary in size from peptides of a few amino acids (with molecular weights of a few hundred) to very large proteins containing over 10 000 amino acids (with molecular weights of over 1 000 000). Most proteins have molecular weights in the range 10 000 to 150 000 (see Fig. 1). Proteins that are part of multisubunit complexes may reach much larger sizes. Proteins are often fractionated on the basis of size (really on the basis of effective radius or Stokes' radius) by passing down a gel-filtration column (or size-exclusion column (SEC). The column is filled with porous beads with characteristic pore sizes. The largest proteins cannot penetrate into the bead and are excluded and elute first in what is called the void or excluded volume. Very small proteins and salts easily pass in and out of the beads and see the entire volume of the column (the column volume). Other intermediate-sized proteins elute between the void and the column volume based on how much time they spend outside and inside the beads.

3.2 Shape

Protein shapes range from approximately spherical (globular) to quite asymmetric. The shape of a protein influences its movement through a solution during centrifugation, through small pores in membranes, into beads during gel filtration, or through gels during electrophoresis. For example, consider two monomeric proteins of the same mass where one is spherical and the other is cigar shaped. During centrifugation through a glycerol gradient, the spherical protein will have a smaller Stokes' radius and, thus, will encounter less friction as it sediments through the solution. It will sediment faster, and thus, appear to be larger than the cigar-shaped protein. On the other hand, during size-exclusion chromatography, the same spherical protein with its smaller Stokes' radius will more readily diffuse into the pores of a gel-filtration bead and will elute later, thus appearing smaller than the cigar-shaped protein.

3.3 Charge

The net charge of a protein is determined by the sum of the positively and negatively charged amino acid residues. If a protein has a preponderance of aspartic and glutamic acid residues, it has a net negative charge at pH 7 and is termed an acidic protein. If it has a preponderance of lysine and arginine residues, it is considered to be a basic protein. The equilibrium between charged and uncharged groups and hence the charge of a protein is determined by the pH of the solution. The charge of the ionizable groups found on unmodified proteins as a function of pH is shown in Table 1. Generally, one used a positively charged resin (an anion-exchange column) to bind a negatively charged protein and a negatively charged resin (a cation-exchange column) to bind a positively charged protein. Bind the protein to the column at low salt (e.g. 0.1 M NaCl) and elute with an increasing salt gradient. At some stage, the ionic attraction of the protein to the column resin will become weak enough to cause the protein to dissociate from the column and elute.

Tab. 1 The charge of the ionizable groups found on unmodified proteins as a function of pH.

Ionizable group	pK_a[a]	pH 2	pH 7	pH 12
C-terminal (COOH)	4.0	0 0 0 0 0	- - - - - - -	- - - - - - - - - - - - - - - - - -
Aspartate (COOH)	4.5	0 0 0 0 0 0	- - - - - -	- - - - - - - - - - - - - - - - - -
Glutamate (COOH)	4.6	0 0 0 0 0 0	- - - - - -	- - - - - - - - - - - - - - - - - -
Histidine (imidazole)	6.2	+ + + + + + + + + +	0 0 0 0 0 0 0	0 0 0 0 0 0 0
N-terminal (amino)	7.3	+ + + + + + + + + + +	0 0 0 0 0 0	0 0 0 0 0 0
Cysteine (SH)	9.3	0 0 0 0 0 0 0 0 0 0 0 0 0 0 0 0	- - - - - - - - - -	
Tyrosine (phenol)	10.1	0 0 0 0 0 0 0 0 0 0 0 0 0 0 0 0	- - - - - - -	
Lysine (amino)	10.4	+ + + + + + + + + + + + + + + + + + +	0 0 0 0	
Arginine (guanido)	12.0	+ +	0	

[a] pK_a is the pH at which the ionizable group is half ionized. The precise pK_a value for a given ionizable group can be influenced by the immediate local environment.

3.4 Isoelectric Point

The isoelectric point (pI) is the pH at which the charge on a protein is zero and is determined by the number and titration curves of the positively and negatively charged amino acid residues on the protein. Protein pI values generally range from 4 to 10 (Fig. 1). An example of a theoretical titration curve and pI determination of E. coli RNA polymerase transcription factor, sigma32 (σ^{32}), is shown in Fig. 3.

3.5 Charge Distribution

The charged amino acid residues may be distributed uniformly on the surface of the protein or they may be clustered such that one region is highly positive while another region is highly negative. Such nonrandom charge distribution can be used to discriminate among proteins. An example is the E. coli σ^{32} protein (Fig. 3). At pH 7.9, σ^{32} has a negative charge of -46 and a positive charge of $+40$, giving a net charge of -6. It is able to bind reasonably tightly to both anion- and cation-exchange columns, apparently because its charged residues are not evenly distributed on the surface. This property can be used to purify this protein because most proteins will not bind to both types of ion-exchange columns under a single solvent condition.

3.6 Hydrophobicity

Most hydrophobic amino acid residues are buried on the inside of a protein, but some are found on the surface. The number and spatial distribution of hydrophobic amino acid residues present on the surface of the protein determine the ability of the protein to bind to hydrophobic column materials (as in hydrophobic interaction chromatography or HIC) and, therefore, can be exploited in fractionation. Generally, load a protein mixture onto a HIC column at high salt (e.g. 1 M ammonium sulfate), where hydrophobic interactions are strongest, and then elute with a decreasing salt gradient to successively elute more and more tightly bound proteins.

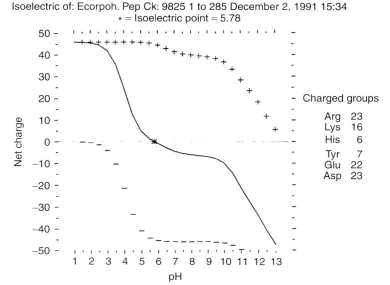

Fig. 3 Titration curve and isoelectric point (pI) of E. coli σ^{32}. This graph shows theoretical plots of the number of positively charged and negatively charged groups and the net charge as a function of pH for the E. coli RNA polymerase transcription factor σ^{32}, based on its amino acid sequence. The pI is indicated by the asterisk and is 5.78. The charged groups are Arg (23), Lys (16), His (6), Tyr (7), Glu (22), and Asp (23). This plot was generated using the Genetics Computer Group Sequence Analysis Software package.

3.7 Solubility

Proteins vary dramatically in their solubility in different solvents, all the way from being essentially insoluble ($<10\ \mu g\ mL^{-1}$) to being very soluble ($>300\ mg\ mL^{-1}$). Key variables that affect the solubility of a protein include pH, ionic strength, the nature of the ions, temperature, and the polarity of the solvent. Proteins are generally less soluble at their isoelectric point where there is less charge repulsion. Proteins are commonly fractionated by adding higher and higher concentrations of the mild salt, ammonium sulfate. Generally, the solubility of a given protein will decrease about 10-fold as the ammonium sulfate increases about 6% in saturation. (It takes about 760 gm of ammonium sulfate added to a liter of water to give a 100% saturated solution, which is about 4.1 M, at 20 °C). Since ammonium sulfate is mild and stabilizing to proteins, is relatively inexpensive and pure, and is highly soluble, it is the most common material used to fractionate proteins on the basis of solubility.

3.8 Density

The density of most proteins is between 1.3 and 1.4 g cm^{-3} and this is not generally a useful property for fractionating proteins. However, proteins containing large amounts of phosphate (e.g. phosvitin, density = 1.8) or lipid moieties (e.g. β-lipoprotein, density = 1.03) are substantially different in density compared with the average protein and may be separated from the bulk of proteins using density methods.

3.9
Ligand Binding

Many enzymes bind substrates, effector molecules, cofactors, or DNA sequences quite tightly. This binding affinity can be used to bind an enzyme to a column to which the appropriate ligand or DNA sequence has been immobilized. For example, the transcription factor AP-1 is purified by binding to a specific DNA affinity column.

3.10
Metal Binding

Many enzymes bind certain chelated metal ions (e.g. Cu^{++}, Zn^{++}, Ca^{++}, Co^{++}, and Ni^{++}) quite tightly, usually through interactions with cysteine or histidine residues. This binding can be used to bind an enzyme to a column to which the appropriate chelated metal ion has been immobilized. See Sect. 3.15 on the use of metal-chelate column for purification of protein tagged by the addition of 6 to 10 terminal histidines.

3.11
Reversible Association

Under certain solution conditions, some enzymes aggregate to form dimers, tetramers, and so on. For example, the ability of *E. coli* RNA polymerase to be a dimer under one condition (0.05 M NaCl) and a monomer under another condition (0.3 M NaCl) can be used if two fractionations based on size are carried out sequentially under those two different conditions.

3.12
Posttranslational Modifications

After protein synthesis, many proteins are modified by the addition of carbohydrates, acyl groups, phosphate groups, or a variety of other moieties. In many cases, these modifications provide handles that can be used in fractionation. For example, proteins containing carbohydrates on their surface can often be bound to columns containing plant lectins, which are molecules capable of binding tightly to certain carbohydrate moieties on glycoproteins. Phosphoproteins will in some cases bind to a chelated Fe^{++} column.

3.13
Specific Sequence or Structure

The precise geometric presentation of amino acid residues on the surface of a protein can be used as the basis of a separation procedure. For example, an antibody that recognizes only a particular site (epitope) on a protein can usually be obtained. An immunoaffinity column can be prepared by attaching a monospecific antibody (which binds only to the protein of interest) to a resin. Immunoaffinity chromatography (IAC) can result in highly selective separation and provides a very effective purification step. One can also immobilize a protein of interest and use it to specifically bind another protein out of a complex protein extract. This process is called *protein affinity chromatography*.

3.14
Unusual Properties

In addition to the types of properties mentioned above, certain proteins have unusual properties that can be exploited during their purification – an example is unusual thermostability. Most proteins unfold and coagulate or precipitate when heated to 95 °C. A protein that remains soluble and active after such heat treatment can be separated easily from the bulk of the other cellular proteins. Another such property is unusual resistance to proteases. These two properties often go hand in hand. An interesting example of a purification involving these properties is that of E. coli alkaline phosphatase. The cellular extract is heated and the insoluble coagulated proteins are removed by centrifugation. The supernatant that contains the phosphatase is then treated with a protease, which digests the remaining contaminating proteins, leaving an essentially pure preparation of alkaline phosphatase.

3.15
Genetically Engineered Purification Handles

With the advent of genetic engineering, it has become relatively easy to clone the cDNA encoding a given protein. It is then possible to construct an overproducing strain of E. coli that can be induced to produce large amounts of a desired gene product. Recently, it has become common to alter the cDNA in such a way as to add a few extra amino acids on the amino terminus or the carboxyl terminus of the protein being expressed. This added "tag" can be used as an effective purification handle. One of the most popular tags is addition of 6 to 10 histidines onto the amino terminus of a protein. One then purifies the protein by its ability to bind tightly to a column containing chelated Ni^{++} or Co^{++} in which it can be washed and then eluted with free imidazole or by lowering the pH to 5.9, where histidine becomes fully protonated and no longer binds to chelated metal.

3.16
What Can Be Learnt from the Amino Acid Sequence of a Protein that is Useful in Purification?

These days it is very common to purify a protein in which the gene has been sequenced. Thus, one can easily deduce the amino acid sequence of the corresponding protein. Can this knowledge help in designing a purification scheme? The answer is that it can be somewhat, but not very, helpful. It is easy to determine the precise molecular weight of the polypeptide chain, but not to predict whether it forms dimers or tetramer or is part of a multisubunit complex. Its charge versus pH and its isoelectric point can be determined as shown in Fig. 3, which gives some idea as to which type of ion-exchange column to use, but again this will only be useful if it is not associated with other proteins. It is possible to calculate its extinction coefficient on the basis of its tryptophan and tyrosine content, which is very useful when it is pure. It is possible to determine if it has membrane-spanning regions or it has potential modification sites. One may be able to deduce that it is a member of a larger family of proteins by sequence alignment or by the presence of conserved sequence motifs that suggest cofactor affinity. However, its shape is not known, since the three-dimensional structure from sequence cannot yet be reliably predicted. Its multisubunit features cannot be predicted. Its ammonium sulfate precipitation properties cannot be predicted. Surface features such as hydrophobic patches, charge distribution, or antigenic sites cannot be predicted. Therefore, one must conclude that protein purification is still an empirical science.

4 Types of Separation Methods

There are a large number of separation processes that can be utilized to fractionate proteins on the basis of the properties listed above. These are summarized in Table 2. The sequential use of several of these separation processes will allow the progressive purification of almost any protein. If the processes are chosen carefully, and if proper attention is paid to separation conditions, and to maintaining the stability of the protein, the purification will result in reasonable efficiency, speed, yield, and purity, while retaining the biological activity and chemical integrity of the polypeptide.

An example of a hypothetical protein fractionation scheme is shown in Fig. 4. This scheme relies on three of the most common fractionation methods, ammonium sulfate precipitation, ion-exchange chromatography, and gel-filtration or size-exclusion chromatography. The purification summary in Table 3 is a typical way of summarizing the yield and specific activity of each of the major steps in a purification scheme.

5 Protein Inactivation and How to Prevent It

A protein purification scheme will generally not be considered successful if the result is a protein that is pure, but inactive. Therefore, one of the key considerations in working with a protein is to prevent it from becoming inactivated.

Tab. 2 Separation processes that can be utilized to fractionate proteins.

Separation process	Basis of separation
Precipitation	
Ammonium sulfate	Solubility
Acetone	Solubility
Polyethyleneimine	Charge, size
Isoelectric	Solubility, pI
Phase partitioning (e.g. with polyethylene glycol)	Solubility
Chromatography	
Ion exchange (IEX)	Charge, charge distribution
Hydrophobic interaction (HIC)	Hydrophobicity
Reverse-phase HPLC	Hydrophobicity, size
Affinity	Ligand-binding site
DNA affinity	DNA-binding site
Lectin affinity	Carbohydrate content and type
Immobilized metal affinity (IMAC)	Metal binding
Immunoaffinity (IAC)	Specific antigenic site
Chromatofocusing (CF)	pI
Gel filtration/size exclusion (SEC)	Size, shape
Electrophoresis	
Gel electrophoresis (PAGE)	Charge, size, shape
Isoelectric focusing (IEF)	pI
Centrifugation	Size, shape, density
Ultrafiltration	Size, shape

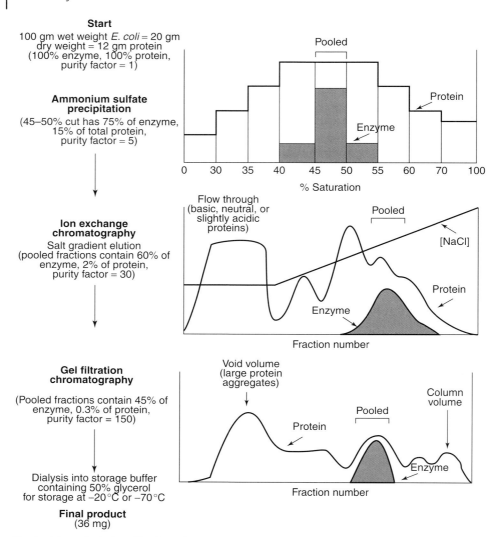

Fig. 4 A hypothetical purification scheme.

Table 4 summarizes some of the main reasons why a protein might become inactivated and what can be done to prevent this inactivation. In general, one works quickly and at low temperature (in a cold room or ice bucket) to avoid proteolytic degradation. One avoids foaming or undue exposure to oxygen and adds a reducing agent to prevent oxidation. One uses a buffer to maintain pH and a chelating agent like EDTA to protect against heavy-metal ions. Addition of 5% glycerol seems to stabilize most proteins and reduce adsorption to the walls of the container. A low salt concentration (e.g. 100 mM) helps increase solubility and prevent ionic adsorption to surfaces. Table 5 gives a good all-purpose buffer that in most

Tab. 3 Summary of hypothetical purification in Fig. 4.

Fraction	Total Protein [mg]	Total Activity [%]	Specific Activity	Step Yield [%]	Overall Yield [%]
Extract	12 000	100	=1		=100
				75	
AS pptn (45–50%)	1800	75	5		75
				80	
IEC (pooled peak)	240	60	30		60
				75	
Gel filtration (peak)	36	45	150 fold purification		45 final yield
Pure standard			150		

Tab. 4 Protein inactivation and ways of preventing it.

Reasons for inactivation	How to prevent it
Oxidation, foaming	Add DTT or TCEP, store under argon
Protease degradation	Add protease inhibitors, cooler, purer
Adsorption to container	Use polypropylene tubes, BSA carrier, glycerol nonionic detergent, protein more concentrated
Aggregation and precipitation	Store less concentrated, add salt, pH away from pI
Heavy metals	Add EDTA, cleaner tube, reagents
Temperature inactivation	Store cooler, add ligand or glycerol to stabilize
Bacterial growth	Use Tris, EDTA, azide, avoid PO_4, OAc^-
Enzymatic reaction (phosphatase)	Cooler, purer, add specific inhibitor
Dissociation of subunits/cofactors	Store more concentrated
pH changed	Avoid CO_2 in room, Tris changes pH with temperature
Inactive/misfolded conformation	Incubate at 37 °C to anneal the structure

Tab. 5 A good all-purpose buffer for keeping proteins happy.

TGED + 0.1 M NaCl

Buffer	50 mM Tris-HCl, pH 7.9 at 20 °C
Stabilizer	5% glycerol
Chelator	0.1 mM EDTA
Reducing agent	0.1 mM DTT (dithiothreitol)
Salt	0.1 M NaCl
Storage buffer – similar to above but has 50% glycerol. Will not freeze at −20 °C. Best stored at −70 °C.	

cases will keep a protein active and happy.

6
Protein Purification Strategy

How one designs a purification scheme will, in large part, determine how successful one will be in achieving the goal of a protein purification: *that of getting a high yield of highly pure and active protein in the minimal number of steps.* Achieving a high final yield requires a high recovery at each step. Four steps at 80% step yield will be $(0.8)^4 = 0.41 = 41\%$ final yield, while 4 steps at 60% step yield will give a 13% final yield. Final purity will be guided by the intended use of the protein, but it should be free of major contaminants and any traces of enzymes that interfere with the intended use. High activity will depend on maintaining the stability of the protein as discussed in Sect. 5. By choosing high-resolution fractionation steps, one can achieve a given fold purification in fewer steps. For example, if the target protein is 0.01% of the total protein in the extract, it will require a 10^4-fold purification. This can be achieved in 4 steps, each giving a 10-fold purification, or 3 steps with a 22-fold purification, or 2 steps capable of 100-fold purification. The fewer the steps the faster the preparation, the lower the protein losses the lower the cost of the purification procedure. Some of the key considerations in designing a purification procedure are (1) to have a convenient assay to follow purification; (2) choose a starting material rich in protein; (3) take precautions to minimize damage, inactivation or loss; (4) use the minimal number of steps; (5) remove the bulk of material quickly; (6) avoid unnecessary duplication, dialysis, and delay; (7) generally use fractionation steps in the order: precipitation, ion exchange, affinity, sizing; and (8) use high-resolution steps where possible.

7
Overproducing Recombinant Proteins

The advent of genetic engineering has given us the ability to routinely clone genes and overproduce their gene products. This has changed the way we think about protein purification. For most protein purifications, one no longer starts with large amounts of naturally occurring material. Instead, clone the gene of interest, insert it into a suitable expression vector, transform the vector into a suitable expression host (e.g. *E. coli*), grow the cells, induce the transcription of the gene of interest, harvest the cells, break open the cells and purify the overproduced recombinant protein. By using an expression vector where the target gene has a strong, inducible promoter, it is possible to express the target to levels as high as 20 to 40% of the total cellular protein.

The most commonly used bacterial expression system was developed by Bill Studier and colleagues at Brookhaven National Laboratory, in the late 1980s. The host strain BL21(DE3) is a derivative of *E. coli* B that is deficient in several proteases to help prevent proteolysis of the recombinant protein and that has an inducible copy of the T7 phage RNA polymerase integrated as a phage lambda lysogen in the bacterial chromosome. The T7 RNA polymerase is kept repressed by the lactose operon repressor due to the placement of a lac operator near the promoter. The gene of interest is inserted into a multicopy expression vector under the control of a T7 RNA polymerase promoter, also with a lac operator to keep it off. The vector also contains an extra copy of the lac repressor gene to enhance repression. The repressor and the presence of an additional plasmid (pLysS, that encodes a T7 lysozyme to inhibit low levels of T7 RNA polymerase activity) help keep uninduced transcription of the target gene low. When the cells have been grown to the desired cell density, the lac operon inducer, IPTG, is added to about 1 mM, causing the repressor to dissociate from the lac operators and allowing expression of the T7 RNA polymerase. The T7 RNA polymerase in turn actively transcribes the many copies of the plasmid-encoded recombinant gene and the resulting mRNA is efficiently translated into the protein of interest. The result can be the production of very large amounts of the recombinant protein, to as much as 20 to 40% of the total cell protein within about 4 h after induction. This means that it is possible to purify as much as 30 to 50 mg of recombinant protein from 1 gm of wet weight bacterial cell paste.

Several refinements have been introduced in the last few years to improve the chances that overproduction will be successful, especially if one is trying to overproduce a mammalian, plant, or archaeal protein. One problem is that human proteins often use codons that are rarely used in *E. coli*. These codons correspond to *E. coli* tRNAs that are very low in abundance in the cell. When one tries to overexpress a gene containing many of these rare codons, the result is that very little, if any, of the protein is produced. This problem was originally solved by changing the DNA sequence of the recombinant gene so that it did not contain rare codons, but it contained the preferred *E. coli* codons. A much easier and more elegant solution has now been developed and is marketed by several biotech research products companies. This involves creation of an improved host bacterial strain that has had 3 to 5 of its rare tRNAs augmented. Many poorly expressed proteins can now be expressed at very high levels. Table 6 lists this and several other problems that have been encountered in protein overexpression in *E. coli* along with how these problems have been solved or alleviated. There are a wide variety of elegantly engineered expression vectors and bacterial expression hosts available from many different biotechnology research

Tab. 6 Problems of poor recombinant protein expression and their solutions.

Problem	Solution
1. Target gene contains rare *E. coli* codons	Supplement host *E. coli* with rare tRNAs
2. Target mRNA is degraded	Use *E. coli* strain deficient in RNase E
3. Target protein is toxic to host cells	Use tighter repression, lower copy plasmid
4. Target protein is a membrane protein	Use a strain that has extra internal membranes
5. Target protein needs to form disulfide bonds	Use strain that is deficient in several key reductases, Gor, TrxB
6. Product normally forms stable heterodimers	Simultaneous coexpression of two different proteins in one *E. coli* strain

products companies worldwide. In addition to many E. coli-based expression hosts, there are also expression hosts such as: Bacillus subtilis, Pichia pastoris, Aspergillus, baculovirus/insect cells, mammalian cells, plants, and animals.

8
Refolding Proteins Solubilized from Inclusion Bodies

One of the most common problems in overexpressing a recombinant protein in E. coli is the fact that while large amounts of the protein are produced, most of it is not soluble, but is found as an insoluble inclusion body. Apparently, the newly synthesized protein, when partially folded into its native structure, exposes some hydrophobic regions and is quite sticky and prone to interaction with other partially folded proteins, leading to aggregation and inclusion body formation. There are two main approaches to dealing with inclusion bodies: (1) try to increase the proportion of the overproduced protein that is soluble; and (2) purify the inclusion body, solubilize it by dissolving it in a protein denaturant, and then refold it into its native structure.

8.1
Increasing Production of Soluble Protein

To purify the soluble material, one will want to increase as much as possible the proportion of the overproduced protein that is soluble. The most common approach is to induce the overproduction in cell growing at 20 to 25 °C. Apparently, the slower growth rate and lower temperatures results in more refolded protein and less aggregation and inclusion body formation. People have tried coexpressing cloned chaperone proteins to facilitate proper folding, but this is not common as it is only effective in some cases. An elegant recent approach, has been to grow the cells at 37 °C, shift the temperature briefly to 42 °C to induce the heat shock response, and then shift the cells to 20 °C for induction. Often if two proteins that normally form stable heterodimers are individually overexpressed they are insoluble, but if coexpressed in the same cell they form soluble, native heterodimers. Finally, many proteins remain soluble when overexpressed as genetic fusions with known proteins that readily fold to form stable native structures (protein fusion partners like NusA, GST, and TrxA).

8.2
Refolding Inclusion Bodies

If one chooses to refold protein solubilized from inclusion bodies, it is common to first wash the inclusion bodies with a nonionic detergent like TritonX-100 to solubilize membranes and break any unbroken cells. The washed inclusion body is then almost pure, and is ready to be solubilized. The real challenge is not the purification, but the refolding. The key to refolding without reaggregation and precipitation is to refold under low protein concentration. Under these conditions, the concentration of sticky, partially refolded material is lower, decreasing the opportunity for interaction and aggregation. However, for larger preparations, the large volumes become a major problem. Usually, one can find refolding conditions that give efficient refolding yields at reasonable protein concentrations.

8.3
A General Procedure for Refolding Proteins from Inclusion Bodies

A general method that the author has found to be quite effective for many proteins is given below.

a. Grow cells and induce overexpression of target protein.
b. Harvest cells, weigh cell pellet, store frozen at -80 °C
c. Break cells by sonication (pLysS cells are easy to break because of the presence of some T7 lysozyme). Otherwise adding lysozyme helps.
d. Centrifuge cell lysate, wash the IB pellet with 1% TritonX-100 to solubilize membranes and membrane proteins, then wash with buffer to remove TritonX-100.
e. Solubilize IB with a denaturant such as 6 M guanidium hydrochloride or 0.3% Sarkosyl to about 1 mg protein mL^{-1}. Difficult-to-refold proteins may need to be diluted to 0.1 mg mL^{-1} in denaturant; 8 M urea can also be used, but there is a risk of carbamylation of the protein.
f. Centrifuge out any undissolved material and slowly drip dilute the solubilized protein into 15 to 60 volumes of suitable refolding buffer. Additives or various buffer variables are often used to improve folding efficiency of a particular protein. These include 0.5 M arginine, 25% glycerol, varying the pH, temperature, presence of divalent ions, and presence of redox buffers.
g. Pass dilute refolded protein over a suitable high-resolution ion-exchange column, wash the column, and then elute with an increasing salt gradient. This step accomplishes several important things: (1) it concentrates the dilute protein; (2) it removes the denaturant; (3) it removes impurities that do not bind to the column or bind weaker or stronger than the target protein; and finally (4) it often separates refolded monomer from soluble multimers that tend to bind tighter and elute later in the gradient.
h. The resulting protein is usually fully active, homogeneous, and capable of forming crystals suitable for three-dimensional structure determination.

See also Proteomics.

Bibliography

Books and Reviews

Burgess, R.R. (1987) Protein Purification, in: Oxender, D., Fox, C.F. (Eds.) *Protein Engineering*, A.R. Liss, New York, pp. 71–82.

Coligan, J.E., Dunn, B.M., Ploegh, H.L., Speicher, D.W., Wingfield, P. (1997) *Current Protocols in Protein Science*, John Wiley & Sons, New York.

Creighton, T.E. (1993) *Proteins: Structures and Molecular Properties*, 2nd edition, W.H. Freeman, San Francisco, CA.

Ford, C.F., Suominen, I., Glatz, C.E. (1991) Fusion tails for the recovery and purification of recombinant proteins, *Protein Expr. Purif.* **2**, 95–107.

Marschak, D., Kadonaga, J., Burgess, R., Knuth, M., Brennan, W., Lin, S.-H. (1996) *Strategies for Protein Purification and Characterization: A Laboratory Manual*, Cold Spring Harbor Press, Cold Spring Harbor.

Neidhardt, F.C., Phillips, T.A. (1985) The Protein Catalog of *E. coli*, in: Celis, J.E., Bravo, R. (Eds.) *Two-Dimensional Gel Electrophoresis of Proteins*, Academic Press, New York, pp. 417–444.

Schein, C.H. (1989) Production of soluble recombinant proteins in bacteria, *Biotechnology* **7**, 1141–1149.

Scopes, R. (1994) *Protein Purification: Principles and Practice*, 3rd Edition, Springer-Verlag, NY.

Simpson, R.J. (Ed.) (2004) *Purifying Proteins for Proteomics: A Laboratory Manual*, Cold Spring Harbor Press, Cold Spring Harbor.

Primary Literature

Bessette, P.H., Aslund, F., Beckwith, J., Georgiou, G. (1999) Efficient folding of proteins with multiple disulfide bonds in the *E. coli* cytoplasm, *Proc. Natl. Acad. Sci. U.S.A.* **96**, 13703–13708.

Burgess, R.R. (1969) A new method for the large-scale purification of *E. coli* DNA-dependent RNA polymerase, *J. Biol. Chem.* **244**, 6160–6167.

Gill, S., von Hippel, P. (1989) Calculation of protein extinction coefficients from amino acid sequence data, *Biochemistry* **182**, 319–326.

Held, D., Yaeger, K., Novy, R. (2003) Coexpression vectors, *Innovations* **18**, 4–6.

Hochuli, E., Bannworth, W., Dobeli, R., Gentz, R., Studber, D. (1988) Genetic approach to facilitate purification of recombinant proteins with a novel metal chelate adsorbent, *Biotechnology* **6**, 1321–1325.

Lopez, P.J., Marchand, I., Joyce, S.A., Dreyfus, M., RNase, E. (1999) The C-terminal half of which organizes the *E. coli* degradosome, participates in mRNA degradation but not rRNA processing in vivo, *Mol. Microbiol.* **33**, 188–199.

Miroux, B., Walker, J. (1996) Over-production of proteins in *E. coli*: mutant hosts that allow synthesis of some membrane proteins and globular proteins at high levels, *J. Mol. Biol.* **260**, 289–298.

Novy, R., Drott, D., Yaeger, K., Mierendorf, R. (2001) Overcoming codon bias of *E. coli* for enhanced protein expression, *Innovations* **12**, 1–3.

Porath, J. (1992) Immobilized metal ion affinity chromatography, *Protein Expr. Purif.* **3**, 206–281.

Studier, W., Rosenberg, A., Dunn, J., Dubendorff, J. (1990) Use of T7 RNA polymerase to direct expression of cloned genes, *Methods Enzymol.* **185**, 60–89.

Protein Repertoire, Evolution of

Christine Vogel[1], Rajkumar Sasidharan[1], and Emma E. Hill[2]
[1] *MRC Laboratory of Molecular Biology, Cambridge*
[2] *University of California, Berkeley, California*

1	General Characteristics of the Protein Repertoire	203
1.1	Sequence, Structure, and Function	203
1.2	Related Proteins Form Families	205
1.3	Most Proteins Fold into Structures	205
1.4	Domains are the Units of Protein Evolution	206
1.4.1	Evolutionary Hierarchical Domain Classification (SCOP)	207
1.4.2	Alternative Domain Classifications	209
1.4.3	Implications and Complications of Domain Definitions	210
1.4.4	Domain Prediction	211
1.5	How Many Different Families, Superfamilies, and Folds Exist in Nature?	212
1.6	The Distribution of Protein and Domain Families Across Different Genomes	212
1.7	Proteins Interact	214
1.8	Proteins are Part of Pathways	215
1.9	Protein and Domain Function	216
1.9.1	Indirect Function Characteristics	216
1.9.2	Functional Annotation of Domains	217
2	Mechanisms in the Evolution of Single Proteins and the Protein Repertoire	217
2.1	Orthology and Paralogy	217
2.2	Duplication – Mechanisms to Create "More"	218
2.2.1	Horizontal Gene Transfer and Transposons	219
2.2.2	Duplication Leads to Protein and Domain Family Expansions	219
2.2.3	Reconstructing the Evolutionary History of Proteins	220
2.3	Divergence after Duplication	221
2.3.1	Changes in Sequence	221

2.3.2	Changes in Function 222
2.3.3	Structural Tolerance to Amino Acid Changes – a Case Study on Globins 223
2.3.4	Changes in Structure 224
2.4	Recombination – Mechanisms to Create "Novel Combinations" 225
2.4.1	Most Proteins Have More than One Domain 226
2.4.2	The Repertoire of Domain Combinations is Limited and Highly Biased 226
2.4.3	The Sequential Order of Domains and Their Geometry are Conserved 226
2.5	Selective and Random Processes 227
3	**Summary and Conclusions – Evolutionary Principles** 228

Bibliography 229
Books and Reviews 229
Primary Literature 230

Keywords

Alternative splicing
Creation of several different protein variants from one gene sequence using different exons.

Convergence
Independent invention of similar features.

Domain
A distinct (evolutionary) unit within a protein – can be defined from a structural, sequence-based, functional, or evolutionary perspective. Here, we concentrate on the structural and evolutionary domains.

Domain accretion
Tendency of eukaryote proteins to have more domains than their prokaryote homologs.

Domain architecture
The total number and type of domains within a protein and their relative orientations to one another.

Domain combination
Two or more structural domains that have combined at some time within a protein generally in tandem, though, domain insertions also exist.

Domain or protein family
Closely related sequences of domains or whole proteins.

Essential gene; essentiality
Gene with a lethal knockout phenotype.

Gene silencing
Upon a gene duplication event, the redundant gene copy is turned nonfunctional.

Homolog; homologous
Inferred evolutionary relationship.

Inserted domains
Protein domain A inserted into another domain B, so that B is interrupted in sequence.

Linker region; linker sequence
Amino acid sequence joining two domains.

Neo-functionalization
Emergence of novel protein function, usually after relaxation of selection upon a gene duplication event.

Nonsynonymous substitutions
Nucleotide substitutions that change the type of resulting amino acid.

Orphan (sequence)
Sequence without detectable homologs. Also called *singleton*.

Orthologue; orthologous
Homologs; homologous sequences related by descent, that is, speciation.

Paralogue; paralogous
Homologs; homologous sequences in one organism, related by gene duplication.

Protein Repertoire
The full complement of proteins encoded within an organism's genome.

Proteome
The fraction of the protein repertoire that is expressed in a particular organism or cell at a particular time.

Pseudogenes
Presumably untranscribed and untranslated, formerly active but now nonfunctional genes.

Sequence–structure–function Relationship
Relationship between the sequence, structure, and function of a protein. Protein evolution is marked by changes in these three characteristics at different rates.

Structural class
Top level of the hierarchical classification of protein domains. Groups domains according to their composition of alpha-helices and beta-strands.

Subfunctionalisation
Modification and divergence protein function, usually after relaxation of selection upon a gene duplication event.

Superfamily
One or more families of related domains or whole proteins.

Superfold
Notion of folds that are populated by many superfamilies.

Synonymous substitutions
Nucleotide substitutions that do not affect the type of resulting amino acid.

Transcriptome
The complete collection of transcribed elements from a genome.

Unifold
Notion of folds that comprise only one or very few superfamilies.

Abbreviations

aa:	amino acid (residues)
BDIM:	birth-death-innovation models
CATH:	class(C), architecture(A), topology(T) and homologous superfamily (H)
DNA:	deoxyribonucleic acid
3D:	3-dimensional
EC:	enzyme commission
GO:	gene ontology
HGT:	horizontal gene transfer
HMM:	hidden Markov model
NCBI:	National Center for Biotechnology Information
NRDB:	non-redundant database
NTP:	nucleotide triphosphate
PDB:	protein data bank
RMSD:	root mean square deviation
RNA:	ribonucleic acid
SCOP:	structural classification of proteins
TIM:	triose phosphate isomerase
WHD:	winged helix domain

Proteins are the key players in our cells. Their entirety is what we call an organism's protein repertoire, encoded in its genome. Proteins are made up of smaller units called *domains*. Although single-domain proteins exist, the majority of proteins consist of at least two domains; the individual domains and nature of their interactions determine the function of the protein. Today, we have information on the sequence and structure of many proteins and domains. These data enable us to identify, describe, and compare the protein repertoires of a wide range of organisms and how they are determined by the tight interplay of sequence, structure, and function. The main evolutionary processes that form the protein repertoire are duplication, recombination, and divergence of the encoding sequences. Here, we describe these processes and how they affect protein structure and function. This allows us to draw general conclusions on the evolution of the protein repertoire on the organismal scale.

1
General Characteristics of the Protein Repertoire

Millions of different species inhabit our planet. One of the most fundamental problems in biology is to describe, organize, and understand the evolution of organisms that have very diverse characteristics. Although natural selection does not necessarily lead to an increase in complexity as organisms evolve, it is apparent that the complexity of certain lineages, such as our own, has increased during evolution.

The anatomy and physiology of an organism and its changes during evolution are determined primarily by its protein repertoire. The modularity of proteins, in turn, reveals the versatility of living organisms and their amazing ability to constantly evolve. One can imagine an increase in complexity by employment of three main mechanisms: (1) the invention of novel *modules*, for example, proteins or cells; (2) an increase in abundance of existing modules and their divergence; or (3) novel interactions between existing modules. The first of these mechanisms, *de novo* protein construction is relatively rare.

Here, we examine the protein repertoire, how it was formed via the duplication and recombination of existing modules, and how this contributed to the multitude of organisms that exist today. This provides, in contrast to single protein analyses, a much broader evolutionary insight into phylogenetic relationships. Understanding how the protein repertoire has evolved to produce functionally diverse and distinct proteins permits an appreciation of how speciation and evolution have occurred at a higher level.

1.1
Sequence, Structure, and Function

Proteins function as enzymes, regulators, messengers, or receptors; they form cellular structures, transporters, and storage devices. Proteins enable cells to grow, divide, and communicate with other cells or to respond to environmental input. They are found in cellular compartments or even outside the cell, attached to the membrane or freely moving around. Proteins bind other molecules, such as DNA, RNA, lipids, metal ions, or carbohydrates, but they also interact with each other, forming large structures or functional complexes.

Proteins are characterized by the intimate relationship between their sequence, structure, and function (Fig. 1). Proteins of similar structure can have an enormous variety of functions, and an example of these are the TIM barrels. In contrast, the same function, for example binding to DNA, can be fulfilled by proteins of most different structures.

Similarly, the sequence and structure of a protein are intrinsically linked, and the chemistry of certain residues along with their surface location is responsible for maintaining the function of binding to the receptor protein. Even though apparently simple, understanding the relationships between these characteristics is crucial for an understanding of protein evolution. The amino acid sequence determines the structure a protein forms. Homologous proteins are of common evolutionary descent, and they often have similar sequences. Such sequence similarity is easy to detect and quantify with tools available today. However, protein structure is more conserved than sequence, and, frequently, the structure of related proteins will be similar despite the sequences having greatly diverged from one another.

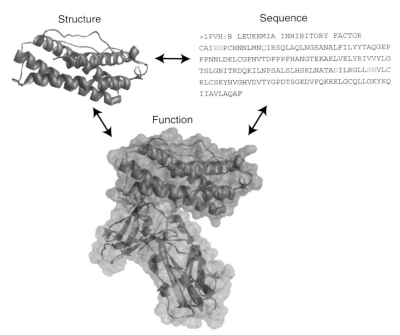

Fig. 1 Proteins are characterized by their sequence, structure, and function. The sequence, individual domain structure (colored purple), and interacting structure (with Gp130 colored blue) of leukemia inhibitory factor are shown (Protein Data Bank (PDB) ID: 1pvh). Within the sequence and interacting domains residues that are colored red form hydrogen bonds with the receptor. (See color plate p. xxxiii.)

In the triangular relationship between sequence, structure, and function (Fig. 1), small changes at the level of the sequence can produce either large or small changes in the structure, which, depending on location, may impact the protein's function. A single nucleotide substitution can produce an amino acid change, which might alter specificity or binding affinity. A classic example of this is the discovery of sickle cell anemia by Vernon Ingram in 1956, the first disease that was documented on a molecular level. Ingram showed that the replacement of glutamic acid by valine at position 6 in hemoglobin β-chain was responsible for the abnormality. This substitution produces alterations in the structure such that haem binding and thus iron transport is compromised, resulting in sickle cell–shaped red blood cells and serious anemia.

In the following sections, we describe some aspects of the sequence, structure, and function of proteins and how they relate to each other.

1.2
Related Proteins Form Families

In order to understand the protein repertoire and how it evolves, we need to know the evolutionary or phylogenetic relationships between the proteins. Margaret Dayhoff was amongst the first to introduce the notion that proteins can be grouped into *families* and *superfamilies* of related molecules and recognized the role of these relationships in protein evolution. These concepts of families and superfamilies have played heavily into construction of our present-day classifications of proteins.

A family was defined as a group of related sequences based on sequence similarity. The evolutionary relationship is obvious from the amino acid sequence alone: if two sequences are highly similar, the proteins are likely to be related. Many methods exist to detect these sequence similarities today, for example, BLAST. As we commented on earlier, the structure of two related proteins can be remarkably similar despite their sequence having diverged beyond recognizable similarity. In superfamilies, which can comprise one or more related families, sequence similarity is often undetectable, and the evolutionary relationship is demonstrated by structural and functional information.

A wonderful example of related proteins with minimal or no sequence similarity is the four-helical cytokine superfamily. These proteins are known to bind to receptors and trigger signaling cascades functioning in immune and circulatory systems. Distinctive structural features reveal, however, that four-helical cytokines can be subdivided into three different families: the interferons that have a fifth helix, the short chain cytokines that have a small two-stranded beta-sheet and subtle folding differences, and finally the long-chain cytokines, which have longer chains. The structure-based family classification was confirmed by analysis of the gene structure, which showed remarkable conservation of intron/exon structure with respect to the secondary structure elements.

1.3
Most Proteins Fold into Structures

Most proteins fold into very distinct three-dimensional globular structures. Protein structures solved by X-ray crystallography or NMR are deposited into the protein data bank (PDB), a central repository for all proteins of known structure. At a basic level, we distinguish alpha-helices, beta-strands that fold into sheets, turns, and loops, as structural elements. These elements are connected to form simple motifs. These structural motifs fold into domains. Polypeptide chains or proteins can consist of several such domains.

A small set of proteins are fibrous proteins that often play structural roles in, for example, keratin in hair. In addition, a significant fraction of proteins do not fold into distinct structures and form so-called *unstructured* or *disordered* proteins. Often, this group of disordered proteins is overlooked when considering the protein repertoire. Machine learning neural network predictors trained to detect regions of sequence that are likely to be disordered predicted that 35 to 51% of eukaryote proteins have at least one disordered region of >50 residues, and that fully disordered proteins correspond to 11% of proteins in Swiss-Prot and 6 to 17% of proteins in various genomes. At this time, most of our protein evolution knowledge is derived from globular proteins – and it is unclear how similar the evolution of unstructured proteins is.

1.4
Domains are the Units of Protein Evolution

We now discuss one of the key concepts for studying the protein repertoire – that of protein *domains*, that is, the structural and functional units that make proteins

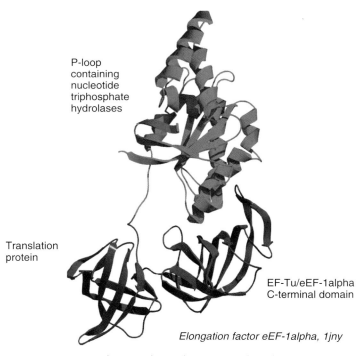

Fig. 2 Domains are the units that make proteins. The eukaryote elongation factor eEF 1-alpha is a beautiful example of a protein that is structured into domains: the P-loop hydrolase domain hydrolases GTP and thus triggers conformational change, the Translation protein domain interacts with the ribosome (not shown) in order to prolong translation.

(Fig. 2). We also discuss the definitions, classification schemes, and properties of domains, and explain how domains allow us to understand protein properties and the evolution of proteins.

The concept of protein domains arose when the first X-ray structures of proteins were determined. These structures, for example, globin or insulin, appeared to be single compact entities. Other protein structures seemed to contain several of these compact entities, and this suggested that proteins are divided into distinct structural units that fold independently, analogous to beads on a string. Typically, domains are regions of compact protein structure with a hydrophobic core. Levitt and Chothia were the first to classify the arrangements of secondary structure elements in a set of proteins available at that time. These representations led to the identification of four distinct protein classes: all-alpha, all-beta, alpha + beta and alpha/beta – these remain the main classes used today.

In 1974, Rossmann et al. realized that domains represented *evolutionary* rather than only structural units. They observed that several nucleotide-binding domains displayed structural similarity, and they speculated that these domains had descended from a domain present in an ancestral protein. Importantly, they also recognized the value of structural information for these analyses: only "a combination of structure and sequence permits a measurement of more distant evolutionary relationships". Later, Murzin et al. constructed the structural classification of proteins database (SCOP), which uses such an evolutionary domain definition (Fig. 3). In SCOP, a structural unit within a protein only qualifies as a domain if it is observed in isolation as a single-domain protein and/or in different combinations with other structural units. Evidence for the evolutionary relationship between domains is taken from detailed examination of protein structure, and, additionally, sequence and function.

Alternative views on domains include the idea that fold space consists of repeated small structural motifs, such as helix-turn-helix motifs, smaller than the compact entities considered domains in other structural approaches. Certainly, folds such as TIM-barrels and other barrel type structures consist of the same small structural motif repeated several times. It is easy to imagine how duplication and assembly of the smaller motifs might lead to the evolution of these larger protein domains. Indeed, some protein fragment matching and assembly-based approaches have been very successful recently when applied to the *ab initio* protein fold prediction problem.

Protein domains are of great diversity, and today there are many databases using their own domain definitions and different protein classifications. The definition of a protein domain remains complicated and can be approached from a structural, sequential, or functional perspective. For answering evolutionary questions, however, it is crucial to define a domain in an evolutionary context – and this is why we use the term *domain* always with reference to the evolutionary unit defined in SCOP.

1.4.1 Evolutionary Hierarchical Domain Classification (SCOP)

SCOP is a completely manual classification for proteins of known structures and is considered by many as the gold-standard classification. The hierarchical classification starts at the level of individual domains or small proteins (Fig. 3). A SCOP domain typically has 100 to 250 amino acid residues, though smaller and

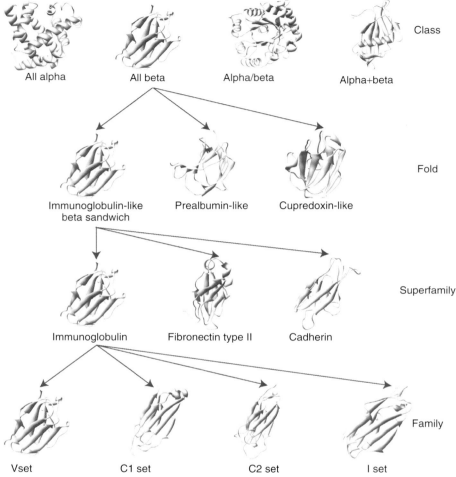

Fig. 3 The SCOP classification. The hierarchical classification is illustrated by the example of the Immunoglobulin domains. One representative structure is given at every level – note that not all folds are shown for this class, and not all superfamilies for this specific fold.

larger domains do occur. Closely related domains are then grouped into families on the basis of having a common evolutionary ancestor evidenced by structural similarity and more than 30% sequence identity. Superfamilies at the next level are made up of families for which the structure and function suggest a common evolutionary relationship, but which have sequence identities below 30%.

The subsequent level of classification is the fold. Superfamilies and families of domains of the same fold have equivalent major secondary structure elements in a similar arrangement with the same topological connections. Superfamilies within the same fold have an unresolved relationship and may be convergently or divergently related – but for which we are currently lacking information either way.

The final level of SCOP is class, which comprises the original all-alpha, all-beta, alpha/beta (a/b), alpha + beta (a + b), and seven additional, smaller classes.

Note that in the literature, the expressions *fold*, *family*, and *superfamily* are often used with different and overlapping meanings and the hierarchical levels are then blurred. For example, fold is sometimes used to describe superfamilies or even single structures; or protein families are not grouped into superfamilies. There are many potential additional relationships between proteins for which we lack evidence as yet, for example, between superfamilies of the same fold. As more protein structures and sequences become available and our methods advance, some of these relationships will be clarified and classifications updated.

1.4.2 Alternative Domain Classifications

In addition to the evolutionary domain definition used in SCOP, there are many other systems of domain classification and they use a variety of definitions and methods. Some databases center their classification completely on structure, others on sequence similarity; the classification can be done manually, completely automatically, or using a combination of both approaches. Some so-called *metasites*, for example, Interpro, combine the output of several different databases.

An alternative and also commonly used protein structural database is the classification architecture topology and homology (CATH) database. CATH uses structural criteria and a domain definition similar to the one used in SCOP. It classifies domains into homologous superfamilies, topologies, architectures, and classes using a combination of manual and automated methods. The main automated approach to domain definition is to identify segments of protein structure with more internal contacts than contacts to other regions of the structure. Additionally, there are many other structural domain classifications, for example, the families of structurally similar proteins (FSSP) or the molecular modeling database (MMDB), both of which are fully automated.

In both SCOP and CATH, the top levels of the hierarchy are defined by the three-dimensional structure, whereas lower levels are identified by sequence similarity. There is an extensive agreement of the classification of the proteins in both databases, although some exact criteria on topological similarity have yet to be determined. There are, however, some striking differences between the domain definitions of these two resources. One such example is that of protein kinases, which are classified as one single domain in SCOP, but is split into two domains in CATH (Fig. 4). In favor of the CATH definition, there are two distinct lobes within the kinase domain. These two lobes, however, are only ever found together and bind ATP at their interface, and this strongly supports the SCOP single-domain definition of kinases as an individual evolutionary domain. Such classifications are to some extent subjective – it is important to remember that alternatives exist and should be used according to the precise context and what question is being looked into.

Some databases rely on sequence rather than structural similarity for clustering domains into families. Examples of these databases are Pfam and SMART, both of which use HMMs as a basis for the alignments they provide. Other databases rely completely on the conservation of sequence stretches in multiple alignments, such as the ones used for TRIBE-MCL or ProtoMap.

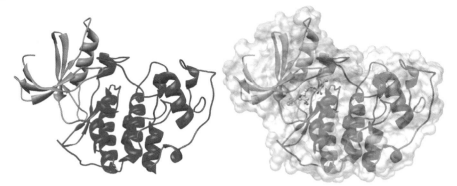

Fig. 4 Protein kinases, as defined in SCOP and CATH. On the left is the kinase domain (PDB ID: 1hck) colored according to the CATH definition of two distinct structural domains (first CATH domain residues 1–84 red, the second CATH domain residues 85–297 blue). On the right, the same coloring is implemented; however, we also show the surface of the entire kinase domain and the location of the bound ATP. This highlights the fact that in terms of surface the entire domain appears as one globular structure, and the location of the ATP binds at the interface of the two CATH defined domains. Functionally, both domains are necessary for ATP to bind and therefore the single-domain SCOP definition seems more accurate in terms of functional, evolutionary domain definition. (See color plate p. xxxiv.)

Sequence-based domain definitions have the advantage that domains of as yet unknown structure and unstructured regions can be included. Homology from sequence similarity, when present, is easier to detect than from structural similarity. We are positive that when significant similarity exists between two sequences, domains are homologous. However, in order to detect all remote relationships, sequence similarity alone is often insufficient and thus it is only of limited use for answering evolutionary questions. In such cases, other conservation of other structural features should be included for consideration. Structure and its finer details, for example, alpha-helix caps, beta-bulges, conserved turns and so on, is a potentially better tool for understanding evolution of the protein repertoire. Currently, such approaches are limited to mostly manual analyses, and we need to develop more automated tools to render their use possible on a genome-wide scale.

1.4.3 Implications and Complications of Domain Definitions

The evolutionary domain definition in SCOP and the knowledge inherent within this classification on the relationships between proteins have important implications. We discuss some of these below.

Firstly, information about the evolutionary relationships between domains is most useful to identify homologous proteins in different genomes. When we find domains of the same superfamily in two different proteins, we know that these proteins, or at least parts of them, are related by descent. The proteins must have arisen by duplication and recombination of the ancestral sequence.

Secondly, the exact definition of a domain in terms of its start and end location within the protein sequence is sometimes complicated. The domains can be joined by long *linker* regions, which sometimes fold into secondary structure elements. Many domains can also have long insertions in the form of loops, alpha helices, or other elements. In the extreme case, whole domains, called *inserted domains*, can be introduced into the middle of another domain, interrupting the sequence contiguity. Thus, despite all efforts to consistently classify protein structures, many exceptions or special cases have to be considered.

Finally, it remains a matter of debate whether individual folds (structural topologies) arose only once during evolution or whether some of them arose by *convergence* from unrelated ancestors. The basic physicochemical interactions and associated structural and sequence motifs are often conserved across different superfamilies within one fold. In contrast, the same activity and/or function can be performed by two or more completely unrelated structures. Anecdotal cases of *convergent* evolution have been reported in the literature where there is currently no evidence of the proteins being derived from a common ancestor. A well-known example are serine proteases: chymotrypsin and subtilisin are members of two different superfamilies, but the three-dimensional arrangement of the catalytic residues and the reaction mechanism are the same. Another example is MutS and topoisomerise II, which share a structural motif that in its dimeric form is suitable for sensing DNA topology and binding recombination intermediates. Such examples might prove wrong once our ability to detect remote homology improves. When we do not have sufficient evidence of different origins of two protein folds, the relationship is really unresolved in terms of evolution rather than a definite case of convergent evolution. This general lack of proof for convergent evolution led to the suggested use of the term *parallel* rather than convergent evolution.

There is, however, very little doubt that domains with clear structure, sequence, and/or functional evidence of homology have one common ancestor, even in the absence of detectable sequence identity. While classification systems such as CATH contain some information on domain homology, to the best of our knowledge, SCOP is the only one, which defines domains from a more hierarchical evolutionary perspective and incorporates several homology levels. Thus, the evolutionary domain definition in SCOP provides the basis for many exciting analyses into the evolution of the protein repertoire. We can examine the changes in structure across homologous proteins, and how this is linked to changes in sequence and function. We can also compare orthologous proteins in different organisms and try and understand how they may contribute to organism-specific features. Finally, knowledge of the relationships between proteins can help method development, such as to establish evaluation procedures for homology recognition algorithms.

1.4.4 Domain Prediction

In order to study the protein repertoire, for example, the number, distribution, or sizes of domain families in different organisms, we need as comprehensive as possible knowledge on the presence or absence of particular domains in those organisms. While we currently know the structure of more than 25 000 proteins, this covers only a small fraction of all proteins in a particular organism. We can increase this coverage by predicting domains and their structures. This is done by inference, that is, exploiting homology with proteins of known structure, or *ab initio*, that is, prediction of the structure of proteins with new folds.

Many classification systems mentioned above include algorithms to predict domains based on sequence and/or structural homology. Today, we have the complete genome sequence for more than 130 organisms, and, for example, with the use of the SUPERFAMILY database, it is possible to make domain predictions for 40% of eukaryote sequences by inferring homology from sequence similarity – and these are the data that we discuss here. Despite some apparent biases in the proteins of known structure, for example, more globular and soluble proteins, and overrepresentation of some proteins of high biological interest, for example, immunoglobulins and globins, we assume that the conclusions drawn from this partial information are still of general relevance and depict a more refined picture of the protein repertoires than that which was previously known – and this picture will continue to be polished as more data becomes available.

1.5
How Many Different Families, Superfamilies, and Folds Exist in Nature?

The first important question that may arise, after consideration of appropriate domain definition and classification, is on the size of the protein universe, that is, how many different families actually exist in nature and how many novel families are yet to be discovered.

There are more than 1200 domain superfamilies in the current version of the SCOP database, and it is likely that these cover the most common superfamilies. Estimates on the total number of folds or superfamilies, however, vary – partly due to the blurred definitions of fold and superfamily. Early speculation about the total number of protein families and superfamilies relied largely on the rate of discovery of new families, and they predicted around 1000 common superfamilies. When other assumptions or methods were used, the estimates of the total number of existing families, superfamilies, or folds vary from a few hundred to several thousand.

In some respect, the exact number of domain superfamilies might be of little importance, since the distribution of the number of proteins per superfamily or the number of superfamilies per fold is extremely uneven: a few folds are very popular and have many superfamilies. These superfamilies, in turn, also occur in many different proteins (Fig. 5). These folds are also called *superfolds*. Superfolds often have very generic functions and contain structures such as the P-loop NTP hydrolases, Rossmann folds or TIM barrels. In contrast, the majority of novel folds that have been discovered recently tend to only have a small number of superfamily members. They are also known as *unifolds*; it remains unclear how many of the unifolds encompass the still unknown proportion of the protein repertoire.

1.6
The Distribution of Protein and Domain Families Across Different Genomes

Once we have an idea of the number of different domain families that exist in nature, we might like to know how they are distributed across different species. We obtain a first and simple measure of similarity between organisms, how they are related and evolved, and how the organisms' complexity is determined by, for example, the number of different domain superfamilies that occur, their kind, and size.

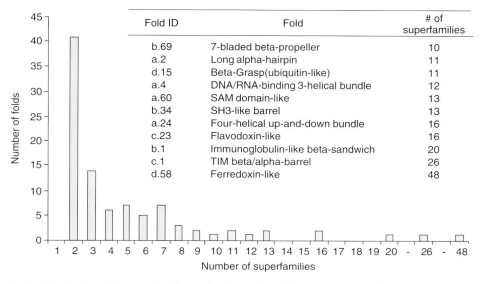

Fig. 5 Distribution of the number of superfamilies within folds in SCOP 1.65. The inset table provides details of the super- or hyper-folds with 10+ superfamilies. There are 602 folds, which have only one superfamily – this is not shown as a bar on the histogram.

Early studies provided a census of genomes in terms of structures, highlighting differences in preferences for secondary structures and folds. Some studies suggest that globular domains occurred early in evolution, with the alpha/beta class being possibly the first.

Later, these analyses were tailored to the distribution of domain families and superfamilies; and again, any conclusions we draw are from those 40% of the sequences to which one or more domains are assigned. For example, in most metazoans, we find about 800 different domain superfamilies out of a total of about 1200 known superfamilies, whereas the unicellular baker's yeast and bacteria have about 650 or even fewer (Fig. 6). Some superfamilies are conserved across different genomes; and the more closely related are two genomes, the larger the fraction of conserved superfamilies. For example, around 400 superfamilies or 35% occur in 14 eukaryote organisms including yeast, invertebrates, and vertebrates. The members of these superfamilies form up to 90% of the domains in these organisms, which makes the protein repertoires between these organisms very similar in terms of domain composition. In contrast, only about 20% of the domain superfamilies or even less occur in all three kingdoms of life. These 20% of all superfamilies comprise, however, around 80% of all domains.

This means that most of the superfamilies evolved long before the split of eukaryotic clades. Sequences encoding domains of these superfamilies have subsequently been duplicated many times and are modified and reused in many different contexts; and we discuss these mechanisms – duplication, modification, and reuse – later in more detail.

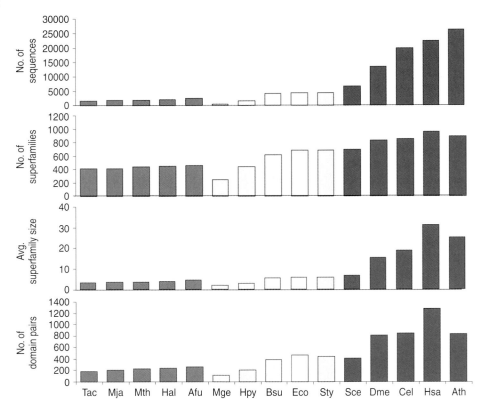

Fig. 6 Different features of the genome and protein repertoire. From top: The total number of sequences (genes) predicted for each organism. The total number of different SCOP domain superfamilies predicted with SUPERFAMILY version 1.65 (http://supfam.mrc-lmb.cam.ac.uk/SUPERFAMILY/). The average number of proteins per domain superfamily that is, number of duplicates per domain superfamily. Finally, the number of different two-domain combinations. Red – Archaea: Tac, *Thermoplasma acidophilum*; Mja, *Methanocaldococcus jannaschii*; Mth, *Methanothermobacter thermautotrophicus* Delta H; Hal, *Halobacterium* sp. NRC-1; Afu, *Archaeoglobus fulgidus* DSM 4304.
Yellow – Bacteria: Mge, *Mycoplasma genitalium*; Hpy, *Helicobacter pylori* 26695; Bsu, *Bacillus subtilis* ssp. subtilis 168; Eco, *Escherichia coli* K12; Sty, *Salmonella typhimurium* LT2. Blue – Eukarya: Sce, *Saccharomyces cerevisiae*; Dme, *Drosophila melanogaster* 3.2; Cel, *Caenorhabditis elegans* WS123; Hsa, *Homo sapiens* 22.34d; Ath, *Arabidopsis thaliana* 5. (See color plate p. xxxv.)

1.7 Proteins Interact

Proteins interact – they are part of complexes, cellular structures, and pathways. Many *stable* protein complexes are crucial for the cell to work: the transcription factor assembly, the splicing machinery. *Transient* interactions of proteins or protein domains have important roles, for example, in signaling cascades or cell cycle regulation. These interactions are part of protein function and hence constrain the evolution of sequence. Indeed, proteins in stable complexes are more conserved than proteins in transient complexes, and, generally, the more interaction partners a protein has, the slower it evolves. Proteins with many interaction partners tend to have well-conserved sequences and frequently have essential functions. Domains within proteins also interact, and we discuss the details of their spatial arrangement in a later section.

Recent years have produced an enormous growth of genome-scale protein interaction data and revealed that, similar to other biological phenomena, protein interaction networks are of a scale-free nature. The hubs in these networks, i.e., proteins with many interaction partners, are either part of stable complexes or are involved in crucial transient interactions mentioned above. With a map of protein interactions, we can better understand the evolution of the protein repertoire with reference to the constraints placed by interactions and the resulting functions.

1.8
Proteins are Part of Pathways

Many proteins, for example enzymes or signaling molecules, are part of pathways, such as in metabolism or signal transduction. In the *Escherichia coli* small-molecular metabolism, about 600 proteins act in about 100 pathways, and one-quarter of the enzymes are involved in multiple pathways. Even this basic set of ancient enzymes seems to have evolved by means of extensive duplication and recombination.

Whether the function of a single protein, or of the entire pathway, is essential decides whether the organism will survive in case of a mutation and thus influences the rate of evolution of the protein. Again, we see many biases: most domain superfamilies occur in only a few pathways, whereas a few domain superfamilies of very generic function occur in many pathways.

The "Horowitz" model assumes that duplicated enzymes tend to be employed in the same pathway so that enzymes in consecutive reaction steps are often homologous. However, despite extensive domain and protein duplication across metabolic enzymes, this is only partially true. For most of the homologous proteins, the catalytic mechanism or cofactor-binding properties are conserved or only slightly modified while the substrate specificity has changed. This suggests that it is much easier to evolve new binding sites than new catalytic mechanisms. Thus, the recruitment of homologous domains in novel pathways primarily occurred on the basis of catalytic mechanism or cofactor binding, and this speaks against the Horowitz model.

The network of biological pathways is amazingly robust. More than half of all genes in yeast have, when knocked out, no, or hardly any, apparent effect on the fitness of the organism, that is, they are nonessential. Many of these genes might, however, be of important function under nonhomeostatic conditions such as when subject to stress. The apparent robustness is not only due to the functional redundancy created by the many duplicate genes but also by alternative routes to producing the same molecules.

The same pathway in different organisms can contain species-specific sets of isozymes. Enzymes responsible for the same reaction in the same pathway but in different organisms can also be completely unrelated. This phenomenon is called *nonorthologous* replacement, and one example is lysyl-tRNA-synthetases, which occurs in two unrelated protein families across bacteria and archaea.

The recruitment of single enzymes from different pathways could be the driving force for pathway evolution, and some enzyme families are capable of changing both reaction and substrate chemistry. Such recruitment seems more prevalent than, for example, whole pathway duplication, enzyme specialization, *de novo* invention of pathways, or retroevolution of pathways. Together, these variations produce "mosaic" network of pathways in different organisms with a widespread plasticity.

1.9
Protein and Domain Function

The function of a protein is usually determined by the interplay of its sequence and three-dimensional structure. However, two proteins of very similar functions may have arisen from different ancestor molecules and/or can have different structures. A well-documented example of functional convergence is the Ser/His/Asp catalytic triad of the serine proteases found in several different protein folds. The converse is also true: proteins that are structurally or evolutionarily related do not necessarily have identical functions such as is seen for the large TIM-barrel fold.

Similar to structure classification, the identification and useful classification of protein function is nontrivial. Several classification schemes for protein function have been developed in recent years, and it is important to choose the scheme according to the question that is to be addressed. Each scheme has its own advantages and disadvantages. One of the oldest systems is that of EC-numbers, classifying chemical reactions of enzymes.

A more recent classification, that comprises both enzymes and nonenzyme proteins, is gene ontology (GO) which classifies protein subcellular localization, molecular function, or the biological process of proteins in three separate ontologies. The distinction between the molecular function of a protein and the biological process in which it acts is very important, and, to the best of our knowledge, GO is the only classification system thus far that makes it. For example, a kinase's molecular function is to transfer phosphate groups. Kinases can, however, also act in processes as different as signal transduction, glycolysis, or muscle movement. A disadvantage or difficulty in GO is its nonhierarchical character and that the depth of description is inconsistent at various levels of the graph.

1.9.1 Indirect Function Characteristics

The function of a protein is tightly coupled to its sequence and structure, but also to many other characteristics, such as its interaction partners, its localization in the cell, position in the genome, and so on. These characteristics co-evolve and, thus, they can be used to predict protein function and to study function evolution.

Firstly, function prediction often relies on sequence similarity, assuming that homologous proteins have related functions. The predictions are reliable if the sequence identity to a homolog of known function is greater than about 40%. If the sequence identity is low, the three-dimensional structure of the protein might still give clues as to the function, but the situation is much less clear.

Secondly, proteins of similar function tend to co-localize on the genome. In prokaryotes, genes that participate in the same process tend to be organized in an operon; but a nonrandom gene order is also observed in eukaryotes. Housekeeping genes, for example, tend to form clusters on the chromosome.

Thirdly, proteins of the same pathway tend to interact. Fourth, a functional link between two proteins is assumed to encourage the physical fusion of the two genes, and this approach has been termed *Rosetta Stone principle*. In the fifth place, two functionally linked proteins also tend to show the same phylogenetic profile. That means two genes that show the same pattern of presence or absence in a number of organisms tend to be functionally linked. Penultimately, functionally related proteins are often co-regulated in the same subcellular compartment or have similar effects on the knockout phenotype.

Last but not least, the existence of the same domains in two proteins, for example, two enzymes from different organisms, and their interactions with other domains can also tell us whether these enzymes carry out a similar function. Thus, the domain composition can be used to predict protein function or localization.

1.9.2 Functional Annotation of Domains

The function annotation of protein domains is even more difficult. GeneOntology annotation is available for some Pfam domains, but only a few hundred of them have equivalents to SCOP domain superfamilies. It is, however, difficult to use a whole-gene-based annotation scheme, such as GO, for domains. Most proteins consist of several domains, which can have different functions. For example, a kinase domain can be found with a lipid-binding and/or protein interaction domain. Such considerations made it necessary to develop a novel domain-based annotation scheme, and Bashton and Chothia have done so. Their scheme is domain-centric and emphasizes domain function in the context of domain neighbors in multidomain proteins, providing functional annotations for a subset of SCOP domains. The annotation is based on detailed examination of the protein structures. In this domain-centric functional classification scheme, domains are classified into seven categories that encompass catalytic activity, cofactor-binding, responsibility for subcellular localization, protein–protein interaction, and so forth.

2 Mechanisms in the Evolution of Single Proteins and the Protein Repertoire

In the previous sections, we described the characteristics of single proteins and of the protein repertoire as it is encoded in an organism's genome. We will now focus on the processes that form the protein repertoire during evolution: firstly, whole sequences or parts thereof are duplicated and therefore the number of copies of a protein or domain is increased in a genome. This can amplify the function of the protein, but also provides material for a second mechanism: sequences, especially of duplicated proteins or domains, diverge by mutations, deletions, and insertions that produce modified structures and functions. Thirdly, whole genes or single exons recombine so that novel protein domain arrangements emerge.

2.1 Orthology and Paralogy

Before discussing the details of these three mechanisms, it is important to introduce an important notion on homology: the distinction between paralogy and orthology.

Paralogs are homologous proteins that arose via gene duplication within one organism (Fig. 7); they are also called *duplicate* genes. The redundancy created by gene duplication relaxes selection

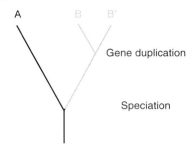

Fig. 7 Orthology and Paralogy. The tree represents the orthologous and paralogous relationships of a gene, protein, or domain. Speciation results in one copy of the gene in each species: A and B are orthologs (gray and black branch, respectively). A gene duplication event in one species (gray branch) results in B′, which is a paralog to B.

pressure and encourages the divergence of protein function. In many cases, however, paralogs still have identical or similar biochemical activities.

Orthologs are protein homologs that arose during a speciation and not a gene duplication event. Thus, orthologous proteins are still under selective pressure and will often have identical functions in different organisms. While we generally observe a core proteome of 1 : 1 orthologs across eukaryote organisms, orthologous genes can, of course, also be duplicated. If the duplication occurs *after* the speciation event, the duplicates are called *in-paralogs* or *co-orthologs*, since the orthology is "shared" between the duplicates. If the duplication occurs *before* the speciation event, the duplicates are called *out-paralogs*. A very useful database collects clusters of orthologous groups of proteins (COGs). Even though these groups of proteins merge paralogs and orthologs, proteins within these clusters often have similar functions.

Finally, "horizontal" gene transfer, as described below, introduces *xenologs* into an organism, for example, proteins that are homologous to proteins in another organism without these two organisms being directly related by descent. In addition, proteins that substitute for function in alternative pathways or in nonorthologous displacement, discussed above, could be considered *analogs*, that is, proteins of functional or structural similarity without homology.

2.2
Duplication – Mechanisms to Create "More"

Duplication of existing building blocks drives evolution and creates paralogs as mentioned above. Haldane (1932) and Muller (1935) were the first to note the importance of duplication, later followed by more exact descriptions and estimates. The genomes of organisms of all three kingdoms of life seem to be formed by many gene duplication events, accompanied by extensive loss of genes. This is also clear at the level of domains: at least 58% of the domains in *Mycoplasma* and 98% of the domains in humans are duplicates. Nice examples of evolution by gene duplication are the HisF and HisA beta/alpha barrels, which have arisen through duplication and fusion of an ancient half-barrel.

Several different mechanisms can be responsible for duplication of existing available coding sequences. First, whole genome duplication takes place as a consequence of nondisjunction during meiosis within species or hybrids. Whole genome duplication has presumably occurred several times at least during the evolution of eukaryotes: once in yeast, later another one in vertebrates. The model plant *Arabidopsis thaliana* shows evidence of at least two rounds.

Second, whole chromosomes or parts of chromosomes can be duplicated. There are well-known human genetic disorders, which are the result of such chromosomal duplications. Perhaps the best known of these is trisomy 21 or Down syndrome in which affected people have a third copy of chromosome 21. These types of chromosomal aberrations are called *aneuploidy*. The majority of chromosome duplications or deletions are deleterious since the duplication of a large number genes causes severe dosage imbalance in the levels of the gene products.

Third, single genes can be duplicated via tandem duplication, retroposition, or transposition as the result of ectopic recombination. Single gene duplication is assumed to be a major mechanism for the creation of recent duplicates. Tandem duplications are the result of unequal crossing-overs during meiosis. Unequal crossing-over between misaligned sequences generates a chromosome with a duplicated region and a chromosome with a deleted region.

Finally, part of a gene or one exon can be duplicated in unequal crossing-overs or replication slippage during DNA replication. Repetitive sequences can increase the chances of unequal crossing-over. Replication slippage has been found to happen for domains, resulting in long stretches of repeated domains in some proteins.

2.2.1 Horizontal Gene Transfer and Transposons

In addition to gene duplication, horizontal gene transfer (HGT) plays an important role in the evolution of prokaryote genomes. HGT is the acquisition of new genes via the movement of genetic material between species, other than by descent. Many eukaryote genes are found in bacteria, interestingly, however, not so in archaea. In eukaryotes, the rates of HGT seem much lower than in prokaryotes. The presence of many genes that arrived through horizontal transfer can seriously distort our picture of the distribution of protein and domain families and impair the reconstruction of the correct evolutionary pathway. In fact, the extensive HGT in prokaryotes led to the depiction of their phylogeny being a "mesh" or network, rather than a tree without a single common bacterial ancestor.

Finally, eukaryote genomes appear to be frequently shaped by jumping genes or *transposons* that move around in the genome, copying and pasting themselves. Estimates suggest that at least half of the human DNA consists of transposons, making a total of around 4 million all together. By their influence on protein structure, for example, via insertion of additional domains, they are of some influence on the shape of the protein repertoire.

2.2.2 Duplication Leads to Protein and Domain Family Expansions

While the earliest evolution of the protein repertoire must have involved the *ab initio* invention of proteins and domains, later on, this may have been replaced by the easier and more efficient reuse of existing building blocks. Thus, it is unsurprising that even the simplest bacterial genomes are the product of extensive gene duplication and recombination.

Lineage-specific expansions of protein families result from relatively recent duplications and are prominent in genomes. Across bacteria, protein, and domain families that function as transcription factors or in signal transduction increase exponentially in number with the size of the bacterial genome. In eukaryotes, we observe the expansion of protein or domain families with very specific functions, and this can give clues on why the organisms have different physiological features. For example, the largest superfamilies in humans are mostly involved in or part of cell surface receptors, protein–protein, or cell–cell interaction, signaling, or cytoskeleton structure, and this explains why we can have so many different cell types forming intricate tissues and organs.

Proteins that function as regulators or cell-adhesion molecules are associated with the advent of multicellularity, and thus the expansion of protein or domain families that have these functions give clues as to how the multicellular organisms evolved. For example, Immunoglobulin and Cadherin domains expanded in the fruit fly as compared to the nematode worm. The functions of these domains in cell-adhesion molecules in the embryonic development of the nervous system can explain why the fly has a much more intricate nervous system than the worm.

The observed extent of domain duplication varies – across different superfamilies and across genomes. In vertebrates, such as human and mouse, each domain superfamily occurs on average in 30 different proteins, while in invertebrates, they occur in only 20 proteins. The numbers are even smaller for bacteria. Generally the duplication results in a distribution of family sizes in genomes that follow a power law, with remarkably constant parameters across organisms. This means that there are only a few highly abundant superfamilies, for example, the P-loop NTP hydrolases, NAD(P)-binding Rossmann domains and certain kinase families. This relationship was first recognized for gene family size, later for domain families or even pseudogene families, and its general implications are discussed in a later section.

2.2.3 Reconstructing the Evolutionary History of Proteins

A classic way to study evolution of a protein repertoire is to reconstruct phylogenetic trees from sequence alignments using one of several methods. One of the most reliable but also the most computationally expensive are maximum likelihood approaches, followed by parsimony, and then distance methods. From the reconstructed trees, it is possible to follow the evolutionary pathway of a protein. This permits an understanding of one protein family – more recent approaches have involved constructing trees for all or most of an organism's proteins in order to obtain an overview of the evolutionary pathway of as many components of the organism as possible. These kinds of analyses can be performed on individual domains, entire sequences, or on concatenated alignments of multiple proteins or domains. Unfortunately, xenologs, that is, homologs in organisms that are not directly related, create difficulties. Therefore, it is advisable to consult several phylogenetic trees since their consensus may reflect the organisms' true evolutionary relationships best.

By reconciling a tree with a species tree, it is then possible to estimate where the duplication and speciation events took place and this differentiates orthologs from paralogs (Fig. 7). This method is vital in phylogenomic-type approaches, which attempt to elucidate function from a proteins phylogenetic location within a tree with respect to homologous proteins of known function.

2.3 Divergence after Duplication

In the absence of duplication, any mutation that occurs is under selective pressure to maintain the protein's structure and function. Once a domain or protein has duplicated, this pressure is relaxed and one copy of the domain can evolve a new or modified function. This can occur by sequence divergence or by combining with other domains to form a multidomain protein with a new series of domains. The positive, negative, or neutral effect on the function determines if the duplicate is retained. Apart from changes in sequence, structure, and function, which we discuss below, duplication also relaxes selection pressure on other characteristics, for example, gene regulation or expression patterns.

Many eukaryotes are diploid, which means they have two copies of each chromosome. Thus, there naturally exists a duplicate of each gene. The diploid nature of many organisms delivers potential for divergence and modification of protein function even without any additional gene duplication. This is exemplified in many well-known cases of dominant and recessive mutations that are carried through populations. One example is the recessive mutation that is the cause of sickle cell anemia, which we discussed earlier.

2.3.1 Changes in Sequence

The first level of changes that occur are at the nucleotide level. Mutations can affect a single nucleotide (point mutations) or several adjacent nucleotides (segmental mutations). Depending on the type of change caused by the mutational event, there can be substitutions, recombinations, deletions, insertions, or inversions. Substitution mutations can be transitions, where one purine is replaced by a different purine or transversions, where a purine is replaced by a pyrimidine or *vice versa*. If these changes happen in protein-coding regions, they can affect the amino acid sequence in two different ways:

Many single nucleotide changes are called *synonymous* substitutions, as they do not alter the resulting amino acid due to the redundancy of the genetic code. Synonymous substitutions are not as much affected by selection, as the protein sequence and structure remains the same, and they display similar rates across different protein families. They can therefore serve as a measure of the overall substitution rate and the absolute evolutionary distance between two sequences.

In contrast, *nonsynonymous substitutions* change the resulting amino acid and are likely to affect the protein and fall under selection. As a consequence of different selection pressures on different protein families, the nonsynonymous substitution rates vary enormously amongst protein families.

We usually distinguish three different effects of sequence changes: deleterious (or negative), neutral, and advantageous (or positive) mutations. The neutral effect of, mostly, synonymous substitutions has a long history in research. One common way to analyze for advantageous mutations is to compare the frequency of nonsynonymous and synonymous substitutions. Examples of advantageous mutations seem relatively rare, and are observed in resistance or pathogen defense genes, for example, in MHC proteins. It is very difficult to estimate the absolute or relative mutation rates, as many different factors affect them: the level of expression of the gene, the function of the resulting protein, the essentiality of the gene, or the relative amino acid cost.

2.3.2 Changes in Function

Once a gene is duplicated within a genome, one of the two copies is free to evolve, while the first can maintain the original function. This evolution may produce modified or new molecular functions that act within the same biological process, or it could create functionality within an unrelated biological process. We distinguish three major effects on protein function after a duplication event. First, the redundant sequence can have a highly negative effect on the fitness of the organism and is consequently silenced. The whole gene is lost or turned into a "dead," untranscribed *pseudogene*. This is a very common process. Second, the redundant protein can evolve a new function that is different from the original function, for example, by sequence divergence or combination with a novel domain partner, and this process is called *neo-functionalization*. Last, if the original gene has multiple functions, they may become split between the two copies, a process called *subfunctionalization*. In eukaryotes, neofunctionalization is much more common than subfunctionalization.

The evolution of domain function is even more complicated, since a domain can either have an independent function, and/or contribute to the function of a multidomain protein in cooperation with other domains. In order to assess domain function evolution, knowledge of the three-dimensional structure is vital. Todd et al. found that for most domains within a superfamily the reaction chemistry is fully or at least largely conserved. The substrate specificity, rather than reaction chemistry, is modulated; this often occurs via changes in the kind of domains adjacent to the catalytic domains, embellishments in the active site, or by variations in the oligomeric state of relatives. Catalytic residues can "migrate" with respect to their position in the amino acid sequence, but their relative position in the three-dimensional structure is mostly preserved. In sum, domain function evolution occurs in two major ways: first, a domain can perform the same function, but in different protein contexts, that is, with different partner domains. Second, a domain may diverge and acquire a novel or modified function.

The way in which an individual protein or protein domain will evolve is very much dependent on the interactions it maintains with other molecules. There is a high level of evolutionary constraint on enzyme binding sites that have chemically specific ligands and restricted binding orientation so that mechanistically the function can still be maintained. Similarly, regulator proteins that bind to specific DNA or RNA sequences also need to maintain a very specific binding site. Conversely, a protein that binds a receptor protein is freer to evolve as the receptor may co-evolve at the same time. Therefore, proteins involved in such interactions can evolve more quickly than proteins that must bind and localize a specific small molecule.

This is exemplified by the fast-evolving four-helical cytokines, specifically human and mouse Interleukin-5, the sequences of which are 70% identical, but they will no longer cross-bind to one another's receptors. Such fast-evolving proteins often seem to have functions with the immune system – a facet of living organisms, which needs to evolve quickly in order to adapt to the many new and changing invasive viruses and bacteria present. For immunoglobulins, crucial for antigen recognition in higher eukaryotes, the ability to evolve quickly is indispensable.

Indeed, they have evolved the ability to switch only one of their domains (the variable domain or V-set) by alternative splicing in order to create more combinations, thus increasing the likelihood of antigen recognition.

In sum, the nature of a protein domain's specific interactions, and the number of interactions it is physically a part of, is likely to impact greatly the rate of evolution for that particular protein or at least the domains within it. The greater the number of interactions and the more constrained structurally the binding partners, the slower the evolution.

2.3.3 Structural Tolerance to Amino Acid Changes – a Case Study on Globins

Implicitly, alterations within the amino acid sequence of a protein will produce changes in the resulting structure. However, the effect on the structure is dependent on where in the protein the change occurs. Traditionally, there have been two approaches to understand tolerance to amino acid changes – theoretical comparison of related sequences and structures from present-day organisms and mutagenesis experiments. In mutagenesis experiments, the roles of specific residues in the folding and stability of a protein are probed by replacing them with alternative residues.

These two approaches convey a similar message on acceptability of mutations in proteins. The hydrophilic exterior of a protein is tolerant to a wide variety of mutations. However, there is a restriction on the choice of mutations in the protein interior; in general, mutations conserve the hydrophobic nature of the core. Thus, the nature and accessibility of the side chains play an important role in the folding and stability of proteins.

Globins represent wonderful examples of how evolutionarily related proteins can be of similar three-dimensional structure and function, but of hardly recognizable sequence similarity. The first two protein structures to be determined, that of sperm whale myoglobin and horse hemoglobin, have only 28% sequence identity. Lesk and Chothia examined the relationship between sequence, structure, and packing in globins in order to understand the effect of sequence divergence on structure and to understand how proteins adapt to mutations.

Although there are approximately 150 residues in globins, the principal determinants of the three-dimensional structure are the 59 residues that are involved in the packing of helices and in the interactions between the helices and the haem group. Only half of these residues are buried. Any mutations of the buried residues retain nonpolar sidechains, although the amino acid shape and size vary. The interface structure between the alpha helices within the globin domains is extensively conserved, except for some shifts and rotations between pairs of helices as a consequence of dense helix packing. If a mutation at the interface changes the volume of the amino acid side chain, the helices move to adjust the packing.

Rare or disallowed changes in globins illustrate the global effect of mutations on the structure and how they are accommodated. If the interface residues switched instead to a completely new set of contacts, the stability of the contact would be lost as the complementarity of the helix surface is compromised beyond allowed limits of variability. For example, a single insertion in the helix interior would turn at least part of the interface by 100°, destroying surface complementarity. Thus, insertions and deletions are never found within a globin helix. Furthermore, mutations in the individual residues in contact with the haem are difficult to accommodate as the haem has comparatively little flexibility. In order to maintain the geometry of haem packing, changes in helix packings are coupled and do not occur independently.

2.3.4 Changes in Structure

How far can sequences diverge given the requirements of structural stability? In order to understand the relationship between divergence of sequence and structure, Chothia and Lesk compared 25 structures from eight different protein families. They determined the common core of a protein family by individually superposing the main-chain atoms of major secondary structure. The core regions undergo structural change as their sequences diverge and accumulate mutations. The further the sequences diverge and the lower the sequence identify, the small is the core common to the proteins. For those pairs of proteins, whose sequence identity is 50% or more, 90% of the residues of the individual structures are part of their common core. In contrast, for pairs whose residue identity drops to about 20%, their common cores contain commonly half of the residues in the individual structures. The structural differences in the common cores of members of individual protein families consist mainly of changes in the relative position and orientation of packed secondary structures and local changes in case of beta-sheets. A more recent study of the four-helical cytokines found around 5% of sites within the proteins that had any kind of minimal hydrophobicity or volume conservation within the hydrophobic cores of the proteins.

This demonstrates how rapid evolution can take place at the sequence level while still producing the same or a similar protein structure. There is, however, also great scope for alterations to protein structure.

First, secondary structure can be changed and additional regions of secondary structure (decorations) can be added around the core. Frequently, N- and C-terminal extensions are added to structural domains. Occasionally, we observe wonderful snapshots of such structural evolution in progress. For example, recent work by Newlove shows how one part of a protein domain in closely related Cro proteins can switch its secondary structure depending on its oligomeric state (Fig. 8). Other fold changes include divergent relationships between proteins with distinct topological isomers such as the catalytic domain of carboxypeptidase G_2

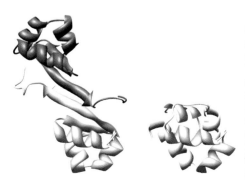

Fig. 8 Cro protein evolution. The dimeric Cro lambda repressor protein is shown on the left-hand side (PDB ID: 1cop with the different domains shaded differently) and on the right is the evolutionarily related Cro 434 protein (PDB ID: 2cro). Secondary structure switching occurs at the dimer interface of Cro lambda, which forms beta-strands, compared to the C-terminal helices in the equivalent position in Cro 434.

and the aminopeptidase from *Aeromonas proteolytica*, and interchangeable parts, that is, topological differences, but similar secondary structures, for example, triabin and the lipocalins.

Second, oligomeric state also contributes to the evolution of the protein repertoire. Paralogs of proteins can have different oligomeric dependencies for stability and functionality. Alteration of oligomeric state can be a mechanism for enormous functional, as well as structural, divergence as seen in the cupin family. Often, within multimeric protein structures we observe *domain-swapping*, that is, individual monomers become intertwined and even more dependent on one another. Indeed domain-swapping has been suggested as a mechanism of protein evolution toward multimerization. Thus, in multimeric proteins, we see again how the modularity of proteins, that is, by combining different numbers of domains, provides a fantastic resource for a variety of evolutionary processes.

Overall, the variety of evolutionary mechanisms, such as insertion, deletion and substitution of sequences, circular permutations, strand invasions and withdrawals, hairpin flips and swaps, can eventually lead to changes in the protein fold, and Grishin et al. detail many wonderful examples. Circular permutations are alterations of protein connectivity evident through ligation of the termini and cleavage at another site. Strand invasion refers to the insertion of a strand into a betasheet such that hydrogen bonds have to be broken in the existing sheet, and reformed with the inserted strand. Strand withdrawal means the removal of a strand. In hairpin flips and swaps, beta-strands are internally flipped, more commonly, this is observed for strands adjacent to one another – this creates or removes crossing loops.

This illustrates how the evolutionary relationship between two proteins, that is, homology, and similarity between their structures are different features and how one does not necessarily determine the other. Rapid evolution can take place at the sequence level and still produce a similar structure and *vice versa*.

2.4
Recombination – Mechanisms to Create "Novel Combinations"

Another way for Nature to diversify the repertoire of proteins is to shuffle their units, the domains, around and create novel combinations. Already 30 years ago, examination of protein structures showed that many proteins are formed of two or more domains; and domains from the same or from different superfamilies combine with each other. Several molecular mechanisms underlie this reshuffling. The major molecular mechanism that leads to multidomain proteins and novel combinations is nonhomologous recombination of domains, sometimes referred to as *domain shuffling*.

There is evidence that in eukaryotes there is a tendency for exon boundaries to coincide with domain boundaries, which suggests that the proteins are formed by intronic recombination.

There is a lot of debate as to whether introns were inserted early on in evolution or late. A detailed study on the conservation of the introns and exons between proteins of known structure revealed that probably some introns were inserted earlier and others later – and, in some cases, their positioning was more conserved within superfamilies than in others. Most of the intron positions showed little or no conservation supporting the view that intron insertion appeared only after the duplication or speciation. In contrast, some extreme examples, for example, the four-helical cytokines, showed an amazing conservation of intron positions with respect to secondary structure elements, and this implied that they were inserted already in the proteins' common ancestor.

Another important recombinatorial mechanism, especially in eukaryotes, is the fusion of genes leading to so-called *Rosetta-Stone sequences*. Gene fusion seems more common than splitting or fission of genes.

2.4.1 Most Proteins Have More than One Domain

Proteins form complex networks of interactions. Thus, even a modest increase in the number of domains in a protein could result in numerous new interactions. Small proteins generally contain just one domain. Large proteins are more often formed by combinations of domains. The total number of domains found in one protein and their relative orientation determines what we call the protein's *domain architecture*. The majority, or at least two-thirds, of proteins consist of two or more domains in prokaryotes and an even larger fraction in eukaryotes. Higher eukaryote proteins also have, on average, more different types of domains than lower organisms. Last, eukaryote proteins acquire domains additional to those seen in their prokaryote orthologs, and this tendency has been termed *domain accretion*. The increasing complexity of domain architectures in higher eukaryotes (Fig. 6) has been correlated with their increased physiological complexity, since more domains in a protein allow a greater variety of interactions and functions.

2.4.2 The Repertoire of Domain Combinations is Limited and Highly Biased

Even though most proteins consist of several domains, the repertoire of combinations of different domains seems very limited. The proteins from more than one hundred different organisms contain only several thousand different combinations of two-domain superfamilies. This is far fewer than would be possible given the total number of about 1200 known superfamilies or the number of multidomain proteins per proteome. This fraction is likely to decrease even more if membrane proteins are included.

There are a few domain superfamilies that are highly versatile and have neighboring domains from many superfamilies, while most superfamilies have only a few different types of adjacent domains. The distribution of the number of partner superfamilies per superfamily follows a power law, like the distribution of superfamily sizes mentioned above. A most versatile superfamily is, for example, the P-loop hydrolase with more than 80 different combination partners in the human genome. We also observe that particular combinations of domains occur in many different proteins, such as the combination of SH3 and SH2 domains is part of many signal transduction proteins. These recurring domain combinations are *supra-domains*, for example, combinations that are duplicated as one unit and reused in different proteins. Thus, supra-domains represent evolutionary units larger than single domains – and they appear to be a major part of the protein repertoire.

2.4.3 The Sequential Order of Domains and Their Geometry are Conserved

Since most proteins consist of multiple domains, and domains determine the function and evolutionary relationships of proteins, it is important to understand the principles of both domain combinations and interactions. If the same domain combination is observed in two different proteins, one possibility is that they have assembled independently by different recombinatorial routes. We assume, however, that most instances of the same two-domain combination or domain architecture have evolved from the same ancestor, and there are several lines of evidence that support this.

First, three-dimensional structural analyses of individual protein families such as the Rossmann domains have shown that proteins with the same domain architecture are related by descent, that is, evolved from one common ancestor. This may even be true for most two-domain protein families of known structures in the current databases. Second, with only a small fraction of exceptions, two domains occur in only one N- to C-terminal order in structural assignments to genome sequences. This conservation of domain order is likely to be historical instead of functional, as a very similar interface and functional sites could be formed by two domains in either order, for instance, given a long linker. Finally, proteins sharing a series of domains also tend to have the same function, which is rarely the case if domain order is switched.

In addition to their conserved sequential order in a protein, the spatial arrangement, or domain *geometry*, is remarkably conserved if the two domains are homologous. This has first been shown for the example of Rossmann domains and their catalytic partner domains, but is possibly true for most of the domain combinations found in SCOP. Studies of small numbers of families of protein complexes, where the interdomain geometry occurs across different polypeptide chains, have also revealed extensive conservation of geometry; and there is a tendency for geometry of interaction of protein domains to be the more conserved, the more similar the domain sequences are.

These results suggested that for proteins of unknown structure, the quarternary structure and protein complex geometry can usually be modeled on the basis of homologous polypeptide(s) of known structure; and impressive examples of such protein models of complexes are the yeast ribosome and exosome.

2.5
Selective and Random Processes

As illustrated above, the sequence, structure and function of proteins are tightly linked and place strong constraints on evolution. In contrast, many evolutionary changes seem to occur in a random fashion. Selection then modifies the outcome of these processes: it favors advantageous changes, rejects disadvantageous changes, and tolerates changes of neutral effect. Some changes may be of neutral effect. This is also true at the level of proteins and domains.

On the one hand, we have seen that many domain or protein families have properties useful in a wide range of contexts; it seems obvious that certain specialized biochemical functions are selected more often than others. Examples are the C2H2 and C2HC zinc-finger domains, which are found in a wide variety of proteins associated with DNA-binding, such as transcription factors. From a structural point of view, there is a tight interplay between the folding kinetics, the stability, and the function of a protein, and the so-called *designability* of a protein, domain, or domain combination could also be an important selection criterion. According to this view, certain folds have topological properties that make them especially *designable* that is stable, modular, and functional. One example is the TIM-barrel fold with a uniquely symmetrical construction.

On the other hand, one of the most striking observations in recent years was that mathematical laws seem to govern many biologically relationships, and it is difficult to imagine how specific selective procedures may have produced them. In particular, power laws describe the distribution of gene family sizes pseudogene families, the lengths of proteins in terms of domains, or the domain superfamilies among folds. Power laws also describe properties of scale-free networks that depict the relationships or interactions between proteins in metabolic pathways and domains within proteins.

Stochastic birth, death, and innovation models (BDIM) deliver one explanation of the power laws. They describe a *preferential attachment* mechanism in which the "fit get fitter" : large protein families tend to get larger; highly versatile domains tend to get even more different combination partners. Innovation comprises both the divergence of sequences and processes that introduce novel sequences into a genome, such as horizontal gene transfer. BDIMs approximate the observed power law distributions remarkably well and support a random mode in evolution. However, they basically ignore the individual properties of proteins and domains – a most crucial abstraction and simplification.

3
Summary and Conclusions – Evolutionary Principles

If life was like a book, written in an unknown language, then the words of this language may be the proteins. These proteins form sentences of pathways. Several sentences in a paragraph might correspond to an organelle; organelles are parts of cells, and cells of organisms. Here, we considered the protein vocabulary, and how the protein words are made of syllables called *domains*. Since the first determination of protein structures some 40 years ago, researchers have been studying these basic units, the domains. Now that the complete genome sequence of many organisms and their domain assignments are available, we can start to understand the rules of how domains form proteins and how our molecular vocabulary evolved.

We have given a description of the repertoire of proteins and their domains, how it has evolved thus far, and the mechanisms involved in bringing about this evolution. Our description revolved around the relationship between sequence, structure, and function of these domains, and how these characteristics determine the rate of protein evolution. We showed how the duplication, recombination, and divergence of protein domains had an enormous impact on the protein repertoire as we see it today.

In a very succinct review, Chothia and colleagues gave two possible reasons why these mechanisms of domain duplication, divergence, and recombination take place so commonly rather than via *ab initio* protein construction. The first sets of proteins created must have arisen via *ab initio* invention. Once an initial set of domains and proteins had evolved that was varied enough to support a basic form of life, it was much faster and more efficient to use duplication and recombination to create new protein combinations with altered functions. In addition, once repair mechanisms present in DNA replication and protein synthesis had been created, they made the *ab initio* process for domain invention too costly and difficult to be used very often.

In addition, many other processes, such as regulation of gene expression, alternative splicing, RNA editing or protein modification, determine how many and which kinds of proteins are active in a particular cell at a particular point in time. They represent mechanisms by which the complexity of the protein repertoire that is active in a cell is increased and delivers one explanation of why higher eukaryotes can do with relatively few genes. They are important mechanisms that deserve separate, more detailed description, and go beyond the scope of this discussion.

For an even more complete understanding of evolution, an integration of most different fields will be necessary: "genomics" analyses will have to join molecular evolution approaches; structural biologists, population geneticists, and experimentalists will have to interact more closely. Today, we already see some fruitful outcomes of such integrative approaches. One example is the recent work by Lynch and Conery. The authors analyze the relationship between the genome complexity and population size: species of large populations, such as bacteria, have small genomes, while for species of small populations, such as humans, we observe the opposite. Another beautiful example examines the link between molecular sequence variation and molecular and macroscopic phenotype; and this is for the chaperone Hsp90, which acts as a buffer and regulator in response to environmental changes. The chaperone assists the maturation, that is, the correct folding, of many key regulatory proteins. Thus, genetic variation can accumulate in genomes and can remain phenotypically silent due to the buffering by the chaperone. Upon environmental changes, however, for example, under stress, the chaperone's function is challenged and it releases novel phenotypes of folded protein structures.

While the groundwork for an understanding of protein structures and their evolution has been done, the exact interaction between selection and random processes is not yet clear. Only future comprehensive, integrative analyses of domain superfamilies and their combinations will reveal how the selection, in combination with random processes, can have produced a most complex protein universe. Thus, eventually, once we fully understand the vocabulary of proteins and grammar of protein evolution, we will be able to read the whole story of the book of life.

See also Genetic Analysis of Populations; Immunoassays; Molecular Systematics and Evolution; Origins of Life, Molecular Basis of; Paleontology, Molecular; Protein Expression by Expansion of the Genetic Code.

Bibliography

Books and Reviews

Branden, C., Tooze, J. (1998) *Introduction to Protein Structure*, Garland Science, New York.
Chothia, C., Gough, J., Vogel, C., Teichmann, S.A. (2003) Evolution of the protein repertoire, *Science* **300**(5626), 1701–1703.
Grishin, N.V. (2001) Fold change in evolution of protein structures, *J. Struct. Biol.* **134**, 167–185.

Koonin, E.V., Galperin, M.Y. (2002) *Sequence–Evolution–Function: Computational Approaches in Comparative Genomics*, Kluwer Academic Publishers, Boston, MA.

Koonin, E.V., Wolf, Y.I., Karev, G.P. (2002) The structure of the protein universe and genome evolution, *Nature* **420**(6912), 218–223.

Li, W.-H., Graur, D. (1999) *Fundamentals of Molecular Evolution*, Sinauer Associates Incorporated, Sunderland, MA.

Patthy, L. (1999) *Protein Evolution*, Blackwell Science, Oxford.

Petsko, G., Ringe, D. (2003) *Protein Structure and Function*, Blackwell Publishing, Oxford.

Ponting, C.P., Russell, R.R. (2002) The natural history of protein domains, *Annu. Rev. Biophys. Biomol. Struct.* **31**, 45–71. Epub 2001 Oct 25.

Rubin, G.M., Yandell, M.D., et al. (2000) Comparative genomics of the eukaryotes, *Science* **287**(5461), 2204–2215.

Todd, A.E., Orengo, C.A., Thornton, J.M. (1999) Evolution of protein function, from a structural perspective, *Curr. Opin. Chem. Biol.* **3**(5), 548–556.

Primary Literature

Alberts, B. (1998) The cell as a collection of protein machines: preparing the next generation of molecular biologists, *Cell* **92**, 291–294.

Aloy, P., Russell, R.B. (2002) Interrogating protein interaction networks through structural biology, *Proc. Natl. Acad. Sci. U.S.A.* **99**, 5896–5901.

Aloy, P., Ceulemans, H., Stark, A., Russell, R.B. (2003) The relationship between sequence and interaction divergence in proteins, *J. Mol. Biol.* **332**, 989–998.

Aloy, P., Ciccarelli, F.D., Leutwein, C., Gavin, A.C., Superti-Furga, G., Bork, P., Bottcher, B., Russell, R.B. (2002) A complex prediction: three-dimensional model of the yeast exosome, *EMBO Rep.* **3**, 628–635.

Alves, R., Chaleil, R.A.G., Sternberg, M.J.E. (2002) Evolution of enzymes in metabolism: a network perspective, *J. Mol. Biol.* **320**, 751–770.

Andrade, M.A., Perez-Iratxeta, C., Ponting, C.P. (2001) Protein repeats: structures, functions, and evolution, *J. Struct. Biol.* **134**, 117–131.

Andreeva, A., Howorth, D., Brenner, S.E., Hubbard, T.J., Chothia, C., Murzin, A.G. (2004) SCOP database in 2004: refinements integrate structure and sequence family data, *Nucleic Acids Res.* **32**, D226–D229.

Anfinsen, C.B., Haber, E., Sela, M., White, F.H. Jr. (1961) The kinetics of formation of native ribonuclease during oxidation of the reduced polypeptide chain, *Proc. Natl. Acad. Sci. U.S.A.* **47**, 1309–1314.

Apic, G., Gough, J., Teichmann, S.A. (2001) Domain combinations in archaeal, eubacterial and eukaryotic proteomes, *J. Mol. Biol.* **310**, 311–325.

Apic, G., Huber, W., Teichmann, S.A. (2003) Multidomain protein families and domain pairs: comparison with known structures and a random model of domain recombination, *J. Struct. Funct. Genomics* **4**, 67–78.

Apweiler, R., Attwood, T.K., Bairoch, A., Bateman, A., Birney, E., Biswas, M., Bucher, P., Cerutti, L., Corpet, F., Croning, M.D.R., et al. (2000) InterPro – an integrated documentation resource for protein families, domains and functional sites, *Bioinformatics* **16**, 1145–1150.

Aravind, L., Mazumder, R., Vasudevan, S., Koonin, E.V. (2002) Trends in protein evolution inferred from sequence and structure analysis, *Curr. Opin. Struct. Biol.* **12**, 392–399.

Aravind, L., Tatusov, R.L., Wolf, Y.I., Walker, D.R., Koonin, E.V. (1998) Evidence for massive gene exchange between archaeal and bacterial hyperthermophiles, *Trends Genet.* **14**, 442–444.

Ashburner, M., Ball, C.A., Blake, J.A., Botstein, D., Butler, H., Cherry, J.M., Davis, A.P., Dolinski, K., Dwight, S.S., Eppig, J.T., et al. (2000) Gene ontology: tool for the unification of biology. The Gene Ontology Consortium, *Nat. Genet.* **25**, 25–29.

Bairoch, A. (2000) The ENZYME database in 2000, *Nucleic Acids Res.* **28**, 304–305.

Bajaj, M., Blundell, T. (1984) Evolution and the tertiary structure of proteins, *Annu. Rev. Biophys. Bioeng.* **13**, 453–492.

Barabasi, A.L., Albert, R. (1999) Emergence of scaling in random networks, *Science* **286**, 509–512.

Barabasi, A.L., Oltvai, Z.N. (2004) Network biology: understanding the cell's functional organization, *Nat. Rev. Genet.* **5**, 101–113.

Bartlett, G.J., Borkakoti, N., Thornton, J.M. (2003) Catalysing new reactions during evolution: economy of residues and mechanism, *J. Mol. Biol.* **331**, 829–860.

Bashton, M., Chothia, C. (2002) The geometry of domain combination in proteins, *J. Mol. Biol.* **315**, 927–939.

Bateman, A., Coin, L., Durbin, R., Finn, R.D., Hollich, V., Griffiths-Jones, S., Khanna, A., Marshall, M., Moxon, S., Sonnhammer, E.L., et al. (2004) The Pfam protein families database, *Nucleic Acids Res.* **32**, D138–D141.

Beckmann, R., Spahn, C.M., Eswar, N., Helmers, J., Penczek, P.A., Sali, A., Frank, J., Blobel, G. (2001) Architecture of the protein-conducting channel associated with the translating 80S ribosome, *Cell* **107**, 361–372.

Berman, H.M., Battistuz, T., Bhat, T.N., Bluhm, W.F., Bourne, P.E., Burkhardt, K., Feng, Z., Gilliland, G.L., Iype, L., Jain, S., et al. (2002) The protein data bank, *Acta Crystallogr. D Biol. Crystallogr.* **58**, 899–907.

Betts, M.J., Guigo, R., Agarwal, P., Russell, R.B. (2001) Exon structure conservation despite low sequence similarity: a relic of dramatic events in evolution? *EMBO J.* **20**, 5354–5360.

Bilke, S., Peterson, C. (2001) Topological properties of citation and metabolic networks, *Phys. Rev. E Stat. Nonlin. Soft Matter Phys.* **64**, 036106.

Bork, P., Jensen, L.J., von Mering, C., Ramani, A.K., Lee, I., Marcotte, E.M. (2004) Protein interaction networks from yeast to human, *Curr. Opin. Struct. Biol.* **14**, 292–299.

Branden, C.I., Tooze, J. (1999) *Introduction to Protein Structure*, Garland Publishing, New York.

Brenner, S.E., Chothia, C., Hubbard, T.J. (1998) Assessing sequence comparison methods with reliable structurally identified distant evolutionary relationships, *Proc. Natl. Acad. Sci. U.S.A.* **95**, 6073–6078.

Brenner, S.E., Chothia, C., Hubbard, T.J.P., Murzin, A.G. (1996) Understanding Protein Structure: Using SCOP for Fold Interpretation, in: Doolittle, R.F. (Ed.) *Computer Methods for Macromolecular Sequence Analysis*, Chap. 37, Methods in Enzymology Vol. 266, Academic Press, Orlando, FL, pp. 635–643.

Caetano-Anolles, G., Caetano-Anolles, D. (2003) An evolutionarily structured universe of protein architecture, *Genome Res.* **13**, 1563–1571.

Castillo-Davis, C.I., Hartl, D.L., Achaz, G. (2004) cis-regulatory and protein evolution in orthologous and duplicate genes, *Genome Res.* **14**, 1530–1536.

Cavalier-Smith, T. (1991) Intron phylogeny: a new hypothesis, *Trends Genet.* **7**, 145–148.

Charlesworth, B., Sniegowski, P., Stephan, W. (1994) The evolutionary dynamics of repetitive DNA in eukaryotes, *Nature* **371**, 215–220.

Chen, J., Anderson, J.B., DeWeese-Scott, C., Fedorova, N.D., Geer, L.Y., He, S., Hurwitz, D.I., Jackson, J.D., Jacobs, A.R., Lanczycki, C.J., et al. (2003) MMDB: Entrez's 3D-structure database, *Nucleic Acids Res.* **31**, 474–477.

Chervitz, S.A., Aravind, L., Sherlock, G., Ball, C.A., Koonin, E.V., Dwight, S.S., Harris, M.A., Dolinski, K., Mohr, S., Smith, T., et al. (1998) Comparison of the complete protein sets of worm and yeast: Orthology and divergence, *Science* **282**, 2022–2028.

Chirgadze, D.Y., Demydchuk, M., Becker, M., Moran, S., Paoli, M. (2004) Snapshot of protein structure evolution reveals conservation of functional dimerization through intertwined folding, *Structure (Camb)* **12**, 1489–1494.

Chothia, C. (1992) Proteins – 1000 families for the molecular biologist, *Nature* **357**, 543–544.

Chothia, C., Lesk, A.M. (1982) Evolution of proteins formed by Beta-sheets. 1. Plastocyanin and Azurin, *J. Mol. Biol.* **160**, 309–323.

Chothia, C., Lesk, A.M. (1986) The relation between the divergence of sequence and structure in proteins, *EMBO J.* **5**, 823–826.

Chothia, C., Gough, J., Vogel, C., Teichmann, S.A. (2003) Evolution of the protein repertoire, *Science* **300**, 1701–1703.

Copley, R.R., Bork, P. (2000) Homology among (beta alpha)(8) barrels: Implications for the evolution of metabolic pathways, *J. Mol. Biol.* **303**, 627–640.

Copley, R.R., Schultz, J., Ponting, C.P., Bork, P. (1999) Protein families in multicellular organisms, *Curr. Opin. Struct. Biol.* **9**, 408–415.

Coulson, A.F., Moult, J. (2002) A unifold, mesofold, and superfold model of protein fold use, *Proteins: Struct., Funct., Genet.* **46**, 61–71.

Dandekar, T., Snel, B., Huynen, M., Bork, P. (1998) Conservation of gene order: a fingerprint of proteins that physically interact, *Trends Biochem. Sci.* **23**, 324–328.

Dandekar, T., Schuster, S., Snel, B., Huynen, M., Bork, P. (1999) Pathway alignment: application to the comparative analysis of glycolytic enzymes, *Biochem. J.* **343**(Pt 1), 115–124.

Darwin, C. (1859) *The Origin of Species by Means of Natural Selection, or the Preservation of Favoured Races in the Struggle for Life*, John Murray, London.

Dayhoff, M.O. (1976) The origin and evolution of protein superfamilies, *Fed. Proc.* **35**, 2132–2138.

Dodson, G., Wlodawer, A. (1998) Catalytic triads and their relatives, *Trends Biochem. Sci.* **23**, 347–352.

Doolittle, W.E. (1998) You are what you eat: a gene transfer ratchet could account for bacterial genes in eukaryotic nuclear genomes, *Trends Genet.* **14**, 307–311.

Doolittle, W.F. (1987) What introns have to tell us – hierarchy in genome evolution, *Cold. Spring Harb. Symp. Quant. Biol.* **52**, 907–913.

Doolittle, W.F. (1999a) Lateral genomics, *Trends Biochem. Sci.* **24**, M5–M8.

Doolittle, W.F. (1999b) Phylogenetic classification and the universal tree, *Science* **284**, 2124–2128.

Doolittle, W.F. (2000) The nature of the universal ancestor and the evolution of the proteome, *Curr. Opin. Struct. Biol.* **10**, 355–358.

Dujon, B., Sherman, D., Fischer, G., Durrens, P., Casaregola, S., Lafontaine, I., De Montigny, J., Marck, C., Neuveglise, C., Talla, E., et al. (2004) Genome evolution in yeasts, *Nature* **430**, 35–44.

Dunker, A.K., Brown, C.J., Lawson, J.D., Iakoucheva, L.M., Obradovic, Z. (2002) Intrinsic disorder and protein function, *Biochemistry* **41**, 6573–6582.

Dunker, A.K., Obradovic, Z., Romero, P., Garner, E.C., Brown, C.J. (2000) Intrinsic protein disorder in complete genomes, *Genome Inform. Ser. Workshop Genome Inform.* **11**, 161–171.

Dunker, A.K., Lawson, J.D., Brown, C.J., Williams, R.M., Romero, P., Oh, J.S., Oldfield, C.J., Campen, A.M., Ratliff, C.M., Hipps, K.W., et al. (2001) Intrinsically disordered protein, *J. Mol. Graph. Model.* **19**, 26–59.

Dunwell, J.M., Purvis, A., Khuri, S. (2004) Cupins: the most functionally diverse protein superfamily? *Phytochemistry* **65**, 7–17.

Eisen, J.A. (1998) A phylogenomic study of the MutS family of proteins, *Nucleic Acids Res.* **26**, 4291–4300.

Enright, A., Van Dongen, S., Ouzounis, C. (2002) An efficient algorithm for large-scale detection of protein families, *Nucleic Acids Res.* **30**, 1575–1584.

Enright, A.J., Iliopoulos, I., Kyrpides, N.C., Ouzounis, C.A. (1999) Protein interaction maps for complete genomes based on gene fusion events, *Nature* **402**, 86–90.

Fitch, W.M. (1970) Distinguishing homologous from analogous proteins, *Syst. Zool.* **19**, 99–113.

Fitch, W.M. (2000) Homology a personal view on some of the problems, *Trends Genet.* **16**, 227–231.

Fraser, H.B., Hirsh, A.E., Steinmetz, L.M., Scharfe, C., Feldman, M.W. (2002) Evolutionary rate in the protein interaction network, *Science* **296**, 750–752.

Galperin, M.Y., Walker, D.R., Koonin, E.V. (1998) Analogous enzymes: Independent inventions in enzyme evolution, *Genome Res.* **8**, 779–790.

Gerstein, M. (1997) A structural census of genomes: comparing bacterial, eukaryotic, and archaeal genomes in terms of protein structure, *J. Mol. Biol.* **274**, 562–576.

Gilbert, W. (1978) Why genes in pieces? *Nature* **271**, 501.

Gilbert, W. (1979) *Introns and Exons: Playgrounds of Evolution*, Academic Press, New York.

Gough, J. (2004) Convergent evolution of domain architectures (is rare), *Bioinformatics*.

Gough, J., Chothia, C. (2002) SUPERFAMILY: HMMs representing all proteins of known structure. SCOP sequence searches, alignments and genome assignments, *Nucleic Acids Res.* **30**, 268–272.

Gough, J., Karplus, K., Hughey, R., Chothia, C. (2001) Assignment of homology to genome sequences using a library of hidden Markov models that represent all proteins of known structure, *J. Mol. Biol.* **313**, 903–919.

Govindarajan, S., Recabarren, R., Goldstein, R.A. (1999) Estimating the total number of protein folds, *Proteins: Struct., Funct., Genet.* **35**, 408–414.

Grishin, N.V. (2001) Fold change in evolution of protein structures, *J. Struct. Biol.* **134**, 167–185.

Gu, Z., Nicolae, D., Lu, H.H., Li, W.H. (2002) Rapid divergence in expression between duplicate genes inferred from microarray data, *Trends Genet.* **18**, 609–613.

Hadley, C., Jones, D.T. (1999) A systematic comparison of protein structure classifications: SCOP, CATH and FSSP, *Structure Fold. Des.* **7**, 1099–1112.

Han, J.D., Bertin, N., Hao, T., Goldberg, D.S., Berriz, G.F., Zhang, L.V., Dupuy, D., Walhout, A.J., Cusick, M.E., Roth, F.P., et al. (2004) Evidence for dynamically organized modularity in the yeast protein-protein interaction network, *Nature* **430**, 88–93.

Harrison, S.C. (2003) Variation on an Src-like theme, *Cell* **112**, 737–740.

Harrison, P.M., Gerstein, M. (2002) Studying genomes through the aeons: Protein families, pseudogenes and proteome evolution, *J. Mol. Biol.* **318**, 1155–1174.

Heger, A., Holm, L. (2000) Towards a covering set of protein family profiles, *Prog. Biophys. Mol. Biol.* **73**, 321–337.

Hegyi, H., Gerstein, M. (1999) The relationship between protein structure and function: a comprehensive survey with application to the yeast genome, *J. Mol. Biol.* **288**, 147–164.

Hegyi, H., Gerstein, M. (2001) Annotation transfer for genomics: measuring functional divergence in multidomain proteins, *Genome Res.* **11**, 1632–1640.

Hill, E., Morea, V., Chothia, C. (2002) Sequence conservation in families whose members have little or no sequence similarity: the four-helical cytokines and cytochromes, *J. Mol. Biol.* **322**, 205–233.

Hill, E., Broadbent, I.D., Chothia, C., Pettitt, J. (2001) Cadherin superfamily proteins in Caenorhabditis elegans and Drosophila melanogaster, *J. Mol. Biol.* **305**, 1011–1024.

Holm, L., Sander, C. (1996) The FSSP database: fold classification based on structure structure alignment of proteins, *Nucleic Acids Res.* **24**, 206–209.

Hughes, A.L., Nei, M. (1988) Pattern of nucleotide substitution at major histocompatibility complex class I loci reveals overdominant selection, *Nature* **335**, 167–170.

Hughes, T.R., Marton, M.J., Jones, A.R., Roberts, C.J., Stoughton, R., Armour, C.D., Bennett, H.A., Coffey, E., Dai, H., He, Y.D., et al. (2000) Functional discovery via a compendium of expression profiles, *Cell* **102**, 109–126.

Hurst, L.D., Smith, N.G. (1999) Do essential genes evolve slowly? *Curr. Biol.* **9**, 747–750.

Hurst, G.D., Werren, J.H. (2001) The role of selfish genetic elements in eukaryotic evolution, *Nat. Rev. Genet.* **2**, 597–606.

Hurst, L.D., Pal, C., Lercher, M.J. (2004) The evolutionary dynamics of eukaryotic gene order, *Nat. Rev. Genet.* **5**, 299–310.

Huynen, M.A., van Nimwegen, E. (1998) The frequency distribution of gene family sizes in complete genomes, *Mol. Biol. Evol.* **15**, 583–589.

Ingram, V.M. (1956) A specific chemical difference between the globins of normal human and sickle cell anaemia haemoglobin, *Nature* **178**, 792–794.

Ingram, V.M. (2004) Sickle cell anemia hemoglobin: the molecular biology of the first "molecular disease"–the crucial importance of serendipity, *Genetics* **167**, 1–7.

Irving, J.A., Whisstock, J.C., Lesk, A.M. (2001) Protein structural alignments and functional genomics, *Proteins: Struct., Funct., Genet.* **42**, 378–382.

Ito, T., Chiba, T., Ozawa, R., Yoshida, M., Hattori, M., Sakaki, Y. (2001) A comprehensive two-hybrid analysis to explore the yeast protein interactome, *Proc. Natl. Acad. Sci. U.S.A.* **98**, 4569–4574.

Janin, J., Chothia, C. (1985) Domains in proteins – definitions, location, and structural principles, *Methods Enzymol.* **115**, 420–430.

Jardine, O., Gough, J., Chothia, C., Teichmann, S.A. (2002) Comparison of the small molecule metabolic enzymes of Escherichia coli and Saccharomyces cerevisiae, *Genome Res.* **12**, 916–929.

Jeong, H., Mason, S.P., Barabasi, A.L., Oltvai, Z.N. (2001) Lethality and centrality in protein networks, *Nature* **411**, 41–42.

Jeong, H., Tombor, B., Albert, R., Oltvai, Z.N., Barabasi, A.L. (2000) The large-scale organization of metabolic networks, *Nature* **407**, 651–654.

Jordan, I.K., Makarova, K.S., Spouge, J.L., Wolf, Y.I., Koonin, E.V. (2001) Lineage-specific gene expansions in bacterial and archaeal genomes, *Genome Res.* **11**, 555–565.

Kaessmann, H., Zollner, S., Nekrutenko, A., Li, W.H. (2002) Signatures of domain shuffling in the human genome, *Genome Res.* **12**, 1642–1650.

Karev, G.P., Wolf, Y.I., Koonin, E.V. (2003) Simple stochastic birth and death models of genome evolution: was there enough time for us to evolve? *Bioinformatics* **19**, 1889–1900.

Karev, G.P., Wolf, Y.I., Rzhetsky, A.Y., Berezovskaya, F.S., Koonin, E.V. (2002) Birth and death of protein domains: a simple model of evolution explains power law behavior, *BMC Evol. Biol.* **2**, 18.

Kendrew, J.C. (1958) A three-dimensional model of the myoglobin molecule obtained by x-ray analysis, *Nature* **181**, 662–666.

Khaitovich, P., Weiss, G., Lachmann, M., Hellmann, I., Enard, W., Muetzel, B., Wirkner, U., Ansorge, W., Paabo, S. (2004) A neutral model of transcriptome evolution, *PLoS Biol.* **2**, E132.

Kidwell, M.G. (2002) Transposable elements and the evolution of genome size in eukaryotes, *Genetica* **115**, 49–63.

Kimura, M. (1983) *The Neutral Theory of Molecular Evolution*, Cambridge University Press, Cambridge, MA.

Kinch, L.N., Grishin, N.V. (2002) Evolution of protein structures and functions, *Curr. Opin. Struct. Biol.* **12**, 400–408.

Koonin, E.V., Aravind, L., Kondrashov, A.S. (2000) The impact of comparative genomics on our understanding of evolution, *Cell* **101**, 573–576.

Koonin, E.V., Makarova, K.S., Aravind, L. (2001) Horizontal gene transfer in prokaryotes: quantification and classification, *Annu. Rev. Microbiol.* **55**, 709–742.

Koonin, E.V., Mushegian, A.R., Bork, P. (1996) Non-orthologous gene displacement, *Trends Genet.* **12**, 334–336.

Koonin, E.V., Wolf, Y.I., Karev, G.P. (2002) The structure of the protein universe and genome evolution, *Nature* **420**, 218–223.

Kuznetsov, V.A. (2002). *Computational and Statistical Approaches to Genomics*, Kluwer, Boston, MA, pp. 125–171.

Lander, E.S., Linton, L.M., Birren, B., Nusbaum, C., Zody, M.C., Baldwin, J., Devon, K., Dewar, K., Doyle, M., FitzHugh, W., et al. (2001) Initial sequencing and analysis of the human genome, *Nature* **409**, 860–921.

Lang, D., Thoma, R., Henn-Sax, M., Sterner, R., Wilmanns, M. (2000) Structural evidence for evolution of the beta/alpha barrel scaffold by gene duplication and fusion, *Science* **289**, 1546–1550.

Lee, D., Grant, A., Buchan, D., Orengo, C. (2003) A structural perspective on genome evolution, *Curr. Opin. Struct. Biol.* **13**, 359–369.

Lercher, M.J., Urrutia, A.O., Hurst, L.D. (2002) Clustering of housekeeping genes provides a unified model of gene order in the human genome, *Nat. Genet.* **31**, 180–183.

Lesk, A.M., Chothia, C. (1980) How different amino acid sequences determine similar protein structures: the structure and evolutionary dynamics of the globins, *J. Mol. Biol.* **136**, 225–270.

Lespinet, O., Wolf, Y.I., Koonin, E.V., Aravind, L. (2002) The role of lineage-specific gene family expansion in the evolution of eukaryotes, *Genome Res.* **12**, 1048–1059.

Letunic, I., Copley, R.R., Schmidt, S., Ciccarelli, F.D., Doerks, T., Schultz, J., Ponting, C.P., Bork, P. (2004) SMART 4.0: towards genomic data integration, *Nucleic Acids Res.* **32**, D142–D144.

Levitt, M., Chothia, C. (1976) Structural patterns in globular proteins, *Nature* **261**, 552–558.

Liu, J., Rost, B. (2003) Domains, motifs and clusters in the protein universe, *Curr. Opin. Chem. Biol.* **7**, 5–11.

Liu, Y., Gerstein, M., Engelman, D.M. (2004) Evolutionary use of domain recombination: a distinction between membrane and soluble proteins, *Proc. Natl. Acad. Sci. U.S.A.* **101**(10), 3495–3497.

Lupas, A.N., Ponting, C.P., Russell, R.B. (2001) On the evolution of protein folds: are similar motifs in different protein folds the result of convergence, insertion, or relics of an ancient peptide world? *J. Struct. Biol.* **134**, 191–203.

Luscombe, N., Qian, J., Zhang, Z., Johnson, T., Gerstein, M. (2002) The dominance of the population by a selected few: power law behavior applies to a wide variety of genomic properties, *Genome Biol.* **3**, RESEARCH0040.

Lynch, M., Conery, J.S. (2000) The evolutionary fate and consequences of duplicate genes, *Science* **290**, 1151–1155.

Lynch, M., Conery, J. (2003) The origins of genome complexity, *Science* **302**, 1401–1404.

Madera, M., Gough, J. (2002) A comparison of profile hidden Markov model procedures for remote homology detection, *Nucleic Acids Res.* **19**, 30.

Madera, M., Vogel, C., Kummerfeld, S.K., Chothia, C., Gough, J. (2004) The SUPERFAMILY database in 2004: additions and improvements, *Nucleic Acids Res.* **32**, D235–D239.

Makalowski, W., Boguski, M.S. (1998) Evolutionary parameters of the transcribed mammalian genome: an analysis of 2,820 orthologous rodent and human sequences, *Proc. Natl. Acad. Sci. U.S.A.* **95**, 9407–9412.

Makarova, K.S., Aravind, L., Daly, M.J., Koonin, E.V. (2000) Specific expansion of protein families in the radioresistant bacterium

Deinococcus radiodurans, *Genetica* **108**, 25–34.

Makarova, K.S., Aravind, L., Galperin, M.Y., Grishin, N.V., Tatusov, R.L., Wolf, Y.I., Koonin, E.V. (1999) Comparative genomics of the archaea (Euryarchaeota): evolution of conserved protein families, the stable core, and the variable shell, *Genome Res.* **9**, 608–628.

Makarova, K.S., Aravind, L., Wolf, Y.I., Tatusov, R.L., Minton, K.W., Koonin, E.V., Daly, M.J. (2001) Genome of the extremely radiation-resistant bacterium Deinococcus radiodurans viewed from the perspective of comparative genomics, *Microbiol. Mol. Biol. Rev.* **65**, 44–79.

Marchler-Bauer, A., Addess, K.J., Chappey, C., Geer, L., Madej, T., Matsuo, Y., Wang, Y., Bryant, S.H. (1999) MMDB: Entrez's 3D structure database, *Nucleic Acids Res.* **27**, 240–243.

Marcotte, E.M., Pellegrini, M., Ng, H.L., Rice, D.W., Yeates, T.O., Eisenberg, D. (1999a) Detecting protein function and protein-protein interactions from genome sequences, *Science* **285**, 751–753.

Marcotte, E.M., Pellegrini, M., Thompson, M.J., Yeates, T.O., Eisenberg, D. (1999b) A combined algorithm for genome-wide prediction of protein function, *Nature* **402**, 83–86.

Martin, A.C., Orengo, C.A., Hutchinson, E.G., Jones, S., Karmirantzou, M., Laskowski, R.A., Mitchell, J.B., Taroni, C., Thornton, J.M. (1998) Protein folds and functions, *Structure Fold. Des.* **6**, 875–884.

Mazet, F., Shimeld, S.M. (2002) Gene duplication and divergence in the early evolution of vertebrates, *Curr. Opin. Genet. Dev.* **12**, 393–396.

McKenzie, A.N., Barry, S.C., Strath, M., Sanderson, C.J. (1991) Structure-function analysis of interleukin-5 utilizing mouse/human chimeric molecules, *EMBO J.* **10**, 1193–1199.

Miles, E.W., Davies, D.R. (2000) Protein evolution. On the ancestry of barrels, *Science* **289**, 1490.

Miyata, T., Yasunaga, T. (1980) Molecular evolution of mRNA: a method for estimating evolutionary rates of synonymous and amino acid substitutions from homologous nucleotide sequences and its application, *J. Mol. Evol.* **16**, 23–36.

Miyata, T., Miyazawa, S., Yasunaga, T. (1979) Two types of amino acid substitutions in protein evolution, *J. Mol. Evol.* **12**, 219–236.

Mott, R., Schultz, J., Bork, P., Ponting, C.P. (2002) Predicting protein cellular localization using a domain projection method, *Genome Res.* **12**, 1168–1174.

Muller, A., MacCallum, R.M., Sternberg, M.J. (2002) Structural characterization of the human proteome, *Genome Res.* **12**, 1625–1641.

Murzin, A.G. (1998) How far divergent evolution goes in proteins, *Curr. Opin. Struct. Biol.* **8**, 380–387.

Murzin, A.G., Brenner, S.E., Hubbard, T., Chothia, C. (1995) Scop – a structural classification of proteins database for the investigation of sequences and structures, *J. Mol. Biol.* **247**, 536–540.

Nagano, N., Orengo, C.A., Thornton, J.M. (2002) One fold with many functions: the evolutionary relationships between TIM-barrel families based on their sequences, structures and functions, *J. Mol. Biol.* **321**, 741–765.

Nagarajan, N., Yona, G. (2004) Automatic prediction of protein domains from sequence information using a hybrid learning system, *Bioinformatics* **20**, 1335–1360.

Nelson, K.E., Clayton, R.A., Gill, S.R., Gwinn, M.L., Dodson, R.J., Haft, D.H., Hickey, E.K., Peterson, J.D., Nelson, W.C., Ketchum, K.A., et al. (1999) Evidence for lateral gene transfer between Archaea and bacteria from genome sequence of Thermotoga maritima, *Nature* **399**, 323–329.

Newlove, T., Konieczka, J.H., Cordes, M.H. (2004) Secondary structure switching in Cro protein evolution, *Structure (Camb)* **12**, 569–581.

Nicola, N.A., Hilton, D.J. (1998) General classes and functions of four-helix bundle cytokines, *Adv. Protein Chem.* **52**, 1–65.

Nobbs, C.L., Watson, H.C., Kendrew, J.C. (1966) Structure of deoxymyoglobin: a crystallographic study, *Nature* **209**, 339–341.

Ogata, H., Audic, S., Barbe, V., Artiguenave, F., Fournier, P.E., Raoult, D., Claverie, J.M. (2000) Selfish DNA in protein-coding genes of Rickettsia, *Science* **290**, 347–350.

Ohno, S. (1970) *Evolution by Gene Duplication*, Springer-Verlag, New York.

Orengo, C.A., Jones, D.T., Thornton, J.M. (1994) Protein superfamilies and domain superfolds, *Nature* **372**, 631–634.

Orengo, C.A., Pearl, F.M., Thornton, J.M. (2003) The CATH domain structure database, *Methods Biochem. Anal.* **44**, 249–271.

Orengo, C.A., Sillitoe, I., Reeves, G., Pearl, F.M.G. (2001) Review: what can structural classifications reveal about protein evolution? *J. Struct. Biol.* **134**, 145–165.

Orengo, C.A., Michie, A.D., Jones, S., Jones, D.T., Swindells, M.B., Thornton, J.M. (1997) CATH – a hierarchic classification of protein domain structures, *Structure* **5**, 1093–1108.

Ouzounis, C.A., Coulson, R.M., Enright, A.J., Kunin, V., Pereira-Leal, J.B. (2003) Classification schemes for protein structure and function, *Nat. Rev. Genet.* **4**, 508–519.

Overbeek, R., Fonstein, M., D'Souza, M., Pusch, G.D., Maltsev, N. (1999) The use of gene clusters to infer functional coupling, *Proc. Natl. Acad. Sci. U.S.A.* **96**, 2896–2901.

Papp, B., Pál, C., Hurst, L. (2003) Dosage sensitivity and the evolution of gene families in yeast, *Nature* **424**, 194–197.

Papp, B., Pal, C., Hurst, L.D. (2004) Metabolic network analysis of the causes and evolution of enzyme dispensability in yeast, *Nature* **429**, 661–664.

Park, J., Lappe, M., Teichmann, S.A. (2001) Mapping protein family interactions: Intramolecular and intermolecular protein family interaction repertoires in the PDB and yeast, *J. Mol. Biol.* **307**, 929–938.

Park, J., Karplus, K., Barrett, C., Hughey, R., Haussler, D., Hubbard, T., Chothia, C. (1998) Sequence comparisons using multiple sequences detect three times as many remote homologues as pairwise methods, *J. Mol. Biol.* **284**, 1201–1210.

Patthy, L. (1994) Exons and introns, *Curr. Opin. Struct. Biol.* **4**, 383–392.

Patthy, L. (1999) Genome evolution and the evolution of exon-shuffling–a review, *Gene* **238**, 103–114.

Patthy, L. (2003) Modular assembly of genes and the evolution of new functions, *Genetica* **118**, 217–231.

Pawson, T., Nash, P. (2003) Assembly of cell regulatory systems through protein interaction domains, *Science* **300**, 445–452.

Pellegrini, M., Marcotte, E.M., Thompson, M.J., Eisenberg, D., Yeates, T.O. (1999) Assigning protein functions by comparative genome analysis: protein phylogenetic profiles, *Proc. Natl. Acad. Sci. U.S.A.* **96**, 4285–4288.

Perutz, M.F., Kendrew, J.C., Watson, H.C. (1965) Structure and function of haemoglobin: II. Some relations between polypeptide chain configuration and amino acid sequence, *J. Mol. Biol.* **13**, 669–678.

Perutz, M., Muirhead, H., Cox, J.M., Goaman, L.C. (1968) Three-dimensional Fourier synthesis of horse oxyhemoglobin at 2.8 A resolution: the atomic model, *Nature* **219**, 131–139.

Ponting, C.P., Russell, R.R. (2002) The natural history of protein domains, *Annu. Rev. Biophys. Biomol. Struct.* **31**, 45–71.

Prabu, M.M., Suguna, K., Vijayan, M. (1999) Variability in quaternary association of proteins with the same tertiary fold: a case study and rationalization involving legume lectins, *Proteins: Struct., Funct., Genet.* **35**, 58–69.

Qian, J., Luscombe, N.M., Gerstein, M. (2001) Protein family and fold occurrence in genomes: Power-law behaviour and evolutionary model, *J. Mol. Biol.* **313**, 673–681.

Ragan, M.A. (2001) Detection of lateral gene transfer among microbial genomes, *Curr. Opin. Genet. Dev.* **11**, 620–626.

Ranea, J.A., Buchan, D.W., Thornton, J.M., Orengo, C.A. (2004) Evolution of protein superfamilies and bacterial genome size, *J. Mol. Biol.* **336**, 871–887.

Reidhaar-Olson, J.F., Sauer, R.T. (1988) Combinatorial cassette mutagenesis as a probe of the informational content of protein sequences, *Science* **241**, 53–57.

Remm, M., Storm, C.E., Sonnhammer, E.L. (2001) Automatic clustering of orthologs and in-paralogs from pairwise species comparisons, *J. Mol. Biol.* **314**, 1041–1052.

Riley, M., Labedan, B. (1996) E.coli Gene Products: Physiological Functions and Common Ancestries, in: Neidhart, F.C. (Ed.) *Escherichia Coli and Salmonella, Cellular and Molecular Biology*, ASM Press, Washington, DC, pp. 2118–2202.

Riley, M., Serres, M.H. (2000) Interim report on genomics of Escherichia coli, *Annu. Rev. Microbiol.* **54**, 341–411.

Rison, S.C., Teichmann, S.A., Thornton, J.M. (2002) Homology, pathway distance and chromosomal localization of the small molecule metabolism enzymes in Escherichia coli, *J. Mol. Biol.* **318**, 911–932.

Rokas, A., Williams, B., King, N., Carroll, S. (2003) Genome-scale approaches to resolving incongruence in molecular phylogenies, *Nature* **425**, 798–804.

Rossmann, M.G., Moras, D., Olsen, K.W. (1974) Chemical and biological evolution of nucleotide-binding protein, *Nature* **250**, 194–199.

Rubin, G.M., Yandell, M.D., Wortman, J.R., Gabor Miklos, G.L., Nelson, C.R., Hariharan, I.K., Fortini, M.E., Li, P.W., Apweiler, R., Fleischmann, W., et al. (2000) Comparative genomics of the eukaryotes, *Science* **287**, 2204–2215.

Russell, R.B. (1994) Domain insertion, *Protein Eng.* **7**, 1407–1410.

Russell, R.B. (1998) Detection of protein three-dimensional side-chain patterns: new examples of convergent evolution, *J. Mol. Biol.* **279**, 1211–1227.

Russell, R.B., Alber, F., Aloy, P., Davis, F.P., Korkin, D., Pichaud, M., Topf, M., Sali, A. (2004) A structural perspective on protein-protein interactions, *Curr. Opin. Struct. Biol.* **14**, 313–324.

Rzhetsky, A., Gomez, S.M. (2001) Birth of scale-free molecular networks and the number of distinct DNA and protein domains per genome, *Bioinformatics* **17**, 988–996.

Sangster, T.A., Lindquist, S., Queitsch, C. (2004) Under cover: causes, effects and implications of Hsp90-mediated genetic capacitance, *Bioessays* **26**, 348–362.

Schmidt, S., Sunyaev, S., Bork, P., Dandekar, T. (2003) Metabolites: a helping hand for pathway evolution?, *Trends Biochem. Sci.* **28**, 336–341.

Schultz, J., Milpetz, F., Bork, P., Ponting, C.P. (1998) SMART, a simple modular architecture research tool: identification of signaling domains, *Proc. Natl. Acad. Sci. U.S.A.* **95**, 5857–5864.

Semple, C., Wolfe, K.H. (1999) Gene duplication and gene conversion in the Caenorhabditis elegans genome, *J. Mol. Evol.* **48**, 555–564.

Serres, M.H., Goswami, S., Riley, M. (2004) GenProtEC: an updated and improved analysis of functions of Escherichia coli K-12 proteins, *Nucleic Acids Res.* **32**, D300–D302.

Sidow, A. (1996) Genome duplications in the evolution of early vertebrates, *Curr. Opin. Genet. Dev.* **6**, 715–722.

Simillion, C., Vandepoele, K., Saeys, Y., Van de Peer, Y. (2004) Building genomic profiles for uncovering segmental homology in the twilight zone, *Genome Res.* **14**, 1095–1106.

Snel, B., Bork, P., Huynen, M. (2000) Genome evolution. Gene fusion versus gene fission, *Trends Genet.* **16**, 9–11.

Snel, B., Bork, P., Huynen, M. (2002) Genomes in flux: the evolution of archaeal and proteobacterial gene content, *Genome Res.* **12**, 17–25.

Sonnhammer, E.L., Koonin, E.V. (2002) Orthology, paralogy and proposed classification for paralog subtypes, *Trends Genet.* **18**, 619–620.

Spahn, C.M., Beckmann, R., Eswar, N., Penczek, P.A., Sali, A., Blobel, G., Frank, J. (2001) Structure of the 80S ribosome from Saccharomyces cerevisiae–tRNA-ribosome and subunit-subunit interactions, *Cell* **107**, 373–386.

Swindells, M.B., Orengo, C.A., Jones, D.T., Hutchinson, E.G., Thornton, J.M. (1998) Contemporary approaches to protein structure classification, *Bioessays* **20**, 884–891.

Tatusov, R.L., Koonin, E.V., Lipman, D.J. (1997) A genomic perspective on protein families, *Science* **278**, 631–637.

Tatusov, R.L., Galperin, M.Y., Natale, D.A., Koonin, E.V. (2000) The COG database: a tool for genome-scale analysis of protein functions and evolution, *Nucleic Acids Res.* **28**, 33–36.

Tatusov, R.L., Fedorova, N.D., Jackson, J.D., Jacobs, A.R., Kiryutin, B., Koonin, E.V., Krylov, D.M., Mazumder, R., Mekhedov, S.L., Nikolskaya, A.N., et al. (2003) The COG database: an updated version includes eukaryotes, *BMC Bioinform.* **4**, 41.

Teichmann, S.A. (2002) The constraints protein-protein interactions place on sequence divergence, *J. Mol. Biol.* **324**, 399–407.

Teichmann, S.A., Park, J., Chothia, C. (1998) Structural assignments to the Mycoplasma genitalium proteins show extensive gene duplications and domain rearrangements, *Proc. Natl. Acad. Sci. U.S.A.* **95**, 14658–14663.

Teichmann, S.A., Rison, S.C., Thornton, J.M., Riley, M., Gough, J., Chothia, C. (2001) Small-molecule metabolism: an enzyme mosaic, *Trends Biotechnol.* **19**, 482–486.

Timsit, Y. (2001) Convergent evolution of MutS and topoisomerase II for clamping DNA crossovers and stacked Holliday junctions, *J. Biomol. Struct. Dyn.* **19**, 215–218.

Todd, A.E., Orengo, C.A., Thornton, J.M. (1999) Evolution of protein function, from a structural perspective, *Curr. Opin. Chem. Biology.* **3**, 548–556.

Todd, A.E., Orengo, C.A., Thornton, J.M. (2001) Evolution of function in protein superfamilies, from a structural perspective, *J. Mol. Biol.* **307**, 1113–1143.

Todd, A.E., Orengo, C.A., Thornton, J.M. (2002) Plasticity of enzyme active sites, *Trends Biochem. Sci.* **27**, 419–426.

Tompa, P. (2002) Intrinsically unstructured proteins, *Trends Biochem. Sci.* **27**, 527–533.

Tsoka, S., Ouzounis, C.A. (2001) Functional versatility and molecular diversity of the metabolic map of Escherichia coli, *Genome Res.* **11**, 1503–1510.

Uetz, P., Giot, L., Cagney, G., Mansfield, T.A., Judson, R.S., Knight, J.R., Lockshon, D., Narayan, V., Srinivasan, M., Pochart, P., et al. (2000) A comprehensive analysis of protein-protein interactions in Saccharomyces cerevisiae, *Nature* **403**, 623–627.

van Nimwegen, E. (2003) Scaling laws in the functional content of genomes, *Trends Genet.* **19**, 479–484.

Vitagliano, L., Masullo, M., Sica, F., Zagari, A., Bocchini, V. (2001) Crystal structure of Sulfolobus solfataricus Elongation factor 1 alpha in complex with GDP, *EMBO J.* **20**, 5305.

Vogel, C., Teichmann, S.A., Chothia, C. (2003) The immunoglobulin superfamily in Drosophila melanogaster and Caenorhabditis elegans and the evolution of complexity, *Development* **130**, 6317–6328.

Vogel, C., Berzuini, C., Bashton, M., Gough, J., Teichmann, S.A. (2004) Supra-domains—evolutionary units larger than single protein domains, *J. Mol. Biol.* **336**, 809–823.

von Mering, C., Krause, R., Snel, B., Cornell, M., Oliver, S.G., Fields, S., Bork, P. (2002) Comparative assessment of large-scale data sets of protein-protein interactions, *Nature* **417**, 399–403.

Wagner, A. (2000a) Decoupled evolution of coding region and mRNA expression patterns after gene duplication: implications for the neutralist-selectionist debate, *Proc. Natl. Acad. Sci. U.S.A.* **97**, 6579–6584.

Wagner, A. (2000b) Robustness against mutations in genetic networks of yeast, *Nat. Genet.* **24**, 355–361.

Wang, Z.X. (1998) A reestimation for the total numbers of protein folds and superfamilies, *Protein Eng.* **11**, 621–626.

Webber, C., Ponting, C.P. (2004) Genes and homology, *Curr. Biol.* **14**, R332–R333.

Wetlaufer, D.B. (1973) Nucleation, rapid folding, and globular intrachain regions in proteins, *Proc. Natl. Acad. Sci. U.S.A.* **70**, 697–701.

Wilson, C.A., Kreychman, J., Gerstein, M. (2000) Assessing annotation transfer for genomics: quantifying the relations between protein sequence, structure and function through traditional and probabilistic scores, *J. Mol. Biol.* **297**, 233–249.

Winzeler, E.A., Liang, H., Shoemaker, D.D., Davis, R.W. (2000) Functional analysis of the yeast genome by precise deletion and parallel phenotypic characterization, *Novartis Found. Symp.* **229**, 105–109; discussion 109–111.

Winzeler, E.A., Shoemaker, D.D., Astromoff, A., Liang, H., Anderson, K., Andre, B., Bangham, R., Benito, R., Boeke, J.D., Bussey, H., et al. (1999) Functional characterization of the S. cerevisiae genome by gene deletion and parallel analysis, *Science* **285**, 901–906.

Wolf, Y.I., Aravind, L., Koonin, E.V. (1999a) Rickettsiae and Chlamydiae – evidence of horizontal gene transfer and gene exchange, *Trends Genet.* **15**, 173–175.

Wolf, Y.I., Grishin, N.V., Koonin, E.V. (2000) Estimating the number of protein folds and families from complete genome data, *J. Mol. Biol.* **299**, 897–905.

Wolf, Y.I., Karev, G., Koonin, E.V. (2002) Scale-free networks in biology: new insights into the fundamentals of evolution? *Bioessays* **24**, 105–109.

Wolf, Y.I., Rogozin, I.B., Koonin, E.V. (2004) Coelomata and not Ecdysozoa: evidence from genome-wide phylogenetic analysis, *Genome Res.* **14**, 29–36.

Wolf, Y.I., Brenner, S.E., Bash, P.A., Koonin, E.V. (1999b) Distribution of protein folds in the three superkingdoms of life, *Genome Res.* **9**, 17–26.

Wolfe, K.H. (2001) Yesterday's polyploids and the mystery of diploidization, *Nat. Rev. Genet.* **2**, 333–341.

Wolfe, K.H., Li, W.H. (2003) Molecular evolution meets the genomics revolution, *Nat. Genet.* **33**, 255–265.

Wolfe, K.H., Shields, D.C. (1997) Molecular evidence for an ancient duplication of the entire yeast genome, *Nature* **387**, 708–713.

Wright, C.S., Alden, R.A., Kraut, J. (1969) Structure of subtilisin BPN' at 2.5 angstrom resolution, *Nature* **221**, 235–242.

Wuchty, S. (2001) Scale-free behavior in protein domain networks, *Mol. Biol. Evol.* **18**, 1694–1702.

Wuchty, S. (2002) Interaction and domain networks of yeast, *Proteomics* **2**, 1715–1723.

Yanai, I., Wolf, Y.I., Koonin, E.V. (2002) Evolution of gene fusions: horizontal transfer versus independent events, *Genome Biol.* **3**, research0024.

Yang, Z., Bielawski, J.P. (2000) Statistical methods for detecting molecular adaptation, *Trends Ecol. Evol.* **15**, 496–503.

Yona, G., Linial, N., Tishby, N., Linial, M. (1998) A map of the protein space–an automatic hierarchical classification of all protein sequences, *Proc. Int. Conf. Intell. Syst. Mol. Biol.* **6**, 212–221.

Zhang, C., DeLisi, C. (1998) Estimating the number of protein folds, *J. Mol. Biol.* **284**, 1301–1305.

Zuckerkandl, E. (1975) The appearance of new structures and functions in proteins during evolution, *J. Mol. Evol.* **7**, 1–57.

Protein Splicing

Kenneth V. Mills
College of the Holy Cross, Worcester, MA

1	**Discovery of Protein Splicing**	243
2	**Chemical Mechanism of Protein Splicing**	243
2.1	N-S (or N-O) Acyl Rearrangement	243
2.2	Transesterification	245
2.3	Asparagine Cyclization Coupled to Peptide Bond Cleavage	246
2.4	Finishing Reactions	247
3	**Intein Structures**	247
3.1	Domain Organization	247
3.2	Three-dimensional Structures	248
4	**Protein Splicing in Trans**	248
5	**Kinetic Studies of Protein Splicing**	249
6	**Relationship of Inteins with Introns and Homing Endonucleases**	250
7	**Relationship of Protein Splicing with Other Forms of Protein Autoprocessing**	250
7.1	Bacterial Intein-like Domains	250
7.2	Hedgehog Proteins	251
7.3	Pyruvoyl Enzyme Autoprocessing	252
7.4	N-terminal Nucleophile Hydrolases	254
7.5	Nucleoporins	254
8	**Inteins in Biotechnology**	255
8.1	Self-cleavable Affinity Tags	255
8.2	Protein Ligation	255

Encyclopedia of Molecular Cell Biology and Molecular Medicine, 2nd Edition. Volume 11
Edited by Robert A. Meyers.
Copyright © 2005 Wiley-VCH Verlag GmbH & Co. KGaA, Weinheim
ISBN: 3-527-30648-X

8.3	Protein Cyclization 258
8.4	Screening Systems Utilizing Inteins 259
9	**Perspective** 259
	Acknowledgments 260
	Bibliography 260
	Books and Reviews 260
	Primary Literature 260

Keywords

Extein
The exteins are the flanking polypeptides interrupted by an intein.

HINT Domain
The HINT domain is the common structural folding pattern shared by **h**edgehog proteins and **int**eins.

Homing Endonuclease
Homing endonucleases recognize and cleave a target allele, facilitating the lateral transfer of the gene encoding an intein (or intron) to a homologous allele lacking the intervening sequence.

Intein
Inteins are intervening protein sequences that facilitate their own posttranslational excision from, as well as the ligation of, the flanking polypeptides.

Protein Splicing
Protein splicing is a posttranslational autoprocessing event by which an intervening protein facilitates its own excision from flanking polypeptides and the specific ligation of those polypeptides.

Trans-splicing
Protein splicing facilitated by a split intein. Intein reassociation must precede protein splicing in trans.

■ Protein splicing is a posttranslational autoprocessing event by which an intervening polypeptide sequence, the intein, facilitates both its own excision from flanking polypeptides, or exteins, as well as the ligation of the exteins. Since the discovery of protein splicing in 1990, significant progress has been made in elucidating the chemical mechanism of protein splicing by both biochemical and structural studies. The more recent discoveries of inteins that can facilitate protein trans-splicing, inteins that splice via noncanonical mechanisms, and bacterial intein-like domains have provided further insight into the mechanism and evolution of protein splicing. The study of protein splicing has also deepened our understanding of related forms of protein autoprocessing, particularly the processing of hedgehog proteins. Inteins have also found wide use in biotechnology applications, including systems for protein purification, ligation, and cyclization.

1
Discovery of Protein Splicing

Protein splicing was discovered in 1990 in *Saccharomyces cerevisiae* (*Sce*). The *Sce VMA1* gene is interrupted by an intervening sequence, which is absent from homologous genes in related organisms. The split gene is transcribed into the corresponding full-length mRNA, with no spliced mRNA product detected. Equal amounts of ATPase and intein are detected using Western blot, indicating that the intervening gene is translated. Protein-splicing elements were also discovered in the *recA* gene of *Mycobacterium tuberculosis* (*Mtu*) and the DNA polymerase gene in *Thermococcus litoralis*.

The experiments that followed defined the essential elements for a functional intein. The fact that splicing occurs if the intein is inserted into a foreign context indicates that the intein itself contains all of the genetic information necessary for protein splicing. Inteins have homology to homing endonuclease domains similar to those found in mobile introns, although it was demonstrated by deletion analysis that the endonuclease activity is not essential for protein splicing. Essential residues were identified by analysis of the modulation of protein splicing by mutation. The nucleophilic cysteine or serine flanking the intein, as well as the conserved asparagine at the C-terminus of the intein, play a role in efficient *in vivo* protein splicing.

2
Chemical Mechanism of Protein Splicing

2.1
N-S (or N-O) Acyl Rearrangement

The first step of protein splicing involves the formation of a linear thioester or ester intermediate at the N-terminal splice junction. (See Fig. 1.) This ester intermediate is the result of an N-S or N-O acyl shift of the peptide bond involving the amino group of the N-terminal cysteine or serine. A linear thioester intermediate was first observed in a *Pyrococcus* sp. GBD Pol intein in which the N-terminal serine was replaced by cysteine. The thioester was detected by reaction with neutral hydroxylamine or ethylenediamine, which results in cleavage at the N-terminal splice junction. The splice junction was identified by analysis of the C-terminal hydroxamate residue of the N-terminal extein after hydroxylamine-induced cleavage. A thioester intermediate was also observed in the *Sce* VMA intein, which naturally contains an N-terminal cysteine.

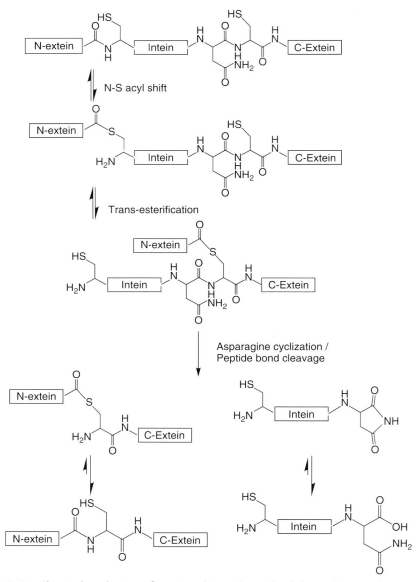

Fig. 1 Chemical mechanism of protein splicing. The nucleophilic residues are shown as cysteine in this example, although the upstream can be cysteine or serine and the downstream nucleophile can be cysteine, serine, or threonine.

Not all inteins initiate protein splicing by an N-S or N-O acyl shift of the scissile peptide bond, as inteins lacking an N-terminal nucleophile have been identified, three of which have been shown to splice. This suggests that these inteins have evolved to splice by bypassing the initial N-S or N-O shift through a direct attack by the downstream nucleophile on the upstream scissile peptide bond. In the *Methanococcus jannaschii* KlbA intein, replacement of the N-terminal alanine with cysteine or serine yields significant amounts of spliced product, and a double mutant of the intein N-terminal alanine to cysteine and the N-terminal cysteine of the C-exteins (The term C-extein will refer to the C-terminal extein segment, and N-extein will refer to the N-terminal extein segment. Amino acid numbering begins with the first residue of the intein, such that the "N-1 residue" is the C-terminal residue of the N-extein and the "C+1" residue is the N-terminal residue of the C-extein.) to alanine can promote N-terminal cleavage by neutral hydroxylamine. The fact that this modified intein is able to facilitate an N-S acyl shift suggests that the replacement of the N-terminal nucleophilic amino acid of the intein by alanine may have occurred recently.

Uncatalyzed N-O or N-S acyl shifts have not been detected, but are known to occur in peptide bonds involving the α-nitrogen of serine or threonine in the presence of strong acid. It has been suggested that catalytic strain may serve in both accelerating the rate of this step of protein splicing as well as in shifting the equilibrium more favorably toward the ester. For instance, in the crystal structure of the *Mycobacterium xenopi* (*Mxe*) GyrA intein, the scissile bond is in a cis conformation. In addition, distortion of the scissile bond in a catalytically active *Mxe* GyrA intein is suggested by NMR data obtained using a segmental isotopically labeled intein. These data include an abnormally low one-bond dipolar coupling constant for the scissile bond and an unusually large upfield proton shift for the amide proton. Experimental evidence suggests that this bond strain is mechanistically relevant, as the identity of the N-1 residue of the *Mxe* GyrA intein has an effect on both splicing yield *in vivo* and DTT-induced N-terminal cleavage *in vitro*. In the crystal structure of the *Sce* VMA intein, the backbone τ angle of the N-1 residue is distorted by $10°$. For this intein as well, a subset of mutations of the N-1 residue prevents protein splicing and DTT-induced N-terminal cleavage, and another set of mutations shifts the equilibrium of the N-S acyl shift to favor the thioester On the other hand, no distortion of the scissile bond is observed in crystal structures of nonsplicing mutants of the *Sce* VMA intein or the *Synechocystis* sp. PCC6803 (*Ssp*) DnaB intein.

2.2
Transesterification

The second step of protein splicing involves the transesterification of the linear ester by the downstream nucleophile (See Fig. 1). The result is a branched ester intermediate, first discovered in the *Psp*-GBD Pol intein. The branched intermediate was identified by reduced mobility in SDS-PAGE and by the results of sequential Edman degradation, in which two amino acids were released at each sequencing cycle, consistent with the N-terminal sequences of the N-extein and intein. Raising the pH shifts the equilibrium to the amide, and the lowering of the pH restores the branched ester, indicating full reversibility. An ester bond between the side chain of the downstream serine and the N-extein was discovered by trapping the branched ester by denaturation and cleaving it under mildly alkaline conditions. Branched thioester intermediates are also observed with inteins with cysteine at the C+1 position.

The upstream and downstream nucleophilic side chains are less than 4 Å apart in a crystal structure of a Sce VMA intein, suggesting facile transesterification. However, in the crystal structures of the Sce VMA and Ssp DnaB inteins, the nucleophilic side chains are separated by 8.5 and 9 Å, which suggests that either these inteins are not in an active form or require a significant conformational change to allow splicing. The downstream cysteine residue in the Mtu RecA intein has an unusually low pKa of 5.8, which may indicate that the transesterification can be facilitated by the unique dielectric environment of the intein. On the other hand, general base catalysis may facilitate transesterification in the Sce VMA intein, as the amino group of the intein N-terminal cysteine produced by the N-S shift may improve the nucleophilicity of the N-terminal thiol of the C-extein.

2.3
Asparagine Cyclization Coupled to Peptide Bond Cleavage

The third step of protein splicing is the cyclization of the C-terminal asparagine of the intein, which is coupled to cleavage of the peptide bond joining the intein and C-extein. (See Fig. 1.) Spontaneous asparagine cyclization in proteins typically results in side-chain deamidation and involves nucleophilic attack of the main-chain amide nitrogen on the side-chain carbonyl carbon to form an amidosuccinimide, followed by hydrolysis to either aspartic acid or iso-aspartic acid. An alternative mode of cyclization involves attack of the side-chain amide nitrogen on the main-chain carbonyl and leads to aminosuccinimide formation coupled to peptide bond cleavage. This may be preceded by deprotonation of the side-chain amide nitrogen, as indicated by studies of model peptides in which the rate of cleavage increases with pH. Spontaneous peptide bond cleavage via succinimide intermediates is observed in some long-lived proteins, such as α-crystallin.

The third step of protein splicing was elucidated by the isolation and analysis of C-terminal aminosuccinimide residues formed during protein splicing mediated by the Psp-GBD Pol intein and the Sce VMA intein. Branched intermediate can accumulate if the conserved asparagine at the C-terminus of the Psp-GBD Pol intein is mutated to aspartic acid or glutamine.

It has been proposed that the conserved penultimate histidine residue plays a role in asparagine cyclization. For instance, protein splicing is arrested and branched intermediate accumulates if the penultimate histidine in the Psp-GBD Pol intein is mutated to alanine. Protein splicing can be enhanced in inteins lacking a penultimate histidine by introducing a histidine residue by mutation. Structural evidence points to a role for the penultimate histidine as well. For instance, in the Mxe GyrA crystal structure, His197 is in a position to serve as a proton donor to the backbone amide nitrogen of the scissile bond. In the Ssp DnaB structure, the penultimate histidine could make a hydrogen bond with the main-chain carbonyl oxygen of the adjacent asparagine. However, at least 27 inteins lacking a penultimate histidine are known. Some of these inteins splice in the absence of the histidine residues, and in some cases splicing is impaired if the penultimate residue is replaced with histidine by mutation. It is possible that this histidine may play a role in coordinating the steps of protein splicing rather than facilitating the asparagine cyclization directly.

Two inteins with C-terminal glutamine can facilitate splicing, as can an intein with a C-terminal aspartic acid. Although it seems likely that these inteins splice via cyclization of the C-terminal residue to a glutarimide or succinanhydride, respectively, excised inteins with cyclized C-terminal residues have not been detected, even if the C-terminal glutamine is changed to asparagine. This is surprising because other inteins with C-terminal aminosuccinimide residues have been isolated, and suggest that the noncanonical inteins may splice via an alternative mechanism.

2.4
Finishing Reactions

Asparagine cyclization leads to the excision of the intein and leaves the exteins linked by an ester. The completion of protein splicing thus requires the hydrolysis of the C-terminal aminosuccinimide residue of the intein and the conversion of the thioester or ester bond ligating the extein segments to a native peptide bond (See Fig. 1). The N-O or N-S acyl rearrangement is probably uncatalyzed because the rearrangement in synthetic depsipeptides occurs in less than a minute, much faster than the measured rates of *in vitro* splicing reactions. However, a crystal structure of a spliced Sce VMA intein shows that the N-terminal amino group of the intein could stabilize the oxythiazolidine intermediate in the S-N rearrangement. Enhancement of the rate of this rearrangement could prevent hydrolysis of the nascent ester-linked extein segments. The hydrolysis of the C-terminal aminosuccinimide is probably uncatalyzed as well, as the timescale of the isolation of inteins containing C-terminal aminosuccinimide is consistent with the 2.3-h half-life of the L-imide in model peptides.

3
Intein Structures

3.1
Domain Organization

Inteins can vary in size from 134 to 608 residues, and many are themselves interrupted by a homing endonuclease domain. In general, inteins contain six conserved motifs related to protein splicing, which are labeled in Fig. 2 according to the revised nomenclature of Pietrokovski. Conserved elements include an N-terminal nucleophilic residue (either Cys or Ser) in block N1, a conserved aspartic acid or glutamic acid in blocks N2 and N4, a TXXH motif in block N3, and a His-Asn sequence in block C1. Inteins are flanked at the C-terminus with a conserved cysteine, serine, or threonine residue. Although inteins can promote protein splicing in a foreign context, the efficiency of protein splicing can

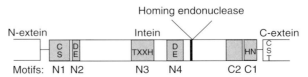

Fig. 2 Conserved elements in a typical intein. The motif nomenclature is that of Pietrokovski. Highly conserved residues are indicated in bold.

be influenced by the extein amino acid sequence. Random mutagenesis and genetic deletion studies have shown that the homing endonuclease domain is not required for protein-splicing function, and about 10% of inteins lack this central endonuclease domain, yet such mini-inteins can still promote splicing. This discovery led to the development of inteins split within the endonuclease domain that could splice in trans both *in vivo* and *in vitro*, and the discovery of a naturally split intein in *Synechocystis* sp. PCC6803.

3.2
Three-dimensional Structures

The crystal structure of the VMA intein from *S. cerevisiae* consists of two separate domains: the protein-splicing domain, which contains both termini of the protein, and the endonuclease domain. The protein-splicing domain is an elongated structure of seven β-sheets in two substructures. A crystal structure of a zinc-inactivated *Sce* VMA intein with the endonuclease domain deleted shows that the N-terminal scissile bond is in a strained trans conformation, although a later crystal structure with short N- and C-terminal exteins has a normal transpeptide bond at both splice junctions. The *M. xenopi* GyrA intein forms a flattened "horseshoe" shape, the edges of which are composed of two long, curved antiparallel β-stands. Two three-stranded β-sheets, related by pseudo twofold symmetry, form a cleft in which the N- and C-terminal residues are separated by less than 3 Å. The N-terminal scissile bond is in a cis conformation. The crystal structure of the *Ssp* DnaB mini-intein has a similar overall folding pattern, but exhibits strain in the C-terminal scissile bond and two separate catalytic sites that may be separately responsible for the reactions at the intein N- and C-termini. The crystal structure of the Hedgehog autoprocessing domain reveals that Hedgehog domains and inteins share a common fold, as is expected from their sequence homology. This common fold is called the *HINT module* (for Hedgehog-INTein).

4
Protein Splicing in Trans

The discovery of minimal inteins provided the impetus for the search for further intein modifications that would allow protein splicing in trans (See Fig. 3). Trans-splicing was first demonstrated *in vivo* in *Escherichia coli* using a plasmid encoding the *Mtu* RecA intein in which the DNA encoding the central homing endonuclease domain was replaced by a linker containing overlapping translation stop and start codons. Protein trans-splicing *in vitro* was demonstrated using separately expressed fragments, one consisting of *E. coli* maltose-binding protein fused to the 105 N-terminal residues of the RecA intein and the other consisting of the 107 C-terminal residues of the intein fused to a domain containing a His-tag. These fragments could be reassociated by denaturation followed by joint renaturation and efficiently mediated *in vitro* protein splicing. This trans-splicing system was improved by the generation of a semisynthetic split intein by replacement of the C-terminal expressed segment of the trans-splicing intein with a synthetic polypeptide. A similar trans-splicing system was derived from the *Psp*-GBD Pol intein, although the homing endonuclease domain could not be deleted without loss of function.

Fig. 3 Protein splicing in trans. The vertical lines are used to suggest interactions between the N- and C-intein segments.

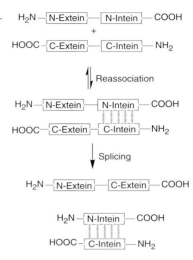

5
Kinetic Studies of Protein Splicing

A naturally split intein was discovered to interrupt the DnaE protein in *Synechocystis* species PCC6803. The N-terminal segment of the DnaE protein is fused to the N-intein, and the C-intein is fused to the C-terminal segment of the DnaE protein. The genes encoding these two proteins are separated by 745 kb of DNA and reside on opposite strands. The split intein can splice *in vivo* if overexpressed in *E. coli*. As a result of this discovery, an *in vitro* trans-splicing system using the DnaE intein was developed that could mediate splicing without the need for joint denaturation and renaturation, which has been used to discover zinc as an inhibitor for protein splicing, allowed for extensive kinetic analysis of protein splicing, and encouraged the development of biotechnological applications and protein-splicing screening systems that will be described below.

Because inteins facilitate an autoprocessing reaction, they serve as both enzyme and substrate and are not true "catalysts." To study the kinetics of the protein-splicing reaction, one must be able to activate an intein from an inactive precursor. This was first accomplished in detail with the split *Ssp* DnaE intein. The rate constant of trans-splicing at 23 °C was 6.6×10^{-5} s^{-1}. The first step of splicing, the N-S acyl shift, was fast in comparison, with a pseudo-first-order rate constant of DTT-induced N-terminal cleavage of 1.0×10^{-3} s^{-1}. Preincubation experiments suggest that intein association is even faster. On the other hand, the rate constant of C-terminal cleavage, that is, asparagine cyclization coupled to peptide bond cleavage, was 1.9×10^{-4} s^{-1}. These data suggest that, for the *Ssp* DnaE intein, the N-S acyl rearrangement proceeds more rapidly than asparagine cyclization, which could serve to ensure that the steps of splicing proceed in order. Further detailed kinetic analysis of inteins and modified inteins could help determine how the intein catalyzes and coordinates the individual chemical steps of splicing.

6
Relationship of Inteins with Introns and Homing Endonucleases

Inteins and introns both interrupt the coding sequence of genes. As with inteins, group I and group II introns can catalyze their own excision from and the religation of flanking sequences. However, although inteins are excised posttranslationally, introns are excised at the RNA level. Like protein splicing, the splicing of introns from precursor RNA consists of a series of transesterification reactions. However, the self-splicing of introns *in vivo* is often inefficient in the absence of protein cofactors. Over 80% of inteins are interrupted by homing endonuclease domains, and many introns contain open reading frames outside of their core domains that encode homing endonucleases.

Homing endonuclease domains interrupt the majority of known inteins and are usually members of the dodecapeptide (LAGLI-DAG) family. The endonuclease domain of the *Sce* VMA intein can mediate intein homing. It cleaves at the intein insertion locus of an intein-less allele of the *Sce VMA* gene during meiosis and converts the allele lacking the intein into one containing an intein. However, the endonuclease is not required for protein-splicing activity. The crystal structure of this intein shows that it is a two-domain protein, with one domain responsible for protein-splicing activity and the other for endonuclease activity. The wide genetic distribution of mini-inteins that lack homing endonucleases suggests that the homing function may have evolved early and that modern mini-inteins lost their endonuclease domains.

The homing ability of inteins due to the endonuclease domain may explain some interesting aspects of intein distribution. For instance, few inteins are found in eubacteria aside from those in the mycobacteria and cyanobacteria. This could be explained by the fact that homing events have not been found to occur in *E. coli*. Another strange case involves the relationship of the DnaB inteins found in *Synechocystis* sp. PCC6803 and in *Rhodothermus marinus*. The DnaB inteins have 54% sequence identity and are inserted at the same site in the host protein, but the codon usage in the *R. marinus* intein is very different from the codon usage in the extein. If this is the result of homing by the same intein, it is curious how it could occur between such unrelated organisms. A third example is that the host proteins can share greater sequence similarity than the inteins. The VMA inteins in *Candida tropicalis* and *S. cerevisiae* are only 37% identical, whereas the host proteins are 88% identical. It would seem unlikely that unrelated inteins would invade the same gene if there were not some role for protein splicing that exerts genetic pressure to maintain the intein.

It is unclear if inteins or introns are more than merely parasitic DNA elements. Inteins usually interrupt conserved domains and introns are usually inserted at domain boundaries, suggesting that they might serve a regulatory function. The most likely case for possible gene regulation by an intein would seem to be the naturally split DnaE inteins, which could play a role in regulating DNA replication.

7
Relationship of Protein Splicing with Other Forms of Protein Autoprocessing

7.1
Bacterial Intein-like Domains

Bacterial intein-like domains (BILs) have been discovered in a wide range of bacterial species. The BILs have more conserved sequence motifs than inteins, and are inserted into variable regions of their host proteins. There are two major categories of BILs: the A-type BILs contain the conserved intein sequence blocks, but mostly lack a C-terminal nucleophile, and the B-type BILs lack intein motifs C1 and C2, but usually have a nucleophilic residue at their penultimate positions.

A-type BILs have been shown to facilitate protein splicing as well as C-terminal cleavage reactions. An interesting case is an A-type BIL from *Clostridium thermocellum*, which facilitates proteins splicing, although it has a C-terminal alanine. This suggests that the BIL might splice via an alternative mechanism in which an N-S acyl rearrangement at the N-terminal splice junction and asparagine cyclization coupled to peptide bond cleavage at the C-terminal splice junction occur independently, and the free amino group of the C-extein's N-terminus then attacks the linear ester, resulting in ligation of the exteins (See Fig. 4). To date, B-type BILs have not been shown to facilitate protein splicing, but some can undergo N- and C-terminal cleavage reactions.

7.2
Hedgehog Proteins

Inteins and hedgehog proteins have many similarities, including amino acid sequence, three-dimensional structure, and catalytic function. Hedgehog proteins are involved in cell patterning during metazoan development, and disruption of the hedgehog-signaling pathway can lead to developmental disorders and tumorigenesis. They are found in almost all metazoans, whereas inteins are found only in unicellular organisms.

Hedgehog proteins, like inteins, undergo posttranslational autoprocessing. The *Drosophila* hedgehog protein was first shown to generate two protein domains from one larger protein precursor. The mechanism of this self-cleavage is similar to that of intein-mediated protein splicing (See Fig. 5). First, an N-S acyl rearrangement of the peptide bond between a conserved Gly-Cys sequence results in a thioester intermediate. Next, a transesterification between the 3β-hydroxyl of cholesterol and the newly formed thioester results in peptide bond cleavage and conjugation of the upstream fragment with cholesterol. In analogy with the role of exteins in protein splicing, the upstream signaling fragment is not required for autoprocessing, as hedgehog self-cleavage can occur in a foreign protein context.

HINT domains also occur in *Caenorhabditis elegans*, and are called *Groundhog* and *Warthog*. Although one of these HINT domains undergoes cleavage near the expected site if expressed in cultured *Drosophila* cells, the polypeptides flanking the HINT domain are dissimilar from those in the hedgehog proteins, which suggests that they may function differently in autoprocessing and cell signaling.

It is likely that the hedgehog proteins and inteins share a common ancestor. There is significant amino acid sequence homology between inteins and the autoprocessing domain of hedgehog proteins. In addition, the *Drosophila* hedgehog protein shares structural homology with intein domains. It has been suggested that inteins evolved from this common ancestor and propagated as a result of the invasion of a homing endonuclease. The hedgehog proteins may have lost the ability to splice

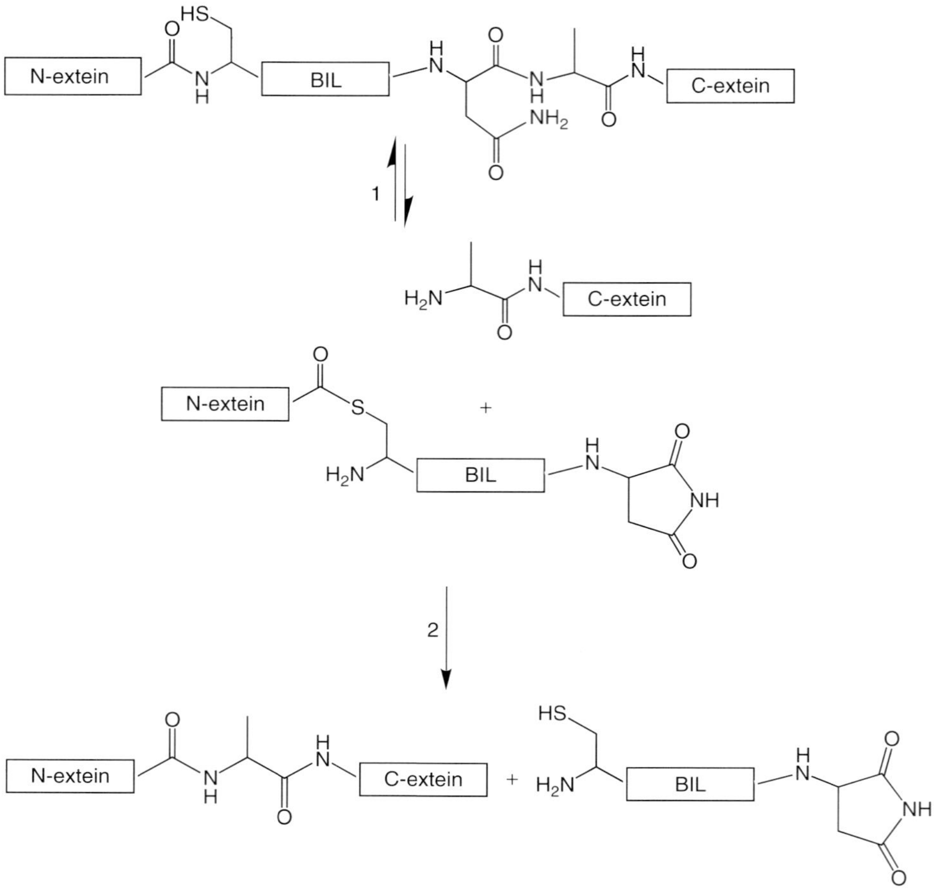

Fig. 4 Proposed alternate mechanism for splicing of a BIL without a C-terminal nucleophile.

while retaining the ability to facilitate an N-S acyl shift.

7.3
Pyruvoyl Enzyme Autoprocessing

A class of decarboxylases and reductases use a covalently linked pyruvoyl moiety as a prosthetic group instead of pyridoxal phosphate cofactors. In both histidine decarboxylase and aspartate decarboxylase, the pyruvoyl moiety is the result of self-catalyzed processing. Although there is no sequence or structural homology to inteins, the reaction is initiated by an N-O acyl rearrangement (See Fig. 5). A vestige of this reaction is seen in the crystal structure of *E. coli* aspartate decarboxylase. The precursor protein, π, is a tetramer, but only three subunits undergo processing. Autoprocessing is initiated by an N-O acyl rearrangement of the peptide bond between Gly24 and Ser25. The presence of an ester intermediate is suggested by the electron density of the unprocessed subunit.

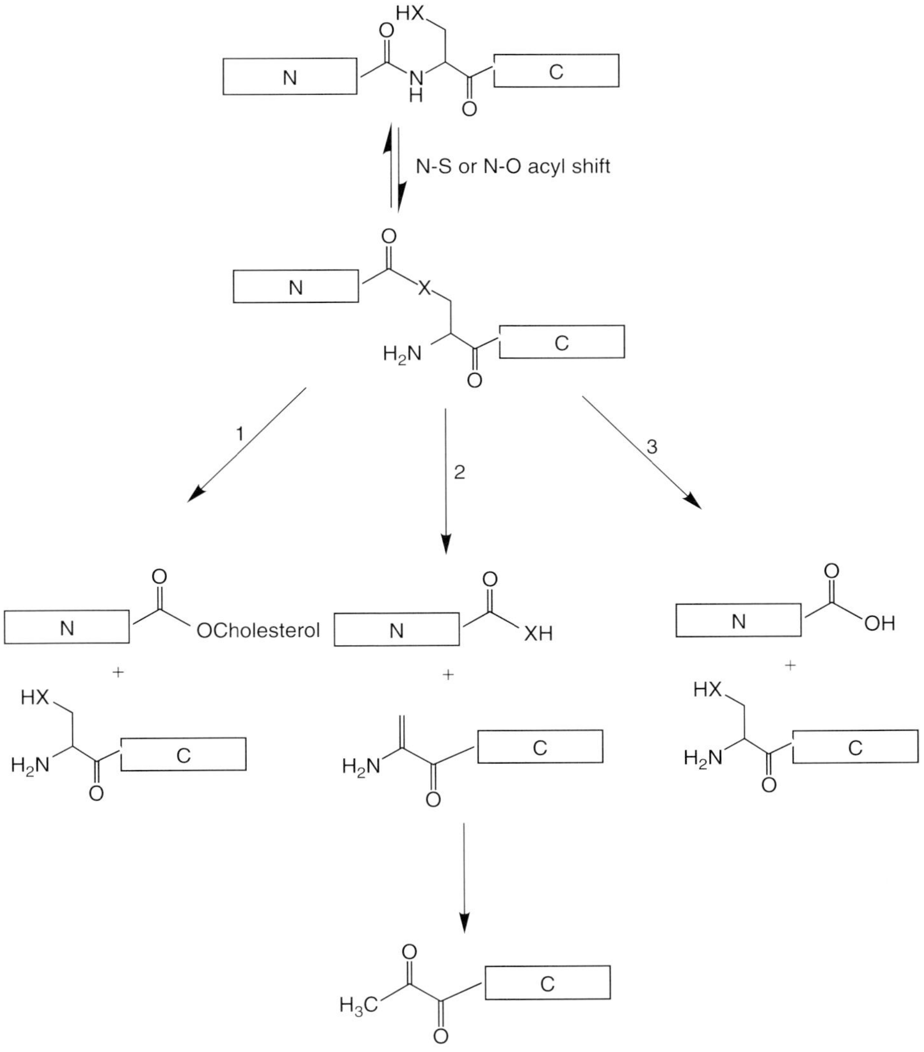

Fig. 5 Related forms of autoprocessing. Following an N-O or N-S acyl shift, the ester can be resolved by (1) transesterification with cholesterol as in hedgehog autoproteolysis; (2) β-elimination following by hydrolysis, as in pyruvoyl enzyme formation; or (3) hydrolysis as in N-terminal nucleophile or nucleoporin autocleavage. "N" and "C" in the boxes indicate the N- and C-terminal domains respectively.

Instead of the second transesterification reaction of inteins or hedgehog proteins, the breakdown of the ester is facilitated by β-elimination, followed by hydration to a carbinolamine and elimination of ammonia to form an N-terminal pyruvoyl group. As with some inteins, the position of the N-O equilibrium may be affected by catalytic strain, in this case by the presence of a tight turn immediately preceding the scissile bond.

7.4
N-terminal Nucleophile Hydrolases

N-terminal nucleophile hydrolases, such as penicillin acylase, glutamine PRPP amidotransferase, and the proteosome are activated by cleavage of the peptide bond upstream of the catalytic residue to generate a catalytic subunit from a larger, inactive precursor. Direct evidence for intramolecular processing in this class of structurally related enzymes exists for glycosylasparaginase from *Flavobacterium meningosepticum* (*Fme*). Active glycosylasparaginase hydrolyzes the β-N-glycosidic bond between asparagine and N-acetylglucosamine in asparagine-linked glycans. Activation of the enzyme initiates with an N-O acyl rearrangement. In place of the transesterification reaction found with inteins, the ester is hydrolyzed to form two protein segments. The discovery that free glycine inhibits autoprocessing allowed the detection of a thioester intermediate by cleavage with neutral hydroxylamine in a Thr1Cys glycosylasparaginase mutant. Although there is no structural or evolutionary relationship between inteins and glycosylasparaginase, it is possible that glycosylasparaginase has also adopted conformational strain as a means to promote an amide to ester rearrangement. The crystal structure of the *Fme* glycosylasparaginase reveals a tight turn involving the scissile bond, resulting in a deviation from planarity of the peptide bond immediately downstream of the scissile bond by more than 20°, as well as deviation of the backbone τ angles by more than 9°. The distortion is stabilized by interaction of the side chain of the N-1 aspartic acid residue (Asp151) with Thr203 and the catalytic Thr152. Perhaps the release of this steric strain favors formation of the ester intermediate.

7.5
Nucleoporins

Another example of protein autoprocessing involves the self-catalyzed cleavage of the yeast nucleoporin Nup145. The nuclear pore complex is responsible for many aspects of cellular function, including the exchange of protein and RNA between the nucleoplasm and the cytosol. Nup145p is a 145-kDa protein that is posttranslationally processed into a 65-kDa N-terminal domain and an 80-kDa C-terminal domain. The C-terminal domain functions as part of a nucleoporin complex that controls mRNA export and nuclear architecture. The function of the N-terminal fragment is not clear, but it is improperly localized if separately expressed from the C-terminal fragment. Thus, Nup145p autoprocessing plays a regulatory role in the cellular localization of essential proteins. The autoprocessing is initiated by an N-O acyl rearrangement that results in hydrolysis of the peptide bond between phenylalanine and serine in a conserved His-Phe-Ser motif (See Fig. 5). Deletion of the first 712 residues of Nup145p does not eliminate autoprocessing, which suggests that the information required for processing is located mostly in the C-terminal domain. In the crystal structure of human nucleoporin Nup98, hydrogen bonding is evident between the histidine and serine of the conserved triad. Although there is no evidence of catalytic strain, there is tight anchoring of the side chain of the conserved phenylalanine of the conserved triad in a hydrophobic pocket as well as β-strand interactions downstream of the scissile bond, which may suggest the use of constrained geometry in the promotion of the amide–ester rearrangement. Although nucleoporin autoprocessing and protein splicing share the same first step, there is no structural or sequence homology between these autoprocessing domains.

8
Inteins in Biotechnology

8.1
Self-cleavable Affinity Tags

One of the first applications developed that takes advantage of protein-splicing elements was the use of inteins modulated by mutation for use in protein purification. Protein purification is often facilitated by the use of affinity tags, but it is sometimes necessary to remove these domains from the protein of interest. Traditionally, this is done using a protease that must then be separated from the target protein. However, if the affinity tag is fused to an intein, it is possible to cleave the intein-affinity tag fusion from the target protein, using either N-terminal or C-terminal cleavage uncoupled from the overall protein-splicing reaction (See Fig. 6).

The first affinity tag purification scheme developed uses a modified form of the *Sce* VMA intein that can be induced to cleave at the intein N-terminus. Precursor protein accumulates if the C-terminal intein asparagine residue is mutated to alanine, preventing the final step of splicing. If the target protein is used as the N-extein and an affinity tag as the C-extein, the precursor can be purified using an affinity resin. The protein target can then be eluted by initiation of N-terminal cleavage by added thiol. If cysteine is used, it can be incorporated into the target protein at its C-terminus by essentially irreversible S-N acyl rearrangement after cleavage. Protocols using mini-inteins can allow purification of larger target proteins. N-terminal cleavage of the *Ssp* DnaB mini-intein can be induced by raising the pH to 8.0 or the temperature to 25 °C, eliminating the need for added thiols.

Alternatively, the target protein can be fused to the intein C-terminus, and the N-extein can be used as the affinity tag. Purification is accomplished by inducing C-terminal cleavage via asparagine cyclization. One would expect higher levels of expression of soluble fusion proteins in such a system, in which a soluble *E. coli* protein such as maltose-binding protein can be used as the N-terminal domain. The first such system described uses a *Sce* VMA intein modified by mutation such that C-terminal cleavage only occurs after N-terminal cleavage is induced by the addition of exogenous thiol. However, in this case the N-extein peptide coelutes with the C-extein target and added thiol. Later systems use modified inteins for which C-terminal cleavage does not require prior cleavage at the intein N-terminus and can be induced by a change in pH or temperature.

8.2
Protein Ligation

Protein semisynthesis in aqueous solution in the absence of enzymes has been a long-standing goal in synthetic chemistry. Two methods of protein semisynthesis are native chemical ligation and expressed protein ligation.

The native chemical ligation scheme developed by Tam and coworkers involves the ligation of a peptide with a C-terminal thioester to a peptide with an N-terminal

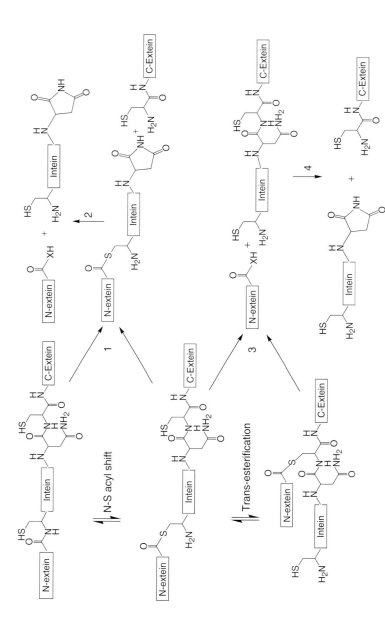

Fig. 6 Side reactions that are manipulated in intein-based protein purification and ligation systems. (1) Premature asparagine cyclization of the precursor or linear ester can lead to C-terminal cleavage, possibly followed by (2) hydrolysis or thiolysis at the N-terminal splice junction. (3) Hydrolysis or thiolysis of the linear or branched esters can lead to N-terminal cleavage, possibly followed by (4) cyclization of asparagine at the C-terminal splice junction coupled to peptide bond cleavage.

cysteine (mediated by a trialkylphosphine and an alkylthiol), or the ligation of a peptide with a C-terminal thiocarboxylic acid to a peptide with an N-terminal β-bromo amino acid at low pH. Another method, developed by Kent and coworkers, ligates a peptide with a C-terminal thioester and a peptide with an N-terminal cysteine and proceeds in the presence of Ellman's reagent. However, the bimolecular nature of both methods results in a relatively slow rate of ligation. The methods also require a cysteine residue at the junction of the ligated peptides as well as the synthesis of the N-terminal segment, which limits the size of the protein target. However, other methods avoid the requirement for the cysteine residue. For instance, in "conformationally assisted protein ligation," the ligation of ribonuclease S with the S-peptide between alanine and serine is facilitated by the prior refolding of the two fragments. Native chemical ligation between a thioester and a 2-mercaptobenzylamine-linked peptide results in a peptide bond without the need for cysteine, albeit with o-benzylmercaptoamine in the product. Another method uses a phosphinobenzenethiol to facilitate native chemical ligation between a peptide with a C-terminal thioester and a peptide with an N-terminal azide, yielding a native peptide with no requirement for cysteine or affinity between the peptide targets. However, this method remains limited to the ligation of small peptides because of the need to synthesize the thioester-containing peptide.

The semisynthesis of proteins with building blocks generated using inteins presents a great advantage over the synthetic peptides used in native chemical ligation procedures. This "expressed protein ligation" was first described using the *Sce* VMA intein with the target protein as the N-extein. The target protein is purified with a C-terminal thioester via N-terminal cleavage of the intein induced by thiophenol. It is then ligated to a peptide with an amino-terminal cysteine. A more efficient procedure was developed using the *Mxe* GyrA mini-intein, using 2-mercaptoethanesulfonic acid to induce N-terminal cleavage. Expressed protein ligation methods of this kind also employ the *Mth* RIR1 mini-intein and the *Ssp* DnaB mini-intein. However, these methods require the use of a synthetic peptide containing an N-terminal cysteine. Two procedures were developed to avoid the need for a synthetic peptide. In one, factor Xa is used to cleave a target peptide upstream of a cysteine residue. The second uses controlled intein C-terminal cleavage to generate a cleaved C-extein with an N-terminal cysteine. Recent applications include intein-mediated biotinylation of proteins, conjugation of an oligonucleotide to a recombinant protein, and semisynthesis of a protein with β-amino acids.

Protein splicing in trans is an essentially unimolecular method of protein ligation. It was originally used to ligate maltose-binding protein to paramyosin or a tetrapeptide using the *Psp*-GBD-Pol intein or to a His-tag-containing domain using the *Mtu* RecA intein. Using a synthetic peptide as the C-terminal fragment of the *Mtu* RecA intein, peptides containing synthetic probes or nonnatural amino acids can be ligated to target proteins. However, these methods are limited by the need for joint denaturation and renaturation. The discovery of a naturally split intein, the *Ssp* DnaE intein, opened the possibility of protein ligation *in vitro* by protein trans-splicing without prior denaturation, and can be applied to proteins that cannot be reversibly denatured. The discovery of "conditional" protein splicing in trans, by which association and therefore subsequent splicing of the split *Sce* VMA intein is induced by the fusion of complementary components of a protein dimerization domain to the intein segments, could allow regulation of protein ligation *in vitro* and *in vivo*.

The discovery of intein-mediated protein ligation in trans has led to a number of interesting applications. For instance, each fragment of a split *Sce* VMA intein was fused to a fragment of green fluorescent protein (GFP). Although the intein fragments interact only weakly, if fused to proteins that interact strongly, they associate properly and splice to yield functional GFP. This system was used *in vivo* to measure protein–protein interactions. In another example, split *Ssp* DnaE fragments, each fused to fragments of 5-enolpyruvylshikimate-3-phosphate synthase (EPSPS), yields a reconstituted, although not spliced, enzyme. The reconstituted enzyme rescues the growth of *E. coli* cells with a depletion of the *EPSPS* gene if cultured in the presence of glyphosate. Herbicide-resistant acetolactate synthase was also split and fused to the two segments of the *Ssp* DnaE intein, which mediates splicing of the acetolactate synthase as well as resistance to valine growth inhibition in *E. coli*. Trans-splicing of EPSPS mediated by the split *Ssp* DnaE intein is also observed in the chloroplast of *Nicotiana tabacum* when a gene encoding for N-terminal intein-EPSPS fusion was integrated into the nuclear DNA and a gene encoding for the C-terminal intein-EPSPS fusion was integrated into the chloroplast genome. These transgenic plants also show an increased resistance to glyphosate. Trans-splicing of split β-glucuronidase in *Arabidopsis* using the *Ssp* DnaE intein has also been observed.

8.3
Protein Cyclization

An interesting use of inteins involves the cyclization of proteins by native chemical ligation, expressed protein ligation, and protein trans-splicing. Cyclized proteins are resistant to exoproteases and useful in the study of protein folding dynamics.

Two methods for protein cyclization have been developed on the basis of native chemical ligation. In one, an unprotected peptide is synthesized with an N-terminal cysteine and a C-terminal thioester on a Boc-Gly-aminomethyl-polystyrene resin with a thioester "handle." The thioester is derivatized with Ellman's reagent and the cyclic peptide is formed in the presence of thiophenol. Another method involves the cyclization of the 47-residue WW domain from human Yes kinase-associated protein. The protein was synthesized with a C-terminal thioester and an N-terminal cysteine, and allowed to fold. The cyclization proceeds whether the protein is folded or not, but the cyclization is an order of magnitude faster when the protein is folded because the N- and C-termini of the protein are in closer proximity.

The size limitations of protein cyclization are largely eliminated by the development of expressed protein ligation protocols. One group fused the N-terminus of a target protein to a modulated *Ssp* DnaB intein and the C-terminus of the target protein to the *Mxe* GyrA intein. Cleavage at the N-terminus of the target protein is initiated by pH-induced asparagine cyclization of the DnaB intein. This is followed by cleavage at the C-terminus of the target protein by addition of 2-mercaptoethanesulfonic acid, resulting in a target protein with an N-terminal cysteine and a C-terminal thioester, which can either cyclize or polymerize. In another method, an SH2 domain is cyclized by fusion to the N-terminus of the *Sce* VMA intein. An N-terminal cysteine is revealed by factor Xa cleavage and reacts with the thioester at the N-terminal splice junction to liberate cyclized protein. A similar method was used to cyclize β-lactamase that is more stable to heat denaturation than the linear form.

Protein cyclization based on transsplicing was first developed using the naturally split *Ssp* DnaE intein *in vivo*. The C-terminal segment of the DnaE intein is fused to the target protein at its N-terminus, and the N-terminal segment of the intein is fused to the target at its C-terminus, such that protein splicing yields a cyclized product. This method is called *SICLOPPS*, or split intein-mediated circular ligation of peptides and proteins. A similar system was used to circularize maltose-binding protein *in vivo* as well as *in vitro*. A split version of the *Pfu* Pol1 intein can cyclize GFP *in vivo*, which denatures at half of the rate of uncyclized GFP. Intein-mediated cyclization can be used to generate cyclic peptide libraries in cells to screen for bioactive compounds.

8.4
Screening Systems Utilizing Inteins

Inteins have been used in the design of genetic screening systems. In one such system, a plasmid-carried gene encoding for aminoglycoside phosphotransferase (aph) was interrupted with the *Mtu* RecA intein, such that cell growth in the presence of kanamycin is dependent on efficient protein splicing. This aph-intein fusion was used to screen for open reading frame insertions into the DNA sequence coding for the intein. Belfort and coworkers created a screening system based on the splicing of thymidylate synthase (TS) interrupted by the *Mtu* RecA intein. In *thyA E. coli*, one can select for protein splicing by growth in the absence of thymine, which would require protein splicing to generate a complementary TS, and select against protein splicing in the presence of thymine and a DHFR inhibitor, which would be lethal in the presence of spliced, complementary TS. This system was used to define the minimal size of a functional intein and to select for an intein that can facilitate pH-dependent C-terminal cleavage.

Temperature-sensitive intein mutants were selected using a system based on the *Mxe* GyrA intein. Splicing of the split GyrA protein results in cell death in the presence of quinoline-related gyrase inhibitors, so this system could be used to select for inhibitors of protein splicing by coupling cell survival to efficient splicing. Protein-splicing inhibitors could also be detected using a system based on the cytotoxicity of the ccdB protein. N-terminal fusion of lacZα to the cell death domain ccdB does not emeliorate the cytotoxicity of ccdB, but the insertion of the *Mtu* RecA intein between lacZα and ccdB does ablate ccdB function. Therefore, inhibition of protein splicing can be coupled to cell survival. Finally, the fluorescence of GFP interrupted by the *Mtu* RecA intein is dependent on removal of the intein by protein splicing, and this could be used as a colorimetric assay for protein-splicing inhibition.

9
Perspective

Protein splicing was first discovered in 1990. Since that time, the chemical steps of this process have been defined, and a variety of clever and useful biotechnology applications have been developed. However, many questions and interesting avenues of inquiry remain. For instance, the means by which inteins facilitate and coordinate each step of splicing remain largely unanswered. It is also unclear if inteins are merely molecular parasites, or if they play a role in the regulation of gene expression. If so, inhibitors of protein splicing could be designed and used to dissect this biological role. Inteins could also serve as drug targets, as one could imagine that the inactivation of such interrupted exteins as the RecA or DnaB proteins of *Mtu* could have antimicrobial effects. The majority of the studies of protein splicing have been done with foreign proteins as the exteins and *in vitro* or in *E. coli*, which does not have inteins, so it will be interesting to study the splicing of inteins in their natural hosts and extein contexts. Finally, it is interesting that inteins and hedgehog domains have considerable sequence and structural homology, but that inteins are found only in unicellular organisms and do not have an identified function, whereas hedgehog proteins are found in most metazoans and play an essential role in embryonic development. Perhaps the recently discovered bacterial intein-like domains will serve as an evolutionary link between these autoprocessing domains.

Acknowledgments

I gratefully acknowledge Professor Henry Paulus of the Boston Biomedical Research Institute for his critical reading of the manuscript.

See also Protein and Nucleic Acid Enzymes; Mass Spectrometry-based Methods of Proteome Analysis; Proteomics.

Bibliography

Books and Reviews

Belfort, M., Reaban, M.E., Coetzee, T., Dalgaard, J.Z. (1995) Prokaryotic introns and inteins: a panoply of form and function, *J. Bacteriol.* **177**, 3897–3903.

Blaschke, U.K., Silberstein, J., Muir, T.W. (2000) Protein engineering by expressed protein ligation, *Methods Enzymol.* **328**, 478–496..

Cotton, G.J., Muir, T.W. (1999) Peptide ligation and its application to protein engineering, *Chem. Biol.* **6**, R247–R256.

Derbyshire, V., Belfort, M. (1998) Lightning strikes twice: intron-intein coincidence, *Proc. Natl. Acad. Sci. U. S. A.* **95**, 1356–1357.

Evans, T.C., Xu, M.Q. Jr. (1999) Intein-mediated protein ligation: harnessing nature's escape artists, *Biopolymers* **51**, 333–342.

Giriat, I., Muir, T.W., Perler, F.B. (2001) Protein splicing and its applications, *Genet. Eng. (N. Y).* **23**, 171–199.

Lambowitz, A.M., Belfort, M. (1993) Introns as mobile genetic elements, *Annu. Rev. Biochem.* **62**, 587–622.

Liu, X.Q. (2000) Protein-splicing intein: Genetic mobility, origin, and evolution, *Annu. Rev. Genet.* **34**, 61–76.

Mann, R.K., Beachy, P.A. (2000) Cholesterol modification of proteins, *Biochim. Biophys. Acta* **1529**, 188–202.

Paulus, H. (2000) Protein splicing and related forms of protein autoprocessing, *Annu. Rev. Biochem.* **69**, 447–496.

Paulus, H. (2003) Inteins as targets for potential antimycobacterial drugs, *Front. Biosci.* **8**, s1157–s1165.

Primary Literature

Adam, E., Perler, F.B. (2002) Development of a positive genetic selection system for inhibition of protein splicing using mycobacterial inteins in Escherichia coli DNA gyrase subunit A, *J. Mol. Microbiol. Biotechnol.* **4**, 479–487.

Albert, A., Dhanaraj, V., Genschel, U., Khan, G., Ramjee, M.K., Pulido, R., Sibanda, B.L., von Delft, F., Witty, M., Blundell, T.L., Smith, A.G., Abell, C. (1998) Crystal structure of aspartate decarboxylase at 2.2 A resolution provides evidence for an ester in protein self-processing, *Nat. Struct. Biol.* **5**, 289–293.

Amitai, G., Dassa, B., Pietrokovski, S. (2004) Protein splicing of inteins with atypical glutamine and aspartate C-terminal residues, *J. Biol. Chem.* **279**, 3121–3131.

Amitai, G., Belenkiy, O., Dassa, B., Shainskaya, A., Pietrokovski, S. (2003) Distribution and function of new bacterial intein-like protein domains, *Mol. Microbiol.* **47**, 61–73.

Arnold, U., Hinderaker, M.P., Nilsson, B.L., Huck, B.R., Gellman, S.H., Raines, R.T. (2002) Protein prosthesis: a semisynthetic enzyme with a beta-peptide reverse turn, *J. Am. Chem. Soc.* **124**, 8522–8523.

Aspock, G., Kagoshima, H., Niklaus, G., Burglin, T.R. (1999) Caenorhabditis elegans has scores of hedgehog-related genes: sequence and expression analysis, *Genome Res.* **9**, 909–923.

Beligere, G.S., Dawson, P.E. (1999) Conformationally Assisted Protein Ligation Using C-Terminal Thioester Peptides, *J. Am. Chem. Soc.* **121**, 6332–6333.

Camarero, J.A., Muir, T.W. (1997) Chemoselective backbone cyclization of unprotected peptides, *Chem. Commun.* **15**, 1369–1370.

Camarero, J.A., Muir, T.W. (1999) Biosynthesis of a head-to-tail cyclized protein with improved biological activity, *J. Am. Chem. Soc.* **121**, 5597–5598.

Camarero, J.A., Pavel, J., Muir, T.W. (1998) Chemical synthesis of a circular protein domain: Evidence for Folding-assisted cyclization, *Angew. Chem. Int. Ed. Engl.* **37**, 5597–5598.

Capasso, S., Mazzarella, L., Sorrentino, G., Balboni, G., Kirby, A.J. (1996) Kinetics and mechanism of the cleavage of the peptide bond next to asparagine, *Peptides* **17**, 1075–1077.

Caspi, J., Amitai, G., Belenkiy, O., Pietrokovski, S. (2003) Distribution of split DnaE inteins in cyanobacteria, *Mol. Microbiol.* **50**, 1569–1577.

Chen, L., Benner, J., Perler, F.B. (2000) Protein splicing in the absence of an intein penultimate histidine, *J. Biol. Chem.* **275**, 20431–20435.

Chen, L., Pradhan, S., Evans, T.C. Jr. (2001) Herbicide resistance from a divided EPSPS protein: the split Synechocystis DnaE intein as an in vivo affinity domain, *Gene* **263**, 39–48.

Chin, H.G., Kim, G.D., Marin, I., Mersha, F., Evans, T.C., Chen, L., Xu, M.Q., Pradhan, S. Jr. (2003) Protein trans-splicing in transgenic plant chloroplast: reconstruction of herbicide resistance from split genes, *Proc. Natl. Acad. Sci. U. S. A.* **100**, 4510–4515.

Chong, S., Xu, M.Q. (1997) Protein splicing of the Saccharomyces cerevisiae VMA intein without the endonuclease motifs, *J. Biol. Chem.* **272**, 15587–15590.

Chong, S., Williams, K.S., Wotkowicz, C., Xu, M.Q. (1998) Modulation of protein splicing of the Saccharomyces cerevisiae vacuolar membrane ATPase intein, *J. Biol. Chem.* **273**, 10567–10577.

Chong, S., Shao, Y., Paulus, H., Benner, J., Perler, F.B., Xu, M.Q. (1996) Protein splicing involving the Saccharomyces cerevisiae VMA intein. The steps in the splicing pathway, side reactions leading to protein cleavage, and establishment of an in vitro splicing system, *J. Biol. Chem.* **271**, 22159–22168.

Chong, S., Montello, G.E., Zhang, A., Cantor, E.J., Liao, W., Xu, M.Q., Benner, J. (1998) Utilizing the C-terminal cleavage activity of a protein splicing element to purify recombinant proteins in a single chromatographic step, *Nucleic Acids Res.* **26**, 5109–5115.

Chong, S., Mersha, F.B., Comb, D.G., Scott, M.E., Landry, D., Vence, L.M., Perler, F.B., Benner, J., Kucera, R.B., Hirvonen, C.A., Pelletier, J.J., Paulus, H., Xu, M.Q. (1997) Single-column purification of free recombinant proteins using a self-cleavable affinity tag derived from a protein splicing element, *Gene* **192**, 271–281.

Clarke, S. (1987) Propensity for spontaneous succinimide formation from aspartyl and asparaginyl residues in cellular proteins, *Int. J. Pept. Protein Res.* **30**, 808–821.

Cooper, A.A., Chen, Y.J., Lindorfer, M.A., Stevens, T.H. (1993) Protein splicing of the yeast TFP1 intervening protein sequence: a model for self-excision, *EMBO J.* **12**, 2575–2583.

Dassa, B., Haviv, H., Amitai, G., Pietrokovski, S. (2004) Protein splicing and auto-cleavage of bacterial intein-like domains lacking a C'-flanking nucleophilic residue, *J. Biol. Chem.* **279**, 32001–32007.

Daugelat, S., Jacobs, W.R. Jr. (1999) The Mycobacterium tuberculosis recA intein can be used in an ORFTRAP to select for open reading frames, Protein Sci. **8**, 644–653.

Davis, E.O., Sedgwick, S.G., Colston, M.J. (1991) Novel structure of the recA locus of Mycobacterium tuberculosis implies processing of the gene product, J. Bacteriol. **173**, 5653–5662.

Davis, E.O., Jenner, P.J., Brooks, P.C., Colston, M.J., Sedgwick, S.G. (1992) Protein splicing in the maturation of M. tuberculosis recA protein: a mechanism for tolerating a novel class of intervening sequence, Cell **71**, 201–210.

Dawson, P.E., Muir, T.W., Clark-Lewis, I., Kent, S.B. (1994) Synthesis of proteins by native chemical ligation, Science **266**, 776–779.

Derbyshire, V., Wood, D.W., Wu, W., Dansereau, J.T., Dalgaard, J.Z., Belfort, M. (1997) Genetic definition of a protein-splicing domain: functional mini-inteins support structure predictions and a model for intein evolution, Proc. Natl. Acad. Sci. U. S. A. **94**, 11466–11471.

Ding, Y., Xu, M.Q., Ghosh, I., Chen, X., Ferrandon, S., Lesage, G., Rao, Z. (2003) Crystal structure of a mini-intein reveals a conserved catalytic module involved in side chain cyclization of asparagine during protein splicing, J. Biol. Chem. **278**, 39133–39142.

Duan, X., Gimble, F.S., Quiocho, F.A. (1997) Crystal structure of PI-SceI, a homing endonuclease with protein splicing activity, Cell **89**, 555–564.

Evans, T.C., Benner, J., Xu, M.Q. Jr. (1998) Semisynthesis of cytotoxic proteins using a modified protein splicing element, Protein Sci. **7**, 2256–2264.

Evans, T.C., Benner, J., Xu, M.Q. Jr. (1999) The cyclization and polymerization of bacterially expressed proteins using modified self-splicing inteins, J. Biol. Chem. **274**, 18359–18363.

Evans, T.C., Benner, J., Xu, M.Q. Jr. (1999) The in vitro ligation of bacterially expressed proteins using an intein from Methanobacterium thermoautotrophicum, J. Biol. Chem. **274**, 3923–3926.

Evans, T.C., Xu, M.Q. Jr. (1999) Intein-mediated protein ligation: harnessing nature's escape artists, Biopolymers **51**, 333–342.

Evans, T.C., Martin, D., Kolly, R., Panne, D., Sun, L., Ghosh, I., Chen, L., Benner, J., Liu, X.Q., Xu, M.Q. Jr. (2000) Protein trans-splicing and cyclization by a naturally split intein from the dnaE gene of Synechocystis species PCC6803, J. Biol. Chem. **275**, 9091–9094.

Gangopadhyay, J.P., Jiang, S.Q., Paulus, H. (2003) An in vitro screening system for protein splicing inhibitors based on green fluorescent protein as an indicator, Anal. Chem. **75**, 2456–2462.

Geiger, T., Clarke, S. (1987) Deamidation, isomerization, and racemization at asparaginyl and aspartyl residues in peptides. Succinimide-linked reactions that contribute to protein degradation, J. Biol. Chem. **262**, 785–794.

Ghosh, I., Sun, L., Xu, M.Q. (2001) Zinc inhibition of protein trans-splicing and identification of regions essential for splicing and association of a split intein*, J. Biol. Chem. **276**, 24051–24058.

Gimble, F.S., Thorner, J. (1992) Homing of a DNA endonuclease gene by meiotic gene conversion in Saccharomyces cerevisiae, Nature **357**, 301–306.

Gorbalenya, A.E. (1998) Noncanonical inteins, Nucleic Acids Res. **26**, 1741–1748.

Gu, H.H., Xu, J., Gallagher, M., Dean, G.E. (1993) Peptide splicing in the vacuolar ATPase subunit A from Candida tropicalis, J. Biol. Chem. **268**, 7372–7381.

Guan, C., Cui, T., Rao, V., Liao, W., Benner, J., Lin, C.L., Comb, D. (1996) Activation of glycosylasparaginase. Formation of active N-terminal threonine by intramolecular autoproteolysis, J. Biol. Chem. **271**, 1732–1737.

Guan, C., Liu, Y., Shao, Y., Cui, T., Liao, W., Ewel, A., Whitaker, R., Paulus, H. (1998) Characterization and functional analysis of the cis-autoproteolysis active center of glycosylasparaginase, J. Biol. Chem. **273**, 9695–9702.

Hall, T.M., Porter, J.A., Young, K.E., Koonin, E.V., Beachy, P.A., Leahy, D.J. (1997) Crystal structure of a Hedgehog autoprocessing domain: homology between Hedgehog and self-splicing proteins, Cell **91**, 85–97.

Hirata, R., Anraku, Y. (1992) Mutations at the putative junction sites of the yeast VMA1 protein, the catalytic subunit of the vacuolar membrane H(+)-ATPase, inhibit its processing by protein splicing, Biochem. Biophys. Res. Commun. **188**, 40–47.

Hirata, R., Ohsumk, Y., Nakano, A., Kawasaki, H., Suzuki, K., Anraku, Y. (1990) Molecular structure of a gene, VMA1, encoding the catalytic subunit of H(+)-translocating adenosine triphosphatase from vacuolar membranes of Saccharomyces cerevisiae, *J. Biol. Chem.* **265**, 6726–6733.

Hodel, A.E., Hodel, M.R., Griffis, E.R., Hennig, K.A., Ratner, G.A., Xu, S., Powers, M.A. (2002) The three-dimensional structure of the autoproteolytic, nuclear pore-targeting domain of the human nucleoporin Nup98, *Mol. Cell.* **10**, 347–358.

Hodges, R.A., Perler, F.B., Noren, C.J., Jack, W.E. (1992) Protein splicing removes intervening sequences in an archaea DNA polymerase, *Nucleic Acids Res.* **20**, 6153–6157.

Iwai, H., Pluckthun, A. (1999) Circular beta-lactamase: stability enhancement by cyclizing the backbone, *FEBS Lett.* **459**, 166–172.

Iwai, H., Lingel, A., Pluckthun, A. (2001) Cyclic green fluorescent protein produced in vivo using an artificially split PI-PfuI intein from Pyrococcus furiosus, *J. Biol. Chem.* **276**, 16548–16554.

Iwai, K., Ando, T. (1967) N-O Acyl Shifts, *Methods Enzymol.* **11**, 263–282.

Kane, P.M., Yamashiro, C.T., Wolczyk, D.F., Neff, N., Goebl, M., Stevens, T.H. (1990) Protein splicing converts the yeast TFP1 gene product to the 69-kD subunit of the vacuolar H(+)-adenosine triphosphatase, *Science* **250**, 651–657.

Kinsella, T.M., Ohashi, C.T., Harder, A.G., Yam, G.C., Li, W., Peelle, B., Pali, E.S., Bennett, M.K., Molineaux, S.M., Anderson, D.A., Masuda, E.S., Payan, D.G. (2002) Retrovirally delivered random cyclic Peptide libraries yield inhibitors of interleukin-4 signaling in human B cells, *J. Biol. Chem.* **277**, 37512–37518.

Klabunde, T., Sharma, S., Telenti, A., Jacobs, W.R., Sacchettini, J.C. Jr. (1998) Crystal structure of GyrA intein from Mycobacterium xenopi reveals structural basis of protein splicing, *Nat. Struct. Biol.* **5**, 31–36.

Lee, J.J., Ekker, S.C., von Kessler, D.P., Porter, J.A., Sun, B.I., Beachy, P.A. (1994) Autoproteolysis in hedgehog protein biogenesis, *Science* **266**, 1528–1537.

Lesaicherre, M.L., Lue, R.Y., Chen, G.Y., Zhu, Q., Yao, S.Q. (2002) Intein-mediated biotinylation of proteins and its application in a protein microarray, *J. Am. Chem. Soc.* **124**, 8768–8769.

Lew, B.M., Paulus, H. (2002) An in vivo screening system against protein splicing useful for the isolation of nonsplicing mutants or inhibitors of the RecA intein of Mycobacterium tuberculosis, *Gene* **282**, 169–177.

Lew, B.M., Mills, K.V., Paulus, H. (1998) Protein splicing in vitro with a semisynthetic two-component minimal intein, *J. Biol. Chem.* **273**, 15887–15890.

Lew, B.M., Mills, K.V., Paulus, H. (1999) Characteristics of protein splicing in trans mediated by a semisynthetic split intein, *Biopolymers* **51**, 355–362.

Liu, X.Q., Hu, Z. (1997) A DnaB intein in Rhodothermus marinus: indication of recent intein homing across remotely related organisms, *Proc. Natl. Acad. Sci. U. S. A.* **94**, 7851–7856.

Liu, X.Q., Yang, J. (2003) Split dnaE genes encoding multiple novel inteins in Trichodesmium erythraeum, *J. Biol. Chem.* **278**, 26315–26318.

Lovrinovic, M., Seidel, R., Wacker, R., Schroeder, H., Seitz, O., Engelhard, M., Goody, R.S., Niemeyer, C.M. (2003) Synthesis of protein-nucleic acid conjugates by expressed protein ligation, *Chem. Commun. (Camb).* **7**, 822–823.

Lum, L., Beachy, P.A. (2004) The Hedgehog response network: sensors, switches, and routers, *Science* **304**, 1755–1759.

Martin, D.D., Xu, M.Q., Evans, T.C. Jr. (2001) Characterization of a naturally occurring trans-splicing intein from Synechocystis sp. PCC6803, *Biochemistry* **40**, 1393–1402.

Mathys, S., Evans, T.C., Chute, I.C., Wu, H., Chong, S., Benner, J., Liu, X.Q., Xu, M.Q. (1999) Characterization of a self-splicing mini-intein and its conversion into autocatalytic N- and C-terminal cleavage elements: facile production of protein building blocks for protein ligation, *Gene* **231**, 1–13.

Mills, K.V., Paulus, H. (2001) Reversible inhibition of protein splicing by zinc ion, *J. Biol. Chem.* **276**, 10832–10838.

Mills, K.V., Lew, B.M., Jiang, S., Paulus, H. (1998) Protein splicing in trans by purified N- and C-terminal fragments of the Mycobacterium tuberculosis RecA intein, *Proc. Natl. Acad. Sci. U. S. A.* **95**, 3543–3548.

Mills, K.V., Manning, J.S., Garcia, A.M., Wuerdeman, L.A. (2004) Protein splicing of a

Pyrococcus abyssi intein with a C-terminal glutamine, *J. Biol. Chem.* **279**, 20685–20691.

Ming, J.E., Muenke, M. (1998) Holoprosencephaly: from Homer to Hedgehog, *Clin. Genet.* **53**, 155–163.

Mizutani, R., Anraku, Y., Satow, Y. (2004) Protein splicing of yeast VMA1-derived endonuclease via thiazolidine intermediates, *J. Synchrotron. Radiat.* **11**, 109–112.

Mizutani, R., Nogami, S., Kawasaki, M., Ohya, Y., Anraku, Y., Satow, Y. (2002) Protein-splicing reaction via a thiazolidine intermediate: crystal structure of the VMA1-derived endonuclease bearing the N and C-terminal propeptides, *J. Mol. Biol.* **316**, 919–929.

Mootz, H.D., Blum, E.S., Tyszkiewicz, A.B., Muir, T.W. (2003) Conditional protein splicing: a new tool to control protein structure and function in vitro and in vivo, *J. Am. Chem. Soc.* **125**, 10561–10569.

Muir, T.W., Sondhi, D., Cole, P.A. (1998) Expressed protein ligation: a general method for protein engineering, *Proc. Natl. Acad. Sci. U. S. A.* **95**, 6705–6710.

Nichols, N.M., Benner, J.S., Martin, D.D., Evans, T.C. Jr. (2003) Zinc ion effects on individual Ssp DnaE intein splicing steps: regulating pathway progression, *Biochemistry* **42**, 5301–5311.

Nilsson, B.L., Kiessling, L.L., Raines, R.T. (2000) Staudinger ligation: a peptide from a thioester and azide, *Org. Lett.* **2**, 1939–1941.

Offer, J., Dawson, P.E. (2000) N^α-2-Mercaptobenzylamine-Assisted Chemical Ligation, *Org. Lett.* **2**, 23–26.

Ozawa, T., Kaihara, A., Sato, M., Tachihara, K., Umezawa, Y. (2001) Split luciferase as an optical probe for detecting protein-protein interactions in mammalian cells based on protein splicing, *Anal. Chem.* **73**, 2516–2521.

Ozawa, T., Takeuchi, T.M., Kaihara, A., Sato, M., Umezawa, Y. (2001) Protein splicing-based reconstitution of split green fluorescent protein for monitoring protein-protein interactions in bacteria: improved sensitivity and reduced screening time, *Anal. Chem.* **73**, 5866–5874.

Perler, F.B. (2002) InBase: the Intein Database, *Nucleic Acids Res.* **30**, 383–384.

Perler, F.B., Comb, D.G., Jack, W.E., Moran, L.S., Qiang, B., Kucera, R.B., Benner, J., Slatko, B.E., Nwankwo, D.O., Hempstead, S.K., Carlow, C.K.S., Jannasch, H. (1992) Intervening sequences in an Archaea DNA polymerase gene, *Proc. Natl. Acad. Sci. U. S. A.* **89**, 5577–5581.

Pietrokovski, S. (1994) Conserved sequence features of inteins (protein introns) and their use in identifying new inteins and related proteins, *Protein Sci.* **3**, 2340–2350.

Pietrokovski, S. (1998) Modular organization of inteins and C-terminal autocatalytic domains, *Protein Sci.* **7**, 64–71.

Poland, B.W., Xu, M.Q., Quiocho, F.A. (2000) Structural insights into the protein splicing mechanism of PI-SceI, *J. Biol. Chem.* **275**, 16408–16413.

Porter, J.A., Young, K.E., Beachy, P.A. (1996) Cholesterol modification of hedgehog signaling proteins in animal development, *Science* **274**, 255–259.

Porter, J.A., Ekker, S.C., Park, W.J., von Kessler, D.P., Young, K.E., Chen, C.H., Ma, Y., Woods, A.S., Cotter, R.J., Koonin, E.V., Beachy, P.A. (1996) Hedgehog patterning activity: role of a lipophilic modification mediated by the carboxy-terminal autoprocessing domain, *Cell* **86**, 21–34.

Romanelli, A., Shekhtman, A., Cowburn, D., Muir, T.W. (2004) Semisynthesis of a segmental isotopically labeled protein splicing precursor: NMR evidence for an unusual peptide bond at the N-extein-intein junction, *Proc. Natl. Acad. Sci. U. S. A.* **101**, 6397–6402.

Rosenblum, J.S., Blobel, G. (1999) Autoproteolysis in nucleoporin biogenesis, *Proc. Natl. Acad. Sci. U. S. A.* **96**, 11370–11375.

Scott, C.P., Abel-Santos, E., Wall, M., Wahnon, D.C., Benkovic, S.J. (1999) Production of cyclic peptides and proteins in vivo, *Proc. Natl. Acad. Sci. U. S. A.* **96**, 13638–13643.

Shao, Y., Paulus, H. (1997) Protein splicing: estimation of the rate of O-N and S-N acyl rearrangements, the last step of the splicing process, *J. Pept. Res.* **50**, 193–198.

Shao, Y., Xu, M.Q., Paulus, H. (1995) Protein splicing: characterization of the aminosuccinimide residue at the carboxyl terminus of the excised intervening sequence, *Biochemistry* **34**, 10844–10850.

Shao, Y., Xu, M.Q., Paulus, H. (1996) Protein splicing: evidence for an N-O acyl rearrangement as the initial step in the splicing process, *Biochemistry* **35**, 3810–3815.

Shingledecker, K., Jiang, S., Paulus, H. (2000) Reactivity of the cysteine residues in the protein splicing active center of the

Mycobacterium tuberculosis RecA intein, *Arch. Biochem. Biophys.* **375**, 138–144.

Shingledecker, K., Jiang, S.Q., Paulus, H. (1998) Molecular dissection of the Mycobacterium tuberculosis RecA intein: design of a minimal intein and of a trans-splicing system involving two intein fragments, *Gene* **207**, 187–195.

Southworth, M.W., Perler, F.B. (2002) Protein splicing of the Deinococcus radiodurans strain R1 Snf2 intein, *J. Bacteriol.* **184**, 6387–6388.

Southworth, M.W., Benner, J., Perler, F.B. (2000) An alternative protein splicing mechanism for inteins lacking an N-terminal nucleophile, *EMBO J.* **19**, 5019–5026.

Southworth, M.W., Amaya, K., Evans, T.C., Xu, M.Q., Perler, F.B. (1999) Purification of proteins fused to either the amino or carboxy terminus of the Mycobacterium xenopi gyrase A intein, *Biotechniques* **27**, 110–114, 116, 118–120.

Southworth, M.W., Adam, E., Panne, D., Byer, R., Kautz, R., Perler, F.B. (1998) Control of protein splicing by intein fragment reassembly, *EMBO J.* **17**, 918–926.

Sun, L., Ghosh, I., Paulus, H., Xu, M.Q. (2001) Protein trans-splicing to produce herbicide-resistant acetolactate synthase, *Appl. Environ. Microbiol.* **67**, 1025–1029.

Takeda, S., Tsukiji, S., Nagamune, T. (2004) Site-specific conjugation of oligonucleotides to the C-terminus of recombinant protein by expressed protein ligation, *Bioorg. Med. Chem. Lett.* **14**, 2407–2410.

Tam, J.P., Lu, Y.A., Liu, C.F., Shao, J. (1995) Peptide synthesis using unprotected peptides through orthogonal coupling methods, *Proc. Natl. Acad. Sci. U. S. A.* **92**, 12485–12489.

Teixeira, M.T., Fabre, E., Dujon, B. (1999) Self-catalyzed cleavage of the yeast nucleoporin Nup145p precursor, *J. Biol. Chem.* **274**, 32439–32444.

Telenti, A., Southworth, M., Alcaide, F., Daugelat, S., Jacobs, W.R., Perler, F.B. Jr. (1997) The Mycobacterium xenopi GyrA protein splicing element: characterization of a minimal intein, *J. Bacteriol.* **179**, 6378–6382.

Vanderslice, P., Copeland, W.C., Robertus, J.D. (1988) Site-directed alteration of serine 82 causes nonproductive chain cleavage in prohistidine decarboxylase, *J. Biol. Chem.* **263**, 10583–10586.

Voorter, C.E., de Haard-Hoekman, W.A., van den Oetelaar, P.J., Bloemendal, H., de Jong, W.W. (1988) Spontaneous peptide bond cleavage in aging alpha-crystallin through a succinimide intermediate, *J. Biol. Chem.* **263**, 19020–19023.

Wang, S., Liu, X.Q. (1997) Identification of an unusual intein in chloroplast ClpP protease of Chlamydomonas eugametos, *J. Biol. Chem.* **272**, 11869–11873.

Wood, D.W., Wu, W., Belfort, G., Derbyshire, V., Belfort, M. (1999) A genetic system yields self-cleaving inteins for bioseparations, *Nat. Biotechnol.* **17**, 889–892.

Wu, H., Hu, Z., Liu, X.Q. (1998) Protein trans-splicing by a split intein encoded in a split DnaE gene of Synechocystis sp. PCC6803, *Proc. Natl. Acad. Sci. U. S. A.* **95**, 9226–9231.

Xu, M.Q., Perler, F.B. (1996) The mechanism of protein splicing and its modulation by mutation, *EMBO J.* **15**, 5146–5153.

Xu, Q., Buckley, D., Guan, C., Guo, H.C. (1999) Structural insights into the mechanism of intramolecular proteolysis, *Cell* **98**, 651–661.

Xu, M.Q., Southworth, M.W., Mersha, F.B., Hornstra, L.J., Perler, F.B. (1993) In vitro protein splicing of purified precursor and the identification of a branched intermediate, *Cell* **75**, 1371–1377.

Xu, M.Q., Comb, D.G., Paulus, H., Noren, C.J., Shao, Y., Perler, F.B. (1994) Protein splicing: an analysis of the branched intermediate and its resolution by succinimide formation, *EMBO J.* **13**, 5517–5522.

Xu, R., Ayers, B., Cowburn, D., Muir, T.W. (1999) Chemical ligation of folded recombinant proteins: segmental isotopic labeling of domains for NMR studies, *Proc. Natl. Acad. Sci. U. S. A.* **96**, 388–393.

Xuan, J., Tarentino, A.L., Grimwood, B.G., Plummer, T.H., Cui, T., Guan, C., Van Roey, P. Jr. (1998) Crystal structure of glycosylasparaginase from Flavobacterium meningosepticum, *Protein Sci.* **7**, 774–781.

Yamamoto, K., Low, B., Rutherford, S.A., Rajagopalan, M., Madiraju, M.V. (2001) The Mycobacterium avium-intracellulare complex dnaB locus and protein intein splicing, *Biochem. Biophys. Res. Commun.* **280**, 898–903.

Yang, J., Fox, G.C., Henry-Smith, T.V. Jr. (2003) Intein-mediated assembly of a functional beta-glucuronidase in transgenic plants, *Proc. Natl. Acad. Sci. U. S. A.* **100**, 3513–3518.

Protein Structure Analysis: High-throughput Approaches

Andrew P. Turnbull and Udo Heinemann
Max Delbrück Center for Molecular Medicine, Berlin, Germany

1	**Bioinformatics** 269	
1.1	Structure-to-function Approaches 269	
1.2	Identification of Disordered Regions in a Protein 272	
1.3	Protein-ligand Complexes 272	
2	**Protein Production** 273	
2.1	Yeast 274	
2.2	Baculovirus–insect Cell 274	
2.3	*Leishmania tarentolae* (Trypanosomatidae) 274	
2.4	Cell-free Expression Systems 275	
3	**Purification** 275	
4	**Structure Determination by X-ray Crystallography** 276	
4.1	Crystallization 276	
4.2	Data Collection 278	
4.3	Phasing 278	
4.4	Automated Structure Determination 279	
4.5	Molecular Replacement 280	
5	**Structure-based Drug Design** 280	
6	**Structure Determination by NMR** 280	
	Bibliography 282	
	Books and Reviews 282	
	Primary Literature 282	

Keywords

Active Site
Part of an enzyme molecule made up of amino acid residues, and involved in substrate binding and catalysis.

Anomalous Diffraction
Diffraction of X-rays near the absorption edge of a scattering atom, where a phase shift occurs and the atomic scattering factor becomes a complex quantity; This phenomenon is exploited in phasing X-ray diffraction data.

Isotope Labeling
Introduction of ^{13}C, ^{15}N and other nonradioactive isotopes into a protein for structure determination by nuclear magnetic resonance (NMR) spectroscopy.

Phasing
Reconstitution of the phase relations of diffracted X-rays as part of a crystal structure analysis, using anomalous diffraction and other techniques.

Posttranslational Modification
Covalent modification of a protein molecule after translation by phosphorylation, glycosylation, acetylation and so on. This is rare in prokaryotic proteins, and common in proteins from eukaryotes.

Protein Disorder
Lack of defined three-dimensional structure in segments of proteins or complete proteins in the absence of stabilizing binding partners.

Protein Domain
Compact folding unit of a protein, recognizable on the sequence and/or three-dimensional structure level.

Protein Fold
Recurrent pattern of three-dimensional structure in a protein or protein domain. There are far fewer protein folds than sequences or sequence families.

Selenomethionine
Nonnatural amino acid that can be incorporated into proteins by gene technology in place of methionine. Its selenium atom is used as an anomalous scatterer for phasing the X-ray diffraction pattern.

Structural Genomics
Large-scale project to determine the shapes of all proteins and other important biomolecules encoded by the genomes of key organisms.

Structural Proteomics
Large-scale project to determine the shapes and functions of all proteins encoded by the genomes of key organisms.

Synchrotron Radiation
Electromagnetic radiation emitted by subatomic particles (electrons or positrons) traveling at high velocity in storage rings. X-rays produced at synchrotrons are used in crystal structure determination.

Developments in the high-throughput analysis of protein structure have been primarily driven by worldwide structural genomics initiatives that are aimed at determining the three-dimensional structures of all proteins and other important biomolecules encoded by the genomes of key organisms. Structural genomics requires a large number of procedural steps in order to convert sequence information into a three-dimensional structure; this has led to new high-throughput methods for protein production, characterization, and structure determination. Over the past decade, the most notable technological advances have been in the fields of X-ray crystallography and NMR – the principal tools of structural genomics – which have facilitated high-throughput, rapid, and cost-effective structure determinations. In recent years, these developments have resulted in an exponential increase in the number of structures being deposited in the Protein Data Bank (PDB; http://www.rcsb.org/pdb/), in which the total number currently exceeds 30 000. Major developments in the fields of bioinformatics, protein production, and structure determination, and the impact of these on high-throughput protein structure analysis will be discussed in what follows.

1
Bioinformatics

Primary sequence analysis, such as similarity searches against protein sequence databases, protein domain architecture determination, identification of specialized local structural motifs, and prediction of protein structure are possible with a variety of homology-based modeling methods (Fig. 1). Sequence database searches are particularly useful in selecting targets for structural genomics initiatives and, where possible, generating homologous probes for determining structures by the molecular replacement method. For example, the Web-based program 3D-PSSM (http://www.sbg.bio.ic.ac.uk/~3dpssm/) is a fast method for predicting the protein fold from the primary amino acid sequence, and SWISS-MODEL (http://swissmodel.expasy.org/) is a fully automated protein structure homology-modeling server.

1.1
Structure-to-function Approaches

There are a number of bioinformatics resources available that are aimed

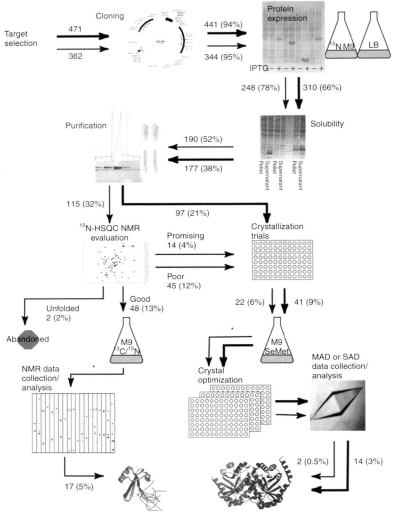

Fig. 1 Schematic flow diagram of the strategies employed in structural genomics initiatives, using *Methanobacterium thermoautotrophicum* as an example. The number of protein targets after each step and the percentage relative to the number of starting targets are indicated in brackets. Thin arrows and italicized numbers are for smaller molecular-weight proteins, and wide arrows and bold numbers are for larger molecular-weight proteins. Diagram taken from Yee, A., Pardee, K., Christendat, D., Savchenko, A., Edwards, A.M., Arrowsmith, C.H. (2003) Structural proteomics: toward high-throughput structural biology as a tool in functional genomics, *Acc. Chem. Res.* **36**, 183–189. Picture with permission from Prof. Cheryl Arrowsmith.

at identifying a protein's biochemical function from its three-dimensional structure (Fig. 2). For example, Dali (http://www2.ebi.ac.uk/dali) and VAST (Vector Alignment Search Tool; http://www.ncbi.nlm.nih.gov:80/Structure/VAST/vastsearch.html) offer Web-based servers for automatically comparing the fold of a newly determined structure against known folds, as represented by the protein structures in the PDB. Such comparisons can often reveal striking similarities between proteins that are not evident from sequence analysis alone, and that can provide important insights into biological function even in the absence of any other biochemical or functional data. However, computer-based approaches fail to assign functions to proteins that adopt novel folds. Enzymes are a notable exception to this rule, because the groupings of residues constituting their active sites tend to be highly conserved in their spatial disposition even in cases where there is no overall similarity in sequence or fold. For example, the relative positioning of the Ser-His-Asp catalytic triad of the serine proteases is highly conserved even when found in protein structures adopting different folds. Hence, screening a new protein structure against a database of enzyme active-site templates such as PROCAT (http://www.biochem.ucl.ac.uk/bsm/PROCAT/PROCAT.html) can be used for detecting key functional residues. The spatial patterns of residues can be automatically generated using various techniques including graph theory (ASSAM) and "fuzzy pattern matching" (RIGOR). Alternative approaches to identifying enzymes

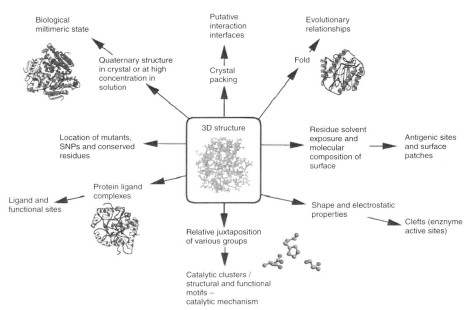

Fig. 2 Summary of the information deriving from the three-dimensional structure of a protein, relating to its biological function. Taken from Thornton, J.M., Todd, A.E., Milburn, D., Borkakoti, N., Orengo, C.A. (2000) From structure to function: approaches and limitations, *Nat. Struct. Biol.* **7**(Suppl.), 991–994.

on the basis of their three-dimensional structure and predicting their functions, have recently been reported. Here, a vector machine-learning algorithm is used, based on the secondary structure of proteins the propensities of amino acids, and surface properties, in order to discriminate enzymes from nonenzymes. Another approach analyses protein surface charges to identify conserved residues that can serve as catalytic sites. Once a general class of biochemical function of a protein has been proposed, experimental screening of enzymatic activity can be used to derive the precise biochemical function. For example, after the structure determination of BioH (an enzyme involved in biotin biosynthesis in *E. coli*) the protein structure was screened against a library of enzyme active sites, a Ser/His/Asp catalytic triad was identified, and subsequent hydrolase assays showed BioH to be a carboxylesterase.

1.2
Identification of Disordered Regions in a Protein

The occurrence of regions in proteins that lack any fixed tertiary structure is increasingly being observed in structural studies. These disordered regions or "random coils" are inherently flexible and are involved in a variety of functions, including the modulation of the specificity/affinity of protein-binding interactions, activation by cleavage, and DNA recognition. During the target selection process, it is important to consider any intrinsic protein disorder, because it can often lead to problems with the expression of protein-coding genes, protein stability, purification, and crystallization. PONDR, DisEMBL, and GlobPlot are useful tools for predicting potential disordered regions within a protein sequence that can be used to help design constructs corresponding to globular proteins or domains. PONDR (Predictor of Naturally Disordered Regions; http://www.pondr.com) and DisEMBL use methods based on artificial neural networks, whereas GlobPlot (http://globplot.embl.de) relies on a novel, propensity-based disorder-prediction algorithm. These methods can also be used to predict inherently flexible regions in protein sequences. For example, PONDR predicted that the linker between the DNA-operator-binding central domain of the transcriptional regulator KorB (KorB-O) and the KorB dimerization domain (KorB-C) is flexible, which was indeed observed in crystal structures and is thought to facilitate complex formation on circular plasmids (Fig. 3).

1.3
Protein-ligand Complexes

Protein-ligand complexes are the most useful in terms of providing functional information, because they reveal the nature of the ligand, the site at which it is bound to the protein, the location of the active site, and of the catalytic machinery (if the protein is an enzyme). There are several examples of structural analyses in which an unexpected protein-bound ligand or cofactor derived from the cloning organism was discovered. For example, the structure of the trimeric human protein p14.5 was found to have picked up benzoate molecules from the crystallization buffer at its inter-subunit tunnels, which most likely mark a hydrolytic active site (Fig. 4). When such data are available at high resolution, proposing a biological function for the protein can be relatively straightforward, because these data identify the nature of the ligand, the

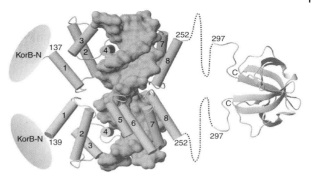

Fig. 3 Natively disordered regions in the bacterial transcriptional regulator and partitioning protein KorB. The KorB DNA-binding domains (KorB-O, center) are connected by flexible linkers to N-terminal domains of unknown structure and function (KorB-N, left), and the KorB dimerization domains (KorB-C, right). Picture taken from Khare et al., 2004.

Fig. 4 Crystal structure of the trimeric human protein, hp14.5. Benzoate molecules picked up from the crystallization buffer bind in the inter-subunit tunnels and mark putative hydrolytic active sites (Manjasetty et al., 2004). Picture with permission from Dr. B.A. Manjasetty.

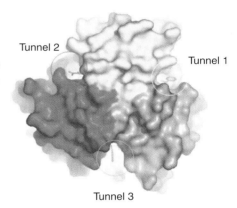

ligand-binding site, and the arrangement of catalytic residues from which a catalytic mechanism can be postulated.

2
Protein Production

Protein expression and purification play a central role in high-throughput protein structure analysis. Cloning using restriction enzymes is impractical for high-throughput approaches, because of the complications of selecting compatible and appropriate restriction enzymes for each cloning procedure, and the multiple steps of experimental refinement and treatment that must be performed. Therefore, high-throughput cloning requires procedures based on the polymerase chain reaction (PCR). High-throughput cloning and expression methods are now being developed in many laboratories, enabling the generation and testing of up to hundreds of DNA constructs for high-level expression in parallel, using rapid and generic protocols. To generate expression vector clones, generic cloning systems can be used; the Gibco/Life Technologies GATEWAY™ system is one such example, streamlining the expression and cloning process by alleviating recloning steps and avoiding the use of restriction enzymes in the cloning and subcloning processes.

High-throughput approaches may rely on both prokaryotic and eukaryotic hosts. *Escherichia coli* expression systems are advantageous for many reasons, most notably because the overproduced protein is usually obtained without any posttranslational modification heterogeneity, and protein expression is cheaper and faster than with eukaryotic systems. However, it is not always possible to obtain soluble protein for many eukaryotic proteins, in particular, human proteins, by using the heterologous expression of eukaryotic genes, in which the codon usage for a cDNA could be suboptimal in *E. coli*. Furthermore, eukaryotic systems are necessary for the expression of proteins that require posttranslational modifications for their correct folding and activity. In situations in which a protein cannot be synthesized in *E. coli*, the eukaryotic yeasts (*Saccharomyces cerevisiae* and *Pichia pastoris*), baculovirus-infected insect cells, *Leishmania tarentolae,* and the cell-free wheat germ systems that have been developed for high-throughput approaches can be used. Furthermore, in these and other eukaryotic expression systems, the codon usage is closer to that in humans. The various merits of each system are discussed below.

2.1
Yeast

The yeasts *S. cerevisiae* and *P. pastoris* can be used for the routine production of recombinant proteins. More recently, *P. pastoris* has emerged as the preferred yeast, because of its strong, highly inducible promoter system resulting in higher yields of recombinant protein, stable genomic integration and posttranslational modifications, such as phosphorylation, which may be important for both the structure and function of some human proteins. In comparison with *S. cerevisiae*, the distribution and chain length of N-linked oligosaccharides are significantly shorter, and therefore this system represents a suitable alternative for the extracellular expression of human proteins. An additional advantage of the methylotrophic *P. pastoris* expression system is that it makes it possible to introduce ^{13}C label into recombinant proteins (by feeding ^{13}C-labeled methanol) for nuclear magnetic resonance (NMR) structural analyses.

2.2
Baculovirus–insect Cell

The recombinant baculovirus–insect cell expression system accomplishes most posttranslational modifications, including phosphorylation, N- and O-linked glycosylation, acylation, disulfide cross-linking, oligomeric assembly, and subcellular targeting, which may all be critical for the accurate production and function of human proteins. In contrast to bacterial expression systems, the recombinant baculovirus expression system usually produces soluble proteins without the need for induction or specific temperature conditions.

2.3
Leishmania tarentolae (Trypanosomatidae)

The use of the parasitic Trypanosomatidae species, *L. tarentolae*, as the host for in vitro protein production has recently been reported, and can achieve high levels of protein expression. The Trypanosomatidae species of parasites naturally produce large amounts of glycoproteins, which is an advantage in the production of sialylated heterologous glycosylated proteins. Furthermore, given the natural auxotrophy of *L. tarentolae* for methionine, it has been suggested that such a system could prove to be useful for the production of selenomethionine-labeled proteins, for high-throughput X-ray crystallographic structure determination using single- or multiple-wavelength anomalous diffraction (SAD and MAD) phasing techniques.

2.4
Cell-free Expression Systems

Cell-free systems facilitate the parallel expression of many protein-coding genes, and are therefore suitable for high-throughput protein production. Obtaining the protein directly in a form suitable for X-ray crystallography or NMR spectroscopy is another prerequisite for high-throughput structure analysis. For the purpose of X-ray analysis, anomalous diffraction techniques (SAD or MAD) necessitate the substitution of methionine residues with selenomethionine. Additionally, NMR structure analysis often requires the labeling of proteins with ^{13}C and/or ^{15}N, which can be introduced through cell growth on media containing these isotopes in the form of ^{13}C-glucose and ^{15}NH$_4$Cl. In this respect, the *E. coli* cell-free protein synthesis system of Kigawa et al. (1999) permits the straightforward incorporation of isotopes for NMR analysis. However, this system is not always suitable for the expression of some eukaryotic proteins and can result in aggregation, formation of insoluble inclusion bodies, and degradation of the expression product. Furthermore, in the case of multidomain proteins, which are found more often in eukaryotes, correct folding occurs more frequently in eukaryotic than in prokaryotic translation systems. Another limitation with the use of *E. coli* systems for high-throughput cell-free expression is that PCR-generated fragments are not transcribed and translated efficiently in these systems: contamination by the mRNA- and DNA-degradation enzymes originating from the cell decreases the stability of the templates and reduces yields. By contrast, in eukaryotic cell-free systems, the added mRNAs are stable for long periods of time, and therefore these systems overcome many of the limitations associated with *E. coli* cell-free systems. Furthermore, cell-free systems can produce high yields of correctly folded proteins and, unlike *in vivo* systems, facilitate the expression of proteins that would otherwise interfere with the host cell physiology. Recently, the synthesis and screening of gene products based on the cell-free system prepared from eukaryotic wheat embryos has been reported; this method bypasses many of the time-consuming cloning steps involved in conventional expression systems and lends itself to the high-throughput expression of proteins, using automated robotic systems.

3
Purification

Protein purification has also seen significant improvements because of the use of affinity tags fused to the protein of interest, so that it can be separated from the host cell proteins rapidly by using standardized purification schemes. N-terminal tags range from large tags, such as glutathione-S-transferase (GST), maltose-binding protein (MBP), thioreductase, and the chitin-binding protein, to fairly small tags such as His_6, and various epitope tags. In some systems, the use of large tags can lead to the production of fusion products that are too large for high-level expression, but the use of such tags can improve the correct folding and stability of the overexpressed protein. His_6 is the most commonly used purification tag, because it is easier to incorporate into expression constructs and it allows a generic one-step purification using automated methods based on nickel-nitrilotriacetic acid (Ni-NTA) or other immobilized metal-affinity chromatography resins. The use of an additional C-terminal StrepII tag enables dual-affinity column purification, which ensures that only full-length gene products are separated. For crystallization to occur, it is generally agreed that the fusion tags must be removed, because they can introduce flexible regions into the protein, which can interfere with crystallization or lead to various forms of microheterogeneity. In a recent study, the compatibility of small peptide affinity tags with protein crystallization was assessed, in which the N-terminus of the chicken spectrin SH3 domain was labeled with a His_6 tag and a StrepII tag, fused to the N- and C-termini, respectively. The resulting protein, His_6-SH3-StrepII, comprised 83 amino acid residues, 23 of which originated from the tags (accounting for 23% of the total fusion protein mass). In contrast to the general consensus that the presence of affinity tags is detrimental in structural studies, this study demonstrated that the fused affinity tags did not interfere with crystallization or structure analysis, and did not change the protein structure. This suggests that, in some cases, protein constructs utilizing both N- and C-terminal peptide tags may lend themselves to structural investigations in high-throughput regimes.

4
Structure Determination by X-ray Crystallography

4.1
Crystallization

Crystallization is regarded as a major bottleneck in structure determination by X-ray crystallography and can be divided into two stages: coarse screening for initial crystallization conditions, followed by optimization of the conditions in order to produce diffraction-quality single crystals. The field of crystallization has recently been revolutionized by significant developments in automation, miniaturization, and process integration. These developments have led to the availability of robotic liquid-handling systems that are capable of rapidly and efficiently screening thousands of crystallization conditions, in which different parameters such as ionic strength, precipitant concentration, additives, pH, and temperature are altered. High-throughput screening begins with the automated preparation and/or reformatting of precipitant solutions into crystallization microplates. The latest nanoliter robotic liquid-dispensing systems are capable of dispensing very small drops (typically containing between 25 and 100 nL of protein), which reduces the amount of protein required for screening conditions and the overhead on protein production. Furthermore, smaller drops equilibrate faster than larger drops, leading to a more rapid appearance of crystals. The crystallization experiments are then regularly monitored using an imaging robot, and the collected images can either be analyzed visually or by using automatic crystal recognition systems. HomeBase™ (The Automation Partnership, UK) represents an integrated system combining high-density plate storage and imaging.

A recent development in protein crystallization has been the use of a microfluidic system for crystallizing proteins, using the free-interface diffusion method at the nanoliter scale. This system is capable of screening hundreds of crystallization conditions in which droplets, each containing solutions of protein, precipitants, and additives in various ratios, are formed in the flow of immiscible fluids inside a polydimethylsiloxane (PDMS)/glass capillary composite microfluidic device (Fig. 5). The system is capable of performing multidimensional screening (mixing 5–10 solutions) and therefore explores more of the crystallization space than the conventional method of vapor diffusion, increasing the likelihood of obtaining crystals. Furthermore, the capillary containing protein crystals can be directly exposed to the synchrotron X-ray beam, eliminating the need to manually manipulate the crystal. These features also mean that it has the potential to serve as the basis for future high-throughput automated crystallization systems.

Membrane proteins represent the most persistent bottleneck for all preparatory procedures and analytical methods, because they are water soluble only in the presence of detergents, and are difficult to overproduce in the quantities required for structural studies. Membrane proteins, which constitute up to 30% of the protein repertoire of an organism, represent the targets for more than 50% of the drugs that are being currently used and tested. An adapted robotic system has recently been reported that enables high-throughput crystallization of membrane proteins using lipidic mesophases.

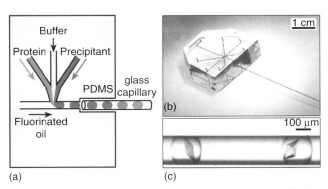

Fig. 5 (a) Schematic illustration of the droplet-based microfluidic system for protein crystallization. (b) Photograph of a polydimethylsiloxane (PDMS)/glass capillary composite microfluidic device. (c) A micrograph of thaumatin crystals grown in droplets, produced by the method outlined in panel (a), within a capillary. Taken from Zheng, B., Tice, J.D., Roach, L.S., Ismagilov, R.F. (2004) A droplet-based, composite PDMS/glass capillary microfluidic system for evaluating protein crystallization conditions by microbatch and vapor-diffusion methods with on-chip X-ray diffraction, *Angew. Chem., Int. Ed. Engl.* **43**, 2508–2511. Picture with permission from Dr. Rustem Ismagilov.

4.2
Data Collection

X-ray data collection has been revolutionized over the past decade, with the development of improved X-ray sources and detectors and the universal adoption of flash-freezing techniques that greatly reduce crystal radiation damage. New third-generation synchrotrons are now available across the world, which provide more intense and stable X-ray beams and, combined with new, faster, larger, and more sensitive X-ray detectors, allow higher-quality data to be collected much more rapidly, leading to a dramatic increase in the success rate of structure determination. Furthermore, specialized, highly collimated microfocus beamlines, such as ID13 at the ESRF (Grenoble, France; http://www.esrf.fr/exp_facilities/ID13/index.html) and the protein crystallography beamline at the Swiss Light Source (SLS), are specifically tailored to study biological crystals with small physical dimensions (microcrystals from 5 to 50 μm in size) or very large unit cell dimensions. Future developments in detector design, such as the MarResearch GmbH (Norderstedt, Germany) flat-panel detector, promise larger continuous active areas, higher spatial resolution, very low noise, and fast readout times (~1 s). High-throughput X-ray data collection has also led to the development of automated robotic sample changers that store and mount crystals sequentially while maintaining the samples at liquid-nitrogen temperatures (100 K). A novel system has recently been described, using a proprietary crystallization plate that facilitates the preliminary investigation of the diffraction properties of crystals *in situ* in the drop, by direct exposure to the X-ray beam. The BIOXHIT project (Biocrystallography (X) on a Highly Integrated Technology Platform for European Structural Genomics; http://www.bioxhit.org), comprising more than 20 research groups from all over Europe, aims to develop an integrated technology platform for synchrotron beamlines by promoting new approaches to crystallization, and fully automating diffraction data collection and structure determination.

Recent developments in technology have not been limited exclusively to synchrotron sources. The latest generation of high-intensity X-ray generators has revolutionized X-ray sources to such an extent that the latest "in-house" systems from Rigaku/MSC (e.g. the FR-E SuperBright; http://www.rigakumsc.com/) and Bruker AXS (http://www.bruker-axs.com) are comparable in intensity to first-generation synchrotron sources and, coupled with automated sample changers, make high-throughput crystallography possible in the laboratory. Furthermore, the development of the Compact Light Source (Lyncean Technologies, Inc., Palo Alto, CA, USA) promises to have a huge impact on protein structure determination, offering the possibility of a "synchrotron beamline" for home laboratory applications. This tunable, tabletop X-ray source combines an electron beam with a laser beam to generate an intense X-ray beam and, as a next-generation X-ray source, directly addresses the increasing demand for high-throughput protein crystallography.

4.3
Phasing

The central problem in X-ray crystallography is the determination of the protein phases. X-ray data collected from a crystal consist of structure factor amplitudes, but there is no way of directly measuring the phase associated with each amplitude. Recent advances in macromolecular phasing have simplified and further automated this crucial stage of X-ray structure determination to such an extent that the eventual success of a project is usually assured, if well-diffracting crystals are available. The techniques of isomorphous replacement, anomalous scattering, molecular replacement, and single (SAD) and multiple (MAD) anomalous dispersion are commonly used to solve the phase problem. However, phase determination has been dramatically facilitated by the widespread adoption of SAD and MAD phasing techniques by the crystallographic community, primarily as a consequence of the availability of stable and tunable synchrotron sources. These allow the optimal exploitation of the anomalous effect as a source of phase information by delivering X-ray energies corresponding to absorption maxima of the anomalous scatterers, very often selenium introduced as selenomethionine through substitution of methionine. Additionally, heavy-atom labels (mercury, platinum, and others) may be bound to crystalline proteins by soaking to yield phase information by anomalous diffraction techniques.

Modern phasing techniques, such as fast halide soaks and sulfur-SAD, hold promise of simpler and faster protein structure determination than traditional methods. Fast halide soaks using bromide and iodide that diffuse rapidly into the crystal and display significant anomalous scattering signals can be used to quickly derivatize protein crystals. Heavy-atom reagents can also be incorporated into the crystal in a relatively short time if a concentration greater than 10 mM is used. These derivatives then display better isomorphism and diffraction qualities than those obtained after a standard, prolonged soak. The availability of stable synchrotron beamlines and improvements in data processing programs make it possible to collect extremely accurate diffraction data, and to determine structures using the very weak anomalous signals from atoms such as sulfur and phosphorus that are inherently present in macromolecules or nucleic acids; several novel structures have been determined by using sulfur-SAD phasing.

4.4
Automated Structure Determination

High-throughput crystallographic structure determination requires software that is automated and designed for minimum user intervention. There has been considerable development in the direct-method programs, SHELXD and SnB, which can automatically determine heavy-atom substructures from a very small signal. HKL2MAP connects several programs from the SHELX suite to guide the user from analyzing scaled diffraction data (SHELXC), through substructure solution (SHELXD) and phasing (SHELXE), to displaying an electron density map (Xfit). There are a number of other automated software systems, such as ACrS (automated crystallographic system) and PHENIX (Python-based hierarchical environment for integrated Xtallography) that are currently being developed to meet the requirements of high-throughput structure determination by combining multiple structure-determination software packages into one intuitive interface.

Finally, new algorithms for interpreting electron density maps and for automated model building, such as, SOLVE/RESOLVE, AUTOSHARP/

SHARP, and ARP/wARP, enable rapid construction of protein models without the need for significant manual intervention. At present, the success rates for these programs are dependent on the resolution of the diffraction data (typically, 2.5-Å resolution or higher is necessary for automatic chain fitting).

4.5 Molecular Replacement

When an approximate structural model of a protein under investigation is available, either from NMR, a homologous X-ray structure, or from homology modeling, initial phases can be obtained using molecular replacement where the homologous probe structure is fitted to the experimental data using three rotational and three translational parameters. The programs AMoRe and MolRep that are integrated into the CCP4i GUI simplify the problem of positioning a molecule in the asymmetric unit by running sequential rotational and translational searches. Advances in molecular replacement include the implementation of the maximum likelihood–based algorithms in BEAST, and the six-dimensional evolutionary search algorithm in EPMR. As the number of protein structures increases, it is anticipated that molecular replacement will become the standard method for structure determination, using a generalized search of all unique protein domains present in the PDB.

compounds bind to their targets accelerates drug development and is more cost-effective. Notable drugs that have been successfully designed using protein three-dimensional structure information include the HIV protease inhibitors, Viracept™ (Agouron, USA and Eli Lilly, USA), and Agenerase™ (Vertex, USA; Kissei, Japan; Glaxo Wellcome, UK). However, it is usually necessary to screen a large number of protein–ligand complex structures in the iterative process of rational structure-based drug design. Hence, high-throughput protein X-ray crystallography offers an unprecedented opportunity for facilitating drug discovery. Recent advances in rapid binding-site analysis of *de novo* targets using virtual ligand (*in silico*) screening and small-molecule cocrystallization methodologies, in combination with the miniaturization and automation of structural biology, enable more rapid lead compound identification and faster optimization, providing a framework for direct integration into the drug discovery process. In the high-throughput structure determination of protein–ligand complexes, it is desirable to use tools that can locate, build, and refine the structure of the bound ligand with minimal human intervention. One such tool is X-LIGAND, part of the QUANTA software package (Accelrys, San Diego, CA, USA) that automatically searches for unoccupied regions of electron density in the structure of the protein–ligand complex in which it tries to fit the ligand.

5 Structure-based Drug Design

Knowledge of the three-dimensional structure of proteins can play a key role in the development of small-molecule drugs, because being able to verify how lead

6 Structure Determination by NMR

Solution-state NMR spectroscopy can serve as a technique complementary to X-ray crystallography in protein structure analysis, particularly in the context of structural genomics initiatives, where many protein targets either do not crystallize or do not form crystals suitable for crystallographic studies (owing to small crystal size or poor diffraction quality). NMR measurements are performed in aqueous solution, obviating the need for growing crystals. This technique is applicable primarily to small proteins (<30 kDa) that are highly soluble (millimolar concentrations), and is particularly useful in the study of proteins that are partially unfolded in the absence of their appropriate binding partners. The additional technique of solid-state NMR is useful in providing structural information for some integral membrane proteins that may not be accessible using crystallographic methods. Furthermore, chemical shift perturbation studies can be used to validate proposed biochemical functions, to map ligand-binding epitopes, and to screen for small-molecule ligands in drug development.

High-throughput, NMR-based structure determination requires rapid and automated data acquisition and analysis methods. The major challenges in realizing this have been those of increasing instrumentation sensitivity (signal-to-noise ratio) and reducing the time required for data collection. These technical issues have been addressed by constructing new high-field magnets and by the recent introduction of cryogenic probes that operate at low temperatures ($\sim$25 K), permitting the investigation of proteins that have either low solubility or low yields from purification. Additionally, the application of TROSY (transverse relaxed optimized spectroscopy), a novel spectroscopic concept based on the selection of slowly relaxing NMR transitions, has provided significant sensitivity enhancements for large proteins.

The most time-consuming aspects of structure determination by NMR are the long data collection times necessary for independently sampling three or more indirect dimensions along with the time taken to interpret the correspondingly large number of spectra from ^{13}C/^{15}N-isotope-labeled samples. Rapid resonance assignment is a prerequisite for high-throughput NMR structure determination; techniques such as reduced-dimensionality ^{13}C,^{15}N,^{1}H-triple resonance NMR also avoid the sampling limited regime through the simultaneous frequency-labeling of two spin types in a single indirect dimension. Heteronuclear multidimensional data reduce complications arising from interspectral variations by maximizing the dimensionality of the spectra and decrease signal overlap of data sets sufficiently for the data to be analyzed automatically. Recent approaches to automated structure elucidation from NMR spectra include NOESY-Jigsaw, in which sparse and unassigned NMR data can be used to reasonably and accurately assess secondary structure and align it. The information thus retrieved is useful for quick structural assays for assessing folds before full structural determination and can therefore assist in fold prediction. Additionally, the program ATNOS (automated NOESY peak picking) enables automated peak picking and NOE signal identification in homonuclear 2D and heteronuclear 3D [^{1}H, ^{1}H]-NOESY spectra during *de novo* protein structure determination.

See also Protein NMR Spectroscopy; Proteomics.

Bibliography

Books and Reviews

Blundell, T.L., Mizuguchi, K. (2000) Structural genomics: an overview, *Prog. Biophys. Mol. Biol.* **73**, 289–295.

Dauter, Z. (2002) New approaches to high-throughput phasing, *Curr. Opin. Struct. Biol.* **12**, 674–678.

Guntert, P. (2004) Automated NMR structure calculation with CYANA, *Methods Mol. Biol.* **278**, 353–378.

Laskowski, R.A., Watson, J.D., Thornton, J.M. (2003) From protein structure to biochemical function? *J. Struct. Funct. Genomics* **4**, 167–177.

Norin, M., Sundstrom, M. (2002) Structural proteomics: developments in structure-to-function predictions, *Trends Biotechnol.* **20**, 79–84.

Stevens, R.C. (2000) Design of high-throughput methods of protein production for structural biology, *Struct. Fold. Des.* **8**, R177–R185.

Stewart, L., Clark, R., Behnke, C. (2002) High-throughput crystallization and structure determination in drug discovery, *Drug Discov. Today* **7**, 187–196.

Tickle, I., Sharff, A., Vinkovic, M., Yon, J., Jhoti, H. (2004) High-throughput protein crystallography and drug discovery, *Chem. Soc. Rev.* **33**, 558–565.

Primary Literature

Abola, E., Kuhn, P., Earnest, T., Stevens, R.C. (2000) Automation of X-ray crystallography, *Nat. Struct. Biol.* **7**, 973–977.

Adams, P., Gopal, k., Grossekunstleve, R.W., Hung, L.-W., Ioerger, T.R., McCoy, A.J., Moriarty, N.W., Pai, R.K., Read, R.J., Romo, T.D., Sacchettini, J.C., Sauter, N.K., Storoni, L.C., Terwilliger, T.C. (2004) Recent developments in the PHENIX software for automated crystallographic structure determination, *J. Synchrotron Radiat.* **11**, 53–55.

Albala, J.S., Franke, K., McConnell, I.R., Pak, K.L., Folta, P.A., Rubinfeld, B., Davies, A.H., Lennon, G.G., Clark, R. (2000) From genes to proteins: high-throughput expression and purification of the human proteome, *J. Cell. Biochem.* **80**, 187–191.

Bailey-Kellogg, C., Widge, A., Kelley, J.J., Berardi, M.J., Bushweller, J.H., Donald, B.R. (2000) The NOESY jigsaw: automated protein secondary structure and main-chain assignment from sparse, unassigned NMR data, *J. Comput. Biol.* **7**, 537–558.

Bate, P., Warwicker, J. (2004) Enzyme/non-enzyme discrimination and prediction of enzyme active site location using charge-based methods, *J. Mol. Biol.* **340**, 263–276.

Breitling, R., Klingner, S., Callewaert, N., Pietrucha, R., Geyer, A., Ehrlich, G., Hartung, R., Muller, A., Contreras, R., Beverley, S.M., Alexandrov, K. (2002) Non-pathogenic trypanosomatid protozoa as a platform for protein research and production, *Protein Expr. Purif.* **25**, 209–218.

Brenner, S.E. (2000) Target selection for structural genomics, *Nat. Struct. Biol.* **7**(Suppl.), 967–969.

Brown, J., Walter, T.S., Carter, L., Abrescia, G.A., Aricescu, A.R., Batuwangala, T.D., Bird, L.E., Brown, N., Chamberlain, P.P., Davis, S.J., Dubinina, E., Endicott, J., Fennelly, J.A., Gilbert, R.J.C., Harkiolaki, M., Hon, W.-C., Kimberley, F., Love, C.A., Mancini, E.J., Manso-Sancho, R., Nichols, C.E., Robinson, R.A., Sutton, G.C., Schueller, N., Sleeman, M.C., Stewart-Jones, G.B., Vuong, M., Welburn, J., Zhang, Z., Stammers, D.K., Owens, R.J., Jones, E.Y., Harlos, K., Stuart, D.I. (2003) A procedure for setting up high-throughput nanoliter crystallization experiments. II. Crystallization results, *J. Appl. Crystallogr.* **36**, 315–318.

Bruel, C., Cha, K., Reeves, P.J., Getmanova, E., Khorana, H.G. (2000) Rhodopsin kinase: expression in mammalian cells and a two-step purification, *Proc. Natl. Acad. Sci. U.S.A.* **97**, 3004–3009.

Brunzelle, J.S., Shafaee, P., Yang, X., Weigand, S., Ren, Z., Anderson, W.F. (2003) Automated crystallographic system for high-throughput protein structure determination, *Acta Crystallogr.* **D59**, 1138–1144.

Cereghino, J.L., Cregg, J.M. (2000) Heterologous protein expression in the methylotrophic yeast *Pichia pastoris*, *FEMS Microbiol. Rev.* **24**, 45–66.

Cherezov, V., Peddi, A., Muthusubramaniam, L., Zheng, Y.F., Caffrey, M. (2004) A robotic system for crystallizing membrane and soluble proteins in lipidic mesophases, *Acta Crystallogr.* **D60**, 1795–1807.

Collaborative Computational Project, N. (1994) The CCP4 suite: programs for protein crystallography, *Acta Crystallogr.* **D50**, 760–763.

Dauter, Z., Dauter, M., Rajashankar, K.R. (2000) Novel approach to phasing proteins: derivatization by short cryo-soaking with halides, *Acta Crystallogr.* **D56**, 232–237.

del Val, C., Mehrle, A., Falkenhahn, M., Seiler, M., Glatting, K.H., Poustka, A., Suhai, S., Wiemann, S. (2004) High-throughput protein analysis integrating bioinformatics and experimental assays, *Nucleic Acids Res.* **32**, 742–748.

Delbruck, H., Ziegelin, G., Lanka, E., Heinemann, U. (2002) An Src homology 3-like domain is responsible for dimerization of the repressor protein KorB encoded by the promiscuous IncP plasmid RP4, *J. Biol. Chem.* **277**, 4191–4198.

Dobson, P.D., Doig, A.J. (2003) Distinguishing enzyme structures from non-enzymes without alignments, *J. Mol. Biol.* **330**, 771–783.

Endo, Y., Sawasaki, T. (2004) High-throughput, genome-scale protein production method based on the wheat germ cell-free expression system, *J. Struct. Funct. Genomics* **5**, 45–57.

Fortelle, Edl., Bricogne, G. (1997) Maximum-likelihood heavy-atom parameter refinement for multiple isomorphous replacement and multiwavelength anomalous diffraction methods, *Methods enzymol.* **276**, 472–494.

Garman, E. (1999) Cool data: quantity AND quality, *Acta Crystallogr.* **D55**, 1641–1653.

Garner, E., Cannon, P., Romero, P., Obradovic, Z., Dunker, A.K. (1998) Predicting disordered regions from amino acid sequence: common themes despite differing structural characterization, *Genome Inf. Ser. Worksh. Genome Inf.* **9**, 201–213.

Garner, E., Romero, P., Dunker, A.K., Brown, C., Obradovic, Z. (1999) Predicting binding regions within disordered proteins, *Genome Inf. Ser. Worksh. Genome Inf.* **10**, 41–50.

Goodwill, K.E., Tennant, M.G., Stevens, R.C. (2001) High-throughput x-ray crystallography for structure-based drug design, *Drug Discov. Today* **15**(Suppl.), 113–118.

Grinna, L.S., Tschopp, J.F. (1989) Size distribution and general structural features of N-linked oligosaccharides from the methylotrophic yeast, *Pichia pastoris*, *Yeast* **5**, 107–115.

Heinemann, U., Frevert, J., Hofmann, K., Illing, G., Maurer, C., Oschkinat, H., Saenger, W. (2000) An integrated approach to structural genomics, *Prog. Biophys. Mol. Biol.* **73**, 347–362.

Herrmann, T., Guntert, P., Wuthrich, K. (2002) Protein NMR structure determination with automated NOE-identification in the NOESY spectra using the new software ATNOS, *J. Biomol. NMR* **24**, 171–189.

Holm, L., Sander, C. (1993) Protein structure comparison by alignment of distance matrices, *J. Mol. Biol.* **233**, 123–138.

Jhoti, H. (2001) High-throughput structural proteomics using X-rays, *Trends Biotechnol.* **19**, S67–S71.

Kaldor, S.W., Kalish, V.J., Davies, J.F. II, Shetty, B.V., Fritz, J.E., Appelt, K., Burgess, J.A., Campanale, K.M., Chirgadze, N.Y., Clawson, D.K., Dressman, B.A., Hatch, S.D., Khalil, D.A., Kosa, M.B., Lubbehusen, P.P., Muesing, M.A., Patick, A.K., Reich, S.H., Su, K.S., Tatlock, J.H. (1997) Viracept (nelfinavir mesylate, AG1343): a potent, orally bioavailable inhibitor of HIV-1 protease, *J. Med. Chem.* **40**, 3979–3985.

Kelley, L.A., MacCallum, R.M., Sternberg, M.J. (2000) Enhanced genome annotation using structural profiles in the program 3D-PSSM, *J. Mol. Biol.* **299**, 499–520.

Khare, D., Ziegelin, G., Lanka, E., Heinemann, U. (2004) Sequence-specific DNA binding determined by contacts outside the helix-turn-helix motif of the ParB homolog KorB, *Nat. Struct. Mol. Biol.* **11**, 656–663.

Kim, E.E., Baker, C.T., Dwyer, M.D., Murcko, M.A., Rao, B.G., Tung, R.D., Navia, M.A. (1995) Crystal structure of HIV-1 protease in complex with VX-478, a potent and orally bioavailable inhibitor of the enzyme, *J. Am. Chem. Soc.* **117**, 1181–1182.

Kissinger, C.R., Gehlhaar, D.K., Fogel, D.B. (1999) Rapid automated molecular replacement by evolutionary search, *Acta Crystallogr.* **D55**, 484–491.

Kleywegt, G.J. (1999) Recognition of spatial motifs in protein structures, *J. Mol. Biol.* **285**, 1887–1897.

Kolb, V.A., Makeyev, E.V., Spirin, A.S. (2000) Co-translational folding of an eukaryotic multidomain protein in a prokaryotic translation system, *J. Biol. Chem.* **275**, 16597–16601.

Kuhn, P., Wilson, K., Patch, M.G., Stevens, R.C. (2002) The genesis of high-throughput structure-based drug discovery using protein crystallography, *Curr. Opin. Chem. Biol.* **6**, 704–710.

Lamzin, V.S., Perrakis, A. (2000) Current state of automated crystallographic data analysis, *Nat. Struct. Biol.* **7**(Suppl.), 978–981.

Li, X., Obradovic, Z., Brown, C.J., Garner, E.C., Dunker, A.K. (2000) Comparing predictors of disordered protein, *Genome Inf. Ser. Worksh. Genome Inf.* **11**, 172–184.

Li, X., Romero, P., Rani, M., Dunker, A.K., Obradovic, Z. (1999) Predicting protein disorder for N-, C-, and internal regions, *Genome Inf. Ser. Worksh. Genome Inf.* **10**, 30–40.

Linding, R., Russell, R.B., Neduva, V., Gibson, T.J. (2003) GlobPlot: exploring protein sequences for globularity and disorder, *Nucleic Acids Res.* **31**, 3701–3708.

Linding, R., Jensen, L.J., Diella, F., Bork, P., Gibson, T.J., Russell, R.B. (2003) Protein disorder prediction: implications for structural proteomics, *Structure* **11**, 1453–1459.

Linial, M., Yona, G. (2000) Methodologies for target selection in structural genomics, *Prog. Biophys. Mol. Biol.* **73**, 297–320.

Madej, T., Gibrat, J.F., Bryant, S.H. (1995) Threading a database of protein cores, *Proteins* **23**, 356–369.

Manjasetty, B.A., Delbruck, H., Pham, D.T., Mueller, U., Fieber-Erdmann, M., Scheich, C., Sievert, V., Bussow, K., Niesen, F.H., Weihofen, W., Loll, B., Saenger, W., Heinemann, U., Neisen, F.H. (2004) Crystal structure of Homo sapiens protein hp14.5, *Proteins* **54**, 797–800.

Montelione, G.T., Zheng, D., Huang, Y.J., Gunsalus, K.C., Szyperski, T. (2000) Protein NMR spectroscopy in structural genomics, *Nat. Struct. Biol.* **7**(Suppl.), 982–985.

Mueller, U., Bussow, K., Diehl, A., Bartl, F.J., Niesen, F.H., Nyarsik, L., Heinemann, U. (2003) Rapid purification and crystal structure analysis of a small protein carrying two terminal affinity tags, *J. Struct. Funct. Genomics* **4**, 217–225.

Navaza, J. (2001) Implementation of molecular replacement in AMoRe, *Acta Crystallogr.* **D57**, 1367–1372.

Netzer, W.J., Hartl, F.U. (1997) Recombination of protein domains facilitated by co-translational folding in eukaryotes, *Nature* **388**, 343–349.

Novotny, M., Madsen, D., Kleywegt, G.J. (2004) Evaluation of protein fold comparison servers, *Proteins* **54**, 260–270.

Oldfield, T.J. (2001) X-LIGAND: an application for the automated addition of flexible ligands into electron density, *Acta Crystallogr.* **D57**, 696–705.

Pandey, N., Ganapathi, M., Kumar, K., Dasgupta, D., Das Sutar, S.K., Dash, D. (2004) Comparative analysis of protein unfoldedness in human housekeeping and non-housekeeping proteins, *Bioinformatics* **20**, 2904–2910.

Pape, T., Schneider, T.R. (2004) HKL2MAP: a graphical user interface for phasing with SHELX programs, *J. Appl. Crystallogr.* **37**, 843–844.

Perrakis, A., Morris, R., Lamzin, V.S. (1999) Automated protein model building combined with iterative structure refinement, *Nat. Struct. Biol.* **6**, 458–463.

Potterton, E., Briggs, P., Turkenburg, M., Dodson, E. (2003) A graphical user interface to the CCP4 program suite, *Acta Crystallogr.* **D59**, 1131–1137.

Prinz, B., Schultchen, J., Rydzewski, R., Holz, C., Boettner, M., Stahl, U., Lang, C. (2004) Establishing a versatile fermentation and purification procedure for human proteins expressed in the yeasts Saccharomyces cerevisiae and Pichia pastoris for structural genomics, *J. Struct. Funct. Genomics* **5**, 29–44.

Read, R.J. (2001) Pushing the boundaries of molecular replacement with maximum likelihood, *Acta Crystallogr.* **D57**, 1373–1382.

Sali, A., Glaeser, R., Earnest, T., Baumeister, W. (2003) From words to literature in structural proteomics, *Nature* **422**, 216–225.

Sanchez, R., Pieper, U., Melo, F., Eswar, N., Marti-Renom, M.A., Madhusudhan, M.S., Mirkovic, N., Sali, A. (2000) Protein structure modeling for structural genomics, *Nat. Struct. Biol.* **7**(Suppl.), 986–990.

Sanishvili, R., Yakunin, A.F., Laskowski, R.A., Skarina, T., Evdokimova, E., Doherty-Kirby, A., Lajoie, G.A., Thornton, J.M., Arrowsmith, C.H., Savchenko, A., Joachimiak, A., Edwards, A.M. (2003) Integrating structure, bioinformatics, and enzymology to discover function: BioH, a new carboxylesterase from *Escherichia coli*, *J. Biol. Chem.* **278**, 26039–26045.

Sawasaki, T., Ogasawara, T., Morishita, R., Endo, Y. (2002) A cell-free protein synthesis system for high-throughput proteomics, *Proc. Natl. Acad. Sci. U.S.A.* **99**, 14652–14657.

Schneider, T.R., Sheldrick, G.M. (2002) Substructure solution with SHELXD, *Acta Crystallogr.* **D58**, 1772–1779.

Schwede, T., Kopp, J., Guex, N., Peitsch, M.C. (2003) SWISS-MODEL: an automated protein homology-modeling server, *Nucleic Acids Res.* **31**, 3381–3385.

Spriggs, R.V., Artymiuk, P.J., Willett, P. (2003) Searching for patterns of amino acids in 3D protein structures, *J. Chem. Inf. Comput. Sci.* **43**, 412–421.

Stevens, R.C., Yokoyama, S., Wilson, I.A. (2001) Global efforts in structural genomics, *Science* **294**, 89–92.

Sun, P.D., Radaev, S. (2002) Generating isomorphous heavy-atom derivatives by a quick-soak method. Part II: phasing of new structures, *Acta Crystallogr.* **D58**, 1099–1103.

Sun, P.D., Radaev, S., Kattah, M. (2002) Generating isomorphous heavy-atom derivatives by a quick-soak method. Part I: test cases, *Acta Crystallogr.* **D58**, 1092–1098.

Szyperski, T., Yeh, D.C., Sukumaran, D.K., Moseley, H.N., Montelione, G.T. (2002) Reduced-dimensionality NMR spectroscopy for high-throughput protein resonance assignment, *Proc. Natl. Acad. Sci. U.S.A.* **99**, 8009–8014.

Terwilliger, T.C. (2000) Maximum-likelihood density modification, *Acta Crystallogr.* **D56**, 965–972.

Terwilliger, T.C., Berendzen, J. (1999) Automated MAD and MIR structure solution, *Acta Crystallogr.* **D55**, 849–861.

Thornton, J.M., Todd, A.E., Milburn, D., Borkakoti, N., Orengo, C.A. (2000) From structure to function: approaches and limitations, *Nat. Struct. Biol.* **7**(Suppl.), 991–994.

Vagin, A., Teplyakov, A. (1997) MOLREP: an automated program for molecular replacement, *J. Appl. Crystallogr.* **30**, 1022–1025.

Voss, S., Skerra, A. (1997) Mutagenesis of a flexible loop in streptavidin leads to higher affinity for the Strep-tag II peptide and improved performance in recombinant protein purification, *Protein Eng.* **10**, 975–982.

Walter, T.S., Diprose, J., Brown, J., Pickford, M., Owens, R.J., Stuart, D.I., Harlos, K. (2003) A procedure for setting up high-throughput nanoliter crystallization experiments. I. Protocol design and validation, *J. Appl. Crystallogr.* **36**, 308–314.

Watanabe, N., Murai, H., Tanaka, I. (2002) Semi-automatic protein crystallization system that allows in situ observation of X-ray diffraction from crystals in the drop, *Acta Crystallogr.* **D58**, 1527–1530.

Watson, J.D., Todd, A.E., Bray, J., Laskowski, R.A., Edwards, A., Joachimiak, A., Orengo, C.A., Thornton, J.M. (2003) Target selection and determination of function in structural genomics, *IUBMB Life* **55**, 249–255.

Xu, H., Hauptman, H., Weeks, C.M. (2002) Sine-enhanced shake-and-bake: the theoretical basis for applications to Se-atom substructures, *Acta Crystallogr.* **D58**, 90–96.

Yee, A., Pardee, K., Christendat, D., Savchenko, A., Edwards, A.M., Arrowsmith, C.H. (2003) Structural proteomics: toward high-throughput structural biology as a tool in functional genomics, *Acc. Chem. Res.* **36**, 183–189.

Zheng, B., Roach, L.S., Ismagilov, R.F. (2003) Screening of protein crystallization conditions on a microfluidic chip using nanoliter-size droplets, *J. Am. Chem. Soc.* **125**, 11170–11171.

Zheng, B., Tice, J.D., Roach, L.S., Ismagilov, R.F. (2004) A droplet-based, composite PDMS/glass capillary microfluidic system for evaluating protein crystallization conditions by microbatch and vapor-diffusion methods with on-chip X-ray diffraction, *Angew. Chem., Int. Ed. Engl.* **43**, 2508–2511.

Protein Translocation Across Membranes

Carla M. Koehler and David K. Hwang
University of California Los Angeles, Los Angeles, CA, USA

1	Overview of Compartmentalization	288
2	Common Principles of Translocation	289
3	Properties of Targeting Sequences	290
4	Signal-gated Translocation Systems	292
4.1	The Sec System	292
4.1.1	The Prokaryotic Sec System	292
4.1.2	The Mammalian Sec System	294
4.2	YidC is Involved in the Assembly of Membrane Proteins	294
4.3	Mitochondrial Protein Import Pathways	295
4.4	Chloroplast Protein Import Pathways	297
5	Signal-assembled Translocation Systems	298
5.1	Peroxisomal Import Pathways	298
5.1.1	Protein Import into the Matrix	299
5.1.2	Protein Import into the Membrane	301
5.2	The Tat/ΔpH Pathway of Bacteria and Thylakoids	301
6	Conclusions	303
	Acknowledgment	303
	Bibliography	303
	Books and Reviews	303
	Primary Literature	304

Encyclopedia of Molecular Cell Biology and Molecular Medicine, 2nd Edition. Volume 11
Edited by Robert A. Meyers.
Copyright © 2005 Wiley-VCH Verlag GmbH & Co. KGaA, Weinheim
ISBN: 3-527-30648-X

Keywords

Molecular Chaperones
Proteins that assist with the correct folding, assembly, or disassembly of other proteins *in vivo*, but are not considered components of the final functional structure.

Receptor
The receptor, usually in association with the translocation channel, recognizes the targeting sequence and directs the substrate to the translocation channel.

Signal or Targeting Sequence
A sequence or motif in a substrate for directing it to a particular location within the cell. The targeting sequence often is a conserved or consensus sequence at the N or C terminus of the protein.

Translocase or Translocon
A system catalyzing the transfer of a substrate across or into a (membrane) barrier.

Translocation Channel
A pore formed by proteins across the membrane that mediates the transport of a protein across or into a membrane.

From prokaryotic to eukaryotic cells, the movement of proteins across membranes as well as the integration into membranes is a fundamental process with shared mechanisms. Protein translocation pathways can be viewed as modular systems with basic components. Transported substrates contain targeting information that is decoded by receptors at docking sites. The substrate is then translocated across the membrane generally through a channel. A source of energy is often required and a translocation motor usually mediates translocation through the channel. In this chapter, we will summarize translocation systems in eukaryotic and prokaryotic systems, emphasizing common themes. Our understanding of protein translocation is continuing to grow with the addition of new components to existing systems and the identification of new translocation machines. The reader is encouraged to use this chapter as a starting point for more detailed excursions into scientific literature.

1
Overview of Compartmentalization

Membranes form important boundaries in cells to regulate selected exchange of molecules and ions between aqueous environments. The plasma membrane is a universal feature of all cell types and contains a wide range of transporters to mediate the exchange between the extracellular milieu and the cytoplasm. Eukaryotic cells also have developed compartmentalization through membrane-bound organelles that specialize in various intracellular biochemical processes.

Experimental analysis using genetics and *in vitro* assays for protein targeting to eukaryotic organelles and the bacterial periplasmic membrane have revealed molecular details of the majority of protein translocation systems. The unifying principles of these systems were first proposed in the signal hypothesis two decades ago, although the components of the systems vary for each pathway. Prokaryotes first developed translocation systems for the translocation of proteins into the periplasmic space and for integration into membranes. The secretion (Sec) system mediates the translocation of proteins to the periplasmic space and has been evolutionarily conserved in the endoplasmic reticulum and chloroplast thylakoid. When the mitochondrion and chloroplast developed in the primitive eukaryotic cell, the progenitor's DNA became integrated into the host nucleus. As genes moved into the nucleus, novel translocation systems developed within the chloroplast and mitochondrion. The peroxisome also developed a translocation system seemingly independent of other organellar systems, which can translocate folded proteins.

2
Common Principles of Translocation

All translocons possess several essential features. A substrate possesses specific targeting information that directs it to the correct location. The targeting information is referred to as a *signal sequence* and is often found at the N terminus of a substrate. Molecular chaperones maintain the precursor in an import-competent state. Receptors are able to decode the information and direct the precursor to the oligomeric membrane complex, termed the *translocon*. Receptors are often bound to the membrane and may transiently associate with the translocation channel. The channel is an aqueous pore in the membrane that is gated, and mediates passage of the substrate across or into the membrane. The precursor generally passes through the channel in an unfolded state. Translocation requires an energetic input such as a membrane potential and/or nucleotide hydrolysis. A translocation motor is required to pull or push the precursor from the cis side (site of synthesis) to the trans side (destination site) of the membrane. The substrate is often processed and assembles into a multi-subunit complex.

Translocation systems can be defined as export or import systems depending upon the difference between the compartments on either side of the membrane. The export system is located in the bacterial plasma membrane, the endoplasmic reticulum, the chloroplast thylakoid membrane, and mitochondrial inner membrane. Proteins are exported from the cytosol into an extracytoplasmic compartment (i.e. the bacterial periplasm, the lumen of the endoplasmic reticulum and thylakoid, and the mitochondrial inner membrane). The import systems are located in the mitochondrial and chloroplast outer and inner membranes, and the peroxisome; this reference is based on the fact that the protein is translocated into a compartment that is functionally equivalent to the cytosol.

The translocons are remarkably flexible, able to accommodate hundreds of distinct proteins substrates while maintaining the permeability barrier of the membrane. The translocation channel must be gated to prevent solutes and ions from passing between compartments. The channel must also recognize "stop-transfer" signals within integral membrane proteins and then gate laterally to allow diffusion of the transmembrane domains into the lipid bilayer. The channels therefore play an active role in translocation.

Two classes of translocons can be defined on the basis of the mechanism by which the substrate is translocated. The most common class, termed *signal-gated translocon* functions like the gated ion channel. The newly synthesized substrate moves through the channel in an unfolded conformation with the assistance of molecular chaperones. Because the precursor is maintained in an unfolded state, the single translocon can accommodate a wide range of substrates. The translocation reaction therefore is similar to that of ion or metabolite transport. In contrast, the second class is able to transport fully folded and/or oligomeric proteins of large dimensions while maintaining a membrane permeability barrier. This class is referred to as *signal-assembled* translocons and is present in peroxisomes, chloroplast thylakoids, and bacteria. Stable translocation channels have not been detected in these systems; thus, it has been proposed that the translocons are assembled in variable dimensions to accommodate the size of the translocated substrate.

3
Properties of Targeting Sequences

Proteins destined for transport into or across a membrane have a signal sequence, usually at the N terminus that is proteolytically cleaved on the trans side of the membrane. Depending upon the system, it is referred to as a *signal-*, *leader-*, *targeting-*, *transit-*, or *presequence*. Generally, export signal sequences are hydrophobic whereas import signal sequences are more hydrophilic. For the prokaryotic and endoplasmic reticulum export systems, the export sequences can generally be interchanged. However, import signal sequences cannot be interchanged between different organellar sequences; swapping the targeting sequence of a resident chloroplast protein to a mitochondrial targeting presequence alters trafficking from the chloroplast to the mitochondrion. Hence, import signals are important for directing a precursor to the correct organelle in the cytosol that contains several organelles including peroxisomes, mitochondria, and chloroplasts.

Signal sequences increase the specific interaction of a precursor with the appropriate transport machinery. The signal sequence may act as a folding inhibitor to maintain the transport competency of the precursor. As a precursor exits from the ribosome in *Escherichia coli*, the signal sequence inhibits folding, thus allowing the molecular chaperone SecB to bind to the precursor. SecB subsequently binds to SecA to deliver the precursor to the translocation channel. The signal sequence then interacts with components in the translocation channel. If the signal sequence is removed, export is severely inhibited, but export can be increased to normal rates by mutations in the translocation channel. Thus, SecB–SecA interactions are important for targeting, rather than the signal sequence. However, in most cases, the signal sequence acts as a true targeting signal because it is specifically recognized by cytosolic chaperones, targeting factors, or receptors.

For the bacterial and endoplasmic reticulum export systems, the targeting sequence is generally a short ($\sim$20 amino acids) sequence consisting of a hydrophobic core and short polar flanking regions located at the N terminus. The targeting sequence is usually cleaved at the trans side of the membrane by a processing protease. If an export signal is not cleaved and possesses a hydrophobic core of 20 amino acids, the sequence can serve as a signal-anchor sequence to anchor the protein in the membrane, generating a transmembrane protein.

A protein with several membrane-spanning domains also can be generated by the sequential coordination of a signal sequence and a downstream hydrophobic stop-transfer sequence. The signal sequence serves as the initiator for translocation and the hydrophobic stop-transfer sequence arrests translocation, probably triggering a lateral opening of the translocation channel. The protein thus escapes from the translocation channel and becomes an integral membrane protein.

Mitochondria and chloroplasts contain an N-terminal bipartite targeting sequence for targeting and subsequent sorting within the organelle. Mitochondrial precursors are sorted to the matrix, inner membrane, and intermembrane space, whereas chloroplast precursors are directed to the thylakoid membrane and lumen, the stroma, and the inner envelope. The mitochondrial targeting sequence consists of an N-terminal amphipathic helix and is approximately 20 to 50 amino acids in length. One face is positively charged while the opposite face is rich in hydrophobic residues. Proteins destined for the intermembrane space contain a bipartite targeting sequence. Specifically, a hydrophobic "stop-transfer" domain to arrest translocation within the translocon is located after the targeting sequence. Import is coordinated by proteolytic cleavage to remove the N-terminal targeting domain in the matrix, followed by precursor arrest in the translocon, and subsequent proteolytic cleavage in the intermembrane space before release. In contrast, mitochondrial precursors targeted for the inner and outer membranes lack an N-terminal targeting presequence and instead, contain targeting information within the mature protein.

Chloroplast precursors contain an N-terminal transit sequence varying in length from 20 to 150 residues. Precursors destined for the thylakoids also contain a bipartite sequence in which the first N-terminal tract designates chloroplast targeting, followed by a thylakoid transfer domain. Transit sequences carry a net positive charge and are rich in hydroxylated amino acids serine and threonine. Interestingly, several transit sequences can be phosphorylated in the cytosol, which enhances import.

Peroxisomal proteins contain three different types of peroxisomal targeting sequences (PTS). For the matrix protein cohort, PTS1 and PTS2 have been characterized: PTS1 is a C-terminal three amino acid sequence (S/A/C)(K/R/H)(L/M) that has been easily identified as genome sequences have become available, whereas PTS2 is a nonapeptide sequence $(R/K)(L/V/I)X_5(H/Q)(L/A)$ near the N terminus. For membrane proteins, a consensus sequence has not been identified, but sequences encompassing the transmembrane domains have been shown to be required for targeting so that both interact with a peroxisome chaperone and mediate insertion into the membrane.

Substrates of the twin-arginine translocation (Tat) system contain an N-terminal cleavable presequence with a twin-arginine motif containing a consensus SRRXFLK. The Tat-type signals are highly conserved throughout plant and bacterial kingdoms. In prokaryotes, the Tat and Sec systems have to discriminate between the two types of precursors. Whereas Tat- and Sec-type targeting signals generally are similar, Tat-type signals are less hydrophobic than Sec-type signals and thus evade the Sec system.

In general, signal sequences for a given membrane transport system lack a strict consensus sequence. 25% of randomly generated peptides can function as signal sequences for various organelles, and roughly 5% of all sequences in the *E. coli* genome can direct a protein to the mitochondrion. Therefore, the secondary structure seems to be important for specific interactions with receptors.

4
Signal-gated Translocation Systems

4.1
The Sec System

The Sec system of bacteria (Fig. 1) and the endoplasmic reticulum (Fig. 2) share common properties. In addition, the chloroplast thylakoid also contains a Sec-related translocon. These translocons contain an oligomeric membrane–protein complex and associate with a translocation motor. The protein-conducting channel of this pathway is evolutionarily conserved in all kingdoms of life.

The bacterial system contains the SecYEG membrane complex that associates with the cytoplasmic ATPase, SecA. SecA functions as both a component of the cytosolic signal receptor complex and as the protein translocation motor. The SecYEG complex associates with the ribosome for cotranslational-membrane-protein insertion and with YidC, for assembly of membrane proteins. YidC has homologs in the chloroplast (Albino3) and mitochondria (Oxa1p).

4.1.1 The Prokaryotic Sec System

The bacterial Sec system utilizes cytosolic chaperones, a translocation channel, and an ATP-dependent translocation motor to push substrates into the periplasm (Fig. 1). The cytosolic SecB tetramer binds to the targeting sequence for precursors destined for export, thus functioning as a molecule chaperone. SecA is a multifunctional protein that acts as a chaperone and ATP-dependent translocation motor. In the cell, SecA is both membrane-associated and soluble. The soluble SecA also functions as a molecule chaperone; in the homodimeric state, SecA binds to the SecB–preprotein complex. When the

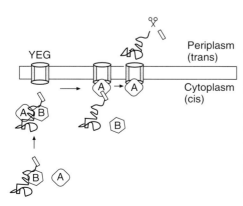

Fig. 1 The Sec translocation pathway for the export of bacterial proteins. Bacterial proteins are synthesized as precursors in the cytosol. SecB binds to the precursor and recruits SecA. The complex moves to the SecYEG translocon. SecB is released and SecA functions as a translocation motor to push the precursor through the SecYEG channel. SecA insertion is coupled with SecG inversion. Note: the channel is in a sealed conformation when not engaged or occupied by SecA.

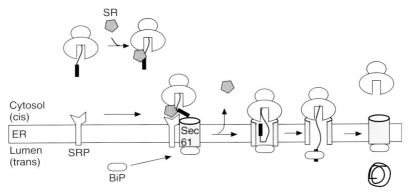

Fig. 2 The Sec cotranslational translocation pathway for the export of mammalian proteins via the endoplasmic reticulum. The soluble signal protein (SR) binds to the targeting signal as it emerges from the ribosome. The ribosome nascent chain complex is then escorted to the signal protein receptor (SRP) at the membrane. SR is released and the ribosome engages the Sec61 translocon, which causes opening of the translocon. The Hsp70 protein BiP gates the channel from the lumenal side and provides a seal to maintain membrane permeability. Translocation proceeds cotranslationally until protein synthesis terminates. The ribosome is released from the translocon and the translocon reverts to its closed state.

SecA–SecB–preprotein complex reaches the membrane, SecA associates with low affinity to negatively charged phospholipids and with high affinity to the translocation channel consisting of SecYEG. The preprotein is then transferred from SecB to SecA, and SecB is released by SecA upon the ATP-dependent initiation of translocation. The preprotein is then translocated through the SecYEG channel by multiple cycles of ATP binding and hydrolysis by SecA. Approximately 2 to 3 kDa segments are translocated per each ATP-dependent insertion reaction. The membrane potential is also required for translocation. SecA "pushing" from the cytosolic side compensates for the lack of a periplasmic energy source that could fuel molecular chaperones in the periplasmic space.

SecYEG forms the core of the protein-conducting channel and along with SecA is termed the *preprotein translocase*. SecY is the largest member of the protein-conducting channel and essential for viability and protein translocation. From experimental approaches and computer modeling, SecY contains 10 membrane-spanning domains with both N and C termini oriented to the cytosol. The fifth cytosolic loop is important for interactions with SecA. In contrast, SecE is a 14-kDa protein with three membrane-spanning domains in *E. coli*; the N terminus localizes to the cytosol while the C terminus faces the periplasm. Interestingly, other bacteria contain a SecE version that is homologous to only the third membrane-spanning domain, thus suggesting that this is an important region in SecE. SecG associates loosely with the protein-conducting channel and contains two membrane-spanning domains. SecG is an unusual protein because it undergoes an inversion in topology upon the membrane insertion of SecA. Although SecG is not essential for viability, it is required for efficient export at low temperatures.

What is the oligomeric state of the protein-conducting channel? The crystal structure from the archaeon *Methanococcus jannaschii* revealed a monomeric complex. However, the diameter of the channel is too narrow to accommodate SecA. Electron microscopy studies with purified SecYEG complex suggested a larger-sized complex of ~85 by ~65 Å with a pore of ~20 Å; the size and mass of these particles may reflect a dimeric or tetrameric form. From experimental studies including analytical ultracentrifugation and blue native gel analysis, it was revealed that the complex forms a dimer. The channel thus seems to have several different oligomeric assemblies. Further structural studies with the channel engaging SecA are required to provide a picture of the translocase in action.

4.1.2 The Mammalian Sec System

The mammalian system contains the SecYEG homologs Sec61α, Sec61β, and Sec61γ subunits, in addition to a unique subunit TRAM that is required for membrane protein integration (Fig. 2). The mammalian translocon forms an aqueous pore that spans the ER (endoplasmic reticulum) membrane, and Sec61α membrane-spanning domains line the channel similar to SecY. Ribosomes synthesizing membrane or secretory proteins are earmarked by a signal sequence at the N terminus of the nascent chain that binds to the signal recognition particle (SRP). SRP directs the ribosome nascent chain complex (RNC) to the signal receptor (SR) at the ER membrane. GTP-dependent interactions then elicit binding of the RNC directly to the translocon, followed by transfer of the nascent chain into the channel.

How is the protein translocated into the channel? The lumenal side of the pore is closed by the Hsp70 chaperone BiP. After RNC docking and elongation of the translocating chain, BiP is released from the pore, allowing passage of the chain into the ER lumen. The RNC then maintains an ion-tight seal on the cytoplasmic side. If a protein contains several membrane-spanning domains, BiP at the lumenal side and the RNC at the cytosolic side alternately gate the pore through a series of orchestrated changes, thus maintaining an ion-tight barrier between the lumen and cytoplasm during translocation of the protein. Specifically, as the nascent chain increases in length, BiP is released from the channel, while the ribosome maintains an ion-tight seal. Upon the detection of a membrane-spanning domain, the luminal end of the pore is closed by BiP, allowing the ribosome to then lift from the translocation channel. Translocated membrane proteins most likely are released laterally into the membrane via parting of the membrane-spanning domains in the Sec channel. The inner diameter of the pore changes dynamically from 9–15 Å in the ribosome-free closed state to 40–60 Å upon ribosome docking. The flexibility of the channel may be required to accommodate multiple membrane-spanning domains, which may leave the translocon in groups.

4.2 YidC is Involved in the Assembly of Membrane Proteins

YidC cooperates with the bacterial Sec system and operates independently to mediate translocation of integral membrane proteins. YidC is an integral membrane protein with six membrane-spanning domains and a large periplasmic domain. YidC interacts with the SecYEG translocon to facilitate integration of membrane proteins during translocation. YidC also acts independently to mediate the insertion of subunits in the energy-transducing complexes such as the F_0 domain of the F_1F_0 ATPase. The chloroplast homolog albino3 is required for protein insertion into the thylakoid membrane and the mitochondrial homolog Oxa1p is required for protein insertion into the inner membrane.

4.3
Mitochondrial Protein Import Pathways

The mitochondrion contains two membranes that separate the mitochondrial matrix from the intermembrane space (Fig. 3). The inner membrane is essentially impermeable to ions because a membrane potential must be maintained for respiration. Most proteins are synthesized on cytosolic ribosomes and imported cotranslationally or posttranslationally into the mitochondrion. Proteins coded on the mitochondrial genome, however, are exported.

The mitochondrion contains a small genome and these proteins are exported to the inner membrane via Oxa1p. Oxa1p is homologous to YidC, but contains five membrane-spanning domains and a large C-terminal domain in the matrix. The ribosome docks on the C-terminal domain to facilitate translocation of newly synthesized hydrophobic membrane-spanning

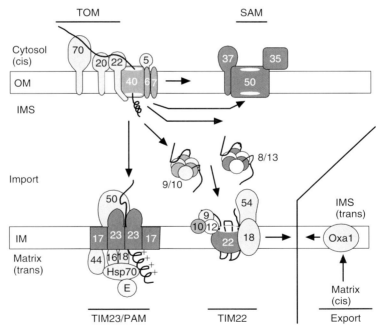

Fig. 3 Translocation pathways of the mitochondrion. Most proteins are imported from the cytosol and pass through the TOM complex. Single pass membranes are released from the TOM channel, whereas β-barrel proteins are sorted to the SAM complex for assembly. Proteins with a typical N-terminal targeting sequence are imported via the TIM23 translocon, whereas inner membrane proteins lacking an N-terminal targeting domain are imported via the TIM22 translocation system. Proteins coded by the mitochondrial genome are exported via Oxa1p.

domains into the membrane. Translocation depends upon the membrane potential. Structural studies suggest that Oxa1p functions as a tetramer, but whether it forms a translocation channel has not been established.

Most proteins are imported from the cytosol. The outer membrane contains the translocase of the outer membrane (TOM) complex through which all precursors are translocated. Cytosolic chaperones such as mitochondrial import stimulation factor and presequence binding factor have been identified in mammals, and the cytosolic chaperones Hsp70 and Hsp90 bind to the Tom70p receptor. Cotranslational import also has been implicated because mRNAs of prokaryotic origin have a signal in the 3′ untranslated region that targets them to the mitochondrion.

The TOM complex contains both, receptors and the translocation pore. The receptor Tom20p binds directly to the N-terminal presequence. In contrast, Tom70p mediates import of precursors that contain targeting information within the mature part of the protein. The precursors are then directed to the translocation pore formed by Tom40p. Small Tom proteins, Tom6p and Tom7p, are thought to gate the pore and allow proteins with one or two membrane-spanning domains to escape laterally into the lipid bilayer.

Proteins such as Tom40p and porin are β-barrel proteins and recently have been shown to be assembled by the sorting and assembly machinery (SAM) complex for outer membrane β-barrel proteins. The SAM complex consists of Sam50p, Sam35p, and Sam37p. Whereas Sam50p is also a β-barrel protein, Sam37p contains one membrane-spanning domain, and Sam35p is peripherally associated with Sam50p on the cytosolic side. The molecular mechanisms of this pathway remain to be elucidated, but preliminary studies suggest that the precursor is passed from the TOM complex via the intermembrane space to the SAM complex. Bacteria contain a homologous protein Omp85, which most likely assembles β-barrel proteins in the outer membrane and may be required for lipid biogenesis. Thus, the SAM pathway seems to be evolutionarily conserved.

For proteins with a typical N-terminal targeting presequence, they are imported from the TOM complex to the TIM23 complex. The TIM23 complex forms the membrane translocon and associates with the protein associated motor (PAM). The TIM23 complex contains the receptors Tim50p and the N-terminal domain of Tim23p, whereas Tim17p and the C-terminal domain of Tim23p form the translocation channel. A membrane potential is required to initiate translocation. The PAM then completes translocation in an ATP-dependent manner. The driving force is generated by the ATPase mHsp70, which is anchored to the membrane by Tim44p. The nucleotide exchange factor Mge1p promotes the reaction cycle of mHsp70, thereby allowing nucleotide release. Pam16p and Pam18p are associated J-proteins that stimulate the ATPase cycle. Upon translocation into the matrix, the targeting sequence may be proteolytically cleaved, and a battery of chaperones facilitates folding of proteins and assembly into functional protein complexes.

Proteins destined for the inner membrane utilize the TIM22 pathway. Precursors of this pathway include the mitochondrial carrier family such as the ADP/ATP carrier, and the phosphate carrier, as well as Tim17p, Tim22p, and Tim23p. In contrast to precursors with an N-terminal targeting presequence, these precursors contain targeting information within the mature part of the protein and cross the

TOM complex as a loop. The estimated size of the TOM translocon is ~20–26 Å, which can accommodate two unfolded polypeptide chains. As the hydrophobic precursors cross the TOM complex and enter the aqueous intermembrane space, the small Tim proteins escort the complexes to the TIM22 insertion complex at the inner membrane. The small Tim proteins are a family of five proteins in yeast. Tim8p partners with Tim13p and Tim9p partners with Tim10p to form 70-kDa complexes in the intermembrane space. The small Tim proteins bind to hydrophobic sequences in the inner membrane substrates and escort them to the insertion complex. The insertion complex contains the membrane proteins Tim22p, Tim54p, and Tim18p and associated Tim12p with a fraction of the Tim9p and Tim10p. Tim22p forms a twin-pore translocase that mediates import into the inner membrane. The mechanism of insertion into the membrane has not been elucidated, but a membrane potential is required. The auxiliary proteins Tim18p and Tim54p most likely play a role in complex assembly and stability.

4.4
Chloroplast Protein Import Pathways

The chloroplast organization is the most complex of all organelles. Proteins can be imported across three distinct membranes, the chloroplast outer and inner envelope membranes and the thylakoid membrane, and into three soluble subcompartments – the space between envelope membranes, the stroma, and the thylakoid lumen (Fig. 4). The translocase of the outer envelope (TOC) and the translocase of the inner envelope (TIC) coordinate the import of chloroplast proteins. Proteins are imported to the thylakoids by one of three different systems, the Sec pathway, the SRP pathway, and the Tat pathway. Additionally, proteins may insert spontaneously into the thylakoid membrane.

Most precursors are imported posttranslationally from the cytosol, except for proteins that are coded on the chloroplast genome. As the precursor emerges from the ribosome, molecular chaperones of the Hsp70 family prevent aggregation and maintain an unfolded import-competent state. The precursor may be phosphorylated by a cytosolic ATP-dependent protein kinase to "fast-track" import. Toc159 and Toc34 are GTPases and form the receptor system for directing the precursor to the membrane. Toc159 cycles between the cytosol and the TOC complex while Toc34 remains associated with the β-barrel channel Toc75. The GTP requirement has not been elucidated, but recent studies suggest a coordinated cycle in which GTP-dependent substrate recognition is coupled to GTP-driven translocation across the outer envelope.

The TOC complex has a molecular mass of approximately 550 kDa consisting of a stoichiometry of 4 Toc34 : 4 Toc75 : 1 Toc159. Toc75 is the most abundant outer envelope protein with a predicted topology of 16 transmembrane β-sheets. Toc75 forms a cation-selective ion channel when reconstituted into planar lipid bilayers. On the basis of electrophysiological measurements, the channel is an estimated 25 Å at the entrance and 15 to 17 Å in the middle, which could accommodate a single polypeptide chain. An accessory protein Toc64 might serve as a docking site for hydrophobic precursors bound to chaperone Hsp70.

The TIC complex is suggested to associate simultaneously with the TOC complex during translocation. In contrast

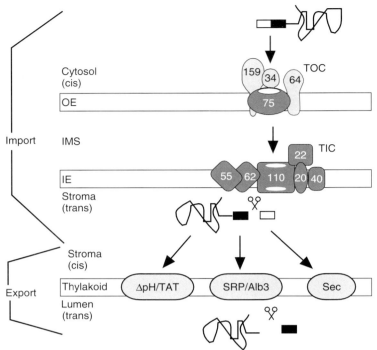

Fig. 4 Translocation pathways of the chloroplast. Precursors are imported from the cytosol and pass through the TOC complex and the TIC complex. In the stroma, the processing peptidase removes the transit sequence; for precursors destined for export to the thylakoid, cleavage of the transit sequences reveals the thylakoid targeting sequence for export via one of three systems.

to the GTP requirement at the TOC complex, ATP hydrolysis is required for translocation across the inner membrane, probably for stromal molecular chaperones that function as a translocation motor. The TIC complex contains several members including Tic110, Tic62, Tic55, Tic40, Tic22, and Tic20 and has a molecular weight of 250 kDa. Of these components, Tic110 is the best characterized and seems to bind to the TOC complex and stromal chaperones Hsp93 and chaperonin 60. Tic62 and Tic55 are redox components that may link the metabolic status of the chloroplast (based on the NAD(P)H:NAD(P) ratio) to the import capacity of the chloroplast.

In *Arabidopsis thaliana*, many of the TIC and TOC components belong to multigene families, allowing complex regulation and reinforcing the importance of import in chloroplast biogenesis.

5 Signal-assembled Translocation Systems

5.1 Peroxisomal Import Pathways

Peroxisomes are single membrane-bound organelles that specialize in β-oxidation of fatty acids, hydrogen peroxide-based respiration, and are a defense against oxidative stress. They are found in virtually every eukaryotic cell, with a typical size of 0.1–1 μm in diameter. In plants, peroxisomes are specialized as glyoxysomes for the glyoxylate cycle, leaf peroxisomes involved in photorespiration pathways, and unspecialized peroxisomes. In mammals, peroxisomes are integral for fatty acid oxidation, and defects in its assembly lead to a class of inherited diseases referred to as *Zellweger's syndrome* or *peroxisome biogenesis disorders* (PBDs). Genetic and biochemical studies in yeast and complementation studies in cells derived from Zellweger's syndrome patients have proved instrumental in identifying many of the proteins involved in peroxisomal biogenesis, termed *peroxins* and coded by PEX genes. Approximately 25 to 30 peroxins have been identified and include both soluble and membrane proteins.

As with other organelles, events in biogenesis include import into the matrix and import of membrane proteins. The matrix protein pathway has been best characterized, but the current mechanistic understanding is not as refined as that of other organelles, because *in vitro* manipulations with peroxisomes are difficult: the organelles are difficult to purify and a robust import assay has not been developed. Biogenesis studies, however, have been advanced by groups working in several experimental systems. The striking differences of peroxisomal biogenesis in comparison with other organelles are that a typical translocation motor and channel have not been identified, and folded proteins can be imported. Additionally, peroxisomes do not possess a membrane potential.

5.1.1 Protein Import into the Matrix

Peroxisomal matrix proteins are synthesized on free ribosomes in the cytosol and imported posttranslationally into peroxisomes, emphasized with PTS1 precursors that have the targeting information at the C terminus. From experimental systems, the simple PTS1 can target a nonresident protein such as green fluorescent protein to the peroxisomal matrix, verifying that the PTS1 sequence is necessary and sufficient for targeting to peroxisomes. The import of matrix proteins depends on cytosolic receptors Pex5p and Pex7p, for PTS1 and PTS2 precursors, respectively (Fig. 5). Pex5p consists of two domains; the C-terminal half contains a high-affinity PTS1-binding site consisting of seven tetratricopeptide repeats (TPR), whereas the N-terminal half mediates peroxisomal targeting and protein–protein interactions at the organelle. Pex7p contains WD-40 motifs and works in conjunction with accessory proteins to guide the precursor to the peroxisome. The cytosolic chaperones thus bind to their cargo and the complexes dock independently at the peroxisome surface.

The docking platform consists of Pex13p, Pex14p, and Pex17p, of which Pex13p and Pex14p are integral peroxisomal membrane proteins (PMP). Both Pex5p and Pex7p converge to this dock, implicating a single translocon for both types of cargo. Peripherally associated Pex17p forms a tight core with PMP Pex14p. Pex13p exposes the N and C termini to the cytosol and recognizes both Pex5p and Pex7p. Pex7p binds to the N terminus, and Pex5p and Pex14p binds to the C terminus, which contains an Src-homology-3 (SH3) domain.

A second group of membrane proteins, Pex2p, Pex10p, and Pex12p (the

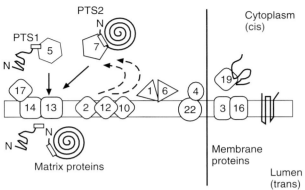

Fig. 5 Translocation pathways of the peroxisome. Pex5p chaperones PTS1 precursors and Pex7p chaperones PTS2 precursors to the docking complex consisting of Pex13p, Pex14p, and Pex17p. The precursor is translocated into the thylakoid lumen and Pex5p and Pex7p are recycled to the cytosol. Peroxisomal membrane proteins are chaperoned to the membrane by Pex19p and associate with Pex3p and Pex16p for insertion into the membrane. Peroxisome translocation mechanisms are not well understood but folded, proteins can be translocated.

RING-finger peroxins), function downstream from the receptor complexes. All three components contain cytoplasmic zinc RING domains, and Pex10p and Pex12p bind to Pex5p. How does Pex5p deliver cargo to the peroxisomal matrix? The mechanism by which Pex5p releases cargo has been under debate because Pex5p has been localized to the cytoplasm, peroxisomal membrane, and peroxisomal matrix. It has been suggested that Pex5p shuttles between the cytoplasm and peroxisomal matrix. In the "extended-shuttle" mechanism, Pex5p undergoes multiple cycles of translocation in and out of the matrix, whereas in the "simple shuttle" mechanism, Pex5p stays at the surface of the membrane and releases the cargo. The consensus is that Pex5p traverses the mammalian peroxisomal membrane through multiple rounds of entry into the matrix and export to the cytosol. Further evidence that Pex5p at least traverses the membrane is confirmed by an interaction between Pex5p and Pex8p, the only known peroxin localized to the trans side of the membrane in *Saccharomycescerevisiae*.

Additional peroxins, Pex4p, Pex22p, Pex1p, and Pex6p, are required to unload the cargo and return the receptor to the cytoplasm. Pex22p is a PMP that serves as a membrane anchor for cytosolic Pex4p, an E2-type ubiquitin conjugating enzyme. The RING-finger domains of Pex10p may interact with Pex4p to link the docking complex. ATP hydrolysis is required for import of matrix proteins, providing the potential for a nucleotide-hydrolysis–driven import cycle. Pex1p and Pex6p are two membrane-associated ATPases of the AAA family (ATPases Associated with diverse cellular Activities). Genetic studies place the requirement for Pex1p and Pex6p between receptor docking/translocation and Pex4p/Pex22p, suggesting that Pex1p and Pex6p are required for unloading the cargo and recycling the receptor.

5.1.2 Protein Import into the Membrane

Matrix protein and PMP import seem to be independent. In contrast to matrix protein import, PMP import does not require ATP hydrolysis, and mutations in genes that abrogate matrix import usually do not affect membrane import. Furthermore, a translocation intermediate that blocked the binding of matrix proteins to isolated glyoxysomes did not affect PMP import.

Thus, of most *pex* mutants in which matrix import is abolished, PMPs are still inserted into peroxisome remnants or "ghosts" because they lack matrix proteins.

However, in cells that are deficient in the PMPs Pex3p, Pex16p, and Pex19p, peroxisome ghosts are not detectable, suggesting that this cohort defines the essential components for PMP import and sorting. Supporting this, mutant cells are seen to lack *pex3* or *pex19* and have unstable PMPs. And these three peroxins interact *in vitro* and *in vivo*.

Like peroxisomal matrix protein import, PMP import seems to share common features (Fig. 5). Pex19p is predominantly cytosolic, but can be found associated with the peroxisomal membrane, possibly via farnesylation at the C terminus. Pex19p thus acts as a chaperone to prevent PMP degradation and to guide PMPs in the cytosol to the peroxisome. As expected, Pex19p binds to several PMPs. In addition to its chaperone role, Pex19p subsequently acts as an import receptor, binding to the docking protein Pex3p. Pex3p thus acts as a docking protein, and as expected, Pex3p inserts independently of Pex19p. A specific role for Pex16p in PMP targeting is not known, but Pex16p may function at an early stage of peroxisomal membrane assembly upstream of Pex3p and Pex19p. Thus, peroxisomal import pathways share the common feature of a cytosolic chaperone, which targets both membrane and matrix proteins to docking sites on the membrane.

5.2 The Tat/ΔpH Pathway of Bacteria and Thylakoids

The Tat system (also referred to as the ΔpH pathway) is conserved in bacteria and chloroplast thylakoids (Fig. 6). The pathway was first characterized in plant chloroplasts in the early 1990s when biochemical studies showed that the pH gradient, rather than a nucleotide hydrolysis,

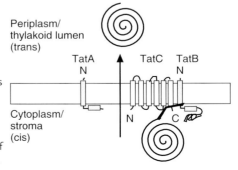

Fig. 6 The Tat/ΔpH translocation pathway for folded proteins. The Tat pathway of the thylakoid lumen and bacteria share common features. TatB and TatC assemble into a complex that binds to the Tat signal sequence. TatA is then recruited and the folded protein is translocated across the membrane in a membrane potential/ΔpH-dependent manner. A nucleotide requirement and chaperones on the cis and trans sides of the membrane have not been identified.

was required for the import of a subset of thylakoid lumenal proteins. The observation of the pathway in prokaryotes led to rapid identification of its important components. The Tat pathway mediates the translocation of large, folded proteins, such as proteins with a range of redox cofactors, including FeS, NiFe, and molybdopterin centers, across the tightly sealed membrane. In some cases, the oligomeric substrates employ a "hitchhiker" mechanism in which only one or a subset of subunits contain targeting signals. In most prokaryotes, the Sec system is the main route for export, but in the halophilic archaeon Halibacterium sp NRC-1, the majority of secretory proteins including traditional Sec-dependent substrates, use the Tat system. The rationale for this adaptation is that the haloarchea has a very high cytoplasmic salt concentration (4–5 M) and newly synthesized proteins must fold quickly to prevent aggregation; hence, the Tat system is required for export of these folded cytoplasmic proteins.

How does the Tat system translocate a folded protein across an energy-transducing membrane? With chloroplast thylakoid import studies, a precursor was constructed with a Tat signal fused to bovine pancreatic trypsin inhibitor (BPTI), which was internally cross-linked to prevent unfolding. This construct imported into isolated chloroplasts and targeted to the thylakoid lumen by the Tat pathway. In a second study, a larger translocation substrate was generated using the reporter dihydrofolate reductase (DHFR), which when folded binds to the folate analog methotrexate (MTX) with very high affinity. This translocation intermediate normally arrests in the mitochondrial TOM or bacterial Sec system because it cannot be unfolded, thus confirming that most translocons must completely unfold proteins prior to translocation. However, the folded DHFR-MTX construct was imported efficiently into the thylakoid lumen, and folding was confirmed because the DHFR–MTX complex was protease resistant. In bacterial systems, GFP (green fluorescent protein) fusions also were used to confirm that folded proteins were exported via the Tat system. Specifically, GFP that is engineered for export via the Sec system does not fold properly in the periplasm and does not fluoresce. In contrast, Tat-targeted GFP exported to the periplasm was folded and detected as a fluorescent halo on the cell periphery, confirming efficient translocation of a folded protein.

Because the Tat pathway is organized in one operon in E. coli, characterization of the translocation components was facilitated by identification of the proteins within the operon. The E. coli system has three Tat proteins (the tatABC operon) and plants have three (Tha4, Hcf106, and cpTatC for tatA, B, and C respectively) homologs. All Tat proteins are located in the membrane. TatA and TatB both contain one transmembrane domain (N terminus facing the periplasm), whereas TatC contains six membrane-spanning domains (N and C termini exposed to the cytoplasm). The TatABC complex is approximately 500 to 600 kDa in size and consists of approximately a 1:11 ratio for the individual components, although TatA is present in higher amounts or in a separate complex. From single particle electron microscope images, the TatABC complex has dimensions of 10 × 13 nm. TatB/Hcf106 and TatC form a stable membrane complex that binds the Tat signal. This then triggers the recruitment of TatA to form the functional translocation system, which requires a membrane potential. For the actual translocation of the substrate, additional factors on the cis and trans side of the membrane seemingly are not required. The current available evidence suggests that the force may be applied within the membrane domain or by structural changes within the Tat translocon itself. After translocation, dissociation of the Tat translocon has been observed, which is consistent with the reversible assembly of the translocon.

Whereas the specific mechanism of transport remains to be elucidated, the transport of large folded proteins is very efficient, but costly. Careful measurements of the prevailing ΔpH during translocation of thylakoid proteins has shown that translocation requires 30 000 protons per protein transported. Moreover, two differently sized substrates required differing levels of ΔpH for translocation, suggesting that the Tat system adapts energetically to the size and/or shape of the substrate. Indeed, future mechanistic studies should provide new ideas about how large folded proteins can be exported across an energy-transducing membrane.

6
Conclusions

Significant progress has been made toward understanding the mechanism of protein translocation across membranes. Yet, there are major questions to be addressed. The events that initiate the translocation reaction have not been deciphered in all translocation systems. The mechanism by which the channel can accommodate precursors of varying sizes, particularly for the signal-assembled translocon, remains a mystery. Moreover, how defects in the translocation systems result in the diseased state in plants and animals are beginning to be elucidated. Eventually, a greater understanding of the mechanisms of translocation will provide information about the broader question of how membrane components assemble.

Acknowledgment

David Hwang was supported by the United States Public Health Service National Research Service Award (GM07185). Carla M. Koehler was supported by grants from the Beckman Foundation, National Institutes of Health, and the Muscular Dystrophy Association.

See also Membrane Traffic: Vesicle Budding and Fusion; Protein Mediated Membrane Fusion.

Bibliography

Books and Reviews

Bauer, J., Hiltbrunner, A., Kessler, F. (2001) Molecular biology of chloroplast biogenesis: gene expression, protein import and intraorganellar sorting, *Cell Mol. Life Sci.* **58**, 420–433.

Blobel, G. (1980) Intracellular protein topogenesis, *Proc. Natl. Acad. Sci. U.S.A.* **77**, 1496–1500.

Eckert, J.H., Erdmann, R. (2003) Peroxisome biogenesis, *Rev. Physiol. Biochem. Pharmacol.* **147**, 75–121.

Johnson, A.E., van Waes, M.A. (1999) The translocon: a dynamic gateway at the ER membrane, *Annu. Rev. Cell Dev. Biol.* **15**, 799–842.

Koehler, C.M. (2004) New developments in mitochondrial assembly, *Annu. Rev. Cell Dev. Biol.* **20**, 309–335.

Robinson, C., Bolhuis, A. (2004) Tat-dependent protein targeting in prokaryotes and chloroplasts, *Biochim. Biophys. Acta* **1694**, 135–147.

Schatz, G., Dobberstein, B. (1996) Common principles of protein translocation across membranes, *Science* **271**, 1519–1526.

Schnell, D.J., Hebert, D.N. (2003) Protein translocons: multifunctional mediators of protein translocation across membranes, *Cell* **112**, 491–505.

Soll, J., Schlieff, E. (2004) Protein import into chloroplasts, *Natl. Rev.* **5**, 198–208.

Walter, P., Johnson, A.E. (1994) Signal sequence recognition and protein targeting to the endoplasmic reticulum membrane, *Annu. Rev. Cell Biol.* **10**, 87–119.

Primary Literature

Abe, Y., Shodai, T., Muto, T., Mihara, K., Nishikawa, S.-I., Endo, T., Kohda, D. (2000) Structural basis of presequence recognition by the mitochondrial presequence receptor Tom20, *Cell* **100**, 551–560.

Akita, M., Shinkai, A., Matsuyama, S., Mizushima, S. (1991) SecA, an essential component of the secretory machinery of *Escherichia coli*, exists as homodimer, *Biochem. Biophys. Res. Commun.* **174**, 211–216.

Alder, N.N., Theg, S.M. (2003) Energetics of protein transport across biological membranes. A study of the thylakoid DeltapH-dependent/cpTat pathway, *Cell* **112**, 231–242.

Attardi, G., Schatz, G. (1988) Biogenesis of mitochondria, *Annu. Rev. Cell Biol.* **4**, 289–333.

Baker, A., Schatz, G. (1987) Sequences from a prokaryotic genome or the mouse dihydrofolate reductase gene can restore the import of a truncated precursor protein into yeast mitochondria, *Proc. Natl. Acad. Sci. U.S.A.* **84**, 3117–3121.

Becker, T., Jelic, M., Vojta, A., Radunz, A., Soll, J., Schlieff, E. (2004) Preprotein recognition by the Toc complex, *EMBO J.* **23**, 520–530.

Berks, B.C., Palmer, T., Sargent, F. (2003) The Tat protein translocation pathway and its role in microbial physiology, *Adv. Microb. Physiol.* **47**, 187–254.

Bessonneau, P., Besson, V., Collinson, I., Duong, F. (2002) The SecYEG preprotein translocation channel is a conformationally dynamic and dimeric structure, *EMBO J.* **21**, 995–1003.

Blobel, G., Dobberstein, B. (1976) Transfer of proteins across membranes. I. Presence of proteolytically processed and unprocessed nascent immunoglobulin light chains on the membrane-bound ribosomes of murine myeloma, *J. Cell Biol.* **67**, 835–851.

Bolhuis, A. (2002) Protein transport in the halophilic archaeon Halobacterium sp. NRC-1: a major role for the twin-arginine translocation pathway? *Microbiology* **142**, 3335–3346.

Brink, S., Bogsch, E.G., Edwards, W.R., Hynds, P.J., Robinson, C. (1998) Targeting of thylakoid proteins by the ΔpH-driven twin-arginine pathway requires a specific signal in the hydrophobic domain in conjunction with the twin-arginine motif, *FEBS Lett.* **434**, 425–430.

Brix, J., Ziegler, G.A., Dietmeier, K., Schneider-Mergener, J., Schulz, G.E., Pfanner, N. (2000) The mitochondrial import receptor Tom70: identification of a 25 kDa core domain with a specific binding site for preproteins, *J. Mol. Biol.* **303**, 479–488.

Caliebe, A., Grimm, R., Kaiser, G., Lubeck, J., Soll, J., Heins, L. (2003) The chloroplastic protein import machinery contains a Rieske-type iron-sulfur cluster and a mononuclear iron-binding protein, *EMBO J.* **278**, 38617–38627.

Chen, X., Schnell, D.J. (1999) Protein import into chloroplasts, *Trends Cell Biol.* **9**, 222–227.

Clark, S.A., Theg, S.M. (1997) A folded protein can be transported across the chloroplast envelope and thylakoid membranes, *Mol. Biol. Cell* **8**, 923–934.

Cline, K., Mori, H. (2001) Thylakoid delta pH-dependent precursor proteins bind to a cpTatC-Hcf106 complex before Tha4-dependent transport, *J. Cell Biol.* **154**, 719–729.

Collins, C.S., Kalish, J.E., Morrell, J.C., McCaffery, J.M., Gould, S.J. (2000) The peroxisome biogenesis factors pex4p, pex22p, pex1p, and pex6p act in the terminal steps of peroxisomal matrix protein import, *Mol. Cell Biol.* **20**, 7516–7526.

Cristobal, S., de Gier, J.W., Nielsen, H., von Heijne, G. (1999) Competition between Sec- and TAT-dependent protein translocation in *Escherichia coli*, *EMBO J.* **18**, 2982–2990.

Crowley, K.S., Liao, S., Worrell, V.E., Reinhart, G.D., Johnson, A.E. (1994) Secretory proteins move through the endoplasmic reticulum membrane via an aqueous, gated pore, *Cell* **78**, 461–471.

Dammai, V., Subramani, S. (2001) The human peroxisomal targeting signal receptor, Pex5p, is translocated into the peroxisomal matrix and recycles to the cytosol, *Cell* **105**, 187–196.

Diestelkotter, P., Just, W.W. (1993) In vitro insertion of the 22 kDa peroxisomal membrane protein into isolated rat liver peroxisomes, *J. Biol. Chem.* **123**, 1717–1725.

Dodt, G., Braverman, N., Valle, D., Gould, S.J. (1996) From expressed sequence tags to peroxisome biogenesis disorder genes, *Ann. N. Y. Acad. Sci.* **804**, 516–523.

Driessen, A.J. (1993) SecA, the peripheral subunit of the *Escherichia coli* precursor protein translocase, is functional as a dimer, *Biochemistry* **32**, 13190–13197.

D'Silva, P.D., Schilke, B., Walter, W., Andrew, A., Craig, E.A. (2003) J protein cochaperone of the mitochondrial inner membrane required for protein import into the mitochondrial matrix, *Proc. Natl. Acad. Sci. U.S.A.* **100**, 13839–13844.

Eichler, J., Duong, F. (2004) Break on through to the other side – the Sec translocon, *Trends Biochem. Sci.* **29**, 221–223.

Endo, T., Yamamoto, H., Esaki, M. (2003) Functional cooperation and separation of translocators in protein import into mitochondria, the double-membrane bounded organelles, *J. Cell. Sci.* **116**, 3259–3267.

Feilmeier, B.J., Iseminger, G., Schroeder, D., Webber, J., Phillips, G.J. (2000) Green fluorescent protein functions as a reporter for protein localization in *Escherichia coli*, *J. Bacteriol.* **182**, 4068–4076.

Fekkes, P., de Wit, J.G., van der Wolk, J.P., Kimsey, H.H., Kumamoto, C.A., Driessen, A.J.M. (1998) Preprotein transfer to the *Escherichia coli* translocase requires the cooperative binding of SecB and the signal sequence to SecA, *Mol. Microbiol.* **29**, 1179–1190.

Flugge, U.I., Hinz, G. (1986) Energy dependence of protein translocation into chloroplasts, *Eur. J. Biochem.* **160**, 563–567.

Fransen, M., Wylin, T., Brees, C., Mannaerts, G.P., Van Veldhoven, P.P. (2001) Human Pex19p binds peroxisomal integral membrane proteins at regions distinct from their sorting sequences, *Mol. Cell. Biol.* **21**, 4413–4424.

Frazier, A.E., Dudek, J., Guiard, B., Voos, W., Li, Y., Lind, M., Meisinger, C., Geissler, A., Sickmann, A., Meyer, H.E., Bilanchone, V., Cumsky, M.G., Truscott, K.N., Pfanner, N., Rehling, P. (2004) Pam16 has an essential role in the mitochondrial protein import motor, *Nat. Struct. Mol. Biol.* **11**, 226–233.

Gatto, G.J. Jr., Geisbrecht, B.V., Gould, S.J., Berg, J.M. (2000) Peroxisomal targeting signal-1 recognition by the TPR domains of human PEX5, *Nat. Struct. Biol.* **7**, 1091–1095.

Ghys, K., Fransen, M., Mannaerts, G.P., Van Veldhoven, P.P. (2002) Functional studies on human Pex7p: subcellular localization and interaction with proteins containing a peroxisome-targeting signal type 2 and other peroxins, *Biochem. J.* **365**, 41–50.

Gould, S.J., Valle, D. (2000) Peroxisome biogenesis disorders: genetics and cell biology, *Trends Genet.* **16**, 340–345.

Gould, S.J., Collins, C.S. (2002) Opinion: peroxisomal protein import: Is it really that complex? *Nat. Rev. Mol. Cell Biol.* **3**, 382–389.

Gray, M.W., Burger, G., Lang, B.F. (1999) Mitochondrial Evolution, *Science* **283**, 1476–1481.

Haigh, N.G., Johnson, A.E. (2002) A new role for BiP: closing the aqueous translocon pore during protein integration into the ER membrane, *J. Cell Biol.* **156**, 261–270.

Hartl, F.U., Lecker, S., Schiebel, E., Hendrick, J.P., Wickner, W. (1990) The binding cascade of SecB to SecA to SecY/E mediates preprotein targeting to the *E. coli* plasma membrane, *Cell* **63**, 269–279.

Hell, K., Neupert, W., Stuart, R.A. (2001) Oxa1p acts as a general membrane insertion machinery for proteins encoded by mitochondrial DNA, *EMBO J.* **20**, 1281–1288.

Hettema, E.H., Girzalsky, W., van Den Berg, M., Erdmann, R., Distel, B. (2000) *Saccharomyces cerevisiae* pex3p and pex19p are required for proper localization and stability of peroxisomal membrane proteins, *EMBO J.* **19**, 223–233.

Hinnah, S.C., Hill, K., Wagner, R., Schlicher, T., Soll, J. (1997) Reconstitution of a chloroplast protein import channel, *EMBO J.* **16**, 7351–7360.

Holroyd, C., Erdmann, R. (2001) Protein translocation machineries of peroxisomes, *FEBS Lett.* **501**, 6–10.

Honsho, M., Hiroshige, T., Fujiki, Y. (2002) The membrane biogenesis of Pex16p – topogenesis and functional roles in peroxisomal membrane assembly, *J. Biol. Chem.* **277**, 44513–44524.

Huhse, B., Rehling, P., Albertini, M., Blank, L., Meller, K., Kunau, W.H. (1998) Pex17p of *Saccharomyces cerevisiae* is a novel peroxin and component of the peroxisomal protein translocation machinery, *J. Cell Biol.* **140**, 49–60.

Imanaka, T., Shinina, Y., Takano, Y., Hashimoto, T., Osumi, T. (1996) Insertion of the 70 kDa peroxisomal membrane protein into peroxisomal membrane *in vivo* and *in vitro*, *J. Biol. Chem.* **271**, 3706–3713.

Jia, L., Dienhart, M., Schramp, M., McCauley, M., Hell, K., Stuart, R.A. (2003) Yeast Oxa1 interacts with mitochondrial ribosomes: the importance of the C-terminal region of Oxa1, *EMBO J.* **22**, 6438–6447.

Jones, J.M., Morrell, J.C., Gould, S.J. (2001) Multiple distinct targeting signals in integral peroxisomal membrane proteins, *J. Cell Biol.* **153**, 1141–1150.

Keegstra, K., Froehlich, J.E. (1999) Protein import into chloroplasts, *Curr. Opin. Plant Biol.* **2**, 471–476.

Kerscher, O., Holder, J., Srinivasan, M., Leung, R.S., Jensen, R.E. (1997) The Tim54p-Tim22p complex mediates insertion of proteins into the mitochondrial inner membrane, *J. Cell Biol.* **139**, 1663–1675.

Kessler, F., Schnell, D. (2004) Chloroplast protein import: solve the GTPase riddle for entry, *Trends Cell Biol.* **14**, 334–338.

Koehler, C.M., Merchant, S., Schatz, G. (1999) How membrane proteins travel across the mitochondrial intermembrane space, *Trends Biochem. Sci.* **24**, 428–432.

Kozjak, V., Wiedemann, N., Milenkovic, D., Lohaus, C., Meyer, H.E., Guiard, B., Meisinger, C., Pfanner, N. (2003) An essential role of Sam50 in the protein sorting and assembly machinery of the mitochondrial outer membrane, *J. Biol. Chem.* **278**, 48520–48523.

Kuhn, A., Stuart, R., Henry, R., Dalbey, R.E. (2003) The Alb3/Oxa1/YidC protein family: membrane-localized chaperones facilitating membrane protein insertion? *Trends Cell Biol.* **13**, 510–516.

Kunau, W.H. (2001) Peroxisomes: the extended shuttle to the peroxisome matrix, *Curr. Biol.* **11**, R659–R662.

Lazarow, P.B., Fujiki, Y. (1985) Biogenesis of peroxisomes, *Annu. Rev. Cell Biol.* **1**, 489–530.

Legakis, J.E., Terlecky, S.R. (2001) PTS2 protein import into mammalian peroxisomes, *Traffic* **2**, 252–260.

Liao, S., Lin, J., Do, H., Johnson, A.E. (1997) Both lumenal and cytosolic gating of the aqueous ER translocon pore are regulated from inside the ribosome during membrane protein integration, *Cell* **90**, 31–41.

Lill, R., Dowhan, W., Wickner, W. (1990) The ATPase activity of SecA is regulated by acidic phospholipids, SecY, and the leader and mature domains of precursor proteins, *Cell* **60**, 271–280.

Martsuzono, Y., Kinoshita, N., Tamura, S., Shimozawa, N., Hamasaki, M., Ghaedi, K., Wanders, R.J., Suzuki, Y., Kondo, N., Fujiki, Y. (1999) Human PEX19: cDNA cloning by functional complementation, mutation analysis in a patient with Zellweger syndrome, and potential role in peroxisomal membrane assembly, *Proc. Natl. Acad. Sci. U.S.A.* **96**, 2116–2121.

Marzioch, M., Erdmann, R., Veenhuis, M., Kunau, W.H. (1994) PAS7 encodes a novel yeast member of the WD-40 protein family essential for import of 3-oxoacyl-CoA thiolase, a PTS2-containing protein, into peroxisomes, *EMBO J.* **13**, 4908–4918.

Mihara, K., Omura, T. (1996) Cytosolic factors in mitochondrial protein import, *Experientia* **52**, 1063–1068.

Milenkovic, D., Kozjak, V., Wiedemann, N., Lohaus, C., Meyer, H.E., Guiard, B., Pfanner, N., Meisinger, C. (2004) Sam35 of the mitochondrial protein sorting and assembly machinery is a peripheral outer membrane protein essential for cell viability, *J. Biol. Chem.* **279**, 22781–22785.

Mori, H., Cline, K. (2002) A twin arginine signal peptide and the pH gradient trigger reversible assembly of the thylakoid delta pH/Tat translocase, *J. Cell Biol.* **157**, 205–210.

Murphy, C.K., Beckwith, J. (1994) Residues essential for the function of SecE, a membrane component of the *Escherichia coli* secretion apparatus, are located in a conserved cytoplasmic region, *Proc. Natl. Acad. Sci. U.S.A.* **91**, 2557–2561.

Neupert, W., Brunner, M. (2002) The protein import motor of mitochondria, *Nat. Rev. Mol. Cell Biol.* **3**, 555–565.

Nishiyama, K., Mizushima, S., Tokuda, H. (1992) The carboxyl-terminal region of SecE interacts with SecY and is functional in the reconstitution of protein translocation activity in *Escherichia coli*, *J. Biol. Chem.* **267**, 7170–7176.

Pfanner, N., Neupert, W. (1987) Distinct steps in the import of ADP/ATP carrier into mitochondria, *J Biol. Chem.* **262**, 7528–7536.

Pfanner, N., Wiedemann, N., Meisinger, C., Lithgow, T. (2004) Assembling the mitochondrial outer membrane, *Nat. Struct. Mol. Biol.* **11**, 1044–1048.

Pohlschroder, M., Prinz, W.A., Hartmann, E., Beckwith, J. (1997) Protein translocation in three domains of life: variations on a theme, *Cell* **81**, 563–566.

Porcelli, I., de Leeuw, E., Wallis, R., van den Brink-van der Laan, E., de Kruijff, B., Wallace, B.A., Palmer, T., Berks, B.C. (2002) Characterization and membrane assembly of the TatA component of the *Escherichia coli* twin-arginine protein transport system, *Biochemistry* **41**, 13690–13697.

Purdue, P.E., Lazarow, P.B. (2001) Peroxisome biogenesis, *Annu. Rev. Cell Dev. Biol.* **17**, 701–752.

Rachubinski, R.A., Subramani, S. (1995) How protein penetrate peroxisomes, *Cell* **83**, 525–528.

Rehling, P., Model, K., Brandner, K., Kovermann, P., Sickmann, A., Meyer, H.E., Kuhlbrandt, W., Wagner, R., Truscott, K.N., Pfanner, N. (2003) Protein insertion into the mitochondrial inner membrane by a twin-pore translocase, *Science* **299**, 1747–1751.

Robinson, C., Bolhuis, A. (2001) Protein targeting by the twin-arginine translocation pathway, *Nat. Rev. Mol. Cell Biol.* **2**, 350–356.

Rodrigue, A., Chanal, A., Beck, K., Muller, M., Wu, L.F. (1999) Co-translocation of a periplasmic enzyme complex by a hitchhiker mechanism through the bacterial tat pathway, *J. Biol. Chem.* **274**, 13223–13228.

Sacksteder, K.A., Gould, S.J. (2000) The genetics of peroxisome biogenesis, *Annu. Rev. Genet.* **34**, 623–652.

Samuelson, J.C., Chen, M., Jiang, F., Moller, I., Wiedmann, M., Kuhn, A., Phillips, G.J., Dalbey, R.E. (2000a) YidC mediates membrane protein insertion in bacteria, *Nature* **406**, 637–641.

Santini, C.L., Bernadac, A., Zhang, M., Chanal, A., Ize, B., Blanco, C., Wu, L.F. (2001) Translocation of jellyfish green fluorescent protein via the Tat system of *Escherichia coli* and change of its periplasmic localization in response to osmotic up-shock, *J. Biol. Chem.* **276**, 8159–8164.

Sargent, F., Bogsch, E.G., Stanley, N.R., Wexler, M., Robinson, C., Berks, B., Palmer, T. (1998) Overlapping functions of components of a bacterial Sec-independent protein export pathway, *EMBO J.* **17**, 3640–3650.

Schatz, P.J., Riggs, P.D., Jacq, A., Fath, M.J., Beckwith, J. (1989) The secE gene encodes an integral membrane protein required for protein export in *Escherichia coli*, *Genes Dev.* **3**, 1035–1044.

Schwartz, M.P., Matouschek, A. (1999) The dimensions of the protein import channel in the outer and inner mitochondrial membranes, *Proc. Natl. Acad. Sci. U.S.A.* **98**, 13086–13090.

Settles, M.A., Yonetani, A., Baron, A., Bush, D.R., Cline, K., Martienssen, R. (1997) Sec-independent protein translocation by the maize Hcf106 protein, *Science* **278**, 1467–1470.

Shimozawa, N., Suzuki, Y., Zhang, Z., Imamura, A., Chaedi, K., Fujiki, Y., Kondo, N. (2000) Identification of PEX3 as the gene mutated in a Zellweger syndrome patient lacking peroxisomal remnant structures, *Hum. Mol. Genet.* **9**, 1995–1999.

Smith, M.D., Schnell, D. (2001) Peroxisomal protein import: the paradigm shifts, *Cell* **105**, 293–296.

Snyder, W.B., Koller, A., Choy, A.J., Subramani, S. (2000) The peroxin Pex19p interacts with multiple, integral membrane proteins at the peroxisomal membrane, *J. Cell Biol.* **149**, 1171–1178.

Sohrt, K., Soll, J. (2000) Toc64, a new component of the protein translocon of chloroplasts, *J. Cell Biol.* **148**, 1213–1221.

Spence, E., Sarcina, M., Ray, N., Moller, S.G., Mullineaux, C.W., Robinson, C. (2003) Membrane-specific targeting of green fluorescent protein by the Tat pathway in the cyanobacterium synechocystis PCC6803, *Mol. Microbiol.* **48**, 1481–1489.

Sylvestre, J., Margeot, A., Jacq, C., Dujardin, G., Corral-Debrinski, M. (2003) The role of the 3′ untranslated region in mRNA sorting to the vicinity of mitochondria is conserved from yeast to human cells, *Mol. Biol. Cell.* **14**, 3848–3856.

Taylor, R.D., Pfanner, N. (2004) The protein import and assembly machinery of the mitochondrial outer membrane, *Biochim. Biophys. Acta* **1658**, 37–43.

Truscott, K.N., Voos, W., Frazier, A.E., Lind, M., Li, Y., Geissler, A., Dudek, J., Muller, H., Sickmann, A., Meyer, H.E., Meisinger, C., Guiard, B., Rehling, P., Pfanner, N. (2003) A J-protein is an essential subunit of the presequence translocase-associated protein import motor of mitochondria, *J. Cell Biol.* **163**, 707–713.

Van den Berg, B., Clemons, W.M. Jr., Collinson, I., Modis, Y., Hartmann, E., Harrison, S.C., Rapoport, T.A. (2004) X-ray structure of a protein-conducting channel, *Nature* **427**, 36–44.

van der Laan, M., Urbanus, M.L., Ten Hagen-Jongman, C.M., Nouwen, N., Oudega, B., Harms, N., Driessen, A.J., Luirink, J. (2003) A conserved function of YidC in the biogenesis of respiratory chain complexes, *Proc. Natl. Acad. Sci. U.S.A.* **100**, 5801–5806.

Waegemann, K., Soll, J. (1996) Phosphorylation of the transit sequence of chloroplast precursor proteins, *J. Biol. Chem.* **271**, 6545–6554.

Waegemann, K., Paulsen, H., Soll, J. (1990) Translocation of proteins into isolated chloroplasts requires cytosolic factors to obtain import competence, *FEBS Lett.* **261**, 89–92.

Walker, M.B., Roy, L.M., Coleman, E., Voelker, R., Barkan, A. (1999) The maize *tha4* gene functions in Sec-independent protein transport chloroplasts and is related to *hcf106*, *tatA*, and *tatB*, *J. Cell Biol.* **147**, 267–276.

Young, J.C., Hoogenraad, N.J., Hartl, F.U. (2003) Molecular chaperones Hsp90 and Hsp70 deliver preproteins to the mitochondrial import receptor Tom70, *Cell* **112**, 41–50.

Zheng, N., Gierasch, L.M. (1996) Signal sequences: the same yet different, *Cell* **86**, 849–852.

Zito, C.R., Oliver, D. (2003) Two-stage binding of SecA to the bacterial translocon regulates ribosome-translocon interaction, *J. Biol. Chem.* **278**, 40640–40646.

Proteomics

Paul Cutler[1], Israel S. Gloger[2], and Christine Debouck[3]
GlaxoSmithKline Pharmaceutical Research & Development,
[1] *Stevenage, Hertfordshire, UK*
[2] *Harlow, Essex, UK*
[3] *Collegeville, PA*

1	**Definitions** 313	
1.1	Proteomics 313	
1.2	Expression Proteomics 313	
1.3	Interaction Proteomics 313	
1.4	Clinical Proteomics and Toxicoproteomics 313	
2	**Technologies** 313	
2.1	2D Gel Electrophoresis (2-DE) 313	
2.2	Mass Spectrometry 315	
2.3	Isotope Coding 318	
2.4	Yeast Two-hybrid Systems 319	
2.5	Protein Arrays 321	
2.6	Bioinformatics 322	
3	**Applications** 323	
3.1	Target Identification and Validation 323	
3.2	Protein–protein Interactions 325	
3.3	Protein Biomarkers in Clinical and Preclinical Research 325	
4	**Future Perspectives** 327	
4.1	Toward a Full Proteome Map 327	
4.2	Biological Validation of Protein Functions 328	
4.3	Clinical and Diagnostic Applications 328	
	Acknowledgments 329	

Encyclopedia of Molecular Cell Biology and Molecular Medicine, 2nd Edition. Volume 11
Edited by Robert A. Meyers.
Copyright © 2005 Wiley-VCH Verlag GmbH & Co. KGaA, Weinheim
ISBN: 3-527-30648-X

Bibliography 329
  Books and Reviews 329
  Primary Literature 329

Keywords

Electrospray Ionization (ESI)
Electrospray is the process of ionization by which fractionated peptides are sprayed in solution through a charged needle toward the mass analyzer.

Isotope-coded Affinity Tag (ICAT)
Isotope-coded affinity tags are reagents used for quantifying the differential expression of peptides by mass spectrometry. ICAT tags contain three key components: a reactive group for labeling peptides (e.g. iodo-group for labeling cysteines), a group for affinity separation (e.g. biotin), and a spacer. The spacer can be derivatized with different isotopes (e.g. hydrogen or deuterium) to discriminate protein expression of peptides from different samples. The ratios of the heavy and light isotopes measured by the mass spectrometer may then be correlated with the relative abundance of the peptide in the two samples.

Liquid Chromatography–Mass Spectrometry (LC-MS)
Liquid chromatography–mass spectrometry is the separation of complex mixtures of peptides or protein samples by liquid chromatography followed by mass spectrometric analysis. Reverse phase liquid chromatography is usually performed on peptides as this utilizes volatile solvents compatible with mass spectrometry.

Mass Spectrometry
Mass spectrometry is a biophysical methodology used to measure the mass-to-charge ratio of proteins or peptides. The method involves the generation of ions in a vacuum from a solid peptide or a peptide in solution, followed by their detection and mass analysis. Mass spectrometry instruments therefore contain a source of ionization to convert the analyte into ions within the gas phase and a mass analyzer for the detection of the ions. Various methods exist for the generation of ions, including electrospray ionization (ESI) and matrix-assisted laser desorption ionization (MALDI). Methods for detecting and analyzing the ions include time of flight (TOF) analyzers, quadrupole electric fields and ion traps.

Matrix-assisted Laser Desorption Ionization (MALDI)
In matrix-assisted laser desorption ionization, ions are formed via the laser excitation of samples embedded in a chromophore matrix. Laser energy excites the chromophore and the energy is transferred to the analyte, which ionizes and enters the gas phase. The mass-to-charge ratios of these ions are usually measured by the "time of flight"

(TOF) from their release until detection. The time of flight can thus be converted into accurate molecular weight values.

Posttranslational Modifications
This is the process by which proteins are structurally modified following translation by the ribosomes, which may lead to altered function or activity. These include deamidation, proteolytic cleavage such as C-terminal truncation, phosphorylation, and glycosylation.

Protein Arrays
These consist in the microarray-based analysis of proteins either in a planar (chip) or suspension format, to qualitatively or quantitatively characterize protein expression or function.

Proteome
The proteome is the complete collection of proteins expressed in a given cell or tissue under a given set of conditions (normal, disease, treated or untreated with stimulants or compounds).

RNA Interference (RNAi)
RNAi (RNA interference) refers to the introduction of homologous double-stranded RNA (dsRNA) to specifically target and thereby silence the expression of one or more genes in a variety of organisms and cell types (e.g. worms, fruit flies, mammalian cells), resulting in null or hypomorphic phenotypes.

Surface-enhanced Laser Desorption Ionization (SELDI™)
In surface-enhanced laser desorption ionization, the analyte(s) are applied in solution to a protein-binding surface or "chip" followed by treatment with a MALDI matrix. The essential element of SELDI is the chemically modified nature of the surface, which facilitates selective adsorption prior to analysis. Laser desorption will release the bound proteins as ions (analogous to MALDI). The ion spectrum can be analyzed and specific patterns compared across different sample groups.

Tandem MS (MS/MS)
Ions detected by mass spectrometry can be selected for further fragmentation into ion products. The fragmented ion spectra contain more detailed information, which can be used to derive *de novo* sequence information about the peptide.

Transcriptomics
Transcriptomics is the study of the transcriptome, the collection of RNAs expressed in a given cell or tissue under given conditions (normal, disease, treated or untreated with stimulants or compounds). The commonly used method to study the transcriptome is microarray, also called *gene chip*, technology that allows the analysis of gene expression to be determined for all genes in a genome.

Two-dimensional Gel Electrophoresis (2-DE)
Two-dimensional gel electrophoresis (2-DE) is the separation of proteins in a polyacrylamide gel matrix, based on two properties. Separation in the first dimension is determined by the isoelectric point of the proteins being analyzed, while the second dimension separates them according to their apparent molecular weight.

Yeast Two-hybrid (Y2H)
This method was developed for the identification of protein–protein interactions using yeast as a host organism. A gene sequence encoding either a protein or part of a protein of interest is fused to the DNA-binding domain of a known, site-specific transcription factor ("bait"), while a single target gene sequence or library of such gene sequences (which may be prepared from a range of tissues etc.) for a potentially interacting protein(s) is fused to the transactivation domain of the same transcription factor ("prey"). Association between these two fusion proteins in yeast facilitates the reconstitution of the DNA binding and transactivation domains as a functional transcription factor, resulting in activation of a selected reporter gene placed under control of a promoter regulated by the transcription factor. Expression of the reporter can be monitored by a variety of methods.

■ The past century has witnessed many impressive advances in molecular biology and medical research, leading to both increased understanding and improvements in the treatment of many diseases. A notable example of this is the impact that the discovery of antibiotics had in treating bacterial infections. One of the most remarkable achievements was the discovery of the structure of DNA by Crick & Watson, an event that subsequently led some 50 years later to the completion of the sequencing of the entire human genome. This milestone has created a wealth of information and novel avenues for a more focused and comprehensive search and understanding of the biochemical, regulatory and structural function of the proteins encoded by the genome – a phase of biological exploration now referred to as the *postgenome era*.

In the postgenome era, there is significant work ongoing not only on the genetic aspects of biology but also the molecular processes guiding gene expression and function at all levels. These include RNA and protein synthesis and their regulation, protein modifications, protein–protein interactions (pathways), and metabolite formation. This review will focus on Proteomics, the study of global gene expression at the protein level. We will describe both currently used and emerging technologies for exploring universal protein expression, including protein fractionation, analysis, and quantification of complex protein solutions. In particular, this review will focus on characterizing the roles of proteins in pharmaceutical research and development, including exploring applications in determining the functional role of proteins and their relevance in human disease, as surrogate biomarkers for predicting/diagnosing disease, disease progression, and for determining compound efficacy and toxicity. Finally, we will assess what are the future prospects of utilizing protein research to reach a better knowledge of gene/protein function in normal and disease biology as well as clinical applications.

1
Definitions

1.1
Proteomics

The rapidly evolving nature of protein analysis has resulted in an ever-expanding definition of the term "proteomics." While it is accepted that proteomics concerns the study and characterization of all proteins, including their synthesis, modifications, function and interactions, the evolution of separation and identification techniques has generated more specific definitions of proteomic activities such as expression proteomics and interaction proteomics. In addition, specific applications have emerged including clinical proteomics and toxicoproteomics.

1.2
Expression Proteomics

Expression proteomics is defined as the global separation, quantification and identification of expressed proteins in organisms, tissues, or cells under normal or other conditions such as disease or following compound treatment. Initially, this area was centered on 2-DE, but the recent development of nongel-based methodologies such as LC-MS and protein arrays have also started to show promise in this area.

1.3
Interaction Proteomics

Interaction proteomics is the study of specific protein–protein or protein–compound interactions and, more generally, the study of the role of protein complexes within biological pathways. Different methodologies such as the yeast two-hybrid system, phage display, protein arrays, and affinity purification of complexes followed by mass spectrometric analysis have been successfully applied to interaction proteomics studies.

1.4
Clinical Proteomics and Toxicoproteomics

Two particularly exciting emerging fields of expression proteomic research relating to clinical applications are clinical proteomics and toxicoproteomics. Clinical proteomics consists in the identification of biomarkers for disease detection or progression, or even regression after medical therapeutic treatment, whereas toxicoproteomics is the elucidation of potentially toxic or pathological effects of chemical agents or environmental factors.

2
Technologies

2.1
2D Gel Electrophoresis (2-DE)

Since the inception of proteomics as a discipline, 2-DE has been the technique most widely used for protein separation and analysis. This is still the case today, with relatively little changes from its original process developed about 30 years ago by O'Farrell, although a number of complementary approaches are being developed. In 2-DE, the separated proteins are visualized by either pre- or postelectrophoretic stains, class-specific stains (e.g. phosphorylation), or by autoradiography for radiolabeled proteins. Typically, between 2000 and 3000 proteins can be resolved in a standard 2-DE gel (Fig. 1). The technique is robust, quantitative, and relatively simple to implement with reasonably high throughput.

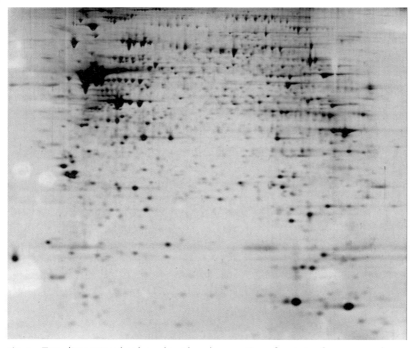

Fig. 1 Two-dimensional polyacrylamide gel separation of proteins from a complex mixture, illustrating separation by isoelectric point in the first (horizontal) dimension and separation by size in the second (vertical) dimension.

Major improvements have enabled greatly increased coverage, reproducibility, and accuracy. These include advances in sample preparation, electrophoresis methodology, as well as sensitivity and quantification of staining. Furthermore, scanning hardware has dramatically improved as have the available algorithms for image analysis and data mining.

Although unparalleled in its resolving power and flexibility, the performance of 2-DE is ultimately finite. Proteins are commonly detected in biological tissues over a dynamic range of $>10^6$. In plasma, the dynamic range has been reported to be on the order of 10^{12}. Because of this huge expression range, high abundance proteins may ultimately mask the visibility and detection of low abundance ones.

There are no "amplification" techniques for proteins analogous to the polymerase chain reaction (PCR) used for transcript identification and quantification. As proteomics is a substrate-limited technique, it is crucial that sufficient sample is available as starting material for analysis. Prefractionation is increasingly becoming a prerequisite for broad proteome coverage. In addition, the technology is biased against proteins with extremes of size, charge, and hydrophobicity. In order to address these "problematic" proteins, a number of novel approaches are being developed. These include multicompartment electrophoresis for enrichment of proteins on the basis of native charge and subcellular fractionation, in particular, isolation of organelle components such as mitochondria and nuclei. As a consequence, initiatives to define the subcellular complement of the nuclear and mitochondrial proteomes are being actively pursued. This type of more focused proteomic analysis has increased sensitivity and has enabled analysis of the nuclear envelope and even the nuclear pore complex. Techniques have also been developed to preenrich proteins by functional class (e.g. proteases) and to specifically remove high abundance proteins. Such approaches include removal of albumin and other highly abundant proteins from plasma and serum prior to clinical proteomic studies.

The original carrier ampholyte-generated pH gradients of the 1970s have now been replaced with immobilized pH gradients (IPG) for isoelectric focusing (IEF). These matrices provide stable and reproducible gradient separations and are available commercially in broad pH range gels (e.g. pH 3–11) and extremely narrow pH range gels (e.g. pH 5–6), enabling extremely high resolution to be attained.

The performance of the second dimension has also been enhanced with the introduction of better instrumentation. In particular, reproducibility and throughput has been improved, while the introduction of the stable gradient gels has increased the resolution and flexibility of the technique. In conjunction with other advances, 2-DE is now capable of producing quantitative and reproducible data at a throughput compatible with biologically significant sample numbers.

The introduction of highly sensitive fluorescent dyes is greatly improving protein detection in gels. Fluorescent stains such as SYPRO Ruby have a better dynamic range, are more quantitative and are more compatible with mass spectrometry than previous stains (e.g. silver stains). A direct consequence of improvements to the performance of the technology is the separation of isoforms of proteins differing by single deamidation or phosphorylation modifications. Subsequently, stains specific for phosphorylation and glycosylation were developed and used successfully for proteins separated by 2-DE. An additional advance is the use of differential gel electrophoresis, incorporating prelabeling protein samples with fluorescent dyes and the ability to run two or even three samples on the same gel for comparative analysis.

Overall, 2-DE remains one of the most reliable and widely used techniques for proteome analysis. This is particularly true for sample profiling and for the analysis of differential protein expression in differing disease states or after treatment of cells or tissues with biologically or medically relevant agents or compounds. The continuous development of new techniques and instrumentation means that the limitations of the technique are constantly being reappraised and 2-DE will remain a very valuable approach for the foreseeable future.

2.2
Mass Spectrometry

One of the fundamental challenges of proteomics work is the requirement for sequence identification of proteins from a given sample following separation. Until a few years ago, the classical protein identification procedure was performed through N-terminal amino acid sequencing, known as *Edman degradation*. The development of mass spectrometry has essentially replaced Edman's method by virtue of providing more sensitive detection and higher sample throughput. These features, coupled with the rapid expansion of genomic databases, have provided a revolutionary approach to protein detection and identification. The key steps describing the use of mass spectrometry to identify proteins from polyacrylamide gels are shown in Fig. 2.

In general, mass spectrometry is performed on tryptic peptides rather than intact proteins as their extraction from the microreticular polyacrylamide matrix of 2-DE is more facile. In addition, peptides are more readily amenable to ionization and analysis. This is not trivial as it necessitates a method for efficiently digesting proteins within the gel and recovering the resulting peptides. The proteases routinely employed, such as trypsin, are presented to the protein substrates, which are present within the gel at concentrations significantly below the Michaelis constant of the enzyme. Nevertheless, methods have been optimized, which enable efficient fragmentation of the proteins in situ.

The main principle of mass spectrometry is the accurate measurement of the masses of the peptide ions and derived peptide fragment ions from the analyses. The first requirement for analyte characterization is the generation of ions in the gas phase. In particular, the so-called "soft ionization" methods of matrix-assisted laser desorption ionization (MALDI) and

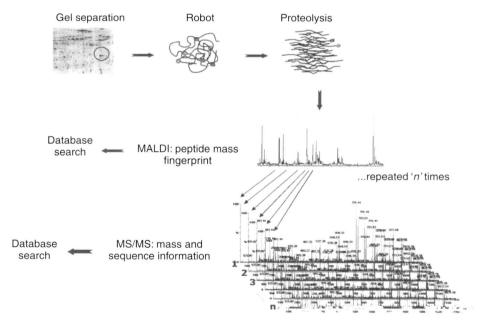

Fig. 2 Qualitative identification of proteins from gels by mass spectrometry. A spot of interest is excised from the gel and digested to peptides by trypsin. Two main methods exist for identifying a protein from the peptides generated. Firstly, the mass of the peptides may be determined by mass fingerprinting and compared to genomic data present in bioinformatics databases. Alternatively, individual peptide ions can be sequentially selected for fragmentation, which generates *de novo* sequence information. A combination of the sequence information and the peptide mass can be used to accurately assign a protein identity by comparison to the genomic database.

electrospray ionization (ESI) have made a dramatic impact on proteomics.

In MALDI, a laser pulse is applied to the sample that is embedded in a matrix (chromophore), which absorbs light at a specific wavelength. The energy in the chromophore is passed to the analyte peptide, which ionizes. The ions created are guided by a strong electrical field into the mass analyzer. MALDI is a particularly "soft" ionization technique resulting in singly charged peptide ions where the mass-to-charge (m/z) ratio reflects the mass of the peptide. This technique is most often used to measure the masses of peptides derived from enzymatic digestion of a protein. This is often referred to as the "fingerprint" of the protein and may be sufficient to identify the protein against a genomic database.

In ESI, ions are formed by electrospraying a peptide solution into the mass spectrometer via a small needle in the presence of a strong electrical field, creating small ionized droplets. The solvent in the droplets is eliminated by applying a counter current flow of gas or heat. Once the ions have entered the mass spectrometer, the mass analyzer separates and measures the ions on the basis of their mass-to-charge ratio (m/z) using electric or magnetic fields with subsequent registration by the detector. In ESI, a series of ions of different charges ($z = 1$, $z = 2$, $z = 3$, etc.) may be formed and these have to be deconvoluted to determine the mass of the parent peptide. This technique provides higher energy ions suitable for fragmentation and *de novo* sequencing. Relative abundance of the peptide will be reflected by the ion current in the analyzer. The ion current intensity will also be a reflection of the propensity of a particular peptide to ionize, which is in turn predicated on the amino acid composition. These confounding influences make absolute quantification without the use of internal standards complex.

In its simplest mode, the mass spectrometer defines the mass of the parent peptide ion. Peptides can then be selected manually or automatically within the mass spectrometer and further fragmented by collision with an inert gas. Peptides undergoing this process (tandem MS or MS/MS) tend to fragment at the amide bond, leading to a mass spectral ladder, whose steps reflect the mass differences between the parent ions sequentially depleted of the terminal amino acid residue (Fig. 2). It is therefore possible to derive *de novo* sequence information and, by reference to genomic databases, unequivocally identify a protein from a single peptide.

In recent years, different variants of mass spectrometry have emerged, encompassing novel approaches for trapping and selecting ions (ion traps, quadrupoles) as well as new mass analyzers such as time of flight (TOF) and Fourier transform ion cyclotron resonance (FT or FTICR-MS). As with the more classical approaches, performance is dependent upon the key parameters of mass accuracy, resolution, and sensitivity.

Biological samples are often extremely complex, and gel-derived digests often contain reagents incompatible with mass spectrometers. To circumvent these issues, separation technologies such as chromatography have been coupled to mass spectrometry. As with 2-DE, methods of fractionation have been developed to increase the proteome coverage or to analyze specific subsets of the proteome. These include methods of fractionation and specific tagging procedures for labeling phosphoproteomes.

Owing to its use of volatile solvents, reverse phase high performance liquid chromatography (RP-HPLC) is particularly compatible with mass spectrometry and is readily interfaced via a standard ESI source (LC-MS). The mass spectrometer is able to sample and analyze the peptides as they elute sequentially from the chromatographic column. While this approach has disadvantages of being slower than MALDI-based approaches, it has the advantage of generating higher quality data.

For extremely high resolution of very complex mixtures, use of two-dimensional chromatography techniques coupled to tandem mass spectrometry (LC/LC-MS/MS), sometimes referred to as *multidimensional protein identification technology* (MudPIT), has been established. The first dimension is often ion exchange chromatography followed by RP-HPLC for the second dimension. Progress in the development of LC/MS-based proteome analysis has raised the realistic prospect of nongel-based global protein expression profiling. Intact proteins are not readily amenable to RP-HPLC in complex mixtures and so it is necessary to predigest the proteins to peptides for analysis. This creates an added dimension of complexity and often precludes adequate analysis of posttranslationally modified proteins. While this technology is still evolving, it offers potential advantages in overcoming the limitations discussed above that are associated with 2-DE approaches. Recent advances have shown that the peptide retention time by LC, the *m/z* value and its intensity may provide sufficient information to both quantitatively and qualitatively identify a protein in a complex mixture.

2.3
Isotope Coding

Mass spectrometry has utilized stable isotopes to quantify relative amounts of individual analytes for many years. Attempts to perform experiments on a global proteomic scale have led to a number of exciting developments. These include the use of stable isotope labeling of amino acids in culture (SILAC) and other metabolic labeling techniques. However, for *in vivo* and larger studies, labeling at the point of analysis is needed. Digestion of proteins with trypsin produces neotermini on the peptides, which provide the opportunity for differential labeling with the isotopes ^{18}O and ^{16}O derived from heavy and light water respectively. The use of isotope-coded affinity tags (ICAT) has gained some acceptance in the last few years, with one notable early study showing differential protein expression in yeast grown under different metabolic conditions.

ICAT probes are a series of labeling reagents containing an affinity tag (biotin), a spacer labeled with a light or heavy isotope, and a thiol-specific reactive group (-iodo) for labeling of cysteine-containing peptides. Labeling is optimized to permit quantitative analysis of samples (Fig. 3). Early ICAT reagents possessed spacers containing either eight deuteriums (D8, the heavy isotope) or eight hydrogens (D0, the light isotope). Two samples are labeled, one with the light reagent and the other with the heavy reagent and the samples are then combined. Once digested, separated, and analyzed by mass spectrometry, the relative abundance of the isotope tags may be used to estimate the relative levels of the individual protein species.

In the earlier incarnation of ICAT, the necessity for high-resolution LC systems to resolve the plethora of peptides resulted in the undesired resolution of the light D0 and heavy D8 form. This is problematic as the mass spectrometric analysis

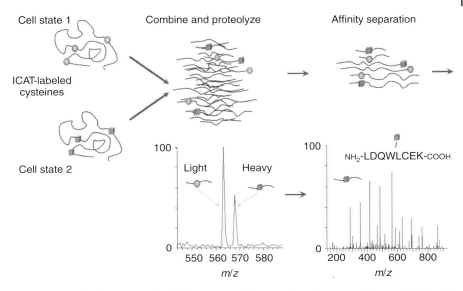

Fig. 3 Principle of isotope-coded affinity tagging. The proteins of two samples are labeled with a tag containing one of two isotopic labels such as eight deuteriums (squares) or eight protons (circles). The samples are combined, affinity separated, and analyzed by mass spectrometry. The relative abundance of the ions with the appropriate mass difference (e.g. 8Da) reflects the relative abundance of the proteins.

is predicated on the two ions entering the mass spectrometer at the same time. The affinity tag specificity for the cysteine residue also limited its applicability only to cysteine-containing proteins. In addition, the reagent itself did not perform ideally in the mass spectrometer, generating heterologous fragmentation, which confounded interpretation. Many new isotopes are being developed today, and it is hoped that future iterations will address these caveats.

2.4
Yeast Two-hybrid Systems

The yeast two-hybrid system identifies putative binary protein interactions, which, if proven, can provide powerful insights into biological pathways. The gene coding region for a protein of interest or part of a protein of interest (the "bait") is fused to the coding region for the DNA-binding domain of a yeast transcription factor (commonly GAL4 or LexA). This is engineered to ensure that the resulting fusion protein is targeted to the nucleus with a selection marker. The coding region for a single protein (the "prey") or a library of proteins from different tissues ("prey" library) is fused to the transcriptional activation domain of the same transcription factor, and also targeted to the nucleus. Screening is performed with reporter genes under the control of a promoter that respond to the reconstituted DNA binding/transcription factor activity, reflecting bait–prey interaction (Fig. 4). Common reporter genes include enzymes, which support cellular growth *(LEU2, HIS3)* and β-galactosidase, for which a simple colorimetric detection is possible.

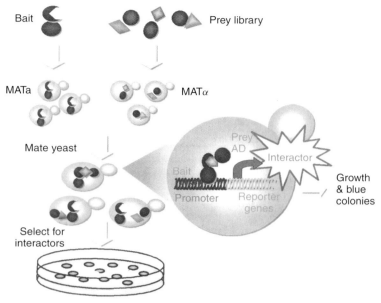

Fig. 4 Principle of the yeast two hybrid system for defining protein–protein interactions. Two chimeras, one containing a DNA-binding domain and the other an activation domain are cotransfected into yeast. If the fusion partners interact, the binding domain and activation domain are brought into proximity and activate transcription of a reporter gene.

In addition to screening prey libraries for interactors, the system can also be used for screening deletion mutants or polymorphisms to characterize the interacting domains. Variants of the yeast two-hybrid technique include yeast one-hybrid systems for protein-DNA interactions and yeast three-hybrid systems for looking at three way protein interactions or protein–drug interactions.

Although there is an association between the strength of the interaction and the level of reporter activity in the yeast two-hybrid system, the interaction is only weakly quantitative and the true strength of the technique is in its qualitative analysis. Although false-positive results due to nonspecific interactions or spontaneous transactivation activity are often likely to confound interpretation, careful experimental design and interpretation as well as follow-up with directed confirmation assays using alternate methods (e.g. coimmunoprecipitation) can help with the identification and validation of true interactors. Often critical protein–protein interactions are specific to a disease state and so the yeast model may generate false-negative data. Other limitations of the technology include aberrant interactions owing to improper folding, difficulty with finding membrane protein interactors, lack of necessary posttranslational modifications, or inappropriate cellular compartmentalization. However, it has the tremendous advantage of being cheap and simple to run, is an *in vivo* system, and can be run on a high-throughput scale. The positive aspects are reflected in its wide utility and role in many important discoveries including the GABA-B receptor heterodimerization and producing genome scale interaction maps for *Saccharomyces cerevisiae* (yeast), *Caenorhabditis elegans* (worm), and *Drosophila* (fruit fly).

2.5
Protein Arrays

Protein arrays constitute an emerging technology, which permits the analysis of protein expression, protein interactions, or protein function. The microarray-like format permits the deposition of various biomolecules onto a solid surface in an ordered array (Fig. 5). Antibody chips allow the detection of protein expression in an analogous method to enzyme-linked immunoassays (ELISAs). Arraying of functional biological molecules will also permit protein interactions as well as protein function to be investigated.

The inherent promise of protein arrays is the capacity to perform routine assays in a multiplexed format. This should greatly improve the comparative nature of a study by reducing assay-to-assay variability. Another benefit in practical terms is the reduction in reagent usage, which has clear relevance for the analysis of scarce clinical samples. Despite these advantages, the performance of protein expression arrays appears to be comparable in terms of sensitivity to ELISAs rather than enhanced. Certain techniques are being developed to try and increase sensitivity, including the use of novel detection systems (e.g. resonance light scattering) and signal amplification systems, but the results to date remain equivocal. Future developments using more advanced detection systems such as mass spectrometry may offer increased sensitivity, although for many applications such as expression analysis, this must be achieved while retaining

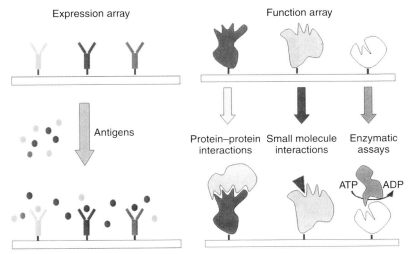

Fig. 5 Protein arrays contain a defined set of proteins arrayed at high density, typically onto a solid surface. Protein arrays can be used for protein profiling by immobilization of antibodies to specific analytes or as functional arrays for detecting protein–protein or protein–drug interactions. Proteins have also been arrayed to detect and characterize the functionality and specificity of both enzymes and substrates.

quantification. Also, as with other "omic-based" technologies, such advances will place demands upon the bioinformatics tools needed to manage and interpret the data.

Currently, the multiplexed protein array assay cannot, in general, compete with individual assays in terms of performance. Although the technology is advancing rapidly with respect to surface chemistry, detection, and data analysis, the limitations of the technology are predominantly in reagent supply. Commercial antibodies with the required sensitivity and selectivity are not available for the vast majority of proteins, and, even where they do exist, only 5% or less are suitable for use in protein arrays, having been produced for other purposes such as immunohistochemistry and western blotting. The production of antibodies is both labor intensive and time consuming. For example, production of monoclonal antibodies requires immunization, generation of hybridoma cell lines, and antibody isolation. The end result is often an antibody that can be thoroughly characterized in terms of sensitivity and selectivity, and a renewable source for producing this antibody over and over again. Although the use of phage libraries and other more novel capture reagents exist, such as aptamers and proteins based on novel scaffolds, the evidence is that maturation is needed to achieve ligands of acceptable performance and assay screening of potential reagents is often rate-limiting. The same challenge is true for functional proteins where the promise of whole genome or near whole genome cloning and expression is still to be realized. Future developments in capture reagents and protein production are likely to control the rate at which omic-scale protein arrays impact the biomedical sciences.

Despite the current limitations, protein arrays for specific applications remain an exciting prospect and are beginning to yield useful and meaningful results.

2.6 Bioinformatics

A robust and expert bioinformatics platform is a key requirement for the capture, organization, analysis, and storage of proteomics data. The aim of both transcriptomics and proteomics studies is to generate high quality quantitative expression data. This needs to be managed in such a way as to enable information to be derived about complex patterns of expression and pathways consistent with pathology, physiology, drug efficacy, toxicology, and so on. By necessity, bioinformatics is a cross-disciplinary activity embracing computer science, software engineering, mathematics, and biology. The massive leap in scale accompanying "omic" technologies has necessitated ever more sophisticated bioinformatic tools to support target selection and validation, and disease understanding. The ability to relate a small piece of information on protein sequence to enable gene identification is critical and fundamental to proteomics, but increasingly important is the ability to place expression changes and interactions into a biological context of pathway and function. A concerted effort has begun to emerge to store, analyze and disseminate proteomic information, although it is still a discipline in its infancy where additional progress is eagerly awaited.

Many searchable databases available on the World Wide Web are being constantly updated to incorporate the increasing volume of data being generated from diverse genomic and proteomic platforms. Placing proteins in their correct functional and biological context is likely to be a complex process with lessons to be learnt from transcriptomics and other disciplines.

3
Applications

3.1
Target Identification and Validation

Since proteins are the ultimate structural and functional components of life, understanding biological systems at the protein level is critical to delineating biological pathways, including those involved in disease as well as determining how drugs exert their beneficial and sometimes toxic effects. These explorations can lead to the identification of new targets (also proteins to be avoided) as well as contributing to validating the involvement of the particular protein involved in the biology of the disease. Proteomics has therefore begun to provide insights into several areas of medical research.

In the field of infectious diseases, proteomics – in conjunction with other disciplines such as transcriptomics and RNA interference (RNAi) – has been used to unravel the mechanism of certain drug activities and to define host–pathogen interactions. This holds the promise of providing new methods for diagnosing and treating infectious diseases. Initial studies focused on the effect of bacterial infection within host cells, including analysis of those proteins that are selectively expressed by the pathogen during infection. These studies also led to increasing our understanding of the inflammatory signaling networks in the process of sepsis. More recent work has focused on the mechanisms of pathogenesis. In addition, work with organisms such as *Plasmodium falciparum*, the malaria parasite, has provided critical insights into major diseases such as malaria, via delineation of the protein expression patterns at distinct stages in the life cycle of the parasite.

Protein changes correlating with disease processes, such as inflammation, have indicated dysregulation of proteins, which may subsequently be modulated by non-steroidal anti-inflammatory drugs. These included a time- and gender-dependent upregulation of acute phase proteins in a rat model. Respiratory diseases, such as chronic obstructive pulmonary disease and asthma, are known to have a strong inflammatory element to their pathology. Genetics has greatly advanced our understanding of the pathobiology underlying complex genetic disorders such as asthma, with numerous genetic loci and individual genes showing a linkage or association. Proteomics has the ability to complement these genetic results by generating biomarkers that may prove invaluable as prognostic tools for predicting asthma susceptibility and disease severity as well as predictors of drug response. As part of the drive to deliver such biomarkers, proteomics has been applied to a number of biofluids, such as bronchoalveolar lavage, together with plasma and serum.

Proteomics has also been deployed to widen the understanding of the therapeutic value of vaccines. For example, in a relatively sophisticated approach to define novel vaccine candidates, 2-DE has been incorporated into studies to define immunogenic proteins, which elicit a host response. This has been reported for therapeutic areas as diverse as infectious diseases and oncology. The latter raises the prospect of developing future cancer vaccines, making provision for a major unmet therapeutic need. Similar approaches have been applied to advancing the diagnosis and treatment of transplant rejection via identification of early events in the process of rejection.

Research in cardiovascular biology in the pregenomic era focused on classical pathophysiological descriptions, with the later incorporation of receptor biology and signal transduction–driven approaches. Cardiovascular research is a discipline for which there are many *in vitro* and *in vivo* model systems available, which provide a vehicle for a more detailed understanding of the molecular basis of disease. In the postgenomic era, proteomics will provide an important role, bridging the gap between genetic and physiological research. Atherosclerosis is a disease of the cardiovascular system, which is now generally accepted to have a large inflammatory component. Several risk factors have been identified including cell adhesion molecules and inflammatory markers such as C reactive protein. Proteomics has recently been instrumental in discovering several putative protein markers associated with atherosclerosis, many of which have a recognized role in inflammation.

Applications of proteomics in the field of oncology are quite diverse, focusing on advancing the understanding of mechanisms, diagnosis, and prognosis, as well as defining possible therapeutic regimes in a number of cancers. One important role has been to advance the definition of mechanisms underlying cellular regulation, including those implicated in the transcriptional regulation of metastasis. Protein expression in hepatocellular carcinoma cell lines, which has proven to be a useful model of persistent viral hepatitis–induced hepatoma, has revealed interesting new proteins and confirmed others previously implicated in the disease. Proteomics has also been used to gain a better understanding of the multifactorial nature of chemoresistance demonstrated by certain cancer cells. The oncology area has been extensively studied using DNA chips (microarrays) for transcript analysis, hence there is a wonderful opportunity for synergy between transcriptomics and proteomics by comparing and contrasting the data for validation of new targets or new disease diagnostic or disease/drug prognostic markers.

In the fields of neurology and psychiatry, the understanding of neurologically relevant systems has been enhanced by significant progress in mapping the human brain proteome. This has been further advanced by the investigation of model organisms such as the mouse and *in vitro* cell systems such as synaptosomes. Alzheimer's disease is a complex genetic disorder in which dysfunction is associated with mutational loci (amyloid precursor protein, presenilin 1 and presenilin 2) as well as susceptibility loci (apolipoprotein E). Progress has been made in this area by detection of oxidatively modified proteins consistent with previously identified peroxynitrite-mediated damage and oxidatively modified RNA. Analysis of human postmortem brains from both Alzheimer's and Down syndrome patients indicates decreased levels of proteins such as synaptosomal-associated protein 25 (snap 25), possibly reflecting impaired synaptogenesis or neuronal loss. Postmortem studies and antemortem analysis of cerebrospinal fluid have also revealed several additional proteins found to be altered in expression in Alzheimer's disease, including apolipoprotein E.

In psychiatric research, the application of genomics and proteomics to disorders as diverse as depression, schizophrenia, and autism is being considered as a major potential contributor in terms of disease diagnosis, understanding, and therapy. New avenues into disease understanding have been illuminated by genetic studies, for example, the discovery of neuregulin, DAOO, and G72 (the latter two associated with glutamate pathways) as schizophrenia susceptibility genes. To date, analysis of cerebrospinal fluid from schizophrenics has produced limited identifiable changes in protein expression and in general, validated markers of psychiatric disorders remain elusive. Despite this difficulty, it is likely that given clinical material of sufficient quality and with detailed disease phenotypic data, proteomics can make a major contribution to both mechanistic understanding and diagnosis.

3.2
Protein–protein Interactions

A key aspect of understanding biological systems is to gain an insight into protein function. Characterizing protein function in a normal state and defining changes in function in the dysregulated state associated with disease can be extremely challenging and while expression proteomics is certainly capable of providing important clues into protein function, investigation of how proteins interact with one another is of additional utility. Association of a protein of unknown function with proteins of known function or linkage to a defined pathway can be invaluable in delineating cellular processes and disease mechanisms.

Various methods exist for defining protein interactions including the yeast two-hybrid system and affinity pull downs or chemoproteomic approaches. In a relatively recent advance in chemoproteomics, a tandem affinity purification (TAP) tag is used in which a fusion of the binding domains of both protein A and calmodulin are employed to enhance selectivity. The two affinity regions are separated by a cleavable tobacco etch virus protease cleavage site. This approach has been applied successfully to create a functionally organized database for yeast–protein interactions, and provide the first insight of how a complex network of proteins may operate in a eukaryotic proteome.

Yeast two-hybrid systems have also been effective in defining new protein targets for pharmaceutical intervention including defining novel nuclear hormone receptors and their cofactors and in characterizing the specific residues or domains within proteins associated with protein function. For instance, the system has also been employed to find peptides, which bind and inhibit key enzymes such as kinases.

As described above, protein arrays are now being developed to define protein function and studies in yeast have already defined and characterized novel binding motifs for proteins such as calmodulin. Perhaps the most immediate benefit of protein interaction studies will be the assembly of large databases of hitherto undefined pathways, which will assist in associating proteins with their functions or, more broadly, with specific pathways and cellular processes.

3.3
Protein Biomarkers in Clinical and Preclinical Research

The discovery and successful application of therapeutic treatments to diseases is a very complicated and lengthy process. Through the development cycle of a new drug candidate, a vast number of hurdles need to be overcome. They mainly involve a clear establishment that the proposed molecule is safe, tolerable, and therapeutically efficacious. During the process, the compound has to be carefully and thoroughly tested in preclinical animal models, and only if those studies are successful in demonstrating the above attributes will the drug progress to tightly controlled human trials. The complexity, scientific rigor, and vast amount of human and financial resources needed during the drug discovery and development process mean that finding ways to predict, diagnose, and understand better the manifestations of disease as well as the outcome of the proposed drug treatment are critical factors to be addressed.

Alongside transcriptomics and metabolomics, proteomics has the potential to dramatically improve the speed of clinical research, discriminating between diseases and disease subtypes. Proteomics applied to biofluids such as plasma offers a relatively noninvasive method of providing bedside/near bedside clinical support. In particular, proteomics has major potential for finding novel biomarkers that can act as predictive signatures of disease onset and progression, act as surrogate markers of therapeutic efficacy, or reveal potential toxicology. The search for novel protein biological biomarkers that can predict at an early stage the efficacy and safety profiles of a drug has benefited from a concerted and vigorous increase in effort in the last decade.

Proteomics technologies are playing a key role in the search for biomarkers, often complementing other approaches such as transcriptomics and genetics. While acknowledging recent advances, initial progress in plasma/serum proteomics was slow owing to the complexity and wide dynamic range of the protein content. For example, while the concentration of serum albumin in normal plasma is in the range of $35-50$ mg·mL^{-1}, the concentration of a cytokine such as interleukin IL6 is 10^9 times lower at $0-5$ pg·mL^{-1}. Methods for removal of albumin and other abundant proteins from plasma have now dramatically improved the scope for clinical proteomics in biofluids. In both preclinical as well as clinical arena, the use of 2-DE gel electrophoresis techniques in conjunction with mass spectrometry as well as the use of immunobased approaches is now providing important and useful data.

Adverse drug reactions are a significant issue both in terms of molecules failing both in preclinical development and in the clinic. Since the estimated cost of bringing a drug to the market is $800 million, any improvement in attrition rates, however small, will accrue significant benefits both to pharmaceutical companies and also to the patients themselves. Initially, the focus of genomic technologies in the area of toxicology was centered on transcriptomics, with the analysis of RNA expression in tissues obtained from animals treated with test compounds. However, the parallel analysis of genes and proteins has brought new insight to both *in vitro* and *in vivo* studies with the goal to predict the toxic effects of potential of new potential drugs much earlier in the development cycle. The aim of predictive toxicology via proteomics is to improve lead and candidate selection and enhance drug efficacy and safety monitoring via profiling of defined surrogate markers from easily obtainable biofluids. Proteomics via 2-DE enables drug-related alterations in expression profiles to be compared and potential biomarkers for further scrutiny to be identified. Several studies have already begun to define protein expression changes, which correlate with drug-mediated toxicity, whereas specific studies have indicated possible mechanisms of toxicity.

Recently, the surface-enhanced laser desorption ionization (SELDI™) time of flight (TOF) technology has been developed. Proteins in a mixture are selectively bound to a surface of defined chemistry and their masses profiled by mass spectrometry. Although the identity of the proteins cannot usually be ascertained from this type of data, a specific mass spectrum can be obtained and used to characterize a specific analyte. This has been applied to define potential biomarkers of ovarian, breast, and prostate cancer.

Application of "omic"-based technologies can offer additional insights over and above the growing area of pharmacogenetics. Pharmacoproteomics represents a more phenotypic approach to the analysis of patient-to-patient variability than is provided by genotyping as it will assess the effects of an individual's genetic makeup as well as the influences of the environment and lifestyle. This makes proteomics a valuable tool in disease diagnosis and the push toward defining subgroups of patients who are more suited to the individual therapies.

4
Future Perspectives

4.1
Toward a Full Proteome Map

The genomes of higher eukaryotes give rise to proteomes of high complexity owing to the predominance of posttranslational modifications and splice variants. Indeed, the level of complexity appears to correlate with that of the organism. It has been estimated that the number of proteins per gene is around 1.1 to 1.3 for bacteria, 3 for yeast and around 10 for humans, although these are estimates and may only hint at the actual levels of complexity.

In addition, the true relevance of a protein function cannot be derived from information about a single protein in isolation. The activity of a given protein will be strongly influenced by its tissue distribution and intracellular localization, as well as its proximity to and interaction with other proteins and metabolites. These factors illustrate the increasing need to move the technology toward total proteome coverage.

Although a laudable aim, currently no method is in sight that will allow the analysis of the entire proteome. Progress toward significantly higher proteome coverage is being made via prefractionation techniques to analyze subproteomes. Several studies have now analyzed proteomes at the organellar level and even at the suborganellar level. This contributes more information about the location, expression, and function of proteins, and these methods for analysis are becoming increasingly sensitive, robust and quantitative. Historically, certain types of proteins have been difficult to study, including highly hydrophobic membrane-bound proteins. Although membrane proteins have been typically underrepresented during analysis by 2-DE, progress in extraction methods for membrane proteins and the introduction of improved solubilization techniques, both in sample preparation and separation, have markedly improved membrane protein detection. In addition, the ability to select individual protein classes such as glycosylated proteins or phosphoproteins via antibodies or specific tags has increased the scope and range of proteome subset analyses. Advances in the technology and, in particular, the recent advances in nongel-based proteomics may herald a major advance toward increased proteome coverage.

4.2 Biological Validation of Protein Functions

As mentioned a number of times in this review, proteomics offers tremendous opportunities to rapidly progress biological and medical research. The ability to analyze and potentially quantify many proteins at once in a multiplexed assay without any preselection makes proteomics a truly open multiplexed discovery platform. However, experience to date has shown that the wealth of information created can be extremely difficult to process and presents a number of novel issues. The multiplexed nature of the assay increases the chances of discovering many proteins associated with a pathology or biological phenotype. It also increases the probability of identifying changes, which occur purely by chance, or as an indirect result of the up- or downregulation of relevant proteins. Thus, the necessity for rigorous validation of protein changes associated with altered phenotypes is of paramount importance and places a large burden on what is currently a potentially resource and rate-limiting step.

While many proteins such as enzymes can be assayed functionally when defined substrates are available, individual immunoassays have long been one of the definitive approaches to measuring individual protein expression. However, as discussed above, the limited availability of high-quality antibody reagents has led to significant effort being expended on provision of antibody reagents for assays and protein arrays. Further, the creation of international consortia is being proposed to undertake and complete the monumental task of producing antibodies against every human protein. Additional validation can also be gleaned from specific pathway expansion, gene silencing and gene knockout experiments.

4.3 Clinical and Diagnostic Applications

A number of issues need to be addressed before proteomics can achieve general and routine applicability in clinical studies. In addition to the caveats of working with plasma and serum, as discussed above, there are issues surrounding the heterogeneous nature of human samples. Biological systems are complex and study populations from clinical cohorts tend to be highly variable. Many clinical studies are derived from populations representing diversity in genetic and environmental and lifestyle factors, as well as in age, disease severity and comorbidity. All of these factors conspire to confound interpretation of clinical proteomic data. The capacity to run proteomic analysis at a scale consistent with large populations is needed to facilitate extraction of biologically and statistically significant information from the confounding elements. The industrialization of 2-DE is not as advanced as is needed to address very large clinical cohorts, although it is now capable of processing thousands of samples. The immediate promise of the emerging techniques such as LC/LC-MS/MS and protein arrays will be to augment information content attainable with 2-DE rather than to provide dramatically increased throughput.

Progress to date in the area of clinical proteomics has been rapid and it will surely continue to improve at an equally impressive rate given the huge demand and benefits. The potential for defining biomarkers for therapeutic or diagnostic use, as well as defining biological mechanisms is immense.

Acknowledgments

The authors wish to thank Jingwen Chen, Bob Hollingsworth, Arthur Moseley, Ian White, and Julia White for their help in preparing the figures, and Mike Trower and Mark Skehel for helpful comments on the manuscript.

See also Genetics, Molecular Basis of; Oncology, Molecular; Protein Modeling; Protein NMR Spectroscopy; Protein Splicing; Serial Analysis of Gene Expression.

Bibliography

Books and Reviews

Aebersold, R., Mann, M. (2003) Mass spectrometry-based proteomics, *Nature* **422**, 198–207.

Anderson, N.L., Anderson, N.G. (2002) The human plasma proteome: history, character and diagnostic prospects, *Mol. Cell. Proteomics* **1**, 845–867.

Cutler, P. (2003) Protein arrays: current state of the art, *Proteomics* **3**, 3–18.

Patterson, S.D., Aebersold, R.H. (2003) Proteomics: the first decade and beyond, *Nat. Genet.* **33**(suppl.), 311–323.

Simpson, R.J. (2002) *Proteins and Proteomics: A Laboratory Manual*, Cold Spring Harbor Press, New York.

Walgren, J.L., Thompson, D.C. (2004) Application of proteomic technologies in the drug development process, *Toxicol. Lett.* **149**, 377–385.

Zerkowski, H.-R., Grussenmeyer, T., Matt, P., Grapow, M., Engelhardt, S., Lefkovits, I. (2004) Proteomics strategies in cardiovascular research, *J. Proteome Res.* **3**, 200–208.

Zhu, H., Snyder, M. (2003) Protein Chip Technology, *Curr. Opin. Chem. Biol.* **7**, 55–63.

Zolg, J.W., Langen, H. (2004) How industry is approaching the search for new diagnostic markers and biomarkers, *Mol. Cell. Proteomics* **3**, 345–354.

Primary Literature

Arrell, D., Neverova, I., van Eyk, J. (2001) Cardiovascular proteomics: evolution and potential, *Circ. Res.* **88**, 763–773.

Baak, J.P.A., Path, F.R.C., Hermsen, M.A.J.A., Meijer, G., Schmidt, J., Janssen, E.A.M. (2003) Genomics and proteomics in cancer, *Eur. J. Cancer* **39**, 1199–1215.

Bailey, W.J., Ulrich, R. (2004) Molecular profiling approaches for identifying novel biomarkers, *Expert Opin. Drug Safety* **3**, 137–151.

Bandara, L.R., Kennedy, S. (2002) Toxicoproteomics – a new preclinical tool, *Drug Discov. Today* **7**, 411–418.

Beranova-Giorgianni, S. (2003) Proteome analysis by two-dimensional gel electrophoresis and mass spectrometry: strengths and limitations, *Trends Anal. Chem.* **22**, 273–281.

Beroza, P., Villar, H.O., Wick, M.M., Martin, G.R. (2002) Chemoproteomics as a basis for postgenomic drug discovery, *Drug Discov. Today* **7**, 807–814.

Blobel, G., Wozniak, R. (2000) Proteomics for the pore, *Nature* **403**, 835–836.

Bodovitz, S. (2003) The protein biochip content problem, *Drug Discov. World* **4**, 50–60.

Cacabelos, R. (2002) Pharmacogenomics for the treatment of dementia, *Ann. Med.* **34**, 357–379.

Chakravati, D., Fiske, M., Fletcher, L., Zagursky, R. (2001) Application of genomics and proteomics for identification of bacterial gene products as potential vaccine candidates, *Vaccine* **19**, 601–612.

Coates, P.J., Hall, P.A. (2004) The yeast two-hybrid system for identifying protein-protein interactions, *J. Pathol.* **199**, 4–7.

Cordwell, S.J., Nouwens, A.S., Walsh, B.J. (2001) Comparative proteomics of bacterial pathogens, *Proteomics* **1**, 461–472.

Davidson, P., Westman-Brinkmalm, A., Nilsson, C., Lindbjer, M., Paulson, L., Andreasen, N., Sjorgen, M., Blennow, K. (2002)

Proteome analysis of cerebrospinal fluid proteins in Alzheimer patients, *Clin. Neurosci. Neuropathol.* **13**, 611–615.

Edman, P. (1970) Sequence determination, *Mol. Biol. Biochem. Biophys.* **8**, 211–255.

Evangelista, C., Lockshon, D., Fields, S. (1996) The yeast two-hybrid system: prospects for protein linkage maps, *Trends Cell Biol.* **6**, 196–199.

Flory, M., Griffin, T., Martin, D., Aebersold, R. (2002) Advances in quantitative proteomics using stable isotope tags, *Trends Biotechnol.* **20**, S23–S29.

Furness, L.M. (2002) Analysis of gene and protein expression for drug mode of toxicity, *Curr. Opin. Drug. Discov. Devel.* **5**, 98–103.

Garfin, D.E. (2003) Two-dimensional gel electrophoresis: an overview, *Trends Anal. Chem.* **22**, 263–272.

Grandi, G. (2001) Antibacterial vaccine design using genomics and proteomics, *Trends Biotechnol.* **19**, 181–188.

Grunenfelder, B., Rummel, G., Vohradsky, J., Roder, D., Langen, H., Jenal, U. (2001) Proteomic analysis of the bacterial cell cycle, *Proc. Natl. Acad. Sci. U.S.A.* **98**, 4681–4686.

Halapi, E., Hakonarson, H. (2004) Recent development in genomic and proteomic research for asthma, *Curr. Opin. Pulm. Med.* **10**, 22–30.

Hegde, P.S., White, I.R., Debouck, C. (2003) Interplay of transcriptomics and proteomics, *Curr. Opin. Biotechnol.* **14**, 647–651.

Herbert, B.R., Harry, J.L., Packer, N.H., Gooley, A.A., Pedersen, S.K., Williams, K.L. (2001) What place for polyacrylamide in proteomics? *Trends Biotechnol.* **19**(Suppl. 10), S3–S9.

Hoving, S., Voshol, H., van Oostrum, J. (2000) Towards high performance two-dimensional gel electrophoresis using ultrazoom gels, *Electrophoresis* **21**, 2617–2621.

Hutter, G., Sinha, P. (2001) Proteomics for studying cancer cells and the development of chemoresistance, *Proteomics* **1**, 1233–1248.

Ideker, T., Thorsson, V., Ranish, J., Christmas, R., Buhler, J., Eng, J., Bumgarner, R., Goodlett, D., Aebersold, R., Hood, L. (2001) Integrated genomic and proteomic analyses of a systematically perturbed metabolic network, *Science* **292**, 929–934.

Jiang, L., Lindpaintner, K., Li, H.-F., Gu, N.-F., Langen, H., Fountoulakis, M. (2003) Proteomic analysis of the cerebrospinal fluid of patients with schizophrenia, *Amino Acids* **25**, 49–57.

Jain, K.K. (2004) Role of pharmacogenomics in the development of personalised medicine, *Pharmacogenomics* **5**, 331–336.

Johnson, J.R., Florens, L., Carucci, D.J., Yates, J.R. (2004) Proteomics in malaria, *J. Proteome Res.* **3**, 296–306.

Jungblut, P., Thiede, B. (1997) Protein identification from 2-DE gels by MALDI mass spectrometry, *Mass Spec. Rev.* **16**, 145–162.

Lander, E.S. et al. (2001) Initial sequencing and analysis of the human genome, *Nature* **409**, 860–921.

Le Naour, F. (2001) Contribution of proteomics to tumor immunology, *Proteomics* **1**, 1295–1302.

Link, A. (2002) Multidimensional peptide separations in proteomics, *Trends Biotechnol.* **20**, S8–S13.

Lubec, G., Krapfenbauer, K., Fountoulakis, M. (2003) Proteomics in brain research: potentials and limitations, *Prog. Neurobiol.* **69**, 193–211.

Luo, J., Isaacs, W.B., Trent, J.M., Duggan, D.J. (2003) Looking beyond morphology: cancer gene expression profiling using DNA microarrays, *Cancer Invest.* **21**, 937–949.

Martin, D., Nelson, P. (2001) From genomics to proteomics: techniques and applications in cancer research, *Trends Cell Biol.* **11**, S60–S65.

O'Farrell, P.H. (1975) High resolution two-dimensional electrophoresis of proteins, *J. Biol. Chem.* **250**, 4007–4021.

Ong, S.-E., Foster, J., Mann, M. (2003) Mass spectrometric-based approaches in quantitative proteomics, *Methods* **29**, 124–130.

Petricoin, E.F., Liotta, L.A. (2002) Proteomic analysis at the bedside: early detection of cancer, *Trends Biotechnol.* **20**, S30–S34.

Petricoin, E.F., Zoon, K.C., Kohn, E.C., Barrett, J.C., Liotta, L.A. (2002) Clinical Proteomics: Translating benchside promise into bedside reality, *Nat. Rev. Drug Discov.* **1**, 683–695.

Pieper, R., Gatlin, C.L., Makusky, A.J., Russo, P.S., Schatz, C.R., Miller, S.S., Su, Q., McGrath, A.M., Estock, M.A., Parmer, P.P., Zhao, M., Huang, S.-T., Zhou, J., Wang, F., Esquer-Biasco, R., Anderson, N.L., Taylor, J., Steiner, S. (2003) The human serum proteome: display of nearly 3700 chromatographically separated protein spots on two-dimensional electrophoresis and identification of 325 distinct proteins, *Proteomics* **3**, 1345–1364.

Rosenblatt, K.P., Bryant-Greewood, P., Killian, J.K., Mehta, A., Geho, D., Espina, V., Perticoin, E.F., Liotta, L.A. (2004) Serum Proteomics in Cancer diagnosis and management, *Annu. Rev. Med.* **55**, 97–112.

Ryan, T., Patterson, S. (2002) Proteomics: drug target discovery on an industrial scale, *Trends Biotechnol.* **20**, S45–S51.

Santoni, V., Molloy, M., Rabilloud, T. (2000) Membrane proteins and proteomics. Un amour impossible? *Electrophoresis* **21**, 1054–1070.

Smith, R., Shen, Y., Tang, K. (2004) Ultrasensitive and quantitative analyses from combined separations – mass spectrometry for the characterisation of proteomics, *Acc. Chem. Res.* **37**, 269–278.

Smolka, M., Zhou, H., Purkayasstha, S., Aebersold, R. (2001) Optimization of the isotope-coded affinity tag-labelling procedure for quantitative proteome analysis, *Anal. Biochem.* **297**, 25–31.

Srinivas, P.T., Verma, M., Zhao, Y., Srivastava, S. (2002) Proteomics for cancer biomarker discovery, *Clin. Chem.* **48**, 1160–1169.

Steiner, S., Anderson, N.L. (2000) Expression profiling in toxicology – potentials and limitations, *Toxicol. Lett.* **112–113**, 467–471.

Tao, W., Aebersold, R. (2003) Advances in quantitative proteomics via stable isotope tagging and mass spectrometry, *Curr. Opin. Biotechnol.* **14**, 110–118.

Taylor, S., Fahy, E., Ghosh, S. (2003) Global organellar proteomics, *Trends Biotechnol.* **21**, 82–88.

Walduck, A., Rudel, R., Meyer, T.F. (2004) Proteomic and gene profiling approaches to study host responses to bacterial infection, *Curr. Opin. Microbiol.* **7**, 33–38.

Washburn, M., Wolters, D., Yates, J. (2001) Large-scale analysis of the yeast proteome by multidimensional protein identification technology, *Nat. Biotechnol.* **19**, 242–247.

Wilson, K.E., Ryan, M.M., Prime, J.E., Pashby, D.P., Orange, P.R., O'Beirne, G., Whateley, J.G., Bahn, S., Morris, C.M. (2004) Functional genomics and proteomics: application in neurosciences, *J. Neurol. Neurosurg. Psych.* **75**, 529–538.

Zhu, H., Klemic, J., Chang, S., Bertone, P., Casamayor, A., Klemic, K., Smith, D., Gerstein, M., Reed, M., Snyder, M. (2000) Analysis of yeast protein kinases using protein chips, *Nat. Genet.* **26**, 283–289.

Zhu, H., Bilgin, M., Bangham, R., Hall, D., Casamayor, A., Bertone, P., Lan, N., Jansen, R., Bidlingmaier, S., Houfek, T., Mitchell, T., Miller, P., Dean, R., Gerstein, M., Snyder, M. (2001) Global analysis of protein activities using proteome chips, *Science* **293**, 2101–2105.

Proton Translocating ATPases

Masamitsu Futai[1], Ge-Hong Sun-Wada[2], and Yoh Wada[2]
[1] *Microbial Chemistry Research Foundation and CREST, Japan Science and Technology Agency, Kamiosaki, Shinagawa, Tokyo, Japan*
[2] *ISIR, Osaka University, Ibaraki, Osaka, Japan*

1	Introduction 335	
2	Catalytic Mechanism of Proton-translocating F-ATPase 337	
3	**Roles of the γ-Subunit: Energy Coupling by Mechanical Rotation** 338	
3.1	Roles of the γ-Subunit in Energy Coupling 338	
3.2	Subunit γ-Rotation 340	
3.3	Mutational Analysis of the γ-Subunit Rotation 342	
4	**Rotational Catalysis of the F-ATPase Holoenzyme** 344	
4.1	Structure of F_0 Sector and Proton Transport Pathway 344	
4.2	Rotational Catalysis of the F-ATPase Holoenzyme 345	
4.3	Rotational Catalysis of F-ATPase in Membranes 347	
5	**Rotational Catalysis of V-ATPASE** 350	
5.1	Catalytic Site and Proton Pathway 350	
5.2	Subunit Rotation of V-ATPase During Catalysis 351	
6	**Epilogue** 353	
	Bibliography 353	
	Books and Reviews 353	
	Primary Literature 354	

Keywords

F_0F_1
ATP synthase or proton-translocating ATPase formed from membrane extrinsic (F_1) and intrinsic (F_0) sectors. Nomenclature from factors of oxidative phosphorylation.

F-ATPase
One of the three classes of ion transport ATPases, mostly proton-translocating ATPase. Nomenclature from F_0F_1 or factors of oxidative phosphorylation.

P-type ATPase
Ion transport ATPase forming acylphosphoenzyme intermediates during catalysis. This group contains proton-translocating ATPases such as H^+/K^+ ATPase and H^+ATPase. They are different from F-ATPase, or V-ATPase, but similar to Na^+/K^+ ATPase.

Rotational Catalysis
Continuous rotation of $\gamma\varepsilon$ C_{10} subunit complex during ATP synthesis or hydrolysis by F-ATPase. Similar mechanism found in V-ATPase.

γ-Subunit
F-ATPase subunit located at the center of the catalytic hexamer ($\alpha_3\beta_3$) formed from the α- and β-subunits, and rotating during catalysis.

V-ATPase
Vacuolar-type proton-translocating ATPase. Nomenclature from a proton pump identified initially in fungal or plant vacuoles, but now found in a wide variety of endomembrane organelles such as lysosomes and endosomes and plasma membrane of specialized cells such as osteoclast or kidney intercalated cell.

Abbreviations

DCCD: dicyclohexylcarbodiimide
Pi: inorganic phosphate

■ Proton-translocating ATPases synthesize or hydrolyze ATP coupling with electrochemical proton gradient, and have important roles in animal and plant physiology. The ion transporting ATPases are classified as F-ATPases (F-type ATPases), V-ATPases (V-type ATPases), and P-ATPases (P-type ATPase). The F-ATPase, also called *ATP synthase*, is found in chloroplasts, mitochondria, or bacterial membranes, and synthesizes most of the ATP, the biological energy currency. The enzyme is formed basically from eight subunits, a-, b-, and c-subunits forming membrane intrinsic F_0 sector with ab_2c_{10} stoichiometry, and α-, β-, γ-, δ-, and

ε-subunits forming peripheral F$_1$ sector with $\alpha_3\beta_3\gamma\delta\varepsilon$ stoichiometry. Three catalytic β-subunits forming a hexamer with α-subunit ($\alpha_3\beta_3$) have catalytic cooperativity. The $\gamma\varepsilon c_{10}$ subunits complex, formed from peripheral ($\gamma\varepsilon$) and intrinsic membrane (c) subunits is a central stalk, rotating during catalysis. The V-ATPase forms inside acidic pH in endomembrane organelles such as lysosomes, endosomes, synaptic vesicles, and so on, and is similar to F-ATPase in subunit organization and structure. The same enzyme is found in plasma membrane of special cells such as osteoclast and kidney intercalated cell. Gastric proton pump H$^+$/K$^+$ATPase or plant plasma membrane H$^+$ATPase belongs to P-ATPases forming acyl phosphate intermediate, and is not discussed in this article.

1
Introduction

The mechanism of ATP synthesis has been a focus of biochemists for more than four decades. ATP synthase was first identified in the mitochondrial inner membrane, and then found successively in chloroplasts and bacterial membranes. This enzyme synthesizes ATP from ADP and phosphate (Pi) coupled with an electrochemical proton gradient, and is also called *proton-translocating ATPase* or F-ATPase (F-type ATPase) because of the reversible proton pumping upon ATP hydrolysis. The name F-ATPase originated from the coupling factor of oxidative phosphorylation sensitive to oligomycin. *Escherichia coli* F-ATPase is composed of a membrane extrinsic F$_1$ sector and a transmembrane F$_0$, formed from five ($\alpha_3\beta_3\gamma\delta\varepsilon$) and three ($ab_2c_{10-14}$) subunit assemblies with different stoichiometries, respectively (Fig. 1). It can also be divided into a catalytic $\alpha_3\beta_3$ hexamer, stalks ($\gamma\varepsilon ab_2$), and membrane (ab_2c_{10-14}) domains. Two stalks (central and peripheral stalk) have been observed by electron microscopy. The mitochondrial enzyme has additional subunits, possibly with regulatory functions.

The higher-ordered X-ray structure of the bovine $\alpha_3\beta_3\gamma$ complex had been solved in 1994 by John Walker and coworkers. Following this breakthrough, the structures of crystals of bovine F$_1$, inhibited by efrapeptin, aurovertin, NBD-Cl (7-chloro-4-nitrobenzo-2-oxa-1,3-diazole), and DCCD (dicyclohexylcarbodiimide) were solved by the same group. Bovine F$_1$ containing 1 mol MgADP trifluoroaluminate and 2 mol MgADP trifluoroaluminate has also been crystallized, and the structures have been determined. These studies, together with biochemical analysis, have contributed greatly to an understanding of the catalytic site and mechanism. The structures of F$_1$ sectors of other origins have also been reported.

The catalytic site is mainly located in the β-subunit of the $\alpha_3\beta_3$ hexamer, and the three sites (one in each β) show strong cooperativity. The amino- and carboxyl-terminal helices of the γ-subunit are located in the center of the $\alpha_3\beta_3$ hexamer, and form a central stalk with the ε-subunit. The peripheral or second stalk is formed from the *a*- and *b*-subunits, and the δ-subunit is located near the top of the α-subunit. The proton pathway is located at the interface between

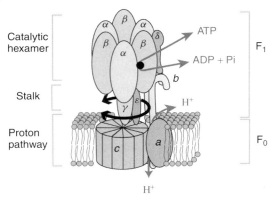

Fig. 1 Schematic model of proton-translocating F-ATPase, showing the subunit structures of the catalytic hexamer ($\alpha_3\beta_3$), stalk and membrane domain. The membrane extrinsic F_1 and transmembrane F_0 sector are shown together with energy coupling between the ATP synthesis–hydrolysis and subunit rotation.

the a- and c-subunits, and the hairpin structure of the purified c-subunit has been solved by Fillingame and coworkers using NMR. A ring structure formed from multiple c-subunits was suggested by early studies involving electron and atomic force microscopy, and was extensively analyzed recently through NMR structure and genetic approaches. The X-ray structure of yeast F_1 with a c-subunit ring has also been reported by Stock et al.

The mechanism of coupling of proton transport and ATP synthesis or hydrolysis has been a major question for this complicated enzyme. The binding change mechanism proposes rotation of the γ-subunit relative to the $\alpha_3\beta_3$ hexamer coupled with the chemistry at the catalytic sites. Rotation of the $\gamma\varepsilon c_{10-14}$ complex relative to $\alpha_3\beta_3\delta ab_2$ upon ATP addition was shown recently, indicating that continuous rotation of the assembly of F_1 and F_0 subunits is involved in the coupling between ATP hydrolysis and proton transport.

Vacuolar-type ATPase (V-ATPase) with significant similarity to F-ATPase was introduced later to a family of proton-translocating ATPases. V-ATPase is apparently different from F-ATPase in its physiological roles, and forms an acidic luminal pH in endomembrane organelles including lysosomes and endosomes, and in extracellular compartments such as resorption lacuna formed between osteoclasts and the bone surface. It has extrinsic membrane V_1 and V_0 transmembrane domains formed from the A, B, C, D, E-, F-, G-, and H-subunits, and a, c, c', c'', and d, respectively. F- and V-ATPase share significant homology, especially in its catalytic subunits and proton pathways, and also share some unique differences, including its membrane sector and stalk region subunit compositions. A series of isoforms have been found for V-ATPase subunits. Thus, comparative studies of the two proton–translocating ATPases are pertinent for understanding the molecular mechanisms of both.

In this chapter, we discuss recent progress made in the understanding of F-ATPase, focusing mainly on the energy coupling between the chemistry and proton transport through subunit rotation, and also discuss V-ATPase from a similar aspect. It should be interesting for readers to follow a series of biochemical studies to show rotational catalysis of the F-ATPase holoenzyme leading to that of the V-ATPase.

2
Catalytic Mechanism of Proton-translocating F-ATPase

As expected from its complicated structure, F-ATPase is not a simple Michaelis–Menten type enzyme. Furthermore, the overall mechanism includes catalysis (chemistry), subunit rotation and proton translocation. X-ray structure and kinetic studies, especially substrate binding analysis with an intrinsic tryptophan probe, revealed that the three catalytic sites are asymmetric. Senior and coworkers have recently discussed and summarized the molecular catalytic mechanism of F-ATPase.

ATP synthesis and hydrolysis could be carried out kinetically at a single site (unisite catalysis), two sites operating together (bisite catalysis), or all three sites working together (trisite catalysis). Unisite catalysis has only been demonstrated for ATP hydrolysis, and can be measured experimentally with an ATP:F_1 ratio of less than 1:3. This rate is 10^5–10^6-fold lower than that of steady state (multisite) catalysis. The enzyme cross-linked chemically and thus could not rotate, but could still carry out unisite catalysis. This catalysis is not a part of the steady state, which includes subunit rotation.

Catalytic residues have been identified by analyzing unisite catalysis of the purified E. coli mutant F_1 sector (Fig. 2). Briefly, βLys155 of the β-subunit is required for binding of the γ-phosphate moiety, as shown by studies involving affinity labeling and mutant enzymes such as βLys155Ala (βLys155 $\rightarrow$ Ala) or βLys155Ser. The βArg182 residue is also involved in the binding. Enzymes with substitutions of βThr156 showed similar properties to those of βLys155. The hydroxyl moiety of βThr156 is essential, possibly for Mg^{2+} binding, since it can only be replaced by a serine residue. βGlu181 is a critical catalytic residue. Its side chain forms a hydrogen bond with a water molecule located near the γ-phosphate of ATP. However, it was shown later that this residue was not involved in nucleotide binding or Mg^{2+} coordination. βTyr331, which is stacked close to the adenine ring, is required for the binding of ADP or ATP. Results of analysis of the tryptophan fluorescence of the βTyr331Trp mutant F_1 are consistent

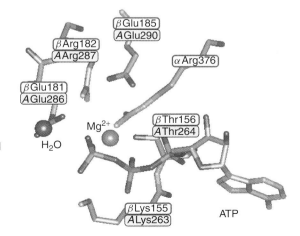

Fig. 2 Catalytic sites of F-ATPase and V-ATPase. The catalytic residues of E. coli F-ATPase are shown together with bound ATP. Their positions are cited according to the bovine crystal structure. Corresponding residues of yeast V-ATPase are also shown.

with the role of the βTyr331 residue. The bovine residues corresponding to those discussed above are located close to the phosphate moiety or the adenine ring of bound ATP or ADP as shown in the X-ray structure (Fig. 2).

Recent results, including the F_1 structure of all three sites filled with nucleotides, support trisite catalysis, in which the three sites are working together during the steady state, and the notion that bisite catalysis does not occur. We leave convincing discussions of these points to the article of Senior and coworkers.

It can be assumed that mutant enzymes defective in steady state catalysis should show impaired unisite catalysis. In this regard, αArg376 mutant enzymes are of interest. The αArg376 residue of the α-subunit is located close to the β- or γ-phosphate of ATP or ADP and Mg at the catalytic site (Fig. 2). However, αAg376 does not directly participate in the chemistry of ATP hydrolysis or synthesis, as shown by the unisite catalysis of mutant F_1 sectors. The αArg376Lys or αArg376Ala mutant enzyme showed 2×10^3-fold lower steady state ATP hydrolysis than the wild type. However, the mutant enzymes showed essentially the same kinetics for unisite catalysis as the wild type, suggesting that they can pass through the transition state. These results indicate that αArg376 is essential for promotion of catalysis to the steady state turnover. This notion is different from the previously suggested roles of αArg376 that was deduced from the structural model and fluoroaluminate binding, an indicator of the formation of the pentacovalent transition state. The βGlu185 residue, located close to the γ-phosphate and Mg at the catalytic site, may have a similar role to αArg376 because the mutant enzymes maintain unisite catalysis, but are defective in multisite catalysis. However, detailed analysis could not be carried out because the mutant F_1 was unstable after solubilization from membranes.

The binding change mechanism of Boyer proposes that the three sites are involved sequentially in ATP synthesis or hydrolysis: at a specific point of time during the steady state, the chemical reaction occurs reversibly at one site and ATP release and/or binding of ATP + Pi occurs at the two other sites with the expenditure of energy. Evidence supporting catalytic cooperativity and the binding change mechanism has accumulated, as reviewed by Boyer in 1997. It includes early kinetic evidence of unisite and steady state catalysis showing three K_m values and three interacting nucleotide binding sites, ^{18}O isotope exchange reactions, inhibitor studies, and so on. These studies suggested that the chemistry, "ADP + Pi $\leftrightarrow$ ATP + H_2O," at the catalytic site is reversible, and provides essentially no energy change. The mechanism is strongly supported by the X-ray structure showing asymmetric catalytic sites and predicting continuous γ-subunit rotation. However, it cannot be concluded that the binding change mechanism was proven at the molecular level, although the mechanism is conceptually accepted and has been extremely useful for understanding the enzyme.

3
Roles of the γ-Subunit: Energy Coupling by Mechanical Rotation

3.1
Roles of the γ-Subunit in Energy Coupling

The essential role of the γ-subunit in catalysis and assembly was shown by early reconstitution experiments: a catalytic core complex exhibiting ATPase activity could be reconstituted from the purified *E. coli* α-, β-, and γ-subunits, but not without the γ-subunit. Consistently, assembly of the F_1 sector is strongly affected by γ-subunit mutations, especially those of residues interacting with the β-subunit. The isolated $\alpha_3\beta_3\gamma$ complex could functionally bind to the F_0 sector only after its assembly with the δ- and ε-subunits. These results indicate that the two minor subunits are required for the functional binding of $\alpha_3\beta_3\gamma$.

Early genetic approaches focused on the γ-subunit carboxyl-terminal region. The amino acid sequences of the γ-subunits are weakly conserved among different species, though their X-ray structures are similar. When the known γ-sequences are aligned, only 28 of the 286 residues of the *E. coli* subunit are conserved, and mostly in the carboxyl- and amino-terminal helices located at the center of the $\alpha_3\beta_3$ hexamer. Seventeen residues are conserved between residues 242 and 286 of the γ-subunit. An enzyme lacking 10 residues at the carboxyl terminus is still capable of *in vivo* ATP synthesis, indicating that the three conserved residues in this region are not required. Structural flexibility of the carboxyl terminus was also demonstrated by a frameshift mutation: the enzyme was still active with the γ-subunit having seven additional residues at its carboxyl terminus together with nine altered residues downstream of γThr277.

The enzyme with the nonsense mutation (γGln269End) was inactive, and substitution of a conserved residue (γGln269, γThr273, or γGlu275) between γGln269 and γLeu276 gave enzymes with reduced ATPase activity and energy coupling. We noticed that mutations resulted in different ratios of ATPase catalysis and proton transport; three mutants (βThr277End, βGln269Leu, and βGlu275Lys) exhibited about 15% of the wild-type ATPase activity, but showed various degrees of ATP-dependent H^+ transport and *in vivo* ATP synthesis. These results suggest active role(s) of the γ-subunit in ATPase activity and energy coupling.

In addition to the carboxyl terminus, the only other conserved region is near the amino terminus. The importance of this region was first suggested by the mutant lacking residues between γLys21 and γAla27, which resulted in failure of assembly of the F_1 complex. We introduced amino acid substitutions systematically into the amino-terminal region of the γ-subunit. Most of the changes between γIle19 and γLys33, γAsp83 and γCys87, or at γAsp65 had no effect, even with a drastic replacement such as hydrophobic to hydrophilic or acidic to basic. Interesting exceptions were the γMet23Arg and γMet23Lys substitutions. These mutants grew only slowly on succinate through oxidative phosphorylation, indicating that they were impaired in ATP synthesis. However, the membranes prepared from the γArg23 and γLys23 strains showed 100 and 65% of the wild-type ATPase activity, but formed only 32 and 17% of the electrochemical proton gradient respectively, indicating that these mutants are defective in energy coupling between ATP hydrolysis and proton transport. In the X-ray structure, the γMet23 residue is located close to the DELSEED (βAsp380–βAsp386) loop of the β-subunit. Thermodynamic and kinetic analyses of the purified γMet23Lys enzyme suggested that the introduced γLys23 residue forms an ionized hydrogen bond with βGlu381 in the loop. Consistent with this interpretation, the phenotype of γMet23Lys was restored by the second mutation, βGlu381Gln. The βGlu381Lys mutation also caused deficient energy coupling. These results suggest that the interaction between the regions around γMet23 and βGlu381, and thus the γ-subunit residue and β-subunit DELSEED loop, are involved in energy coupling.

The γMet23Lys mutation was suppressed by substitution of carboxyl-terminal residues, including γArg242, γGlu269, γAla270, γIle272, γThr273, γGlu278, γIle279k, and γVal280. From these results, Nakamoto et al. suggested that γMet23, γArg242, and the region between γGlu269 and γVal280 are three interacting domains that are required for efficient energy coupling. The X-ray structure shows that γMet23 located in the amino-terminal helix is near γArg242, but γGly269 and γVal280 in the carboxyl-terminal helix are near the top of the $\alpha_3\beta_3$ hexamer. Furthermore, second-site mutations mapped to the amino (residues 18, 34, and 35) and carboxyl (residues 236, 238, 242, and 262) termini suppressed γGln269Glu and γThr273Val mutations. The higher-ordered structure clearly shows that γGlu269 or γThr273 does not interact directly with the residues of the second mutations. The occurrence of suppression at a distance may suggest that the two α-helices of the γ-subunit located at the center of the $\alpha_3\beta_3$ complex undergo long-range conformational changes during catalysis. As expected, the relative orientations of the γ-subunit to the three β-subunits (β_E, β_{DP}, and β_{TP}) are different in the crystal structure, strongly supporting γ-subunit rotation during ATP hydrolysis or synthesis.

3.2 Subunit γ-Rotation

The higher-ordered structure of $\alpha_3\beta_3\gamma$ indicated that α and β are arranged alternately around the amino- and carboxyl-terminal α-helices of the γ-subunit. The binding change mechanism predicts that the catalytic sites in the three β-subunits participate sequentially in ATP synthesis or hydrolysis via conformation transmission through the γ-subunit rotation. This rotation was suggested by experiments on chemical cross-linking between γCys87 and βCys380 (originally βAsp380) in the DELSEED loop, and analysis of polarized absorption recovery after photobleaching of a probe linked to the carboxyl terminus of the chloroplast γ-subunit.

The continuous unidirectional γ-subunit rotation in F_1 was recorded directly by Noji, Yoshida, and their colleagues (Fig. 3). They immobilized the Bacillus $\alpha_3\beta_3\gamma$ complex on a glass surface through histidine residues introduced into the β-subunit. The fluorescent actin filament connected to the γ-subunit rotated continuously in an anticlockwise direction during ATP hydrolysis. The rotation became slower with an increase in the filament length, and generated a frictional torque of $\sim$40 pN nm. The ε-subunit rotation was also shown using the same approach, consistent with the tight association of γ and ε. A 120° step rotation was shown in the presence of a dilute ATP concentration, indicating that the γ-subunit rotates, interacting with the three β-subunits successively. Furthermore, a refined measurement system involving gold beads revealed that the 120° step could be divided into 90 and 30° steps. These two steps were proposed to correspond to ATP binding and ADP release, respectively.

The rotation of an actin filament connected to the γ-subunit of E. coli F_1 was observed using a similar system to that described for the Bacillus $\alpha_3\beta_3\gamma$ complex. The rotation was anticlockwise, inhibited by azide (F-ATPase inhibitor),

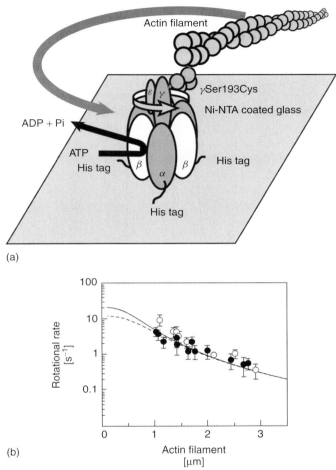

(a)

(b)

Fig. 3 Rotation of the γ-subunit of F-ATPase. F_1 was immobilized through a histidine tag introduced to the α- or β-subunit, and an actin filament was connected to the carboxyl terminus of the γ-subunit. ATP-dependent rotation of the wild-type (open circles) and γMet23Lys mutant (filled circles) F_1 sector are shown. Taken from Omote, H., Sambonmatsu, N., Saito, K., Sambongi, Y., Iwamoto-Kihara, A., Yanagida, T., Wada, Y., Futai, M. (1999) The γ-subunit rotation and torque generation in F_1-ATPase from wild-type or uncoupled mutant Esche, *Proc. Natl. Acad. Sci. U.S.A.* **96**, 7780–7784. Experimental procedure developed by Noji et al. (1997).

and generated a frictional torque of ~40 pN·nm, as reported for the Bacillus complex. The F_1 sector, proven to be a chemically driven motor, should be connected functionally to the F_0 sector to complete ATP-driven proton transport. Conversely, proton transport through F_0 should be coupled to the γ-subunit rotation and chemistry at the catalytic sites. Questions on the energy coupling among chemistry, rotation, and proton transport will be answered with the E. coli enzyme by taking advantage of the accumulated genetic and biochemical information.

3.3
Mutational Analysis of the γ-Subunit Rotation

We were interested in characterizing the γ-rotation of a series of mutant enzymes defective in energy coupling and catalytic cooperativity. As described above, the γMet23Lys mutant is defective in energy coupling between catalysis and proton transport. We thought that the γ-rotation in the mutant F_1 may be defective since the γ-residue was replaced. Similar to the original mutant, the γMe23Lys enzyme engineered for rotation observation exhibited essentially the same ATPase activity as the engineered wild type. However, the enzyme could not form an electrochemical proton gradient in membrane vesicles, or carry out *in vivo* oxidative phosphorylation. Unexpectedly, an actin filament connected to the γ-subunit of γMet23Lys rotated, and generated essentially the same torque as that of the wild type (Fig. 3). These results suggest that the γMet23Lys mutant F-ATPase could couple between chemistry and rotation, but was defective in transforming mechanical work into proton translocation or vice versa. In this regard, Al-Shawi et al. showed that the γMet23Lys mutant is defective in communication between F_1 and F_0. These results also suggested that analysis of F_1 sector rotation is not enough to understand F-ATPase catalysis. It became imperative to determine which subunit complex is rotating in the F-ATPase holo enzyme purified or embedded in the membrane.

Mutant F_1 sectors with substitution of the βSer174 residue and their suppressors have been useful for understanding the rotation mechanism (Fig. 4). Replacing βSer174 with other residues altered the ATPase activity to between 150 and 10% of the wild-type level, and the larger the side chain of the residue introduced, the lower the ATPase activity observed. Both the βSer174Leu and βSer174Phe enzymes retained about 10% of the wild-type ATPase activity. However, the two mutants showed a difference in energy coupling: the βSer174Leu mutant could still grow on succinate by oxidative phosphorylation and transport protons into the isolated membrane vesicles, whereas the βSer174Phe mutant could not grow and showed no proton transport. Consistent with these observations, their F_1 sectors differed in γ-subunit rotation. The F_1 sector with βSer174Phe showed apparently slower rotation than the wild type, and generated significantly lower frictional torque ($\sim$17 pN nm), whereas the F_1 with βSer174Leu was similar to the wild type. These results suggest that the rotation or torque generation is closely related to the energy coupling with proton transport.

Biochemical defects of the βSer174Phe mutation were suppressed by a second-site mutation, βGly149 to Ser, Cys, or Ala. Double mutants such as βSer174Phe/βGly149 Ser showed essentially the same ATPase activity and proton transport as the wild type. As expected from these results, the double mutant F_1 generated the wild-type torque ($\sim$40 pN nm). The high-resolution structure of the bovine F_1 predicts that βSer174 is located on the β-subunit surface within the loop between an α-helix (helix B) and a β-sheet (β-sheet 4) (Fig. 4). βGly149, the first residue of the phosphate-binding P-loop connected to helix B, is

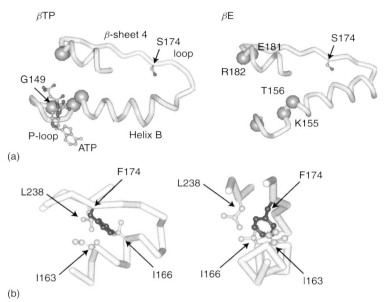

Fig. 4 Models of the *E. coli* β-subunit domain including βGly149 and βSer174. (a) Domain structure for the ATP-bound (β_{TP}) and empty (β_E) β-subunit. The bovine structure was used to model *E. coli* domains. Positions of ATP and residues discussed in the text are indicated. Nomenclatures for the α-helix and β-sheet are those of Abrahams et al. (1994). (b) Models of the βSer174Phe mutant domain structure. The positions of the residues discussed in the text are shown. Taken from Iko, Y., Sambongi, Y., Tanabe, M., Iwamoto-Kihara, A., Saito, K., Ueda, I., Wada, Y., Futai, M. (2001) ATP synthase F_1 sector rotation. Defective torque generation in the β subunit Ser-174 to Phe mutant and its suppression by second mutations, *J. Biol. Chem.* **276**, 47508–47511.

close to the catalytic site. The structure of the βGly149–βSer174 region is significantly different between the nucleotide-bound and empty β-subunits. Thus, it can be assumed that the conformational change of the catalytic site, followed by that of the βGly149–βSer174 domain, leads to the γ-subunit rotation for the energy coupling to proton transport. The βSer174Phe mutation was also suppressed by αArg296Cys of the α-subunit. It is of interest to analyze the rotation of the double mutant and related strains.

Energy minimization with a simple potential function of the modeled βSer174Phe mutant F_1 predicts that the side chain of βPhe174 interacts with that of βIle163 or βIle166 of the β-subunit with no nucleotide bound. We substituted βIle163 or βIle166 with a less bulky Ala residue in the βSer174Phe mutant. As expected, the F_1 with the βIle163Ala/βSer174Phe or βIle166Ala/βSer174Phe double mutant could rotate and generate almost similar torque to the wild type. These results suggest that the βGly149–βSer174 domain plays important roles in the rotation and torque generation essential for energy coupling.

The role of the DELSEED loop has been focused on because its conformation is significantly different among the three β-subunits ($β_E$ and $β_T$ or $β_D$), consistent with the roles of $γ$Met23 and $β$Glu381 (the second residue of the DELSEED loop) in energy coupling. However, mutagenesis studies involving replacement of residues in the DELSEED loop of thermophilic Bacillus $α_3β_3γ$ did not reveal significant effects on the torque generation. The negative results may suggest that the loop is not related to the conformation change driving the $β$-subunit rotation. However, this is difficult to conclude from the directed mutations without structural analysis. Furthermore, the experimental system with actin filaments gives rotation rates and torque values with high deviations. Extensive substitution of related residues and their second mutations should be analyzed for the final conclusion, as discussed earlier, for a region around $β$Ser174. Analysis of rotation in the F-ATPase holoenzyme may be necessary, as pointed out, for the $γ$Met23Lys mutation.

4
Rotational Catalysis of the F-ATPase Holoenzyme

4.1
Structure of F_0 Sector and Proton Transport Pathway

As discussed earlier, the F_0 membrane sector is composed of a-, b-, and c-subunits, whose stoichiometry (1 : 2 : 10 ± 1, for a : b : c) was first determined from the stained bands on polyacrylamide gel electrophoresis. On the basis of structural prediction, Cox et al. proposed a model of F_0 in which the a- and b-subunit helices are surrounded by a ring of c-subunits. However, this model was not consistent with the electron and atomic force microscopic images. In the model derived from the images, the a- and b-subunits are attached to one side of the symmetric ring structure formed by the c-subunits. An atomic force microscope image indicated <12 c-subunits in the ring. The ab_2 complex was purified recently after solubilization of F_0 from membranes utilizing a histidine tag introduced at the a-subunit amino terminus. The proton pathway (F_0) was reconstituted from the ab_2 complex and c-subunit, confirming that ab_2 and the c-ring are two functional subcomplexes of the F_0 sector.

Subunit a spans the membrane with five helices, and its fourth helix could be cross-linked to the carboxyl-terminal helix of a c-subunit when Cys residues were introduced at appropriate positions, aArg210 in the fourth helix is involved in proton transport, and all the substitutions, even aArg210 to Lys, gave an F_0 sector with no ability to transport protons. The structure of the amino terminus (residues 1–34) of the b-subunit, including the transmembrane helix, was solved by NMR in a chloroform–methanol–water mixture. Extramembrane helices interact with the $δ$- and $α$-subunits at the top of the F_1 sector, which is consistent with the model proposing that the b-subunit closely interacts with an $α_3β_3$ hexamer. The b-subunit dimer interaction with the F_1 sector is necessary for a second stalk. The carboxyl-terminal helix of a c-subunit can be cross-linked to the membrane helix of the b-subunit. Fillingame and coworkers solved the structure of the E. coli c-subunit by NMR: monomeric c (in a mixture of chloroform, methanol, and water, pH 5) gave a hairpin-like structure formed from the two helices connected by a polar region that interacts with the $γ$- and $ε$-subunits. Structural models of F_0

and the c-ring were reviewed recently by Fillingame and coworkers.

The front face of one c-subunit packs with the back face of a second c-subunit to form a dimer, consistent with the results of mutant studies. Fillingame and coworkers proposed that dimers form a ring of 12 monomers, with the amino and carboxyl α-helices of each c-subunit in the interior and at the periphery, respectively, and the ring was modeled from the results of molecular dynamic calculations. The model was supported by cross-linking analysis. However, a ring containing more than 10 monomers was not found in purified F_0F_1, and recent recalculation gave essentially the same, but a slightly smaller ring, formed from 10 copies of the c-subunit with the two helices in similar orientations. In an alternative model, the carboxyl- and amino-terminal α-helices form inner and outer rings, respectively. However, this model is not supported by cross-linking experiments. The X-ray structure indicated an yeast F_1 tightly bound with a c-ring formed from 10 monomers.

These results established that the c-subunits form a ring of 10 monomers. However, the number of monomers forming the ring is still controversial: atomic force microscopy of chloroplast and bacterial F_0 demonstrated rings of 14 and 11 c-subunits, respectively. The difference in the copy number may be due to the difference in species, loss of a part of the monomers, or reorganization of the ring during purification. Structural studies on F-ATPase or the F_0 sector will eventually explain the discrepancy.

cAsp61, in the middle of the second transmembrane α-helix, is responsible for proton transport, and close to cAla24 and cIle28, of which substitutions by other residues reduced the DCCD reactivity of cAsp61. The stoichiometry of the a- and c-subunits (1 : 10) indicates that one aArg210 and multiple cAsp61 are required for proton translocation in F-ATPase. As the pK_a of the cAsp61 carboxyl moiety is 7.1, the NMR c-subunit structure at pH 5 is at the fully protonated stage. The interaction between cAsp61 and aArg210 may lower the pK_a to facilitate proton release into the proton pathway. Rastogi and Girvin solved the c-subunit structure at pH 8, with cAsp61 being in an almost completely deprotonated form. The difference between the c structures at pH 5 and 8 is that the carboxyl-terminal α-helix was rotated by 140° with respect to the amino-terminal helix. From the two structures and the results of cross-linking experiments, Fillingame suggested that the carboxyl-terminal α-helix rotates during proton transport, interacts with aArg210, and deprotonates cAsp61. This c-subunit structural change possibly drives stepwise rotation of the c-ring.

4.2
Rotational Catalysis of the F-ATPase Holoenzyme

To complete ATP hydrolysis-dependent proton transport, the γ-rotation should be transmitted to the F_0 membrane sector. Conversely, for ATP synthesis, proton transport through F_0 should generate torque to drive the γ-subunit rotation coupled to chemistry. The γ-rotation should be coupled to protonation–deprotonation of cAsp61 in either direction. Depending on the modeling of F_0 (a- and b-subunit inside the c-ring), rotation of a, b, γ, δ, and ε relative to the complex of α, β, and the c-ring once has been suggested. This speculation led to experimental tests, although an alternative model in which the c-subunit ring attached by the a- and

b-subunit helices is supported experimentally, as discussed above.

Different mechanisms could be proposed for the coupling between γ-rotation within the $\alpha_3\beta_3$ hexamer and proton transport through the F_0 sector: (1) the γ-subunit rotates on the surface of the c-ring; and (2) the γ-subunit and the c-ring rotate as a single complex. Models should be consistent with the proton transport continuously utilizing one aArg210 and 10–14 cAsp61 residues of the a-subunit and the c-ring, respectively. Consistent with mechanism (2), energy transduction of F-ATPase with a counterrotating rotor ($\gamma\varepsilon c$ ring) and stator has been proposed. Convincing evidence supporting this mechanism includes the results of chemical cross-linking. Cross-linking between the γ- and c-subunits did not affect F-ATPase activity, indicating that sequential interaction of the rotating γ-subunits with multiple c-subunits is not necessary during ATP synthesis–hydrolysis. Similarly, cross-linking between the γ- and ε-subunits did not affect ATP hydrolysis, supporting the rotation of an actin filament connected to either subunit. However, $\alpha-\gamma$, $\alpha-\varepsilon$, $\beta-\gamma$, and $\beta-\varepsilon$ cross-linking resulted in loss of the activity, confirming that $\varepsilon\gamma$ rotates against $\alpha_3\beta_3$. These results suggest that a complex formed from γ-, ε-, and c-subunits is a mechanical unit for relative rotation to the $\alpha_3\beta_3$ hexamer.

We obtained direct evidence of continuous rotation of the E. coli $\varepsilon\gamma c_{10-14}$ complex during ATP hydrolysis (Fig. 5). F-ATPase engineered for rotation was solubilized from membranes, purified, and immobilized through histidine tags introduced into the α-subunit (Fig. 5a). Upon the addition of ATP, the filament connected to the c-ring rotated anticlockwise continuously, and generated similar torque to that observed for the γ-rotation in the F_1 sector. The rotation was inhibited by venturicidin, a specific inhibitor for F-ATPase but not for ATPase activity of the F_1 sector. The rotation and ATPase activity became less sensitive to the antibiotic when the cIle38Thr mutation was introduced into the c-subunit, indicating that antibiotic binding to the c-ring inhibited rotation.

Pänke et al. also observed c-subunit rotation during ATP hydrolysis. They immobilized F_0F_1 through a histidine tag introduced into the β-subunit, and an actin filament was connected to the c-subunit through a streptag and streptavidin. The characteristics of the rotation observed were the same as those with the above systems. These results are consistent with the recent experiments showing that complete cross-linking of the γ-, ε-, and c-subunit had no effect on ATP hydrolysis, proton translocation, or ATP synthesis.

The important conclusion drawn from these experiments is that the c-subunit ring rotates when F-ATPase is immobilized through the α- or β-subunit. As discussed by Pänke et al. and Wada et al., it is not easy to prove that all the subunits were integrated into F-ATPase rotating under the microscope. However, indirect evidence supports the intactness of the enzyme used for rotation: reconstitution studies indicated that the a-, b-, and c-subunit are required to form F_0 capable of F_1 binding, and all F_1 subunits are required for binding to the F_0 sector. Early genetic and biochemical studies led to the conclusion that all three F_0 subunits are required for F_1 binding. The intactness of F-ATPase in these observations was also supported by the results for the membrane-bound enzyme, which rotated in essentially a similar manner to the purified F-ATPase.

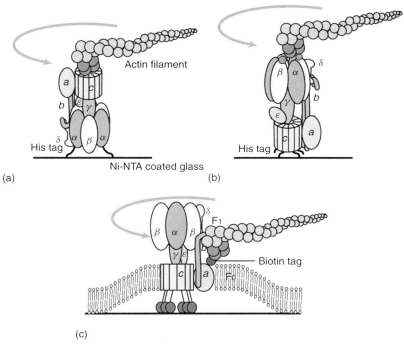

Fig. 5 Rotational catalysis of F-ATPase. F_0F_1 was immobilized through a histidine tag attached to the α-subunit (a) or c-ring (b, c), and a fluorescent actin filament was connected as a probe to the c (a), α (b), or a (c) subunit to observe rotational catalysis. Continuous rotation of the probe connected to the purified F_0F_1 (a or b) or membrane-embedded F_0F_1 (c) was observed upon addition of ATP.

We further addressed the basic question of whether the rotor and stator are interchangeable in F-ATPase (Fig. 5b). Thus, F-ATPase was immobilized on a glass surface through a histidine tag introduced into the c-subunit, and an actin filament was connected to the β-subunit through the biotin-binding domain of transcarboxylase (biotin tag). ATP-dependent filament rotation generated the same frictional torque as observed above, similar to the case of F-ATPase immobilized through the α- or β-subunit and with an actin filament connected to the c-subunit. Thus, either of the two subcomplexes ($\varepsilon\gamma c_{10-14}$ or $\alpha_3\beta_3\delta ab_2$) could be a rotor or a stator. Their actual roles *in vivo* will depend on the viscous drag due to the cytosol and membranes.

4.3
Rotational Catalysis of F-ATPase in Membranes

Although rotation of the $\gamma\varepsilon C_{10-14}$ complex has been shown using purified F-ATPase, a more favorable experimental system is the isolated membrane because the original integrity of the enzyme is maintained. The membrane could be used to answer one of the obvious questions of whether the c-ring rotates relative to the a-subunit. Such rotation would support the model of proton transport through the interface of the two subunits (Fig. 5c). It may be difficult to test the rotation of a probe connected to the c-ring embedded in membranes by immobilizing F-ATPase through the α- or β-subunit. As described earlier, purified F-ATPase can be immobilized through a histidine tag connected to the c-ring, and an actin filament can be connected to the α- or β-subunit through a biotin tag. This experimental system was modified to test the rotation of F-ATPase in the membrane. As histidine and biotin tags face the periplasm and cytoplasm of the intact E. coli cell, respectively, membrane preparation for rotation should be carefully considered. Right-side out or everted membrane vesicles cannot be used for testing rotation, because only one of the two tags in these vesicles is accessible from the medium: the histidine tag faces the medium for right-side out vesicles, but the biotin tag is inside the same vesicles. Similarly, the histidine tag is inside everted vesicles, and thus cannot be used for immobilizing F-ATPase in these vesicles. Therefore, F-ATPase in planar membranes should be used to test for the rotation.

We prepared membrane fragments by passing E. coli cells through a French press by a slight modification of the procedure used to prepare everted vesicles. A labeling experiment with streptavidine-conjugated gold particles indicated that the preparation contained a significant number of planar membranes. Using this preparation, we observed ATP-dependent rotation of an actin filament connected to the β-subunit. As expected from the previous studies, the rotation was counterclockwise and sensitive to DCCD.

To complete a model of proton translocation through the rotation in F-ATPase, it is essential to show rotation of the c-ring relative to the a-subunit (Fig. 5c). The c-subunit (cMet65Cys) cross-linked with the a-subunit (αAsn214Cys) was highly protected from labeling with ^{14}C-DCCD. However, when F-ATPase was labeled with ^{14}C-DCCD and subjected to the ATPase reaction before cross-linking, there was significantly increased labeling of the a–c dimer. This experiment supports the rotational catalysis in F_0, although no

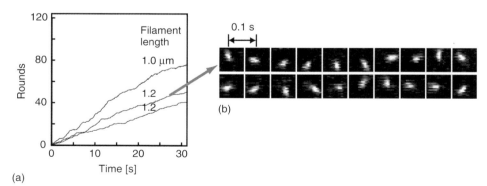

Fig. 6 Relative rotation of the a-subunit in membrane-embedded F-ATPase. Time courses of rotating filaments connected to the a-subunit of membrane F-ATPase immobilized through the c-subunit are shown, along with time courses of rotating filaments of varying length (a) and typical sequential video images (video interval, 100 ms) (b). Taken from Nishio, K., Iwamoto-Kihara, A., Yamamoto, A., Wada, Y., Futai, M. (2002) Subunit rotation of ATP synthase embedded in membranes: a or β subunit rotation relative to the c subunit ring, Proc. Natl. Acad. Sci. U.S.A. **99**, 13448–13452.

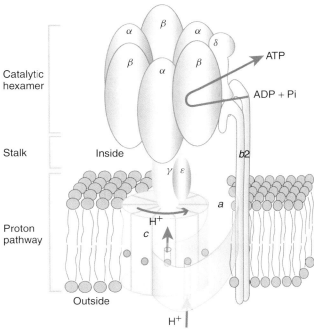

Fig. 7 Schematic mechanism of rotational catalysis by F-ATPase. For ATP synthesis, electrochemical proton transport changes the c-subunit conformation that drives the rotational movement of the c-ring together with the γ- and ε-subunits. Conversely, ATP hydrolysis drives the γ- and ε-subunit rotation together with the c-ring to transport protons. Modified from Wang, H., Oster, G. (1998) Energy transduction in the F_1 motor of ATP synthase, *Nature* **396**, 279–282; Fillingame, R.H., Angevine, C.M., Dmitriev. O.Y. (2002) Coupling proton movement to c-ring rotation in F_0F_1 ATP synthase: aqueous access channel and helix rotations at a–c interface, *Biochim. Biophys. Acta* **1555**, 29–36; Junge, W., Lill, H., Engelbrecht, S. (1997) ATP synthase: an electrochemical transducer with rotatory mechanics, *Trends Biochem. Sci.* **22**, 420–423.

information on the mechanism could be obtained. We could show rotation of the actin filament connected to the a-subunit relative to the c-ring using membrane fragments (Fig. 6).

These results and those obtained with purified F_0F_1 or the F_1 sector indicate that $\gamma\varepsilon c_{10-14}$ and $\alpha_3\beta_3\delta ab_2$ are mechanical units, that is, an interchangeable rotor and stator, respectively. Notably, the results obtained with membrane-embedded F-ATPase and those with the purified one are similar. Furthermore, the experimental system will be extremely valuable for studying the mechanism of coupling of rotation and proton transport, although further modification(s) may be necessary.

These studies established rotational catalysis by F-ATPase (Fig. 7). For ATP hydrolysis, the F-ATPase is a chemically driven motor rotating $\gamma\varepsilon c_{10-14}$ to drive proton transport. Studies on the mechanism of the rotation in the F_0 sector were started only recently. The rotation is inhibited by venturicidin or DCCD, suggesting that the tight or covalent binding of these bulky chemicals to the c-subunit inhibits mechanical rotation. Junge and coworkers showed that the c-subunit ring with the cAsp61Asn mutation can still rotate during ATP hydrolysis, indicating that the proton transport is not obligatory for the chemically driven rotation (ATP hydrolysis-dependent c-ring rotation). For ATP synthesis, the same enzyme is a potential-driven motor rotating $\gamma\varepsilon c_{10-14}$ through an electrochemical proton gradient. This membrane system will be useful for studying the rotation in this direction.

5
Rotational Catalysis of V-ATPASE

5.1
Catalytic Site and Proton Pathway

V-ATPase (vacuolar-type ATPase) acidifies the lumens of endomembrane organelles such as lysosomes, endosomes, and synaptic vesicles. The same enzyme in the plasma membranes of specialized cells, including osteoclasts and renal intercalated cells, pumps protons into extracellular compartments such as resorption lacunae and collecting ducts, respectively. Despite significant physiological differences, V-ATPase exhibits similarities with F-ATPase. The catalytic A-subunit of V-ATPase is homologous to F-ATPase β. The homology is striking in the phosphate-binding P-loop and other sequences, including the F-ATPase catalytic residues discussed above. E. coli F-ATPase residues at the catalytic site, βLys155, βThr156, βGlu181, βArg182, and βGlu185, correspond to ALys263, AThr264, AGlu266, AArg267, and AGlu290 of the yeast V-ATPase A-subunit, respectively (Fig. 2). Mutational studies of yeast V-ATPase supported the notion that they are catalytic residues. The kinetics indicate that V-ATPase has three catalytic sites exhibiting cooperativity. The cysteine residue in the V-ATPase P-loop may be involved in regulation. F-ATPase with Cys introduced at the corresponding position became sensitive to sulfhydryl reagents, similar to V-ATPase.

The membrane V_0 sector of V-ATPase has a more complicated subunit composition: yeast V_0 is formed from c-, c'-, c''-, a-, and d-subunits. Of the three proteolipid subunits, the c- and c'- having four transmembrane helices are a duplicated form of the F-ATPase c-subunit, and show 56% identity in amino acid residues with each other. The carboxyl moieties (cGlu137 and c'Glu145) for proton transport are located on the fourth helix of the c- and c'-subunits, respectively. The c''-subunit, exhibiting some homology with c and c', is also required in proton transport: c''Glu188 in the middle of the third α-helix is also implicated for proton transport. The c''-subunit is conserved in mammalians, whereas the c'-subunit is only found in yeast. The stoichiometry of the $c : c' : c''$ subunits in yeast is $n:1:1$. The three proteolipids may form a ring structure similar to the F_0c-ring. Assuming that 4, 1, and 1 molecules of the c, c' and c'' subunits, respectively, form a ring, V_0 has ~ 6 proton–translocating residues altogether. This is in contrast with 10–14 residues for F_0. Mutation of V-ATPase aArg735 abolished proton translocation similar to that of F-ATPase aArg210, although the subunit a of V_0 and F_0 exhibit little homology. Thus, V_0 has multiple carboxyl moieties and one arginine essential for proton translocation.

The similarities between the two ATPases raised the interesting question of whether V-ATPase can synthesize ATP. Unlike F-ATPase localized with the respiratory chain in the mitochondrial membrane, mammalian or yeast membranes containing V-ATPase do not have a system for generating an electrochemical proton gradient. Does V-ATPase synthesize ATP when a membrane potential or proton gradient is generated? To answer this question, we expressed a plant proton–translocating pyrophosphatase in yeast vacuoles. Upon hydrolysis of pyrophosphate, the vacuolar membrane vesicles could form an electrochemical proton gradient, which was lowered by the addition of ADP + Pi. As expected from this result, we observed ATP synthesis sensitive to the V-ATPase specific inhibitor bafilomycin. The results further support the similarities between the two ATPases. Plant vacuoles having both V-ATPase and pyrophosphatase may synthesize ATP similar to the yeast chimeric system.

5.2
Subunit Rotation of V-ATPase During Catalysis

Despite the similarities of V- and F-ATPase discussed earlier, they are also significantly different. Typically, F-ATPase of *E. coli* has eight subunits, whereas yeast V-ATPase has 13. Most of the subunits unique to V-ATPase are located in the stalk region. Furthermore, the presence of isoforms of subunits B, E, G, d, and a has been found for V-ATPase, whereas the information on isoforms is limited for F-ATPase. As pointed out above, V_0, possibly its ring structure, may be different from F_0. These differences and similarities led us to examine the rotational catalysis in V-ATPase.

As described above, rotation of an actin filament connected to the a-, α-, or β-subunit was observed when F-ATPase was immobilized through the c-subunit ring. The correspondence of minor subunits in the stalk region between the two AT-Pases was difficult to determine (Fig. 8).

```
Yeast G    1 ---------------------------------------------------------MSQKN       5
E.coli b   1 VNLNATILGQAIAFVLFVLFCMKYVWPPLMAAIEKRQKEIADGLASAERAHKDLDLAKAS     60

           6 GIATLLQAEKEAHEIVSKARKYRQDKLKQAKTDAAKEIDSYKIQKDKELKEFEQKNAGGV     65
          61 ATDQLKKAKAEAQVIIEQANKRRSQILDEAKAEAEQERTKIVAQAQAEIEA-ERKRARE-    118

          66 GELEKKAEAG-VQGELAEIKKIAEKKKD-DVVKILIETVIKPSAEVHINAL              114
         119 -ELRKQVAILAVAGAEKIIERSVDEAANSDIVDKLVAEL------------              156


Yeast G    1 --------------------------------------MSQKNGIATLLQAEKEAHEIVSK    23
E.coli δ   1 MSEFITVARPYAKAAFDFAVEHQSVERWQDMLAFAAEVTKNEQMAELLSGALAPETLAES     60

          24 ARKYRQDKLKQAKTDAAKEIDSYK-IQKDKELK-EFEQKNAGGVG--ELEKKAEAGVQGE    79
          61 FIAVCGEQLDENGQNLIRVMAENGRLNALPDVLEQFIHLRAVSEATAEVDVISAAALSEQ    120

          80 -LAEIKKIAEKKKDDVVKIL--IETVIKPSAEVHINAL-----------------        114
         121 QLAKISAAMEKRLSRKVKLNCKIDKSVMAGVIIRAGDMVIDGSVRGRLERLADVLQS     177
```

Fig. 8 Similarities of the V-ATPase *G*-subunit with the F-ATPase *b*- or *δ*-subunit. The subunits are aligned to obtain maximal homology. Identical amino acid residues are boxed.

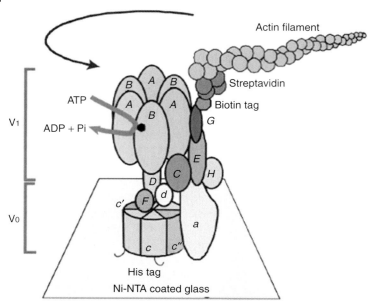

Fig. 9 Rotation of an actin filament connected to the G-subunit of V-ATPase. Solubilized V-ATPase was immobilized on a glass surface through the c-subunit, and an actin filament was connected to the G-subunit. Rotation was observed upon the addition of ATP, and inhibited by nitrate and concanamycin. Taken from Hirata, T., Iwamoto-Kihara, A., Sun-Wada, G.-H., Okajima, T., Wada, Y., Futai, M. (2003) Subunit rotation of vacuolar-type proton pumping ATPase: relative rotation of the G and C subunits, *J. Biol. Chem.* **278**, 23741–23719.

The G-subunit was suggested to exist in the stalk domain of a recent model, and to be accessible from the cytosol. The G-subunit and F-ATPase b exhibit ~24% homology, although the G-subunit lacks a transmembrane region. However, recent cross-linking studies suggested that the G-subunit is located near the top of V_1, similar to the F-ATPase δ-subunit. Thus, the G-subunit may correspond to the b or δ-subunit of F_1, and could be a candidate for connecting a probe to observe rotation, although the homology is quite limited.

On the basis of these considerations, we engineered the yeast chromosome (*VMA3* and *VMA10* for the c- and G-subunit, respectively), solubilized V-ATPase from vacuolar membranes, and tested rotational catalysis. An actin filament connected to the G-subunit rotated upon ATP hydrolysis (Fig. 9). The rotation was inhibited by nitrate and concanamycin, similar to their effects on ATPase activity. Concanamycin inhibition was striking: the rotation terminated within a few seconds after the addition, which is consistent with its tight affinity. This antibiotic is similar to bafilomycin, which binds to the V_0 sector, possibly to its c-subunit, indicating that concanamycin, possibly bound to the rotor–stator interface, terminated the rotation. Torque generated by the rotation was similar to that with F-ATPase.

These results suggest that the A_3B_3 hexamer rotates together with the G-subunit when the c-subunit ring is immobilized. An interesting question is "Which part of the V-ATPase rotates *in vivo*?" In this regard, Holiday and coworkers reported that the B-subunit interacts with the cytoskeleton, possibly actin. Thus, the c-subunit ring together with the subunit corresponding to the γ- (possibly D) subunit may be rotating *in vivo* if the V-ATPase holo enzyme is immobilized with the cytoskeleton. However, it is also possible that the rotor–stator is determined only by the slight difference in viscous drag applied to each subcomplex.

The rotational catalysis of the peripheral membrane sector of *Thermus thermophilus* ATPase was shown recently. This Archaean ATPase is more similar to F-ATPase than V-ATPase, as pointed out by the authors. In this regard, the ATPase of Archae (Archaean bacterial) plasma membranes has been classified as an A-type ATPase. Thus, three related proton–translocating ATPases carry out common rotational catalysis.

6
Epilogue

Proton-translocating F-ATPase (ATP synthase) has been a fascinating membrane enzyme for generations of biochemists. As Paul Boyer called it *A splendid molecular machine* it is more than an ordinary enzyme. F-ATPase synthesizes or hydrolyzes ATP coupling between proton translocation and chemistry, and has become a real molecular machine in this sense, since mechanical rotation of its subunit complex was included recently in its mechanism. For the engineering aspect of the mechanism, the F_1 sector immobilized on a metal surface can rotate a metal plate.

As described above, experimental systems have been established for further studies. Details of the rotation mechanism will be revealed by studies involving *E. coli* F-ATPase together with the progress regarding physical methods and the higher-ordered X-ray structure determinations of the F_0 sector.

It was interesting to learn that rotational catalysis has been expanded to V-ATPase and also distantly related Archaean ATPase. The basic mechanism of V-ATPase rotation may be similar to that of F-ATPase, as expected from the structural similarities between the two enzymes. Thus, the rotational mechanism of *E. coli* F-ATPase should be studied extensively, using its genetic and biochemical advantages. V-ATPase may have a more fascinating regulatory mechanism that is supported by a unique V_0 structure and a series of isoforms in the stalk region. The regulatory role of stalk subunit(s) including subunit E in energy coupling has already been discussed.

See also Biological Regulation by Protein Phosphorylation; Metabolic Basis of Cellular Energy; Protein and Nucleic Acid Enzymes; Protein Modeling.

Bibliography

Books and Reviews

Boyer, P.D. (1997) The ATP synthase – a splendid molecular machine, *Annu. Rev. Biochem.* **66**, 717–749.

Fillingame, R.H. (1996) Membrane sectors of F- and V-type H^+-transporting ATPases, *Curr. Opin. Struct. Biol.* **6**, 491–498.

Fillingame, R.H., Dmitriev, O.Y. (2002) Structural model of the transmembrane Fo rotary sector of H$^+$-transporting ATP synthase derived by solution NMR and intersubunit cross-linking in situ, *Biochim. Biophys. Acta* **1565**, 232–245.

Fillingame, R.H., Angevine, C.M., Dmitriev, O.Y. (2002) Coupling proton movements to c-ring rotation in F$_1$F$_0$ATP synthase: aqueous access channels and helix rotations at the a-c interface, *Biochim. Biophys. Acta* **1555**, 29–36.

Futai, M., Noumi, T., Maeda, M. (1989) ATP synthase (H$^+$-ATPase): results by combined biochemical and molecular biological approaches, *Annu. Rev. Biochem.* **58**, 111–136.

Futai, M., Wada, Y., Kaplan, J. (2004) *Handbook of ATPases*, Wiley-VCH Verlag Gmbh & Co., KGaA, Weinheim.

Nelson, N., Harvey, W.R. (1999) Vacuolar and plasma membrane proton-adenosinetriphosphatases, *Physiol. Rev.* **79**, 361–385.

Nishi, T., Forgac, M. (2002) The vacuolar (H$^+$)-ATPases – nature's most versatile proton pumps, *Nat. Rev. Mol. Cell Biol.* **3**, 94–103.

Panefsky, H.S., Cross, R.L. (1991) Structure and mechanism of F$_0$F$_1$-type ATP synthases and ATPases, *Adv. Enzymol. Relat. Areas Mol. Biol.* **64**, 173–214.

Racker, E. (1976) *A New Look at Mechanisms in Bioenergetics*, Academic Press, New York.

Senior, A.E., Nadanaciva, S., Weber, J. (2002) The molecular mechanism of ATP synthesis by F$_1$F$_0$-ATP synthase, *Biochim. Biophys. Acta* **1553**, 188–211.

Stock, D., Gibbons, C., Arechaga, I., Leslie, A.G.W., Walker, J.E. (2000) The rotary mechanism of ATP synthase, *Curr. Opin. Struct. Biol.* **10**, 672–679.

Weber, J., Senior, A.E. (1997) Catalytic mechanism of F$_1$-ATPase, *Biochim. Biophys. Acta* **1319**, 19–58.

Primary Literature

Abrahams, J.P., Leslie, A.G.W., Lutter, R., Walker, J.E. (1994) Structure at 2.8 Å resolution of F$_1$-ATPase from bovine heart mitochondria, *Nature* **370**, 621–628.

Abrahams, J.P., Buchanan, S.K., Raaji Van, M.J., Fearnley, I.M., Leslie, A.G.W., Walker, J.E. (1996) The structure of bovine F$_1$-ATPase complexed with the peptide antibiotic efrapeptin, *Proc. Natl. Acad. Sci. U.S.A.* **93**, 9420–9424.

Aggeler, R., Haughton, M.A., Capaldi, R.A. (1995) Disulfide bond formation between the COOH-terminal domain of the β subunits and the γ and ε subunits of the *Escherichia coli* F$_1$-ATPase. Structural implications and functional consequences. *J. Biol. Chem.* **270**, 9185–9191.

Al-Shawi, M.K., Ketchum, C.J., Nakamoto, R.K. (1997) Energy coupling, turnover, and stability of the F$_0$F$_1$ ATP synthase are dependent on the energy of interaction between γ and β subunits, *J. Biol. Chem.* **272**, 2300–2306.

Arata, Y., Baleja, J.D., Forgac, M. (2002a) Localization of subunits D, E, and G in the yeast V-ATPase complex using cysteine-mediated cross-linking to subunit B, *Biochemistry* **41**, 11301–11307.

Arata, Y., Baleja, J.D., Forgac, M. (2002b) Cysteine-directed cross-linking to subunit B suggests that subunit E forms part of the peripheral stalk of the vacuolar H$^+$-ATPase, *J. Biol. Chem.* **277**, 3357–3363.

Bianchet, M.A., Hullihen, J., Pederson, P.L., Amzel, L.M. (1998) The 2.8-Å structure of rat liver F$_1$-ATPase: configuration of a critical intermediate in ATP synthesis/hydrolysis, *Proc. Natl. Acad. Sci. U.S.A.* **95**, 11065–11070.

Birkenhäger, R., Hoppert, M., Deckers-Hebestreit, G., Mayer, F., Altendorf, K. (1995) The F$_0$ complex of the *Escherichia coli* ATP synthase. Investigation by electron spectroscopic imaging and immunoelectron microscopy, *Eur. J. Biochem.* **230**, 58–67.

Braig, K., Menz, R.I., Montgomery, M.G., Leslie, A.G.W., Walker, J.E. (2000) Structure of bovine mitochondrial F$_1$-ATPase inhibited by Mg^{2+} ADP and aluminium fluoride, *Structure* **8**, 567–573.

Cox, G.B., Fimmel, A.L., Gibson, F., Hatch, L. (1986) The mechanism of ATP synthase: a reassessment of the functions of the b and a subunits, *Biochim. Biophys. Acta* **849**, 62–69.

Dancan, T.M., Bulygin, V.V., Zou, Y., Hutcheon, M.L., Cross, R.L. (1995) Rotation of subunits during catalysis by *Escherichia coli* F$_1$-ATPase, *Proc. Natl. Acad. Sci. U.S.A.* **92**, 10964–10968.

Dmitriev, O.Y., Jones, P.C., Fillingame, R.H. (1999) Structure of the subunit c oligomer in the F$_1$F$_0$ ATP synthase: model derived from solution structure of the monomer and cross-linking in the native enzyme, *Proc. Natl. Acad. Sci. U.S.A.* **96**, 7785–7790.

Dmitriev, O.Y., Jones, P.C., Jiang, W.P., Fillingame, R.H. (1999) Structure of the membrane domain of subunit b of the *Escherichia coli* F_0F_1 ATP synthase, *J. Biol. Chem.* **274**, 15598–15604.

Dunn, S.D. (1982) The isolated γ subunits of *Escherichia coli* F_1 ATPase binds the ε subunit, *J. Biol. Chem.* **257**, 7354–7359.

Dunn, S.D., Futai, M. (1980) Reconstitution of a functional coupling factor from the isolated subunits of *Escherichia coli* F_1 ATPase, *J. Biol. Chem.* **255**, 113–118.

Elston, T., Wang, H., Oster, G. (1998) Energy transduction in ATP synthase, *Nature* **391**, 510–513.

Eya, S., Maeda, M., Futai, M. (1991) Role of the carboxyl terminal region of H^+-ATPase F_0F_1 a subunit from *Escherichia coli*, *Arch. Biochem. Biophys.* **284**, 71–77.

Fillingame, R.H., Jiang, W.P., Dmitriev, O.Y. (2000) The oligomeric subunit crotor in the F_0 sector of ATP synthase: unresolved questions in our understanding of function, *J. Bioenerg. Biomembr.* **32**, 433–439.

Foster, D.L., Fillingame, R.H. (1982) Stoichiometry of subunits in the H^+-ATPase complex of *Escherichia coli*, *J. Biol. Chem.* **257**, 2009–2015.

Futai, M. (1977) Reconstitution of ATPase activity from the isolated α, β, and γ subunits of the coupling factor, F_1, of *Escherichia coli*, *Biochem. Biophys. Res. Commun.* **79**, 1231–1237.

Futai, M., Omote, H., Sambongi, Y., Wada, Y. (2000) Synthase (H^+ ATPase): coupling between catalysis, mechanical work, and proton translocation, *Biochim. Biophys. Acta* **1458**, 276–288.

Futai, M., Oka, T., Sun-Wada, G.-H., Moriyama, Y., Kanazawa, H., Wada, Y. (2000) Luminal acidification of diverse organelles by V-ATPase in animal cells, *J. Exp. Biol.* **203**, 107–116.

Garcia, J.J., Capaldi, R.A. (1998) Unisite catalysis without rotation of the γ-ε domain in *Escherichia coli* F_1-ATPase, *J. Biol. Chem.* **273**, 15940–15945.

Gibbons, C., Montgomery, M.G., Leslie, A.G.W., Walker, J.E. (2000) The structure of the central stalk in bovine F_1-ATPase at 2.4 Å resolution, *Nat. Struct. Biol.* **7**, 1055–1061.

Girvin, M.E., Rastogi, V.K., Abildgaard, F., Markley, J.L., Fillingame, R.H. (1998) Solution structure of the transmembrane H^+-transporting subunit c of the F_1F_0 ATP synthase, *Biochemistry* **37**, 8817–8824.

Groth, G., Walker, J.E. (1997) Model of the c-subunit oligomer in the membrane domain of F-ATPases, *FEBS Lett.* **410**, 117–123.

Groth, G., Pohl, E. (2001) The structure of the chloroplast F_1-ATPase at 3.2 Å resolution, *J. Biol. Chem.* **276**, 1345–1352.

Gumbiowski, K., Pänke, O., Junge, W., Engelbrecht, S. (2002) Rotation of the c subunit oligomer in EF_0EF_1 mutant cD61N, *J. Biol. Chem.* **277**, 31287–31290.

Hanada, H., Moriyama, Y., Maeda, M., Futai, M. (1990) Kinetic studies of chromaffin granule H^+-ATPase and effects of bafilomycin A1, *Biochem. Biophys. Res. Commun.* **170**, 873–878.

Hara, K.Y., Noji, H., Bald, D., Yasuda, R., Kinoshita, K. Jr., Yoshida, M. (2000) The role of the DELSEED motif of the β subunit in rotation of F_1-ATPase, *J. Biol. Chem.* **275**, 14260–14263.

Hausrath, A.C., Gruber, G., Matthews, B.W., Capaldi, R.A. (1999) Structural features of the γ subunit of the *Escherichia coli* F_1 ATPase revealed by a 4.4-Å resolution map obtained by X-ray crystallography, *Proc. Natl. Acad. Sci. U.S.A.* **96**, 13697–13702.

Hirata, R., Graham, L.A., Takatsuki, A., Stevens, T.H., Anraku, Y. (1997) VMA11 and VMA16 encode second and third proteolipid subunits of the *Saccharomyces cerevisiae* vacuolar membrane H^+-ATPase, *J. Biol. Chem.* **272**, 4975–4803.

Hirata, H., Nakamura, N., Omote, H., Wada, Y., Futai, M. (2000) Regulation and reversibility of vacuolar H^+-ATPase, *J. Biol. Chem.* **275**, 386–389.

Hirata, T., Iwamoto-Kihara, A., Sun-Wada, G.-H., Okajima, T., Wada, Y., Futai, M. (2003) Subunit rotation of vacuolar-type proton pumping ATPase: relative rotation of the G and c subunits, *J. Biol. Chem.* **278**, 23741–23719.

Holiday, L.S., Lu, M., Lee, B.S., Nelson, R.D., Solivan, S., Zhang, L., Gluck, S.L. (2000) The amino-terminal domain of the B subunit of vacuolar H^+-ATPase contains a filamentous actin binding site, *J. Biol. Chem.* **275**, 32331–32337.

Huss, M., Ingenhorst, G., Konig, S., Gassel, M., Dorse, S., Zeeck, A., Altendorf, K., Wieczorek, H. (2002) Concanamycin A, the specific inhibitor of V-ATPases, binds to the V_0 subunit c, *J. Biol. Chem.* **277**, 40544–40548.

Hutcheon, M.L., Duncan, T.M., Ngai, H., Cross, R.L. (2001) Energy-driven subunit rotation at the interface between subunit a and the c oligomer in the F_0 sector of *Escherichia coli* ATP synthase, *Proc. Natl. Acad. Sci. U.S.A.* **98**, 8519–8524.

Ida, K., Noumi, T., Maeda, M., Fukui, T., Futai, M. (1991) Catalytic site of F_1-ATPase of *Escherichia coli* Lys-155 and Lys-201 of the β subunit are located near the γ-phosphate group of ATP in the presence of Mg^{2+}, *J. Biol. Chem.* **266**, 5424–5429.

Ihara, K., Abe, T., Sugimura, K.I., Mukohata, Y. (1992) Halobacterial A-ATP synthase in relation to V-ATPase, *J. Exp. Biol.* **172**, 475–485.

Iko, Y., Sambongi, Y., Tanabe, M., Iwamoto-Kihara, A., Saito, K., Ueda, I., Wada, Y., Futai, M. (2001) ATP synthase F_1 sector rotation. Defective torque generation in the β subunit Ser-174 to Phe mutant and its suppression by second mutations, *J. Biol. Chem.* **276**, 47508–47511.

Imamura, H., Nakano, M., Noji, H., Mureyuki, E., Ohkuma, S., Yoshida, M., Yokoyama, K. (2003) Evidence for rotation of V_1-ATPase, *Proc. Natl. Acad. Sci. U.S.A.* **100**, 2312–2315.

Iwamoto, A., Miki, J., Maeda, M., Futai, M. (1990) H^+-ATPase γ subunit of *Escherichia coli*. Role of the conserved carboxyl-terminal region, *J. Biol. Chem.* **265**, 5043–5048.

Iwamoto, A., Omote, H., Hanada, H., Tomioka, N., Itai, A., Maeda, M., Futai, M. (1991) Mutations in Ser174 and the glycine-rich sequence (Gly149, Gly150, and Thr156) in the β subunit of *Escherichia coli* H^+-ATPase. *J. Biol. Chem.* **266**, 16350–16355.

Jeanteur-De Beukelaer, C., Omote, H., Iwamoto-Kihara, A., Maeda, M., Futai, M. (1995) β-γ subunit interaction is required for catalysis by H^+-ATPase (ATP synthase). β subunit amino acid replacements suppress a γ subunit mutation having a long unrelated carboxyl terminus, *J. Biol. Chem.* **270**, 22850–22854.

Jiang, W.P., Hermolin, J., Fillingame, R.H. (2001) The preferred stoichiometry of c subunits in the rotary motor sector of *Escherichia coli* ATP synthase is 10, *Proc. Natl. Acad. Sci. U.S.A.* **98**, 4966–4971.

Jones, P.C., Fillingame, R.H. (1998) Genetic fusions of subunit c in the F_0 sector of H^+-transporting ATP synthase. Functional dimers and trimers and determination of stoichiometry by cross-linking analysis, *J. Biol. Chem.* **273**, 29701–29705.

Jones, P.C., Hermolin, J., Jiang, W.P., Fillingame, R.H. (2000) Insights into the rotary catalytic mechanism of F_0F_1 ATP synthase from the cross-linking of subunits b and c in the *Escherichia coli* enzyme, *J. Biol. Chem.* **275**, 31340–31346.

Junge, W., Lill, H., Engelbrecht, S. (1997) ATP synthase: an electrochemical transducer with rotatory mechanics, *Trends Biochem. Sci.* **22**, 420–423.

Kato-Yamada, K., Noji, H., Yasuda, R., Kinoshita, K. Jr., Yoshida, M. (1998) Direct observation of the rotation of epsilon subunit in F_1-ATPase, *J. Biol. Chem.* **273**, 19375–19377.

Kawasaki-Nishi, S., Nishi, T., Forgac, M. (2001) Arg-735 of the 100-kDa subunit a of the yeast V-ATPase is essential for proton translocation, *Proc. Natl. Acad Sci. U.S.A.* **98**, 12397–12402.

Ketchum, C.J., Al-Shawi, M.K., Nakamoto, K.K. (1998) Intergenic suppression of the γ M23K uncoupling mutation in F_0F_1 ATP synthase by β Glu-381 substitutions: the role of the β 380DELSEED386 segment in energy coupling. *Biochem. J.* **330**, 707–712.

Le, N.P., Omote, H., Wada, Y., Al-Shawi, M.K., Nakamoto, R.K., Futai, M. (2000) *Escherichia coli* ATP synthase alpha subunit Arg-376: the catalytic site arginine does not participate in the hydrolysis/synthesis reaction but is required for promotion to the steady state, *Biochemistry* **39**, 2778–2783.

Liu, Q., Kane, P.M., Newman, P.R., Forgac, M. (1996) Site-directed mutagenesis of the yeast V-ATPase B subunit (Vma2p), *J. Biol. Chem.* **271**, 2018–2022.

Liu, Q., Leng, X.H., Newman, P.R., Vasilyeva, E., Kane, P.M., Forgac, M. (1997) Site-directed mutagenesis of the yeast V-ATPase A subunit, *J. Biol. Chem.* **272**, 11750–11756.

Löbau, S., Weber, J., Wilke-Mounts, S., Senior, A.E. (1997) F_1-ATPase, roles of three catalytic site residues, *J. Biol. Chem.* **272**, 3648–3456.

Long, J.C., Wang, S., Vik, S.B. (1998) Membrane topology of subunit a of the F_1F_0 ATP synthase as determined by labeling of unique cysteine residues, *J. Biol. Chem.* **273**, 16235–16240.

MacLeod, K.J., Vasilyeva, E., Baleja, J.D., Forgac, M. (1998) Mutational analysis of the nucleotide binding sites of the yeast vacuolar proton-translocating ATPase, *J. Biol. Chem.* **273**, 150–156.

Mc Lachlin, D.T., Dunn, S.D. (2000) Disulfide linkage of the b and Δ subunits does not affect the function of the *Escherichia coli* ATP synthase *Biochemistry* **39**, 3486–3490.

Menz, R.I., Walker, J.E., Leslie, A.G.W. (2001) Structure of bovine mitochondrial F_1-ATPase with nucleotide bound to all three catalytic sites: implications for the mechanism of rotary catalysis, *Cell* **106**, 331–341.

Nadanaciva, S., Webe, J., Senior, A.E. (1999) The role of β-Arg-182, an essential catalytic site residue in *Escherichia coli* F_1-ATPase, *Biochemistry* **38**, 7670–7677.

Nadanacivam, S., Weber, J., Wilke-Mounts, S., Senior, A.E. (1999) Importance of F_1-ATPase residue αArg-376 for catalytic transition state stabilization, *Biochemistry* **38**, 15493–15499.

Nakamoto, R.K., Maeda, M., Futai, M. (1993) The γ subunit of the *Escherichia coli* ATP synthase. Mutations in the carboxyl-terminal region restore energy couplin γ to the amino-terminal mutant γ Met-23–> Lys, *J. Biol. Chem.* **268**, 867–872.

Nakamoto, R.K., Al-Shawi, M.K., Futai, M. (1995) The ATP synthase γ subunit. Suppressor mutagenesis reveals three helical regions involved in energy coupling, *J. Biol. Chem.* **270**, 14042–14046.

Nakamoto, R.K., Shin, K., Iwamoto, A., Omote, H., Maeda, M., Futai, M. (1992) *Escherichia coli* F_0F_1-ATPase. Residues involved in catalysis and coupling, *Ann. N. Y. Acad. Sci.* **671**, 335–344.

Nishio, K., Iwamoto-Kihara, A., Yamamoto, A., Wada, Y., Futai, M. (2002) Subunit rotation of ATP synthase embedded in membranes: a or β subunit rotation relative to the c subunit ring, *Proc. Natl. Acad. Sci. U.S.A.* **99**, 13448–13452.

Noji, H., Yasuda, R., Yoshida, H., Kinoshita, K. Jr. (1997) Direct observation of the rotation of F_1-ATPase, *Nature* **386**, 299–302.

Omote, H., Maeda, M., Futai, M. (1992) Effects of mutations of conserved Lys-155 and Thr-156 residues in the phosphate-binding glycine-rich sequence of the F_1-ATPase β subunit of *Escherichia coli*, *J. Biol. Chem.* **267**, 20571–20576.

Omote, H., Le, N.P., Park, M.-Y., Maeda, M., Futai, M. (1995) β subunit Glu-185 of *Escherichia coli* H^+-ATPase (ATP synthase) is an essential residue for cooperative catalysis, *J. Biol. Chem.* **270**, 25656–25660.

Omote, H., Tainaka, K., Iwamoto-Kihara, A., Wada, Y., Futai, M. (1998) Stability of the *Escherichia coli* ATP synthase F_0F_1 complex is dependent on interactions between γGln-269 and the β subunit loop β Asp-301-β Asp-305, *Arch. Biochem. Biophys.* **308**, 277–282.

Omote, H., Sambonmatsu, N., Saito, K., Sambongi, Y., Iwamoto-Kihara, A., Yanagida, T., Wada, Y., Futai, M. (1999) The γ-subunit rotation and torque generation in F_1-ATPase from wild-type or uncoupled mutant *Escherichia coli*, *Proc. Natl. Acad. Sci. U.S.A.* **96**, 7780–7784.

Orriss, G.L., Leslie, A.G.W., Braig, K., Walker, J.E. (1998) Bovine F_1-ATPase covalently inhibited with 4-chloro-7-nitrobenzofurazan: the structure provides further support for a rotary catalytic mechanism, *Structure* **6**, 831–837.

Pänke, O., Gumbiowski, K., Junge, W., Engelbrecht, S. (2000) F-ATPase: specific observation of the rotating c subunit oligomer of EF_0EF_1, *FEBS Lett.* **472**, 34–38.

Park, M.Y., Omote, H., Maeda, M., Futai, M. (1994) Conserved Glu-181 and Arg-182 residues of *Escherichia coli* H^{+-} ATPase (ATP synthase) β subunit are essential for catalysis: properties of 33 mutants between β Glu-161 and β Lys-201 residues, *J. Biochem.* **116**, 1139–1145.

Powell, B., Graham, L.A., Stevens, T.H. (2000) Molecular characterization of the yeast vacuolar H^+-ATPase proton pore, *J. Biol. Chem.* **275**, 26354–23660.

Rastogi, V.K., Girvin, M.E. (1999) Structural changes linked to proton translocation by subunit c of the ATP synthase, *Nature* **402**, 263–268.

Rodgers, A.J.W., Capaldi, R.A. (1998) The second stalk composed of the β- and Δ-subunits connects F_0 to F_1 via an α-subunit in the *Escherichia coli* ATP synthase, *J. Biol. Chem.* **273**, 29406–29410.

Sabbert, D., Engelbrecht, S., Junge, W. (1996) Functional and idling rotatory motion within F_1-ATPase, *Nature* **381**, 623–625.

Sambongi, Y., Iko, Y., Tanabe, M., Omote, H., Iwamoto-Kihara, A., Ueda, I., Yanagida, T., Wada, Y., Futai, M. (1999) Mechanical rotation of the c subunit oligomer in ATP synthase (F_0F_1): direct observation, *Science* **286**, 1722–1724.

Schneider, E., Altendorf, K. (1985) All three subunits are required for the reconstitution of an active proton channel (F_0) of *Escherichia coli* ATP synthase (F_1F_0), *EMBO J.* **4**, 515–518.

Seelert, H., Poetsch, A., Dencher, N.A., Engel, A., Stahlberg, H., Müller, D.J. (2000) Structural biology. Proton-powered turbine of a plant motor, *Nature* **405**, 418–419.

Senior, A.E., Al-Shawi, M.K. (1992) Further examination of seventeen mutations in *Escherichia coli* F_1-ATPase β-subunit, *J. Biol. Chem.* **267**, 21471–21478.

Senior, A.E., Nadanaciva, S., Weber, J. (2000) Rate acceleration of ATP hydrolysis by F_1F_0-ATP synthase, *J. Exp. Biol.* **203**, 35–40.

Shin, K., Nakamoto, R.K., Maeda, M., Futai, M. (1992) F_0F_1-ATPase γ subunit mutations perturb the coupling between catalysis and transport, *J. Biol. Chem.* **267**, 20835–20839.

Shirakihara, Y., Leslie, A.G.W., Abrahams, J.P., Walker, J.E., Ueda, T., Sekimoto, Y., Kambara, M., Saika, K., Kagawa, Y., Yoshida, M. (1997) The crystal structure of the nucleotide-free $\alpha 3\beta 3$ subcomplex of F_1-ATPase from the thermophilic *Bacillus* PS3 is a symmetric trimer, *Structure* **5**, 825–836.

Singh, S., Turina, P., Bustamante, C.J., Keller, D.J., Capaldi, R.A. (1996) Topographical structure of membrane-bound *Escherichia coli* F_1F_0 ATP synthase in aqueous buffer, *FEBS Lett.* **397**, 30–34.

Soong, R.K., Bachand, G.D., Neves, H.P., Olkhovets, A.G., Craighead, H.G., Montemagno, C.D. (2000) Powering an inorganic nanodevice with a biomolecular motor, *Science* **290**, 1555–1558.

Stahlberg, H., Müller, D.J., Suda, K., Fotiadis, D., Engel, A., Meier, T., Matthey, U., Dimroth, P. (2001) Bacterial Na^+-ATP synthase has an undecameric rotor, *EMBO Rep.* **2**, 229–233.

Stalz, W.-D., Greie, J.-C., Deckers-Hebestreit, G., Altendorf, K. (2003) Direct interaction of subunits a and b of the F_0 complex of *Escherichia coli* ATP synthase by forming an ab_2 subcomplex, *J. Biol. Chem.* **278**, 27068–27071.

Stock, D., Leslie, A., Walker, J.E. (1999) Molecular architecture of the rotary motor in ATP synthase, *Science* **286**, 1700–1705.

Sun-Wada, G.-H., Wada, Y., Futai, M. (2003) Vacuolar H^+ pumping ATPases in luminal acidic organelles and extracellular compartments: common rotational mechanism and diverse physiological roles, *J. Bioenerg. Biomembr.* **35**, 347–358.

Sun-Wada, G.-H., Murakami, H., Nakai, H., Wada, Y., Futai, M. (2001) Mouse Atp6f, the gene encoding the 23-kDa proteolipid of vacuolar proton translocating ATPase, *Gene* **274**, 93–99.

Sun-Wada, G.-H., Imai-Senga, Y., Yamamoto, A., Murata, Y., Hirata, T., Wada, Y., Futai, M. (2002) A proton pump ATPase with testis-specific E1-subunit isoform required for acrosome acidification, *J. Biol. Chem.* **277**, 18098–18105. Y.

Takeyasu, K., Omote, H., Nettikadan, S., Tokumasu, F., Iwamoto-Kihara, A., Futai, M. (1996) Molecular imaging of *Escherichia coli* F_0F_1-ATPase in reconstituted membranes using atomic force microscopy, *FEBS Lett.* **392**, 110–113.

Tanabe, M., Nishio, K., Iko, Y., Sambongi, Y., Iwamoto-Kihara, A., Wada, Y., Futai, M. (2001) Rotation of a complex of the γ subunit and c ring of *Escherichia coli* ATP synthase. The rotor and stator are interchangeable, *J. Biol. Chem.* **276**, 15269–15274.

Tsunoda, S.P., Aggeler, R., Yoshida, M., Capaldi, R.A. (2001) Rotation of the c subunit oligomer in fully functional F_1F_0 ATP synthase, *Proc. Natl. Acad. Sci. U.S.A.* **98**, 898–902.

Valiyaveetil, F.I., Fillingame, R.H. (1998) Transmembrane topography of subunit a in the *Escherichia coli* F_1F_0 ATP synthase, *J. Biol. Chem.* **273**, 16241–16247.

Van Raaji, M.J., Abrahams, J.P., Leslie, A.G.W., Walker, J.E. (1996) The structure of bovine F_1-ATPase complexed with the antibiotic inhibitor aurovertin B, *Proc. Natl. Acad. Sci. U.S.A.* **93**, 6913–6917.

Wada, Y., Sambongi, Y., Futai, M. (2000) Biological nano motor, ATP synthase F_0F_1: from catalysis to $\gamma \varepsilon c_{10-12}$ subunit assembly rotation, *Biochim. Biophys. Acta* **1459**, 499–505.

Watts, S.D., Capaldi, R.A. (1997) Interactions between the F_1 and F_0 parts in the *Escherichia coli* ATP synthase. Associations involving the loop region of c subunits, *J. Biol. Chem.* **272**, 15065–15068.

Weber, J., Senior, A.E. (2000) ATP synthase: what we know about ATP hydrolysis and what we do not know about ATP synthesis, *Biochim. Biophys. Acta* **1458**, 300–309.

Weber, J., Senior, A.E. (2001) Bi-site catalysis in F_1-ATPase: does it exist? *J. Biol. Chem.* **276**, 35422–35428.

Weber, J., Hammond, S.T., Wike-Mounts, S., Senior, A.E. (1998) Mg^{2+} coordination in catalytic sites of F_1-ATPase, *Biochemistry* **37**, 608–614.

Weber, J., Lee, R.S.V., Grell, E., Wise, J.G., Senior, A.E. (1992) On the location and function of tyrosine $\beta 331$ in the catalytic site of *Escherichia coli* F_1-ATPase, *J. Biol. Chem.* **267**, 1712–1718.

Wikens, S., Capaldi, R.A. (1998) ATP synthase's second stalk comes into focus, *Nature* **393**, 29.

Yasuda, R., Noji, H., Kinoshita, K. Jr., Yoshida, M. (1998) F_1-ATPase is a highly efficient molecular motor that rotates with discrete 120 degree steps, *Cell* **93**, 1117–1124.

Yasuda, R., Noji, H., Yoshida, H., Kinoshita, K. Jr. (2001) Resolution of distinct rotational substeps by submillisecond kinetic analysis of F_1-ATPase, *Nature* **410**, 898–904.

Protusions: Cilia and Flagella: *see* Mobile Structures: Cilia and Flagella

Pufferfish Genomes: *Takifugu* and *Tetraodon*

Melody S. Clark[1] and Hugues Roest Crollius[2]
[1] *Biological Sciences Division, British Antarctic Survey Cambridge, UK*
[2] *Laboratoire Dynamique et Organisation des Genomes (LDOG), Department of Biology, Ecole Normale Superiere, Paris, France*

1	**Introduction** 363	
2	**Classification, Habitat, and Morphology of Pufferfish**	**364**
2.1	Classification and Geographical Distribution of Pufferfish	364
2.2	Pufferfish Morphology 365	
3	**Pufferfish Genomes** 365	
3.1	Genome Size 365	
3.2	Repetitive Elements 366	
3.3	Gene Number 367	
3.4	Pufferfish Chromosomes 368	
3.5	Pufferfish Genetics 369	
4	**The Genome Projects** 370	
4.1	Sequencing Projects 370	
4.2	Physical maps 371	
4.3	Resources 371	
5	**Pufferfish as a Genome Tool** 371	
5.1	Comparative Genome Analysis in Vertebrates 371	
5.2	Functional Analyses Including Detection of Regulatory Elements	375

Encyclopedia of Molecular Cell Biology and Molecular Medicine, 2nd Edition. Volume 11
Edited by Robert A. Meyers.
Copyright © 2005 Wiley-VCH Verlag GmbH & Co. KGaA, Weinheim
ISBN: 3-527-30648-X

6   Prospects   377

        Bibliography   377
        Books and Reviews   377
        Primary Literature   378

Keywords

Conserved Order
Demonstration that three or more homologous genes lie on one chromosome in the same order in two separate species.

Conserved Segment
The syntenic association of two or more homologous genes in two separate species that are contiguous (not interrupted by different chromosome segments) in both species.

Conserved Synteny
The linked association of two or more homologous genes in two separate species regardless of gene order or interspersion of noncontiguous segments between two markers.

Karyotype
The chromosome complement of an organism.

Synteny
This term simply means linkage, that is, if two genes are syntenic, they are both present on the same chromosome.

Takifugu
While this is a species of pufferfish, this is generally the abbreviated name used in this text for *Takifugu rubripes*. This species is also known as *Fugu*, *Fugu rubripes*, and *Torafugu*. Its common name is the Japanese pufferfish.

Tetraodon
While this is also a species of pufferfish, this is generally the abbreviated name used in this text for *Tetraodon nigroviridis*. In the past, this species had been misnamed as *Tetraodon fluviatilis*, which is a different species of the *Tetraodon* family. *T. nigroviridis* is currently designated as *Chelonodon nigroviridis* in Fishbase. Its common name is the spotted green pufferfish.

The pufferfish have some of the smallest vertebrate genomes known (350–500 Mb). This lead to two species, the Japanese pufferfish and the spotted green pufferfish (*Takifugu rubripes* and *Tetraodon nigroviridis* respectively), being adopted as model vertebrates for in-depth genome-sequencing programs. These data have provided a wealth of information for comparative genomics studies in both fish and other vertebrate species, particularly human. This article describes the generalized features of the pufferfish genomes, the genome projects and resources available, in addition to the development of the pufferfish as genome tools.

1
Introduction

The raison d'être behind the pufferfish genome projects has to be viewed in a historical context. When the initial pufferfish genome project (using *Takifugu*) was proposed by Sydney Brenner in the early 1980s, sequencing technologies were still in their infancy. The sequencing programs of *Escherichia coli* and *Caenorhabditis elegans* had been started and it was these that provided the impetus behind the development of the high-throughput technologies, which are taken for granted today. Having said that, it did not seem conceivable at that time that they could be improved to such an extent as to allow the sequencing of the human genome. Therefore, Sydney Brenner proposed the need for an economical vertebrate model genome to bridge the gap between the data arising from the bacterial and invertebrate projects and human (Table 1). It had been known since 1968, from the work of Hinegardner, that several species of pufferfish had small genomes of approximately 0.4 pg per haploid nucleus (400 Mb). This equates to approximately one-eighth the size of the human genome. After preliminary sequencing scanning experiments showed a similar gene content to human, *T. rubripes* became accepted as

Tab. 1 Genome sizes of selected organisms. When the pufferfish was proposed as a model genome, sequencing technologies were relatively limited, with work started only on the bacteria and *Saccharomyces cerevisiae* and *C. elegans*.

Organism	Common name	Genome size [Mb]
Escherichia coli	Bacteria	4.64
Saccharomyces cerevisiae	Yeast	14
Caenorhabditis elegans	nematode	97
Takifugu rubripes	Japanese pufferfish	365
Tetraodon nigroviridis	spotted green pufferfish	350
Danio rerio	zebrafish	1900
Xenopus laevis	African clawed toad	3100
Gallus gallus	chicken	1200
Bos taurus	cattle	3651
Rattus norvegicus	rat	3000
Mus musculus	mouse	2500
Homo sapien	human	3000

a universal model genome. However, *Takifugu* was in some ways rather limited; it was only available from Japan and was therefore not universally accessible to researchers; it also grows rapidly to 1 kg within the first year and so cannot be kept in small laboratory fish tanks. Therefore, an alternative pufferfish species was proposed by Hans Lehrach – the spotted green pufferfish or *T. nigroviridis*. This is a small freshwater pufferfish that is available from many aquarium shops worldwide and so can be easily kept in most laboratories, providing a readily available source of DNA and tissue-specific RNA. Paradoxically, with the massive improvement in sequencing technologies, the human genome has been sequenced in advance of the pufferfish and the excess capacity of the large sequencing factories has since enabled the sequencing of both pufferfish genomes. The data from these is proving extremely valuable in comparative genomics and in the functional characterization of genes, as will be discussed Sect. 5.

2
Classification, Habitat, and Morphology of Pufferfish

2.1
Classification and Geographical Distribution of Pufferfish

Both pufferfish species (*T. rubripes* and *T. nigroviridis*) are members of the series Percomorpha, order Tetraodontiformes, family Tetraodontidae. The Tetraodontidae family comprises 19 genera with approximately 121 species. They are largely marine species, with several of them entering and occurring in brackish and freshwater. They live in tropical and subtropical regions, in the Atlantic, Indian, and Pacific oceans. Approximately 12 species of *Cariotetraodon*, *Chonerhinos*, and *Tetraodon* occur only in freshwater, primarily in the Congo river and southern Asia. The geographical locations of the two pufferfish species subject to genome sequencing are detailed further:

- *T. rubripes* is found in the Northwest Pacific: western part of the Sea of Japan and the East China and Yellow Seas northward to Muroran, Hokkaido, Japan. It is a temperate marine species, which occasionally enters brackish waters. It grows rapidly to a large size (70 cm is thought to be the maximum size it attains) and is not thought to breed until three-years old. It is a very important commercial species in the Japanese fisheries/aquaculture industry.
- *T. nigroviridis* is found in Asia: Sri Lanka to Indonesia, Malaysia, and north to China. It is a tropical (24–28 °C) freshwater species. It is a relatively small pufferfish species (maximum size 14 cm) and is sold in the aquarium trade worldwide.

The family Tetraodontidae is further split into two subfamilies: the Canthigastrinae (sharpnose puffers), which only comprises one genus and 26 species, while the rest are consigned to the Tetraodontinae. Paleontological evidence indicates that the families of the Tetraodontiform fishes diverged from each other between 90 and 50 million years ago, the fossil records for the Tetraodontiformes going back at least to the early Eocene (58 million years ago), indicating a long evolutionary history. Phylogenetic trees built from mitochondrial and nuclear genes showed that *Tetraodon* and *Takifugu* shared a common ancestor between 18 and 30 million years ago.

With two pufferfish genomes available for data mining, biologists are increasingly interested in related species as a means of exploiting these as biological models. While the order Tetraodontiformes is well defined (Fig. 1), there is relatively little information on the relationships between the different family members.

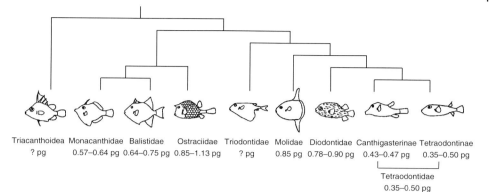

Fig. 1 Phylogeny and haploid genome sizes for sampled members of seven Tetraodontiform families. The phylogeny was based on morphological characters by Winterbottom, 1974. Figure reproduced from Brainerd, E.L., Slutz, S.S., Hall, E.K., Phillis, R.W. (2001) Patterns of genome size evolution in tetraodontiform fishes, *Evolution Int. J. Org. Evolution* **55**, 2363–2368, with permission.

2.2
Pufferfish Morphology

Pufferfish are generally recognized as the most advanced order of the teleosts, having a highly derived body form. The name Tetraodontiformes literally means four-toothed, which refers to the common pattern of four teeth in the outer jaws, although it has to be said that, on first glance, these often resemble more of a beak structure. Pufferfish are characterized by a high degree of fusion or loss of numerous bones in both the head and body. They usually have no lower ribs and have a reduced number of vertebrae when compared to most other fish species. The scales are usually modified as spines, shields, or bony plates over a thick leathery skin. The pufferfish have modified stomachs, which allows them to inflate to an enormous size by gulping down water, or in some cases air, when annoyed or threatened.

Some species, including *T. rubripes*, contain the alkaloid poison tetraodotoxin in certain tissues of their body. This is one of the most potent poisons known and acts by selectively blocking the voltage-sensitive sodium channels, which are essential components in cellular signaling pathways. The pufferfish have a point mutation in the protein sequence of the sodium-channel pump and so are resistant to the poison. In *Takifugu*, tetraodotoxin is present in high concentrations in the liver and ovaries and at lower concentrations in the intestines. The flesh, skin, and testes are not poisonous. However, despite the presence of a potentially lethal poison in certain tissues, *Takifugu* is highly prized item on the sushi menu. Hence the need for specially trained chefs to prepare *Takifugu*, ensuring that none of the toxic tissues is included in the sushi.

3
Pufferfish Genomes

3.1
Genome Size

The two pufferfish have similar genome characteristics, which will be described in general terms here. Measurements of the nuclear DNA content showed that both contain approximately 0.35 to 0.4 pg per haploid genome, equivalent to a genome size between 350 and 400 megabases (Mb). The alignment of syntenic regions between *Tetraodon* and *Takifugu* indicates that the latter is larger by approximately 10%, although it will be necessary to wait until both genome sequences are complete for a final estimate. Both genomes are equivalent to approximately one-eighth the size of the human genome.

These compact genomes are characterized by smaller intergenic regions and introns and fewer repeat sequences (less than 10% of the genome). Data from the *Takifugu* genome shows that 75% of the introns are less than 425 bp, with a modal value of 79 bp. Only 500 introns are greater than 10 kb. The comparable figures for human are a modal value of 87 bp, which is not significantly different, but the 75% threshold is raised to 2609 bp, and over 12 000 introns are greater than 10 kb. The pufferfish genomes are prized by molecular biologists for their small genes, and, in general, pufferfish genes are scaled down compared to their human orthologs; however, a note of warning: a significant number are 1.3× larger.

Questions always arise as to why the genomes are so small and about the mechanisms involved. There is great diversity in fish genome sizes and this is also true of the Tetraodontiformes. Members of the Diodontidae (the spiny pufferfish), the sister family to the Tetraodontidae (smooth pufferfish), have genomes twice as large (Fig. 1). Studies indicate that the Diodontidae genomes have reached equilibrium, that is, are stable, but that this state of equilibrium was disturbed in the smooth pufferfish, following their divergence from the spiny puffers. The smaller genomes are thought to be the result of a decline in the rate of large-scale insertions, probably as a result of decreased transposable element activity. These are often found in the genome as middle repeat sequences (approximately 20% of the *Diodon hystrix* genome), but these are rare in *Takifugu*, a conclusion that presumably extends to the other smooth pufferfish with their compact genomes.

3.2
Repetitive Elements

Aside from a reduced number of middle repeat transposable elements, other types of such elements are also present in much lower quantities (approximately 1% of the total genome) compared to the other larger vertebrate genomes. It is the number rather than the types of transposable elements that are reduced as pufferfish contain sequences homologous to all known families of transposable elements including Ty3-gypsy, Ty1-copia, LINES, DNA transposons, and retroviruses. In general, these elements are dispersed throughout the genome but not in an even manner. *In situ* localization studies in *Tetraodon* show that they are compartmentalized in specific regions that correspond to the short arms of 10 pairs of small subtelocentric chromosomes (Fig. 2). These chromosome arms are heterochromatic and also contain a large number of pseudogenes that have arisen through duplication and retrotransposition. Characterization of such genes (iSET and Trapeze) has been carried out in *Tetraodon*, but it is not known whether they are present in *Takifugu* as the heterochromatic regions have been left out of the genome assembly to facilitate contiguation.

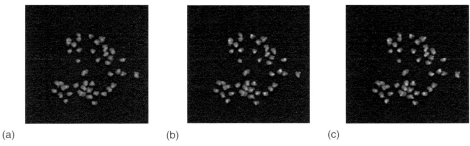

Fig. 2 *In situ* hybridization of two different non-LTR retrotransposons to *Tetraodon* chromosomes. (a) Green signals of Zebulon retrotransposon; (b) red signals of Barbar retrotransposon; (c) combined signal (yellow) of both retrotransposons. Both of these elements share the same sites in the genome, at the tip of some subtelocentric chromosomes. Babar is more abundant (in the shotgun data) and has more signals here, as seen by the additional red hybridizations in the combined data. (See color plate p. xxvii).

In addition to these transposable elements, the pufferfish genomes also contain two satellite sequences. The first is a 118-bp repeat, which maps to the centromeres. Interestingly, and despite their exact same size in both species, the two centromeric repeats show no noticeable sequence similarity. The second satellite is a subtelocentric repeat (10-mer in the case of *Tetraodon* (GGCGTCTGAG) and 20-mer in the case of *Takifugu* (GGCATCTGATC-CTGGTAGCT)). Minisatellites are present to the extent of approximately 41% in *Tetraodon* (unknown in *Takifugu*) and microsatellites comprise between 2 to 3.5% of the pufferfish genomes. In total, repeat sequences comprise less than 10% of the genome.

The G + C content is higher in the pufferfish (46.2% in *Tetraodon* and 47.7% in *Takifugu*) compared to human (40.3%), which is thought to be due to the relative increased coding content. In *Takifugu*, the G + C heterogeneity is also less marked compared to mammals but is similar in *Tetraodon*, albeit shifted to higher G + C percentage. The reasons why *Tetraodon* contains more regions higher in G + C content than the closely related *Takifugu* are still unknown. In fact, a more homogeneous distribution in *Tetraodon* would be expected since it is a general condition associated with poikilothermic animals. The heterogeneous distribution of G + C content in mammals is associated with gene-rich and gene-poor regions, and the pufferfish also have gene-dense and gene-sparse regions. Overall, the average gene density in the pufferfish is one gene for every 10 kb, compared to 40 kb in human.

3.3
Gene Number

As regards the actual number of genes present in the pufferfish, there are only current estimates based on gene prediction programs. The number of predicted genes is greater than that of human (35 180 compared to 24 847), but it is suspected that fish have more genes than mammals owing to additional duplication events. This is supported by results from whole-genome comparisons between *Takifugu* and *Tetraodon* on the one hand, and human and mouse on the other hand, which show that the two fish contain a slight excess of regions conserved during evolution between tetrapods and teleosts. However, the number of transcripts does not seem to vary much between the two species (38 510 and 37 347). It should be noted that no estimate has been made of the rate of alternative transcription in the pufferfish and this could dramatically alter the number of transcripts. In human, 60% of genes are thought to be alternatively transcribed.

When the *Takifugu* draft genome was compared to the human sequence, three-quarters of human predicted genes showed a strong match on BLAST sequence similarity searching. One-quarter of human predicted proteins did not show a match and, therefore, are either highly diverged or not present. Six thousand *Takifugu* predicted genes showed no match to human predicted proteins. Although some of these genes may be fish-specific, it is unlikely that this is the case for the majority. Of the nonmatching human proteins, there were many cell surface receptor–ligand system proteins of the immune system, hemopoietic system, and energy/metabolism of homotherms. Immune cytokines, in general, were either not detected in *Takifugu* or showed very distant similarity. However, a focused effort to identify the class I and II cytokine receptors and their ligands in *Tetraodon* showed that the complete repertoire of these genes was present. This work required a careful examination of gene-structure conservation and a systematic cloning of transcripts for these genes, and highlights the fact that automatic annotation of genomes may leave out complete functional families, especially those submitted to accelerated sequence divergence.

3.4
Pufferfish Chromosomes

In depth, karyotypic analysis of most pufferfish chromosomes is difficult because of their small size (2 µm, on average). Therefore, most analysis is in the form of a basic giesma-stained karyotype. Cytogenetic information is available for 53 out of the 339 marine Tetraodontiformes. The Triacanthidae are considered the most basal clade of the Tetraodontiformes (based on morphological and osteologic criteria) and have a diploid chromosome complement of $2n = 2x = 48$. This mirrors the generalized Percomorpha complement from which they were derived and also represents the karyotypic base for all Tetraodontiformes (Fig. 3). Karyotypes of the more specialized species range from $2n = 2x = 28$ to 52, which are thought to represent derivations of this karyotype. This inference is supported by the occurrence of bi-armed chromosomes in this group, which is considered a derived condition in fish karyotypes.

Takifugu rubripes. $2n = 2x = 44$. Analysis of six different *Takifugu* species karyotypes shows that they are virtually identical, and the occurrence of natural hybrid species suggests that their morphology is indeed very similar. There is a large pair of submetacentric chromosomes within the complement, which is the characteristic marker chromosome pair of the group.

Tetraodon nigroviridis. $2n = 2x = 42$. In addition to the larger submetacentric chromosome pair, characteristic of the *Takifugu* species, *Tetraodon* also contains a large

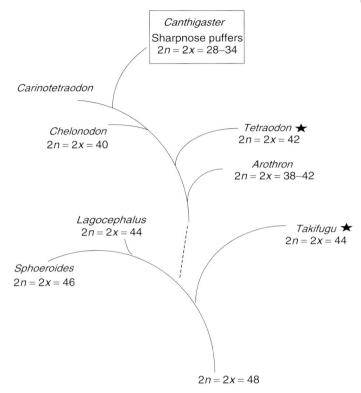

Fig. 3 Chromosome numbers of sampled Tetraodontinae overlaid on a restricted phylogeny on the basis of anatomical diversity, derived from J.C. (1980) *Osteology, phylogeny and higher classification of the fishes of the order Plectognathi (Tetraodontiformes)*. NOAA Technical Report NMFS Circular 434. US Department of Commerce, Washington, USA. Not all species of Tetraodontidae are shown. *Sphoeroides* is regarded as the most basal, having the most generalized osteology. $2n = 2x = 48$ is regarded as the ancestral karyotype. The stars indicate the species to which the sequenced pufferfish genomes belong. Virtually nothing is known about the evolutionary timescales between species in this family, although it has been estimated that *Takifugu* and *Tetraodon* diverged approximately 18 to 30 million years ago.

metacentric chromosome, presumably the result of a centric fusion between two of the smaller acrocentric chromosome pairs. This species is by far the best studied, in cytogenetic and physical mapping terms. Silver staining shows that there are two pairs of nucleolar organizer (NOR) chromosomes that are transcriptionally active.

3.5
Pufferfish Genetics

Owing to the problems in breeding these species, there are no linkage maps for pufferfish. In fact, no morphological characters enable the distinction between males and females in *Tetraodon*. Evidence suggests that the inability to breed pufferfish may be due to some form of captive-induced stress. Histological studies of gonads have been performed in *Tetraodon* and show that both females and males are sexually mature when imported from Southeast Asia to Europe, but a high level of occlusions in female in vitellogenesis suggest that the last stages of sexual maturation are blocked because of some external factors that may be due to the stress of transportation. Breeding problems also arise with *Takifugu*, even when attempted in their native Japan, without being subjected to lengthy air-freight transportation.

4
The Genome Projects

4.1
Sequencing Projects

Both pufferfish genome projects have been generated using a Whole-Genome Shotgun (WGS) approach, where single read sequences are generated from the ends of cloned genomic DNA fragments. Both are currently at the stage of a "working draft," which means that, several times, genome coverage of clones have been assembled into a series of contigs (termed *scaffolds*). These vary dramatically in size from several 100 kb to 1 to 2 kb and are gradually being assembled into increasingly larger contigs. If both pufferfish were to be completely sequenced, then there would eventually be 21 contigs for *Tetraodon* ($2n = 2x = 42$) and 22 contigs for *Takifugu* ($2n = 2x = 44$), with each contig representing one of each chromosome pair. However, it is not known whether this will be the case; some regions, such as the telomeres and centromeres, are particularly difficult to assemble, and it is more likely that each genome will remain as a draft. Because of the nature of these projects, data are continually changing and, therefore, specific details are not given here and the reader is referred to the relevant web pages (below).

Tetraodon. This sequencing project was started at Genoscope in 1997 and they were joined by the Whitehead Institute for Genomic Research (WIGR) in June 2001. The approach taken was to end sequence a number of libraries containing different insert sizes: BAC (120–160 kb), cosmid (40 kb), and plasmid (4, 2.5, and 2 kb) libraries. The *Tetraodon* draft-sequence data is available on two sites, with an option on the French language version at Genoscope.

Genoscope: www.genoscope.cns.fr/externe/English/Projets/Projet_C/C.html

WIGR: www-genome.wi.mit.edu/annotation/Tetraodon

Takifugu. October 2000 saw the announcement by the US department of Energy that a consortium led by the Joint Genome Institute (JGI) was to completely sequence *Takifugu*. The consortium comprised groups from within the JGI, Molecular Sciences Institute, and Institute for Systems Biology in the US, along with MRC HGMP Resource Centre in the UK and the Institute of Molecular and Cell Biology in Singapore. This project mainly used end sequencing from a 2-kb plasmid library, but with additional sequence from a 5.5-kb plasmid library and BACs and cosmids. The *Takifugu* sequence information is available on four sites at the moment, all with different sequence viewers, although the final official site is expected to be Ensembl.

Ensembl: www.ensembl.org/Fugu_rubripes

HGMP: http://www.fugu.hgmp.mrc.ac.uk

Institute of Molecular and Cellular Biology: http://www.fugu-sg.org

JGI: http://genome.jgi-psf.org/fugu6/fugu6.home.html

4.2
Physical maps

The requirement for physical mapping has, to a large extent, been superseded by the genome information in which large areas of the genome are represented by sequence contigs. These contigs are available in several databases (detailed in Sect. 4.1) and are generally well annotated with gene homology information. Genome maps are however useful for confirming the integrity of whole-genome sequence assemblies and providing sequencing material in regions where there are gaps. Because pufferfish cannot be bred to produce genetic linkage maps, the only map prior to the sequence data was generated by physically linking genomic clones across chromosomes. In *Tetraodon*, this was performed in several complementary ways. Firstly, single copy probes derived from the ends of BAC clones were hybridized onto high-density colony membranes carrying 10-fold redundant coverage of the genome in BAC clones. This generated nearly 3000 clone contigs spanning about 60% of the *Tetraodon* genome. Then, the same BAC clones were digested with restriction enzymes and their electrophoretic profile compared (the "fingerprinting" technique) to examine the extent of their overlap. This approach produced approximately 2000 contigs independently from the hybridization method, and both sets were then merged to provide a structure onto which the assembled genome sequence could be layered and confirmed. This mapping information, together with added paired clone end sequences, was used to generate "ultracontigs" by linking together sequence scaffolds. Several hundred double-color *in situ* hybridizations on metaphase chromosomes were finally necessary to anchor these ultracontigs (generally 1 to 10 megabases in size) to the *Tetraodon* genome. A similar mapping program is currently ongoing for the *Takifugu* genome, using the clone fingerprinting technique.

4.3
Resources

There are quite a few genomic resources available for both pufferfish species, with additional cDNA libraries available for both *Takifugu* and *Tetraodon*. There is also a P1 artificial chromosome library available for a related pufferfish species, *Sphoeroides nephelus* (the Southern pufferfish). This comprises 30 000 clones, which is equivalent to 7 to 8 genome equivalents. These are all detailed below in Table 2.

5
Pufferfish as a Genome Tool

5.1
Comparative Genome Analysis in Vertebrates

With the current abundance of whole-genome sequence in both the prokaryote and eukaryote kingdoms, "comparative genomics" is a term increasingly used to

Tab. 2 Resources (genomic and cDNA) for pufferfish.

Species	Library	Type	Average insert size [kb]	Available
Takifugu	genomic	cosmid	40	http://www.hgmp.mrc.ac.uk/geneservice/
	genomic	BAC	120	http://www.hgmp.mrc.ac.uk/geneservice/
	cDNA	plasmid		http://www.hgmp.mrc.ac.uk/ geneservice/info@image.llnl.gov
Tetraodon	genomic (A)	BAC	126	www.genoscope.cns.fr/
	genomic (B)	BAC	153	www.genoscope.cns.fr/
	genomic (G)	plasmid	4	www.genoscope.cns.fr/
	cDNA	plasmid		www.genoscope.cns.fr/
Sphoeroides	genomic	P1	125–150	replicas: camemiya@vmrc.org pools: jpostle@oregon.uoregon.edu replicas and filters: http://www.rzpd.de

describe any approach where the sequence of two genomes (whole or partial) are compared to extract some knowledge about functional elements and the evolutionary processes that shape their sequences. Such studies were slow to take off in vertebrates because of the large genome size of the few species that were used as models or that were of interest to genome biologists (human, mouse, frog, zebrafish, chicken, etc.). The advent of sequencing projects in pufferfish changed this situation by providing a relatively economical and fast access to large regions of vertebrate genomes to be used in such comparative analyses.

The position of *Takifugu* and *Tetraodon* relative to mammals in the evolutionary tree is crucial to their importance as genome tools. The long divergence time since the split with tetrapods (approximately 400 million years) implies that functionally inert sequences that are submitted to a neutral rate of mutation have had time to diverge sufficiently to provide a good contrast with those that are functionally important and are thus submitted to purifying selection. This was empirically evidenced by the many small-scale analyses of regions sampled from the *Takifugu* genome and compared to their orthologous regions in human and sometimes mouse.

Initially, one of the reasons behind proposing the pufferfish as a model was the hypothesis that gene finding in human would be facilitated by a comparative positional cloning approach in a compact genome. For example, if the candidate disease region was 4 Mb in human, then by taking flanking markers to this region and mapping them to the pufferfish, this would equate to only 500 kb in the pufferfish. This left a much smaller region of DNA, in which to search, identify, and sequence the putative candidate disease gene. This relied on there being large-scale conservation of synteny between the pufferfish and human (the main reference organism). At the beginning of the 1980s, very little was known about gene-order conservation between vertebrates, so early pufferfish studies concentrated on these small-scale mapping studies. In retrospect, not surprisingly, there were many examples of conservation of synteny between the pufferfish and human; however, many of these studies only encompassed two or three genes and often did not consider conservation of gene order. As larger data sets became available, it was clear that conservation of gene order over larger regions was rare (Fig. 4). Analysis of the *Takifugu* genome data showed that only 3.8% of the genome is in segments with no intervening sequences (perfect conservation of gene order between the pufferfish and human), while 5% is in segments with 1 intervening gene (mapping from a completely different human chromosome), 7% has up to 5 intervening genes, and 9.3% is in segments with up to 15 intervening genes. This is substantiated by mapping data in the zebrafish, which demonstrates large-scale conservation of synteny between zebrafish and human, at the whole chromosome level, but massive rearrangement of gene order between the two species. So the chance of finding a human candidate disease gene in the pufferfish by positional cloning is rare.

	Map position	Gene	*Tetraodon* chromosome
9p	24.3	DMRT1	8
	24.1	AK3	1
	24.1	GLDC	9
	23	TYRP1	11
	13.3	CNTFR	8
Centromere			
9q	21.13	ANXA1	10
	21.13	GCNT1	8
	21.32	HNRPK	12
	22.32	HSD17B3	7
	22.33	TMOD	13
	33	EPB72	14
	33.2	PTGS1	8
	33.3	PBX3	15
	34.11	AK1	16
	34.11	ENG	17
	34.11	ASS	10
	34.12	ABL1	10

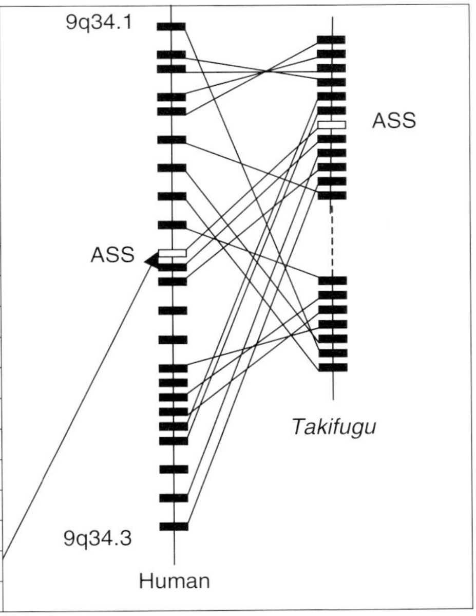

Fig. 4 Comparative mapping data from human chromosome 9, compared with *in situ* hybridization mapping data onto *Tetraodon* chromosomes (data taken from Gruetzner, F., Roest Crollius, H., Lutjens, G., Jaillon, O., Weissenbach, J., Ropers, H.H., Haaf, T. (2002) Four-hundred million years of conserved synteny of human Xp and Xq genes on three *Tetraodon* chromosomes, *Genome Res.* **12**, 1316–1322, with human map positions updated with Ensembl v.16.33.1) and also *Takifugu* BAC mapping data taken from Bouchireb, N., Gruetzner, F., Haaf, T., Stephens, R.J., Elgar, G., Green, A.J., Clark, M.S. (2001) Comparative mapping of the human 9q34 region in *Fugu rubripes*, *Cytogenet. Cell Genet.* **94**, 173–179. The data clearly shows that genes from human chromosome 9 are scattered across 12 different *Tetraodon* chromosomes. ASS, which maps to human 9q34.11, was mapped to a single *Takifugu* chromosome along with 2 BAC contigs containing 19 other human 9q34 genes. The diagram shows the rearrangements in gene order between the two species, which, in human, covers a region of 10 Mb.

There are exceptions to this general rule. At the gene level, there are the Hox complexes, and this is probably because of their coordinated expression in development processes. At the chromosomal level, there is some conservation of synteny of genes found on the human X. However, the X was almost certainly subjected to a special selection process during vertebrate development when compared with the rest of the chromosome complement. So, although one of the original reasons for studying the pufferfish is no longer robust, the pufferfish genomes have contributed significantly to our understanding of gene and chromosome evolution. Comparative analyses of human and pufferfish are still very useful for gene finding and also for promoter identification, as will be described in the following sections. As regards conservation of gene order between the two pufferfish, preliminary data indicates that they are nearly identical, which is not surprising when considering their close evolutionary relatedness.

Another way to exploit pufferfish genome sequence is to identify genes in other species. This became possible on a genome scale when the first samples of the *Tetraodon* WGS sequence became available. First, it was necessary to optimize sequence comparison algorithms so that the computing time required to compare a large fraction of vertebrate genome could be counted in days rather than months or years. Such parameters were thus adjusted using the BLAST sequence comparison algorithm so that about 40% of the human genome and 30% of the *Tetraodon* genome (the respective fraction of the two genomes available at the time) could be compared in about 5 days on a 100 CPU Unix cluster. The alignments produced were then filtered, on the basis of their length and their percent identity, so that only those that were thought to align over protein coding regions were retained. The criteria used in this selection were calibrated on known pufferfish and human genes. When several regions or sequence reads from *Tetraodon* overlapped on a given human region, this stretch of human DNA was dubbed an evolutionary conserved region (ecore). This first large-scale comparison between significant fractions of two vertebrate genomes produced a surprising result: the number of ecores estimated in the entire genome (once extrapolated from the 40% used in the comparison) was much lower than what would be expected if humans were to have the 80 000 to 100 000 genes given by previous estimates. Results from Exofish, the tool developed around the Human–*Tetraodon* comparison, suggested that the human genome contained between 28 000 and 34 000 genes, which is similar to the plant *Arabidopsis thaliana* and only five times more than the unicellular yeast. At the same time, this number was supported by a careful reanalysis of human EST sequence clusters, which suggested that the human genome contained no more than 35 000 genes. Today, with a finished sequence at hand, automatic annotation programs such as the Ensembl project have identified approximately 24 000 human genes and it is thought that few genes remain to be discovered. In fact, a comparison between the *Takifugu* draft genome sequence and an earlier version of the human genome suggested that less than 1000 human genes conserved between fish and mammals remained to be annotated.

Pufferfish genomes still have much to contribute to genome research. For instance, the human and mouse genomes are notorious for their abundance of transposable elements (approximately 45% of the genome). Conversely, probably the main reason for the small pufferfish genome size is their paucity in repeats. This difference could be exploited to address a key question: how much DNA is under evolutionary selection in a vertebrate genome? Because this fraction is likely to be small, it cannot be easily measured against the background of neutrally evolving DNA represented by the repeats. In fact, a rough estimate indicated that approximately 5% of the mammalian genome could be under selection, when comparing the mouse draft sequence to the human. But aligning the two pufferfish genomes together may provide a much better answer, and, moreover, may indicate where these regions could be.

5.2
Functional Analyses Including Detection of Regulatory Elements

The "classic" gene functional analysis tools of transgenics, knock-ins, and knockouts are not possible in the pufferfish because of problems with breeding and size (in the case of *Takifugu*). Additionally, there has been only one pufferfish cell line made and so cell culture expression studies in "same species" is also not generally available. However, this will not hinder the exploitation of the pufferfish genomes. While characterization of pufferfish gene structure is valuable for gene identification (see Sect. 5.1), identifying conserved functional domains, understanding gene evolution processes, and to fish biologists working on other species, the main exploitation will be for the identification and characterization of conserved noncoding regulatory elements. Understanding gene function is the next stage for all genome projects and activity is obviously particularly strong in the mammalian field. Comparative analyses of mammalian genomes with those of the pufferfish can greatly advance this area of research.

The true extent of a gene, or rather functional unit, comprising coding regions, 5 and 3′ UTRs (untranslated regions), and regulatory elements can span many 100 kb in mammals. An example of this is the human *Sox9* gene, which is surrounded by 2 Mb 5′ and 500 kb 3′ of noncoding DNA. In *Takifugu*, these regions are reduced to 68 kb and 97 kb respectively, representing a considerable compaction of sequence. What exists in these noncoding regions between genes is only just being investigated; although, homology between species is generally very low when considering UTRs and these can be determined relatively easily by sequencing of full-length cDNAs or by RACE experiments. Regulatory regions are much more difficult to define, they are very small, often less than 10 bp in length, and may be at a considerable distance from the coding region of the gene they control. This is the real use of the pufferfish; because of their small genome size, the essential regulatory elements may also be squeezed much closer to the genes they regulate. Despite the evolutionary distance between the pufferfish and mammals, some of the regulatory sequences remain similar enough to those of mammals to be recognized by sequence comparisons and also to be functional in transgenic organisms. To identify and characterize a regulatory element, there are two requirements: that the regulatory elements are present in the DNA and that there is an assay system.

Recognition of regulatory regions can be determined using a variety of methods, such as BLAST sequence similarity searching, Clustal multiple sequence alignments, or purpose-built programs: Pipmaker (http://bio.cse.psu.edu/pipmaker), Vista (http://www-gsd.lbl.gov/vista), or

Dotter. Having determined that there are conserved regions between the pufferfish and mammals in a gene of interest, these then require experimental testing.

So far, use of pufferfish DNA in transgenics has had a mixed history, particularly with regard to the use of whole pufferfish genes. Success was achieved in rat for the *Takifugu* isotocin and oxytocin genes, but other examples, such as the HD and Wt1 genes, have resulted in aberrant splicing and nonfunctional products. The sequencing of the zebrafish genome and the increasing number of zebrafish facilities has opened up a new and potentially very powerful avenue for pufferfish research. While pufferfish genes may not generally splice correctly in mammals or mammalian cell lines, there appear to be no such problems in zebrafish, indicating that the aberrant splicing of pufferfish genes that occurs in mammals is related to evolutionary differences in splicing efficiency between fish and mammals. Having said that, it is the regulatory regions that are the real source of exploitation, rather than the actual genes themselves.

Pufferfish promoter regions have been inserted upstream of reporter constructs and expressed in a range of cell lines (mammalian and piscine). They have enabled the narrowing of regions responsible for the control of gene expression in genes as varied as the *Mx* protein (involved in the innate immune system), *Ick* (a lymphocyte-specific protein tyrosine kinase), *Pax 6* (involved in eye development), *tyrosinase* (pigmentation gene), and *SCL* (stem cell leukemia gene involved in hemopoiesis and vasculogenesis). This number of examples, currently rather limited, will dramatically increase over the next few years, as more groups access and mine the pufferfish data.

Pufferfish promoters have also been assayed, to a lesser extent, in transgenic animals. The *Takifugu* SCL locus directed appropriate expression to hemopoietic and neural tissue in transgenic zebrafish embryos and the Wt1 gene from the same species was faithfully expressed in the developing zebrafish kidney. Ick and Pax 6 *Takifugu* noncoding regions were successfully tested in transgenic mice. In the human *Pax 6* gene, there is a 3′ cis-acting regulatory element termed C1170Box123. This contains three modules in human and mouse, but only one in *Takifugu*. However, this one conserved module was assayed in mice, using both *Takifugu* and human 3′ constructs and both produced similar reporter gene-expression patterns. This latter example clearly demonstrates the point that not all regulatory regions will be conserved between fish and mammals. Indeed, this cannot be expected as there are considerable differences in morphology, physiology, and biochemistry between the two sets of species. Also, just because no homology is detected between mammalian and pufferfish promoter sequences, this does not mean that pufferfish 5′ regions cannot direct expression of mammalian genes. Although the *Takifugu* CRABP-1 promoter showed no homology to that of mouse, expression of the mouse gene was still obtained, but in a much more tissue-restricted manner.

Analysis of promoter elements using comparisons between species does rely on similarity of expression of the gene under study. Clearly there will be many examples between mammals and fish where this is not so, but fish also have an additional use, as regards dissection of promoter elements. It is well documented that fish have extra genes. It is thought that these duplicate pairs survive via the processes of nonfunctionalization, neofunctionalization, and subfunctionalization. This duplication, which in many quarters is seen as a drawback of fish models, is actually very much an advantage. Comparison of duplicated genes between fish species and also with other vertebrates allows subfunction partitioning between the two genes, a situation that is not possible in mammalian models. Analysis of promoters between duplicated genes in fish can assign tissue- or developmental-specific functions to either or both of the pair of genes and can break up pleiotropic gene function and regulation. Few examples are available so far, but analysis of zebrafish, the two pufferfish, and *Xiphophorus*-duplicated *mitf* genes demonstrated subfunctionalization as a result of degeneration of alternative exons. This again is an area of investigation set to expand, particularly with regard to genes, which present in single copies in mammals are lethal when mutated. Mutations of such genes in fish are often still viable (as there are two copies) if only one of each duplicated pair is treated at any one time.

6
Prospects

The completion of the *Takifugu* and *Tetraodon* draft genome sequences mark a new step in the field of whole-genome sequencing in eukaryotes. Indeed, the first eukaryotes to be sequenced other than human (yeast, worm, fly, mustard weed, and mouse) were natural first choices: decades of laboratory experiments had accumulated a wealth of genetic and functional information, and their genome sequence was thought of as a conceptual framework in which to unify this diverse and heterogeneous data. In contrast, very little was known on the biology and life cycle of the two pufferfish described in this article. The reason for sequencing them lay exclusively in their position in the phylogenetic tree with respect to human and in their unusually small genome size. Their genome sequence has been a platform for gene discovery and genome comparisons.

But the *Takifugu* and *Tetraodon* genomes have not said their last word. Today, with the upcoming availability of the zebrafish genome sequence, it is expected that much will be gained by adequately combining functional tools developed in this species and comparative genomics tools represented by the pufferfish genomes.

See also Genetics, Molecular Basis of; Genomic Sequencing (Core Article); Immunoassays; Shotgun Sequencing (SGS); Whole Genome Human Chromosome Physical Mapping.

Bibliography

Books and Reviews

Ebert, E. (2001) *The Puffers of Fresh and Brackish Waters*, Aqualog verlag GmbH, Rodgau.

Helfman, G.S., Collette, B.B., Facey, D.E. (1997) *The Diversity of Fishes*, Blackwell Science, Malden, Massachusetts, USA.

Muller, F., Blader, P., Strahle, U. (2002) Search for enhancers: teleost models in comparative genomic and transgenic analysis of *cis* regulatory elements, *Bioessays* 24, 564–572.

Nelson, J.S. (1994) *Fishes of the World*, J. Wiley & Sons, New York, USA.

Rothenberg, E.V. (2001) Mapping of complex regulatory elements by pufferfish/zebrafish transgenesis, *Proc. Natl. Acad. Sci. U.S.A.* 98, 6540–6542.

Thomas, J.W., Touchman, J.W. (2002) Vertebrate genome sequencing: building a backbone for comparative genomics, *Trends Genet.* **18**, 104–108.

Tyler, J.C. (1980) *Osteology, phylogeny and higher classification of the fishes of the order Plectognathi (Tetraodontiformes)*. NOAA Technical Report NMFS Circular 434. US Department of Commerce, Washington, USA.

Winterbottom, R. (1974) *The Familial Phylogeny of the Tetraodontiformes (Acanthopterygii: Pisces) as Evidenced by their Comparative Mycology*, Smithsonian Contributions to Zoology, Number 155. Smithsonian Institution Press, Washington, DC.

Primary Literature

Altschmied, J., Delfgaauw, J., Wilde, B., Duschl, J., Bouneau, L., Volff, J.N., Schartl, M. (2002) Subfunctionalization of duplicate mitf genes associated with differential degeneration of alternative exons in fish, *Genetics* **161**, 259–267.

Aparicio, S., Hawker, K., Cottage, A., Mikawa, Y., Zuo, L., Venkatesh, B., Chen, E., Krumlauf, R., Brenner, S. (1997) Organization of the *Fugu rubripes* Hox clusters: evidence for continuing evolution of vertebrate Hox complexes, *Nat. Genet.* **16**, 79–83.

Aparicio, S., Chapman, J., Stupka, E., Putnam, N., Chia, J.M., Dehal, P., Christoffels, A., Rash, S., Hoon, S., Smit, A., et al. (2002) Whole-genome shotgun assembly and analysis of the genome of *Fugu rubripes*, *Science* **297**, 1301–1310.

Bagheri-Fam, S., Ferraz, C., Demaille, J., Scherer, G., Pfeifer, D. (2001) Comparative genomics of the SOX9 region in human and *Fugu rubripes*: conservation of short regulatory sequence elements within large intergenic regions, *Genomics* **78**, 73–82.

Barton, L.M., Gottgens, B., Gering, M., Gilbert, J.G., Grafham, D., Rogers, J., Bentley, D., Patient, R., Green, A.R. (2001) Regulation of the stem cell leukemia (SCL) gene: a tale of two fishes, *Proc. Natl. Acad. Sci. U.S.A.* **98**, 6747–6752.

Baxendale, S., Abdulla, S., Elgar, G., Buck, D., Berks, M., Micklem, G., Durbin, R., Bates, G., Brenner, S., Beck, S. (1995) Comparative sequence analysis of the human and pufferfish Huntington's disease genes, *Nat. Genet.* **10**, 67–76.

Bouchireb, N., Gruetzner, F., Haaf, T., Stephens, R.J., Elgar, G., Green, A.J., Clark, M.S. (2001) Comparative mapping of the human 9q34 region in *Fugu rubripes*, *Cytogenet. Cell Genet.* **94**, 173–179.

Brainerd, E.L., Slutz, S.S., Hall, E.K., Phillis, R.W. (2001) Patterns of genome size evolution in tetraodontiform fishes, *Evolution Int. J. Org. Evolution* **55**, 2363–2368.

Brenner, S., Elgar, G., Sandford, R., Macrae, A., Venkatesh, B., Aparicio, S. (1993) Characterization of the pufferfish (*Fugu*) genome as a compact model vertebrate genome, *Nature* **366**, 265–268.

Brenner, S., Venkatesh, B., Yap, W.H., Chou, C.F., Tay, A., Ponniah, S., Wang, Y., Tan, Y.H. (2002) Conserved regulation of the lymphocyte-specific expression of lck in the *Fugu* and mammals, *Proc. Natl. Acad. Sci. U.S.A.* **99**, 2936–2941.

Brum, M.J.I., Galetti, Jr, P.M. (1997) Teleostei ground plan karyotype, *J. Comp. Biol.* **2**, 91–102.

Camacho-Hubner, A., Richard, C., Beermann, F. (2002) Genomic structure and evolutionary conservation of the tyrosinase gene family from *Fugu*, *Gene* **285**, 59–68.

Camacho-Hubner, A., Rossier, A., Beermann, F. (2000) The *Fugu* rubripes tyrosinase gene promoter targets transgene expression to pigment cells in the mouse, *Genesis* **28**, 99–105.

Crnogorac-Jurcevic, T., Brown, J.R., Lehrach, H., Schalkwyk, L.C. (1997) *Tetraodon fluviatilis*, a new pufferfish model for genome studies, *Genomics* **41**, 177–184.

Dalle Nogare, D.E., Clark, M.S., Elgar, G., Frame, I.G., Poulter, R.T. (2002) Xena, a full-length basal retroelement from tetraodontid fish, *Mol. Biol. Evol.* **19**, 247–255.

Dasilva, C., Hadji, H., Ozouf-Costaz, C., Nicaud, S., Jaillon, O., Weissenbach, J., Roest Crollius, H. (2002) Remarkable compartmentalization of transposable elements and pseudogenes in the heterochromatin of the *Tetraodon nigroviridis* genome, *Proc. Natl. Acad. Sci. U.S.A.* **99**, 13636–13641.

Fischer, C., Ozouf-Costaz, C., Roest Crollius, H., Dasilva, C., Jaillon, O., Bouneau, L., Bonillo, C., Weissenbach, J., Bernot, A. (2000) Karyotype and chromosomal localization of characteristic tandem repeats in the pufferfish

Tetraodon nigroviridis, *Cytogenet. Cell genet.* **88**, 50–55.

Force, A., Lynch, M., Pickett, F.B., Amores, A., Yan, Y.L., Postlethwait, J. (1999) Preservation of duplicate genes by complementary, degenerative mutations, *Genetics* **151**, 1531–1545.

Gilligan, P., Brenner, S., Venkatesh, B. (2002) *Fugu* and human sequence comparison identifies novel human genes and conserved non-coding sequences, *Gene* **294**, 35–44.

Griffin, C., Kleinjan, D.A., Doe, B., van Heyningen, V. (2002) New 3' elements control Pax6 expression in the developing pretectum, neural retina and olfactory region, *Mech. Dev.* **112**, 89–100.

Gruetzner, F., Lutjens, G., Rovira, C., Barnes, D.W., Ropers, H.H., Haaf, T. (1999) Classical and molecular cytogenetics of the pufferfish *Tetraodon nigroviridis*, *Chromosome Res.* **7**, 655–662.

Gruetzner, F., Roest Crollius, H., Lutjens, G., Jaillon, O., Weissenbach, J., Ropers, H.H., Haaf, T. (2002) Four-hundred million years of conserved synteny of human Xp and Xq genes on three *Tetraodon* chromosomes, *Genome Res.* **12**, 1316–1322.

Hinegardner, R. (1972) Cellular DNA content and the evolution of teleostean fishes, *Am. Nat.* **106**, 621–644.

Jaillon, O., et al. (2003) The *Tetraodon nigroviridis* genome sequence (in preparation).

Kleinjan, D.A., Dekker, S., Guy, J.A., Grosveld, F.G. (1998) Cloning and sequencing of the CRABP-I locus from chicken and pufferfish: analysis of the promoter regions in transgenic mice, *Transgenic Res.* **7**, 85–94.

Lamatsch, D.K., Steinlein, C., Schmid, M., Schartl, M. (2000) Noninvasive determination of genome size and ploidy level in fishes by flow cytometry: detection of triploid *Poecilia formosa*, *Cytometry* **39**, 91–95.

Miles, C.G., Rankin, L., Smith, S.I., Niksic, M., Elgar, G., Hastie, N.D. (2003) Faithful expression of a tagged *Fugu* WT1 protein from a genomic transgene in zebrafish: efficient splicing of pufferfish genes in zebrafish but not mice, *Nucleic Acids Res.* **31**, 2795–2802.

Miyaki, K., Tabeta, O., Kayano, H. (1995) Karyotypes in six species of pufferfishes genus *Takifugu* (Tetraodontidae, Tetraodontiformes), *Fish. Sci.* **61**, 594–598.

Neafsey, D.E., Palumbi, S.R. (2003) Genome size evolution in pufferfish: a comparative analysis of diodontid and tetraodontid pufferfish genomes, *Genome Res.* **13**, 821–830.

Postlethwait, J.H., Woods, I.G., Ngo-Hazelett, P., Yan, Y.L., Kelly, P.D., Chu, F., Huang, H., Hill-Force, A., Talbot, W.S. (2000) Zebrafish comparative genomics and the origins of vertebrate chromosomes, *Genome Res.* **10**, 1890–1902.

Poulter, R., Butler, M. (1998) A retrotransposon family from the pufferfish (*fugu*) *Fugu rubripes*, *Gene* **215**, 241–249.

Poulter, R., Butler, M., Ormandy, J. (1999) A LINE element from the pufferfish (*fugu*) *Fugu rubripes* which shows similarity to the CR1 family of non-LTR retrotransposons, *Gene* **227**, 169–179.

Roest Crollius, H., Jaillon, O., Bernot, A., Dasilva, C., Bouneau, L., Fizames, C., Wincker, P., Brottier, P., Quetier, F., Saurin, W., Weissenbach, J. (2000) Human gene number estimate provided by genome wide analysis using *Tetraodon nigroviridis* genomic DNA, *Nat. Genet.* **25**, 235–238.

Roest Crollius, H., Jaillon, O., Dasilva, C., Ozouf-Costaz, C., Fizames, C., Fischer, C., Bouneau, L., Billault, A., Quetier, F., Saurin, W., Bernot, A., Weissenbach, J. (2000) Characterization and repeat analysis of the compact genome of the freshwater pufferfish *Tetraodon nigroviridis*, *Genome Res.* **10**, 939–949.

Sathasivam, K., Baxendale, S., Mangiarini, L., Bertaux, F., Hetherington, C., Kanazawa, I., Lehrach, H., Bates, G.P. (1997) Aberrant processing of the *Fugu* HD (FrHD) mRNA in mouse cells and in transgenic mice, *Hum. Mol. Genet.* **6**, 2141–2149.

Schuddekopf, K., Schorpp, M., Boehm, T. (1996) The *whn* transcription factor encoded by the nude locus contains an evolutionarily conserved and functionally indispensable activation domain, *Proc. Natl. Acad. Sci. U.S.A.* **93**, 9661–9664.

Thomas, J.W., Touchman, J.W., Blakesley, R.W., Bouffard, G.G., Beckstrom-Sternberg, S.M., Margulies, E.H., Blanchette, M., Siepel, A.C., Thomas, P.J., McDowell, J.C., et al. (2003) Comparative analyses of multi-species sequences from targeted genomic regions, *Nature* **424**, 788–793.

Venkatesh, B., Si-Hoe, S.L., Murphy, D., Brenner, S. (1997) Transgenic rats reveal functional conservation of regulatory controls between the *Fugu* isotocin and rat oxytocin genes, *Proc. Natl. Acad. Sci. U.S.A.* **94**, 12462–12466.

Yap, W.H., Tay, A., Brenner, S., Venkatesh, B. (2003) Molecular cloning of the pufferfish (*Takifugu rubripes*) Mx gene and functional characterization of its promoter, *Immunogenetics* **54**, 705–713.

Quantitative Analysis of Biochemical Data

Albert Jeltsch[1], *Jim Hoggett*[2], *and Claus Urbanke*[3]
[1] *International University Bremen, Bremen, Germany*
[2] *University of York, York, UK*
[3] *Medizinische Hochschule Hannover, Zentrale Einrichtung Biophysikalisch-Biochemische Verfahren, Hannover, Germany*

1	**Quantitative Data** 383	
2	**Numeric Modeling of Experiments** 383	
2.1	Formal Definition of the Mathematical Problem	383
2.2	Fitting 386	
2.3	Experimental Systems 387	
2.4	Measurement and Signals 388	
2.5	Models 388	
2.6	Selection of Appropriate Models 389	
2.7	Parameters 389	
2.8	Essential Steps in the Analysis 390	
2.9	Fitting Data by the Method of Least Squares 390	
2.10	Global Fitting of Multiple Data Sets 392	
2.11	Introduction to Error Estimation 393	
2.12	Introduction to Numerical Integration 394	
3	**Applications** 396	
3.1	Linear Regression 396	
3.2	Michaelis–Menten Kinetics 396	
3.3	Analysis of Circular Dichroism Experiments to Determine the Secondary Structure Composition of Proteins 397	
3.4	Dissociation Kinetics 397	
3.5	Binding Data 398	
3.6	Independent Identical Binding Sites 398	
3.6.1	Analysis of Simple Binding Data 399	
3.6.2	Independent Nonidentical Binding Sites 399	

Encyclopedia of Molecular Cell Biology and Molecular Medicine, 2nd Edition. Volume 11
Edited by Robert A. Meyers.
Copyright © 2005 Wiley-VCH Verlag GmbH & Co. KGaA, Weinheim
ISBN: 3-527-30648-X

3.6.3	Cooperative Binding 400
3.7	pH Dependence of Enzyme Catalyzed Reactions 401
3.8	Pre-steady State Kinetics 402
3.9	Association Kinetics Analyzed by Numerical Integration 403
3.10	Surface Binding Reactions 404
3.11	Analysis of Competition Experiments 404
3.12	Regularization – Application to Analytical Ultracentrifugation 406

Bibliography 408
Books and Reviews 408
Primary Literature 408

Keywords

Algorithm
A defined rule for calculating output numbers from input numbers, usually carried out by a computer.

Gaussian (or Normal) Distribution
A distribution of values with the probability of the distribution being proportional to the distance squared from the mean.

Global Fitting
Fitting experimental data from various data sets to a single model function.

Least-squares Fit
Method of fitting experimental data to a function by minimization of the squared deviations of the experimental and theoretical points.

Model
An abstraction (usually a simplification) of the processes that are happening in a real system that allows the processes to be described in mathematical language.

Numerical Integration
Obtaining the value of an integral or the solution of a differential equation in a stepwise procedure using only numerical values.

Occam's Razor
Also known as the principle of parsimony, this means that other things being equal, the model or hypothesis with the smaller number of assumptions or parameters is to be preferred.

Parameters
Constant elements of a function.

> Biochemistry is an experimental science characterized by an analytical and quantitative approach. Methods for the analysis of quantitative data are essential tools for handling the results of many biochemical studies. Some approaches are relatively straightforward and familiar to the biochemical community; others are more specialized, and their use is consequently less common. This chapter provides an account of two principal approaches to quantitative data analysis: least-squares fitting and regularization methods. A short summary of the theoretical background of these two approaches is followed by extensive coverage of applications taken from diverse areas including: enzyme kinetics, ligand binding, analytical ultracentrifugation, circular dichroism (CD), and the pH dependence of biochemical reactions. The discussion covers the nature of mathematical modeling of systems, and emphasizes the importance of selection of an appropriate physical model to describe the data.

1
Quantitative Data

The outcome of some biochemical experiments can be expressed as simple descriptive statements, such as "the desired band or peak was observed." In others, such as studies of the rates of processes or the affinity of a ligand for its target, the results need to be given in quantitative form as numerical values of one or more parameters. Such quantitative results are derived by mathematical analysis of the raw experimental data. As an example, consider the time course of an enzymatic reaction. The raw data are the dependence of product concentration (the dependent variable, conventionally shown on the y-axis) on the time (the independent variable, conventionally on the x-axis). The mathematical methods used to handle these and similar quantitative data can be of greater or lesser complexity depending on the processes involved and on the detail or depth of analysis required.

2
Numeric Modeling of Experiments

2.1
Formal Definition of the Mathematical Problem

This paragraph gives a formal introduction to the mathematical approaches most relevant to quantitative biochemical data analysis. For details on practical application, please refer to Sects. 2.2 and 3. In quantitative biochemical experiments, the outcome R can be described by a set of values usually expressed as real numbers. The elements of this set usually have all the mathematical properties of a function f, assigning the observable quantities (concentrations, spectroscopic signals etc.) to the independent variables $(x_i^1 \cdots x_i^n)$

(place, time, concentration of reactants ...) and constant parameters ($p_1 \cdots p_m$) (temperature, pH, buffer composition ...):

$$R = \{f_1(x_1^1 \ldots x_1^n; p_1 \ldots p_m),$$
$$f_2(x_2^1 \ldots x_2^n; p_1 \ldots p_m), \ldots,$$
$$f_k(x_k^1 \ldots x_k^n; p_1 \ldots p_m)\} \quad (1)$$

This general treatment can be simplified if one considers only a single observable. Then the result can be represented as an n-dimensional surface in $(n+1)$-dimensional space, created by the n-dimensions of the independent variables and one additional dimension for the experimental result. For further simplification, if there is only a single independent variable, this hypersurface is reduced to a situation that is much more familiar to a biochemist, that is, a set of points in a two-dimensional plot. One should, however, keep in mind that simply connecting these points with a continuous line introduces an assumption about the order of the functions and implies that one expects "in-between" functions that have not been measured. This procedure is only valid if there are good reasons to support this assumption, but fortunately this is more often than not true.

To illustrate this general function further, let us consider an enzyme kinetic measurement where the concentration of the product formed is measured as a function of time, while the enzyme concentration, initial substrate concentration, and temperature, as well as buffer composition are kept constant. In formal mathematical language this is expressed as follows:

$$c_{\text{Product}}(t) = f(t; c_{\text{Enzyme}}, c_{\text{Substrate}}^{\text{initial}},$$
$$pH, T, c_{\text{Buffer}})$$

In the experiment that we have just envisaged, time is the only independent variable, but repeating the experiment for different enzyme concentrations or initial substrate concentrations may introduce additional independent variables such that the complete experiment R consists of a set of time series of product concentration at different enzyme and initial substrate concentrations. Depending on the scientific question that is being posed, one also might vary the pH, temperature, or buffer composition. While in most experiments it is assumed that the independent variables and the constant parameters are known *a priori*, the values of these functions, which are the experimental results, are of course subject to experimental uncertainties, in other words – they are noisy. In obtaining such an experimental value, one is effectively taking a sample from a potentially infinitely large set of possible readings whose distribution reflects the experimental uncertainty. It follows, therefore, that statistical factors must be taken into consideration in any quantitative interpretation of results, a statement that will be all too familiar to anyone who has actually been involved in carrying out experiments and analyzing data.

Evaluation of such quantitative data is thus concerned with finding a suitable algebraic description for the earlier-mentioned functions to interpret it properly. The algebraic description that was used historically was a linear function, because then the only way of carrying out a regression analysis of data was fitting to a straight line. Data that were not themselves linear were transformed in such a way that they satisfied a linear relationship. Well-known examples of such transformations are the Scatchard analysis of binding data or Lineweaver–Burk analysis of enzyme kinetics. Unfortunately, all such linearization procedures inevitably distort the experimental data by nonlinear transformation of the noise, which is one of the major sources of error in data interpretation.

Since computing power is now not a limiting factor and linearization of data is no longer necessary, or indeed desirable, algorithms have been developed that allow the direct analysis of the raw experimental data without the necessity of transformation. Basically, an algebraic model is devised that can simulate the unknown function that is being measured experimentally. Such a model should be based on physical considerations about the experiment, but it may also be partly or completely empirical. In addition to the independent variables $(x_i^1 \cdots x_i^n)$ and constant parameters $(c_1 \cdots c_m)$ of the experiment, such a model will also include some model parameters $(p_1 \cdots p_o)$. These model parameters include physical constants, like rate constants or equilibrium binding constants whose determination usually is the aim of the experiment. The result of such modeling then is a set R_{Model} of numerical functions, ideally similar to those functions measured experimentally:

$$R_{\text{Model}} = \{f_1^{\text{Model}}(x_1^1 \cdots x_1^n; c_1 \cdots c_m; p_1 \cdots p_o), \cdots,$$
$$f_k^{\text{Model}}(x_k^1 \cdots x_k^n; c_1 \cdots c_m; p_1 \cdots p_o)\} \quad (2)$$

There are essentially two strategies that can be used to find an optimal description of an experimental result:

In *fitting algorithms*, R_{Model} is used to describe the measurement directly and optimization procedures are used to find the set of model parameters $(p_1 \cdots p_o)$ that best describes the experiment. The methods to find this optimal set will be discussed in Sect. 2.2.

Regularization methods can be exploited in circumstances where the fitting procedure is used to generate a best-fit function rather than values of individual parameters. In general, the fitting is carried out to a so-called describing function, which is of the following form $\int a(r) \cdot R_{\text{Model}}(r) dr$.

$R_{\text{Model}}(r)$ is a function that depends on the nature of the experiment being undertaken. For example, for kinetic data it may be appropriate to take an exponential function of the form $R = a \cdot e^{-t/\tau}$ where a (the amplitude of the function) and τ (its characteristic time dependence) are parameters used in the fitting algorithm. For ultracentrifuge data, the $R_{\text{Model}}(r)$ function is generally based on Lamm's differential equation expressing the concentration c as a function of distance (x) and time (t), the independent variables, and s and D are the sedimentation and diffusion coefficients, which are time and position independent parameters.

The other element of the describing function $a(r)$ is an arbitrary function whose form is part of the fitting procedure. Different criteria can be used to decide on the most appropriate form of the function $a(r)$. One that is commonly used is the maximum entropy principle, which selects the solution with the highest informational entropy, or alternatively expresses, the minimal information expressed as $\int (a \ln a) dr$. The rationale for this criterion is that, according to Occam's razor, the solution with the highest parsimony is to be preferred, a view that can be justified by Bayesian statistical arguments. In operational terms, it is important that the quality of the fit to the data should be balanced against any constraints such as maximum entropy or parsimony requirements.

As an example, one might envisage an exponential function $R = a \cdot e^{-t/\tau}$. In this function, both a and τ are the parameters that would be used in a fitting algorithm. However, a describes the height of the function (or its intensity) and τ describes the shape (or its steepness). The describing function is defined as $\int a(\tau) \cdot e^{-t/\tau} d\tau$. Since τ will now take all possible values (in practice the range will be limited) only $a(\tau)$ is left as an arbitrary function to be calculated to describe the experiment best. It is interesting to note that this procedure represents a projection of the experimental data onto $a(\tau)$. For example, a sum of three exponential functions $R = \sum_{i=1}^{3} a_i e^{-t/\tau_i}$ would result in an $a(\tau)$, which has the value of a_i at τ_i and zero everywhere else. Provencher in 1982 had given a full description of this method, and the same is also illustrated later in this chapter (Sect. 3.12) in an analysis of the sedimentation behavior of protein mixtures in the analytical ultracentrifuge that was first described by Schuck in 2000.

2.2
Fitting

Regardless of the strategy used, data evaluation by modeling involves searching for a numerical model that best describes the experimental data according to some suitable criterion. The questions that arise when data are being analyzed quantitatively are essentially the following:

- How well does the model under consideration, which is usually proposed on the basis of previous experience, perform in explaining the experimental data, bearing in mind the accuracy of that data? Is the model satisfactory, or is it necessary to consider alternatives?
- What values of the parameters characterizing the system (rate constants, equilibrium binding constants etc.) are most consistent with the experimental data?
- How accurate are these parameters, and what are the limits of error?

There are several important criteria that all procedures for data analysis should satisfy, chiefly that:

- the experimenter should be able to see the results of the analysis graphically to check whether they are reasonable, and get a feel for the accuracy;
- however the results are manipulated, the original raw data should not be lost;
- there should be no hidden error propagation in the operations, for example, by using transformations involving $1/x$, y^x and similar functions.

The most widely used criterion for judging the quality of description or fit of the data by a model is the least-squares criterion where the sum of squared deviations (SSD) or the root mean square deviation (RMSD) is a minimum for a given parameter set.

When having o parameters and k data points we obtain:

$$SSD(p_1 \cdots p_o) = \sum_{i=1}^{k} \left[f_i^{Model} - f_i^{measured} \right]^2$$

$$RMSD(p_1 \cdots p_o) = \sqrt{\frac{1}{k} \sum_{i=1}^{k} \left[f_i^{Model} - f_i^{measured} \right]^2} \quad (3)$$

If the original experimental measurements are normally (Gaussian) distributed around the "true" value, then such a minimization will lead to a parameter set of maximum likelihood. This set of parameters can be interpreted as the result of the experiment.

However, it should be borne in mind that the least-squares criterion is only one possible criterion of fit, and other criteria exist, which can be expressed by using the following more general definition of model deviation D:

$$D(p_1 \ldots p_o) = \sum_{i=1}^{k} |(f_i^{\text{Model}} - f_i^{\text{measured}})|^n \quad (4)$$

where n can be any real number. In case of $n = 1$ the fitting procedure is called *robust fitting*. Robust fitting can be of advantage whenever the sampled measurements are not distributed in a Gaussian manner, but contain some "outliers" (Johnson, 2000).

The basic concepts underlying the methods of data analysis discussed here are illustrated in Fig. 1. The results of an experiment are data. A model is a description of the processes taking place in the experimental system being observed, which defines a mathematical relationship between the independent variables and the results. The model also defines physical parameters as variables to be fitted or used in regularization. With plausible initial values of the parameters, the mathematical relationships are used to obtain simulated data, which are compared with the experimental data. The values of the parameters are then varied until an optimal fit is obtained of the simulated and experimental results.

In the following sections, the basic concepts of quantitative data analysis are discussed, together with the terms used in Fig. 1.

2.3
Experimental Systems

The system is made up of various components and species. "Components" are molecules that differ in their covalent structures, for example, enzyme, substrate, and product; components can interact to form complexes, for example, an enzyme–substrate complex. "Species" are all the entities present in the solution that differ in either their covalent or

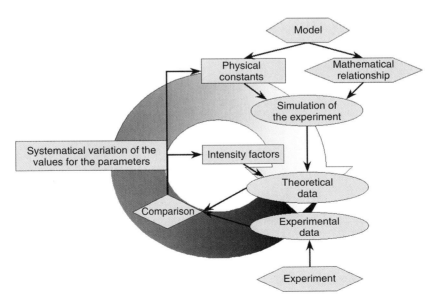

Fig. 1 Schematic illustration of fitting procedures used for quantitative data analysis.

noncovalent structures; this will include components, complexes, and where relevant, different conformations of these.

2.4
Measurement and Signals

In our analysis, we consider a general relationship between the measurement or "signal" and the composition of the solution; the signal is the experimental quantity being measured, which gives information about the processes taking place. It is assumed that the total measured signal is additive in terms of the contributions of all of the species (i) in the solution, and furthermore, that the signal from each species is proportional to its concentration (c_i). Different species contribute in differing extents to the total signal, and the proportionality constant (f_i) is termed *the intensity factor* of the species i. This intensity factor defines the relationship between the concentration of the species and the measured signal (e.g. CPM, absorbance, fluorescence intensity, etc.). It is usual to have to take account of a nonspecific but constant background signal, which we here define as the *baseline* (BL). Thus, the observed signal is given by:

$$S = BL + \sum_i f_i c_i \quad (5)$$

In many cases, this expression simplifies by using the mass conservation laws. So, for example, if we have the species A, B, that form AB in solution, then the signal is given by the expression

$$\begin{aligned} S &= BL + f_A c_A + f_B c_B + f_{AB} c_{AB} \\ &= BL + f_A c_A^0 + f_B c_B^0 + (f_{AB} - f_A - f_B) c_{AB} \\ &= C + f c_{AB} \end{aligned} \quad (6)$$

with only a single intensity factor and a constant term. To simplify the analysis, and to make the numerical analyses more stable, it is important that realistic assumptions are made about which species contribute to the signal, for example, in the case of radioactive detection, only those species that are labeled.

2.5
Models

As described in the mathematical definition section, a model represents an abstraction of the processes that are happening, or could be happening, during the experiment. From this model, we can derive a mathematical relationship between the experimental results and the independent variables. Consider the simple case of the two species A and B forming a complex AB in a time-dependent process:

$$c_{AB} = f(t, c_A, c_B)$$

The experimental results would be the concentrations c_{AB}, and the independent variables would be c_A, c_B, and the time t.

The model provides specific parameters for the fitting process, enabling theoretical data to be evaluated. These can be calculated either analytically or numerically. If the mathematical relationship between the signal (observation) and the parameters is sufficiently simple, it may be possible to obtain analytical solutions and calculate the theoretical signal directly, that is, in the present example to obtain values of c_{AB} knowing the initial concentrations of the concentrations c_A^0, and c_B^0, and the time t.

However, in many cases, the mathematical relationships are not that simple, and analytical solutions may either not be possible in principle, or be too difficult and cumbersome in practice. In such cases, the theoretical data can be simulated by numerical methods.

A model is, of course, only a working hypothesis, whose validity is judged by its success in accounting for the data. If its performance is not satisfactory, then alternative models should be sought or devised. However, if a model is to be replaced by a more complicated one, then it is important to check that the data really warrant this. More complicated models generally have more parameters, and more parameters will always lead to better fitting of the data. One should be guided here by the Principle of Parsimony, that other things being equal, the preferred model is the simplest one with the fewest parameters.

2.6
Selection of Appropriate Models

The choice of the right model to be used to describe experimental results is one of the trickiest and most interesting tasks in scientific work, and this is a subject that can only be touched on here. As discussed earlier, we are guided by the Principle of Parsimony that in science one should seek the simplest explanation for phenomena. In the present context, it means that we should define models with as few parameters as possible, consistent with obtaining a satisfactory description of the data. This is a sensible approach, because if a simple model fits the data adequately, then so necessarily must more complicated versions of that model. It follows that experimental observations can only serve to rule out models often, but not always, because they are oversimplified; the data can never prove that a model is correct. The question naturally arises at this stage about how one can establish whether a model is successful in accounting for the data or not. There are several criteria for assessing the quality of a model.

- The absolute magnitude of the deviations between the theoretical and experimental data. Does the theoretical curve lie in the region of experimental uncertainty of the data points (taking particular care not to overestimate the accuracy of the data)?
- The direction of the deviations between the theoretical and experimental data. Are the deviations randomly distributed, sometimes above and sometimes below the curve, or are they clustered, above the curve in one region and below in another? If the deviations are not randomly distributed, this indicates that the theoretical curve is not a satisfactory fit to the experimental data. One reason for this is that the model is wrong and is not an adequate description of the situation; another is that systematic errors have been made in carrying out the experiment.
- Whether alternative models are available, which can account for the data more satisfactorily.

A good model should also have predictive power and suggest additional experiments that can be carried out to test the model further.

2.7
Parameters

Depending on the model under consideration, one obtains a set of parameters that establish the relationship between the experimental data and the assumptions underlying the model. It is important to distinguish two kinds of parameters: global and local. This distinction is important when several data sets are being considered jointly in the analysis; the values of the global parameters must be the same in all cases, whereas, those of the local parameters may vary from one data set to another.

- Global parameters: the values of the global parameters are the same for all of the data sets that are being considered in the analysis. We are dealing here with physical quantities such as binding or rate constants whose values we wish to determine.
- Local parameters: the values of the intensity factors discussed earlier can differ from experiment to experiment. Examples of intensity factors are: radioactivity ($CPM = f_i \cdot c_i$), fluorescence intensity ($signal = f_i \cdot c_i$), absorbance spectroscopy ($OD = f_i \cdot c_i$, where f_i is the extinction coefficient of species i), ELISA ($signal = f_i \cdot c_i$), and so on. Although the precise values of these factors, which are local parameters, often are not particularly interesting in understanding the system, they are needed for the analysis.

2.8
Essential Steps in the Analysis

There are three basic steps in every data analysis (cf. Fig. 1):

- arbitrary initial values of the parameters are introduced into the model to calculate theoretical concentrations for all of the species of interest in the system.
- these theoretical concentrations are combined with initial values for the intensity factors to obtain theoretical values for the measurement or signal.
- the values of the parameters and intensity factors are varied to obtain the best fit of the theoretical values of the signal to the experimental values; the combination of parameters that best fits the data is the result of the analysis.

2.9
Fitting Data by the Method of Least Squares

The classical method for fitting data to theoretical curves is linear regression. This procedure allows the equation of the best straight line fitting a set of experimental data to be calculated directly:

$$y = a + bx$$

$$\text{slope } b = \frac{n \sum xy - \left(\sum x\right)\left(\sum y\right)}{n \sum x^2 - \left(\sum x\right)^2} \quad (7)$$

y intercept $a =$

$$\frac{\left(\sum x^2\right)\left(\sum y\right) - \left(\sum x\right)\left(\sum xy\right)}{n \sum x^2 - \left(\sum x\right)^2}$$

Until relatively recently, this was the only method that could be used conveniently to fit data by regression. This is the reason why so many classical approaches for evaluating biochemical data depended on linearizing data, sometimes by quite complex transformations. The best-known examples are the use of the Lineweaver–Burk transformation of the Michaelis–Menten model to derive enzyme kinetic data, and of the Scatchard plot to analyze ligand binding equilibria. These linearization procedures are generally no longer recommended, or necessary.

In contrast to the explicit analytical solution of "least-squares fit" used in linear regression, our present treatment of data analysis relies on an iterative optimization, which is a completely different approach: as a result of the operations discussed in the previous section theoretical data are calculated, dependent on the model and choice of parameters, which can be compared with the experimental results. The deviation between theoretical and experimental data is often expressed as the sum of the errors squared for all the data points, alternatively called the *sum of squared deviations* (SSD):

$$SSD = \sum_i (S_{i,\text{exp}} - S_{i,\text{theo}})^2 \quad (8)$$

This deviation is now minimized by variation of the parameters. The combination of parameter values that best fit the experimental data – using this deviation as the criterion of best fit is the desired solution of the analysis. This process of finding a solution is termed *iteration* because the solution is located by trying out many possible combinations of parameters; since the equations being fitted are in general nonlinear, the process is more specifically one of the iterative nonlinear least-squares fitting.

The process is essentially as follows. All possible combinations of the parameters (physical constants and intensity factors), of number N, define an $(N+1)$-dimensional error space. Every point in that space has a characteristic value of the sum of the squared deviations or SSD, which thus generates an error surface in $(N+1)$-dimensional space. If, for simplicity, we consider a model with only two parameters, these can be represented on the x- and y-axes, and the value of SSD on the z-axis. The error surface is now simply a surface in conventional three-dimensional space. An even simpler example with one parameter is illustrated in Fig. 2, where the parameter is shown on the x-axis and the SSD is on the y-axis. The task in the fitting procedure is to locate the minimum value in the SSD curve (region A in Fig. 2). It is impracticable to try out all possible values of the combined set of parameters, particularly when there are many of them. The procedure adopted in most computer programs is, starting from initial values of the parameters (provided by the user) calculations are made of the slope (or derivative) of the error surface in $(N+1)$-dimensional space. This is done by making a small variation in each of the parameters in turn, and calculating the SSD. The program then locates the region where the slope is steepest (downwards) and it alters the parameters by a small step in that direction to generate a new set of parameters, which fit the data better. From this new set of parameters, the program repeats the operation in a second iterative cycle to locate the direction of steepest descent, and hence a new set of parameters.

This procedure depends on certain features that merit comment.

- The step length in the iteration is critical: if it is too short, then the process of

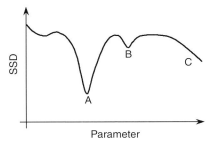

Fig. 2 Two-dimensional representation of an error surface. Region A is the location of the global minimum, region B is a local minimum, and region C represents an area where the model is no longer valid and the slope of the error surface is directed away from the minimum.

locating the minimum takes too long, whereas, if the step length is too long the algorithm used in the program can miss the target area, and thus never locate the minimum. The SOLVER algorithm used in Excel selects the step length automatically depending on the slope of the error surface and the result of the previous round of iteration.
- The "result" located by the program can be a local minimum (e.g. region B in Fig. 2). Locating the global minimum is often not straightforward, particularly when the error surface is complex, and the program can find itself trapped in a local minimum. The best means of avoiding this, or at least detecting when it is happening, is to begin the iteration process from different initial parameter estimates, and check whether the same solution is found in every case. If this does not happen, the solution with the lowest SSD corresponds to the best solution, although it should be noted that in some cases alternative solutions might be equally good in terms of their SSD values, bearing in mind the accuracy of the experimental data.
- To avoid local minima, most algorithms also test randomly selected points in the error surface. The extent to which a program carries out these tests determines the speed of locating the minimum and the tendency of the algorithm to become trapped in local minima.

All models have limits to their region of validity; for example, negative values of rate or binding constants do not correspond to physically meaningful situations. In such regions, mathematical errors will arise, such as attempting to find the square root of a negative number, even though all of the equations have been correctly programmed.

The slope of the error surface can lead the iterations into regions that are remote from the minimum. This situation can readily lead to failure to locate the minimum when the initial parameter estimates are not very good. To remedy this, a fresh set of initial estimates should be selected, which fit the data better. In Fig. 2, for example, it would be difficult to locate the minimum if the program started in region C since the slope in the error surface is pointing in the wrong direction.

The usual criterion of "best fit" is the sum of errors squared (the SSD discussed earlier) rather than the absolute magnitude of the errors. This procedure is mathematically justified when the errors in the data follow the Gaussian (or normal) distribution. Under these conditions, the error distribution function is given by the following equation in which x is the measurement, μ the mean, and σ the standard deviation:

$$f(x) = \frac{1}{\sigma\sqrt{2\pi}} e^{-\frac{1}{2}\left(\frac{x-\mu}{\sigma}\right)^2} \qquad (9)$$

When the data are distributed according to this function, the frequency of occurrence of data falls according to the square of the deviation. In practice, the sum of error square (SSD) criterion is also used in cases where it has not been explicitly established that the errors are normally distributed, and it appears to function quite well.

2.10
Global Fitting of Multiple Data Sets

If different sets of experiments have been carried out under circumstances in which the observations depend on a common set of parameters, then it is sensible to attempt a global fitting of the data sets to obtain best estimates of the parameters. It is important here to distinguish clearly between the global and local parameters discussed earlier (Sect. 2.7). The global parameters are valid for all of the data sets and are fitted to all of the data, whereas the local parameters may assume different values for the various data sets. For example, if one were investigating a thermodynamic equilibrium, and monitoring the process using radioactive detection, the value of the equilibrium constant must be the same under the same conditions, whereas the specific activities of the reaction participants could well be different. In global data fitting, it is particularly important to keep the number of parameters as small as possible. There are two reasons for this. First, the general consideration that following the Principle of Parsimony one should seek to account for experimental data using the smallest number of variables. Second, that iterative fitting of the data becomes much more difficult (in fact, exponentially so) as the number of variables increases; the process becomes much slower, and there is an increasing risk that local minima will interfere with the fitting. To keep the number of parameters as small as possible, it is important to check, in particular, whether all of the local parameters are needed. For example, in the general case, it is assumed that all of the reaction participants contribute to the experimental signal or measurement, but if this is not in fact true, then it is better to set the intensity factors of as many species as possible to 0, and only allow the minimum necessary number of species to contribute to the signal. For example, in studies based on fluorescence detection, only species containing a fluorophore need to be assigned intensity factors.

One difficulty that can arise in global data analysis is that the signal intensities of different data sets can be very different. If the data are treated equally, this can lead to the situation that data sets or curves with high intensities completely dominate those with lower intensities, simply because their error squared parameters (SSD) are so much larger. The most effective way of dealing with this situation is to weight the SSDs of the different data sets or curves by a suitable factor, for example, by the mean value of the data set, or by the relevant intensity factor. It should be emphasized that weighting factors must never be treated as variables in the fitting process.

2.11
Introduction to Error Estimation

One of the most difficult tasks in day-to-day scientific activity is making reliable estimates of the errors and uncertainties in the data. How reliable are my data? How accurate are the parameters calculated from them? Can I, or should I, exclude particular models for explaining my data? These are examples of the sort of questions that need to be asked. We have already discussed the question of judging how well models perform in accounting for data; we now turn to the question of assessing accuracy, on the basis that the model used is an appropriate one. It should be noted that we are dealing here with statistical errors and the treatment of outliers; systematic errors cannot be detected by these approaches.

Statistical analysis of repeated measurements. If a very large number of data are available, then it is sensible to consider carrying out a rigorous statistical analysis. The simplest procedure is to do many replicates of the same experiment (or series of experiments, if more than one data set is needed for the analysis) and then analyze them independently. This is a very good way of assessing the error range (determined as standard deviations, maximal range etc.) of the individual parameters; the problem is the amount of work involved.

The accuracy of individual measurements. We are concerned here with the problem of assessing accuracy when the number of available data is limited. One means of gauging error is to remove individual data points from the fitting process to get a feel for the "robustness" of the data. In effect, what this process does is to analyze the data on the assumption that the single experiment removed had not been carried out. It is possible in this way to assess how reliable the data are, and specifically to determine whether the outcome was highly dependent on the single result, implying that one would need to be very sure about it. This form of analysis is straightforward and revealing, and it ought to be a part of every data evaluation.

Analysis of error surfaces. To conclude this section, we consider a more quantitative approach to error estimation. The first step is to estimate the accuracy of the individual data points; this can either be done by analysis of the variability of replicate measurements, or from the variation of the fitted result. From that, one can assess the shape of the error surface in the region of the minimum. The procedure is straightforward: the square root of the error, defined as the SSD, is taken as a measure of the quality of the fit. A maximum allowed error is defined, which depends on the reliability of the individual points, for example, 30% more than with the best fit, if the points are scattered by about 30%. Then each variable (not the SSD as before) is minimized and also maximized. A further condition is imposed so that the sum of errors squared (SSD) should not increase by more than the fraction defined earlier. This method allows good estimates to be made of the different accuracy of the component variables, and also enables accuracy to be estimated reliably even in complex analyses. Finally, it reveals whether parameters are correlated. This is an important matter since it happens often, and in some extreme cases where parameters are tightly correlated, it leads to situations where individual constants are effectively not defined at all, merely their products or quotients. Correlations can also occur between global and local parameters. This issue is illustrated in Fig. 3, where a set of initial slopes of an enzyme reaction are analyzed using the Michaelis–Menten model (see Sect. 3.2, Eq. 12). Since K_m is defined using v_{max}, depending on the shape of the curve any change of v_{max} will cause a certain change of K_m as well.

2.12
Introduction to Numerical Integration

Kinetic processes can be described by differential equations; for example,

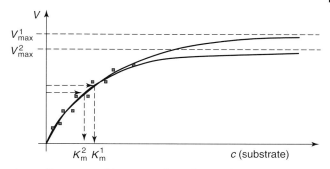

Fig. 3 Illustration of the origin of correlation of parameters with the example of a Michaelis–Menten analysis of steady state kinetics data. In this example, there is a strong correlation of k_{cat} and K_m.

for a reversible bimolecular association reaction:

$$A + B \underset{k_{21}}{\overset{k_{12}}{\rightleftharpoons}} AB$$

is described by

$$\frac{\partial c_{AB}}{\partial t} = k_1 c_A c_B - k_{-1} c_{AB} \quad (10)$$

This equation defines directly the change in concentration of the species AB with given concentrations of the reactants A and B, and the product AB. This is a differential equation whose solution is an expression of the form $c_{AB} = f(t, c_A^0, c_B^0)$. The solution involves a process of integration, which is often difficult, and sometimes impossible, at least analytically. In such cases, numerical integration can be used to simulate the time-dependent variation of c_{AB} in an experiment, enabling theoretical data to be obtained even for complex systems.

The procedure for numerical integration is as follows. Initial conditions are first selected: c_A^0, c_B^0, c_{AB}^0 and from this initial state the concentrations of the three component species are altered stepwise using "fluxes" defined from the differential equation given earlier, with a finite time increment Δt.

Two different fluxes exist:

F_1: "association," $A + B$

$\longrightarrow AB$ for which $F_1 = k_1 c_A c_B \Delta t$

F_{-1}: "dissociation," AB

$\longrightarrow A + B$ for which $F_{-1} = k_{-1} c_{AB} \Delta t$

(11)

The concentration changes are defined in terms of these fluxes as follows:

$$\Delta c_A = -F_1 + F_{-1}$$
$$\Delta c_B = -F_1 + F_{-1}$$
$$\Delta c_{AB} = -F_{-1} + F_1$$

from which new concentrations are obtained using the following general expression, where $c_{i,\text{old}}$ is the "old" concentration of the species (i) before the incremental change Δc_i:

$$c_i = c_{i,\text{old}} + \Delta c_i$$

The formulae given in Eq. (11) are prototypes for bimolecular (F_1) and monomolecular (F_{-1}) elementary reactions respectively. By combining these prototype equations, kinetic schemes of any desired complexity can be described and analyzed. The time interval is a critical parameter in the integration process, if it is chosen too short, the process takes too long, if it is too long, the process is inaccurate.

3 Applications

3.1 Linear Regression

Situations arise very often where data need to be fitted to linear equations. Linear regression is one of the classical procedures in general regression analysis, and before the advent of accessible nonlinear fitting methods it was the only one that could be readily used. For n data pairs in the form (x, y) where y is a function of x, the linear equation of the form $y = a + bx$ that minimizes the sum of errors squared (SSD) is given by Eq. (7).

3.2 Michaelis–Menten Kinetics

The Michaelis–Menten model shown below is the simplest mechanism for describing the kinetics of enzyme catalyzed reactions:

$$S + E \underset{k_{-1}}{\overset{k_1}{\rightleftharpoons}} ES \longrightarrow E + P$$

According to this mechanism, the rate constant for the reaction k (dimension: time^{-1}) depends on two parameters: K_m and k_{cat}. In the simple mechanism shown earlier and with the assumption that ES is in a steady state, $K_m = (k_2 + k_{-1})/k_1$ and $k_{cat} = k_2$. The dimensions of K_m and k_{cat} are concentration and (time)$^{-1}$ respectively. The rate of the reaction v (dimension: concentration/time) is given by the expression $v = kc_{E,total}$, and v_{max} is equal to $k_{cat}c_{E,total}$. The dependence of the reaction rate constant (k) on substrate concentration is given by Eq. (12), from which it can be seen that the K_m value is the concentration of substrate that gives half of the maximum rate k_{cat}.

$$k(c_S) = k_{cat} \frac{c_S}{c_S + K_m} \quad (12)$$

To evaluate K_m and k_{cat}, the rate of reaction is measured as a function of substrate concentration and the two kinetic parameters are determined using Eq. (12). The classical method of doing this is by fitting the data to a linearized form of Eq. (12) such as the Lineweaver–Burk plot shown in Eq. (13):

$$\frac{1}{k(c_S)} = \frac{1}{k_{cat}} + \frac{K_m}{k_{cat}} \frac{1}{c_S} \quad (13)$$

From this it follows that a plot of $1/k$ against $1/c_S$ should give a straight line with an x-intercept of $-1/K_m$ and a y-intercept of $1/k_{cat}$. The Lineweaver–Burk analysis illustrates very clearly the sort of problem that can arise when dealing with linearized data. An assumption that underlies simple linear regression following the procedure discussed in the previous section is that all of the data points have the same error, or specifically, standard deviation. This assumption is no longer valid when the data are transformed as is shown in the following diagrams.

Figure 4(a) illustrates a series of measurements where all the data have the same error; in Fig. 4(b), the same data are shown after transformation for Lineweaver–Burk analysis. It can be seen that the data points at low concentration (i.e. at high values of $1/k$ and $1/c_S$) have a much higher error than the other points, and the situation

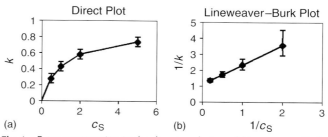

Fig. 4 Error propagation in the direct analysis and Lineweaver–Burk analysis of Michaelis–Menten kinetics.

is made worse because these inaccurate points are also the ones that exert the most leverage on the linear regression, and hence on the derived kinetic parameters. A very comprehensive analysis of the kinetics of enzyme catalyzed reactions has been presented by Bisswanger in 2000.

3.3
Analysis of Circular Dichroism Experiments to Determine the Secondary Structure Composition of Proteins

Circular dichroism (CD) is the spectroscopic method of choice to investigate the secondary structure of proteins. The CD spectrum of a protein is given by the sum of the contributions of its α-helical, β-sheet, β-turn, and disordered ("random") structural elements. The results of CD experiments can be analyzed very simply to provide information about the fraction of α-helical, β-sheet, β-turn, and random regions in a protein. The analysis depends on comparison of the CD spectrum with reference spectra for the respective secondary structure elements, which are derived from the CD spectra of a large set of proteins of known structure. Then, the CD signal is given by:

$$\theta(\lambda) = c_P(f_\alpha \varepsilon_\alpha^{\text{ref}} + f_\beta \varepsilon_\beta^{\text{ref}} + f_t \varepsilon_t^{\text{ref}} + f_r \varepsilon_r^{\text{ref}}) \tag{14}$$

where c_P is the concentration of the protein, f_α, f_β, f_t, and f_r denote the fractions of the protein in α-helical, β-sheet, turn and random conformation, and $\varepsilon_i^{\text{ref}}$ is the extinction coefficients of the respective secondary structure element taken from the reference spectra. For analysis, $\theta(\lambda)^{\text{theo}}$ is fitted to $\theta(\lambda)^{\text{exp}}$ by variation of f_α, f_β, f_t, and f_r.

3.4
Dissociation Kinetics

Dissociation reactions of the general form $AB \rightarrow A + B$ are monomolecular processes, where the rate of decay of the complex is proportional to its concentration. The concentration dependence of c_{AB} is given by the following differential equation:

$$\frac{dc_{AB}}{dt} = k_{-1} c_{AB}$$

which on integration yields Eq. (15), where $c_{AB}(t)$ is the concentration of complex at any time t, and c_{AB}^0 is the initial concentration at time $t = 0$:

$$c_{AB}(t) = c_{AB}^0 e^{-k_{-1}t} \tag{15}$$

Analysis of dissociation processes yields values for the rate constant k_{-1}, whose dimensions are $(time)^{-1}$. This rate constant is related to the lifetime (τ) of the complex AB by the expression $\tau = (k_{-1})^{-1}$, and to the half-life ($t_{1/2}$) by the expression $t_{1/2} = (\ln 2/k_{-1})$.

3.5
Binding Data

The equilibrium constant for a simple bimolecular association process

$$A + B \underset{}{\overset{K_{Ass}}{\rightleftharpoons}} AB$$

is defined by the expression:

$$K_{Ass} = \frac{c_{AB}}{c_A c_B} \qquad (16)$$

This equilibrium constant is expressed as the association constant, which has dimensions (concentration)$^{-1}$, in molar terms M^{-1}. The dissociation constant K_{Diss} is the reciprocal of K_{Ass} and has dimensions of concentration (M). The objective of the following derivation is to obtain an equation of the form $c_{AB} = f(c_{A,tot}, c_{B,tot}, K_{Ass})$, where $c_{i,tot}$ is the total or stoichiometric concentration of the component i (which is known), in contrast to the quantity c_i in Eq. (16), which is the free concentration of the species in solution, which is not known. An equation of this form will enable us to calculate the theoretical data.

Using the conservation conditions: $c_{A,tot} = c_A + c_{AB}$ and $c_{B,tot} = c_B + c_{AB}$ Eq. (16) can be written in the form:

$$K_{Ass} = \frac{c_{AB}}{[(c_{A,tot} - c_{AB})(c_{B,tot} - c_{AB})]} \qquad (17)$$

The only unknown in this equation is the term c_{AB}. Expanding and rearranging Eq. (17) yields the following quadratic equation:

$$c_{AB}^2 - \left(c_{A,tot} + c_{B,tot} + \frac{1}{K_{Ass}}\right) c_{AB} + c_{A,tot} c_{B,tot} = 0$$

The solutions of a quadratic equation of the general form $x^2 + px + q = 0$ are given by the two roots x_1 and x_2:

$$x_{1,2} = -\frac{p}{2} \pm \sqrt{\left(\frac{p}{2}\right)^2 - q}$$

In the present case, only the negative square root term is physically meaningful, so the concentration of AB is given by the following equation:

$$c_{AB} = \frac{c_{A,tot} + c_{B,tot} + 1/K_{Ass}}{2}$$
$$- \sqrt{\left(\frac{c_{A,tot} + c_{B,tot} + 1/K_{Ass}}{2}\right)^2 - c_{A,tot} c_{B,tot}}$$
(18)

To determine values of K_{Ass} binding data are needed, where the total concentrations of either A or B are comparable in magnitude to $1/K_{Ass}$.

3.6
Independent Identical Binding Sites

The earlier-mentioned model and equations have to be modified if one of the species (say A) has several binding sites for the other species B. If the binding sites are independent and do not interact, then binding to each site on A can be described by Eq. (19). Taking all of the binding sites into account yields a hyperbolic binding curve whose binding equation only differs from Eq. (18), where for every molecule of A, n binding sites exit such that the total concentration of binding sites is $nc_{A,tot}$:

$$c_{B,bound} = \frac{nc_{A,tot} + c_{B,tot} + 1/K_{Ass}}{2}$$
$$- \sqrt{\left(\frac{nc_{A,tot} + c_{B,tot} + 1/K_{Ass}}{2}\right)^2 - nc_{A,tot} c_{B,tot}}$$
(19)

To obtain accurate estimates of the number of binding sites (n), binding experiments (usually titrations) need to be performed under conditions where the total concentration of A is relatively high, specifically that $c_{A,tot} \gg 1/K_{Ass}$; these conditions define a "stoichiometric titration" where, effectively all of the B added is bound until the sites on A are saturated (see Fig. 5). Titrations under these conditions are insensitive to the value of the association constant, so to obtain reliable estimates of K_{Ass}, data are needed from titrations at much lower concentrations, where $c_{A,tot} \leq 1/K_{Ass}$. It should be clear from this discussion that it is not easy to evaluate both n and K_{Ass} accurately, and it is usually necessary to do a global analysis of several data sets, obtained under different concentration conditions.

3.6.1 Analysis of Simple Binding Data

The equation for n identical, noninteracting binding sites (Eq. 19) is in principle soluble, although the solution is not straightforward. When binding is more complex and the sites are of different affinity and interacting, then analytical solutions cannot be obtained. However, analysis of the binding can be simplified by carrying out experiments under conditions where one of the interacting partners (say A) is present at a much lower concentration than the other. The concentration of the partner in excess (B) is varied, and the proportion of available binding sites on A, which are occupied ($c_{AB}/c_{A,tot}$) is measured. The simplification in the analysis arises because the free concentration of B can be taken to be the same as the total concentration (since $c_{AB} \ll c_B$). Equation (16) can be simplified considerably yielding, after inserting the conservation condition for A and rearrangement:

$$\frac{c_{AB}}{c_{A,tot}} = \frac{c_B K_{Ass}}{1 + c_B K_{Ass}} \quad (20)$$

3.6.2 Independent Nonidentical Binding Sites

Consider a macromolecule A that can bind several molecules of B. In the simplest case, where A possesses two binding sites for B, there are four possible species, A, AB, BA, and BAB, whose concentrations depend on four binding constants:

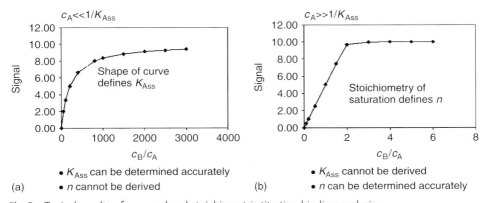

Fig. 5 Typical results of a normal and stoichiometric titration binding analysis.

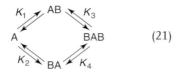

$$(21)$$

Under conditions where $c_{A,\text{tot}} \ll c_{B,\text{tot}}$, terms involving the total concentration of A do not occur in the analysis (as mentioned earlier), and it is therefore not possible to use Eq. (20) to analyze the stoichiometry of the binding equilibrium. However, even under these experimental conditions, it is possible to obtain information about the number of binding sites, provided the binding constants of the two processes are sufficiently different in magnitude.

Information about the minimum number of binding sites for B on the macromolecular species A can also be obtained if a signal can be measured, which specifically monitors the concentration of A fully saturated with B (BAB in our scheme). For example, the enzyme DNA polymerase has two binding sites for metal ions, and both need to be occupied for the enzyme to be active. If it is assumed that the two sites are independent, and hence $K_1 = K_4$ and $K_2 = K_3$ in Eq. (21), the following expression can be derived for the occupancy of the two sites (designated 1 & 2):

$$\theta_1 = \frac{c_{AB}}{c_{A,\text{tot}}} = \frac{c_B K_1}{1 + c_B K_1}$$

$$\theta_2 = \frac{c_{BA}}{c_{A,\text{tot}}} = \frac{c_B K_2}{1 + c_B K_2} \qquad (22)$$

and the proportion of A where both sites are occupied is given by:

$$\theta_{1,2} = \frac{c_{BAB}}{c_{A,\text{tot}}} = \theta_1 \theta_2$$

In the special case of identical binding sites ($K_{\text{Ass},1} = K_{\text{Ass},2}$), the dependence of $\theta_{1,2}$ on the total concentration of A ($c_{A,\text{tot}}$) is weakly sigmoidal at low concentrations of B, and not hyperbolic; this is a direct indication that A can bind more than one B. The total concentration of bound ligand ($= \theta_1 + \theta_2$) follows a hyperbolic dependence, as expected since the sites are independent.

3.6.3 Cooperative Binding

In the previous section, we discussed the case where the various binding sites were noninteracting; in this section we consider the other limiting case where A is either free, or fully occupied by B as the species AB_n, and the intermediate states $AB, AB_2, \ldots, AB_{n-2}$, and AB_{n-1} are not populated. This behavior arises because of positive interactions between the sites resulting in cooperative binding; according to Eq. (21), cooperative binding occurs when $K_3 \gg K_1$ and $K_4 \gg K_2$. The model considered here represents "all or none" behavior, which is not just a theoretical model but also one that does actually occur with biopolymers.

In cooperative binding following the "all or none" model

$$A + nB \rightleftharpoons AB_n$$

the association constant is defined by the expression:

$$K_{\text{Ass}}^{\text{app}} = \frac{c_{AB_n}}{c_A (c_B)^n}$$

Introducing the conservation condition for A with the further assumption that $c_{AB} \ll c_B$ yields the following equation:

$$\frac{c_{AB_n}}{c_{A,\text{tot}}} = \frac{(c_{B,\text{tot}})^n K_{\text{Ass}}^{\text{app}}}{1 + (c_{B,\text{tot}})^n K_{\text{Ass}}^{\text{app}}} \qquad (23)$$

This equation describes a sigmoidal binding curve, where the degree of sigmoidal behavior depends on the magnitude of n. The intrinsic binding constant of B for A (K_{Ass}) can be determined from the apparent binding constant (K_{Ass}^{app}) using the following relationship:

$$K_{Ass}^{app} = (K_{Ass})^n$$

3.7
pH Dependence of Enzyme Catalyzed Reactions

The rate of an enzyme catalyzed reaction is not only dependent on the concentrations of enzyme and substrate, but also on the conditions of the reaction. An important parameter affecting rate is the pH, defined as $-\log_{10} c(H^+)$, and it is very common that enzymes have a pH optimum. The pH can have several effects: (1) protons may participate in the catalytic reaction itself; (2) the protonation state of substrates and cosubstrates may alter, with consequent effects on rate; (3) the protonation state of the enzyme itself may alter. In our example here, we deal with the last case. Proteins contain many groups that can undergo protonation–deprotonation reactions, including the N-terminal amino and C-terminal carboxyl groups, and the side chains of the following amino acids: Asp, Glu, His, Cys, Tyr, Lys, and Arg. Protonation/deprotonation reactions are examples of ligand binding equilibria, which are simplified by the experimental approach of using a buffered solution such that the concentration of H^+ in the solution remains constant throughout the experiment. The state of protonation of a group is conveniently represented by its pK_a value, which is the negative decadic logarithm of the dissociation constant for the protonation reaction:

$$K_a = \frac{c(H^+)c(A^-)}{c(HA)}$$

$$pK_a = -\log_{10} K_a$$

The proportions of a group A in the protonated $\theta(H)$ and deprotonated $\theta(-)$ states can be evaluated as follows:

$$\theta(-) = \frac{A^-}{A_{tot}} = \frac{K_a}{c(H^+) + K_a}$$

$$= \frac{10^{-pK_a}}{10^{-pH} + 10^{-pK_a}}$$

$$\theta(H) = 1 - \theta(-) \quad (24)$$

The following questions are important for analyzing the pH dependence of enzyme catalyzed reactions:

– how many protonation reactions participate?
– which protonation state must the pH-sensitive groups on the enzyme be in?
– what are the pK_a values of these groups?

We consider a general model to analyze the protonation equilibria. If the enzyme possesses n groups that can participate in protonation–deprotonation equilibria, then in principle 2^n different species can be formed. For example, if $n = 3$, all three groups can be protonated (HHH), two (HH–, H–H, and –HH), one (H–, –H–, and –H) or none (–). The probability (P) of occurrence of these species, and hence their relative concentrations, depends on the product of the probabilities that each individual group is in a particular state:

$P(HHH) = P(\text{group 1 is protonated})$
  $\times P(\text{group 2 is protonated})$
  $\times P(\text{group 3 is protonated})$

$P(HH-) = P(\text{group 1 is protonated})$
  $\times P(\text{group 2 is protonated})$
  $\times P(\text{group 3 is unprotonated})$ and so on.

The probability of a group being in a particular protonation state is given by

Eq. (24), and combination of these probabilities multiplied by the total concentration of enzyme yields the concentrations of the different species.

The turnover rate is used as an "effect" or signal to monitor the protonation, and thus the observed rates can be used to analyze the thermodynamic protonation equilibria. In the general case, every species would be assigned an intensity factor, and the signal (observed rate of the reaction) would be the sum of all of these factors. For many analyses, one makes the simplifying assumption that only one species is catalytically active.

3.8
Pre-steady State Kinetics

Enzyme reactions proceed, in general, via several intermediate states. A simple model incorporating multiple states is shown below: enzyme and substrate associate to form an enzyme–substrate complex, which undergoes a conformational change to ES# before breaking down into enzyme and product.

$$E + S \underset{k_{-1}}{\overset{k_1}{\rightleftharpoons}} ES \underset{k_{-2}}{\overset{k_1}{\rightleftharpoons}} ES\# \longrightarrow E + P$$

Since the concentrations of all the intermediate states are constant under steady state conditions, all of these states can, at least formally, be incorporated into a single kinetic intermediate state. It follows that under steady state conditions, kinetic data can provide no information about the existence and kinetic properties of intermediate enzyme–substrate complexes. An understanding of the mechanism of an enzyme catalyzed reaction needs information about these intermediate states, which is therefore usually obtained from kinetic studies before steady state has been established, usually by rapid reaction methods. Comprehensive coverage of the techniques and methods of analysis of pre-steady state kinetics is beyond the scope of this chapter, but we discuss here methods for analyzing simple exponential processes. Two approaches are used. In the first, the observed signal $S(t)$ is fitted to an exponential function of the following form:

$$S(t) = Ae^{-t}/\tau, \text{ for decreasing signals.}$$
$$S(t) = A[1 - e^{-t/\tau}] \text{ for increasing signals.}$$
(25)

A is the amplitude of the reaction and τ the time constant, with the dimension of (time). If the kinetic mechanism of the observed process is known, then rate constants can be derived from the time constant. For example, for a simple dissociation process, the rate constant (k_{-1}) is given by $1/\tau$. In this case, the value of τ is independent of reactant concentration.

If both forward and back reactions can take place, then $1/\tau$ depends on both k_1 and k_{-1}. In the special case when the concentration of S is much greater than that of E, then the association rate constant is given by the equation $1/\tau = k_1 c_S + k_{-1}$. Values of the two rate constants can be determined from the dependence of τ on the substrate concentration c_S; a linear regression of $1/\tau$ versus c_S yields k_1 as the slope of the plot and k_{-1} as the y-intercept. For this analysis to be valid, it is important to be sure that the observed reaction represents a single exponential process. If the reaction involves more than one exponential phase, then more complex models need to be considered, since the minimal number of reaction steps is given by the number of exponential processes.

This method of analysis has several disadvantages, one of which is that intermediate parameters (τ) are evaluated from the data, which then form the basis for global fitting of the data; consequently, the global fitting is not carried out on the raw data directly. A second drawback is that the predictive power of this analysis as regards mechanism is rather limited.

A more powerful method for the analysis of such systems is to use direct integration of the differential equations that describe the mechanism of the reaction. An advantage of this procedure is that the fitting is carried out directly to the raw data. Since in most cases, particularly those of any kinetic complexity, the resulting systems of differential equations cannot be integrated analytically, numerical integration has to be used, which is not a problem in the currently given computer performance.

3.9
Association Kinetics Analyzed by Numerical Integration

The rate of a bimolecular association process $A + B \rightarrow AB$ is given by Eq. (26):

$$\frac{dc_{AB}}{dt} = k_1 c_A c_B \quad (26)$$

The rate constants for bimolecular association reactions have dimensions (concentration)$^{-1}$ (time)$^{-1}$. Although this differential equation has a very simple form, it does not have a very straightforward analytical solution. For this reason we use numerical integration methods to simulate theoretical data. This is a general approach that can be used to obtain solutions of complex kinetic processes. Although it is always easy to formulate differential equations like Eq. (26), which

express the time dependence of the various concentrations, solving the equations is another matter; it is often impossible to obtain explicit analytical solutions of the form $c = f(t)$, from which concentrations of the reaction participants can be directly determined. What can, however, be evaluated is the concentration change (or "flux") for a species in a given time interval Δt under given conditions:

$$F_1 = k_1 c_A(t) c_B(t) \Delta t$$

The solution can be obtained by proceeding stepwise (using small values of Δt) and calculating $c_{AB}(t)$ using the expression:

$$c_{AB}(t + \Delta t) = c_{AB}(t) + F_1$$

This procedure is called *numerical integration*.

If we consider the following equilibrium:

$$E + S \underset{k_{-1}}{\overset{k_1}{\rightleftharpoons}} ES$$

There are two different fluxes:

$$E + S \longrightarrow ES, \text{ where } F_1 = k_1 c_E c_S \Delta t$$

$$ES \longrightarrow E + S, \text{ where } F_{-1} = k_{-1} c_{ES} \Delta t$$

The concentration changes are defined as follows:

$$\Delta c_E = -F_1 + F_{-1}$$
$$\Delta c_S = -F_1 + F_{-1}$$
$$\Delta c_{ES} = -F_{-1} + F_1$$

from which new concentrations can be derived using the following expression where, as before, $c_{i,\text{old}}$ is the old concentration of the species i before the new increment Δc_i:

$$c_i = c_{i,\text{old}} + \Delta c_i$$

3.10
Surface Binding Reactions

With the development and widespread use of the surface plasmon resonance (SPR) approach to study biomolecular interactions, analysis of binding reactions at surfaces is becoming increasingly important. Assuming that only one class of binding sites is present, one only has to consider that the surface has a certain binding capacity C. Since in a typical SPR experiment, the analyte is flushed along the surface in a continuous buffer flow, the concentration of analyte (c_A) available for binding is constant. Therefore, the flux (F_1) of the analyte to the surface is given by the following expression, where k_1 is the binding rate constant and Δt is the time increment used for numerical integration:

$$F_1 = k_1 c_A (C - c_{A,bound}) \Delta t \quad (27)$$

Release of bound analyte is given by a second flux, just as described in Eq. (27):

$$F_{-1} = k_{-1} c_{A,bound} \Delta t$$

3.11
Analysis of Competition Experiments

Competition experiments are widely used in the biosciences, particularly in studies of binding interactions. A simple example is shown in the following:

$$AB + C \xrightleftharpoons{K_1} A + B + C \xrightleftharpoons{K_2} AC + B \quad (28)$$

In this example, the equilibrium between A, B, and AB is affected by the addition of C. The popularity of the competition technique is due to the fact that it can be used to investigate interactions (in this case the binding of $A + C = AC$) without having to detect the participating species (free C and the complex AC). The method relies on using one interaction (here $A + B = AB$) as a reporter to monitor the other. This assumes, of course, that a suitable signal is available to follow the formation of AB. Figure 6 illustrates the formation of AB, and the effect of adding C to a system containing A, B, and AB: on addition of C the species AC is formed at the expense of AB whose concentration falls, with a concomitant decrease in the observed signal.

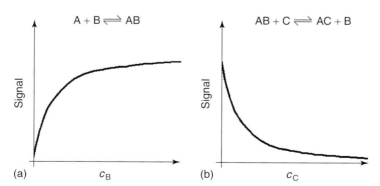

Fig. 6 Indirect analysis of molecular interaction of A and C by competition of the interaction of A with B.

It is also a desirable feature of competition experiments that they allow more precise comparison of the binding of different species (in this case B and C) to a common target (A) than is possible in separate binding experiments. It is also possible to use this approach with a single experimental set-up to test the binding of many different ligands to A, on the basis that all these ligands compete with B for the same binding site.

The analysis of coupled equilibria is the most complex problem that is considered in this chapter. It may seem surprising that such apparently straightforward systems like those shown in Eqs. (21) and (28) should present such great difficulties in analysis, the more so because it is a trivial matter to calculate the equilibrium constants, if the concentrations of the various species are known. However, that situation arises very rarely for several reasons:

- in most investigations, only some of the species can be detected;
- it is usually the case that only one "signal" is measured, whose dependence on the concentration of reaction participants may be complex, and must be derived from the model;
- experimental error.

Proceeding as we have done before with complex systems, we calculate theoretical data to deal with these systems. Since straightforward analytical solutions are not available, even for such simple cases as Eq. (28), numerical methods are used to simulate solutions of the equilibria. From an arbitrary initial state (e.g. only free A, B, and C present) we calculate for both isolated equilibria the final equilibrium concentrations, and the concentration changes needed to establish equilibrium. We then use these fluxes. Since, in every step, there are several fluxes, whose coupling is not taken into account, this approximate procedure takes us towards the equilibrium, but not directly to it. So the equilibrium has to be located iteratively.

We may write:

$$K_1 = \frac{c_{AB, \text{Equil.}}}{c_{A, \text{Equil.}} \cdot c_{B, \text{Equil.}}}$$

or:

$$c_{AB, \text{Equil.}} = K_1 c_{A, \text{Equil.}} c_{B, \text{Equil.}}$$

where, $c_{i, \text{Equil.}}$ is the equilibrium concentration of i.

We are looking for the flux F, necessary to take us from the present concentrations to the equilibrium concentrations:

$$c_{AB, \text{Equil.}} = c_{AB} + F$$
$$c_{A, \text{Equil.}} = c_A - F$$
$$c_{B, \text{Equil.}} = c_B - F$$

Thus, we have:

$$(c_{AB} + F) = K_1(c_A - F)(c_B - F)$$

and

$$F^2 - \left(c_A + c_B + \frac{1}{K_1}\right)F + \frac{c_A c_B - c_{AB}}{K_1} = 0$$

We calculate these fluxes for the two equilibria, and use these. We reach a new concentration state, which is not the true equilibrium state, since we have not taken the coupling of the equilibria into account. So the operation is repeated with newly calculated fluxes, and after ca. 50 iterations, we reach a stable equilibrium. The concentrations correspond to the joint equilibrium specified by the two thermodynamic association constants K_1 and K_2.

For practical applications, it is important that the calculated fluxes are not too large so that calculation does not overshoot the target. So, for example, if K_1 and K_2 are very large, and $c_A > c_B$ and $c_C > c_B$, our initial analysis of the first equilibrium will show that almost all of the B must flow to AB. At the same time, analysis of the second equilibrium will show that all of the B should flow to BC. When we take account of the two fluxes, we obtain negative concentrations in the first iteration. Since in this example, there are at most two fluxes for a single species (free B), this complication can be avoided by reducing every flux by 50%.

3.12
Regularization – Application to Analytical Ultracentrifugation

As outlined in the mathematical introduction (Sect 2.1), any model that describes an experiment can be used either as a discretely calculated function to describe an experiment or in a regularizing fashion to calculate the distribution of model parameters that best describe the experiment. In biochemistry, the latter approach was first applied to the analysis of relaxation kinetics or circular dichroism data. Here, we show how the approach can be used to analyze sedimentation velocity experiments in the analytical ultracentrifuge.

In an ultracentrifuge, dissolved macromolecules (e.g. proteins) are subjected to large gravitational fields, which causes them to sediment with a velocity that is described by the sedimentation coefficient s:

$$v = \frac{dx}{dt} = s\omega^2 x$$

where v is the velocity of the molecule, ω is the angular speed of the rotor, and x is the distance from the center of rotation. The sedimentation coefficient has the dimension of time and it is usually given in units of S (for Svedberg) with $1\,S = 10^{-13}$ s. Larger molecules sediment faster and consequently have larger values of sedimentation coefficient than smaller ones. These differences in sedimentation can be exploited to detect interactions of macromolecules. In an analytical ultracentrifuge, the local concentration of particles in the centrifuge cell can be observed by optical absorption or refractive index measurements.

To illustrate such an experiment Fig. 7(a) shows the sedimentation of a mixture of two interacting proteins, N-acetyl-glutamate-kinase (NAGK) and the signal transducing protein PII from *Synechococcus* sp. PCC7942. NAGK is a hexamer with a molecular mass of 192 kg mol^{-1} and a sedimentation constant of 8 S, while PII is a tetramer with a molecular mass of 42 kg and sedimentation coefficient of 3 S. Two clearly separated boundaries can be observed, the faster sedimenting boundary (10 S) is formed by the complex of these two proteins while the slower one represents PII protein whose concentration in this experiment was chosen to be much larger than that of NAGK (Maheswaran et al. 2004). It is clear that the two boundaries are spread due to diffusion and that a straightforward analysis of the amounts of free and complex bound protein is difficult to estimate from the data directly. However, the movement of such boundaries in the analytical ultracentrifuge can be described using the Lamm equation derived by Lamm in 1929:

$$\frac{\partial c}{\partial t} = \frac{1}{x}\frac{\partial}{\partial x}\left[\frac{1}{x}\left(D\frac{\partial c}{\partial x} - s\omega^2 x^2\right)\right]$$

This equation can be solved by numerical methods and represents a model

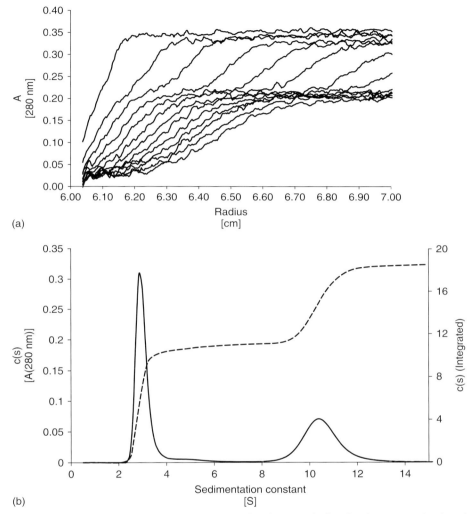

Fig. 7 Sedimentation of a mixture of 11.2 μM NAGK and 2.24 μM PII proteins (for details cf. text.) (a) Sedimentation profiles in the analytical ultracentrifuge taken every 19 min at a speed of 38 000 rpm. (b) Sedimentation coefficient distribution calculated with SEDFIT. The dotted line is the integral over the distribution and can be used to calculate the absorptions of free and bound proteins.

function of the type $c = f(x, t; s, D)$ where x and t are the independent variables and, s and D, the diffusion coefficient, are parameters that are constant throughout an experiment. Regularization algorithm given by Schuck in 2000 is similar to the following description using Lamm's differential equation. Thus, s and D are related to molecular parameters by the equations

$$s = \frac{M(1 - \bar{v}\sigma)}{N_A f} \text{ and } D = \frac{RT}{N_A f}$$

where M represents the molar mass (kg mol^{-1}), $\bar{v}$ the partial specific volume (m^3 kg^{-1}) and the f frictional coefficient (kg s^{-1}) of the particle; N_A and R are Avogadro's number and the gas constant, respectively. For a sphere of radius R_0, the frictional coefficient is given by Stokes law $f = 6 \cdot \pi \cdot \eta \cdot R_0$. Using $\bar{v}$ and the frictional ratio f/f_0, which is the ratio of the actual frictional coefficient of the particle to the frictional coefficient of a perfect sphere with the same volume, the diffusion coefficient can be related to the sedimentation coefficient by

$$D = \frac{RT}{6\pi\eta N_A} \sqrt[3]{\frac{2(1-\bar{v}\rho)}{9\bar{v}s\eta(f/f_0)^4}}$$

To analyze the sedimentation data using the regularizing procedure, we write the describing function in the form $\int c(s)f(x, t; s)ds$ from which the sedimentation coefficient distribution $c(s)$ can be calculated from the original sedimentation data. The program to do this (SEDFIT) is available from Peter Schuck (http://www.analyticalultracentrifugation.com/default.htm).

Figure 7(b) shows the sedimentation profiles represented as the sedimentation coefficient distribution and from the integral of the two peaks the absorption and thus relative amounts of free and bound protein can be evaluated; from this it was calculated that the proteins form a 1:1 (NAGK hexamer: PII trimer) complex.

Bibliography

The present text is a modified and expanded version of Chapter 9 published in Pingoud, A., Urbanke, C., Hoggett, J. & Jeltsch, A. (2002) Biochemical Methods, Wiley-VCH. In that book, detailed step-by-step instructions can be found on how to implement many of the procedures described here in Excel worksheets.

Books and Reviews

Bisswanger, H. (2002) *Enzyme Kinetics Principles and Methods*, Wiley-VCH, Weinheim, Germany.

Brand, L. (Ed.) (1992) *Methods in Enzymology, Vol. 210: Numerical Computer Methods*, Elsevier, Amsterdam, Netherlands.

Brand, L. (Ed.) (2004) *Methods in Enzymology, Vol. 383: Numerical Computer Methods*, Part D, Elsevier, Amsterdam, Netherlands.

Johnson, M. (Ed.) (1994) *Methods in Enzymology, Vol. 240: Numerical Computer Methods*, Part B, Elsevier, Amsterdam, Netherlands.

Johnson, M. (Ed.) (2000) *Methods in Enzymology, Vol. 321: Numerical Computer Methods*, Part C, Elsevier, Amsterdam, Netherlands.

Johnson, M. (Ed.) (2004) *Methods in Enzymology, Vol. 384: Numerical Computer Methods*, Part E, Elsevier, Amsterdam, Netherlands.

Purich, D. (Ed.) (1995) *Methods in Enzymology, Vol. 249: Enzyme Kinetics and Mechanism*, Part D, Elsevier, Amsterdam, Netherlands.

Primary Literature

Johnson, M.L. (2000) Outliers and robust parameter estimation, *Methods Enzymol.* **321**, 417–424.

Lamm, O. (1929) Die differentialgleichung der ultrazentrifugierung, *Ark. Matematik, Astron. Ffys.* **21B**(2), 1–4.

Maheswaran, M., Urbanke, C., Forchhammer, K. (2004) Complex formation and catalytic activation by the PII signaling protein of N-Acetyl-L-glutamate kinase from *Synechococcus elongatus* strain PCC 7942, *J. Biol. Chem.* **279**(53), 55202–55210, Epub 2004 Oct 22.

Provencher, S.W. (1982a) A constrained regularization method for inverting data represented by linear algebraic or integral functions, *Comput. Phys. Commun.* **27**, 213–227.

Provencher, S.W. (1982b) CONTIN: a general purpose constrained regularization program for inverting noisy linear algebraic and integral equations, *Comput. Phys. Commun.* **27**, 229–242.

Rusling, J.F., Kumosinski, T.F. (1996) *Nonlinear Computer Modeling of Chemical and Biochemical Data*, Academic Press, New York.

Schuck, P. (2000) Size-distribution analysis of macromolecules by sedimentation velocity ultracentrifugation and Lamm equation modeling, *Biophys. J.* **78**(3), 1606–1619.

Radioisotopes in Molecular Biology

Robert James Slater
University of Hertfordshire, Hatfield, UK

1	**Principles** 413	
1.1	Radioactive Decay 413	
1.2	Kinetics of Decay 413	
1.3	Units 415	
2	**Radiation Protection** 416	
2.1	General Principles 416	
2.2	Unpacking and Dispensing Radioactive Solutions 419	
3	**The Choice of Radionuclide** 420	
4	**Nucleic Acid Labeling** 422	
5	**Protein Labeling** 424	
6	**Detection** 427	
6.1	Ionization Monitors 427	
6.2	Scintillation Counters 427	
6.3	Cerenkov Counting 429	
6.4	Autoradiography 429	
6.5	Personal Dosimeters 431	
	Bibliography 431	
	Books and Reviews 431	
	Primary Literature 432	

Keywords

Autoradiography
The detection of radioactivity using photographic emulsions.

Fluorography
A sensitive form of autoradiography in which radioactivity is detected by light emanating from a scintillator (or "fluor") in close contact with the sample.

Half-life
The time taken for the activity of a radionuclide to lose half its value by decay.

Quenching
Any process that reduces the efficiency of detection of radioactivity.

Radioactivity
The emission of ionizing radiations by matter.

Radioisotopes
Atoms with the same atomic number (protons) but differing mass numbers (protons plus neutrons), and having unstable nuclei that decay to a stable state by the emission of ionizing radiations; for example, ^{14}C is a radioisotope of ^{12}C.

Radionuclide
An atomic species that is radioactive, for example, ^{3}H, ^{14}C and so on.

Specific Activity
The rate of decay per unit mass, for example dpm g^{-1}, Ci $mole^{-1}$, Bq $mmol^{-1}$.

■ Radioisotopes are used extensively in molecular biology. They can be incorporated into DNA, RNA, and protein molecules, both *in vivo* and *in vitro*. As a consequence, the presence or metabolism of macromolecules can be investigated or "traced." The incorporation of radioisotopes allows the detection of minute quantities, thereby facilitating experimental techniques that require high sensitivity. Labeling *in vitro* is much more efficient than labeling *in vivo*; specific activities regularly reach 10^8–10^9 dpm μ g^{-1} for *in vitro* ^{32}P–labeling of DNA, for example. High-specific activity leads to greater sensitivity. Specific activity is inversely related to the half-life of the radioisotope. Low-energy emitters, such as ^3H and ^{35}S provide high resolution in autoradiography. High-energy, high-specific activity emitters such as ^{32}P for nucleic acid labeling, provide great sensitivity but low resolution in autoradiography. The radionuclides ^3H, ^{14}C, ^{35}S, and ^{125}I are most commonly used for protein labeling; ^{32}P and ^{33}P are most commonly used for nucleic acid labeling.

1
Principles

1.1
Radioactive Decay

Atomic nuclei contain a number of protons (referred to as the atomic number, Z) and, except in the case of hydrogen, neutrons (N) also. The total number of these nuclear atomic particles (or nucleons) is the mass number (A), that is $A = Z + N$. The atomic number determines the element (e.g. when $Z = 6$ the element is carbon). Isotopes are elements that have the same atomic number but differing mass numbers. For example, ^{12}C and ^{14}C are different isotopes of carbon, both with atomic number 6, but differing mass numbers, 12 and 14 respectively, thereby indicating that the isotopes have either 6 or 8 neutrons. Some atoms are unstable: they have the incorrect number of neutrons for nuclear stability and are referred to as radionuclides or, more commonly, radioisotopes. Radioactivity is the means by which unstable atoms reach a stable state. The process can occur in one step, as is the case with most radioisotopes used in biological experiments, or in a number of steps, such as in the case of uranium, which decays through several radionuclides to a stable isotope of lead. There are many ways in which this decay can occur.

Alpha decay is loss of a helium nucleus, ^4_2He, for example:

$$^{226}_{88}Ra \longrightarrow {}^{222}_{86}Rn + {}^4_2He$$

Alpha particles are heavy and slow moving, have poor penetration (3–9 cm), and a net double positive charge. Alpha emitters are found mostly amongst the heavier elements and are rarely used by biologists.

Beta decay results from the conversion of a neutron to a proton (negatron emission) or a proton to a neutron (positron emission). Of most relevance to biologists is the negatron emission form of β decay, involving the loss of an electron, which forms the ionizing radiation. Many commonly used radionuclides decay by this mechanism (e.g. ^3H, ^{14}C, ^{35}S, ^{33}P, ^{32}P). For example, ^{14}C decays as follows:

$$^{14}_6C \longrightarrow {}^{14}_7N + \beta^-$$

Beta particles are less charged, lighter, and faster moving than alpha particles; their energy and therefore the range of the radiation produced, varies with the source. This has significant implications for their use and safe handling, and this is discussed later (Sect. 2).

Gamma decay is the emission of electromagnetic radiation, with properties identical to X rays. It often accompanies β-decay. For example:

$$^{131}_{63}I \longrightarrow {}^{131}_{64}Xe + \beta^- + \gamma$$

Electron capture is a process by which a proton is combined with an electron from the innermost K shell, forming a neutron; following a complex series of events, an X ray is emitted. An example of this kind of decay is shown by ^{125}I:

$$^{125}_{53}I \longrightarrow {}^{125}_{52}Te + X\ ray$$

Decay of ^{125}I is also accompanied by emission of Auger electrons, radiation of low energy, providing high-resolution detection in autoradiography (see Sect. 3).

1.2
Kinetics of Decay

It is impossible to predict when an individual unstable nucleus will decay, but when observing the decay of a large number of unstable atoms there is a statistical probability that a certain number will decay within a given time. The number that decay per unit time is proportional to the number of unstable atoms present. Mathematically this is represented as:

$$\frac{-dN}{dt} \alpha N \quad \text{or} \quad \frac{-dN}{dt} = \lambda N \quad (1)$$

where N = number of unstable nuclei, t = time and λ = the decay constant $\sec^{-1}$. Therefore the amount of radioactivity (A, the count rate) is:

$$A = \frac{dN}{dt} = \lambda N \quad (2)$$

As a consequence, radioisotopes exhibit exponential decay as is illustrated by the graphs in Fig. 1, the rate at which a radioisotope decays being determined by its decay constant, λ. This constant is a fundamental property of a radioisotope, and is not influenced by temperature, pressure, and so on. Integration of Eq. (1) with respect to time (letting $N = N_0$ at $t = 0$) gives the following expression:

$$N = N_0 e^{-\lambda t} \quad (3)$$

and

$$\ln \frac{Nt}{N_0} = -\lambda t \quad (4)$$

(where $N = Nt$ at time t)

or:

$$\ln \text{count rate(time, } t_2)$$
$$= \ln \text{count rate(time, } t_1) - \lambda t \quad (5)$$

Thus, if the decay constant is known, the count rate at any time can be calculated from a known count rate and time. In practice, decay constants are rarely quoted;

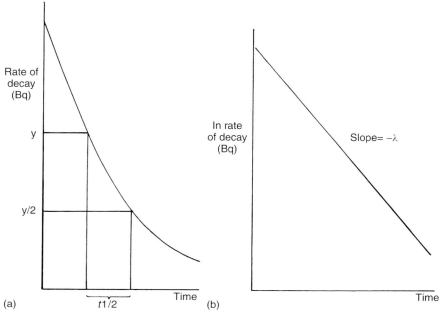

Fig. 1 The rate of decay with time on a linear (a) and semilogarithmic (b) scale.

half-life ($t_{1/2}$) is in many ways a more convenient way to express decay kinetics. The half-life is the time taken for the count rate to drop to half of a given value, and is depicted in Fig. 1.

From Eq. (4), letting $N_t = (1/2)N_0$:

$$\ln\tfrac{1}{2} = -\lambda t_{1/2}$$

or

$$\ln 2 = \lambda t_{1/2} \qquad (6)$$

or

$$\lambda = \frac{0.693}{t_{1/2}} \qquad (7)$$

So, if the half-life is known, a combination of Eqs. (7 and 5) can be used to determine future or past levels of radioactivity. This is particularly valuable when working with short half-life radioisotopes such as ^{32}P. Decay tables based on this type of mathematics are available in the reference literature and suppliers' catalogues.

1.3
Units

The International System of Units (SI System) uses the becquerel (Bq) as the unit of radioactivity, 1Bq being one disintegration per second (1d.p.s.). However, a frequently used unit is the curie (Ci). This unit originates from the number of disintegrations from a gram of pure radium: 1 Ci, therefore, being 3.7×10^{10} d.p.s. (or 37 GBq). This is a very large amount of radioactivity, and most molecular biologists use quantities measured in microcuries (μCi); 1 μCi is 2.22×10^6 d.p.m. (disintegrations per minute) or 37 kBq. The becquerel is the internationally recognized unit of radioactivity and has been adopted by the International Commission for Radiological Protection (ICRP). As a consequence, most countries have incorporated the unit into their legislation regarding use and disposal of radioactivity; suppliers of radioisotopes, however, frequently still use the Ci for historical reasons.

The units Bq, Ci, d.p.s., and d.p.m. all refer to the actual number of nuclear disintegrations. The rate of decay detected by an instrument is measured in counts per minute (c.p.m.). This is usually less than the true rate, d.p.m., because counting is rarely 100% efficient. Consequently, it is normal practice to calibrate the detector and calculate d.p.m. as c.p.m. × 100/efficiency, as a percentage value. Surprisingly, it is sometimes possible for the recorded count rate (c.p.m.) to exceed the actual rate of decay. This can occur in liquid scintillation counters and is discussed in Sect. 6.2.

The term *specific activity* is important in relation to experiments with radioactivity. The term refers to the amount of radioactivity per unit mass; thus d.p.m. g^{-1}, c.p.m. g^{-1}, Ci mole^{-1}, Bq mole^{-1} are all units of specific activity. Frequently, molecular biologists require the highest specific activity possible; for example, DNA labeling with high-specific activity ^{32}P-labeled nucleotides results in probes with the maximum possible sensitivity. The highest specific activities are obtained from radioisotopes with the shortest half-lives (see Sect. 3). Sometimes it is necessary or desirable to change the specific activity by addition of nonradioactive material (a "cold" carrier). This can be achieved by applying the following formula:

$$W = Ma\left(\frac{1}{A'} - \frac{1}{A}\right) \qquad (8)$$

where W = mass of cold carrier required, expressed in mg; a = amount of radioactivity present, expressed in MBq; M = molecular weight of the compound; A = the original specific activity MBq mmol^{-1};

Tab. 1   A summary of units and their definitions.

Unit	Abbreviation	Definition
Counts per minute or second	c.p.m. c.p.s.	The recorded rate of decay
Disintegration per minute or second	d.p.m. d.p.s.	The actual rate of decay
Curie	Ci	The number of d.p.s. equivalent to 1 g of radium (3.7×10^{10} d.p.s.)
Millicurie	mCi	Ci $\times 10^{-3}$ or 2.22×10^{9} d.p.m.
Microcurie	mCi	Ci $\times 10^{-6}$ or 2.22×10^{6} d.p.m.
Becquerel (SI Unit)	Bq	1 d.p.s.
Gigabecquerel (SI Unit)	GBq	10^{9} Bq or 27.027 mCi
Megabecquerel (SI Unit)	MBq	10^{6} Bq or 27.027 µCi
Electron volt	eV	The energy attained by an electron accelerated through a potential difference of 1 V. Equivalent to 1.6×10^{-19} joules.
Roentgen	R	The amount of radiation, which produces 1.61×10^{15} ion pairs per kg of air (2.58×10^{-4} coulombs kg^{-1})
Rad	rad	That dose which gives an energy absorption of 0.01 joule kg^{-1} (J kg^{-1})
Gray (SI Unit)	Gy	That dose which gives an energy absorption of 1 joule kg^{-1}. Thus, 1 Gy = 100 rad.
Rem	rem	That amount of radiation which gives a dose in human equivalent to 1 rad of X rays.
Sievert	Sv	That amount of radiation, which gives a dose in human equivalent to 1 gray of X rays. Thus, 1 Sv = 100 rem.

A' = the required specific activity MBq mmol^{-1}.

Units and definitions are summarized in Table 1.

2
Radiation Protection

2.1
General Principles

Toxicity is the greatest practical disadvantage of using radioisotopes. They produce ionizing radiations, which, when absorbed by cells, cause ionization and the production of free radicals. These radicals, in turn, cause mutation of DNA, hydrolysis of proteins, and ultimately cell death. The toxicity of radiation is dependent not simply on the amount present but on the amount absorbed, the energy of the radiation and its nature, and therefore the biological effect. The international unit to describe absorbed dose, taking into account its biological effectiveness, is the sievert (Sv), known as the equivalent dose. This is calculated from the absorbed dose (gray, Gy), which is a measure of the energy absorbed per unit mass (joule kg^{-1}), multiplied by a radiation-weighting factor (W_R). For X rays, γ-rays and ß-radiations this weighting factor is one, that is the equivalent dose = absorbed dose = joule kg^{-1}. Other forms of radiation such as neutrons and α-particles are more toxic, with a W_R of up to 20.

Radiation is emitted from a source in all directions, so the level of irradiation, and therefore the dose, is inversely related to the square of the distance from the source. A commonly used parameter is dose rate (Sv per minute or hour, etc.). If this is known, either from quoted figures or measurements with a dose rate meter, total doses can be calculated. For example, if a source is delivering 1mSv per hour and you work with the source for 15 min the dose will be 250 μSv. To calculate dose rates at any distance, use the formula:

$$\text{Dose rate}_1 \times \text{distance}_1^2 = \text{dose rate}_2 \times \text{distance}_2^2 \quad (9)$$

There are official dose limits for workers using radiation: 20 mSv year^{-1} for the whole body. To put this in perspective, it represents about 10 times the background radiation. There are also annual dose limits for specific organs, for example, 150 mSv for the lens of the eye, and 500 mSv for the hand; however, a fundamental principle behind radiation protection is that doses must be as low as reasonably practical (the ALARP principle). In practice this means that workers should not work up to a dose limit, but rather always strive to reduce their exposure. For example, radioisotopes should only be used if there is a net benefit (see the end of Sect. 6.4); and when radioisotopes are used, all practical steps must be taken to reduce dose to workers, (for example, by choosing a low-energy emitter).

So far the discussion has been primarily concerned with *external radiation*, that is, radioisotopes outside the body. However, a potential hazard in working with radioisotopes in a biology laboratory is via *internal radiation*: radiation that has entered the body by, for example, ingestion or inhalation. The annual limit on intake (ALI) provides a guide to relevant dosimetry. ALIs for radioisotopes used in molecular biology are included in Table 2, but it should be noted that the figures are for the elemental isotopes, and not particular formulations. The ALI for ^{3}H-thymidine is lower, for example, than that for ^{3}H$_2$O, because it is more toxic due to its longer clearance time from the body. Exhaustive lists of ALI can be found in the detailed radiation protection literature.

The various radioisotopes used in molecular biology fall into two broad categories of hazard: those that present an external radiation hazard (^{32}P and ^{125}I) and those that present only a negligible (^{14}C, ^{35}S, ^{33}P) or zero (^{3}H) external radiation risk. Clearly, therefore, it is incumbent on the laboratory worker to avoid ^{32}P or ^{125}I wherever possible. All radioisotopes present an internal radiation risk, particularly when handled as liquids or gases, but again, the penetrating, higher energy radiations such as from ^{32}P and ^{125}I present the greatest risk. The figures for annual limit on intake shown in Table 2 give an indication of the relative risk from ingestion. Shielding, other than that routinely used for storage, is not usually required for ^{14}C, ^{35}S, ^{33}P or ^{3}H when used at the levels appropriate for most molecular biology experiments. Here the potential risk is through ingestion, which can be avoided by adherence to good laboratory protocols that are generally applicable to work with all radionuclides: wearing of laboratory coat, gloves, and safety spectacles, working in spill trays lined with absorbent paper, double containment of stocks for storage

Tab. 2 Properties and radiation protection data of radioisotopes commonly used in the molecular biological sciences. Radioisotopes vary considerably with respect to their properties and potential hazards. The table below summarizes some generally useful information about isotopes used in biological experiments.

Property/radiation protection criteria	^3H	^{14}C	^{35}S	^{33}P	^{32}P	^{125}I	^{131}I
$t_{1/2}$	12.3 years	5730 years	87.4 days	25.4 days	14.3 days	59.6 days	8.04 days
Mode of decay	β^-	β^-	β^-	β^-	β^-	X (EC)	γ and β
Max β energy (MeV)	0.019	0.156	0.167	0.249	1.709	Auger electrons	0.806
Monitor	Swabs counted by liquid scintillation	β-counter	β-counter	β-counter	β-counter	γ-probe	β-counter
Biological monitoring	Urine	Urine, breath	Urine	Urine	Urine	Thyroid	Thyroid
ALI[a]	480[b]	34	15	14	6.3	1.3[c]	0.9[c]
Critical organ	Whole body	Whole body/fat	Whole body/testis	Bone	Bone	Thyroid	Thyroid
Maximum range in air	6 mm	24 cm	26 cm	49 cm	790 cm	>10 m	>10 m
Shielding required	None	Perspex 1 cm	Perspex 1 cm	Perspex 1 cm	Perspex 1 cm (β-dose rate 210 mSv at surface of 1 MBq ml^{-1})	Lead 0.25 mm or lead	Lead 13 mm
γ dose rate (μSv h^{-1} from 1GBq at 1 m)	–	–	–	–	–	34	51
Special considerations	Monitoring difficulties lead to potential internal hazard. DNA precursors more toxic than tritiated water.	Avoid generation of CO_2	Avoid generation of SO_2		Potential high source of external radiation. Lead shielding and finger dosimeters for quantities greater than 300 and 30 MBq respectively.	Iodine sublimes, work in fume hood. Spills should be treated with sodium thiosulphate solution prior to decontamination. Iodine compounds may penetrate rubber gloves – wear two pairs.	

[a] Based on a dose limit of 20 mSv using the most restrictive dose coefficients for inhalation or ingestion; some figures differ in Germany.
[b] Organically bound H-3.
[c] Based on dose equivalent limit of 500 mSv to thyroid.
Source: Reproduced with permission from Oxford University Press.

and transport, regular and routine monitoring, no eating and drinking (or sucking pencils etc.), in the laboratory, no mouth pipetting, clear delineation of areas used for radioisotope work and washing hands when leaving the laboratory.

Dose rates from λ sources are quoted as the specific dose rate constant. For ^{125}I, this figure is 3.4×10^{-2} μSv h^{-1} MBq^{-1} at 1 m.

For ^{32}P the following formula can be used:

$$\text{dose rate}(\mu\text{Sv h}^{-1}) \text{at 30 cm} = 54 \times \text{activity (MBq)} \quad (10)$$

Using Eqs. (9 and 10), a 1 MBq source of ^{32}P (approx 27 μCi) has a dose rate of 0.9 μSv min^{-1} at 30 cm and 0.81 Sv min^{-1} at 1 cm.

Clearly, shielding is required for work with ^{32}P and ^{125}I; the most practical is 1 cm acrylic for ^{32}P and 1 cm lead-impregnated acrylic for ^{125}I. High-energy β-emitters generate secondary X rays from absorbers, known as bremsstrahlung radiation. The generation of this radiation increases with the atomic weight of the absorber. Hence lead-impregnated acrylic can generate bremsstrahlung radiation, and is therefore not ideally suited for use with ^{32}P.

Vials or test tubes containing ^{32}P or ^{125}I must not be handled directly even when using a body shield, as doses can potentially be very high. They must be manipulated with forceps or some other appropriate device, or kept in acrylic holders such as those specifically marketed by radioisotope and radiation protection equipment suppliers. Automatic pipettes should carry small acrylic shields. Workers who frequently use ^{32}P may be advised to wear fingertip and body dosemeters. In all cases, the member of the staff who is responsible for radiation protection (e.g. Radiation Protection Supervisor or Radiation Protection Advisor) should be consulted. He/she will advise on the laboratory requirements and the type of laboratory required. In the United Kingdom, this is most likely to be a "Supervised Laboratory" (i.e. one in which it is theoretically possible to exceed 1/10 but not more than 3/10 of the recommended dose limit). When dealing with penetrating radiations, remember the basic rules of radiation protection:

maximize distance
minimize time of exposure
use shielding

Do dose estimates on your experiments using the formula described above (Eqs. 9 and 10) and the data in Table 2. Monitor regularly. A sensible precaution when using radioactivity in a protocol for the first time is to do a dummy run with ink in place of the radioisotope solution. A further simple precaution is to always add radioisotopes to mixtures last where possible; this reduces the time of exposure and risk of contamination.

2.2 Unpacking and Dispensing Radioactive Solutions

The following advice applies to unpacking and dispensing solutions emitting penetrating radiations such as from ^{32}P or ^{125}I.

1. Study the data sheet supplied with the radioisotope, do all necessary calculations (including dose assessment on the unpacking and dispensing procedure) and prepare a receptacle containing nonradioactive carrier as necessary, securely attach a label to this receptacle

that states your name, date, radionuclide and radioactive concentration. Manipulation of the stock radioactive solution should be carried out only when everything else is ready.
2. Wear laboratory coat, personal dosimeters (body and fingertip, see Sect. 6.5) safety spectacles, one or two pairs of disposable gloves, lay out all your equipment behind a body screen in a lined spill tray ensuring that no receptacles for liquids can be knocked over. Switch on a monitor and check the battery.
3. Open the sealed container, wipe test the interior with a piece of tissue paper held by forceps, and check the paper against the monitor. Regularly repeat this wipe test throughout the procedure to check for contamination outside the sample vial.
4. Remove lid of inner container surrounding the sample vial. Wipe test the top of the vial. Dispense sample without removing vial if possible; alternatively, if necessary, transfer the inner vial, with forceps, to an acrylic holder.
5. If dispensing with a syringe through a Teflon seal, insert a second needle in the seal as an air bleed. Surround the needle on the syringe with tissue to absorb any tiny drops of contamination. Handle the syringe for as short a time as possible but without reducing dexterity.
6. If dispensing with an automatic pipette, remove lid or seal with care and use an acrylic pipette shield.
7. Place contaminated tissues in a shielded container, rinse syringe/pipette tips in detergent solution for aqueous disposal (this minimizes the radioactivity in solid waste), and place disposable syringe or pipette tip in an appropriate shielded waste disposal container.
8. Repack the stock container, and wipe test all surfaces.
9. Monitor gloves, cuffs, all working surfaces, forceps and so on. Return stock to storage area. Upon completion of work remove laboratory coat, wash and monitor hands, and complete all necessary records in the radioisotope log.

If you are dispensing low-energy radioisotopes, it may be unrealistic to apply all aspects of the above procedure: ^{3}H for example, is not detected by bench monitors and the risk from external radiation is negligible, and therefore shielding is not required. However, the same **principles** apply, in that the risk of contamination should be reduced wherever possible, by, for example, working in a spill tray and preparing everything prior to dispensing the radioisotope.

3
The Choice of Radionuclide

The relative order of increasing radiation hazard is: ^{3}H, ^{14}C, ^{35}S, ^{33}P, ^{32}P, ^{131}I, and ^{125}I. Choose the radioisotope with lowest toxicity where possible. Remember that the shorter the half-life the greater the potential specific activity. The high-energy emitters give high counting efficiencies and strong signals in autoradiography. The lower-energy emitters give lower detection efficiencies but high resolution in autoradiography. The relative merits of radionuclides used in molecular biology are provided in Table 3. As a consequence, ^{32}P provides the highest sensitivity for labeling of probes and filter hybridization for genomic southern blots and northern analysis; ^{35}S and ^{33}P are preferred for DNA sequencing, and ^{35}S and ^{3}H are preferred for *in situ* hybridization and detection of unscheduled DNA synthesis. Phosphorus-33 is a good compromise for most DNA experiments as it combines the features of

Tab. 3 The relative merits of commonly used radionuclides.

	Advantages	Disadvantages
^3H	Safety High specific activity Wide choice of labeling position in organic compounds High resolution in autoradiography	Low efficiency of detection ^3H is large compared with ^1H and may cause an isotope effect Isotope exchange with the environment
^{14}C	Safety Wide choice of labeling position in organic compounds High resolution in autoradiography	Low specific activity significantly reduces value in molecular biology
^{35}S	High specific activity Good resolution in autoradiography	Relatively long biological half-life
^{33}P	High specific activity Safety relative to ^{32}P Longer half-life than ^{32}P Good resolution in autoradiography	Short half-life relative to ^{35}S Cost
^{32}P	Sensitivity of detection (10 fg or 50 dpm cm^{-2} in filter hybridization) Short half-life simplified disposal High specific activity Cerenkov counting	Short half-life effects cost and experimental design External radiation hazard Poor resolution in autoradiography
^{125}I	Ease of detection High specific activity Efficient *in vitro* labeling of proteins	Safety: radiation is penetrating and ^{125}I accumulates in thyroid, ALI is very low

relatively high-specific activity, and lower toxicity than ^{32}P.

4
Nucleic Acid Labeling

Radiolabels are most commonly introduced into nucleic acids by addition of single- or multiple-labeled nucleotides using enzyme-catalyzed reactions (Table 4 and Figs. 2 and 3). Most DNA and RNA polymerases are specific for particular pre-existing nucleic acid templates (e.g. DNA-dependent RNA polymerase) and synthesize a polynucleotide complementary to that template. There are two polymerization reactions that do not create polynucleotides complementary to a template: those catalyzed by terminal deoxynucleotidyl transferase from calf intestine, which adds deoxynucleotides to a 3′ end of DNA or RNA; and poly (A) polymerase, which adds adenosine monophosphates to 3′ ends of RNA molecules. All DNA polymerases require a nucleotide with a free 3′-hydroxyl hydrogen bonded to the template (i.e. a primer). RNA polymerases require a promoter, a specific DNA sequence on a double stranded-DNA molecule, in order to initiate transcription (RNA synthesis).

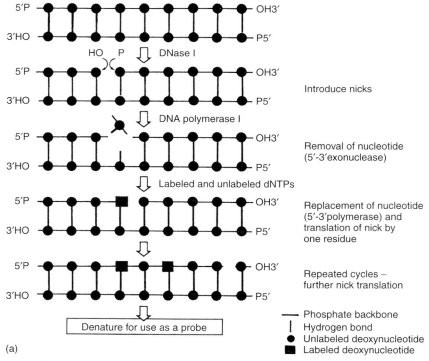

Fig. 2 Some methods for *in vitro* labeling of DNA are as follows: (a) the nick-translation reaction; (b) the random primer labeling reaction; (c) the 5′-end-labeling reaction using T4 polynucleotide kinase; (d) the 3′-end-labeling reaction using terminal deoxynucleotidyl transferase; (e) in-filling reaction using Klenow polymerase. (Reprinted with permission from Oxford University Press.)

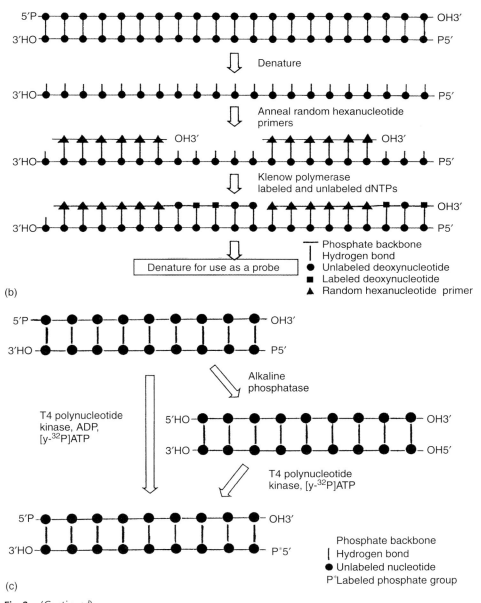

Fig. 2 (Continued)

Modification reactions can be used to incorporate labeling, for example, the addition of a ^{32}P phosphate to the 5'-end of a nucleic acid molecule by the enzyme T4 polynucleotide kinase using $[^{32}P\text{-}\gamma]$ ATP as substrate.

When a labeling reaction is complete it is necessary to estimate the extent of

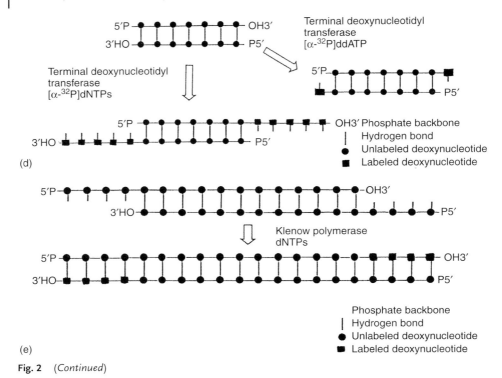

Fig. 2 (*Continued*)

incorporation. To do this, a small aliquot (2 μl) is removed from the reaction and diluted to a known volume (say 20 μl) with water. Two aliquots of this are counted in a scintillation counter. Two further aliquots are transferred to small tubes containing a few μg of carrier DNA or RNA in solution. The tubes are then filled with 10% trichloroacetic acid solution and left on ice for 30 min. The nucleic acids precipitate and can be collected on glass fiber filters. These are also counted in a scintillation counter. Comparison of these counts with the unprecipitated samples gives an indication of the percentage of incorporation, which should be at least 50% for most labeling protocols.

It is not always necessary to remove free, unincorporated labeled nucleotides from labeled probes; but if it is thought to be necessary, free label can be removed by ethanol precipitation of the probe in the presence of 0.1 M ammonium acetate or by the use of a 'spin column': Sephadex G25 swollen in buffer in a disposable syringe or pipette tip plugged with siliconized glass wool. The sample is applied to the column and, following centrifugation, the labeled probe is recovered from the centrifuge tube. Free label remains in the column. Alternatively, commercially made kits for purification of nucleic acids can be used, but these are usually more expensive.

5
Protein Labeling

Proteins are most frequently radiolabeled with ^{3}H, ^{14}C, ^{35}S, or ^{125}I. Tritium and

Tab. 4 Properties of nucleic acid–labeling reactions.

Method	Labeling density	Template	Amount of template (typical)	Reaction time	Incorporation efficiency	Nature of probe	Amount of probe	Specific activity of probe (dpm μg^{-1})
Random prime	Uniform	ssDNA (and denatured dsDNA)	25 ng	5 min–3 h	~75%	DNA	40–50 ng	5×10^9
Nick translation	Uniform	dsDNA	0.5–1 µg	~2 h	~60%	DNA	0.5–1 µg	5×10^8
Transcription labeling	Uniform	dsDNA	0.5–1 µg	1 h	~75%	RNA	~250 ng	5×10^9
3′-end labeling	End	Oligo or DNA fragment	~10 pmol ends	30–60 min	Variable	Oligo or DNA	10 pmols	5×10^6
End repair	End	Oligo or dsDNA fragment	~100 pmols ends	15 min	Variable	Oligo or DNA	100 pmols	5×10^5
5′-end labeling	End	Oligo or DNA fragment (dephosphorylated for forward reaction)	~10 pmol ends	~1 h	Variable	Oligo or DNA	10 pmols	5×10^6

Source: Printed with permission from Oxford University Press.

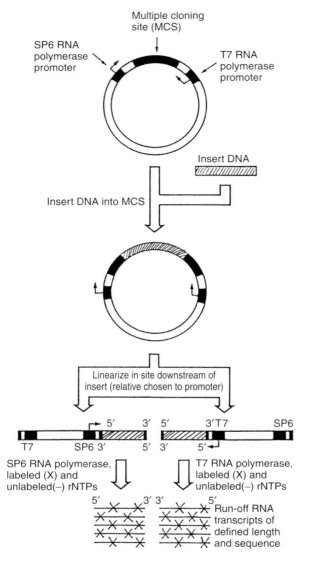

Fig. 3 Preparation of RNA probes using a plasmid containing SP6 and T7 RNA polymerase promoters. (Reprinted with permission from Oxford University Press.)

^{14}C are used for reasons of safety and practicality (because of their long half-life) whenever it is not essential to have proteins labeled to high-specific activity. Proteins can be labeled *in vivo* by incubating cells with labeled amino-acids. Alternatively they can be labeled *in vitro* with N-succinimidyl [2,3-^{3}H] propionate or by reductive methylation with [^{14}C] formaldehyde. Both techniques normally result in proteins that retain their biological activity. Methionine labeled with ^{35}S is used for *in vivo* labeling and *in vitro* protein synthesis as it provides a higher specific activity and good resolution in autoradiography of polyacrylamide gels used for analysis of products.

Radioiodination is used when proteins with a high-specific activity are required, for example, in radioimmunoassay and related techniques. Activities of up to 10^8 dpm µg^{-1} can be achieved. There are many methods that rely on either direct oxidative iodination or indirect conjugation iodination. Probably the most commonly used technique is the chloramine T procedure, as it is relatively simple and inexpensive (for protocol see Slater, Ed., 2002).

Most experiments that involve iodinated proteins require a preparation that is free of unincorporated radioactive iodide. This is most commonly removed by either gel filtration or high performance liquid chromatography (HPLC).

Iodination of proteins is potentially one of the most hazardous techniques in molecular biology. The levels of ^{125}I employed are often in the region of 18.5 MBq (500 µCi). Rigid application of the principles of radiation protection is required, labeling should be carried out only by highly trained personnel with the permission of the radiation protection advisor, and a controlled facility is usually required (for example, if the level of ^{125}I used exceeds 10 MBq).

6
Detection

6.1
Ionization Monitors

An ionization monitor consists of a small gas chamber containing two electrodes, a voltage supply, and a scaler. Ionizing radiation enters through a thin end window, the gas is ionized and a pulse of current is recorded on the scaler, usually displayed as counts per second, counts per minute or Bq cm^{-2}. Gas ionization monitors are suitable for detecting ^{32}P, and they will pick up ^{14}C, ^{35}S and ^{33}P if a sensitive instrument with a very thin end window is used. They do not detect ^{3}H, because the β-particles cannot penetrate the end window. The instruments detect ^{125}I, but not particularly effectively, since electromagnetic radiation passes through the gas and results in relatively poor ionization.

The monitors must be regularly checked and officially calibrated annually. If you use ^{32}P, it is also worthwhile to do a simple calibration on your instrument for your own information. Place a known volume of ^{32}P in a plastic tube with the lid off and record the decay rate at various distances above the tube. Monitoring 0.1 MBq ^{32}P at 10 cm (equivalent to a dose rate of approximately 0.5 µSv min^{-1}) should give a reading in excess of 50 c.p.s. All personnel working with radioisotopes (except ^{3}H) should have constant access to a monitor while at the bench.

6.2
Scintillation Counters

These instruments work on the principle that certain substances (termed *scintillators* or *fluors*) emit light upon absorption of ionizing radiations. The scintillators can be solids (e.g. sodium iodide) or liquids (e.g. toluene). The light is detected by one or more photomultiplier tubes that give an electrical signal to a scaler. The number of electrical signals relates to the number of disintegrations in the sample, and the strength of the signal relates to the energy of the absorbed radiation.

Scintillation probes are small handheld instruments useful for monitoring electromagnetic radiation (X or γ), say from ^{125}I. They either display c.p.s. values or, for dosimetry purposes, μSv h^{-1}. One of these instruments should be available whenever ^{125}I is used. As with ionization monitors, they should be calibrated officially once per year. You can calibrate them approximately yourself as an aid to monitoring (see Sect. 6.1).

Bench scintillation counters are designed for the analysis of large numbers of experimental samples. They either contain solid scintillators for detection of X or γ radiation or are designed for liquid scintillation counting. The latter instruments are found in most biochemical laboratories and are designed particularly for detecting β-emitters. The sample, usually in liquid form, is pipetted into a small container (or vial) to which a scintillation fluid is added.

There are many recipes for scintillation fluids (often referred to as cocktails). They are based on an organic solvent and one or more solutes termed *primary fluors* (when one solute is used) or *primary and secondary fluors* (when two solutes are used). The purpose of these fluors is to enhance the light output and shift the wavelength to the optimum for photomultiplier tubes. Most laboratories buy their cocktails ready prepared, and the manufacturer's catalogue should be carefully consulted. Essentially there are three forms of cocktail: a relatively straightforward formulation for counting organic samples, detergent-containing fluids for counting aqueous samples, and fluids that form gels for counting dispersed powders. The most commonly used ingredients are toluene or xylene as solvent, PPO (2,5-diphenyloxazole) or butyl-PBD (2-(4-t-butylphenyl)-5-(4-biphenyl)-1,3,4-oxadiazole) as primary solute, POPOP (l,4-bis(5-phenyloxazol-2-yl)benzene) as secondary solute and Triton-X100 as detergent.

An important concept in liquid scintillation counting is *quench*. This is anything that reduces counting efficiency, for example, colored compounds or chemicals such as chloroform that interfere with the scintillation process. Because a range of experimental samples may be subject to differing degrees of quench, it is necessary to calculate the counting efficiency of every sample individually. This can be done by re-counting samples following the addition of known amounts of radioactivity (an internal standard). However, this is tedious where there are a large number of samples, and contributes to the problem of waste disposal. Consequently, instruments have in-built means for automatically determining the counting efficiency. This works by counting a solid source of radioactivity within the machine that irradiates the sample vial. When quenching occurs, the energy spectrum of the radiation, as detected by the photomultipliers, appears to shift to lower-energy values. Analysis of this shift results in an assessment of counting efficiency within the sample vial. The instrument then corrects for the efficiency and prints out data as d.p.m. This system works well in most cases. However, it is essential to realize that it can be relied upon only when the sample is in solution in the vial, and is not appropriate for heterologous systems such as counting filter papers placed in scintillation fluid. This is because the filter paper will absorb some radiation, while the automatic quench correction mechanism will be recording only the properties of the solution. To accurately assess counting efficiency on filter papers, the paper has to be removed after counting, a known aliquot of radioisotope applied, the filter re-dried, and counted again. The increase in counts divided by the known added radioactivity provides the counting efficiency.

The ability of scintillation counters to analyze the energy of radiations is of great value. Radioisotopes have characteristic energy spectra. If these are sufficiently different, a scintillation counter can detect radioisotopes separately, even when they are present in the same vial. Beta spectra are all similar in shape but different radioisotopes have differing maximum energies (see Table 2); thus ^{3}H, ^{14}C, and ^{32}P for example, can be distinguished by their energy spectra. This means that dual labeling experiments with these radionuclides is feasible.

The only way to monitor for ^{3}H is to perform wipe tests with damp tissues, and count in a liquid scintillation counter, but if other forms of contamination were present, they would be detected by the counter on the basis of spectral analysis. The fact that different isotopes have differing energies means that scintillation counters can be used to some extent to determine types of contamination (e.g. to determine whether the contamination is ^{3}H or ^{14}C).

Liquid scintillation counting sometimes presents two problems, *photoluminescence* and *chemiluminescence*, that result in light output from scintillation fluid that is not the result of radioactivity. That is, the c.p.m. will exceed the true d.p.m. It is caused by storage of scintillation fluid in bright sunlight, or by certain chemicals. Most modern counters are able to detect it and make a note to that effect on the printout. If in doubt, samples can be stored in the dark for a few hours, and then recounted.

6.3
Cerenkov Counting

High-energy β-particles travel through water faster than light; they cause polarization of molecules along their path, and these emit photons of light as they return to a ground state. The energy of radiation from ^{32}P is high enough to cause the Cerenkov effect. Consequently water can be used in place of scintillation fluid for this radioisotope. The counting efficiency is relatively low (about 30% compared with >90% in a scintillation cocktail) and the counter should be specifically calibrated for it, although an acceptable shortcut is to count the samples as if they were ^{3}H. The big advantages of Cerenkov counting are: aqueous samples can be counted and recovered for experimental use, and there is no accumulation of organic radioactive waste. It is also very inexpensive.

6.4
Autoradiography

Many experiments in molecular biology rely on this technique, in particular, analysis of nucleic acid samples separated on gels (southern and northern blotting) or detection of particular clones or plaques by hybridization screening. The basic principle underlying autoradiography is that light, a β-particle, X ray, or γ-ray can produce silver atoms in the crystals of silver halide that are present in photographic emulsions. This process is reversible if only a single atom of silver produced from one photon, forms in a crystal. However, silver atoms catalyze the conversion of silver halide to silver by the chemicals in the developer; this stabilizes the image.

Tab. 5 Use and sensitivity of autoradiography and fluorography.

Isotope	Method	Example applications	Preflashed film	Inensifying screen	Temp. of exposure	Approximate detection limit after 24 h exposure [dpm/cm^{-2}]
^{3}H	Direct	Whenever very high resolution is required, e.g. in situ hybridization Acrylamide gels	No	No	Room temp.	8×10^6
	Fluorography		Yes	No	$-70\,°C$	8×10^3
^{14}C, ^{35}S or ^{33}P	Direct	DNA sequencing	No	No	Room temp.	6×10^3
	Indirect	SDS-polyacrylamide gels	Yes	Yes	$-70\,°C$	4×10^2
^{32}P	Direct	DNA sequencing	No	No	Room temp.	5×10^2
	Indirect	Southern analysis	Yes	Yes	$-70\,°C$	5×10^1
^{125}I	Direct	Subcellular localization	No	No	Room temp.	2×10^3
	Indirect	Western blotting	Yes	Yes	$-70\,°C$	1×10^2

If the radiation is of very low energy (e.g. from ^{3}H) it cannot penetrate very far and would, for example, be absorbed by the matrix of an agarose or polyacrylamide gel. Conversely, if radiation is very penetrating (e.g. from ^{32}P or ^{125}I) it passes through film without giving up much of its energy and is less efficiently detected than might be expected. To combat these difficulties there are techniques called *fluorography*, where samples such as gels are impregnated with fluors such as PPO (see Sect. 6.2), and *indirect autoradiography*, where intensifying screens partly absorb penetrating radiation and emit light. These adaptations increase sensitivity but reduce resolution. In both of these cases, exposure at low temperature is required to stabilize the latent image. In addition, a brief (<1ms) flash of light is required to preexpose the film and make it more responsive to low signals. Specific films are sold for direct or indirect autoradiography. A summary of some of the key features of autoradiography and fluorography (including use of intensifying screens) is provided in Table 5.

As discussed in Sect. 2, it is a requirement that if alternatives to radioisotopes are available, they should be used wherever possible. Nonradioactive labeling methods for DNA and proteins are improving all the time. Light-based detection techniques seem particularly successful in western blotting, for example. In the case of ^{32}P, a general rule at present is that if acceptable results are obtained within a 24-hour exposure to autoradiography, then a nonradioactive alternative should be sought. If longer exposures are required, particularly with intensifying screens, then it is likely that ^{32}P is still required for meeting the necessary sensitivity parameters.

6.5
Personal Dosimeters

These are either film badges or thermoluminescent detectors (TLDs). The latter contain a phosphor such as LiF that becomes and remains excited once exposed to radiation. Heat treatment results in light emission, the intensity of which relates to the dose received. Small TLDs can be worn on the fingers, under gloves, for hand dosimetry.

Personal dosimeters are not appropriate for workers using low-energy emitters such as ^{14}C, ^{35}S or ^{3}H but they are of value to users of ^{32}P or ^{125}I. Their issue should beon the advice of senior staff with appropriate training and responsibility for radiation protection, after dosimeter calculations and estimates of exposure levels to individuals.

See also Labeling, Biophysical; Mass Spectrometry-based Methods of Proteome Analysis; Nucleic Acid Hybrids, Formation and Structure of.

Bibliography

Books and Reviews

Ballance, P.E., Day, R.L., Morgan, J. (1992) *Phosphorus – 32: Practical Radiation Protection H and H Scientific Consultants*, Leeds, England.

Mundy, C.R., Cunningham, M.W., Read, C.A. (1991) Nucleic Acid Labelling and Detection, in: Brown, T.A. (Ed.) *Essential Molecular Biology a Practical Approach*, Oxford University Press, Oxford and New York.

Slater, R.J. (Ed.) (2002) *Radioisotopes in Biology*, 2nd edition, Oxford University Press, Oxford and New York.

Primary Literature

International Commission on Radiological Protection. (1991) *Recommendations of the International Commission on Radiological Protection*, ICRP Publication 60, Annals of the ICRP 21 Nos 1–3, Pergamon Press, Oxford.

International Commission on Radiological Protection. (1994) *Dose Coefficients for Intakes of Radionuclides by Workers*, ICRP Publication 68, Annals of the ICRP 24 No 4, Pergamon Press, Oxford.

Rat Genome (*Rattus norvegicus*)

Kim C. Worley and Preethi Gunaratne
Human Genome Sequencing Center, Baylor College of Medicine, Houston, TX, USA

1	The Rat and the Rat Genome	436
2	The Assembly Strategy and Results	437
3	**Features of the Rat Genome**	441
3.1	Genome Size	441
3.2	Telomeres and Centromeres	441
3.3	Orthologous Chromosomal Segments and Large-scale Rearrangements	443
3.4	Segmental Duplications	443
3.5	Gains and Losses of DNA	445
3.6	Substitution Rates	448
3.7	G + C Content and CpG Islands	449
3.8	Shift in Substitution Spectra between Mouse and Rat	449
3.9	Evolutionary Hotspots	450
3.10	Covariation of Evolutionary and Genomic Features	452
4	**Evolution of Genes**	452
4.1	Construction of Gene Set and Determination of Orthology	452
4.2	Properties of Orthologous Genes	453
4.3	Indels and Repeats in Protein-coding Sequences	454
4.4	Transcription-associated Substitution and Asymmetry	455
4.5	Conservation of Intronic Splice Signals	456
4.6	Gene Duplications	456
4.7	Conservation of Gene-regulatory Regions	459
4.8	Pseudogenes and Gene Loss	459
4.9	*In Situ* Loss of Rat Genes	460
4.10	Noncoding RNA Genes	462

5	**Evolution of Transposable Elements** 462
5.1	LINE-1 Activity in the Rat Lineage 462
5.2	Different Activity of SINEs in the Rat and Mouse Lineage 464
5.3	Colocalization of SINEs in Rat and Mouse 465
5.4	Endogenous Retroviruses and Derivatives 466
5.5	Simple Repeats 466
5.6	Prevalent, Medium-length Duplications in Rodents 467

6	**Rat-specific Biology** 467
6.1	Chemosensation 467
6.2	α_{2u} Globulin Pheromones 468
6.3	Detoxification 469
6.4	Proteolysis 469

7	**Human Disease Gene Orthologs in the Rat Genome** 469

8	**Summary** 474
	Acknowledgment 475
	Bibliography 475
	Books and Reviews 475
	Primary Literature 475

Keywords

BAC
A bacterial artificial chromosome, a large-insert (~200 kb) cloning vector for genomic sequences.

Contig
A contiguous set of overlapping segments of DNA.

Finished sequence
Complete, contiguous sequence generated with an accuracy of 1 error per 10 000 bp.

Genome
The entire complement of nuclear DNA in an individual or the representative sequence composite from several individuals.

Nonsynonymous
Nucleotide changes in the coding region of a gene that change the amino acid sequence of the translated protein. Some amino acid changes are more deleterious to

the protein function than others. The nonsynonymous substitution rate is the number of nonsynonymous substitutions per nonsynonymous site K_A.

ORF
Open reading frame – a sequence that translates without internal stop codons into a protein sequence.

Orthologous
Sequence regions in different organisms that originated from the same sequence in the last common ancestor of the organisms.

Scaffold
A set of contigs with sequence gaps between them that are linked by mate-pair information or marker information that may or may not give an estimated gap size.

Synonymous
Nucleotide changes in the coding region of a gene that do not change the amino acid sequence of the translated protein. The synonymous substitution rate is the number of synonymous substitutions per synonymous site K_S.

Whole-genome Shotgun (WGS)
Sequence generated randomly from a genome. Usually sheared genomic DNA is subcloned into a library in plasmid vectors, the library is then sampled and clones sequenced from both ends of the inserts.

▪ The Brown Norway rat was the third mammalian genome to be sequenced. The three-way comparison of the human, rat, and mouse sequences resolves details of mammalian evolution and allows divergence events to be placed on different branches of the evolutionary tree. The comparison of the human to invertebrate and rodent genomes highlights the consequences of evolution over 1000 million years and 75 million years, while the comparison of the two rodent genomes describes changes that occurred in the 12 to 24 million years since the common ancestor of the rat and mouse.

A number of insights came from these comparisons:

The rat genome is 2.75 gigabases (Gb), smaller than human (2.9 Gb), and slightly larger than mouse (2.6 Gb). The three genomes encode similar numbers of genes. The majority of the genes have persisted without deletion or duplication and with well-conserved intron–exon structures. The exceptions are members of gene families that have expanded through gene duplication. Genes found in rat, but not mouse included genes producing pheromones, or involved in immunity, chemosensation, detoxification, or proteolysis.

Human genes known to be associated with disease have orthologs in the rat, but their rates of synonymous substitutions are significantly different from other genes.

Three percent of the rat genome is in large segmental duplication, primarily located near the centromeres. Expansions of major gene families are due to these genomic duplications.

About 40% of the rat genome aligns orthologously to mouse and human; this eutherian core of the genome contains the vast majority of exons and known regulatory elements (which comprise 1–2% of the genome). Only a portion of this core (5–6%) appears to be under selective constraint in rodents and primates, while the remainder appears to be evolving neutrally. Outside the eutherian core, the majority of the 30% of the rat genome that aligns only with mouse is rodent-specific repeats. More than half of the nonaligning sequence is rat-specific repeats.

More genomic changes have occurred in the rodent lineages than in the primate. Large rearrangements include approximately 250 rearrangements between a murid ancestor and human (the majority between the eutherian ancestor and the murid ancestor, and approximately 50 each from the murid ancestor to the rat and mouse). There is a threefold higher base substitution rate in neutral DNA along the rodent lineage than along the human lineage, with the rate on the rat branch 5 to 10% higher than along the mouse branch. Microdeletions are more frequent than microinsertions in both the rat and mouse branches. Most interestingly, there is a strong correlation between local rates of microinsertions and microdeletions, nucleotide substitutions, and transposable element insertions in the rat and mouse lineages, although the events occurred independently since the divergence of the two branches.

1
The Rat and the Rat Genome

The rat, although revered as the first sign in the Chinese zodiac and bearer of the Hindu god Ganesh, is a known carrier of over 70 diseases. Rats are involved in the transmission of several infectious diseases to man, including cholera, bubonic plague, typhus, leptospirosis, cowpox, and hantavirus infections. A major agricultural pest, rats and other rodents consume approximately one-fifth of the annual food harvest.

The laboratory rat (*Rattus norvegicus*) has contributed to human health by testing new drugs, and improving the understanding of essential nutrients and the pathobiology of human disease. It was the first mammal domesticated for scientific research (1828). The rat has been the model of choice for physiologists and nutritionists, and there are over 234 inbred strains developed to study genetic diseases.

The rat genome-sequencing project produced a draft sequence that, unlike mouse and human, would not ultimately be finished to remove all sequence gaps and produce a high base accuracy. For this reason, the quality of the draft was important. Although gaps remained, the overall sequence quality supported detailed analyses.

The sequence was generated from two inbred females, (BN/SsNHsd/Mcwi) by a network of centers led by the Human Genome Sequencing Center, Baylor College of Medicine. An international team representing over 20 groups contributed to the analysis reported and summarized here.

2
The Assembly Strategy and Results

The rat genome project used a combined WGS (whole-genome shotgun) with BAC clone strategy (Fig. 1). The project benefited from the logistically simpler WGS sequence generation and the local sequence assembly to resolve duplications afforded by BAC clones. The Atlas assembly suite, designed to combine these data sets, provided a BAC fisher to present localized combined BAC + WGS data to the public prior to the availability of the complete assembly.

Over 44 million DNA sequence reads were generated (Table 1). After removal of low-quality reads and vector contaminants, 36 million reads were used in the assembly where 34 million reads were retained. This was a sevenfold sequence coverage, with 60% of the reads from the WGS. Coverage estimates range from 7.3x (when estimated from the entire "trimmed" length of the sequence data) to 6.9x (when estimated from the sequence with quality of Phred20 or higher).

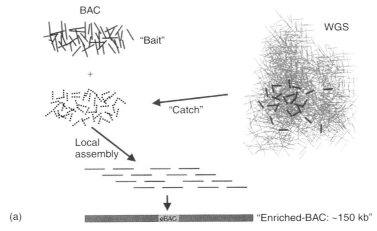

Fig. 1 The new "combined" sequence strategy and Atlas software.
(a) Formation of "eBACs." The strategy combined the advantages of both BAC and WGS sequence data. Modest sequence coverage (~1.8-fold) from a BAC is used as "bait" to "catch" WGS reads from the same region of the genome. These reads, and their mate pairs, are assembled using Phrap to form an enriched BAC or "eBAC." This stringent local assembly retains 95% of the "catch." (b) Creation of higher-order structures. Multiple eBACs are assembled into bactigs on the basis of sequence overlaps. The bactigs are joined into superbactigs by large clone mate-pair information (at least 2 links), extended into ultrabactigs using additional information (single links, FPC contigs, synteny, markers), and ultimately aligned to genome mapping data (RH and physical maps) to form the complete assembly. Used with permission from *Nature* **428**, 493–521 (2004).

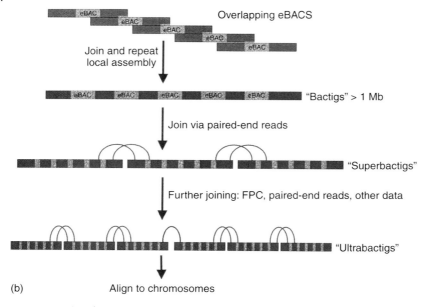

(b)

Fig. 1 (Continued)

Simultaneous to the sequencing, a "fingerprint contig" (FPC) map was developed, which was used in combination with the ongoing sequencing to identify BACs for sequence skimming. The parallel development of mapping and sequencing resources permitted the data-gathering phase of the project to be completed in less than two years.

The statistics of the rat draft assembly (v. 3.1) are given in Table 2. The current assembly (v. 3.4) splices in a number of finished BAC sequences to the v. 3.1 assembly. Much of the gene and protein feature analysis was developed using earlier assemblies (v. 2.0 and 2.1), while the genome description is based on v. 3.1.

The majority of the genome is in contigs larger than the expected mammalian gene (N50 = 38 kb). These contigs are linked into 783 larger scaffolds; those anchored to the radiation hybrid map had an N50 of 5.4 Mb, while the smaller scaffolds that could not be anchored had an N50 of 1.2 Mb.

The quality of the assembled sequence was assessed using sequence from finished BACs, the comparison showed that the bases within contigs were of high quality (1.32 mismatches per 10 kb), similar to the finished sequence. The majority of mismatches occurred at the ends of contigs in regions that average 750 bp and total <0.9% of the genome. Only six mismatches (insertions or deletions) were found within contigs when compared to 13 Mb of finished sequence or one per 2.2 Mb.

The assembly accuracy was judged in comparison to linkage and radiation hybrid maps. The majority of the genetic markers (13/3824) and sequence-tagged sites (96.9%) had consistent chromosome placement (Fig. 2). The maps are congruent with the assembly except for possible mismapped markers.

Tab. 1 Clones and reads used in the RGSP.

Insert size [kb][a]	Source or vector	Reads [millions]				Bases [billions]		Sequence coverage[b]		Clone coverage[c]
		All[d]	Used	Paired	Assembled	Trimmed	≥Phred20	Trimmed	≥Phred20	
2–4	Plasmid	9.6	8.6	7.4	7.9	4.8	4.5	1.8	1.6	3.70
4.5–7.5	Plasmid	4.5	4.3	3.6	3.6	2.4	2.3	0.87	0.82	2.96
10	Plasmid	8.4	7.2	6.4	6.4	4.1	3.8	1.5	1.4	11.63
50	Plasmid	1.7	1.3	1.0	1.1	0.69	0.65	0.25	0.24	9.47
150–250	BAC	0.32	0.31	0.26	0.26	0.18	0.16	0.07	0.06	9.26
Total WGS		24.5	21.7	18.7	19.2	12.1	11.3	4.4	4.1	37.0
2–5	BAC skims	19.6	14.6	13.2	14.5	8.0	7.7	2.9	2.8	4.8[e]
Total		44.1	36.3	31.9	33.7	20.2	19.0	7.3	6.9	41.8

[a] Grouped in ranges of sizes for individual libraries tracked to specific multiples of 0.5 kb.
[b] Total bases in used reads divided by sampled genome size including all cloned and sequenced euchromatic or heterochromatic regions.
[c] Estimated as sum of insert sizes divided by sampled genome size.
[d] WGS reads available on the NCBI Trace Archive as of March 21, 2003; BAC skim reads attempted at BCM–HGSC as of May 12, 2003; BAC end reads obtained directly from TIGR.
[e] Refers to coverage from 2–5 kb subclones from BACs. The BACs that were skimmed amounted to 1.58x clone coverage.

Tab. 2 Statistics of the RGSP draft sequence assembly.

Features[a]	Number	N50 Length [kb]	Bases [Gb]	Bases plus gaps [Gb][b]	Percentage of genome[c]			
					Sampled [2.78 Gb]		Assembled [2.75 Gb]	
					Bases	Bases + Gaps	Bases	Bases + Gaps
Anchored contigs	127 810	38	2.476	2.481	89.1	89.2	90.0	90.2
Anchored superbactig scaffolds	783	5402	2.476	2.509	89.1	90.3	90.0	91.2
Anchored ultrabactigs	291	18 985	2.476	2.687	89.1	96.6	90.0	97.7
Unanchored superbactigs, main scaffolds	134	1210	0.056	0.062	2.0	2.2	2.0	2.3
Unanchored ultrabactigs	128	1529	0.056	0.069	2.0	2.5	2.0	2.5
All superbactigs, main scaffolds	917	5301	2.533	2.571	91.1	92.5	92.1	93.5
Minor scaffolds	4345	8	0.033	0.038	1.2	1.4	1.2	1.4

[a] Anchored sequences are those that can be placed on chromosomes because they contain known markers. The main scaffold for each superbactig is the largest set of contigs (in terms of total contig sequence), which can be ordered and oriented using mate-pair links and ordering of BACs. Scaffolds, which cannot be ordered and oriented with respect to the main scaffold, are termed *minor scaffolds*.
[b] Ambiguous bases (Ns) are counted in the gap sizes, and excluded in the base counts.
[c] Computed as bases plus gaps divided by estimated genome size. Sampled genome size is based on oligonucleotide frequency statistics of unassembled WGS reads. Assembled genome size is based on cumulative contig sequence following assembly.

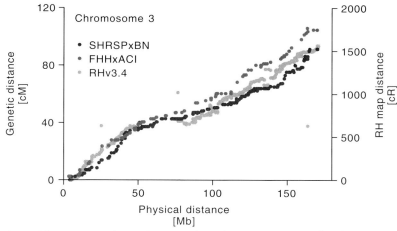

Fig. 2 Map correspondence. Correspondence between positions of markers on two genetic maps of the rat (SHRSPxBN intercross and FHHxACI intercross), on the rat radiation hybrid map, and their position on the rat genome assembly (Rnor3.1). Used with permission from *Nature* **428**, 493–521 (2004). (See color plate p. xxxix.)

3 Features of the Rat Genome

3.1 Genome Size

Genome assemblies are usually smaller than the actual genome size owing to underrepresentation of sequences due to cloning bias and sequencing and assembly difficulties. However, equating assembled genome size with the euchromatic, clonable portion of the genome (CpG) does not take into account the heterochromatic sequence included in the assembly. The rat genome size was estimated by two methods: scaling the assembled genome size by the fraction of features found in the assembly, and measuring the clonable or sampled genome size on the basis of the distribution of short oligomers in the WGS reads before assembly and in the assembly. Both estimates gave a relatively consistent measure of estimated genome size of 2.75 Gb. This conservative estimate was still considerably higher than the size estimated for the mouse draft genome sequence, which has different repeat content and appears to have underrepresented segmental duplications because of technical reasons.

3.2 Telomeres and Centromeres

The rat has both metacentric and telocentric chromosomes, unlike the wholly telocentric mouse chromosomes. As expected, the draft sequence does not contain complete telomere and centromere sequences. The approximate physical location of the centromeres relative to the genomic sequence is shown in Fig. 3. Several of the putative centromere positions coincide with segmental duplication blocks and classical satellite repeat clusters, consistent with enrichment of these sequence features in rat pericentromeric DNA. Human subtelomere regions are characterized by an abundance of segmentally duplicated DNA and an enrichment of internal (TTAGGG)n-like sequence islands.

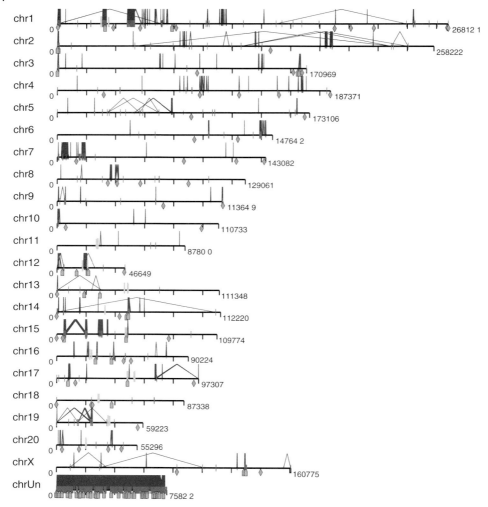

Fig. 3 Distribution of segmental duplications in the rat genome. Interchromosomal duplications (red) and intrachromosomal duplications (blue) are depicted for all duplications with $\geq$90% sequence identity and $\geq$20 kb length. The intrachromosomal duplications are drawn with connecting blue line segments; those with no apparent connectors are local duplications very closely spaced on the chromosome (below the resolution limit for the figure). P arms on the left and the q arms on the right. Chromosomes 2, 4–10, and X are telocentric; the assemblies begin with pericentric sequences of the q arms, and no centromeres are indicated. For the remaining chromosomes, the approximate centromere positions were estimated from the most proximal STS/gene marker to the p and q arm as determined by fluorescent *in situ* hybridization (FISH) (cyan vertical lines; no chromosome 3 data). The chrUn sequence is contigs not incorporated into any chromosomes. Green arrows indicate 1-Mb intervals with more than tenfold enrichment of classic rat satellite repeats within the assembly. Orange diamonds indicate 1-Mb intervals with more than tenfold enrichment of internal (TTAGGG)n-like sequences. For more details, see http://ratparalogy.cwru.edu. Used with permission from *Nature* **428**, 493–521 (2004). (See color plate p. xlvi.)

Approximately one-third of the euchromatic rat subtelomeric regions contain similar features, suggesting that Rnor3.1 might extend very close to the chromosome ends.

3.3
Orthologous Chromosomal Segments and Large-scale Rearrangements

Multimegabase regions of chromosomes have been passed from the primate-rodent ancestor to human and murid rodent descendants with minimal rearrangements of gene order. These regions, bounded by the breaks that occurred during ancient large-scale chromosomal rearrangements are referred to as *orthologous chromosomal segments*. The sequence of these rearrangements was tentatively reconstructed using the human genome and other outgroup data (see Fig. 4). Inspection shows events that preceded and follow the rat–mouse divergence are interleaved. At 1-Mb resolution, multiple methods report virtually indistinguishable sets of orthologous chromosomes segments: 278 between human and rat, 280 between human and mouse, and 105 between rat and mouse. The larger numbers of breaks in orthologous segments between the human and the rodents is expected because of the greater evolutionary distance.

Understanding the number and timing of rearrangement events that have occurred in each of the three individual lineages (see tree in Fig. 5a) since the common primate-rodent ancestor required a more detailed analysis. The X chromosome is presented here as an example; its history is easier to trace completely since rearrangements between the X and the autosomes are rare. There are 16 human–mouse–rat orthologous segments of at least 300 kb in size (Fig. 6a). The most parsimonious scenario requires 15 inversions in the descent from the primate-rodent ancestor. Outgroup data from cat, cow, and dog resolve the timing of these rearrangements more precisely. Most of these events occurred in the rodent lineage: five (or four) before the divergence of rat and mouse, five in the rat lineage and five in the mouse lineage. At most one rearrangement occurred in the human lineage since divergence from the common ancestor with rodents. The analysis of the whole genome showed similar results. The assignment of the considerable rearrangement activity to the rodent branch following primate-rodent divergence is consistent with previous, lower resolution studies.

3.4
Segmental Duplications

Segmental duplications are defined here as genomic regions of at least 5 kb in length that are repeated with >90% identity remaining between the copies. The rat has approximately 2.9% of its bases in these duplicated regions, whereas the human has 5 to 6%, and the mouse has 1.0 to 2.0%. These duplicated structures are particularly challenging to assemble, so some of the mouse–rat difference is attributable to the BAC-based approach used in the rat assembly compared to the WGS-mouse approach. Most of these regions have less than 99.5% identity and are, therefore, not simply overlapping sequences that were not joined by the assembly program. Nearly 44% of these blocks of segmental duplications are mapped to the "unplaced" chromosome in Rnor3.1 indicating the difficulty of anchoring these elements to the genome.

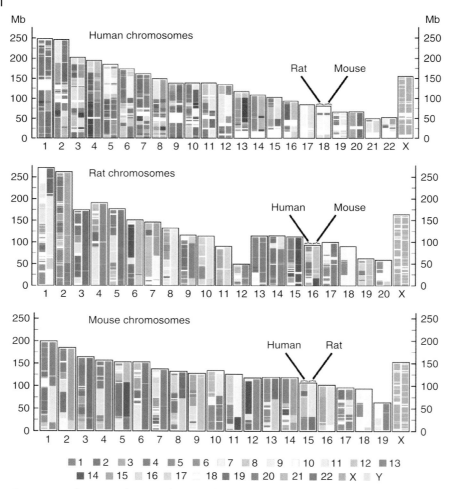

Fig. 4 Map of conserved synteny between the human, mouse, and rat genomes: For each species, each chromosome (x-axis) is a two column boxed pane (p-arm at the bottom) colored according to conserved synteny to chromosomes of the other two species. The same chromosome color code is used for all species (indicated below). For example, the first 30 Mb of mouse chromosome 15 is shown to be similar to part of human chromosome 5 (by the red in left column) and part of rat chromosome 2 (by the olive in right column). An interactive version is accessible (http://www.genboree.org). Used with permission from *Nature* **428**, 493–521 (2004). (See color plate p. xlvii.)

Intrachromosomal duplications are three times more common than interchromosomal duplications and are significantly enriched near the telomeric and centromeric regions (Fig. 3). The pericentromeric accumulation of segmental duplications in the rat seems to be a general property of mammalian chromosome architecture.

There is considerable clustering of segmental duplications. For many of the largest clusters, the underlying sequence

Fig. 5 Substitutions and microindels (1–10 bp) in the evolution of the human, mouse, and rat genomes: (a) The lengths of the labeled branches in the tree (top panel) are proportional to the number of substitutions per site inferred from all sites with aligned bases in all three genomes. (b) The table shows the midpoint and variation in these branch length estimates when estimated from different sequence alignment programs and different neutral sites, including sites from ancestral repeats, fourfold degenerate sites in codons, and rodent-specific sites ("in neutral sites only" row). Other rows give midpoints and variation for microindels on each branch of the tree. Used with permission from *Nature* **428**, 493–521 (2004).

alignments show a wide range of sequence identity, suggesting that duplication events have occurred continuously over millions of years. In contrast, an analysis of all duplicated regions showed a bimodal distribution consistent with bursts of segmental duplication (particularly intrachromosomal duplication) that occurred approximately 5 and 8 Myr ago.

The segmental duplications of the rat genome were of considerable interest because they represent an important mechanism for the generation of new genes. Sixty-three (of 4532 total) NCBI reference sequence genes were completely or partially located within duplicated regions. As discussed in the following, many of these genes present in multiple copies, belong to recently duplicated gene families, and contribute to distinctive elements of rat biology.

3.5 Gains and Losses of DNA

In addition to large rearrangements and segmental duplications, genome architecture is strongly influenced by insertion and deletion events that add and remove DNA over evolutionary time. To characterize the origins and losses of sequence elements in the human, mouse, and rat genomes, the nucleotide bases were categorized using alignment data and annotations of the insertions of repetitive elements (Fig. 7).

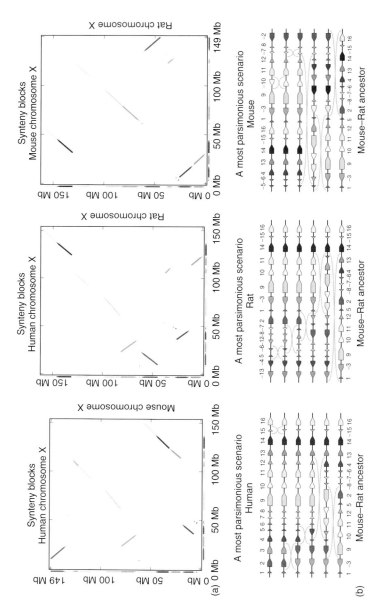

Fig. 6 X chromosome in each pair of species: (a) GRIMM–Synteny computes 16 three-way orthologous segments (≥300 kb) on the X chromosome of human, mouse, and rat, shown for each pair of species, using consistent colors. (b) The arrangement (order and orientation) of the 16 blocks implies that at least 15 rearrangement events occurred during X chromosome evolution of these species. The program MGR determined that evolutionary scenarios with 15 events are achievable and all have the same median ancestor (located at the last common mouse–rat ancestor). Shown is a possible (not unique) most parsimonious inversion scenario from each species to that ancestor. Note the last common ancestor of human, mouse, and rat should be on the evolutionary path between this median ancestor and human. Used with permission from *Nature* **428**, 493–521 (2004). (See color plate p. lv.)

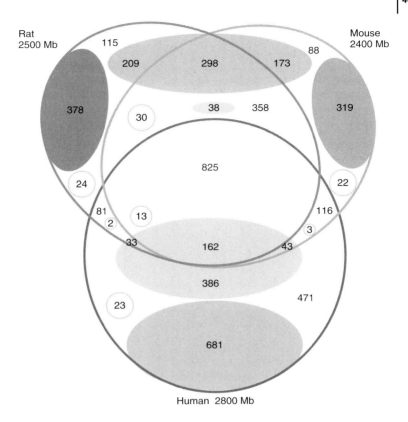

Fig. 7 Aligning portions and origins of sequences in rat, mouse, and human genomes. Each outlined ellipse is a genome, and the overlapping areas indicate the amount of sequence that aligns in all three species (rat, mouse, and human) or in only two species. Nonoverlapping regions represent sequences that do not align. Types of repeats classified by ancestry: those that predate the human–rodent divergence (gray), those that arose on the rodent lineage before the rat–mouse divergence (lavender), species-specific (orange for rat, green for mouse, blue for human), and simple (yellow), placed to illustrate the approximate amount of each type in each alignment category. Uncolored areas are nonrepetitive DNA – the bulk is assumed to be ancestral to the human–rodent divergence. Numbers of nucleotides (in Mb) are given for each sector (type of sequence and alignment category). Used with permission from *Nature* **428**, 493–521 (2004). (See color plate p. xlviii.)

The estimates of the amount of repeats represent lower bounds because some repeats, especially the older ones, are not recognized even though the rodent repeat database was greatly expanded by analyzing the rat and mouse genomes.

About a billion nucleotides (39% of the euchromatic rat genome) align in all three species, constituting an "ancestral core," which is retained in these genomes. The ancestral core contains 94 to 95% of the known coding exons and regulatory regions. The levels of three-way conservation of "mammalian ancestral repeats" (transposon relics retained in all three species) confirm estimates that 5 to 6% of the human genome is accumulating substitutions more slowly than the neutral rate in the three lineages, and hence may be under purifying selection. In this constrained fraction, non-coding regions outnumber coding regions regardless of the constrain, an observation that supports comparative analyses on limited subsets of the genome. The preponderance of noncoding elements in the most constrained fraction of the genome underscores the likelihood that they play critical roles in mammalian biology.

About 700 Mb (28%) of the rat euchromatic genome aligns only with the mouse. At least 40% of this is rodent-specific repeats that are inserted on the branch from the primate-rodent ancestor to the murid ancestor. Some of the remainder is ancestral mammalian repeats or single copy DNA deleted in the human lineage but retained in rodents (Fig. 7). Although this 700 Mb of rodent-specific DNA is primarily neutral, it may also contain some functional elements lost in the human lineage in addition to sequences representing gains of rodent-specific functions, including some coding exons.

3.6
Substitution Rates

The alignment data allow relatively precise estimates of the rates of neutral substitutions and microindel events ($<=10$ bp). As in previous studies comparing human and mouse, both synonymous fourfold degenerate ("4D") sites in protein-coding regions and sites in mammalian ancestral repeats were used in this analysis. In addition, the primarily neutral rodent-specific sites that did not align to human discussed earlier were used.

Estimates for the neutral substitution level between the two rodents range from 0.15 to 0.20 substitutions per site, while estimates for the entire tree of human, mouse, and rat range from 0.52 to 0.65 substitutions per site (Fig. 5). This difference was predictable because of the evolutionary closeness of the two rodents. Less predictable was the finding that for all classes of neutral sites analyzed, the branch connecting the rat to the common rodent ancestor is 5 to 10% longer than the mouse branch. Thus, for as yet unknown reasons, the rat lineage has accumulated substantially more point substitutions than the mouse lineage since their last common ancestor.

Four-way alignments with sequence from orthologous ancestral repeats in the three genomes and the repeat consensus sequences (to approximate the progenitor repeat sequence) were used to distinguish substitutions on the branches from the primate-rodent ancestor to either the human or rodent ancestor. This revealed a three-to-one speed-up in rodent substitution rates relative to human, larger than estimated in the mouse analysis but consistent with more recent studies, which also use multiple sequence alignments.

Estimates for rates of microdeletion events are, for all branches, approximately twofold higher than rates of microinsertion (Fig. 5b), suggesting a fundamental difference in the mechanisms that generate these mutations. Furthermore, there are substantial lineage specific rate differences. The rat lineage has accumulated microdeletions more rapidly than the mouse, while the opposite holds true for microinsertions. As with substitutions, both microinsertions and microdeletion rates are substantially slower in the human lineage. The size distributions of microindels (1–10 bp) were heavily weighted toward the smallest indels and similar for insertions, deletions, and both the mouse and rat branches.

3.7
G + C Content and CpG Islands

The G + C content of the rat varies significantly across the genome (Fig. 8a) in a distribution that more closely resembles that of mouse than human. The variation of G + C content is coupled with differences in the distribution of CpG islands – short regions that are associated with the 5′ ends of genes and gene regulation and that escape the depletion of CpG dinucleotides owing to deamination of methylated cytosine. The 2.6-Gb rat genome assembly contains 15 975 CpG islands in nonrepetitive sequences of the genome. This is similar to the 15 500 CpG islands reported in the 2.5-Gb mouse genome, but far fewer than the 27 000 reported in the human genome. The distribution of CpG islands by chromosome is shown in Fig. 8b.

The changes in CpG island content predate the rat-mouse split and are consistent with the accelerated loss of CpG dinucleotides in rodents compared with humans. It remains possible, however, that the greater number of human regions with extremely high G + C content are due to distributational changes occurring in the primate, rather than the rodent lineage.

3.8
Shift in Substitution Spectra between Mouse and Rat

The nonrepetitive fraction of the rat genome is enriched for G + C content relative to the mouse genome by ∼0.35% over 1.3 billion nucleotides. This is a subtle but substantial difference that may be explained in part by differences in the spectra of mutation events that have accumulated in the mouse and rat lineages. The small minorities of substitutions where events can be assigned to either the mouse or the rat lineage by virtue of a nucleotide match in the alignment column between human and only one rodent were studied. Of the ∼117 million alignment columns meeting this criteria, ∼60 million involve a change in the rat lineage versus ∼57 million in the mouse, reflecting the increase in the rates of point substitution in the rat lineage (Fig. 5b). Fifty percent of these events in rat involve a substitution from an A/T to a G/C compared to only 47% of all mouse changes. The complementary change, G/C to A/T, is more common in the mouse lineage (38% vs. 35% in rat). There is not a similar substantial difference between changes that do not alter G + C content. In addition, this bias is not confined to particular transition or transversion events, nor can it be explained simply as a result of divergent substitution rates of CpG dinucleotides. Thus, this shift appears to be a general

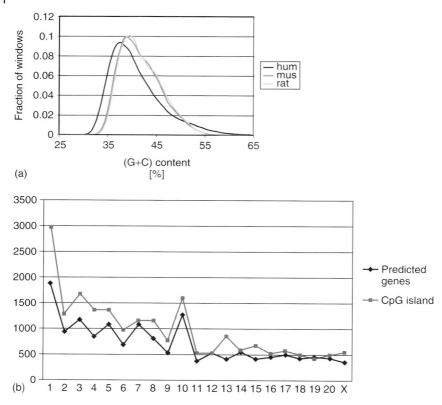

Fig. 8 Base composition distribution analysis. (a) The fraction of 20 kb nonoverlapping windows with a given G + C content is shown for human, mouse, and rat. (b) The number of Ensembl predicted genes per chromosome and the number of CpG islands per chromosome. The density of CpG islands averages 5.9 islands per Mb across chromosomes and 5.7 islands per Mb across the genome. Chromosome 1 has more CpG islands than other chromosomes, yet neither the island density nor ratio to predicted genes exceeds the normal distribution. The number of CpG islands per chromosome and the number of predicted genes are correlated ($R^2 = 0.96$). Used with permission from Nature **428**, 493–521 (2004). (See color plate p. xlix.)

change that results in an increase in G + C content in the rat genome. Biochemical changes in the repair or replication enzymes might be responsible, and the observation that recombination rates are slightly higher in rat than in mouse may suggest a role for G + C biased mismatch repair. However, population genetic factors, such as selection, cannot be ruled out.

3.9 Evolutionary Hotspots

Comparison of the two rodent genomes, using human as outgroup, reveals regions that are conserved yet under different levels of constraint in mouse and rat. These regions may have distinct functional roles and contribute to species-specific differences. In 5055 regions >=100 bp, with at least a tenfold difference in the estimated number of substitutions per site on the mouse and rat branches and <0.25 substitutions per site on the human branch (to avoid fast-evolving regions), analysis found them enriched twofold in transcribed regions. Thirty-nine percent of mouse hot spots were found in 18% of the

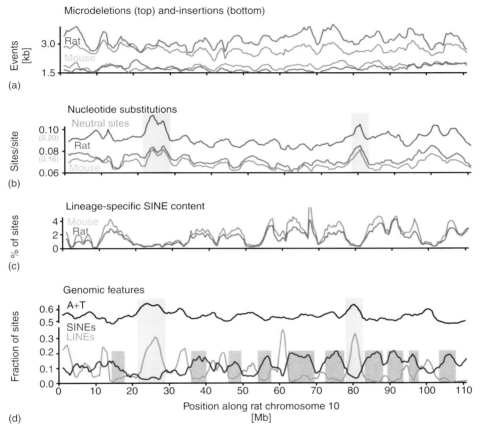

Fig. 9 Variability of several evolutionary and genomic features along rat chromosome 10. (a) Rates of microdeletion and microinsertion events (less than 11 bp) in the mouse and rat lineages since their last common ancestor, revealing regional correlations. (b) Rates of point substitution in the mouse and rat lineages. Red and green lines represent rates of substitution within each lineage estimated from sites common to human, mouse, and rat. Blue represents the neutral distance separating the rodents, as estimated from rodent-specific sites. Note the regional correlation among all three plots, despite being estimated in different lineages (mouse and rat) and from different sites (mammalian vs rodent-specific). (c) Density of SINEs inserted independently into the rat or mouse genomes after their last common ancestor. (d) A + T content of the rat, and density in the rat genome of LINEs and SINEs that originated since the last common ancestor of human, mouse, and rat. Pink boxes highlight regions of the chromosome in which substitution rates, AT content, and LINE density are correlated. Blue boxes highlight regions in which SINE density is high but LINE density is low. Used with permission from *Nature* **428**, 493–521 (2004). (See color plate p. l.)

mouse genome covered by RefSeq genes; and 17% of rat hotspots were found in 8% of the rat genome covered by RefSeq genes. Similar numbers were also observed with coding exons and EST regions. Half of all hotspots in the mouse genome lie totally in noncoding regions.

3.10
Covariation of Evolutionary and Genomic Features

A high-resolution analysis of the genomic and evolutionary landscape of rat chromosome 10 uncovered strong correlations between certain microevolutionary features. Strongly correlated, in particular, are the local rates of microdeletion ($R2 = 0.71$; Fig. 9a), microinsertion ($R2 = 0.56$; Fig. 9a), and point substitution ($R2 = 0.86$; Fig. 9b) between the two independent lineages of mouse and rat. In addition, microinsertion rates are correlated with microdeletion rates ($R2 = 0.55$; Fig. 9a). These strong correlations are also observed in an independent genome-wide analysis, both on the original data and after factoring out the effects of G + C content. Perhaps surprisingly, substantially less correlation is seen between microindel and point substitution rates (compare Figs. 9a and b).

The local point substitution rate in sites common to human, mouse, and rat strongly correlates with that in rodent-specific sites ($R2 = 0.57$; Fig. 9b, blue line versus red/green). These two classes of sites, while interdigitated at the level of tens to thousands of bases, constitute sites that are otherwise evolutionarily independent. This result confirms that local rate variation is not solely determined by stochastic effects and extends, at high resolution, the previously documented regional correlation in rate between 4D sites and ancestral repeat sites.

4
Evolution of Genes

A substantial motivation for sequencing the rat genome was to study protein-coding genes. Besides being the first step in accurately defining the rat proteome, this fundamental data set yields insights into differences between the rat and other mammalian species with a complete genome sequence. Estimation of the rat gene content is possible because of relatively mature gene-prediction programs and rodent-transcript data. Mouse and human genome sequences also allow characterization of mutational events in proteins such as amino acid repeats and codon insertions and deletions. The quality of the rat sequence also allows us to distinguish between the functional genes and pseudogenes.

An estimated 90% of rat genes possess strict orthologs in both mouse and human genomes. Studies also identified genes arising from recent duplication events occurring only in rat, and not in mouse or human. These genes contribute characteristic features to rat-specific biology, including aspects of reproduction, immunity, and toxin metabolism. In contrast, almost all human "disease genes" have rat orthologs. This underscores the importance of the rat as a model organism in experimental science.

4.1
Construction of Gene Set and Determination of Orthology

The Ensembl gene-prediction pipeline has predicted 20 973 genes with 28 516 transcripts, and 205 623 exons in rat. These genes contain an average of 9.7 exons, with a median exon number of 6.0. At least 20% of the genes are alternatively spliced, with an average of 1.3 transcripts predicted per gene. Of the 17% single exon transcripts, 1355 contain frameshifts relative to the predicted protein and 1176 are likely processed pseudogenes. The majority of transcripts are supported by rodent (61%) or vertebrate (72%) transcript evidence, and have at least one untranslated region predicted (60%). The coding densities ranged from 1.2 to 2.2% and the coding fraction of RefSeq genes covered by these predictions ranged from 82 to 98%. The number of coding exons per gene and average exon length were similar in the three species. Differences were observed in intron length, with an average of 5338 bp in human, 4212 bp in mouse, and 5002 bp in rat.

4.2 Properties of Orthologous Genes

Orthology relationships were conservatively predicted and 12 440 rat genes showed clear, unambiguous 1:1 correspondence with a gene in the mouse genome. Accounting for potential errors 86 to 94% of the rat genes have 1:1 mouse orthologs. The remaining genes were associated with lineage-specific gene family expansions or contractions. Surprisingly, a similar proportion (89–90%) of rat genes possessed a single ortholog in the human genome, perhaps because of less resolution in the draft genomes. The majority of nucleotide changes within protein-coding regions that reflected synonymous or non-synonymous substitutions yielded K_A/K_S ratios of less than 0.25 indicating purifying selection. Values of 1 suggest neutral evolution, and values greater than 1 indicate positive selection. Examination of ortholog pairs in orthologous genomic segments (Table 3) showed a slight increase in median K_S values, indicating that the rat lineage has more neutral substitutions in gene coding regions than the mouse lineage.

Rat genes shared with mouse, but with no counterparts in human are expected to reflect either a rapidly evolving gene set, or genes that may have arisen from noncoding DNA, or been converted to pseudogenes in the human lineage. Thirty-one ENSEMBL rat genes were collected that have no nonrodent homolog in current databases. These are twofold over-represented among genes in paralogous gene clusters, and threefold overrepresented among genes whose proteins are likely to be secreted. This is consistent with observations that clusters of paralogous genes, and secreted proteins, evolve relatively rapidly.

Tab. 3 1:1 orthologous genes in human, mouse, and rat genomes[b].

	Human/Mouse	Human/Rat	Mouse/Rat
1:1 ortholog relationships	11 084	10 066	11 503
Median K_S values[a]	0.56 (0.39–0.80)	0.57 (0.40–0.82)	0.19 (0.13–0.26)
Median K_A/K_S values[a]	0.10 (0.03–0.24)	0.09 (0.03–0.21)	0.11 (0.03–0.28)
Median % amino acid identity[a]	88.0% (74.4–96.3%)	88.3% (75.9–96.4%)	95.0% (88.0–98.7%)
Median % nucleotide identity[a]	85.1% (77.4–90.0%)	85.1% (77.8–89.9%)	93.4% (89.2-–95.7%)

[a] Numbers in parentheses represent the 16th and 83rd percentiles.
[b] Data obtained from Ensembl, *Homo sapiens* version 11.31 (24 841 genes), *Mus musculus* version 10.3 (22 345 genes), *Rattus norvegicus* version 11.2 (21 022 genes).

The paucity of rodent-specific genes indicates that *de novo* invention of complete genes in rodents is rare. This is not unexpected, since the majority of eukaryotic protein-coding genes are modular structures containing coding and noncoding exons, splicing signals, and regulatory sequences; and the chances of independent evolution and successful assembly of these elements into a functional gene are small, given the relatively short evolutionary time available since the mouse–rat split. However, individual rodent-specific exons may arise more frequently, particularly if the exon is alternatively spliced. Of the 2302 potential novel rodent-specific exons, with transcript support, none matched human transcripts but approximately half (1116) appear to be present in alternative splice forms found in rodents. These exons are speculated to contain the few successful lineage-specific survivors of the constant process of gene evolution, by birth and death of individual exons.

4.3 Indels and Repeats in Protein-coding Sequences

In contrast to small indels occurring in the bulk of the genome, indels within protein-coding regions are likely lethal, or deleterious, and rapidly removed from the population by purifying selection. Indel rates within rat coding sequences were 50-fold lower than in bulk genomic DNA. The whole genome excess of deletions compared to insertions (Fig. 5b) was also evident in coding sequences. Deletions are ~16% more likely than insertions to be removed from coding sequences by selection.

Owing to the triplet nature of the genetic code, indels of multiples of 3 nucleotides in length (3n indels) are less likely to be deleterious. Direct comparison of 3n indel rates between bulk DNA (0.77 indel/kb for mouse, 0.83 indel/kb for rat) and coding sequence (0.087 indel/kb for mouse and 0.084 indel/kb for rat) showed that 3n indels were ninefold underrepresented in coding sequences. At least 44% of indels were duplicative insertion or deletion of a tandemly duplicated sequence, collectively termed *sequence slippage*. Sequence slippage contributed approximately equally to observed insertions and deletions, the overall excess of deletions could be attributed specifically to an excess of nonslippage deletion over nonslippage insertion in both mouse and rat lineages. Of slippage indels, 13% were trinucleotide repeats known to be particularly prone to sequence slippage and encode homopolymeric amino acid tracts.

Other characteristics of amino acid repeat variations were searched to gain better understanding of dynamic changes in length of homopolymeric amino acid tracts on gene evolution and disease susceptibility. Most species-specific amino acid repeats (80–90%) were found in indel regions and regions encoding species-specific repeats were more likely to contain tandem trinucleotide repeats than those encoding conserved repeats. This was consistent with involvement of slippage in the generation of novel repeats in proteins and extended previous observations for glutamine repeats in a more limited human–mouse dataset.

The percentage of proteins containing amino acid repeats was 13.7% in rat, 14.9% in mouse, and 17.6% in human. The most frequently occurring tandem amino acid repeats were glutamic acid, proline, alanine, leucine, serine, glycine, glutamine, and lysine. Tandem trinucleotide repeats were significantly more abundant in human than in rodent coding sequences, in striking contrast to the frequencies observed in bulk genomic sequences (29 trinucleotide repeats/Mb in rat, 32 repeats/Mb in mouse and 13 repeats/Mb in human, Sect. 5.5). The conservation of human repeats was higher in mouse (52%) than in rat (46.5%), suggesting a higher rate of repeat loss in the rat lineage than in the mouse lineage.

Functional consequences of these in-frame changes in rat, mouse, and human were investigated through clustering of proteins based on annotation of function and cellular localization, and mapping indels onto protein structural and sequence features. The rate that indels accumulated in secreted (3.9×10^{-4} indel/aa) and nuclear (4.0×10^{-4}) proteins is approximately twice that of cytoplasmic (2.4×10^{-4}) and mitochondrial (1.4×10^{-4}) proteins. Likewise, ligand-binding proteins acquire indels (3.1×10^{-4}) at a higher rate than enzymes (2.1×10^{-4}). These trends exactly mirror those observed for amino acid substitution rates, suggesting tight coupling of selective constraints between indels and substitutions. Transcription regulators showed the highest rate of indels (4.3×10^{-4}), a finding that may relate to the overrepresentation of homopolymorphic amino acid tracts in these proteins.

Known protein domains exhibited 3.3-fold fewer indels than expected by chance, again paralleling nucleotide substitution rate differences between domains and nondomain sequences. Transmembrane regions were refractory to accumulating indels, exhibiting a sixfold reduction compared to that expected by chance. Low-complexity regions were 3.1-fold enriched, reflecting their relatively unstructured nature and enrichment for indel prone trinucleotide repeats. Mapping of indels onto groups of known structures revealed indels are 21% more likely to be tolerated in loop regions than the structural core of the protein.

An interesting observation was indel frequency and amino acid repeat occurrence both correlated positively with G + C coding sequence content of the local sequence environment. This may in part be explained by the correlation of polymerase slippage-prone trinucleotide repeat sequences and G + C content. There is also a positive correlation between CpG dinucleotide frequency and coding sequence insertions, but not deletions. This effect diminishes rapidly with increasing distance from the site of the insertion.

4.4
Transcription-associated Substitution and Asymmetry

A significant strand asymmetry for neutral substitutions in transcribed regions has been reported. Within an intron the higher rate of A $\rightarrow$ G substitutions over that of T $\rightarrow$ C substitutions, together with a smaller excess of G $\rightarrow$ A over C $\rightarrow$ T substitutions, leads to an excess of G + T over C + A on the coding strand. The asymmetries are hypothesized to be a by-product of transcription-coupled repair in germline cells. Examining the 3-way alignments of rat, mouse, and human verified that the strand asymmetries for neutral substitutions exist in introns across the rat genome (Table 4). These asymmetries are also seen if the study is limited to ancestral repeat sites, excludes ancestral repeat sites, excludes CpG dinucleotides, is limited to positions flanked by sites that are identical in the aligned sequences (in the case of observations 2 and 3 in Table 4), or considers introns of RefSeq genes for human or mouse. Thus, it appears that strand asymmetry of substitution events

Tab. 4 Strand asymmetry of substitutions in introns of rat genes.

1	Base frequencies on coding strand $(G + T)/(C + A)$	Rat genome 1.060	
2	Ratio of purine transitions to pyrimidine transitions $\text{rate}(A \leftrightarrow G)/\text{rate}(C \leftrightarrow T)$	Rat–Mouse 1.036	Rat–Human 1.036
3	Rate of transitions $\text{rate}(A \rightarrow G)/\text{rate}(T \rightarrow C)$ $\text{rate}(G \rightarrow A)/\text{rate}(C \rightarrow T)$	Rat 1.058 1.017	Mouse 1.091 1.00*

Data in (1) were computed from the rat genome, those in (2) were computed from pairwise alignments, and data in (3) were computed from 3-way alignments. All values except * were highly significant (p-values $<10^{-4}$).

within transcribed regions of the genome is a robust genome-wide phenomenon.

4.5
Conservation of Intronic Splice Signals

The dynamics of evolution of consensus splice signals in mammalian genes was examined using 6352 human–mouse–rat orthologous introns from 976 genes. Intron class is extremely well conserved: no conversion between U2 and U12 introns, nor switching within U12 introns between the major AT–AC and GT–AG subtypes was observed, although U12 switching has been documented at larger evolutionary distances. In contrast, conversions between canonical GT–AG and noncanonical GC–AG subtypes of U2 introns are not uncommon. Only ~70% of GC–AG introns are conserved between human and mouse/rat, and only 90% are conserved between mouse and rat. Using human as outgroup, we detected 9 GT to GC conversions after divergence of mouse and rat (from 6282 introns likely to be GT–AG prior to human and rodents split), and 2 GC to GT conversions (from 34 GC–AG introns likely to predate human and rodent split). Given the higher rate of conversion from GT to GC than the reverse, these results give some indication of the degree to which mutation from T to C is tolerated in donor sites. This substitution appears to be better tolerated in introns with very strong donor sites, since in these introns the proportion of GC donor sites is ~11%, which is much higher than the 0.7% overall frequency of GC donor sites in U2 introns. Very few other noncanonical configurations in U2 introns are conserved, suggesting that most correspond to transient, evolutionarily unstable states, pseudogenes, or misannotations.

4.6
Gene Duplications

Duplication of genomic segments represents a frequent and robust mechanism for generating new genes. Since there were no compelling data showing rat-specific genes arising directly from noncoding sequences, gene duplications were examined to measure their potential contribution to rat-specific biology. A previous study showed such gene clusters in mouse without counterparts in human are subject to rapid, adaptive evolution. Using methods that directly identified paralogous clusters found 784 rat paralog clusters containing 3089 genes. This was lower than in mouse (910 clusters per 3784 genes), but the difference probably reflects the larger number of gene predictions from the mouse assembly. Using methods that analyzed genomic segmental duplications, it appears that the timing of expansion of these individual families, is reflected in local gene duplication and retention within clusters. Neutral substitution rate varies among orthologs by approximately twofold (Fig. 10). Rates of change among ancestral gene duplications (those that predate the mouse/rat split) were relatively constant. Mouse-specific and rat-specific duplications occurred at similar rates, except for those with $K_S < 0.04$ that are reduced in mouse-specific duplications (Fig. 10), though this may be accounted for by different assembly methods.

The rat paralog pairs that probably arose after the rat/mouse split (12–24 Myr ago) have a K_S value of ≥ 0.2 (Table 3). Six hundred and forty nine $K_S < 0.2$ gene duplication events were found in rat, a lower number than found in mouse (755). For both rodents, this represents a likelihood of a gene duplicating between 1.3×10^{-3} and 2.6×10^{-3} every Myr. This is consistent with a previous estimate for Drosophila genes, and an order of magnitude lower than an estimate for *Caenorhabditis elegans* genes.

Immunoglobulin, T-cell receptor α-chain, and α_{2u} globulin genes appear to be duplicating at the fastest rates in the rat genome (Table 5). Since divergence with mouse, these rat clusters have increased gene content several-fold. This recapitulates previous observations that rapidly evolving and duplicating genes are overrepresented in olfaction and odorant-detection, antigen

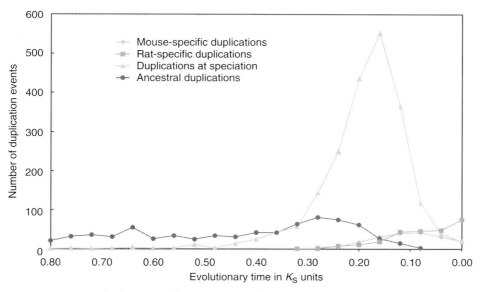

Fig. 10 Variation in the frequency of gene duplications during the evolutionary histories of the rat and mouse. The sequence of gene duplication events was inferred from phylogenetic trees determined from pairwise estimates of genetic divergence under neutral selection (K_S). The median K_S value for mouse:rat 1:1 orthologs is 0.19; this value corresponds to the divergence time of the mouse and rat lineages. Used with permission from *Nature* **428**, 493–521 (2004).

Tab. 5 Recent gene duplications ($K_S < 0.2$) in the rat lineage. Duplications involving retroviral genes, fragmented genes with internal repeats, and likely pseudogene clusters were removed from this list. Only gene clusters exhibiting at least 3 duplications are shown.

Cluster ID/ Chrom.	Recent duplication events	Numbers of genes involved	Extant/ Ancestral cluster size	Annotation	Process
249/4	38	53	60/22	Immunoglobulin κ-chain V	Immunity
640/15	38	47	53/15	TCR α-chain V	Immunity
346/6	25	35	44/15	Immunoglobulin heavy chain V	Immunity
190/3	22	42	168/146	Olfactory receptor	Chemosensation
578/13	16	28	59/43	Olfactory receptor	Chemosensation
400/8	15	26	82/67	Olfactory receptor	Chemosensation
743/20	15	21	37/22	Olfactory receptor	Chemosensation
72/1	12	22	102/90	Olfactory receptor	Chemosensation
500/10	12	18	32/20	Olfactory receptor	Chemosensation
51/1	6	7	16/10	Glandular kallikrein	Reproduction?
256/4	6	8	10/4	Vomeronasal receptor V1R	Chemosensation
488/10	6	10	11/5	Olfactory receptor	Chemosensation
644/15	6	10	14/8	Granzyme serine protease	Immunity
4/1	5	6	9/4	Trace amine receptor, GPCR	Neuropeptide receptors?
248/4	5	9	15/10	Vomeronasal receptor V1R	Chemosensation
393/8	5	10	31/26	Olfactory receptor	Chemosensation
522/10	5	8	19/14	Keratin-associated protein	Epithelial cell function
550/11	5	8	17/12	Olfactory receptor	Chemosensation
635/15	5	9	20/15	Olfactory receptor	Chemosensation
79/1	4	8	38/34	Olfactory receptor	Chemosensation
88/1	4	6	11/7	Olfactory receptor	Chemosensation
109/1	4	7	43/39	Olfactory receptor	Chemosensation
294/5	4	5	5/1	α_{2u} globulin	Chemosensation
310/5	4	5	11/7	Olfactory receptor	Chemosensation
353/7	4	7	13/9	Olfactory receptor	Chemosensation
399/8	4	5	6/2	Ly6-like urinary protein	Chemosensation?
638/15	4	6	6/2	Ribonuclease A	Immunity
690/17	4	6	21/17	Prolactin paralog	Reproduction
239/4	3	6	6/3	Prolactin-induced protein	Reproduction
253/4	3	4	5/2	Camello-like N-acetyltransferase	Developmental regulator
274/4	3	6	20/17	Ly-49 lectin natural killer cell protein	Immunity
297/5	3	4	5/2	Interferon-α	Immunity
523/10	3	4	6/3	Keratin-associated protein	Epithelial cell function
746/20	3	5	6/3	MHC class 1b (M10)	Chemosensation

recognition, and reproduction. An examination of duplicated genomic segments showed this enrichment for most of the same genes and also elements involved in foreign compound detoxification (cytochrome P450 and carboxylesterase genes). Together these are exciting findings because each of these categories can easily be associated with a familiar feature of rat-specific biology, and further investigation could explain some differences between rats and their evolutionary neighbors.

4.7
Conservation of Gene-regulatory Regions

As the third mammal to be fully sequenced, the rat can add significantly to the utility of nucleotide alignments for identifying conserved noncoding sequences. This power increases roughly as a function of total amount of neutral substitution represented in the alignment, and rat adds about 15% to the human-mouse comparison (Fig. 5). Many conserved mammalian noncoding sequences are expected to have regulatory function, and can be predicted using further analyses based upon these alignments.

Typical genome-wide human–mouse–rat alignments show strong conservation for a coding exon, as well as for several noncoding regions (Fig. 11). For example, the intronic region, in Fig. 11 contains 504 bp that are highly conserved in human, mouse, and rat. The last 100 bp of this alignment block are identical in all three species. Peaks in regulatory potential score are correlated with conservation score, and in the highly conserved intronic segment, they are higher for the three-way regulatory potential score than for the two-way scores using human and just one rodent. These data are illustrative, but form the foundation of ongoing efforts to identify genome sequences involved in gene regulation.

Requiring conservation among mammalian genomes greatly increases specificity of predictions of transcription factor binding sites. Transcription factor databases such as TRANSFAC contain known transcription factor binding sites and some knowledge of their distribution, but simply searching a sequence with these motifs provides little discriminatory power. Using a set of 164 weight matrices for 109 transcription factors extracted from TRANSFAC finds 186 792 933 matches in the April 2003 reference human genome sequence, but this was reduced to only 4 188 229 by demanding conservation in the human–mouse–rat three-way alignments. This is a 44-fold increase in specificity.

4.8
Pseudogenes and Gene Loss

Pseudogenes found in rat were classified according to whether they arose from retrotransposition, in which case they integrated into the genome randomly, or whether they arose from tandem duplication and neutral sequence substitution. Using human–rat synteny, 80% of pseudogenes exhibited no significant similarity to the corresponding human orthologous region, and therefore were considered to be retrotransposed, processed pseudogenes. The total pseudogene count and processed pseudogene proportion are consistent with those found for human. Pseudogenes are normally not subjected to selective constraint and therefore accumulate sequence modifications neutrally. Indeed, nearly all of our identified pseudogenes (97 ± 3%) evolved under neutrality according to a

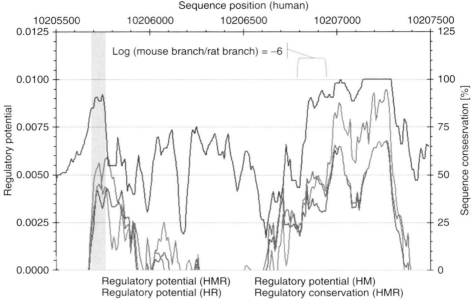

Fig. 11 Close-up of PEX14 (peroxisomal membrane protein) locus on human chromosome 1 (with homologous mouse chromosome 4 and rat chromosome 5). Conservation score computed on 3-way human–mouse–rat alignments presents a clear coding exon peak (gray bar) and very high values in a 504 bp noncoding, intronic segment (right; last 100 bp of alignment are identical in all three organisms). The latter segment showed a striking difference between the inferred mouse and rat branch lengths: the gray bracket corresponds to a phylogenetic tree where the logarithm of mouse to rat branch length ratio is −6. Regulatory potential (RP) scores that discriminate between conserved regulatory elements and neutrally evolving DNA are calculated from 3-way (human–mouse–rat) and 2-way (human-rodent) alignments. Here the 3-way regulatory potential scores are enhanced over the 2-way scores. Used with permission from *Nature* **428**, 493–521 (2004). (See color plate p. li.)

K_A/K_S test, and therefore, are consistent with being pseudogenic.

As with the human genome, the largest group of rat pseudogenes (totaling 2188) consists of ribosomal protein genes. Other large rat pseudogene families arose from olfactory receptors (552, see Sect. 6.1), glyceraldehyde 3-phosphate dehydrogenase (251), protein kinases (177), and RNA binding RNP-1 proteins (174). Pseudogenes homologous to a meiotic spindle-associated protein spindlin are particularly numerous in rat (at least 53 copies), compared to mouse (approximately 3 copies). This suggests that spindlin pseudogenes may have distributed rapidly by a recently active transposable element.

4.9
In Situ Loss of Rat Genes

As an organism evolves, its need for certain genes may be reduced, or lost, owing to changes in its ecological niche. Loss of selective constraints leads to accumulation of nonsense and/or frameshift mutations without retrotransposition or duplication. These nonprocessed pseudogenes are

Tab. 6 Candidate rat pseudogenes, orthologous to mouse and human functional genes.

Mouse gene	Human gene	Strand	Rat genome Coordinates[a]	Frameshifts stops[b]	Annotation
ENSMUSG00000013611	ENSG00000174226	+	7:92752590–92807556	1/0	Sorting nexin
ENSMUSG00000024364	ENSG00000158402	+	18:62742414–62770427	2/0	Dual specificity phosphatase CDC25c
ENSMUSG00000026293	ENSG00000077044	+	9:9563484?–95692601	1/0	Diacylglycerol kinase δ
ENSMUSG00000026785	ENSG00000160447	+	3:9210762–9229984	5/0	Protein kinase PKNβ
ENSMUSG00000026829	ENSG00000148288	+	3:7662414–7664521	2/2	Forssman glycolipid synthetase
ENSMUSG00000027426	ENSG00000125846	+	3:125918806–125924149	1/1	Zinc finger protein 133
ENSMUSG00000028000	ENSG00000138799	−	2:221272797–221304350	1/0	Complement factor I
ENSMUSG00000029203	ENSG00000078140	−	14:44385206–44441888	1/0	Ubiquitin-protein ligase E2 (HIP2)
ENSMUSG00000030270	ENSG00000144550	−	20:8332585–8362331	3/0	Copine (membrane trafficking)
ENSMUSG00000035449	ENSG00000167646	+	1:67374986–67381472	1/0	Cardiac troponin I
ENSMUSG00000037029	ENSG00000105261	−	1:82728049–82730272	1/0	Zinc finger protein 146
ENSMUSG00000037432	ENSG00000167137	−	9:42465695–42498651	1/1	Dysferlin-like protein
ENSMUSG00000039660	ENSG00000167137	+	3:9320401–9326997	4/0	Similar to yeast YMR310c RNA binding protein
ENSMUSG00000042653	ENSG00000137634	+	8:49938446–49939091	1/0	Brush border 61.9 kDa-like protein

[a] Coordinates from rat v2.0.
[b] Mouse genes were used as templates for predicting rat pseudogenes.

interesting since they link environmental changes to genomic mutation events. However, predicted pseudogenes with disrupted reading frames might also be indicative of errors in genome sequence or assembly. By constraining the search to orthologous genomic regions, we identified 14 rat putative nonprocessed pseudogenes (Table 6) with apparently functional, single human, and mouse orthologs.

4.10
Noncoding RNA Genes

The abundance and distribution of noncoding RNAs was investigated in rat. Cytoplasmic tRNA gene identification in rodents is complicated by tRNA derived ID SINE elements (B2 and ID). TRNAscan−SE predicted 175 943 tRNAs (genes and pseudogenes); however, the majority (175 285) were SINEs identified by RepeatMasker. This is far greater than found in mouse (24 402/25 078) or human (25/636). Of the remaining 666 predictions, 163 were annotated as tRNA pseudogenes and 4 were annotated as undetermined by tRNAscan−SE. An additional 68 predictions were removed because their best database match in either human, mouse, or rat tRNA databases matched tRNAs with either a different amino acid or anticodon (violating the wobble rules that specify the distinct anticodons expected). The total of 431 tRNAs (including a single selenocysteine tRNA) identified in the rat genome is comparable to that for mouse (435) and human (492). These three species share a core set of approximately 300 tRNAs using a cutoff of $\geq 95\%$ sequence identity and $\geq 95\%$ sequence length.

A total of 454 noncodingRNAs (other than tRNAs) were identified by sequence comparison to known noncoding RNAs. These include 113 micro (mi) RNAs, 5 ribosomal RNAs, 287 small nucleolar (sno) RNAs, and small nuclear (sn) RNAs; 49 various other ncRNAs such as signal recognition particle (SRP) RNA, 7SK RNA, telomerase RNA, Rnase P RNA, brain-specific repetitive (Bsr) RNA, noncoding transcript abundantly expressed in brain (Ntab) RNA, small cytoplasmic (sc) RNA, and 626 pseudogenes. Complete 18S and 28S rRNA genes and more rRNAs were not identified presumably owing to assembly issues.

5
Evolution of Transposable Elements

Most interspersed repeats are immobilized copies of transposable elements that have accrued substitutions in proportion to their time spent fixed in the genome (for introduction see references in Nature paper). About 40% of the rat genome draft is identified as interspersed repetitive DNA derived from transposable elements, similar to mouse (Table 7) and lower than for the human (almost 50%). The latter difference is mainly because of the lower substitution rate in the human lineage, which allows us to recognize much older (Mesozoic) sequences as interspersed repeats. Almost all repeats are derived from retroposons, elements that procreate via reverse transcription of their transcripts. As in mouse, there is no evidence for activity of DNA transposons since the rat−mouse split. Many aspects of the rat and the mouse genomes' repeat structure are shared; here we focus on the differences.

5.1
LINE-1 Activity in the Rat Lineage

The long interspersed nucleotide element (LINE-1 or L1) is an autonomous

Tab. 7 Composition of interspersed repeats in the rat genome.

	Rat				Mouse	
	Copies [×10³]	Total length [Mb]	Fraction of genome [%]	Lineage specific [%]	Fraction of genome [%]	Lineage specific [%]
LINEs:	657	594.0	23.11	11.70	20.10	9.74
LINE-1	597	584.2	22.73	11.70	19.65	9.74
LINE-2	48	8.4	0.33	–	0.38	–
L3/CR1	11	1.4	0.06	–	0.06	–
SINEs:	1360	181.3	7.05	1.52	7.78	1.80
B1(Alu)	384	42.3	1.65	0.16	2.53	0.92
B4(ID_B1)	359	55.4	2.15	0.00	2.25	0.00
ID	225	19.6	0.76	0.54	0.20	0.00
B2	328	55.2	2.15	0.68	2.29	0.74
MIR	109	13.0	0.51	–	0.56	–
LTR elements:	556	232.2	9.04	1.84	10.28	2.85
ERV_classI	40	24.9	0.97	0.56	0.79	0.36
ERV_classII	141	83.4	3.24	1.02	4.13	1.73
ERV-L (III)	74	21.6	0.84	0.04	1.08	0.23
MaLRs	302	102.5	3.99	0.22	4.27	0.53
DNA elements:	108	20.9	0.81	–	0.86	–
Charlie(hAT)	80	14.8	0.58	–	0.60	–
Tigger(Tc1)	18	4.0	0.16	–	0.17	–
Unclassified:	14	7.3	0.28	–	0.37	–
Total	2690	1036	40.31	14.90	39.45	14.26
Small RNAs:	8	0.6	0.03	0.01	0.03	0.01
Satellites:	14	6.4	0.25	?	0.31	?
Simple repeats:	897	61.1	2.38	?	2.41	?

Data for Rnor3.1 and October 2003 mouse (MM4), excluding Y chromosome, using the December 17 2003 version of RepeatMasker. To highlight the differences between rat and mouse repeat content, column 5 and 6 show the fraction of the genomes comprised of lineage specific repeats. The LINE-1 numbers include all HAL1 copies, while all BC1 scRNA and >10% diverged tRNA–Ala matches, far more common than other small RNA pseudogenes and closely related to ID, have been counted as ID matches.

retroelement, with an internal RNA polymerase II promoter and two open reading frames (ORFs). ORF1 is an RNA binding protein with chaperone-like activity suggesting a role in mediating nucleic acid strand transfer steps during L1 reverse transcription. ORF2 encodes a protein with both reverse transcriptase and DNA endonuclease activity. LINEs are usually 5' truncated so that only a small subset extends to include the promoter region and can function as a source for more copies.

Many classes of LINE-like elements exist, but only L1 has been active in rodents. Over half a million copies, in variable stages of decay, comprise 22% of the rat genome. Although over 10% of the human genome is L1 copies introduced before the rodent–primate split, only 2% of the rat genome could be recognized as such because of the fast substitution rate in the rodent lineage. Thus, probably well over one quarter of all rat DNA is derived directly from the L1 gene.

Following the mouse/rat split, L1 activity appears to have increased in rat. The 6 rat-specific L1 subfamilies, represented by 150 000 copies, cover 12% of the rat genome. Mouse-specific L1 copies accumulated over the same period cover only 10% of the genome (Table 7). This greater accumulation of L1 copies could explain some of the size difference of the rat and mouse genome.

In addition to the traditional L1 elements, there are 7500 copies (10 Mb) of a nonautonomous element that is derived from L1 by deletion of most of its ORF2. On the basis of their low divergence, the presently identified HAL1-like (HAL1 for Half-a-LINE) elements operated only a few million years ago in the mouse lineage (MusHAL1) and still propagate in the rat genome (RNHAL1). RNHAL1 only contains an ORF1, while MusHAL1 encoded an endonuclease as well, though no reverse transcriptase. The repeated origin and high copy number of HAL1s suggests that the ORF1 product, which binds strongly to its mRNA, may render this transcript a superior target for L1 mediated reverse transcription. In this way, HAL1 resembles the nonautonomous, endogenous retrovirus-derived MaLR elements (in the following), which, for over 100 million years, retained only the retroviral gag ORF that encodes an RNA binding protein. A potential advantage of HAL1 over L1 is its shorter length, which, considering the usual 5′ truncation of copies, increases the chance that a copy includes the internal promoter elements and can become a source gene.

5.2
Different Activity of SINEs in the Rat and Mouse Lineage

The most successful usurpers of the L1 retrotransposition machinery are SINEs. These are small RNA-derived sequences with an internal RNA polymerase III promoter. The human Alu SINE was shown to be transposed by L1. L1 lacks sequence specificity and rodent and primate SINE sequences are unrelated to L1. Though any transcript can be retroposed, as evidenced by the numerous mammalian processed pseudogenes, L1-dependent SINEs probably have features that make them especially efficient targets of the L1 reverse transcriptase.

While before the radiation of most mammalian orders L1 was at least as active as L2, the L2 dependent MIR was the only known (and very abundant) SINE of that time. All of the currently active SINEs in different mammalian orders appear to have arisen after the demise of L2 (and consequently MIR), as if an opportunity (or necessity) arose for the creation and expansion of other SINEs.

Four different SINEs are distinguished in rat and mouse. The B1 element and the primate Alu originated from a 7SLRNA gene, probably just before the rodent–primate split and after the speciation from most other eutherians, where Alu/B1 elements are not known. The other SINEs (B4, B2, ID) are rodent-specific and have tRNA-like internal promoter regions.

The fortunes of these SINEs during mouse and rat evolution have been different (Fig. 12). B4 probably became extinct before the mouse–rat speciation, while B2 has remained productive in both lineages, scattering >100 000 copies in each genome after this time. Interestingly, the fate of the B1 and ID SINEs has been opposite in rat and mouse. While B1 is still active in mouse, having left over 200 000 mouse-specific copies in its trail, the youngest of the 40 000 rat-specific B1 copies are 6 to 7% diverged from their source, indicating a relatively early extinction in the rat lineage.

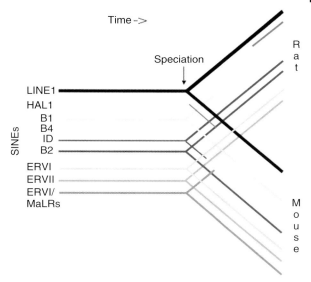

Fig. 12 Historical view of rodent repeated sequences. Relationships of the major families of interspersed repeats (Table 7) are shown for the rat and mouse genomes, indicating losses and gains of repeat families after speciation. The lines indicate activity as a function of time. Note that HAL1-like elements appear to have arisen both in the mouse and rat lineage. Used with permission from *Nature* **428**, 493–521 (2004). (See color plate p. lii.)

On the other hand, after the mouse–rat split only a few hundred ID copies may have inserted in mouse, while this heretofore minor SINE (~60 000 copies predate the speciation) picked up activity in rat to produce 160 000 ID copies.

5.3 Colocalization of SINEs in Rat and Mouse

Although the fates of SINE families differ, the number of SINEs inserted after speciation in each lineage is remarkably similar, ~300 000 copies. As MIR was replaced by L1 driven SINEs, it appears that the demise of B1 in rat allowed the expansion of IDs. Moreover, these independently inserted and unrelated SINEs (ID and B1 only share a mechanism of retroposition) accumulated at orthologous sites: the density of rat-specific SINEs in 14 243 ~100 kb windows in the rat genome is highly correlated ($R^2 = 0.83$) with the density of mouse-specific SINEs in orthologous regions in mouse (Fig. 5). These data corroborate and refine the observation of a strong correlation between the location of primate and rodent-specific SINEs in 1-Mb windows. At 100 kb, no correlation is seen for interspersed repeats other than SINEs.

Insertions of SINEs at the same location have been reported, and the correlation could reflect the existence of conserved hotspots for SINE insertions. However, primate data does not support this. Likewise, gene conversions do not significantly contribute to the observed correlation.

Figure 9(c) displays the lineage-specific SINE densities on rat chromosome 10 and in the mouse orthologous blocks, showing a stronger correlation than any other feature. The cause of the unusual distribution patterns of SINEs, accumulating in gene-rich regions where other interspersed repeats are scarce, apparently is a conserved feature independent of the primary sequence of the SINE and effective over smaller than isochore-sized regions.

In the human genome, the most recent (unfixed) Alus are distributed similar to L1, whereas older copies gradually take on the opposite distribution of SINEs. However, this temporal shift in SINE distribution pattern is not observed in mouse or rat.

Some regions of high LINE content coincide with regions that exhibit both higher AT content and an elevated rate of point substitution (Fig. 9, pink rectangles). In a genome-wide analysis, LINE content correlates strongly with substitution rates, and about 80% of this correlation is explained by higher rates in AT-rich regions. SINE density shows the opposite correlation both on chromosome 10 (Fig. 9) and genome-wide.

These phenomena, in conjunction with an overall trend in substitution rates toward AT-richness, suggest a model by which quickly evolving regions accumulate a higher-than-average AT content, which attracts LINE elements. While distinct cause–effect relationships such as this remain largely speculative, these results reinforce the idea that local genomic context strongly shapes local genomic features and rates of evolution.

5.4
Endogenous Retroviruses and Derivatives

Retrovirus-like elements are the other major contributors to interspersed repeats in the rodent genome. These have several 100-bp long terminal repeats (LTRs) with transcriptional regulatory sequences that flank an internal sequence that, in autonomous elements, encodes all proteins necessary for retrotransposition. All mammalian LTR elements are endogenous retroviruses (ERVs; classes I, II, or III) or their nonautonomous derivatives.

The most productive retrovirus in mammals has been the class III element ERV-L, primarily through its ancient nonautonomous derivatives called "MaLRs" with 350 000 copies occupying ∼5% of the rat genome (Table 7). Human ERV-L and MaLR copies are >6% diverged from their reconstructed source genes and must have died out around the time of our speciation from new world monkeys. In mouse, several thousand almost identical MaLR and ERV-L copies suggest sustained activity. In contrast, rat ERV-L activity was silenced a few million years ago given that the least diverged MaLR and ERV-L copies differ by >4% from each other. Other class III ERVs were active earlier in rodent evolution, before the mouse–rat speciation. Class I and class II elements still thrive in rat, where 2 of 4 class I and 4 of 9 class II autonomous ERVs appear to be active.

5.5
Simple Repeats

Interspersed simple sequence repeats (SSRs), regions of tandemly repeated (1–6 bp) units that probably arise from slippage during DNA replication and can expand and compress by unequal crossing over. Remarkable differences were noted between the human and mouse genomes' SSR content with threefold to fivefold more base pairs contained in near (>90%) perfect SSRs in mouse. SSRs are both more frequent and on average longer in mouse. Polypurine (or polypyrimidine) repeats are especially (10-fold) overrepresented in the mouse genome. As discussed in Sect. 4.3, this contrasts sharply with the greater frequency of triplet repeats coding for amino acids in human relative to the rodents.

Rat and mouse SSR content are more similar, with genome coverage of ~1.4% (for >90% perfect elements compared to 0.45% in human) and are of similar average length. For example, the most common mammalian SSR, the (CA)n repeat, averages 42 bp long in mouse and 44 bp in rat. Some potentially significant differences are that polypurine SSRs are of similar average length but 1.2-fold more common in mouse, while the rare SSRs containing CG dimers are 1.5-fold more frequently observed in rat.

5.6
Prevalent, Medium-length Duplications in Rodents

In addition to the transpositionally derived and simple repeats, the rat and mouse genomes contain a substantial amount of medium-length unclassified duplications (typically 100–5000 bp). These are readily seen in self-comparisons and in intrarodent comparisons after masking the known repeats, but they are much less prevalent in comparisons with the human genome. A substantial fraction of the rodent genomes consists of currently unexplained repeats, which may include (1) novel families of low-copy rodent interspersed repeats; (2) extensions of known but not fully characterized rodent repeats; and (3) duplications generated by a mechanism different from transposition.

6
Rat-specific Biology

The rat genome sequence and predicted gene set reveals genetic differences between rats and mice that might specify their differences in physiology and behavior. In particular, recently duplicated genes are enriched in elements involved in chemosensation and functional aspects of reproduction (Table 5). The differences in the gene complements of rat and mouse are illustrated by in-depth analyses of olfactory receptors (ORs), pheromones, cytochromes P450, proteases, and protease inhibitors.

6.1
Chemosensation

The ability to emit and sense specific smells is a key feature of survival for most animals in the wild. Rat and mouse pheromones, vomeronasal receptors, and ORs genes were duplicated frequently during the time since the common ancestor of rats and mice (Table 5). The rapid evolution of these genes is attributed to conspecific competition, in particular sexual selection.

The rat genome has 1866 ORs in 113 locations: 69 multigene clusters, and 49 single genes. Extrapolating for the 9.8% of the genome not in the assembly, there are ~2070 OR genes and pseudogenes. The rat therefore has ~37% more OR genes and pseudogenes than the ~1510 ORs of the mouse, assuming similar representation of recently duplicated sequences in the two genome assemblies used. Of the 1774 OR sequences that are not interrupted by assembly gaps, 1216 (69%) encode intact proteins, while the remaining 547 (31%) sequences are likely pseudogenes with in-frame stop codons, frameshifts, and/or interspersed repeat elements. Fewer mouse OR homologs are pseudogenes (~20%), but the larger family size in rat still leaves it with substantially more intact ORs than the mouse (~1430 vs ~1210). Striking rat-specific expansions of two ancestral clusters account for much of the difference in OR family size and pseudogene content between rat and mouse, although many other clusters exhibit more subtle changes. Significant differences between human and mouse in OR families have also been reported, but the functional implications of OR repertoire size on the ability of different species to detect and discriminate odorants are not yet known.

6.2
α_{2u} Globulin Pheromones

The α_{2u} globulin genes are odorant-binding proteins that also contribute to essential survival functions in animals. α_{2u} globulin homologs are likely to be highly heterogeneous among murid species. Distinct homologs and genomic arrangement differences have been observed among mouse strains. The evolution of α_{2u} globulin genes on rat chromosome 5 has remodeled this genomic region (Fig. 13). The orthologous human genomic region contains a single homolog, suggesting that the common ancestor of rodents and human possessed one gene. The genome of C57BL/6J mice contains 4 homologous genes, and 7 pseudogenes, while the rat genome contains 10 α_{2u} globulin genes and 12 pseudogenes in a single region.

Phylogenetic trees constructed using amino acid, and noncoding DNA, sequences show that, surprisingly, the rat α_{2u} globulin gene clusters appear to have arisen recently via a rapid burst of gene duplication since the rat/mouse split (Table 5). This is consistent with the Rfp37-like zinc finger-like pseudogene having accompanied virtually all of the rat-specific α_{2u} globulin gene duplications (Fig. 13). The sequences of these genes are

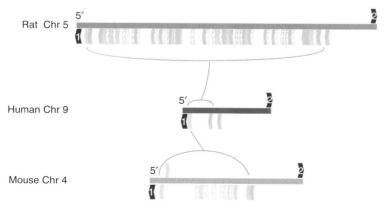

Fig. 13 Adaptive remodeling of genomes and genes. Orthologous regions of rat, human and mouse genomes encoding pheromone-carrier proteins of the lipocalin family (α_{2u} globulins in rat and major urinary proteins in mouse) shown in brown. Zfp37-like zinc finger genes are shown in blue. Filled arrows represent likely genes, whereas striped arrows represent likely pseudogenes. Gene expansions are bracketed. Arrow head orientation represents transcriptional direction. Flanking genes 1 and 2 are TSCOT and CTR1 respectively. (See color plate p. xliv).

also evolving rapidly with median K_A/K_S values of 0.77 and 1.06, for rat and mouse genes respectively. Amino acid sites that appear to have been subject to adaptive evolution are situated both within the ligand-binding cavity, and on the solvent-exposed periphery of the α_{2u} globulin structure. This demonstrates how genome analysis can reveal the imprint of adaptive evolution from megabase to single base levels.

The rapid evolution of these genes, and the remodeling of their genomic regions, can be attributed to the known roles of rat α_{2u} globulins and mouse homologs in conspecific competition and sexual selection. These proteins are pheromones and pheromone carriers that are present in large quantities in rodent urine, and act as scent markers indicating dominance and subspecies identity.

6.3
Detoxification

Cytochrome P450 is a well-recognized participant in metabolic detoxification and rapid evolution is observed within this family. These enzymes are particularly relevant to clinical and pharmacological studies in humans as they metabolize many toxic and endogenous compounds. Rodents are important model organisms for understanding human drug metabolism, so it is important to identify 1:1 orthologs and species-specific expansions and losses. Compared to human genes, there are clear expansions of several rodent P450 subfamilies, but there are also significant differences between rat and mouse subfamilies (Fig. 14a). The fastest evolving subfamily seems to be CYP2J containing a single gene in human, but at least 4 in rat and 8 in mouse (Fig. 14b,c). CYP2J enzymes catalyze the NADPH-dependent oxidation of arachidonic acid to various eicosanoids, which in turn possess numerous biological activities including modulation of ion transport, control of bronchial and vascular smooth muscle tone, and stimulation of peptide hormone secretion. The genomic ordering of genes and their phylogenetic tree indicate an ongoing expansion in the rodents (Fig. 14b,c). This suggests that adaptive evolution has been involved in diversifying their functions.

6.4
Proteolysis

Protease genes also represent an example of rapid evolution in the rat genome. The rat contains 626 protease genes (~1.7% of the genes), similar to mouse (641) and more than human (561). One hundred and two rat protease genes are not found in human, and 42 are absent from mouse. Several rat gene families have expanded; most are involved in reproductive or immunological functions, and have evolved independently in the rat and mouse lineages.

These gene family expansions dramatically illustrate how large-scale genomic changes have accompanied species-specific innovation. Positive selection of duplicated genes has afforded the rat an enhanced repertoire of precisely those genes that allow reproductive success despite severe competition from both within its own, and with other species. This serves as a general illustration of the importance of chemosensation, detoxification, and proteolysis in innovation and adaptation.

7
Human Disease Gene Orthologs in the Rat Genome

The rat is recognized as the premier model for studying the physiological aspects of

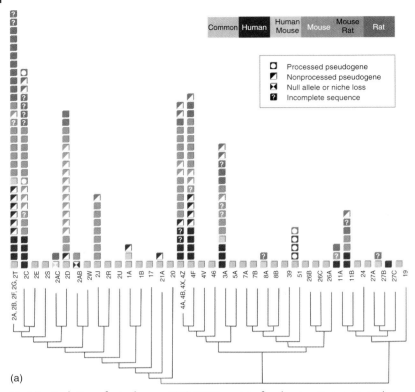

Fig. 14 Evolution of cytochrome P450 (CYP) protein families in rat, mouse, and human. (a) Dendrogram of topology from 234 full-length sequences. The 279 sequences of 300 amino acids; subfamily names and chromosome numbers are shown. Black branches have >70% bootstrap support. Incomplete sequences (they contain Ns) are included in counts of functional genes (84 rat, 87 mouse, and 57 human) and pseudogenes (including fragments not shown; 77 rat, 121 mouse, and 52 human). Thus, 64 rat genes and 12 pseudogenes were in predicted gene sets. Human CYP4F is a null allele due to an in-frame STOP codon in the genome, although a full-length translation exists (SwissProt P98187). Rat CYP27B, missing in the genome, is "incomplete" since there is a RefSeq entry (NP_446215). Grouped subfamilies CYP2A, 2B, 2F, 2G, 2T, and CYP4A, 4B, 4X, 4Z, occur in gene clusters; thus nine loci contain multiple functional genes in a species. One (CYP1A) has fewer rat genes than human, seven have more rodent than human, and all nine have different copy numbers. CYP2AC is a rat-specific subfamily (orthologs are pseudogenes). CYP27C has no rodent counterpart. Rodent-specific expansion, rat CYP2J is illustrated below. (b) The neighbor-joining tree, with the single human gene, contains clear mouse (Mm) and rat (Rn) orthologous pairs (bootstrap values >700/1000 trials shown). Bar indicates 0.1 substitutions per site. (c) All rat genes have a single mouse counterpart except for CYP2J 3, which has further expanded in mouse (mouse CYP2J 3a, 3b, and 3c) by two consecutive single duplications. The genes flanking the CYP2J orthologous regions (rat chromosome 5, 126.9–127.3 Mb; mouse chromosome 4, 94.0–94.6 Mb; human chromosome 1, 54.7–54.8 Mb) are hook1 (HOOK1; pink) and nuclear factor I/A (NFIA; cyan). Genes (solid) and gene fragments (dashed boxes) are shown above (forward strand) and below (reverse strand) the horizontal line. No orthology relation could be concluded for most of these cases. Used with permission from *Nature* **428**, 493–521 (2004). (See color plate p. liii.)

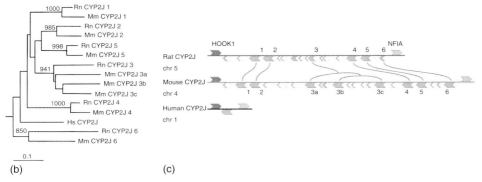

Fig. 14 (*Continued*)

many human diseases, but it has not had as prominent a role in the study of simple genetic disease traits. The rat genome provides an opportunity to use the more than 1000 human Mendelian disorders that have associated loci and alleles to link the rat to human disease examples. The precise identification of the rat orthologs of human genes that are mutated in disease creates further opportunities to discover and develop rat models for biomedical research.

Predicted rat genes were compared with 1112 well-characterized human disease genes that were verified and classified on the basis of pathophysiology. So, 844 (76%) have 1 : 1 orthologs in the rat as predicted by Ensembl. These predictions are likely to be of high quality since 97.4% of the 11 422 rat:human 1 : 1 orthologs predicted by Ensembl were found in orthologous genomic regions.

The proportion of human disease genes with single orthologs in the rat (76%) was higher than the proportion for all human genes (46%). Careful analysis of the remaining 268 human genes not predicted to show 1 : 1 orthology indicated that only six of the human disease genes lack likely rat orthologs among genome, cDNA, EST, and protein sequences. Thus, it appears that, in general, genes involved in human disease are unlikely to have diverged, or to have become duplicated, deleted, or lost as pseudogenes, between rat and human divergence.

Comparisons of K_S, K_A, and the K_A/K_S ratio values of human disease orthologs with those of all remaining ortholog pairs found only the K_S distributions differed significantly indicating that coding regions of human disease genes and their rat counterparts have mutated more rapidly than the nondisease genes. Factors influencing the specific loci could cause this result, or the disease genes may characteristically reside in genomic regions that exhibit higher mutation rates.

The disease gene set was next grouped into 16 disease-system categories and analyzed (Fig. 15). Only five disease systems exhibited significant K_A/K_S differences with respect to the remaining samples. Neurological and malformation-syndrome disease categories manifested K_A/K_S ratios consistent with purifying selection acting on these gene sets. In contrast, the pulmonary, hematological, and immune categories manifested the highest median K_A/K_S ratios, consistent with a role for more positive selection, or reduced selective constraints, among these genes.

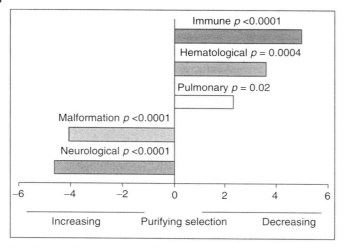

Fig. 15 Selective constraints differ for human disease systems in the rat genome. Human disease-system categories showing significant differences ($p < 0.05$) in a nonparametric test [Mann–Whitney–Wilcoxon] comparing K_A/K_S (human:rat) ratios. P-values from two-level tests between genes from one disease system and the remaining genes. (Mean−Mean0)/Std0 values from multilevel tests from sixteen categorized disease systems. Negative values for neurological (−4.63) and malformation-syndrome (−4.04) categories were observed to be consistent with K_A/K_S ranges where purifying selection predominates. Immune, hematological, and pulmonary categories show positive values of 4.98, 3.59, and 2.34, respectively. Used with permission from *Nature* **428**, 493–521 (2004).

Orthologs of more diverse phyla demonstrated consistent results.

These results demonstrate that various disease systems exhibit significantly different average evolutionary rates. The higher rates noted for the immune system disease genes are consistent with rapid diversification of the functions of the immune systems of rodents and humans. This is expected for genes involved in controlling species-restricted infectious agents if strong adaptive pressure acts during host–pathogen coevolution. Thus, results of studies of these rodent genes may be less directly relevant to our understanding of human immune system diseases than results obtained for other pathophysiology disease systems where conservation is greater and purifying selection is stronger.

A number of genes were examined that harbor triplet nucleotide repeats, and are involved in human neurological disorders, such as Huntington disease, a condition caused by CAG triplet repeat expansion producing abnormally long polyglutamine tracts in an otherwise normal protein. Analysis of the rat:human orthologs of these disease genes indicated that repeat length is substantially shorter in all cases in the rat than in the normal human gene (Fig. 16). To date, there are no naturally occurring rat strains described that exhibit neurological disease associated with repeat expansion mechanisms. The rat may lack repeat

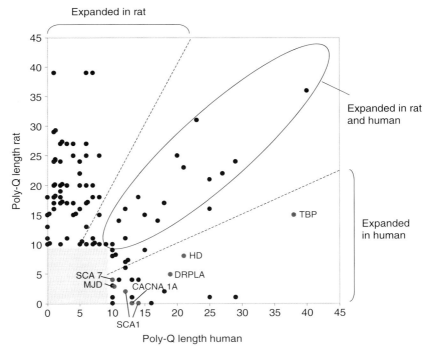

Fig. 16 Polyglutamine repeat length comparison between human and rat. Points represent protein poly-Q length for rat and human. Red points correspond to repeats in genes associated with human disease: SCA1, spinocerebellar ataxia 1 protein, or ataxin1; SCA7, spinocerebellar ataxia 7 protein; MJD, Machado–Joseph disease protein; CACNA1A, spinocerebellar ataxia 6 protein, or calcium channel α-1A subunit isoform 1; DRPLA, dentatorubro–pallidoluysian atrophy protein; HD, Huntington's disease protein, or huntingtin; TBP, TATA binding protein or spinocerebellar ataxia 17 protein. Repeat lengths over 10 were examined; green shading delineates the range not included in the analysis. Also noted are a set that are expanded in rat and human (black circle) and a set where repeats are expanded in the rat. Used with permission from *Nature* **428**, 493–521 (2004). (See color plate p. liv.)

expansion mutational mechanisms or these orthologs may fail to reach a "critical repeat length" susceptible to such mutational mechanisms. Other human genes without known disease association also contain glutamine repeats that are much shorter in the rat orthologs, and thus may be susceptible to mutations that arise through repeat expansion mechanisms. In Fig. 16, it may also be observed that a relatively high proportion of repeats are significantly longer in the rat than in their corresponding human ortholog.

In addition to enabling the direct comparison of rat/human disease orthologs, the rat genome sequence itself is an invaluable aid for discovery of other rat genes to study as disease models. First, genes underlying disease phenotypes with simple inheritance that have been mapped to chromosomal regions can be more easily pursued in both species. Indeed, the rearrangements of conserved segments between the two species have significant value since they tighten the boundaries of the mapped disease regions and reduce the number of genes potentially associated with the disease phenotype. Second, the identification of multiple alleles contributing to quantitative and complex trait differences that are involved in disease processes can be pursued with more accuracy, both in the initial association phases, and in subsequent efforts to detect causative alleles.

8
Summary

As the third mammalian genome to be sequenced, the rat genome has provided both predictable and surprising information about mammalian species. Although clear at the outset that ongoing rat research would benefit from the resource of a genome sequence, there was uncertainty about how many new insights would be found, especially considering the superficial similarities between the rat and the already sequenced mouse. Instead, the results of the sequencing and analysis have generated some deep insights into the evolutionary processes that have given rise to these different species. In addition, the project was invaluable for further developing the methods for the generation and analysis of large genome sequence data sets.

The generation of the rat draft tested the new "combined approach" for large genome sequencing. The high quality of the overall attests of this overall strategy and the supporting software, provide a suitable approach to this problem. With a BAC "skimming" component in the underlying data set, the assembly recovered a fraction of the genome that was expected, by analogy to the mouse project to be difficult to assemble from pure WGS data. The BAC skimming component also allowed progressive generation of high quality local assemblies that were of use to the rat community as the project developed.

The rat genome data have improved the utility of the rat model enormously. The rat gene content provides a parts list that can be explored with the high degree of confidence and precision that is appropriate for biomedical research. The resources for physical and genetic mapping have also been improved, since the relative position of individual markers is now known with confidence and there are computational resources to bridge the process of genetic association with gene modeling and experimental investigation.

The expected benefit of a third mammalian sequence providing an outgroup by which to discriminate the timing of events that had already been noted between mouse and human was fully realized. Using the three sequences and other partial data sets from additional organisms, it was possible to measure some of the overall faster rate of evolutionary change in the rodent lineage shared by mice and rats, as well as the peculiar acceleration of some aspects of rat-specific evolution. The observation of specific expanded gene families in the rat should provide material for targeted studies for some time.

The sequence of the rat genome represents the beginning of the full analysis of the mammalian genome and its complex evolutionary history. Much of the additional data required to complete this story will be from other genomes, distantly related to rat. Nevertheless, the prospect of a finished rat genome, polymorphism data from analysis of other rat strains, and targeted cDNA clone collections will provide rat-specific reagents for routine use in research. Together with the ongoing efforts to fully develop methods to genetically manipulate whole rats and provide effective "gene knockouts" the current and future rat genome resources will ensure a place for this organism in genomic and biomedical research for some time.

Acknowledgment

This chapter is based largely on the publication of the rat genome sequence by the Rat Genome Sequencing Project Consortium (*Nature* **428**, 493–521 (2004)). We thank E. Eichler, K. Fetchel, R. Hardison, P. Havlak, K. Kalafus, M. Krzywinski, A. Milosavljevic, L. Pachter, P. Pevzner, C. Ponting, K. Roskin, A. Sidow, A. Smit, G. Tessler, and D. Torrents, for the figures. The rat genome-sequencing project was funded primarily by the NHGRI and NHLBI of the NIH.

See also Genetics, Molecular Basis of; Genomic Sequencing (Core Article).

Bibliography

Books and Reviews

Chiaromonte, F. et al. (2003) The share of human genomic DNA under selection estimated from human–mouse genomic alignments, *Cold Spring Harb. Symp. Quant. Biol.* **68**, 245–254.

Danielson, P.B. (2002) The cytochrome P450 superfamily: biochemistry, evolution and drug metabolism in humans, *Curr. Drug Metab.* **3**, 561–597.

Hedrich, H.J. (2000) in: Krinke, G.J. (Ed.) *History, Strains, and Models in the Laboratory Rat*, Academic Press, San Diego, CA, pp. 3–16.

Lopez-Otin, C., Overall, C.M. (2002) Protease degradomics: a new challenge for proteomics, *Nat. Rev. Mol. Cell Biol.* **3**, 509–519.

Lynch, M., Conery, J.S. (2000) The evolutionary fate and consequences of duplicate genes, *Science* **290**, 1151–1155.

Montoya-Burgos, J.I., Boursot, P., Galtier, N. (2003) Recombination explains isochores in mammalian genomes, *Trends Genet.* **19**, 128–130.

Ohno, S. (1970) *Evolution by Gene Duplication*, Springer, Berlin, Germany.

Prak, E.T., Kazazian, H.H. Jr. (2000) Mobile elements and the human genome, *Nat. Rev. Genet.* **1**, 134–144.

Reddy, P.S., Housman, D.E. (1997) The complex pathology of trinucleotide repeats, *Curr. Opin. Cell Biol.* **9**, 364–372.

Scarborough, P.E., Ma, J., Qu, W., Zeldin, D.C. (1999) P450 subfamily CYP2J and their role in the bioactivation of arachidonic acid in extrahepatic tissues, *Drug Metab. Rev.* **31**, 205–234.

Smit, A.F. (1999) Interspersed repeats and other mementos of transposable elements in mammalian genomes, *Curr. Opin. Genet. Dev.* **9**, 657–663.

Weiner, A.M. (2002) SINEs and LINEs: the art of biting the hand that feeds you, *Curr. Opin. Cell Biol.* **14**, 343–350.

Willson, T.M., Kliewer, S.A. (2002) PXR, CAR and drug metabolism, *Nat. Rev. Drug Discov.* **1**, 259–266.

Primary Literature

Adams, M.D. et al. (2000) The genome sequence of *Drosophila melanogaster*, *Science* **287**, 2185–2195.

Adkins, R.M., Gelke, E.L., Rowe, D., Honeycutt, R.L. (2001) Molecular phylogeny and divergence time estimates for major rodent groups: evidence from multiple genes, *Mol. Biol. Evol.* **18**, 777–791.

Alba, M.M., Guigo, R. (2004) Comparative analysis of amino acid repeats in rodents and humans, *Genome Res.* **14**, 549–554.

Alba, M.M., Santibanez-Koref, M.F., Hancock, J.M. (1999) Conservation of polyglutamine tract size between mice and humans depends on codon interruption, *Mol. Biol. Evol.* **16**, 1641–1644.

Altschul, S.F., Lipman, D.J. (1990) Protein database searches for multiple alignments, *Proc. Natl. Acad. Sci. U.S.A.* **87**, 5509–5513.

Antequera, F., Bird, A. (1993) Number of CpG islands and genes in human and mouse, *Proc. Natl. Acad. Sci. U.S.A.* **90**, 11995–11999.

Aparicio, S., et al. (2002) Whole-genome shotgun assembly and analysis of the genome of *Fugu rubripes*, *Science* **297**, 1301–1310.

Bailey, J.A., et al. (2002) Recent segmental duplications in the human genome, *Science* **297**, 1003–1007.

Batzoglou, S., et al. (2002) ARACHNE: a whole-genome shotgun assembler, *Genome Res.* **12**, 177–189.

Benit, L., et al. (1997) Cloning of a new murine endogenous retrovirus, MuERV-L, with strong similarity to the human HERV-L element and with a gag coding sequence closely related to the Fv1 restriction gene, *J. Virol.* **71**, 5652–5657.

Bird, A.P. (1980) DNA methylation and the frequency of CpG in animal DNA, *Nucleic Acids Res.* **8**, 1499–1504.

Birdsell, J.A. (2002) Integrating genomics, bioinformatics, and classical genetics to study the effects of recombination on genome evolution, *Mol. Biol. Evol.* **19**, 1181–1197.

Boffelli, D., et al. (2003) Phylogenetic shadowing of primate sequences to find functional regions of the human genome, *Science* **299**, 1391–1394.

Bourque, G., Pevzner, P.A. (2002) Genome-scale evolution: reconstructing gene orders in the ancestral species, *Genome Res.* **12**, 26–36.

Bourque, G., Pevzner, P.A., Tesler, G. (2004) Reconstructing the genomic architecture of ancestral mammals: lessons from human, mouse, and rat genomes, *Genome Res.* **14**, 507–516.

Bray, N., Pachter, L. (2004) MAVID Constrained ancestral alignment of multiple sequence, *Genome Res.* **14**, 693–699.

Burge, C.B., Padgett, R.A., Sharp, P.A. (1998) Evolutionary fates and origins of U12-type introns, *Mol. Cell* **2**, 773–785.

Cai, L., et al. (1997) Construction and characterization of a 10-genome equivalent yeast artificial chromosome library for the laboratory rat, *Rattus norvegicus*, *Genomics* **39**, 385–392.

Cantrell, M.A., et al. (2001) An ancient retrovirus-like element contains hot spots for SINE insertion, *Genetics* **158**, 769–777.

Cavaggioni, A., Mucignat-Caretta, C. (2000) Major urinary proteins, 2U-globulins and aphrodisin, *Biochim. Biophys. Acta* **1482**, 218–228.

Chakrabarti, K., Pachter, L. (2004) Visualization of multiple genome annotations and alignments with the K-BROWSER, *Genome Res.* **14**, 716–720.

Chen, R., Sodergren, E., Gibbs, R., Weinstock, G.M. (2004) Dynamic building of a BAC clone tiling path for genome sequencing project, *Genome Res.* **14**, 679–684.

Cheung, J., et al. (2003) Recent segmental and gene duplications in the mouse genome, *Genome Biol.* **4**, R47 [online].

Clark, A.J., Hickman, J., Bishop, J. (1984) A 45-kb DNA domain with two divergently orientated genes is the unit of organisation of the murine major urinary protein genes, *EMBO J.* **3**, 2055–2064.

Cooper, G.M., et al. (2004) Characterization of evolutionary rates and constraints in three mammalian genomes, *Genome Res.* **14**, 539–548.

Cooper, G.M., Brudno, M., Green, E.D., Batzoglou, S., Sidow, A. (2003) Quantitative estimates of sequence divergence for comparative analyses of mammalian genomes, *Genome Res.* **13**, 813–820.

Costas, J. (2003) Molecular characterization of the recent intragenomic spread of the murine endogenous retrovirus MuERV-L, *J. Mol. Evol.* **56**, 181–186.

Dehal, P., et al. (2002) The draft genome of *Ciona intestinalis*: insights into chordate and vertebrate origins, *Science* **298**, 2157–2167.

Dermitzakis, E.T., et al. (2002) Numerous potentially functional but non-genic conserved sequences on human chromosome 21, *Nature* **420**, 578–582.

Dermitzakis, E.T., et al. (2003) Evolutionary discrimination of mammalian conserved non-genic sequences (CNGs), *Science* **302**, 1033–1035.

Dewannieux, M., Esnault, C., Heidmann, T. (2003) LINE-mediated retrotransposition of marked Alu sequences, *Nat. Genet.* **35**, 41–48.

Duret, L., Mouchiroud, D. (2000) Determinants of substitution rates in mammalian genes: expression pattern affects selection intensity but not mutation rate, *Mol. Biol. Evol.* **17**, 68–74.

Eichler, E.E. (1998) Masquerading repeats: paralogous pitfalls of the human genome, *Genome Res.* **8**, 758–762.

Eichler, E.E. (2001) Segmental duplications: what's missing, misassigned, and misassembled – and should we care? *Genome Res.* **11**, 653–656.

Elnitski, L., et al. (2003) Distinguishing regulatory DNA from neutral sites, *Genome Res.* **13**, 64–72.

Emes, R.D., Goodstadt, L., Winter, E.E., Ponting, C.P. (2003) Comparison of the genomes of human and mouse lays the foundation of genome zoology, *Hum. Mol. Genet.* **12**, 701–709.

Emes, R.D., Beatson, S.A., Ponting, C.P., Goodstadt, L. (2004) Evolution and comparative genomics of odorant- and pheromone-associated genes in rodents, *Genome Res.* **14**, 591–602.

Felsenfeld, A., Peterson, J., Schloss, J., Guyer, M. (1999) Assessing the quality of the DNA sequence from the human genome project, *Genome Res.* **9**, 1–4.

Goff, S.A., et al. (2002) A draft sequence of the rice genome (*Oryza sativa* L. ssp. *japonica*), *Science* **296**, 92–100.

Graves, J.A., Gecz, J., Hameister, H. (2002) Evolution of the human X – a smart and sexy chromosome that controls speciation and development, *Cytogenet. Genome Res.* **99**, 141–145.

Green, P. (2002) Whole-genome disassembly, *Proc. Natl. Acad. Sci. U.S.A.* **99**, 4143–4144.

Green, H., Wang, N. (1994) Codon reiteration and the evolution of proteins, *Proc. Natl. Acad. Sci. U.S.A.* **91**, 4298–4302.

Gumucio, D.L., et al. (1992) Phylogenetic footprinting reveals a nuclear protein which binds to silencer sequences in the human and globin genes, *Mol. Cell. Biol.* **12**, 4919–4929.

Gurates, B., et al. (2002) WT1 and DAX-1 inhibit aromatase P450 expression in human endometrial and endometriotic stromal cells, *J. Clin. Endocrinol. Metab.* **87**, 4369–4377.

Guy, J., et al. (2003) Genomic sequence and transcriptional profile of the boundary between pericentromeric satellites and genes on human chromosome arm 10p, *Genome Res.* **13**, 159–172.

Haldi, M.L., et al. (1997) Construction of a large-insert yeast artificial chromosome library of the rat genome, *Mamm. Genome* **8**, 284.

Hardison, R., et al. (1993) Comparative analysis of the locus control region of the rabbit-like gene cluster: HS3 increases transient expression of an embryonic-globin gene, *Nucleic Acids Res.* **21**, 1265–1272.

Hardison, R.C., et al. (2003) Covariation in frequencies of substitution, deletion, transposition, and recombination during eutherian evolution, *Genome Res.* **13**, 13–26.

Havlak, P., et al. (2004) The *Atlas* genome assembly system, *Genome Res.* **14**, 721–732.

Hayward, B.E., Zavanelli, M., Furano, A.V. (1997) Recombination creates novel L1 (LINE-1) elements in *Rattus norvegicus*, *Genetics* **146**, 641–654.

Hillier, L.W., et al. (2003) The DNA sequence of human chromosome 7, *Nature* **424**, 157–164.

Horvath, J.E., et al. (2003) Using a pericentromeric interspersed repeat to recapitulate the phylogeny and expansion of human centromeric segmental duplications, *Mol. Biol. Evol.* **20**, 1463–1479.

Huang, H., et al. Evolutionary conservation of human disease gene orthologs in the rat and mouse genomes. *Genome Biol.* (submitted).

Hubbard, T., et al. (2002) The Ensembl genome database project, *Nucleic Acids Res.* **30**, 38–41.

Hurst, L.D. (2002) The Ka/Ks ratio: diagnosing the form of sequence evolution, *Trends Genet.* **18**, 486–487.

Hurst, J.L., et al. (2001) Individual recognition in mice mediated by major urinary proteins, *Nature* **414**, 631–634.

International Human Genome Sequencing Consortium. (2001) Initial sequencing and analysis of the human genome. *Nature* **409**, 860–921.

Jaffe, D.B., et al. (2003) Whole-genome sequence assembly for mammalian genomes: arachne 2, *Genome Res.* **13**, 91–96.

Jensen-Seaman, M.I., et al. (2004) Comparative recombination rates in the rat, mouse, and human genomes, *Genome Res.* **14**, 528–538.

Kalafus, K.J., Jackson, A.R., Milosavljevic, A. (2004) Pash: efficient genome-scale sequence anchoring by positional hashing, *Genome Res.* **14**, 672–678.

Kent, W.J., Baertsch, R., Hinrichs, A., Miller, W., Haussler, D. (2003) Evolution's cauldron: duplication, deletion, and rearrangement in the mouse and human genomes, *Proc. Natl. Acad. Sci. U.S.A.* **100**, 11484–11489.

Kirkness, E.F., et al. (2003) The dog genome: survey sequencing and comparative analysis, *Science* **301**, 1898–1903.

Kolbe, D., et al. (2004) Regulatory potential scores from genome-wide 3-way alignments of human, mouse and rat, *Genome Res.* **14**, 700–707.

Krzywinski, M., et al. (2004) Integrated and sequence-ordered BAC and YAC-based physical maps for the rat genome, *Genome Res.* **14**, 766–779.

Kwitek, A.E., et al. (2004) High density rat radiation hybrid maps containing over 24,000 SSLPs, genes, and ESTs provide a direct link to the rat genome sequence, *Genome Res.* **14**, 750–757.

Levinson, G., Gutman, G.A. (1987) Slipped-strand mispairing: a major mechanism for DNA sequence evolution, *Mol. Biol. Evol.* **4**, 203–221.

Li, X., Waterman, M.S. (2003) Estimating the repeat structure and length of DNA sequences using L-tuples, *Genome Res.* **13**, 1916–1922.

Loots, G.G., Ovcharenko, I., Pachter, L., Dubchak, I., Rubin, E.M. (2002) rVista for comparative sequence-based discovery of functional transcription factor binding sites, *Genome Res.* **12**, 832–839.

Ma, B., Tromp, J., Li, M. (2002) PatternHunter: faster and more sensitive homology search, *Bioinformatics* **18**, 440–445.

Makalowski, W., Boguski, M.S. (1998) Evolutionary parameters of the transcribed mammalian genome: an analysis of 2,820 orthologous rodent and human sequences, *Proc. Natl. Acad. Sci. U.S.A.* **95**, 9407–9412.

Margulies, E.H., Blanchette, M., Haussler, D., Green, E. (2003) Identification and characterization of multi-species conserved sequences, *Genome Res.* **13**, 2507–2518.

Marra, M.A., et al. (1997) High throughput fingerprint analysis of large-insert clones, *Genome Res.* **7**, 1072–1084.

Martin, S.L., Bushman, F.D. (2001) Nucleic acid chaperone activity of the ORF1 protein from the mouse LINE-1 retrotransposon, *Mol. Cell. Biol.* **21**, 467–475.

Misra, S., et al. (2002) Annotation of the *Drosophila melanogaster* euchromatic genome: a systematic review. *Genome Biol.* **3**(12), RESEARCH0083. Epub 2002 Dec. 31.

Modrek, B., Lee, C.J. (2003) Alternative splicing in the human, mouse and rat genomes is associated with an increased frequency of exon creation and/or loss, *Nat. Genet.* **34**, 177–180.

Mouse Genome Sequencing Consortium. (2002) Initial sequencing and comparative analysis of the mouse genome. *Nature* **420**, 520–562.

Mural, R.J., et al. (2002) A comparison of whole-genome shotgun-derived mouse chromosome 16 and the human genome, *Science* **296**, 1661–1671.

Murphy, W.J., Fronicke, L., O'Brien, S.J., Stanyon, R. (2003) The origin of human chromosome 1 and its homologs in placental mammals, *Genome Res.* **13**, 1880–1888.

Murphy, W.J., Sun, S., Chen, Z.Q., Pecon-Slattery, J., O'Brien, S.J. (1999) Extensive conservation of sex chromosome organization between cat and human revealed by parallel radiation hybrid mapping, *Genome Res.* **9**, 1223–1230.

Murphy, W.J., Bourque, G., Tesler, G., Pevzner, P., O'Brien, S.J. (2003) Reconstructing the genomic architecture of mammalian ancestors using multispecies comparative maps, *Hum. Genomics* **1**, 30–40.

Myers, E.W., et al. (2000) A whole-genome assembly of *Drosophila*, *Science* **287**, 2196–2204.

Myers, E.W., Sutton, G.G., Smith, H.O., Adams, M.D., Venter, J.C. (2002) On the sequencing and assembly of the human genome, *Proc. Natl. Acad. Sci. U.S.A.* **99**, 4145–4146.

Nadeau, J.H., Taylor, B.A. (1984) Lengths of chromosomal segments conserved since divergence of man and mouse, *Proc. Natl. Acad. Sci. U.S.A.* **81**, 814–818.

Nekrutenko, A. (2004) Identification of novel exons from rat-mouse comparisons, *J. Mol. Evol.* **59**(5), 703–708.

Nekrutenko, A., Makova, K.D., Li, W.H. (2002) The KA/KS ratio test for assessing the protein-coding potential of genomic regions: an empirical and simulation study, *Genome Res.* **12**, 198–202.

Nekrutenko, A., Chung, W.Y., Li, W.H. (2003) An evolutionary approach reveals a high protein-coding capacity of the human genome, *Trends Genet.* **19**, 306–310.

Nelson, D.R. (1999) Cytochrome P450 and the individuality of species, *Arch. Biochem. Biophys.* **369**, 1–10.

Osoegawa, K., et al. (2004) BAC Resources for the rat genome project, *Genome Res.* **14**, 780–785.

Ostertag, E.M., Kazazian, H.H. Jr. (2001) Biology of mammalian L1 retrotransposons, *Annu. Rev. Genet.* **35**, 501–538.

Pennacchio, L.A., Rubin, E.M. (2001) Genomic strategies to identify mammalian regulatory sequences, *Nat. Rev. Genet.* **2**, 100–109.

Pevzner, P., Tesler, G. (2003a) Human and mouse genomic sequences reveal extensive breakpoint reuse in mammalian evolution, *Proc. Natl. Acad. Sci. U.S.A.* **100**, 7672–7677.

Pevzner, P., Tesler, G. (2003b) Genome rearrangements in mammalian evolution: lessons from human and mouse genomes, *Genome Res.* **13**, 37–45.

Pruitt, K.D., Maglott, D.R. (2001) RefSeq and LocusLink: NCBI gene-centered resources, *Nucleic Acids Res.* **29**, 137–140.

Puente, X.S., Lopez-Otin, C.A. (2004) A genomic analysis of rat proteases and protease inhibitors, *Genome Res.* **14**, 609–622.

Puente, X.S., Sanchez, L.M., Overall, C.M., Lopez-Otin, C. (2003) Human and mouse proteases: a comparative genomic approach, *Nat. Rev. Genet.* **4**, 544–558.

Quentin, Y. (1994) A master sequence related to a free left Alu monomer (FLAM) at the origin of the B1 family in rodent genomes, *Nucleic Acids Res.* **22**, 2222–2227.

Rat Genome Sequencing Project Consortium. (2004) Genome sequence of the Brown Norway rat yields insights into mammalian evolution, *Nature* **428**, 493–521.

Riethman, H., et al. (2004) Mapping and initial analysis of human subtelomeric sequence assemblies, *Genome Res.* **14**, 18–28.

Roskin, K.M., Diekhans, M., Haussler, D. (2003) in: Vingron, M., Istrail, S., Pevzner, P., Waterman, M. (Eds.) *Proceedings of the 7th Annual International Conference on Research in Computational Biology (RECOMB 2003)*, ACM Press, New York, pp. 257–266, doi:10.1145/640075.640109.

Rothenburg, S., Eiben, M., Koch-Nolte, F., Haag, F. (2002) Independent integration of rodent identifier (ID) elements into orthologous sites of some RT6 alleles of *Rattus norvegicus* and *Rattus rattus*, *J. Mol. Evol.* **55**, 251–259.

Rouquier, S., Blancher, A., Giorgi, D. (2000) The olfactory receptor gene repertoire in primates and mouse: evidence for reduction of the functional fraction in primates, *Proc. Natl. Acad. Sci. U.S.A.* **97**, 2870–2874.

Roy-Engel, A.M., et al. (2002) Non-traditional Alu evolution and primate genomic diversity, *J. Mol. Biol.* **316**, 1033–1040.

Saitou, N., Nei, M. (1987) The neighbor-joining method: a new method for reconstructing phylogenetic trees, *Mol. Biol. Evol.* **4**, 406–425.

Salem, A.H., et al. (2003) Alu elements and hominid phylogenetics, *Proc. Natl. Acad. Sci. U.S.A.* **100**, 12787–127891.

Salem, A.H., Kilroy, G.E., Watkins, W.S., Jorde, L.B., Batzer, M.A. (2003) Recently integrated Alu elements and human genomic diversity, *Mol. Biol. Evol.* **20**, 1349–1361.

Schwartz, S., et al. (2003) Human–mouse alignments with BLASTZ, *Genome Res.* **13**, 103–107.

Smit, A.F. (1993) Identification of a new, abundant superfamily of mammalian LTR-transposons, *Nucleic Acids Res.* **21**, 1863–1872.

Springer, M.S., Murphy, W.J., Eizirik, E., O'Brien, S.J. (2003) Placental mammal diversification and the Cretaceous-Tertiary boundary, *Proc. Natl. Acad. Sci. U.S.A.* **100**, 1056–1061.

Stanyon, R., Stone, G., Garcia, M., Froenicke, L. (2003) Reciprocal chromosome painting shows that squirrels, unlike murid rodents, have a highly conserved genome organization, *Genomics* **82**, 245–249.

Steen, R.G., et al. (1999) A high-density integrated genetic linkage and radiation hybrid map of the laboratory rat, *Genome Res.* **9**, (insert), AP1–AP8.

Stenson, P.D., et al. (2003) Human Gene Mutation Database (HGMD): 2003 update, *Hum. Mutat.* **21**, 577–581.

Tagle, D.A., et al. (1988) Embryonic epsilon and globin genes of a prosimian primate (*Galago crassicaudatus*). Nucleotide and amino acid sequences, developmental regulation and phylogenetic footprints, *J. Mol. Biol.* **203**, 439–455.

Taylor, M.S., Ponting, C.P., Copley, R.R. (2004) Occurrence and consequences of coding sequence insertions and deletions in mammalian genomes, *Genome Res.* **14**, 555–566.

The International Human Genome Mapping Consortium. (2001) A physical map of the human genome, *Nature* **409**, 934–941.

Thomas, J.W., et al. (2003a) Pericentromeric duplications in the laboratory mouse, *Genome Res.* **13**, 55–63.

Thomas, J.W., et al. (2003b) Comparative analyses of multi-species sequences from targeted genomic regions, *Nature* **424**, 788–793.

Torrents, D., Suyama, M., Bork, P. (2003) A genome-wide survey of human pseudogenes, *Genome Res.* **13**, 2559–2567.

Trinklein, N.D., Aldred, S.J., Saldanha, A.J., Myers, R.M. (2003) Identification and functional analysis of human transcriptional promoters, *Genome Res.* **13**, 308–312.

Tuzun, E., Bailey, J.A., Eichler, E.E. (2004) Recent segmental duplications in the working draft assembly of the brown Norway rat, *Genome Res.* **14**, 493–506.

Venter, J.C., et al. (2001) The sequence of the human genome, *Science* **291**, 1304–1351.

Ventura, M., Archidiacono, N., Rocchi, M. (2001) Centromere emergence in evolution, *Genome Res.* **11**, 595–599.

Vitt, U., et al. (2004) Identification of candidate disease genes by EST alignments, synteny and expression and verification of Ensembl genes on rat chromosome 1q43-54, *Genome Res.* **14**, 640–650.

Waterston, R.H., Lander, E.S., Sulston, J.E. (2002) On the sequencing of the human genome, *Proc. Natl. Acad. Sci. U.S.A.* **99**, 3712–3716.

Waterston, R.H., Lander, E.S., Sulston, J.E. (2003) More on the sequencing of the human genome. *Proc. Natl. Acad. Sci. U.S.A.* **100**, 3022–3024; author reply (2003) **100**, 3025–3026.

Wingender, E., et al. (2001) The TRANSFAC system on gene expression regulation, *Nucleic Acids Res.* **29**, 281–283.

Wolfe, K.H., Sharp, P.M. (1993) Mammalian gene evolution: nucleotide sequence divergence between mouse and rat, *J. Mol. Evol.* **37**, 441–456.

Yang, S., et al. (2004) Patterns of insertions and their covariation with substitutions in the rat, mouse and human genomes, *Genome Res.* **14**, 517–527.

Yang, Z., Goldman, N., Friday, A. (1994) Comparison of models for nucleotide substitution used in maximum-likelihood phylogenetic estimation, *Mol. Biol. Evol.* **11**, 316–324.

Yap, V.B., Pachter, L. (2004) Identification of evolutionary hotspots in the rodent genomes, *Genome Res.* **14**, 574–579.

Young, J.M., et al. (2002) Different evolutionary processes shaped the mouse and human olfactory receptor gene families, *Hum. Mol. Genet.* **11**, 535–546.

Yu, J., et al. (2002) A draft sequence of the rice genome (*Oryza sativa* L. ssp. *indica*), *Science* **296**, 79–92.

Yunis, J.J., Prakash, O. (1982) The origin of man: a chromosomal pictorial legacy, *Science* **215**, 1525–1530.

Zhang, X., Firestein, S. (2002) The olfactory receptor gene superfamily of the mouse, *Nat. Neurosci.* **5**, 124–133.

Zhang, Z., Harrison, P., Gerstein, M. (2002) Identification and analysis of over 2000 ribosomal protein pseudogenes in the human genome, *Genome Res.* **12**, 1466–1482.

Real-Time Quantitative PCR: Theory and Practice

Gregory L. Shipley
The University of Texas Health Science Center, Houston, TX

1	**Real-time PCR Basics** 483	
1.1	Principles of Real-time PCR 483	
1.2	The Reverse Transcription Reaction 485	
1.3	The Polymerase Chain Reaction 487	
2	**Fluorescent Dyes, Assay Chemistries, and Real-time Instruments** 488	
2.1	Fluorescence in Real-time PCR 488	
2.2	Real-time Instruments 490	
2.3	Selected Assay Chemistries 492	
3	**RNA and DNA Isolation for Real-time PCR** 499	
3.1	Isolation of DNA 499	
3.2	Isolation of RNA 499	
4	**Standards Used for Real-time Quantitative PCR** 500	
4.1	RNA Standards 500	
4.2	DNA Standards 501	
5	**Real-time PCR Assay Design** 503	
5.1	Available Software 503	
5.2	Basic Rules for Real-time Assay Development 504	
5.3	Key Factors in Assay Development 506	
6	**Running a Real-time PCR Experiment** 509	
6.1	Practical Considerations 509	
6.2	Reagents 510	
6.3	Controls 512	

Encyclopedia of Molecular Cell Biology and Molecular Medicine, 2nd Edition. Volume 11
Edited by Robert A. Meyers.
Copyright © 2005 Wiley-VCH Verlag GmbH & Co. KGaA, Weinheim
ISBN: 3-527-30648-X

7		Data Analysis 513	
7.1		Baseline Boundaries and Threshold Levels	514
7.2		Fluorescent Signal and Sample Normalization	514
7.3		Quantification Using a Standard Curve	516
7.4		Quantification Using the ddCt Method	518
8		Future Directions 520	
		Bibliography 521	
		Books and Reviews 521	
		Primary Literature 521	

Keywords

Amplicon
The region of a target sequence that is amplified during the PCR, encompassing the sequence bounded by and including the two primer sequences.

Baseline
The range of early PCR cycles that contain low levels of increasing fluorescent signal that is undetectable over the baseline fluorescence of the reporter dye.

Fluorescence Resonance Energy Transfer
Light energy transferred from an excited dye to a neighboring dye in close proximity, resulting in the quenching of the visible emission of light from the primary dye.

Polymerase Chain Reaction
Primer-directed geometric increase in a DNA template using a thermostable DNA polymerase, necessary reagents, and a succession of heating and cooling steps performed in a thermocycling device.

Real-time PCR
This is the continuous collection of fluorescent signals from one or more polymerase chain reactions. Quantitative real-time PCR is the conversion of the fluorescent signal from each reaction to a numerical value for each sample. This is accomplished by calculation from a known reference sample (ddCt) or interpolation from a quantified standard curve. Real-time PCR is the most sensitive and accurate method for the detection and accurate quantification of nucleic acids from any biological source. The term real-time PCR in this article refers to both PCR and RT-PCR.

Reverse Transcription
The primer-directed synthesis of a DNA copy (cDNA) using RNA as the template and reverse transcriptase.

Threshold
A level of fluorescent signal that is significantly above the baseline fluorescence level. The threshold is set by the software or manually by the user within the geometric phase of the amplification curve and is used to define the cycle threshold (Ct) or crossing point (Cp).

Transcript quantification methodology has evolved from Northern blots to RNase protection to competitive RT-PCR. Each of these methods made strides toward greater sensitivity and accuracy in quantification. Since the introduction of the 7700 Sequence Detection System and Taqman® chemistry by Applied Biosystems in 1996, quantitative real-time PCR has become widely accepted as the most sensitive and accurate method for the quantification of both RNA and DNA from a wide variety of sources. Today, there are a number of real-time instruments from a variety of vendors and several detection methodologies and kit chemistries available. At the same time, prices for real-time platforms are falling dramatically. This trend means that the technology is moving from one found primarily in core laboratories to the bench tops of individual investigators.

There are several ways to interrogate a cell for changes induced by artificial or natural agents during a biological process. One way is to look for changes in cellular transcript levels that may indicate downstream changes in the corresponding protein. In another instance, the focus may be on the presence or absence of a viral or bacterial pathogen. In this case, detecting not only the presence but also the degree of infection provides valuable information. Alternatively, looking for the presence or the level of expression from a transgene or the inhibition of expression of an endogenous gene by an RNAi agent may be the question of interest. In all cases, quantitative real-time PCR technology can be utilized to provide the required results. However, successful implementation of the technology requires the user to have a basic background in the theoretical principles of real-time PCR as well as their practical application. The goal of this review is to provide this information.

It will not be possible to cover all aspects of real-time PCR in this review. For a more comprehensive overview of real-time PCR, "A-Z of Quantitative PCR," edited by Stephen Bustin, is highly recommended.

1
Real-time PCR Basics

Before any meaningful discussion of real-time PCR can be initiated, it will be useful to discuss the theoretical principles governing and basic chemistries utilized in real-time PCR.

1.1
Principles of Real-time PCR

Prior to real-time PCR, the effect of the primers and other components of the reagent chemistry on the amplification process during the polymerase chain reaction was something of a black box.

Tubes bearing assay chemistry, primers, and template were exposed to 30 or more cycles of heating and cooling on a thermocycler. The reaction product(s) were then separated on a gel. The presence of a single band of the correct size with any magnitude was deemed a successful experiment. With the advent of real-time PCR, the ability to observe the status of the reaction products during every cycle has brought new insight into the kinetics of template amplification. Most real-time PCR experiments are run for 40 cycles to capture the complete amplification profile from every tube or well. In an ideal PCR, the increase in template occurs geometrically throughout the entire 40 cycles. However, we now know that this is not the case. The amplification of any template can be defined by four phases: (1) the baseline; (2) the true geometric phase; (3) a linear phase as the efficiency of the PCR starts to fall in each successive cycle; and (4) a plateau as amplification arrests (Fig. 1). All three phases can be seen in the amplification plots from any real-time PCR reaction. An example of amplification curves from a 5-log dilution series of an oligonucleotide standard can be see in Fig. 2. Why the PCR slows down and then stops is not clear, but may be due to the build up of products that are inhibitory to the reaction. One thing that is clear is that the concentration of all reactants is in vast excess and none of them should be depleted during 40 cycles of PCR.

Samples with very high amounts of template have been observed, using real-time PCR, to stop amplification sooner and with fewer end products than those with lower initial template concentrations. The result is that the one with higher template will appear to have less band intensity on a gel compared to the one with

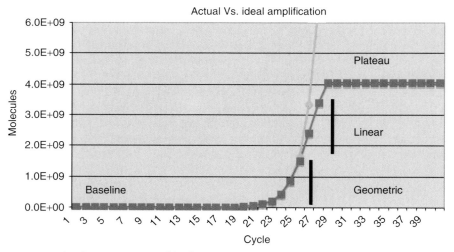

Fig. 1 Graphical representation of the fluorescent signals from an ideal versus an actual reaction over 40 cycles of real-time PCR. In an ideal PCR, there are two phases, a baseline where the signal is below the level of detection followed by a geometric increase in fluorescence that continues over the remaining cycles. However, in an actual reaction, there are four phases. As in the ideal reaction, there is a baseline followed by a geometric phase. However, the amplification becomes less than ideal, leading to a linear phase and finally a plateau where no further increase in signal occurs over the remaining signals.

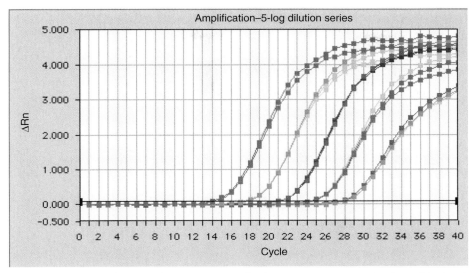

Fig. 2 Dilution curves from a real-time PCR experiment illustrating the four phases of a polymerase chain reaction. Amplification curves from a 5-log standard curve illustrating the four phases of signal generation in real-time PCR: baseline, signal being made, but not detectable by the instrument; geometric, detectable signal with maximal PCR efficiency; linear, postgeometric with slowly declining PCR efficiency; and plateau, no or very little new product made.

significantly lower template concentration after 30 or more cycles of PCR. Therefore, the only way to use the polymerase chain reaction for quantitative measurements is to compare samples in the early, geometric part of their amplification curves where the true differences in the samples can be detected. The ability to do so is what allows real-time PCR to be a *bona fide* quantitative technique.

1.2
The Reverse Transcription Reaction

Quantification of RNA begins with the reverse transcription reaction, which converts the RNA into a complementary DNA sequence or cDNA (also known as "copy DNA"). Two viral reverse transcriptase (RT) enzymes, used to make cDNA, are commercially available: AMV-RT from avian myeloblastosis virus and MMLV-RT from Moloney murine leukemia virus. AMV is a dimeric protein, while MMLV is a monomer. Both enzymes have RNase H activity, the ability to cleave RNA in a RNA/DNA hybrid. However, the activity in AMV-RT is much higher than that in MMLV-RT. This activity can be separated from the reverse transcriptase activity by site-directed mutagenesis. AMV-RT is a more processive enzyme (incorporates more nucleotides per unit time on the template) and is active at higher temperatures than MMLV-RT (42 °C vs. 37 °C for wild-type enzymes). Cloned variants of these enzymes have pushed the temperature limits to 58 °C for AMV-RT and 55 °C for MMLV-RT. Although both have been used for real-time PCR, MMLV-RT works best in practice, which is not consistent with the above characteristics. Both enzymes are inhibitory to the PCR. The inhibition is enzyme concentration dependent. For this reason, it is important to keep the amount of reverse transcriptase as low as possible and still produce cDNA with 100% efficiency. The exact mechanism for the inhibition is not known. It has been proposed that although the high melting temperature of the PCR kills the reverse transcriptase activity, the DNA binding capacity of the enzyme may persist, forming a physical barrier that affects the activity of *Taq* polymerase. The higher temperature stability, dimeric structure, and greater RNase H activity of AMV-RT, may contribute to greater PCR inhibition than the MMLV-RT enzyme.

One major issue for RT-PCR is what primer to use in the reverse transcriptase reaction? The classic primer for cDNA library and standard RT-PCR has been oligo-dT, an oligonucleotide with 18 to 20 dT residues. Since most (roughly 95%) messenger RNA found in higher eukaryotes have poly-A tails at their 3′ ends, the oligo-dT will initiate cDNA synthesis utilizing mRNA molecules, but not rRNA or tRNAs. There are, however, two flaws in this approach for real-time quantitative PCR. The first is that long poly-A sequence stretches exist within mRNAs as well within some of the other RNA classes and these sites will initiate cDNA synthesis as well. Most important, however, is the fact that not all real-time PCR assays are targeted at sequences near the 3′ end of the transcript. Some mRNAs are over 15 Kb in length with 3′ untranslated regions (UTRs) that are 4 Kb or more. In a situation where you need to make enough cDNA to make a recombinant bacterial clone, this is not a problem. However, real-time PCR is a quantitative process where every molecule present should be detected for proper comparisons to be made. Priming with oligo-dT alone will not ensure true quantification for every transcript for every assay unless this caveat has been taken into consideration at the time of assay design. Another approach is the use of random primers that can be found as hexamers up to decamers. Random primers are short oligonucleotides that, as a population, contain every possible oligo sequence. Thus, there will be a subpopulation of the correct primer sequence for every 6-mer up to a 10-mer in every target sequence. The frequency with which a random primer will find a complementary sequence in any template is given by the formula 4^x, where x is the length of the primer sequence. For a 6-mer, a perfect match will occur every 4096 bases, while a 10-mer would be approximately every 1.05×10^6 bases. Using random primers, more than one cDNA may be synthesized for each transcript molecule resulting in the whole transcript being made into multiple cDNAs. The rationale is that every transcript in the total RNA population will be converted to cDNA with equal efficiency. However, there is no data to confirm that premise. Further, especially with random hexamers, there is the possibility that a primer will bind within the amplicon sequence resulting in only a partial cDNA, thus making that molecule undetectable in that assay. Keeping these caveats in mind, in practice, random primers can work well and are the method of choice if multiplexed assays are being performed where more than one transcript or gene will be amplified and detected simultaneously within the same reaction vessel. Some investigators hedge their bets and use a combination of both oligo-dT and random primers. For single-assay reactions, however, there is a third choice. That is to use an assay-specific primer. That is, the same primer that will be used as the reverse primer in the following real-time PCR. This method will make, in theory, cDNA from every possible RNA template for that assay in precisely the correct location required for the following PCR. Unlike when oligo-dT or random primers have been used, the complexity of the resulting cDNA population is significantly lower when using an assay-specific primer. This means that the number of inappropriate priming events that could potentially occur during the PCR will be greatly reduced. Unlike when oligo-dT is used as a primer restricting the assayable region to the 3′-end of the message, use of assay-specific primers allows the amplified region of the assay to lie in any part of the transcript. As an aside, it should be mentioned that RNA can form complex structures resulting in self-priming and extra-assay cDNA synthesis even in the absence of any exogenous primer.

1.3
The Polymerase Chain Reaction

It is hard to think of another laboratory technique that has come along in our time that has had a more dramatic impact on scientific knowledge over such a broad range of subjects than the PCR. Real-time PCR has extended the sensitivity and specificity of nucleic acid detection greatly beyond what was available previously. The component that makes the polymerase chain reaction possible was the discovery of thermostable DNA polymerases that could withstand the multiple cycles of heating required to melt long DNA duplexes and cooling to temperature optimal for the succeeding polymerization steps. The first enzyme, and still the most commonly used, is *Taq* DNA polymerase from *Thermus aquaticus*. *Taq* is still one of the most processive thermostable enzymes commercially available. Most members of *Thermus* sp. also have a 5′-exonuclease activity that is required for cleavage of certain dual-labeled probes required for signaling.

There are two functional types of *Taq* commercially available, wild type and hot start. Hot-start enzymes are similar to their non-hot-start counterparts, but they have been reversibly inactivated by, (1) a heat activated mutation; (2) a small molecule; or (3) one or more antibodies directed at the active site of the enzyme. As the name implies, it is necessary to heat hot-start enzymes for their activation. The time varies greatly for enzyme activation with a minute at 94 to 95 °C being sufficient for the antibody inactivated enzymes while, 10 or more minutes at high temperature are necessary for mutated or chemically inactivated enzymes. All of these transient inhibition mechanisms are effective in repressing enzymatic activity until the initial melting step.

To fully appreciate the difference between using a hot-start *Taq* and non-hot-start *Taq*, you have to understand what happens to the targeted DNA sequence during the PCR. The most important part of the PCR is the first two cycles. During the first cycle, single-stranded cDNA or viral DNA is made into a double-stranded DNA molecule facilitated by new strand synthesis initiated from the forward primer in the reaction mix. Therefore, true amplification does not begin until the second PCR cycle for single-stranded templates. For a double-stranded DNA template, amplification begins during the first PCR cycle because both primers have a template. Nonspecific bands due to inappropriate priming events (false priming) occur during the first cycle of the PCR. This is true for single- or double-stranded templates. This can occur because the reaction starts at a low temperature prior to reaching the first melting step of the PCR cycle, usually 94 to 95 °C. It is during this initial warming of the reaction mixture that false priming and extension by the DNA polymerase can occur. Thus, the use of a hot-start *Taq* that requires 1 to 10 min at 94 to 95 °C for activation will not be active in the lower temperature range during the first cycle of the PCR. For the remaining cycles of the PCR, the lowest temperature is the annealing temperature, usually 50 to 60 °C, depending on the T_m (melting temperature) of the primers. At this elevated temperature, false priming is much less likely to occur.

2
Fluorescent Dyes, Assay Chemistries, and Real-time Instruments

Real-time PCR has evolved from a single instrument that utilized one detection assay and two dyes for signaling to a large number of instruments, with a variety of assay chemistries and fluorescent dyes available. To fully comprehend how real-time PCR works, it is necessary to have a basic understanding of fluorescence and how different assay chemistries and real-time instruments work.

2.1
Fluorescence in Real-time PCR

All real-time instruments are based on the real-time detection of an increasing fluorescent signal over the length of an experiment. The increase in fluorescent signal detected is directly proportional to the corresponding increase in the number of template molecules made during amplification. It is this principle that makes fluorescence useful for quantitative real-time PCR as a detection mechanism.

Fluorescent compounds absorb light as photons within a narrow, optimal wavelength. Light at the optimal wavelength is called the excitation wavelength. When a fluorescent molecule absorbs a quantum of light, the molecule is boosted to a higher energy state. This higher energy state is transient and short lived. This higher energy state decays, falls back to the ground state, and light is given off. This form of emission is called fluorescence. The emitted light is necessarily at a longer wavelength than the initial excitation wavelength. This shift in wavelength is called a Stoke's shift. For every fluorescent dye, there is an optimal absorbance or excitation wavelength and an optimal, shifted emission wavelength used for detection.

Having a steady state fluorescent signal throughout the PCR would not be useful. What is required is a very low initial signal that increases in proportion to the products made during the PCR to a high level that is easily detectable. One way to achieve this kind of signal production is to take advantage of the principle of fluorescence resonance energy transfer or FRET. During FRET, one dye acts as a donor and the other as the acceptor. The excited donor dye transfers its emission energy to the second dye, and that energy is used to excite the electrons in the second dye. The excited acceptor dye then emits light at its emission wavelength which can then be detected. Under conditions of FRET, the light emitted from the donor dye is quenched. The efficiency of energy transfer between the two dyes falls as the 6th power base 10 of the distance between the dye molecules. Therefore, the donor and acceptor molecules have to be in close proximity (10–100 Å). Table 1 shows a list

Tab. 1 Common dyes used in real-time PCR.

Dye name	Max. Ab [nM]	Max. Em [nM]
DABCYL	453	None
Fluorescein™	492	520
6-FAM™	494	518
JOE™	520	548
TET™	521	536
Yakima Yellow™	526	448
BHQ1™	534	None
HEX™	535	556
VIC™	538	554
Cy3™	552	570
TAMRA™	565	580
ROX™	585	605
LightCycler640™	625	640
Cy5™	643	667
LightCycler705™	685	705

of the most common dyes used in real-time PCR.

There are three classes of fluorescent dyes used in real-time PCR based on their function: (1) the reporter, which is the dye monitored during the experiment; (2) the quencher, the dye responsible for quenching the reporter signal in FRET-based assays; and (3) the reference, a dye common to all reaction vessels used to normalize the signal.

- *Reporter dyes:* the most used reporter dye is 6-FAM, a fluorescein derivative with a green emission. This dye can be efficiently excited by the argon laser found in the initial real-time machines (see the following) and has remained as the reporter dye of choice of many chemistries in use today. The first multiplexing experiments, running two or more assays simultaneously in the same reaction, were done with laser-based instruments. For this reason, other fluorescein derivatives with progressively longer excitation wavelengths were made available. The instruments available today can utilize a much wider variety of reporter dyes, from the green through the red spectrum. Another reporter dye in wide use is SYBR Green I that has very low fluorescence as a free dye, but increases dramatically when bound to double-stranded DNA. The number of dyes now on the market that are suitable for use as reporters is continuing to increase.

- *Quencher dyes:* when FRET is used as the signaling mechanism, the fluorescence of the reporter dye is quenched by a neighboring dye. There is at least one exception to this rule and that will be discussed below. There are two classes of quencher dyes; those that emit light and those that do not. The initial FRET pair was 6-FAM and TAMRA, a rhodamine derivative that emits light in the red spectrum. The FAM/TAMRA combination works very well, but TAMRA is sensitive to the alkaline hydrolysis step used to remove oligonucleotides from the solid support used for their synthesis. The result is a comparatively low yield of functional FAM/TAMRA molecules at the same or higher cost compared to a FAM/dark dye oligonucleotide. The dark dyes represent the second class of quenchers. As the name implies, no light is emitted from these dyes although they work very well as quenchers and are available for a large spectrum of reporter dyes. The initial dark dye was DABSYL and is used today when a high degree of quenching is not desired when the reporter and quencher are transiently separated. An example is a molecular beacon (see the following). A second generation of dark dyes is the black hole quenchers or BHQs. As the name implies, these dyes are very efficient quenchers and emit no light. There are three different variants of these dyes available, BHQ1, 2, and 3.

The higher the number, the more red-shifted the reporter dye used in a FRET pair. BHQ dyes are particularly effective when maximal quenching is desired. An example would be Taqman® probes (see the following).

- *Reference dyes:* All real-time PCR instruments have the capability of monitoring multiple reactions during the course of an experiment. It is important that the signal from each reaction vessel be the same so that the results can be compared. One way to achieve this is to include a reference dye in the reaction mix. The most commonly used reference dye is ROX, a red dye that is well shifted from 6-FAM. In practice, any dye can be used as a reference dye as long as it doesn't significantly spectrally overlap the reporter dye(s) in the reaction. Not every instrument requires the presence of a reference dye in the reaction. This will be discussed in the following text. However, signal normalization is a part of the data analysis of all real-time PCR software packages.

2.2
Real-time Instruments

There are three major components in every real-time PCR instrument: (1) the light source; (2) the light gathering and detection system; and (3) the thermocycler. Although each real-time platform is designed to measure a fluorescent signal from the same fluorescent dyes, the way each instrument excites and detects those dyes can be quite distinct. The mechanism driving the heating and cooling cycles can also vary. These differences are driven partially by patent issues and also by an on-going desire to build a more compact, less expensive instrument that delivers comparable or better performance when compared to earlier models. A basic understanding of how real-time PCR instruments work is important for fully understanding how the data is produced.

Light Sources: All the assays currently in use utilize fluorescent dyes as a read out. Therefore, the light source has to illuminate the fluorescent dye in use near its optimum excitation wavelength. There are three different types of light sources in use today: lasers, quartz halogen-tungsten lamps, and light emitting diodes (LEDs).

Lasers emit light that is coherent, monochromatic, and directional. The only two real-time platforms to use a laser are the Applied Biosystems 7700, still widely used but no longer manufactured, and the 7900. Both use an argon laser that emits light at 488 nm. The advantage of a laser is the light is monochromatic and intense. The disadvantage is the reporter dyes used have to have an excitation maximum near 488 nM. Argon lasers work well for fluorescein dyes attached to oligonucleotides such as 6-FAM or the free dye SYBR Green I. However, as the excitation wavelength for a reporter dye moves toward the red spectrum argon lasers lose their ability to excite the dye.

Quartz halogen-tungsten lamps are capable of emitting a steady beam of light from 350 nm to over 2000 nm and are sometimes referred to as "white light" sources. For this reason, a filter must be employed that only allows a narrow band of light at or close to the optimal excitation wavelength for a particular fluorescent dye. Most machines with quartz halogen-tungsten light sources have 4 or 5 filter sets to include a broad range of dyes with excitation maxima distributed over the green–red spectrum.

Light emitting diodes or LEDs are semiconductors that emit light within a narrow bandwidth, usually 30 to 40 nm, when stimulated by an electric current. Rather than using filters, as mentioned earlier, one or more LEDs are used that emit light at increasingly longer wavelengths to match the desired excitation maxima of the reporter dye. Thus, for every dye that has a significantly different excitation wavelength, a matching LED for the new wavelength must be present. The light from most LEDs does not have the light intensity of a laser or even a quartz-halogen bulb, but there is very little heat generated during their use and they require less power to operate.

Real-time Instruments: – The following is a description of the light path in three different real-time instruments. Each one is an example of a machine that utilizes one of the three light sources described above. Although the majority of the new instruments on the market use quartz halogen-tungsten lamps, there are still those on the market that use LEDs or a laser as a light source.

The Applied Biosystems 7900HT – This is the only machine available in the market today that uses a laser light source. However, it is also the only machine that can use either 96- or 384-well plates. It may well be that the high intensity of the laser light source is required for the excitation of reporters in 384 wells. The laser emits light at 488 nm that is directed at a beam splitter that redirects the light over a scan head with 16 lenses, sitting just above the plate in the heated cover. The lenses focus the light into either 8 or 16 wells at a time, depending on which plate is in the instrument. The emitted light comes back up through the lenses and passes through a dichroic mirror that deflects the individual beams of emitted light onto a CCD camera for detection. The scan head goes down each column of the plate, using a stepper motor, and collects the emitted light from the entire plate. The scan head returns to the first column and a new scan ensues until all the cycles have been completed. Multiple scans are made for each column during the melting and annealing/extension phases of each cycle. The plate rests in a metal block that is heated and cooled by a Peltier thermocycler. Thermocyclers, based on the Peltier principle, utilize electron flow to heat and cool the metal block rather than a conventional heater unit opposed by a cooling device.

The Roche Lightcycler and Lightcycler2 – These instruments use a single blue LED light source in an optical range similar to the argon laser. The Lightcyclers have a 32-sample capacity. Each reaction chemistry is contained within a thin capillary tube, which results in a high surface to volume ratio. Cycling times in the Lightcycler units are much reduced because heated air is used instead of a metal block for heating and cooling. This is due to the excellent heat transfer possible using capillary tubes. The samples spin past the light source like tubes in a centrifuge using a stepper motor. The excitation light goes through each tube in sequence and the emitted light is focused on the capillary tip, goes through a dichroic mirror, and is then read by either 3 (Lightcycler) or 6 (Lightcycler 2) photo detection diodes, each specific for an increasing and narrow wavelength range.

The Stratagene Mx3000p – This instrument utilizes a quartz tungsten-halogen (white light) source. The excitation range is from 350 to 750 nm with a similar detection range. Excitation and emission wavelengths are selected using filter pairs specific for each reporter dye to be detected. Light passes through an excitation filter into a fiberoptic cable and into the well of a 96-well plate. The emitted light passes back up the fiberoptic cable through an emission band pass filter and onto a photomultiplier tube (PMT) for detection. The plate is scanned once for each fluorescent dye with the filter wheels moving in tandem to bring the new filter pair into the optical path. The thermocycler is a combination of Peltier, convection, and resistive technologies unique to the Stratagene machines that results in uniform temperatures across the block. The Stratagene Mx3000p is just one example of many instruments on the market that use a quartz halogen-tungsten light source.

2.3
Selected Assay Chemistries

Just as understanding the mechanical workings of the real-time detector being used are important, knowing the principles underlying the fluorescent signaling system utilized in an assay are equally important. There are three mechanisms used to generate a fluorescent signal utilized in the large variety of assay chemistries available for real-time PCR. First is the binding of a free fluorescent dye to the newly amplified double-stranded DNA, second is a fluorescent dye that is associated with one of the PCR primers that generates a signal when incorporated into the amplified product, and third are fluorescently labeled oligonucleotides, independent of the primers called probes, that bear the fluorescent signal. There are three probe types and the method of signaling is different for each of them. In all of these assay types, the signaling molecule has to have a very low initial fluorescence that increases in direct proportion to the amount of DNA made during the PCR. Except for the intercalating dyes, all the other chemistries available depend on FRET.

SYBR Green I – The simplest assay mechanism involves the incorporation of a free dye into double-stranded DNA as it is synthesized (Fig. 3). SYBR Green I

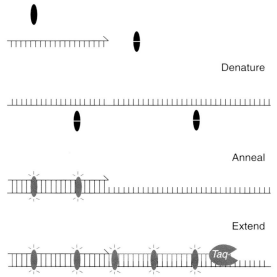

Fig. 3 Graphical representation of SYBR Green I in fluorescent signal generation during a PCR. Free dye has very low fluorescence and will not bind to single-stranded, denatured DNA. During primer annealing, a double-stranded structure is formed and SYBR green dye begins to bind and emit a fluorescent signal. During primer extension by *Taq* DNA polymerase, the number of SYBR green dye molecules bound per double-stranded molecule and the fluorescent signal increases proportionally. The process is repeated in each cycle with increasing total fluorescence (see color plate p. xxxiv).

has a very low fluorescent signal when in solution as a free dye and a low affinity for single-stranded nucleic acids at the concentration used for real-time PCR. The latter characteristic is important, as a high affinity for single-stranded DNA would be inhibitory to the progression of the PCR. Following the synthesis of double-stranded DNA, the dye binds to the minor groove and a dramatic increase in fluorescent signal can be detected. Initially, dyes such as ethidium bromide were used for this purpose. However, ethidium bromide has a relatively high signal as free dye, has a higher affinity for single-stranded DNA, and the increase in fluorescence over background following binding is fivefold lower than SYBR Green I. The positive aspects of using a dye-based real-time assay are as follows: (1) SYBR Green I is a relatively low cost signaling molecule; (2) only one dye is required for all assays; and (3) assay design is limited to two simple primers. The negative aspects of a dye-based assay are as follows: (1) lower specificity as all double-stranded DNA molecules made during the PCR will make a signal; (2) setting up a new assay requires running the initial products on a real-time instrument as well as a DNA acrylamide gel to check for nonspecific amplification products and primer dimers that can add a false signal, in the case of the former, and increased assay background noise from the latter; and (3) the use of a more expensive hot-start *Taq* is a must to reduce inappropriate priming events that can occur at sites with similar complementary sequences during the initial PCR cycle.

Primer-based signaling assays – As the title implies, the signal is generated by a reporter dye on one of the two primers required for the PCR. There are many imaginative reporter–dye primer combinations available that can be used in real-time PCR. Only two will be presented here. These two assay types represent the extremes in the level of complexity found in primer-based signally systems.

1. LUX primers (light upon extension) have a 4 to 6 base extension on the 5′ end of the primer complementary to the internal sequence of the primer that is complementary to the target sequence. A single reporter dye is covalently linked to an internal base near the 3′ end of the primer. The 5′ extended sequence forms a hairpin loop and stem structure within the primer sequence (Fig. 4). When the reporter dye is in close proximity to the double-stranded stem, particularly to guanidines, the signal is quenched without the need for another dye. Guaninidine residues are able to act as quenchers, similar to but not as efficient as an appropriate FRET dye pair. When the primer becomes incorporated into a PCR product, the stem structure is broken and this relieves the quenching of the fluorescent signal. The maximum increase in signal is about 10-fold over background. The advantages of this system are as follows: (1) relatively low cost, only a single dye is required on one of the primers; (2) relatively simple assay design; and (3) can use multiple, assay-specific reporter dyes. The disadvantages are as follows: (1) an acrylamide gel is necessary to determine assay specificity; (2) a hot-start *Taq* is necessary for optimal primer assay fidelity; and (3) it has the lowest signal to background fluorescent signal of any assay.

2. Scorpion primers are named after their mode of binding during the signaling phase and represent the most complex primer-based signaling system. Their advantage is that they have the highest

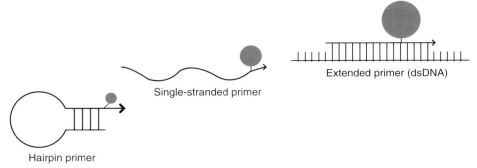

Fig. 4 Structure and mode of action of a LUX primer (light upon extension). LUX primers are inherently quenched because of a designed, short self-complementary sequence added to the 5′ end of the primer. The fluorescent moiety on the 3′ end of the primer is quenched in this confirmation. Following the melting and subsequent annealing phase, the primer binds to complementary sequence on the template and becomes incorporated into the new strand, leading to a significant increase in fluorescence.

template specificity of any primer-based assay. Two variations of the scorpion primer assay have been described, one based on a stem and loop and a newer one based on a linear duplex. Both result in efficient quenching of the reporter signal. The loop version is shown in Fig. 5 and will be described here, but they both have identical fluorescent reporter mechanisms. The basic components of scorpion primers are (1) the actual primer sequence; (2) a PCR stopper moiety to prevent PCR read through; (3) a probe-specific sequence; and (4) a fluorescence detection system comprised of reporter and quencher dyes. The primer structure begins with a standard complementary sequence for the transcript of interest on the 3′ end. Upstream from the primer sequence is the PCR stopper that prevents the upstream sequence from becoming double stranded followed by an internal DABSYL quencher dye. Next comes a stretch of oligonucleotides that can anneal to a complementary internal sequence in the PCR product, followed by 5 to 6 bases of complementary sequence to an internal region just upstream of the DABSYL dye that can form a stem structure bringing the reporter dye at the 5′ end into close proximity to the DABSYL quencher (Fig. 5). The reporter signal is quenched by FRET between the two dyes at the annealing temperature. Detection begins when the 3′ end of the scorpion primer binds to a complementary region of the single-stranded template and is extended during the extension phase. In the next PCR cycle, the resulting DNA duplex and internal stem of the scorpion primer are melted, resulting in a single-stranded molecule with the scorpion reporter incorporated. At the annealing temperature, the internal loop portion of the scorpion primer complementary to a stretch of sequence within the newly synthesized strand can now anneal to the newly made template sequence. The reporter dye is no longer quenched by FRET because the reporter dye is physically distant from the DABSYL and reporter signal

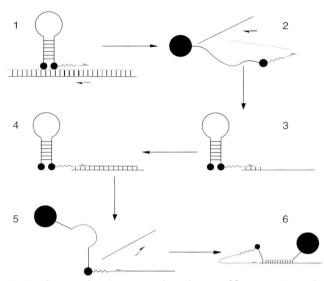

Fig. 5 The complex structure and mechanics of fluorescent signal generation by a scorpion primer. (1) dsDNA template and complete scorpion primer with a FRET quenched reporter dye; (2) melting of template and primer structure transiently releasing FRET quench of reporter dye; (3) template-complementary region of the 3′ end of the primer anneals to ssDNA template; (4) primer extension by *Taq* DNA polymerase; (5) second melting step making the newly made DNA single stranded; (6) internal complementary sequence of the scorpion primer anneals to an internal region of the newly made template releasing the reporter dye to generate a signal.

is detected. The scorpion primer is the only primer-based assay system that has a comparable template specificity to a probe-based assay. It is also the most complex and costly to make.

Probe-based signaling assays – Probe-based assays introduce one or two new oligonucleotides whose sole purpose is to provide the fluorescent signal for the assay. These extra, fluorescently labeled oligonucleotides anneal to a region in the amplicon sequence that lies between, and does not overlap, the two primers. The primers are still the driving force for amplification and they determine the efficiency and sensitivity of the assay. However, with the introduction of the intervening oligonucleotide probe, the primers are no longer the sole determinants of assay specificity. As in primer-based assays, the amplified sequence is determined by the primers alone. However, unlike primer-based assays, fluorescent signal will be realized solely from those amplicons that also have a complementary sequence for the fluorescently labeled probe.

There are three different probe designs in common use and each relies on FRET to either quench or induce the reporter signal. All three of these assay formats have increased specificity over SYBR Green I or simple primer-based systems. This is because both primers and the probe have to anneal to the correct target sequence and work together as a unit for signaling to occur.

1. Taqman® assays are the original chemistry for real-time PCR first commercialized by Applied Biosystems in 1996 (Fig. 6). The probe consists of a dual-labeled oligonucleotide with a reporter dye on the 5′ end and a strong quencher dye such as a BHQ, on the 3′ end. The probe is designed with a higher T_m than the two primers, usually 8 to 10 °C, to ensure it will bind to the template before the primers as the temperature is lowered from the melting to the annealing setting. When free in solution during the annealing step, the probe is quenched. However, when the probe is able to anneal to a single-stranded complementary region that lies just 3′ to one of the primer binding sites, it is partially dequenched because the reporter and quencher dyes are separated and FRET is much reduced. It is critical that the probe bind to every available template prior to the initiation of primer extension to ensure that they all are detected. As *Taq* DNA polymerase synthesizes a new strand, it displaces and cleaves the probe molecule utilizing an inherent 5′-exonuclease activity within the enzyme that is separate from the polymerase activity. For the most part, only DNA polymerases from *Thermus* sp. (such as *Taq* from *T. aquaticus*) have this 5′

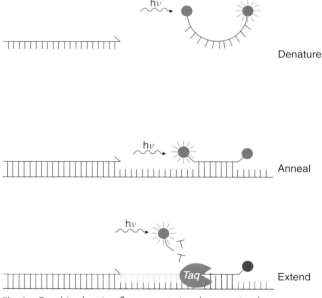

Fig. 6 Graphic showing fluorescent signal generation by a Taqman® probe. When free in solution, the reporter fluorescent dye is quenched by FRET. Owing to a higher T_m, the probe anneals to a complementary template sequence before the primer on the same strand. During primer extension by *Taq* DNA polymerase, the probe is displaced and degraded by a nuclease present in *Taq* DNA polymerase. Release of the reporter dye from the quencher dye allows the full signal to be realized.

nucleotidase activity. Following probe cleavage, the reporter dye is now free in solution and permanently dequenched after being physically removed from the quencher dye. The Taqman name stems from this cleavage step, from the computer game Pac Man.

2. Molecular beacons are similar to Taqman® probes in that they are dual-labeled oligonucleotides having a reporter dye and a DABSYL quencher. Once again, the T_m value of the probe is higher than the two primers to ensure probe binding prior to primer extension. The difference in probe structure, compared to Taqman® probes, is the addition of 4 to 6 extra self-complementary bases on both ends of the probe sequence that do not anneal to the target sequence (Fig. 7). At the annealing temperature of the PCR cycle, this forces the two dyes to come into close proximity with one another when a short stem is formed between the two short complementary sequences, maximizing the quenching of the reporter by the quencher dye via FRET when the probe is free in solution. A fluorescent reporter dye signal is generated following the annealing of the probe to a complementary sequence on the template strand. The annealing of the molecular beacon to an intervening sequence between the two primers physically separates the reporter dye from the quencher dye and FRET is no longer possible. Unlike a Taqman® probe, however, molecular beacons do not depend on cleavage by the DNA polymerase to generate a signal. The probe is simply displaced during new strand synthesis. Displacement is facilitated by keeping the T_m differences between the probe and the two primers at a lower temperature, such as 5 °C instead of 10 °C for Taqman® probes. Overall,

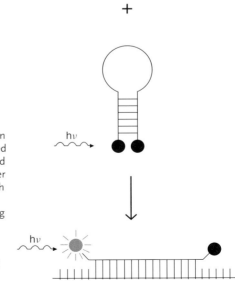

Fig. 7 Mechanism of signal generation for a molecular beacon probe. Designed complementary sequences at the 5′ and 3′ ends bring the reporter and quencher dyes in close proximity leading to much reduced reporter signal via FRET. Following template and beacon melting and annealing, full reporter signal is realized owing to loss of FRET. Unlike Taqman® probes, molecular beacons are not degraded and produce a signal only when annealed to the template.

when compared with Taqman® probes, molecular beacons have a higher signal to noise ratio due to optimal FRET between the reporter and quencher dyes. They have also been reported to work better on high G/C templates and show greater sensitivity to single-base mismatches, which makes them ideal for SNP (single nucleotide polymorphism) detection studies. However, they can be more challenging to design than Taqman® probes.

3. Hybridization probes are used primarily in conjunction with the Roche Lightcycler real-time detector, although they could be used on any real-time machine that can detect the reporter signal. For this assay system, two oligonucleotides, both with equal and higher T_m values than the primers, are made that again anneal to complementary sequences between the two PCR primers (Fig. 8). One is labeled with a donor dye on the 3′ end and the other has a signaling acceptor dye on the 5′ end. Unlike the Taqman® and molecular beacon probes that utilize FRET to quench the reporter signal, hybridization probes use the fluorescent signal induced by FRET as the reporter signal. The fluorescent signal is made when the two probes anneal to a complementary single-stranded DNA template. The excitation light stimulates the acceptor dye (green) and that emission energy is transferred to the acceptor dye (red) by FRET. The light emitted by the acceptor dye is monitored and recorded by the real-time

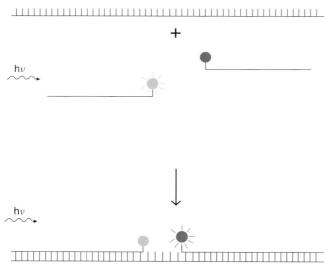

Fig. 8 Graphical representation showing the mechanism of action for a hybridization probe. Two oligo probes bearing a single dye each, one with a fluoroscein dye at the 3′ end and the other with a rhodamine dye at the 5′ end. When the two oligos anneal to a complementary template, the fluoroscein dye is excited by the light source in the instrument and transfers its energy to the rhodamine dye via FRET. FRET can only occur when the two dyes are in close proximity. The instrument is set to detect the rhodamine signal. The signal from the fluoroscein dye is not monitored.

instrument as a positive signal. While free in solution, no signal is generated by the two probe molecules at the correct wavelength as the acceptor dye emission wavelength is not monitored and the reporter dye will not absorb the wavelength of the light source. As with molecular beacons, the probe molecules do not have to be cleaved. The advantage of this system is a low background. Similar to the two probe-based assays described above, this probe-based system is highly template specific. However, designing two tandem oligonucleotide probes that anneal between a pair of primers that will work with high efficiency can be even more challenging than the other two probe systems.

3
RNA and DNA Isolation for Real-time PCR

The key factor for isolating good DNA or RNA is how the sample is treated prior to nucleic acid isolation. If the sample is not treated properly and the macromolecules are degraded *in situ*, no amount of care during the nucleic acid isolation phase can rescue the sample. For animal, plant, or microbial sources, it is imperative that they are properly treated following harvest by either initiating isolation of nucleic acids immediately or processing the sample for storage that will preserve the RNA and/or DNA *in situ*. Quick-freezing of the sample in LN_2 followed by storage at $-80\,^\circ C$ works well to stabilize the nucleic acids in most samples. A relatively new option for sample storage from Ambion, RNALater, can be used to store samples for short times at room temperature or $-20\,^\circ C$ for extended periods of time.

Following sample collection, the next most important thing in the isolation process is to make sure the tissues or cells are disrupted rapidly in the isolation medium. For this purpose, a rotor-stator homogenizer or a bead-beater works very well for most samples.

3.1
Isolation of DNA

Procedures for the isolation of DNA from animal tissue or cells involve incubating the homogenate in buffers containing EDTA, SDS, and proteinase K at elevated temperatures ($50\,^\circ$–$60\,^\circ C$). This will destroy any remaining cellular structure and contaminating proteins while preserving the DNA. Ribonucleases A and T1 are added to degrade the RNA. The digest is then extracted with phenol, phenol-chloroform, and chloroform followed by ethanol precipitation of the DNA. Isolation of DNA from some eukaryotic microbes and most plants is more involved as they have carbohydrates that copurify with the DNA. For these samples, an extra purification step utilizing CTAB (cetyltrimethylammonium bromide) will free the DNA from the contaminants. It is not necessary to worry about shearing the DNA during isolation, as the amplified region for real-time PCR is quite small, in the range of 75 to 250 bases. A final purification over a silicon-based column will result in DNA of high purity suitable for PCR.

3.2
Isolation of RNA

The isolation of RNA can be more problematic than DNA primarily because RNA is much more susceptible to degradation prior to and during isolation. The primary reason for measuring transcripts by quantitative real-time RT-PCR is to compare mRNA levels between control and experimental samples. To ensure transcript levels that reflect the original status of the source, it is important to stop any further changes in the RNA population as rapidly as possible after harvesting the sample. The best method for RNA isolation can depend on the origin of the sample. In general, a combination of thorough homogenization in a guanidinium/phenol-based buffer followed by purification on any of the commercially available silicon-based columns results in excellent yields and the highest quality RNA for real-time PCR. This approach will work for the vast majority of samples from microbes to mammalian tissue. As with DNA isolation, those sources that present problems for DNA isolation will present the same problems for RNA isolation. Methods are available in the literature for those particular tissues or organisms. One exception is isolating RNA from fixed and embedded tissue. We have found that a detergent-based isolation medium followed by proteinase K digestion of deparafinized tissue have proven to work best with limited amounts of starting material where extractions by organic solvents are not practical for maximal yields.

Two excellent sources describing multiple methods for the extraction of nucleic acids are "Molecular Cloning" and "Current Protocols in Molecular Biology."

4
Standards Used for Real-time Quantitative PCR

The term "standard," when used in the context of real-time PCR, means any nucleic acid template that bears the amplicon sequence of the gene or transcript being interrogated. Standards are used to quantify the gene or transcript levels in unknown samples. This is usually accomplished by making a dilution series, 10-fold decrements are common, over a 5 to 6 log range of standard concentration. By assaying the dilution series of a standard along with samples of unknown target levels, it is possible to determine the level of the target sequence in the unknown samples relative to the standards.

The term "absolute" has been used in conjunction with real-time quantitative PCR. This is a misnomer as no set of universally used standard assays with independently measured and verified standards are currently available. The ideal situation would be a single, quantified standard from a central agency such as NIST (National Institute of Standards and Technology, USA) for every nucleic acid target used in a real-time PCR assay. However, it is currently not possible to measure the number of molecules in any standard preparation that will be functional in the PCR, other than by real-time PCR itself. In the absence of this ideal situation, the best that can be achieved is a measurement of target molecules in any sample relative to the nucleic acid standard being used. Standards currently in use are quantified on the basis of their total nucleic acid concentration using absorbance at A_{260} or fluorescent dye–binding assays such as picogreen, ribogreen, or oligreen from Molecular Probes/Invitrogen.

4.1
RNA Standards

Unquestionably, the best standard to use for quantitative real-time RT-PCR would be an RNA molecule bearing the target region of the transcript being measured. This is because both the standard and the sample RNA target have to be converted into cDNA by reverse transcriptase as well as amplified by *Taq* DNA polymerase. Thus, the efficiency of both synthetic processes will be similar in both reactions for the RNA standard and RNA target sequences in the unknowns.

The simplest way to make an RNA standard is to dilute a concentrated, total RNA preparation. The problem with using total RNA to make a standard dilution curve is the highly variable amount of different transcripts in any RNA population. Thus, the number of dilution points detected will be variable for any given assay. Although it would be impractical to quantify unknown RNA samples using dilutions of a total RNA preparation, it does allow a measure of PCR efficiency (from the slope) and a determination of the relative low-end sensitivity of the assay.

The *in vitro* production of RNA of the size of an entire RNA molecule (some mRNAs are over 15 Kb in length) is not practical. However, it is possible to synthesize full-length RNA molecules *in vitro* that are <500 bases in length, in large quantities. To do so requires a double-stranded DNA template bearing the sequence that will be amplified during the PCR. The DNA template can then be cloned such that it is physically placed 3′ of a bacterial viral promoter sequence, usually T7 RNA polymerase. Alternatively, the T7 promoter can be added to the DNA target using a forward PCR primer with the promoter site added to the 5′ end of the primer. In either case, large quantities of RNA can be made per engineered DNA template. This is because RNA polymerases can use the same DNA template multiple times during a synthesis reaction. The full-length RNA is then separated and isolated from any shorter molecules in the total synthesis product utilizing an acrylamide gel. RNA standards are very stable over many years if stored at $-80\,°C$ and treated with care to avoid ribonuclease contamination. RNA standards can be quantified using an A_{260} measurement or, more specifically, by using the ribogreen fluorescent assay from Molecular Probes/Invitrogen.

The easiest RNA standard would be one that was synthesized *de novo* in the laboratory or by a company. The cost of RNA oligonucleotide synthesis has dropped dramatically in the last few years, mostly due to the increasing use of RNAi in research clinical applications. However, synthetic RNAs, the length of even the shortest assay amplicons (60–70 bases), are still very expensive compared to their DNA counterparts.

4.2
DNA Standards

From a practical point of view, DNA standards are much easier to use (no reverse transcriptase step required) and can be made in multiple ways inexpensively. The simplest DNA standard is the purified PCR product from one or more RT-PCR or PCR reactions. Once it has been determined that there is a single amplification product by polyacrylamide gel electrophoresis and excess primers and other reaction components have been removed, the PCR product can be quantified by A_{260} or using the picogreen fluorescent assay (Molecular Probes/Invitrogen). This method is easy, inexpensive, and a quick way to make a standard curve over a broad template concentration to test the efficiency and low-end sensitivity of a new real-time PCR assay.

Another approach to making a DNA standard is to clone an amplified double-stranded DNA product made from cDNA or a DNA template into a plasmid. Once a recombinant bacterial clone is established, a permanent resource for that DNA template is available. Making the DNA standard is as simple as a plasmid preparation and quantification of the purified DNA by A_{260} or picogreen assay.

As a standard for DNA samples, both of these methods are excellent. As standards for RT-PCR, however, both DNA standard types have two drawbacks: (1) DNA standards are not linked to the reverse transcription reaction and (2) cDNA is single stranded and DNA is double stranded, which means that when a single-stranded cDNA is being made into a double-stranded molecule during PCR cycle 1, a double-stranded standard is already being amplified. This must be kept in mind when using dsDNA standards to quantify RNA samples. A more detailed discussion of how standard curves are used to quantify samples is given in Sect. 6.2.

The easiest way to make a DNA standard that is also single stranded is to simply synthesize it yourself or have it made by a company on a DNA synthesizer. For this strategy to work, however, the assay amplicon has to be no longer than 100 bases, preferably less than 80. The ability of modern DNA synthesizers to make long oligonucleotides with high fidelity is quite good. However, there can be a subset of molecules in the oligonucleotide synthesis that are not completely deprotected (still have a blocking group attached), that have a

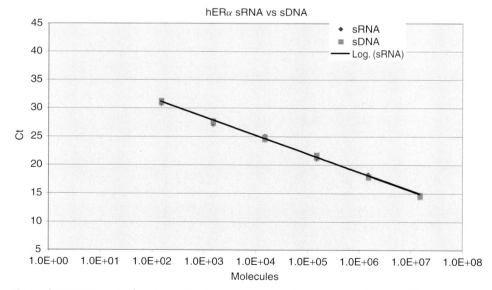

Fig. 9 A comparison made using an *in vitro* transcribed RNA standard to a synthetically manufactured oligo DNA standard for the human estrogen receptor α-(ERα) assay. Both standards were run on the same 96-well plate with identical two-step chemistry. RNA and DNA standards can work identically under the same assay conditions.

missing or inappropriate base in one or more positions and thus cannot serve as a fully functional template in the polymerase chain reaction. Yields of full-length molecules, following isolation on a polyacrylamide gel, are such that one synthesis will easily last the lifetime of the investigator running the assay.

The real question is, do DNA oligo standards yield equivalent results compared to RNA standards? Our experience has been that they do make comparable standards. One example is the comparison of standard curves generated from a single-stranded DNA oligonucleotide versus an *in vitro* transcribed RNA for the human estrogen receptor α-transcript (Fig. 9). The slopes and values for the two standards are nearly identical, showing that the RNA has been converted to cDNA and both templates amplified equally over a 6-log range of template concentrations. One caveat is that assays with amplicons over 100 bases have been difficult to test in the same way because of problems in synthesizing longer oligonucleotides. Nonetheless, in multiple assays, single-stranded DNA oligonucleotides under 100 bases have been shown to work just as well as their RNA counterparts. This means that for the relatively short amplicons used in real-time PCR assays, the RT efficiency has to be near 100% when assay-specific primers are used.

5
Real-time PCR Assay Design

The hallmarks of real-time PCR are the specificity, sensitivity, and efficiency of the assays. An assay is specific if it can be demonstrated that the detected target sequence is the only one being measured in the real-time PCR. Sensitivity is the lowest detectable amount of template that still maintains a linear relationship in a dilution series of standard template. A detectable signal for an assay that no longer falls on the standard curve is not usable for quantification and is below the sensitivity of the assay. The efficiency of an assay is 100% when every molecule in the reaction is exactly doubled. The primary factor that determines if an assay will have high levels of these three characteristics is determined by how well it is designed.

5.1
Available Software

Many multifaceted molecular biology software programs available today, such as MacVector, Vector NTI, Oligo, and others, have subroutines for designing compatible primer pairs for primer-based real-time PCR. Designing assays with an intervening oligonucleotide probe, however, requires a different software algorithm. In all software programs, primers are selected on the basis of criteria set by the investigator, the target specified, and with the goal that the oligonucleotide components not interact with one another during the PCR. Such interactions can lead to the formation of primer dimers during the PCR that increase the baseline fluorescent signal, called *noise*, of the assay. When designing a probe-based assay, the intervening probe sequence has to be taken into account as well. Only a few programs on the market are designed to find primer pairs and a compatible probe together as an assay set.

Table 2 lists the most prevalently used of the available commercial software and those that can be accessed and used over the Web.

Tab. 2 Real-time PCR assay design software.

Primer-based assays[a]		
Commercial software	Company	Web site
MacVector or DS Gene	Accelrys	http://www.accelrys.com/
Vector NTI	InforMax/Invitrogen	http://www.invitrogen.com/
Oligo	Mol. Biol. Insights	http://www.oligo.net/
DNAStar	DNASTAR, Inc.	www.dnastar.com
Probe-based Assays		
Commercial Software	Company	Web site
Primer Express	ABI	http://www.appliedbiosystems.com/
Beacon Designer	Premier Biosoft	http://www.premierbiosoft.com/
Lightcycler Software	Roche	http://www.roche-applied-science.com/
Webware	Company	Web site
Primer3	MIT	
LUX Designer	Invitrogen	
PrimeQuest	IDT	http://scitools.idtdna.com/Primerquest/

[a] Not intended to be an all inclusive list.

5.2 Basic Rules for Real-time Assay Development

Guidelines for designing primer- or probe-based assays are outlined in Table 3. By following these guidelines, we have designed assays that have worked well experimentally for many hundreds of transcripts, including viral, prokaryotic, and eukaryotic gene sequences. What has been proven empirically is that the efficiency of an assay is determined solely by how well the primers work as a pair. The probe, if present, is a signaling molecule and adds greatly to assay specificity.

SYBR Green I dye-based and simple fluorescent-labeled primer assays follow the same rules of assay design as for non-fluorescent PCR reactions. Primer concentrations should be kept relatively low, 50 to 100 nM, to discourage nonspecific priming. In contrast, when using probe-based assays and Taqman® assays, in particular, higher primer concentrations are required for optimal performance, 300 to 400 nM.

It is best to keep the T_m (melting temperature) of both primers the same. This simplifies assay quality control (QC) as changes in annealing temperature will affect the annealing of both primers equally. One way to avoid nonspecific binding is by raising the annealing temperature slightly during the PCR. Conversely, when little to no PCR product is formed, lowering the temperature may lead to increased primer binding to the template. Keeping the percent G/C content similar for both primers has also been found to be best for optimal assay performance. Pairing a short primer with a high G/C content and a long primer with low G/C content will not result in as efficient an assay as when the %G/C content is more balanced. Primers work best if they have 2 to 3 G- or C-residues in the last five bases of the primer sequence at the 3′ end. The binding specificity of PCR primers resides in the last 5 to 6 bases at their 3′ end; the rest of the primer sequence determines the annealing temperature. Many investigators insist on putting a G or C residue on

Tab. 3   Rules for the design of real-time PCR assays.

Rules for primer design

Topic	Rule	Comment
Primer concentrations	Set primer concentrations to 50–100 nM for SYBR green and simple fluorescent primer–based assays; 300–400 nM for Taqman probe–based assays	For SYBR green assays, a low primer concentration will discourage primer dimers and false priming; fluorescently labeled primers may require higher concentrations depending on signal strength
Primer T_m	The primer T_m values should be equal	Good for assay optimization; raising or lowering the annealing temperature will have an equal effect on both primers
Primer % G/C	%G/C of a primer pair should be within 25% of one another	Primers with very disparate %G/C values work satisfactorily but keeping the %G/C close, and thus primer lengths, works best empirically
Primer structure	Primers work best with 2–3 G/Cs in the last 5 bases	Too many Gs or Cs in the last 5 bases will cause false priming and low efficiencies; too few will not prime well
3′ Clamp	G/C clamp on 3′ end works well but is not necessary	Primers with 3′ G- or C-residues work well but primers that end in an A or T can work equally well; 2 or more 3′-Ts do not work well
G residues	No more than 3 Gs in a row for primers or probes; causes a bend in the oligo structure	Multiple G-residues cause the DNA structure to bend which is not conducive to good binding for a short oligo, not such an issue for a long DNA or RNA template
Amplicon T_m	Increase T_m of the amplicon from 85 to 95 °C for high G/C regions or organisms	Often, software will set the optimal amplicon T_m at 85 °C but higher G/C sequences will require a higher setting

Rules for probe design

Topic	Rule	Comment
5′-G residue	Reporter dye should not be linked to a G-residue	Guanine will act as a FRET pair with many fluorescein dyes and quench the signal (e.g. FAM, TET, VIC, etc.)
Probe T_m	Probes should have an 8–10 °C higher T_m than the primers	Probe T_m should be 8–10 °C above the primer T_m to ensure they can bind before the primers and generate a signal by probe cleavage or unfolding
Probe length	Taqman probes should not be over 30 bases or will not quench properly	Empirical experiments have shown that cleavage probes over 30 bases in length do not quench well and thus give a poor signal to noise ratio. High A/T rich sequences lead to long probes, lowering the T_m to 55 or 50 °C allows reasonable probe length with standard reagents, assay will still work at 55° or 60 °C
More Cs than Gs	Probes should have more Cs than Gs, particularly if there is a large imbalance between the two	Multiple G-residues cause the DNA structure to bend which is not conducive to binding in a short oligonucleotide probes that have many more G- than C-residues should be made from the complementary sequence

the 3′ end of the primer. This is called "clamping" the primer. Although primers work well when "clamped," it is not an absolute requirement for maximal PCR efficiency. Long homogeneous runs of any base should be avoided, but, in particular, G-residues. Tandem runs of G-residues over three bases in length tend to cause a bend in the oligonucleotide structure and this deviation from linear structure will impair the ability of the primer (or probe) to anneal properly.

There are rules for the design of fluorescently labeled probes as well. Probes labeled on the 5′ end with a fluorescein dye (e.g. 6-FAM, TET, VIC, etc.) should not begin with a G-residue. Guanine has the ability to effectively diminish the fluorescein signal, as mentioned previously for LUX primers. It is important that probes have a 5 to 10 °C higher T_m than the primers to ensure they bind to the template prior to the primers. This is particularly important for Taqman® probes that require cleavage of the probe for signal production. Similarly, Taqman® probes should be no longer than 30 bases. Probes longer than 30 bases will not quench well prior to cleavage because of poor FRET between the reporter and quencher dyes and generate little increase in fluorescent signal following cleavage compared to shorter, more efficiently quenched probes. When making assays for organisms with a low G/C content (lower than 40%), keeping the probe shorter than 30 bases can be problematic. An easy solution to this problem is simply to lower the optimal T_m setting during assay design. Taqman® assays work very well utilizing 50° or 55 °C annealing temperatures instead of the more common 60 °C. Probe length is not an issue for molecular beacons because of the forced close proximity of the reporter and quencher dyes in the probe design, or dual probe assays designed for the Lightcycler, where again the fluorescent signal comes from the close proximity of the dyes following binding. In general, probes work better if they have more C- than G-residues, particularly if there is a large imbalance between the two bases. This is a general rule applicable to all oligonucleotides used in real-time PCR.

5.3
Key Factors in Assay Development

The most important part of assay design is selecting the correct sequence before you begin. It is critical to have confidence in the accuracy and specificity of the sequence being investigated. There are two key issues, first to know that the sequence in question is accurate and second that the assay will be specific for the intended target. Mismatches between the actual sequence and those in the primers or probe can have a negative effect on assay performance. Genes or transcripts with regions of highly conserved sequence among family members can be problematic. Splice variants, and pseudogenes are some of the main issues to keep in mind when selecting a particular region within the sequence for assay design. For sequenced genomes with information available in public databases, accuracy usually can be determined *in silico*. However, for other organisms, it is important that the target sequence be validated prior to assay design. In this case, it may be necessary to sequence the final PCR product during assay development to ensure the intended target is being amplified.

When possible, it can be very useful to download sequences from other species for the same transcript or gene. When a human assay is the goal, for example, retrieving the human, mouse, and rat or other mammalian sequences can yield helpful and sometimes unexpected information. An alignment of all species will sometimes illuminate conserved regions that may be useful in making a multi-species assay or point out a region to be avoided if a transgenic mouse with a human gene is the target. Multiple species alignments also may point out splice variants present in the rat or mouse sequence in this example, not known in the human transcript. In these instances, avoid exons involved in any known splice variants. It may be the same splice variant that exists in the human as well, but has not yet been observed and/or reported.

Another issue is making a transcript-specific assay for one family member of a multigene family when all members share a large amount of sequence homology. Again, alignments will allow you to find unique places to place an assay. As mentioned above, making mouse- and human-specific assays for the same transcript in a transgenic environment can be a challenge. In this case, it is critical there be no cross-detection of the transcripts from the two species. Should the sequence homology be highly conserved throughout the coding region, an assay can be placed in the 5′ or 3′ untranslated region (UTR). However, these regions of the transcript can contain a sequence that is not well balanced among the four bases and, thus, not very PCR friendly.

There are some philosophical issues to be decided prior to designing a real-time PCR assay. The first is whether to design assays across intron/exon junction regions of a transcript. The primary reason for spanning a junction region is to avoid the detection of contaminating DNA in the RNA preparation. The sequence surrounding junctions is not always compatible with efficient real-time PCR assay design. This is not a problem for primer-based assay with larger amplicons (i.e., 250 bases), but can be a problem for probe-based assays with shorter amplicons (i.e., 65–80 bases). Regardless of assay type, one thing to keep in mind is that there are a lot of pseudogenes in the mammalian genome. Pseudogenes are spliced versions of transcripts that have been introduced back into the DNA sequence, most likely by retroviruses. One estimate is the human genome contains 30 000 genes and 20 000 pseudogenes. Some genes have multiple pseudogenes and others have none. The problem is that it is impossible to predict which have them and which do not. Therefore, crossing a junction does not guarantee you will not detect contaminating DNA in your RNA sample. However, in practice, it does reduce background from DNA contamination for many transcripts. Philosophically, it is more important to have an efficient assay than to insist on crossing a junction. However, for assays that do not cross an intron/exon junction, this means that all the RNA samples have to be DNase I treated to ensure that any contaminating DNA does not contribute a significant signal. DNase I treatment is very straightforward, has little deleterious effect on the RNA, and takes only a short time. Three key factors to keep in mind when DNase I treating RNA samples: (1) keep the DNase I concentration low, a 1/10 dilution of the stock enzyme using a volume of diluted enzyme dependent on the RNA concentration; (2) DNase I is a divalent cation-requiring enzyme, 1-mM $MgCl_2$ is sufficient for the reaction; and (3) divalent cations and high heat will degrade RNA, keep the DNase I heat killing temperature and time low and brief; 75 °C and 10 min are sufficient.

Before an assay is put to use, it is important that it be tested for three important criteria: specificity, efficiency, and sensitivity. We call the empirical determination of these three assay characteristics, quality control (QC). Quality control procedures are slightly different for SYBR Green I and primer-based assays, compared to probe-based and scorpion primer assays. Determining assay specificity for primer-based assays involves running the first real-time PCR products on a DNA acrylamide gel to look closely for extra-assay bands and primer dimers. As mentioned earlier, primer dimers will increase the fluorescent background of the assay and reduce the signal to noise ratio. The latter can have a deleterious effect on assay sensitivity. A melt-curve should be run at the end of these assays, as well, to make sure the product produced makes a single, sharp melt peak. This is also true for every run done with primer-based assays. However, a word of caution with regard to the DNA profiles generated during melt-curve analysis. This method is a quick analysis of the homogeneity of the PCR products generated in every well on the plate. However, it is not a very sensitive assay. PCR products that are not that far apart in size will merely make the melt peak slightly wider, you will not get two distinct peaks unless the bands are quite disparate in size. For this assay type, the amplicon should be 200 to 250 bases in length so that primer dimers, less than 30 mers, can be readily seen as a separately resolved band on the resulting graph. The determination of assay sensitivity and efficiency is the same as for probe-based assays and will be discussed in the following text.

Probe-based assays and scorpion primers have a higher level of template specificity. For this reason, all QC analyses can be done on the real-time machine without the need for an acrylamide gel. The reason is that extra bands or primer dimers, should they be made, will not be detected because the intervening probe oligonucleotide(s) or the tail of the scorpion primer will bind solely to the central region of the correct amplicon and that binding is essential for the generation of a fluorescent signal. Therefore, for these assay types, all components must work in concert to obtain a fluorescent signal. Further, unlike SYBR Green I or fluorescent primer–based assays, the shortest amplicon possible for probe-based assays is the most desirable. The shorter the amplicon, the easier it is to achieve maximal efficiency of the PCR.

Once assay specificity has been established, the overriding theme that should be observed during assay QC is the maximization of assay efficiency and sensitivity. A proven strategy for achieving this goal utilized in designing hundreds of Taqman® probe–based assays is to order multiple primers (two forward and two reverse) around a single probe. This same design principle of multiple primers should be utilized for primer-based assays as well to minimize possible extra-assay bands and primer dimers. Testing the primers in all possible combinations will empirically determine the best primer pair for that assay. The current design software is not sophisticated enough, at present, to ensure an ideal assay every time. The primer pair that results in the lowest number of cycles necessary for a detectable signal and the highest probe dequench at a fixed template concentration is the best primer pair. In some cases, the four primer combinations may not differ greatly, but one pair will fit the optimal criteria better than the other three. In other cases, a larger 3 by 4 primer matrix (12 unique combinations) resulted in only two acceptable primer combinations and only one of those resulted in an optimal assay. There are other examples where moving a primer one base up or down the sequence had a dramatic effect on assay performance. In all of these examples, all the primers fit the design rules outlined previously. In our experience, only by empirical experimentation can the best primer pair for an assay be determined.

Once an optimal primer pair with high template specificity has been selected, the sensitivity and efficiency of the assay must be empirically determined. Both can be found from running a dilution series of an assay standard. A standard curve should be run during assay QC, even if one is not going to be used with unknown samples, to determine assay efficiency and sensitivity. From the slope of the assay standard dilution series, assay efficiency can be determined. Acceptable slopes are -3.2 to -3.5 (see Table 7). In the real world, assays having this range of slopes are seen often and will yield excellent results. The lowest point on the standard curve that is no longer linear with the higher dilutions (falls off the standard curve) is no longer quantifiable in that assay. This is just past the low-end sensitivity for the assay. It is critical that this point be determined during assay QC so that incorrect values are not reported for the assay. This is particularly important for assays that will not include a standard dilution series on every plate.

Multiplexing is having the components for more than one assay in a single reaction. What are the design criteria for a good multiplexed assay? An acceptable multiplexed set of assays should have the same efficiency and sensitivity for their respective target sequences together that they have when run in isolation. This can only be achieved when the primers or primers and probe from one assay are able to amplify their target sequence without interfering with the other assay components in the reaction and visa versa. This means that two assays that work perfectly in isolation may very well not work with the same efficiency when combined. Thus, multiplexed assays have to be designed together as a super set to ensure operational autonomy. However, currently, there is no software available to the end user to assist in this process. Thus, beyond checking for pair-wise interactions among the reactants using available software, the optimization of a multiplex assay is an empirical one.

6
Running a Real-time PCR Experiment

It would not be practical in this review to discuss every nuance of performing a real-time PCR experiment. However, there are some important points that should be taken into consideration before any experiment is initiated.

6.1
Practical Considerations

When working with RNA, standard laboratory procedures should be followed: gloves, virgin plasticware whenever possible (RNase-free), and use of reagents and enzymes that are RNase-free. The most important component is the water. It is worth the expense to purchase commercially available nuclease-free water. DEPC-treated water made in the laboratory is not good for real-time PCR use because residual DEPC (diethyl pyrocarbonate) is a PCR inhibitor. This residue is not present in commercially available nuclease-free water.

Any lab that practices PCR will experience contamination problems, primarily from amplification products from earlier amplification reactions. Problems can be minimized by following a few, easy precautions. First, what is amplified in the tubes or plate wells should stay in the tubes or plate wells. If you have to use PCR products for a gel, for example, open the tube or plate and run the gel in a different room than the one you use to set up your RT and/or PCR reactions. Secondly, use a different set of pipettes in that room. It is best to make small aliquots of water in disposable labware for daily use and dispose of these at the end of the day. Use a different aliquot of water in each room. In time, everything that is exposed to concentrated DNA template for an assay will become contaminated by DNA-containing aerosols that are generated during pipetting. For amplicons of 70 to 250 bases, real-time PCR has the sensitivity to accurately detect template molecules in the high attogram (10^{-18} grams) range. Wipe the barrels of your pipetters down with 70% ethanol regularly to help eliminate this problem.

In an attempt to circumvent the contamination problem, many commercial kits substitute dUTP for dTTP and come with the enzyme uracil N-glycosylase (from the *Escherichia coli ung* gene). This enzyme will remove dU from any DNA molecule that contains it, degrading the template. Use of dUTP and uracil N-glycosylase prior to beginning the PCR is advertised by many companies as a way of eliminating potential contamination by amplified DNA. There are two problems with this approach. First, dUTP has to be used at twice the concentration of dTTP because it is not incorporated as efficiently by *Taq*. Secondly, other potentially contaminating templates in the laboratory (e.g. cDNAs made without dU, DNA standards and genomic DNA) will not contain dU and those will cause contamination anyway. In the end, using this strategy costs a lot more, increases the time of every instrument run, and adds nothing to the amplification of the final PCR product.

In practice, amplification of a target starting from an RNA substrate is more difficult than from DNA. Most failed RT-PCRs are due to a failed RT reaction. There are several reasons why the reverse transcriptase step does not succeed: (1) a very low abundance of the target transcript; (2) nonspecific binding of the primers and thus inefficient production of the correct cDNA template; (3) the inability to make enough complete templates of a long cDNA for the ensuing PCR; (4) a badly degraded RNA template or; (5) the presence of reverse transcriptase inhibitors in the RNA preparation. In contrast, the PCR is a comparatively robust reaction primarily because the DNA template is already present. Most failed PCRs stem from a bad component such as degraded dNTPs, improper or degraded primers, not enough $MgCl_2$, PCR inhibitors in the sample, and most often degraded or incorrect template.

6.2
Reagents

Most manufacturers make two kinds of reagent kits for real-time RT-PCR, a one-step and a two-step kit. In a two-step procedure, cDNA is made in one tube under ideal reaction conditions utilizing a thermocycler or water bath at the optimal temperature followed by the use of some or all of that cDNA for the succeeding PCR, also done under ideal reaction conditions, in a thermocycler or a real-time PCR machine. In a one-step protocol, RNA template is added to a combined reagent mix containing both reverse transcriptase and a hot-start *Taq*. The RT and PCR reactions are performed sequentially in a real-time instrument. Hot-start *Taq* has made one-step protocols possible as the RT reaction is run at a lower temperature than that required for activation of the hot-start *Taq*. This method saves time and requires fewer manipulations. However, it has been shown empirically that one-step procedures are not as sensitive to low RNA template concentrations as two-step protocols. This is most likely because the components of the reaction mixture are a compromise to accommodate both enzymatic reactions. In those instances where transcript levels could be quite low and/or highly variable from assay to assay, use of a two-step protocol gives the best template sensitivity. However, if the same assays are being run repeatedly utilizing similar samples that have a known template range, a time-saving one-step procedure is more efficient.

Purchasing assay reagents from a company saves the time required to put the individual reagents together and provides a consistent reagent base for laboratory operations. When an investigator is new to real-time PCR, it is best to purchase a reagent kit to eliminate that variable should other problems arise. However, there is a financial cost to be paid for this convenience. Over time, the cost savings of constructing your own reagent mixes can be substantial. Further, knowing exactly what components are in each reaction mixture and their laboratory history makes tracking down reagent problems much easier.

The most important concept in making your own assay buffers is that of the master mix. A master mix is a large volume of all the reaction components common to each reaction, that is, for PCR that would be everything but the template, for a single assay. This will ensure that every reaction is as close to the same as possible, except for the one variable being measured.

An example of a master mix for a reverse transcriptase reaction using an assay-specific reverse primer and the reaction conditions are shown in Table 4. The RT master mix is for a single assay and is designed to easily continue on to the following real-time PCR. For wells where a no amplification control (no reverse transcriptase) reaction is required, some of this master mix can be added to a second tube just prior to the addition of the reverse transcriptase.

For SYBR Green I, or a fluorescent primer–based assay, it is important to use a hot-start *Taq* to minimize the number of nonspecific bands and primer dimers formed. An example of a master mix for the PCR we have used successively utilizing SYBR Green I can be seen in Table 5. For best results, we have found that a three-step PCR cycle will eliminate most primer dimmers during data collection at 72 °C.

Tab. 4 Example reverse transcriptase reaction.

Stock	Component	Final	Volume [μL]
*****	DEPC-H2O	*****	2.75
10X	RT Bfr	1X	1.00
2.5 mM	dNTP Mix[a]	500 μM	2.00
20 μM	Reverse Primer	100 nM	0.05
50 U/μL	Rtase	10 U/10 μL	0.20
			6.00

6 μL of RTMaster mix + 4 μL Sample
Thermocycler: 50 °C – 30 min; 72 °C – 5 min; 20 °C – soak

[a] Equal volumes 10 mM individual dNTPs = 2.5 mM each.

Tab. 5 Example SYBR green PCR reaction mix.

Stock	Component	Final	Volume [µL]
*****	PCR-H$_2$O[a]	*****	17.35
10X	PCR Buffer	1X	4.00
50X	ROX[b]	1X	1.00
40%	Glycerol[c]	8%	10.00
10%	Tween 20[c]	0.10%	0.50
50 mM	MgCl$_2$	5 mM	4.00
2.5 mM	dNTP Mix	200 µM	2.00
20 µM	Forward Primer (+)	100 nM	0.25
20 µM	Reverse Primer (−)	100 nM	0.20
100X	SYBR Green I[d]	1/40 000	0.50
5 U/µL	Hot-start Taq[e]	1.25 U	0.20
			40.00

PCR cycle:
95 °C − 1–10 min; 40 cycles: 95 °C − 15 s, 60 °C − 2 s, 72 °C − 30 s

[a] Any nuclease-free water.
[b] Invitrogen, Carlsbad, CA.
[c] High quality, nuclease-free.
[d] Molecular Probes (Invitrogen), Eugene, OR.
[e] Hot-start Taq required.

This buffer and cycling protocol would most likely work well for fluorescently labeled primers as well. For probe-based assays, a simpler PCR master mix is shown in Table 6. Note that with probe-based assays, a faster two-step PCR cycle can be used. In both cases, the master mix is designed to be added directly to the 10-µL post-RT cDNA reaction and inserted directly into a real-time PCR instrument for analysis.

6.3
Controls

One thing that is often not stressed enough is the importance of running the proper controls in a real-time PCR experiment. There are two essential controls to be considered when running a real-time PCR experiment. The first is a no-template control, sometimes called the "NTC". It is critical to know if there is low-level contamination present in the reaction mixture, especially if the ddCt method is being used (Sect. 7.4). One might argue that using commercial master mixes would make this control superfluous. However, the most likely source of contamination comes from the reagents added by the investigator. A second important control is for the reverse transcriptase reaction to check for signal from DNA contamination. This is done by running a reaction without reverse transcriptase, also known as a "no amplification control" or "NAC." As stated above, designing assays across intron/exon junctions is no guarantee that there will be no signal from DNA contamination. Running this control on junction-spanning assays once during QC using 20 to 50 ng of genomic DNA, much higher than any

Tab. 6 Example probe-based PCR reaction mix.

Stock	Component	Final	Volume [μL]
*****	PCR-H2O[a]	*****	23.00
10X	PCR Buffer	1X	5.00
100X	ROX[b]	1X	0.50
50 mM	MgCl2	5 mM	5.00
2.5 mM	dNTP Mix	200 μM	4.00
20 μM	Forward Primer(+)	400 nM	1.00
20 μM	Reverse Primer(−)	400 nM	1.00
20 μM	Probe(+)FAM	100 nM	0.25
5 U/μL	Taq[c]	1.25 U	0.25
			40.00

PCR cycle:
95 °C − 1–10 min; 40 cycles: 95 °C − 12 s, 60 °C − 30 s

[a] Any nuclease-free water.
[b] Invitrogen, Carlsbad, CA.
[c] Hot-start Taq unless assay-specific primers used for cDNA synthesis.

contamination level, will determine the sensitivity of the assay to extraneous DNA. For assays that do not cross a junction, we run one NAC for every sample on each 96-well assay plate. Even though each sample has been treated with DNase I, some assays are much more sensitive to DNA contamination than others, and uniform elimination of DNA contamination is not guaranteed. This control is particularly important for low-level transcripts where it will take 25 to 30 cycles for template detection. Control information for each assay should be included in all manuscripts that report data from real-time PCR experiments.

7
Data Analysis

In reality, it is nearly impossible for the person analyzing real-time PCR data to change the relationship of the samples with one another. That is because of the way the fluorescent signals are detected and the data recorded in the software. Having said that, data analysis can be performed incorrectly and have an impact on the final values obtained for the samples.

As with all quantitative techniques, it is important to compensate for differences in nucleic acid concentration from sample to sample. This process is called *sample normalization*. The most common method of normalizing sample values is to divide those values by values from an invariant transcript or gene, depending on whether the samples are RNA or DNA. If both the assay of interest and the one used to normalize the data have been incorrectly analyzed, the resulting ratio can be far from the true value, which can lead to an incorrect interpretation of the final data. Assay normalization will be discussed in Sect. 7.2.

7.1
Baseline Boundaries and Threshold Levels

The two most important settings for data analysis are the range of cycles that will define the baseline and the threshold. The baseline range represents a number of early cycles where only background signal can be detected. Thus, the signal is steady within this set of cycles. The rule of thumb for setting the upper end of this range is two cycles below the cycle where the highest signal on the plate or in a set of tubes is detected. Some software programs calculate the values for each individual sample on the basis of the baseline setting of the whole sample set. Others base their calculations on each individual well or tube. The low-end setting of the background defaults to cycle 3 in most software programs. This is because the first two cycles can have slightly higher overall signals as the PCR begins. This phenomenon is strictly related to the quenching efficiency of the fluorescent signaling mechanism for all, but SYBR Green I assays.

The cycle threshold (Ct) or crossing point (Cp) is the number of cycles it takes the fluorescent signal intensity from the sample to reach an arbitrary set value. In practical terms, the threshold or crossing point is represented by a horizontal line set across the rising amplification curves. The threshold setting is at a fluorescence intensity level significantly above the background signal. The minimum threshold is usually set at 10 times the standard deviation of the mean of the fluorescent signal(s) within the background range. The threshold is either preset in the software to a fixed value or determined by an algorithm in the software. In either case, it can be changed by the investigator. However, it is important that the threshold stay within the geometric part of the amplification curves of all samples being analyzed. The Ct or Cp is calculated from the fluorescence intensity values measured by the real-time instrument for all the samples in the experiment. Figure 10 shows a magnified view of the low end of the fluorescence scale from a real-time experiment to emphasize the baseline and threshold settings in relation to the amplification curves over a 6-log standard dilution series. In this case, the baseline is set from 3 to 10 as the highest signal on the plate has a Ct of just over 12 cycles. The threshold is set at a fluorescent intensity (deltaRn) of 0.1, which is well within the geometric phase of this assay. This data was collected on an ABI 7700. Different software will have different relative scales for the y-axis. Thus, these absolute values will not be applicable to all real-time machines. However, the principles of how to set the baseline and threshold will be the same for all real-time PCR instruments.

7.2
Fluorescent Signal and Sample Normalization

The fluorescent signal values from a real-time PCR run can be reported in up to three different ways. The first is the raw fluorescence, which will have a scale commensurate with the sensitivity of the detector in the machine and the readout of the software. This is a measure of the raw signal coming from each well or tube and is often labeled R. The second value is called the deltaR (ΔR), which is calculated by subtracting the background signal in the well or tube from the raw fluorescence R and equals the net positive signal in the well or tube. This normalizes the signal for each individual

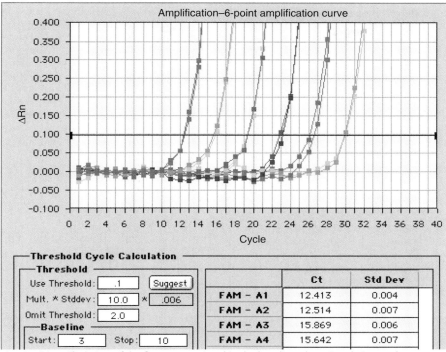

Fig. 10 Magnified view of the fluorescent signal in the baseline phase in relation to the threshold setting. A demonstration of setting the threshold high enough to be above the fluorescent signals (noise) in the baseline, but low enough to be within the geometric phase for each amplification curve. The threshold is the black horizontal line at 0.1 ΔRn (see color plate p. xxxvi).

well or tube for background differences. In machines that use ROX as an internal reference dye within the master mix, the intensity can be reported in a third way, as the deltaRn (ΔRn), which equals the ΔR/ROX signal. The ROX signal detected by the real-time instrument in each well or tube should be equivalent in theory, but not always in practice. This signal will vary with the mechanical light path and sensitivity of the instrument. By dividing by the ROX signal, the net fluorescence from each sample is normalized for any differences in signal intensity from well to well. The deltaRn calculation is the first normalization process in data analysis.

As mentioned earlier, the way to level the playing field for all samples in an experiment is to divide the primary data by some value that is a measure of some invariant component within the sample. The ratio between the invariant gene or transcript and the gene or transcript of interest corrects for differences in the amount of DNA or RNA added into the assay. For DNA experiments, the genes selected for normalization are usually present in a single copy per genome such as ApoB100 for mammalian samples. For transcript normalization, however, this can be more problematic. The critical criterion for using a transcript for data normalization is that it be at the same level in all the samples (e.g. transcripts/cell), regardless of the experimental procedure. For many experiments, transcripts such as β-actin remain constant over the course of an experiment. However, in some experiments, the treatment protocol for the experimental samples, such as the addition of drugs or changes in the physical conditions compared to controls, can cause a change in the relative level of housekeeping genes. In this case, housekeeping genes have been observed to change in tandem with the transcripts of interest. Dividing sample values by those from housekeeping genes that mirror the changes seen in the target transcripts will result in a ratio showing no change with treatment. It is critical, therefore, to make sure the transcript being used is biologically invariant throughout the experiment for all samples.

The question of what transcript to use for data normalization is one with no stock answer. Every experiment has to be monitored for changes in the transcript chosen to normalize the data. A paper looking at different concentrations of housekeeping genes commonly used to normalize real-time PCR data from many tissue types found that the amount of each housekeeping transcript was variable from tissue to tissue and that no single transcript was optimal for all tissues investigated. They concluded that all transcripts can change slightly with treatment so the best way to combat this effect was to use a minimum of three housekeeping genes for data normalization. Although this was a very a good study, the economics and time constraints of this approach have kept it from being widely adopted.

A way to skirt the problem is to use 18SrRNA or 28SrRNA. These are the two most abundant RNA species in any eukaryotic cell and are usually stable to most experimental perturbations. Some would argue that using RNA polymerase III transcripts (ribosomal RNA) to normalize RNA polymerase II transcripts (messenger RNA) is not appropriate because they are at much different levels in the cell and under different regulatory pressures within the cell. For most experimental protocols, however, the two ribosomal RNAs work well for data normalization. However, in certain experiments, it has been shown that even the ribosomal transcripts can be affected by the experimental conditions.

There are other instances where any transcript would not be appropriate for data normalization. They include studies involving apoptosis, measuring DNA or RNA in the serum, actinomycin D studies of mRNA half-life, and instances where the experimental sample has more cell types than the control, for example, sites of inflammation. In these cases, the best alternative is to measure total RNA or DNA using a fluorescent assay such as the ribogreen or picogreen assays respectively from Molecular Probes. Measuring the total nucleic acid present in a sample in this manner is far superior to estimating their concentration using absorbance at 260 nM because absorbance will detect free nucleotides, very short oligonucleotides, contaminating DNA or RNA, and impurities that absorb light at this wavelength. None of these would be detected in a real-time PCR assay, but would have an inconsistent effect on the value for each sample and, thus, on their final normalized values.

7.3
Quantification Using a Standard Curve

There are experiments where it is not necessary to quantify the results. In these cases, a determination of the presence or absence of the target sequence is sufficient. Examples are the presence or absence of a translocation in a tumor population or a pathogen in a tissue. However, the majority of studies require an accurate measure of the number of target molecules detected.

One method for accomplishing this goal is to run a dilution series of a quantified assay template alongside the unknown samples. The kinds of standards that can be used for this purpose were discussed in Sect. 4. In practice, a 10-fold dilution series spanning a 5-log concentration range of standard template is commonly used for quantification of unknown samples. However, during QC, a 6- to 7-log curve is used to determine the efficiency and lowest endpoint (sensitivity) of the assay. An example of a 6-log standard curve is shown in Fig. 11. In this example, the lowest dilution falls on the linear line and thus defines the low end of this assay, 200 molecules, because the last dilution, 7 logs, did not fall on the standard curve (not shown). The lowest detectable level for any assay, and thus its sensitivity, is the last dilution that is still on the standard curve. Unknown samples with Ct values below this level cannot be accurately quantified by the assay. It is extremely important to determine the low-end value for each assay in use. Otherwise, values may be reported that are not within the linear range of the assay and will not represent the true measure of the target. This is because the software is written with the assumption that the Ct or Cp value for the sample falls on the standard curve. If this assumption is not true and the Ct is off the curve, the value assigned to that sample by the software will be incorrect because the Ct value will be forced to fit the curve and an overestimate will result.

The standard curve has a negative slope because the more molecules there are, the fewer PCR cycles it takes for the signal in the reaction to reach the threshold (Ct). In Fig. 11, there are three values associated with the quality of the standard curve, which is a refection of the quality of the assay. The first is the slope from which you can calculate the PCR efficiency of the reaction. The slope (rise over run) is a measure of how many PCR cycles it takes to amplify the template 10-fold. The efficiency is derived from the amplification calculation. The relationship of assay slope to PCR efficiency and template amplification per cycle can be seen in Table 7. From Table 7, a slope of -3.33 equals 100% efficiency and an ideal amplification of template per cycle of 1 to 2. Therefore, the difference between each point in the standard curve is 3.33 cycles (y-axis) for every 10-fold dilution of standard template (Fig. 11). The second value is the y-intercept, which is the theoretical number of cycles required to generate a measurable signal (Ct) from a single template molecule. It is theoretical because most real-time quantitative PCR assays are not sensitive enough to maintain linearity down to one molecule. For RT-PCR, linearity to the low hundreds of molecules is very good due to the inhibition of *Taq* DNA polymerase by reverse transcriptase. The practical utility of this number is it tells you when your standard has been calibrated properly. Values from 36 to 38 cycles are acceptable due to variations in the slope from assay to assay. The third number is r^2 or correlation coefficient, which tells you how well the values making up the standard curve fall on a straight line. For real-time PCR, this value should not be below 0.990 for a good assay and is generally higher than 0.995.

Tab. 7 PCR efficiency versus assay slope.

Assay slope	PCR efficiency	Product amplification
−3.60	89.6%	1.90
−3.55	91.3%	1.91
−0.60	93.1%	1.93
−3.45	94.9%	1.95
−3.40	96.8%	1.97
−3.35	98.8%	1.99
−3.33	100.0%	2.00
−3.30	101.0%	2.01
−0.12	103.1%	2.03
−3.20	105.4%	2.05
−3.15	107.7%	2.08
−3.10	110.2%	2.10

Exponential amplification = 10E(−1/slope).
Efficiency = [10E(−1/slope)]−1.

Molecule numbers are determined for each unknown sample by interpolation from the standard curve. Each unknown is measured in duplicate or triplicate and the mean of the measurements made by the real-time instrument, the Ct values, are averaged in software. Using the mean Ct, the numbers of molecules are determined for each unknown. Following the red lines in the example seen in Fig. 11, a Ct of 29.5 would equate to 200 molecules of target sequence in the sample, relative to this standard preparation. The software determines a formula for the standard curve using a perturbation of $y = mx + b$, where m = slope and b is the y-intercept. The investigator defines how many molecules are assigned to each dilution of standard on the basis of the best measurement possible. The number of molecules, relative to the standard being used, can be calculated for a defined mass using the length of the standard and taking into account whether it is single or double-stranded. The software then uses the formula for the standard curve to calculate the number of molecules for every unknown sample based on their mean Ct values.

The advantages of using a standard curve are as follows: (1) you have a direct measure of the assay quality for each plate or group of samples and (2) data can be compared from plate to plate with confidence. Disadvantages are as follows: (1) care must be taken to ensure the standard curves for each assay are reproducibly made from run to run and (2) standard curves take up wells on the plate that could be used for unknown samples.

There are many examples where using a standard curve to determine a value for unknown samples is preferred. One would be in screening patients for a pathogen. Knowing the pathogen is present early in an infection would be very useful. Equally useful would be to know how many pathogens are present. This added knowledge would be very helpful in determining how aggressively the patient should be treated. Another example would be a large study with many animals or tissue culture plates involving controls, multiple test agents, and multiple transcript assays. Having a quantitative value for each transcript provides great flexibility in how data will be analyzed utilizing a variety of statistical methods.

7.4
Quantification Using the ddCt Method

Not all studies require that a specific value, interpolated from a standard curve, be determined for each sample. Rather, the goal of the experiment is to simply calculate a fold difference between one sample against the remaining samples on the plate or a group of similar control samples compared to one or more

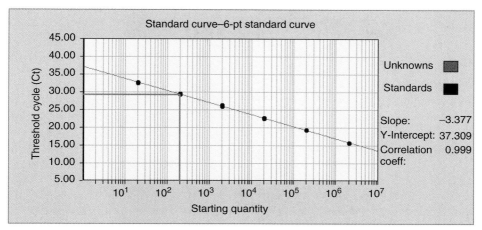

Fig. 11 A 6-log standard curve with a graphic demonstration of how the software determines molecules from the Ct values measured by real-time instruments. A Ct (threshold cycle) value of approximately 29.5 (y-axis) would result in an interpolated value of 200 molecules (x-axis) using this standard curve. A change in the slope of the curve or the measured Ct values of the standards would change the number of molecules obtained from the same Ct reading (see color plate p. xxxvi).

experimental groups on the plate. For these studies, there is a way to calculate a fold difference directly from the Ct values obtained from the real-time PCR data. A second assay, using an invariant housekeeping transcript or one of the ribosomal transcripts, must be performed to normalize the data for differences in sample loading. Armed with the Ct values from the transcript of interest and one used for normalization, a method called the delta delta Ct determination or ddCt can be employed.

As the name implies, the deltas involve subtraction steps. The first is to subtract the Ct of the assay used for data normalization from the Ct of the assay of interest. This now gives a normalized Ct value for each sample on the plate as follows:

Housekeeping gene assay
  Sample 1 — Ct = 21.5
  Sample 2 — Ct = 20.9
  Sample 3 — Ct = 21.2

Gene of interest
  Sample 1 — Ct = 25.2
  Sample 2 — Ct = 22.7
  Sample 3 — Ct = 21.4

Therefore, the deltaCt, normalized for loading for each sample is as follows:

  Sample 1 = 3.7
  Sample 2 = 1.8
  Sample 3 = 0.2

Let us assume that, in this experiment, we want to know the fold difference of Samples 2 and 3 compared to Sample 1. We then have to subtract again, for the second delta:

  deltaCt S2 − deltaCt S1 = 1.8 − 3.7
                        = −1.9

  deltaCt S3 − deltaCt S1 = 0.2 − 3.7
                        = −3.7

Relative fold difference can be determined by the formula, 2^{-ddCt}

Sample 2 relative to Sample 1 = $2^{-(-1.9)}$
$$= 3.73 \text{ fold}$$
Sample 3 relative to Sample 1 = $2^{-(-3.7)}$
$$= 13.00 \text{ fold}$$

Advantages of using the ddCt method are as follows: (1) you do not have to run a standard curve on each plate and (2) the calculation is easy to perform. Disadvantages are as follows: (1) there is no direct evidence each plate has run as expected (slope); (2) comparing data from plate to plate must be controlled by a common sample(s); and (3) final data can only be expressed as a fold difference.

The ddCt method works very well for large studies where there are many samples being compared to a common control. One example of this kind of assay could be the screening of a small molecule library where assaying a maximal number of compounds in each assay set is critical for throughput. The purpose of such a screen could be to find the most active compounds affecting a target transcript from a biological pathway. Comparing this data with another screen utilizing another target transcript from a second pathway might aid in the selection of a compound that has selective activity; high in one assay and low in the second. Those that fit the criteria of the screen would then be moved on to other assays to confirm and elaborate the findings of the real-time PCR screen.

8
Future Directions

Since the inception of real-time PCR by ABI in 1996 with the 7700, the footprint of newer instruments has been steadily getting smaller, the cost has fallen from about $100 000 per machine to about one-quarter of that price, and, in some ways, the newer machines are better than the original. Not only can you get high quality data for less money but the software is also more sophisticated in assisting the user in analyzing and presenting the data. The result of the smaller sizes and lower cost points is that these instruments are migrating from the domain of the core laboratory into individual research laboratories at an increasing rate. The only problem with this trend is that the companies selling real-time PCR instrumentation are not keeping up with training and/or the training is inadequate. The rapid increase in instruments in individual laboratories will put even more pressure on companies selling real-time PCR instruments. Hopefully, this review has provided enough of an introduction to the technique to at least make the reader aware of the major issues involved in performing real-time PCR correctly.

One thing the newer real-time instruments excel at is multiplexing. The hardware has evolved to the point that selective collection of fluorescent light over the entire green–red spectrum is now possible. There are multiple fluorescent dyes available within this broad range capable of being used as reporters covalently attached to fluorescently labeled primers or probes. However, assay design software currently on the market is only capable of designing assays that work in isolation. One way of keeping down costs and saving time would be to obtain information on the assay of interest as well as information for a second transcript useful for data normalization, design two compatible assays, and then run them together in the same, multiplexed reaction. The ability of the investigator to design good multiplexed assays in the same way that single assays can be made, presently, has not kept up with the hardware. Assay design software and the commercial sector will have to catch up with the currently available instrumentation before multiplexing becomes a true advantage to the consumer. Fortunately, there are promising signs of new software for multiplexing on the horizon.

The real value of real-time PCR instruments is their ability to collect data over time. It is this temporal component that adds an invaluable dimension to the data set. It would seem logical then that these instruments could be used for more than studies utilizing nucleic acids or, at the least, more than just PCR. There may be a number of assays that are currently read at end point, that would be amenable to the detection capabilities of a real-time instrument, and could benefit from the addition of a temporal component. One challenge yet unmet by manufacturers of real-time instruments is to expand the use of these instruments to other facets of investigation beyond PCR.

See also Immuno-PCR; Oligonucleotides; RNA Methodologies.

Bibliography

Books and Reviews

Ausubel, F.M. (2001) *Current Protocols in Molecular Biology*, Wiley InterScience, New York.

Bustin, S.A. (2000) Absolute quantification of mRNA using real-time reverse transcription polymerase chain reaction assays, *J. Mol. Endocrinol.* **25**, 169–193.

Bustin, S.A. (2004) *A-Z of Quantitative PCR*, International University Line, La Jolla, CA.

Chomczynski, P. (1993) A reagent for the single-step simultaneous isolation of RNA, DNA and proteins from cell and tissue samples, *Bio Techniques* **15**, 532–534, 536–537.

Gibson, U.E., Heid, C.A., Williams, P.M. (1996) A novel method for real time quantitative RT-PCR, *Genome Res.* **6**, 995–1001.

Heid, C.A., Stevens, J., Livak, K.J., Williams, P.M. (1996) Real time quantitative PCR, *Genome Res.* **6**, 986–994.

Herdewijn, P. (2004) *Oligonucleotide Synthesis: Methods and Applications*, Humana Press, Totowa, NJ

Sambrook, J., Russell, D.W. (2001) *Molecular Cloning: A Laboratory Manual*, Cold Spring Harbor Laboratory Press, Cold Spring Harbor, NY.

Selvin, P.R. (1995) Fluorescence resonance energy transfer, *Methods Enzymol.* **246**, 300–334.

Tricarico, C., Pinzani, P., Bianchi, S., Paglierani, M., Distante, V., Pazzagli, M., Bustin, S.A., Orlando, C. (2002) Quantitative real-time reverse transcription polymerase chain reaction: normalization to rRNA or single housekeeping genes is inappropriate for human tissue biopsies, *Anal. Biochem.* **309**, 293–300.

Primary Literature

Al-Robaiy, S., Rupf, S., Eschrich, K. (2001) Rapid competitive PCR using melting curve analysis for DNA quantification, *Bio Techniques* **31**, 1382–1386, 1388.

Barragan-Gonzalez, E., Lopez-Guerrero, J.A., Bolufer-Gilabert, P., Sanz-Alonso, M., De la Rubia-Comos, J., Sempere-Talens, A. (1997) The type of reverse transcriptase affects the sensitivity of some reverse transcription PCR methods, *Clin. Chim. Acta* **260**, 73–83.

Berney, C., Danuser, G. (2003) FRET or no FRET: a quantitative comparison, *Biophys. J.* **84**, 3992–4010.

Bustamante, L.Y., Crooke, A., Martinez, J., Diez, A., Bautista, J.M. (2004) Dual-function stem molecular beacons to assess mRNA expression in AT-rich transcripts of Plasmodium falciparum, *Bio Techniques* **36**, 488–492, 494.

Bustin, S.A. (2002) Quantification of mRNA using real-time reverse transcription PCR (RT-PCR): trends and problems, *J. Mol. Endocrinol.* **29**, 23–39.

Bustin, S.A., Nolan, T. (2004) Pitfalls of quantitative real-time reverse-transcription polymerase chain reaction, *J Biomol. Tech.* **15**, 155–166.

Calogero, A., Hospers, G.A., Timmer-Bosscha, H., Koops, H.S., Mulder, N.H. (2000) Effect of specific or random c-DNA priming on sensitivity of tyrosinase nested RT-PCR: potential clinical relevance, *Anticancer Res.* **20**, 3545–3548.

Casabianca, A., Orlandi, C., Fraternale, A., Magnani, M. (2004) Development of a real-time PCR assay using SYBR Green I for provirus load quantification in a murine model of AIDS, *J Clin. Microbiol.* **42**, 4361–4364.

Chen, R., Huang, W., Lin, Z., Zhou, Z., Yu, H., Zhu, D. (2004) Development of a novel real-time RT-PCR assay with LUX primer for the detection of swine transmissible gastroenteritis virus, *J. Virol. Methods* **122**, 57–61.

Chomczynski, P., Mackey, K., Drews, R., Wilfinger, W. (1997) DNAzol: a reagent for the rapid isolation of genomic DNA, *Bio Techniques* **22**, 550–553.

Eleaume, H., Jabbouri, S. (2004) Comparison of two standardisation methods in real-time quantitative RT-PCR to follow Staphylococcus aureus genes expression during in vitro growth, *J. Microbiol. Methods* **59**, 363–370.

Fehlmann, C., Krapf, R., Solioz, M. (1993) Reverse transcriptase can block polymerase chain reaction, *Clin. Chem.* **39**, 368–369.

Gerard, G.F., Fox, D.K., Nathan, M., D'Alessio, J.M. (1997) Reverse transcriptase. The use of cloned Moloney murine leukemia virus reverse transcriptase to synthesize DNA from RNA, *Mol. Biotechnol.* **8**, 61–77.

Gilliland, G., Perrin, S., Blanchard, K., Bunn, H.F. (1990) Analysis of cytokine mRNA and DNA: detection and quantitation by competitive polymerase chain reaction, *Proc. Natl. Acad. Sci. U.S.A.* **87**, 2725–2729.

Gjerdrum, L.M., Sorensen, B.S., Kjeldsen, E., Sorensen, F.B., Nexo, E., Hamilton-Dutoit, S. (2004) Real-time quantitative PCR of microdissected paraffin-embedded breast carcinoma: an alternative method for HER-2/neu analysis, *J. Mol. Diagn.* **6**, 42–51.

Huang, Y., Kong, D., Yang, Y., Niu, R., Shen, H., Mi, H. (2004) Real-time quantitative assay of telomerase activity using the duplex scorpion primer, *Biotechnol. Lett.* **26**, 891–895.

Kellogg, D.E., Sninsky, J.J., Kwok, S. (1990) Quantitation of HIV-1 proviral DNA relative to cellular DNA by the polymerase chain reaction, *Anal. Biochem.* **189**, 202–208.

Maaroufi, Y., Ahariz, N., Husson, M., Crokaert, F. (2004) Comparison of different methods of isolation of DNA of commonly encountered Candida species and its quantitation by using a real-time PCR-based assay, *J. Clin. Microbiol.* **42**, 3159–3163.

Mallet, F., Oriol, G., Mary, C., Verrier, B., Mandrand, B. (1995) Continuous RT-PCR using AMV-RT and *Taq* DNA polymerase: characterization and comparison to uncoupled procedures, *Bio Techniques* **18**, 678–687.

Margraf, R.L., Page, S., Erali, M., Wittwer, C.T. (2004) Single-tube method for nucleic acid extraction, amplification, purification, and sequencing, *Clin. Chem.* **50**, 1755–1761.

Martins, T.B., Hillyard, D.R., Litwin, C.M., Taggart, E.W., Jaskowski, T.D., Hill, H.R. (2000) Evaluation of a PCR probe capture assay for the detection of Toxoplasma gondii. Incorporation of uracil N-glycosylase for contamination control, *Am. J. Clin. Pathol.* **113**, 714–721.

Mhlanga, M.M., Malmberg, L. (2001) Using molecular beacons to detect single nucleotide polymorphisms with real-time PCR, *Methods* **25**, 463–471.

Myers, T.W., Gelfand, D.H. (1991) Reverse transcription and DNA amplification by a Thermus thermophilus DNA polymerase, *Biochemistry* **30**, 7661–7666.

Parks, S.B., Popovich, B.W., Press, R.D. (2001) Real-time polymerase chain reaction with fluorescent hybridization probes for the detection of prevalent mutations causing common thrombophilic and iron overload phenotypes, *Am. J. Clin. Pathol.* **115**, 439–447.

Pawlowski, V., Revillion, F., Hornez, L., Peyrat, J.P. (2000) A real-time one-step reverse transcriptase-polymerase chain reaction method to quantify c-erbB-2 expression in human breast cancer, *Cancer. Detect. Prev.* **24**, 212–223.

Peters, I.R., Helps, C.R., Hall, E.J., Day, M.J. (2004) Real-time RT-PCR: considerations for efficient and sensitive assay design, *J. Immunol. Methods* **286**, 203–217.

Robert, C., McGraw, S., Massicotte, L., Pravetoni, M., Gandolfi, F., Sirard, M.A. (2002) Quantification of housekeeping transcript levels during the development of bovine preimplantation embryos, *Biol. Reprod.* **67**, 1465–1472.

Rutledge, R.G., Cote, C. (2003) Mathematics of quantitative kinetic PCR and the application of standard curves, *Nucleic Acids Res.* **31**, e93.

Sears, J.F., Khan, A.S. (2003) Single-tube fluorescent product-enhanced reverse transcriptase assay with Ampliwax (STF-PERT) for retrovirus quantitation, *J. Virol. Methods* **108**, 139–142.

Sellner, L.N., Coelen, R.J., Mackenzie, J.S. (1992) Reverse transcriptase inhibits *Taq* polymerase activity, *Nucleic Acids Res.* **20**, 1487–1490.

Szabo, A., Perou, C.M., Karaca, M., Perreard, L., Quackenbush, J.F., Bernard, P.S. (2004) Statistical modeling for selecting housekeeper genes, *Genome Biol.* **5**, R59.

Telesnitsky, A., Goff, S.P. (1993) RNase H domain mutations affect the interaction between Moloney murine leukemia virus reverse transcriptase and its primer-template, *Proc. Natl. Acad. Sci. U. S. A.* **90**, 1276–1280.

Tosh, C., Hemadri, D., Sanyal, A., Pattnaik, B., Venkataramanan, R. (1997) One-tube and one-buffer system of RT-PCR amplification of 1D gene of foot-and-mouth disease virus field isolates, *Acta Virol.* **41**, 153–155.

Vandesompele, J., De Preter, K., Pattyn, F., Poppe, B., Van Roy, N., De Paepe, A., Speleman, F. (2002) Accurate normalization of real-time quantitative RT-PCR data by geometric averaging of multiple internal control genes, *Genome Biol.* **3**, RESEARCH0034.1–0034.11.

Vu, H.L., Troubetzkoy, S., Nguyen, H.H., Russell, M.W., Mestecky, J. (2000) A method for quantification of absolute amounts of nucleic acids by (RT)-PCR and a new mathematical model for data analysis, *Nucleic Acids Res.* **28**, E18.

Wang, C., Gao, D., Vaglenov, A., Kaltenboeck, B. (2004) One-step real-time duplex reverse transcription PCRs simultaneously quantify analyte and housekeeping gene mRNAs, *Bio Techniques* **36**, 508–516, 518–509.

Whelan, J.A., Russell, N.B., Whelan, M.A. (2003) A method for the absolute quantification of cDNA using real-time PCR, *J Immunol. Methods* **278**, 261–269.

Wilhelm, J., Pingoud, A., Hahn, M. (2003) Validation of an algorithm for automatic quantification of nucleic acid copy numbers by real-time polymerase chain reaction, *Anal. Biochem.* **317**, 218–225.

Wilkening, S., Bader, A. (2004) Quantitative real-time polymerase chain reaction: methodical analysis and mathematical model, *J. Biomol. Tech.* **15**, 107–111.

Yin, J.L., Shackel, N.A., Zekry, A., McGuinness, P.H., Richards, C., Putten, K.V., McCaughan, G.W., Eris, J.M., Bishop, G.A. (2001) Real-time reverse transcriptase-polymerase chain reaction (RT-PCR) for measurement of cytokine and growth factor mRNA expression with fluorogenic probes or SYBR Green I, *Immunol. Cell. Biol.* **79**, 213–221.

RecA Superfamily Proteins

Dharia A. McGrew and Kendall L. Knight
University of Massachusetts Medical School, Worcester, MA

1	**RecA: Function in Homologous Recombination**	**528**
1.1	RecA: Structure and History	528
2	**ATP Binding and Hydrolysis**	**531**
2.1	The Walker A motif	531
2.2	The Walker B motif	531
2.3	The ATP Cap	532
2.4	ATP Hydrolysis	533
3	**DNA Binding Domains**	**534**
3.1	Loop1 and Loop2	534
3.2	C-terminal Gateway	534
3.3	Eukaryotic and Archaeal Uncertainty	534
3.3.1	Helix-hairpin-helix	536
4	**Oligomeric Interface Domains**	**536**
4.1	N-terminal Polymerization Motif	536
4.2	Other Cross-subunit Contacts	537
5	**Allosteric Mechanism**	**538**
5.1	Allosteric Switch: Gln194	538
5.2	ATP Sensing: Phe217	538
6	**Proteins that interact with RecA**	**540**
6.1	RecBCD	540
6.2	RecF, O, and R	542
6.3	RecX and DinI	542

Encyclopedia of Molecular Cell Biology and Molecular Medicine, 2nd Edition. Volume 11
Edited by Robert A. Meyers.
Copyright © 2005 Wiley-VCH Verlag GmbH & Co. KGaA, Weinheim
ISBN: 3-527-30648-X

6.4	Coprotease Substrates	543
6.5	PolV	543
6.6	Proteins that Interact with Rad51	543
7	**Future Directions**	544
	Acknowledgments	545
	Bibliography	**545**
	Books and Reviews	545
	Primary Literature	545

Keywords

Allosteric Switch
A glutamine residue conserved in all RecA superfamily proteins that interacts with the γ-phosphate of bound ATP and triggers conformational changes that contribute to formation of the active form of the protein filament.

ATP Cap
A loop formed by 6 to 7 residues that extends into the ATP binding site of the neighboring subunit to make contacts that are likely involved in stabilizing the structure of the filament and perhaps transmitting ATP-induced allosteric information across the subunit interface.

C-terminal Gateway
A domain formed by the C-terminal region of RecA proteins involved in regulating access of double-stranded DNA into the core of active RecA/ssDNA/ATP filaments.

Helix-hairpin-helix motif
A domain typically involved in non-sequence-specific DNA binding found in the N-terminal region of RadA and Rad51 proteins that may regulate binding of double-stranded DNA in a manner similar to the C-terminal gateway in RecA proteins.

Holliday Junction
The X-shaped crossover structure resulting from the action of RecA-like proteins during the "homologous pairing and strand exchange" step in the HR pathway. This type of structure had been proposed as an intermediate in the HR pathway by Robin Holliday in 1964.

Homologous Recombination (HR)
The transfer of genetic information between two double-stranded DNA molecules of similar sequence that results from a physical exchange of DNA strands.

N-terminal Polymerization Motif
A short β-strand that forms a section of the oligomeric interface by interacting with another β-strand in the core region of the neighboring subunit. This motif is conserved in all RecA superfamily structures solved to date, as well as in the human BRCA2 protein where it likely serves to regulate assembly of Rad51 filaments at the site of a DNA break.

Rad51-mediator Activity
First described for the yeast Rad52 protein regarding its ability to assist Rad51 filament formation, this function now appears to have been conserved in all organisms in which RecA/Rad51-mediated recombination occurs. Examples include RecFOR/RecA and BRCA2/human Rad51 interactions.

Translesion Bypass
DNA replication catalyzed by an error-prone polymerase across a damaged region of template DNA. Specific protein–protein interactions between RecA and the polymerase at the site of damage are important to optimal polymerase function.

■ The bacterial RecA protein is the central component in both the catalysis of homologous recombinational DNA repair and the regulation of expression of a family of DNA repair genes. It functions in a number of mechanistically distinct processes, all of which contribute to the maintenance of genomic integrity, and all of which require formation of an active RecA/ATP/DNA complex. As cellular, biochemical, and structural studies of the eukaryotic and archaeal RecA homologs progress (Rad51 and RadA, respectively), it is clear that despite fundamental similarities in the mechanism of DNA strand exchange catalyzed by both proteins, there are distinct differences in their catalytic mechanisms as well as dramatic differences in the cellular pathways in which each protein plays important roles. For example, both RecA and the human Rad51 protein require ATP binding and hydrolysis for the catalysis of DNA strand exchange, but the use of ATP is distinctly different for each protein. Additionally, whereas RecA plays a specific role in transcriptional regulation of DNA repair genes, no such activity is seen for Rad51. In this review, we offer a comparative view of the biochemical functions of RecA and Rad51, with a particular focus on descriptions of our current understanding of the molecular design, catalytic organization, and allosteric regulation of enzyme function.

1
RecA: Function in Homologous Recombination

The bacterial RecA protein participates in a remarkably diverse set of biological functions, all of which contribute to the maintenance of genomic integrity during the lifetime of the organism as well as to ensure that genetic information is transmitted to subsequent generations with high fidelity. Bacterial RecA is directly responsible for (1) regulation of expression of a set of DNA repair genes referred to as the SOS regulon; (2) catalysis of DNA strand exchange; and (3) promoting bypass of DNA lesions by a specialized DNA polymerase during a cell's recovery from DNA damage. There are RecA-like proteins in all free living organisms, and, in all cases, their recombination function plays an important role in the repair of damaged DNA, whether the damage results from inherent errors in the machinery responsible for replicating DNA, or from exposure to exogenous mutagens. In fact, in their search for mutants that showed defects in homologous recombination (HR), Clark and Margulies discovered in 1965 that defects in *recA* greatly increased the sensitivity of bacteria to the DNA-damaging effects of UV light, thus establishing the link between homologous recombination and DNA repair. The primary enzymatic function of RecA-like proteins is the catalysis of HR. This involves exchange of genetic information between DNA molecules of similar sequence, a process resulting from the physical transfer of DNA strands. The HR pathway is traditionally broken down into four steps as seen in Fig. 1, and RecA-like proteins specifically catalyze the "homologous pairing and strand exchange" step, thus creating crossover structures known as *Holliday junctions* (named after Robin Holliday). In the scheme depicted in Fig. 1, the "initiation" step involves processing double-strand DNA breaks by exonucleases. In bacteria, this is accomplished by the RecBCD enzyme, but the mammalian counterpart(s) have not yet been identified. This leaves an ssDNA tail that serves as substrate for RecA proteins. "Pairing and strand exchange" results when the RecA/ssDNA complex invades a homologous duplex DNA, for example, a sister chromatid. The RecFOR complex is indicated at this step, but current work indicates that this complex regulates assembly and disassembly of the RecA filament specifically during repair of DNA gaps, as described below. The RecA protein itself can perform "branch migration," but, *in vivo*, this process is carried by specialized motor proteins, for example, the bacterial RuvAB helicase complex. Again, although there are a number of candidate helicases in mammalian cells, the one(s) responsible for branch migration of Holliday crossovers have yet to be identified. "Resolution" is achieved by the action of structure-specific endonucleases ("resolvases"), and, in bacteria, this function is performed by the RuvC protein. In mammalian cells, the Rad51C/Xrcc3 protein complex has been shown to be associated with Holliday junction resolvase activity, but it is currently unclear if one of these proteins or an associated nuclease is responsible for junction cleavage.

1.1
RecA: Structure and History

The first X-ray crystal structure of the *Escherichia coli* RecA protein was solved in 1992 by Story et al. In this work, the structures solved were that of protein alone and the protein–ADP complex. In both cases, RecA formed a helical filament

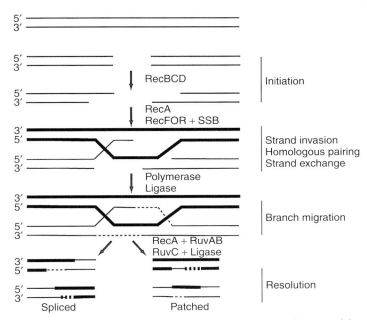

Fig. 1 Steps in the homologous recombination pathway. A basic model for homologous recombinational repair of DNA double-strand breaks (DSBs) in *E. coli* that can be separated into four major steps and are described in the text. The function of RecA in "homologous pairing and strand exchange" is the primary focus of this review.

with six subunits per turn and a distance between turns (pitch) of approximately 83 Å. Previous electron microscopic investigations had clearly established that the RecA filament could exist in two distinct oligomeric states, an "inactive" and "active" filament. The inactive filament is formed by RecA in the absence of nucleotide or by a RecA/DNA/ADP complex and has 6.2 subunits per turn with a pitch of approximately 68 Å. The active RecA filament forms in the presence of DNA and ATP, also has 6.2 subunits per turn but has an extended pitch between 95 and 100 Å. Because the RecA crystal structure contained ADP and has a pitch intermediate between the inactive and active filament, it was not clear what aspects reflected the structure of an active RecA filament. Since that time, many groups have attempted to solve the structure of a catalytically active RecA/DNA/ATP complex, but with no success. Therefore, the original RecA structure has served for more than a decade as a model for the design of experiments aimed at characterizing the molecular mechanism of HR. During this time, and now with the availability of new structures of archaeal RadA and eukaryotic Rad51 proteins, a more complete picture is emerging regarding the structural dynamics of RecA-like filaments, and the relationship of filament structure to its catalytic function.

This review focuses on our current understanding of the structural design of the bacterial RecA protein, the functional coordination of its various domains, and

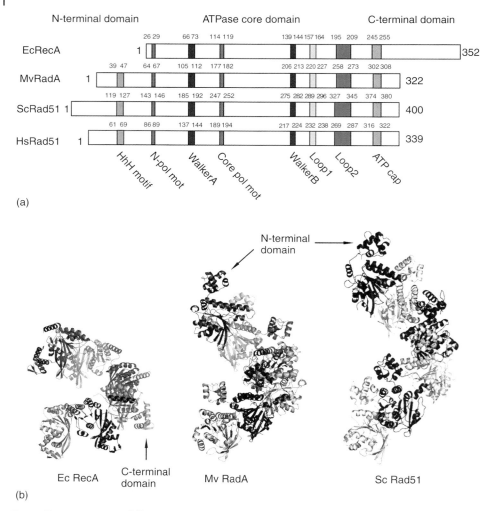

Fig. 2 Domain maps and filament structures of RecA superfamily proteins. (a) Domain map alignment of EcRecA, MvRad51, ScRad51, and HsRad51. HhH, helix-hairpin-helix; N-pol mot, N-terminal polymerization motif (see text). (b) One helical turn (6 subunits) of EcRecA, MvRadA, and ScRad51 protein filament with alternating subunits shown in dark and light colors. These filaments are arranged so that the N-terminal domain (MvRadA and ScRad51) is shown on top, and C-terminal domain (RecA) is positioned at the bottom, as indicated by arrows. Images are made from PDB files: 1REA, 1T4G, and 1SZP.

how new structures of RecA-like proteins now provide important insights into the catalytic organization of these recombinase enzymes. The alignment in Fig. 2(a) indicates some of the conserved domains within RecA, RadA, and Rad51 proteins that will be discussed below. Each member of the RecA superfamily contains the consensus Walker A and B ATP binding site motifs, and recent structural analysis of

RadA and Rad51 proteins reveal a new motif referred to as the *ATP cap* that appears to play important roles in ATP binding and cross-subunit communication. The L1 and L2 domains, as well as the extended C-terminus of RecA and N-terminus of RadA and Rad51, are discussed in terms of their roles in DNA binding. Given the fact that all of these proteins form head-to-tail polymers, large areas of the protein surface are involved in formation of the subunit interface. However, one region, in particular, the N-terminal polymerization motif, deserves mention because of its structural conservation across all members of the superfamily, and especially because of its potential role in specifically regulating Rad51 assembly in human cells.

2
ATP Binding and Hydrolysis

2.1
The Walker A motif

RecA is a DNA-dependant ATPase and contains the traditional Walker A and Walker B sequences that are found in most NTP-requiring enzymes. The Walker A motif, whose consensus sequence is defined as GXXXXGK(T/S), is also known as the P-loop because it contains residues that interact with the phosphate groups of bound NTP. In *E. coli* RecA, this motif extends from residues 66 to 73, and the highly conserved GPESSGKT sequence is identical in 61 of 64 bacterial RecA proteins. Despite this high degree of sequence identity, only four residues have been shown to be absolutely critical to RecA function: Gly66, Gly71, Lys72, and Thr73. In all currently available X-ray crystal structures of RecA proteins, the P-loop lies on the inside surface of the protein filament and near, but not directly within, the subunit–subunit interface. Although numerous mutant RecA proteins have been created with substitutions in the P-loop, work with one mutant, in particular, (Lys72Arg) has provided significant insight into important aspects of the catalytic mechanism of RecA. Although this substitution inhibits ATP hydrolysis, but not binding, and blocks recombination activities *in vivo*, the purified protein still catalyzes strand exchange up to 1.5 kb pairs. This established the idea that ATP binding, rather than ATP hydrolysis, induced a conformationally active form of RecA. Later studies using this same mutant protein showed that ATP turnover is important for coupling the exchange of subunits between free and bound states with further progression of strand exchange.

2.2
The Walker B motif

The Walker B motif in *E. coli* RecA extends from residues 139 to 144 with Asp139 and Asp144 flanking four hydrophobic residues. Although the function of Walker B motifs are less well defined than Walker A and may differ depending on the specific protein, they are most frequently thought to assist in coordinating Mg^{2+} in the bound Mg^{2+}–NTP complex via an Asp residue, play roles in transmitting ATP-induced allosteric information between proteins domains, or, in some cases, may be directly involved in ATP-mediated DNA binding. In the RecA crystal structure, Asp144 binds the Mg^{2+} in the ATP site and is identical in 64 of 64 RecA sequences. Asp139 lies within the subunit interface and is seen in the RecA structure to make a cross-subunit ionic contact with Lys6. However, the relevance of this interaction to the active form of RecA has been questioned because substitution of various polar and nonpolar amino acids for Asp139 has no deleterious effects on RecA function. A five-residue sequence following the Walker B motif, residues 145 to 149, is identical in 63 of 64 RecA sequences, and analysis of *Mycobacterium tuberculosis* RecA in complex with various NTPs suggests that this "connector" segment serves to link the L2 and L1 regions, both physically and functionally upon binding of NTP, thus playing a role in coordinating the primary and secondary DNA binding sites of RecA.

2.3
The ATP Cap

Recently solved structures of archaeal RadA proteins and yeast Rad51 now provide new and interesting insights into the arrangement of residues at the subunit interface that may be important for stabilizing the binding of ATP and regulating ATP-induced allosteric effects within the protein filaments. The archaeal *Methanococcus voltae* RadA structures were solved in the presence of the nonhydrolyzable ATP analog (AMP-PNP) and form a filament with six subunits per turn and a pitch of approximately 104 to 107 Å. This pitch approximates what has been determined to be the pitch of the active form of RecA. In these structures, the subunit interface near the ATP binding site shows significant differences relative to the arrangement of residues in currently published RecA structures. These MvRadA structures reveal a previously uncharacterized structural region, referred to as the "ATP cap," in which residues from one subunit make specific contacts with the ATP in the binding site of the neighboring subunit. The ATP cap in MvRadA spans residues 302 to 308 and one particularly notable feature is Pro307, which stacks directly against the purine base of AMP-PNP (Fig. 3). Remarkably, this Pro is conserved in all RecA superfamily proteins, including all bacterial RecAs currently sequenced. Other residues in the ATP cap are in close contact with the neighbor's AMP-PNP. For example, Asp302 forms a hydrogen bond with the hydroxyl proton at the 3'position of ribose as well as several water-mediated hydrogen bonds with phosphate oxygen at the α and β positions of AMP-PNP. Interestingly, the equivalent residue in *E. coli* RecA is Lys248, which could make similar bonding interactions with the neighboring ATP. A Lys248Ala inhibits RecA function by disrupting the stability of the protein filament. The new yeast Rad51 structure forms a filament with a pitch of 130 Å, which is within the limits seen by electron microscopy for active RecA/ATPγS/dsDNA filaments. Although crystallization solutions contained ATPγS and a 96-base poly-dT oligonucleotide, only a sulfate ion was observed in the ATP binding site in a position that would be occupied by the β-phosphate of bound ATPγS. Importantly, a region equivalent to the ATP cap in MvRadA is seen to extend into the ATP binding site of the neighboring subunit, and Pro379 is in virtually the same position as the equivalent Pro307 in the MvRadA structure (Fig. 3).

When considering how these new RadA and Rad51 structures may relate to RecA, it is important to note the model proposed by VanLoock et al. In this study, the authors performed a rigid body rotation of a subunit in the RecA crystal structure and manually docked the subunit volume into a reconstruction of the active form of

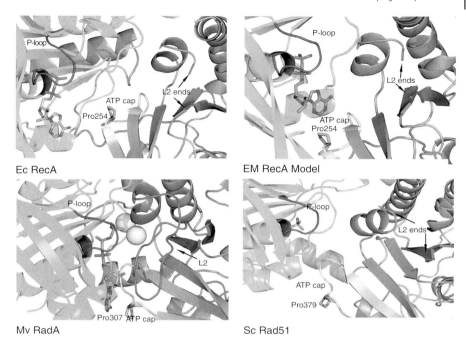

Fig. 3 ATP binding region and the "ATP cap." This view of subunit interface is seen from the inside of the filament. The ribbon diagram of the α-carbon backbone is blue for one subunit and green for the neighbor. In the green subunit, the Walker A motif, or P-loop, is in red, and the following ligands are seen bound to this motif: ADP in EcRecA and EcRecA model; AMP-PNP in MvRadA; a sulfate ion in ScRad51. In the blue subunit, the L2 region (MvRadA) or residues that flank L2 (other 3 panels) are in pink, the "ATP cap" is in yellow, and its universally conserved Proline is in orange. The yellow spheres in the MvRadA structure correspond to two K^+ atoms that are bound to the AMP-PNP. Images are made from PDB files: 1REA, 1N03, 1XU4, and 1SZP (see color plate p. xxxvii).

the RecA filament derived from analyses of electron micrographs. The resulting structure showed an arrangement of subunit–subunit interactions, very similar to those seen in the new RadA and Rad51 structures, and the region that would be the RecA ATP cap is shown in Fig. 3.

2.4
ATP Hydrolysis

While the binding of ATP induces a conformationally active form of RecA, the mechanistic requirement for ATP hydrolysis is still the subject of ongoing studies. Two important features of RecA-mediated strand exchange that require ATP hydrolysis are the ability to progress in a directional fashion and to bypass a variety of DNA lesions or certain lengths of heterology. A recent review of the motor activity of RecA describes models for the coupling of ATP hydrolysis and strand exchange. In the structures of RecA, Glu96 is positioned within the ATP binding site such that it can promote hydrolysis of ATP to ADP, and mutational analysis supports this role. Although both the yeast and human Rad51 proteins are DNA-dependent ATPases, the mechanistic steps required for ATP binding and hydrolysis are not yet well understood (see Sect. 5: Allosteric Mechanism).

3 DNA Binding Domains

3.1 Loop1 and Loop2

To carry out the process of DNA strand exchange, RecA and Rad51 filaments must bind both, an ssDNA substrate and the homologous duplex DNA. Despite the availability of several new crystal structures for RecA, RadA, and Rad51 proteins, none has included a DNA substrate. Therefore, detail regarding the specific interactions between protein and DNA is still lacking. However, a number of mutagenesis and biochemical efforts have provided a good understanding of domains within the RecA protein that bind DNA and their positions within the filament structure. On the basis of the original RecA structure, Story et al. proposed that two disordered regions were likely to be the primary (or ssDNA) and secondary (or dsDNA) binding sites. Loop 2 (L2) extends from residues 195 to 207 and is considered to be the primary DNA binding site, while loop 1 (residues 157–164) is considered to be part of the secondary DNA binding site (Fig. 4). L2 lies within the inner surface of helical filament structure, while L1 is seen on the top surface of the filament (Fig. 4). Some studies suggest that the L2 and L1 DNA binding domains may interact both physically and functionally, and a "connector" segment extending from residues 145 to 149 may be specifically involved in transmitting NTP-induced structural changes in L2 to L1. The extensive number of biochemical, mutational, and structural studies giving rise to the current model of DNA binding by the RecA protein have been reviewed recently. Interestingly, the MvRadA structures reveal a much closer physical approach of the L2 and L1 regions than is seen in the RecA structures (Fig. 4). Therefore, if the MvRadA structures do reflect the active form of the protein filament, it would support the very reasonable idea that the primary and secondary DNA binding sites are likely to interact both physically and functionally during the catalysis of DNA strand exchange.

3.2 C-terminal Gateway

In the currently available RecA structures, the C-terminal 25 residues are unresolved, but biochemical studies suggest that this region of the protein serves as a "gateway" that regulates the accessibility of dsDNA to its DNA binding site. Further work suggests that a network of salt bridges in this region precludes access of dsDNA to the filament interior, and increasing Mg^{2+} disrupts these interactions thereby opening the "gateway." How this might be accomplished *in vivo* remains to be determined.

3.3 Eukaryotic and Archaeal Uncertainty

While the identity and position of DNA binding domains within various RecA proteins is now fairly well-modeled, the DNA binding domains within the eukaryotic RadA and Rad51 proteins are currently not well defined. The L2 region of HsRad51 includes residues 270 to 287 and lies near the filament axis. However, mutagenesis studies suggest that L2 is not involved in

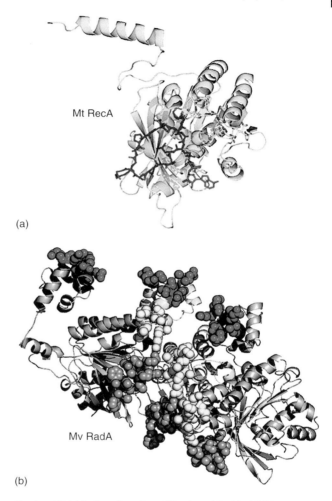

Fig. 4 DNA binding domains of RecA and RadA. (a) Monomer of MtRecA, with the L1 and L2 DNA binding regions shown in yellow and blue respectively. (b) A trimer of MvRadA as seen from the inside of the filament, showing amino acid side chains of L1 (yellow), L2 (blue) and the N-terminal Helix-hairpin-Helix domain (red) as space-filling spheres. Images are made from PDB file: 1G18/1G19 and 1XU4. The green MtRecA subunit (a) is in approximately the same orientation as the green MvRadA subunit (b) (see color plate p. xxxviii).

binding to ssDNA. We have recently created mutant HsRad51 proteins carrying either multiple or single Ala substitutions throughout the L2 region and find that all mutant proteins bind ssDNA with an affinity similar to wild-type HsRad51. Therefore, it seems that the L2 region in HsRad51 does not serve the same function as L2 in RecA. In contrast, several mutant HsRad51 proteins with Ala substitutions in the L1 region (residues 231–239) show significant decreases in binding to ssDNA.

3.3.1 Helix-hairpin-helix

Relative to the bacterial RecA proteins the eukaryotic and archaeal Rad51 and RadA proteins have an extended N-terminus that contains a helix-hairpin-helix motif ("HhH motif" in Fig. 2a; HsRad51 residues 61–70). NMR chemical shift perturbation mapping of this region suggests that it has a role in directly binding dsDNA in HsRad51. Crystal structures of Rad51 and RadA filaments show that this region is positioned along the upper surface of the filament and could serve as the dsDNA binding domain or be involved in regulating access of dsDNA to the filament interior similar to the C-terminal region of RecA (Fig. 2b, Fig. 4).

4
Oligomeric Interface Domains

The oligomeric form of RecA-like proteins required for the catalysis of DNA strand exchange is an extended helical polymer that wraps around ss- and dsDNA, making contact with the nucleic acid polyphosphate backbone, but not the bases. Formation of the nucleoprotein filament extends the helical pitch of dsDNA from an axial rise per base pair of 3.4 Å (standard B-form dsDNA) to 5.2 Å. These DNA binding properties of RecA led to the idea that one of the primary functions of RecA-like proteins was to reveal the DNA bases in a way that optimizes strand switching and formation of new Watson–Crick bases pairs during the catalysis of strand exchange.

4.1
N-terminal Polymerization Motif

The first crystal structure of E. coli RecA revealed an extensive surface on each side of the subunits that formed the oligomeric interface. Earlier genetic and biochemical studies had predicted that a domain in the extreme N-terminus of RecA played an important role in oligomer formation. Comparison of all RecA, RadA, and Rad51 structures now reveals a small conserved structural domain recently termed the *polymerization motif* by Shin et al. Not only is this motif important for the oligomeric stability of RecA-like proteins, in the case of human Rad51, it appears to play an important role in the BRCA2-mediated regulation of Rad51 nucleoprotein filament assembly at the site of DNA damage. The motif forms a short β-strand ("N-pol mot" in Fig. 2a) that interacts in antiparallel fashion with a β-strand in the ATPase core ("Core pol mot" in Fig. 2a) of the neighboring subunit (Fig. 5). Alignment of this conserved region in E. coli RecA and several eukaryotic RadA and Rad51 proteins is shown in Table 1. This region is highly conserved among 64 bacterial RecA sequences, and, although not well conserved with the equivalent sequences in eukaryotic RadA and Rad51 proteins, the structure is conserved among all RecA-like proteins. In the PfRad51 structure Phe97 and Ala100 interact with the neighboring subunit through contacts with

Tab. 1 Alignment of N-terminal Polymerization Motif across RecA Superfamily Proteins.

MvRadA	63-	G		F	K	S	G	I	-68
HsRad51	85-	G		F	T	T	A	T	-90
PfRad51	95-	G	T	F	M	R	A	D	-101
ScRad51	143-	G		F	V	T	A	A	-148
HsBRC4	1523-	G		F	H	T	A	S	-1528
EcRecA	24	G	S	I	M	R	L	G	-30

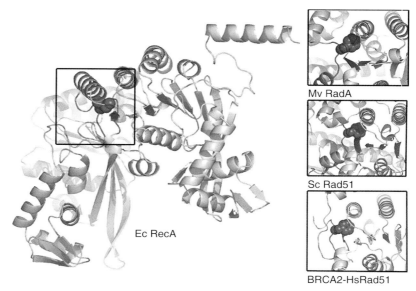

Fig. 5 N-terminal Polymerization motif. The left-hand image is two subunits of the EcRecA filaments (viewed from the outside surface of the filament), showing the N-terminal polymerization motif (green) and the Core Polymerization Motif (red). Isoleucine 26 is shown in space-filling spheres. The boxes on the right show the equivalent interface region in MvRadA (Phe64), ScRad51 (Phe144), and the BRC4-HsRad51 fusion protein (Phe1524) with indicated phenylalanine residues interacting with the neighboring subunit. Images were made from PDB files: 1REA, 1XU4, 1SZP and 1NOW (see color plate p. xxxix).

hydrophobic regions, and similar cross-subunit interactions are observed for the conserved residues in the ScRad51 and MvRadA. Mutational analysis of the equivalent residues in HsRad51, Phe86 and Ala89, supports the idea that both are important for oligomer stability. Further support for the idea that this region is an important part of the oligomeric interface comes from fluorescence studies of Xenopus Rad51. The structure of an HsRad51-BRC4 fusion protein gave rise to the idea that the BRC repeats in BRCA2 interact with Rad51 by mimicking the N-terminal polymerization motif surface. The alignment in Table 1 shows that Phe1524 and Ala1527 in the BRC4 repeat are conserved in this motif. Examination of the E. coli RecA, MvRadA, ScRad51, and HsRad51-BRC4 protein structures shows a striking similarity in cross-subunit interactions in this region of the interface, highlighted by insertion of a conserved Phe or Ile into a hydrophobic area of the neighboring subunit (Fig. 5). Further details of the mechanism by which the BRCA2 protein regulates assembly of HsRad51 nucleoprotein filaments at the site of DNA damage is provided in a recent review by Pellegrini and Venkitaraman.

4.2
Other Cross-subunit Contacts

Residues involved in making cross-subunit contacts account for a high percentage of the conserved residues seen in 64 bacterial RecA sequences. Roca and Cox noted a highly conserved region corresponding to residues 214 to 222 in *E. coli* RecA that they termed a *RecA signature sequence*. In this stretch, 5 of the 9 residues are identical in 64 bacterial RecA sequences (Ala214, Leu215, Lys216, Phe217, Arg222) and the other positions show a high degree of chemical conservation. Modeling studies of *E. coli* RecA and the crystal structures of the RadA and Rad51 proteins support the idea that this region of the interface is more extensively involved with direct contacts with the ATP binding site in the neighboring subunit than the original RecA structure revealed. This region will be discussed in greater detail in the Sect. 5: Allosteric Mechanism.

5 Allosteric Mechanism

5.1 Allosteric Switch: Gln194

Early biochemical studies of RecA showed that ATP serves as an allosteric effector of enzyme function, that is, its binding causes RecA to become catalytically active. For example, although RecA binds with low affinity to ssDNA in the absence of ATP, its DNA binding affinity is increased dramatically in the presence of ATP or a nonhydrolyzable analog such as ATPγS. The RecA crystal structure shows that Gln194 is positioned within the ATP site near to where the γ-phosphate of bound ATP would be, and, therefore, may serve as the sensor of ATP binding. Mutational and biochemical studies confirmed that Gln194 is indeed an "allosteric switch" that triggers conformational changes within the protein in response to ATP binding, an action that is critical for RecA function (see below). Subsequent studies suggested that Gln194 also plays a more direct role in ATP hydrolysis in a manner similar to G-proteins.

5.2 ATP Sensing: Phe217

Large-scale mutagenesis efforts identified Phe217 as a critical subunit interface residue involved in transmitting ATP-mediated allosteric information throughout the oligomeric filament structure. Of 15 different substitutions at Phe217, only Tyr maintained wild-type-like activity *in vivo*, while all others were significantly inhibited. Biochemical studies showed that the Phe217Tyr substitution resulted in a hypercooperativity mutant that increased the kinetics of ATP-induced filament assembly by more than 250-fold. Together with other mutagenesis studies, these data suggested that the following molecular events occur within the RecA structure upon ATP binding. The interaction of Gln194 with the γ-phosphate of ATP triggers a conformational change in the L2 region (residues 195–209). Glycine residues at positions 211 and 212 are nonmutable, and, undoubtedly, are critical to propagating the conformational change into a small helix immediately downstream of L2 (residues 213–218). This results in a specific movement of Phe217 toward the neighboring subunit. However, in the original RecA crystal structure, it was not apparent what residues in the neighbor might interact with Phe217. Subsequent modeling studies using image analyses of electron micrographs suggested that subunits in the active RecA filament are reoriented such that Phe217 and other nearby side chains, for example, Lys216 and Arg222, lie very close to the ATP binding site of the neighboring subunit. In fact, the structures of MvRadA and ScRad51 show that this is likely the case (Fig. 6). The MvRadA/AMP-PNP structure, solved in potassium-rich media, shows that the NTP sits directly within the interface between two subunits. His280, which aligns with *E. coli* RecA Phe217, makes a hydrogen bond with the γ-phosphate of AMP-PNP in the neighboring subunit (Fig. 6). A similar arrangement is seen in the ScRad51 structure in which His352 (equivalent to RecA Phe217 and MvRadA His280) lies close to a sulfate ion in the ATP binding site of the neighboring subunit (sulfate results from slow hydrolysis of ATPγS; Fig. 6). In this study, a His352Ala mutant was shown to be inhibited for ssDNA binding, thereby supporting the model in which His352 is important for formation of a filament that is competent for Rad51 function. Interestingly, when MvRadA was crystallized in low potassium media, the details of side chain conformations at the interface differ, in that access to the neighboring ATP binding site by His280 is blocked by Phe107 (Fig. 6). Similarly,

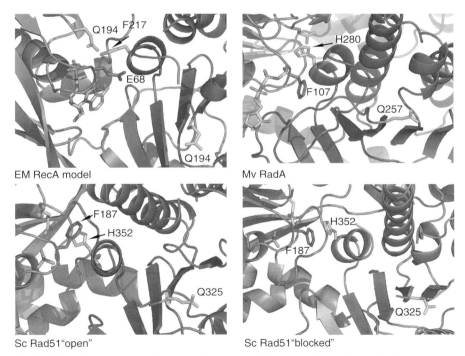

Fig. 6 Allosteric mechanism. Four panels showing different conformations of residues important in sensing ATP binding and conformational change of the filament structure. Glutamine 194 (EcRecA) and its homologous residues in MvRadA (Gln257) and ScRad51 (Gln325) are shown in yellow, while Phe217 (EcRecA) and the homologous histidine residues (MvRadA, His280; ScRad51, His352) are shown in purple. Glutamate 68 (EcRecA) and equivalent phenylalanine residues are in red. The region between Q194 and Phe217 (loop2) is unstructured in all but MvRadA. Images are made from PDB files: 1N03, 1XU4, and 1SZP (see color plate p. xl).

alternating interfaces in the ScRad51 structure show that His352 is blocked from interacting with the neighbor's ATP site by Phe187 (equivalent to MvRadA Phe107). These results may be indicative of the inherent flexibility within a RecA or Rad51 filament as it progresses through the cycle of ATP binding and hydrolysis during the catalysis of DNA strand exchange (see below). Therefore, given these new archaeal RadA and yeast Rad51 structures it seems likely that the positioning of bound ATP within the subunit interface contributes directly to subunit communication and to the cooperative nature of many aspects of recombinase function. A mechanistic similarity to other oligomeric ATPases, for example, DNA helicases, regarding the positioning of ATP binding sites has recently been described by Egelman. Although the contribution made by His280 in MvRadA and His352 in ScRad51 seems to involve direct hydrogen bonding with ATP in the neighboring subunit, it is currently not clear what kind of cross-subunit interactions are made by Phe217 in RecA. It is possible that Phe217 stacks with the purine ring of bound ATP, but further studies will be required to understand the specific molecular interactions of Phe217 in the active form of the RecA filament.

Despite a very similar arrangement of equivalent side chains in the ATP binding site and subunit interface, the human Rad51 protein shows a distinct difference regarding its use of ATP as an allosteric effector compared to RecA and yeast Rad51. Whereas RecA and ScRad51 require ATP to achieve a high-affinity ssDNA binding conformation, HsRad51 does not. It binds ssDNA cooperatively and with high affinity in the absence of ATP. This can be seen in Fig. 7, which shows biosensor DNA binding traces for *E. coli* RecA, ScRad51, and HsRad51. In this assay, protein is added directly to a cuvette containing immobilized ssDNA and binding of protein is measured in real time. Additionally, whereas a Gln194Asn or Gln194Ala mutation in RecA prevents high-affinity binding to ssDNA, equivalent mutations in HsRad51, for example, Gln268Asn, show no impairment of ssDNA binding. Although catalysis of DNA strand exchange by HsRad51 *in vitro* appears to be dependent on ATP hydrolysis, conflicting reports have appeared in the literature regarding the requirement for ATP hydrolysis for HsRad51 function *in vivo*. There are clear differences in the mechanisms by which various recombinase enzymes use NTP cofactors to drive catalysis of DNA strand exchange and further work will be required to fully understand these distinct requirements.

6
Proteins that interact with RecA

Although RecA shows proficient strand exchange activity *in vitro*, its activities *in vivo* depend on specific interactions with a variety of other proteins that are either directly involved in the catalysis of DNA repair or in the regulation of expression of DNA repair genes. Recent efforts have improved our understanding of both specific regions within the RecA protein structure responsible for these interactions as well as the functional consequences of these interactions.

6.1
RecBCD

The RecBCD enzyme is a highly processive helicase and nuclease whose activities undergo significant modification upon encountering a recombination hotspot

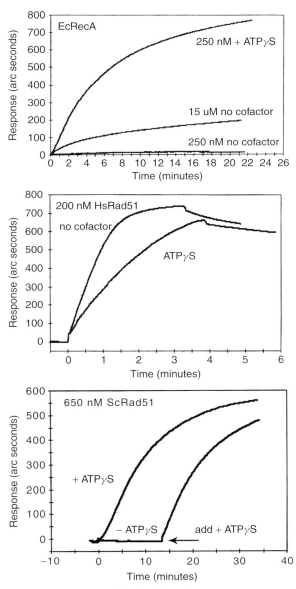

Fig. 7 Biosensor DNA binding data. These curves represent real-time binding of the indicated protein to ssDNA immobilized in an IAsys biosensor cuvette. The concentration of protein, as well as the presence or absence of nucleotide cofactor (ATPγS), is indicated in each panel.

known as *chi*. RecBCD generates 3'single-stranded ends that serve as a substrate for RecA binding and strand exchange. RecBCD also promotes loading of RecA onto ssDNA by virtue of a direct protein–protein interaction mediated by the C-terminal domain of RecB and some region of the core domain of RecA. The recently published X-ray structure of RecBCD now provides a clearer view of how the many activities of this enzyme are coordinated, as well as how the RecA loading site in RecB is likely revealed following RecBCDs encounter with *chi*.

6.2
RecF, O, and R

While early genetic studies suggested that the RecF, RecO, and RecR proteins play subtle, but important, roles in RecA-mediated recombination, recent studies show that they are specifically involved in RecA-mediated repair of gapped DNA lesions that frequently result when a replication complex stalls at a DNA lesion. It appears that these proteins can function both as subcomplexes as well as a heterotrimeric complex. RecFR and RecOR regulate the loading of RecA and the degree of RecA filament extension, respectively, within a gapped region. The RecFOR complex has been reported to serve as a structure-specific mediator of RecA by targeting recombinational repair to the ssDNA–dsDNA junctions in gapped DNA. Other studies support a role for RecFOR in clearing RecA from recombination intermediates to allow replication restart. Although no specific protein–protein interactions have been demonstrated between RecA and RecF, RecO, or RecR, studies by Bork et al. suggest that RecR specifically contacts RecA.

6.3
RecX and DinI

The *recX* and *dinI* genes are part of the SOS regulon and, through direct interaction with the RecA filament, serve to regulate RecA activity. In many bacterial species, the RecX protein is a negative modulator of RecA activity, inhibiting RecA ATPase, strand exchange, and coprotease activities, and serves to prevent harmful overexpression of RecA protein caused by DNA damage. However, in *Neisseria gonorrhoeae* RecX appears to be necessary for normal RecA function. Three-dimensional reconstructions of electron micrographs show that RecX binds to RecA nucleoprotein filaments by spanning the subunit–subunit interface, contacting the C-terminal domain of one subunit and the core domain of the neighboring subunit. However, recent studies show that the maximal inhibitory effect of RecX occurs at low, substoichiometric amounts of RecX relative to RecA. Drees et al. also show that RecX blocks extension of RecA nucleoprotein filaments through a proposed end-capping mechanism, resulting in a net filament disassembly. The interaction between RecX and RecA is modulated by the negatively charged C-terminal region of RecA, and is also enhanced by increasing Mg^+ concentration.

Until very recently, DinI had been thought to be a negative regulator of RecA function, serving to attenuate the SOS response following DNA repair. However, new studies support the idea that the action of DinI opposes that of RecX and serves to stabilize active RecA nucleoprotein filaments. The inhibitory effects of DinI on RecA function occur only at elevated concentrations of DinI, whereas lower concentrations of DinI promote RecA activities. These studies have shown the opposing effects of RecX and DinI to occur both *in vitro* and *in vivo*. Interestingly, at concentrations of DinI that stabilize RecA nucleoprotein filaments, RecA-mediated cleavage of UmuD is inhibited, whereas cleavage of LexA is not. DinI binds to the core domain of RecA and may specifically compete for UmuD binding, but not LexA. Additionally, the C-terminal domain of RecA modulates DinI function and provides regulated access to the core domain of RecA, perhaps in the same way that the C-terminal domain regulates access of dsDNA to the interior of the RecA filament.

6.4
Coprotease Substrates

The active complex of RecA/ATP/ssDNA also catalyzes the autoproteolytic inactivation of several repressor proteins, for example, LexA, phage λ cI and the phage ϕ80 repressors as well as the autoproteolytic activation of the UmuD protein. This activity has no mechanistic resemblance to the strand exchange activity of RecA, nor does it require extended nucleoprotein filaments. The original assumption was that RecA itself had a protease activity, but with the discovery by Little in 1984 that LexA and the phage λ cI repressors undergo a specific autodigestion reaction, it became clear that binding to RecA induced a conformation in these substrate proteins that was competent for autocleavage. Although binding sites for coprotease substrates within the RecA structure have yet to be specifically defined, a large number of mutagenesis, biochemical, and biophysical studies have provided strong evidence that the coprotease binding site includes specific positions within the groove of the RecA filament as well as positions on the upper surface of the filament that overlap with the L1 DNA binding domain. This subject and all of the studies leading up to this model have been reviewed recently by McGrew and Knight.

6.5
PolV

In addition to its activities as a strand exchange and coprotease enzyme, RecA participates directly in an important process referred to "translesion bypass DNA synthesis." The UmuD$'_2$C complex (pol V) is an error-prone DNA polymerases in *E. coli*, and, as part of the recovery from DNA damage, pol V is activated to ensure that some of the unrepaired DNA lesions will not block DNA synthesis to an extent that kills the cell. Owing to its low fidelity, pol V gives rise to an increased level of mutations, and this process in bacteria has been termed *SOS mutagenesis* or *UV-induced mutagenesis*. RecA has several distinct roles in this process: (1) it serves to induce all SOS genes, including the *umuC* and *umuD* genes, by catalyzing autodigestion of the LexA repressor; (2) it catalyzes the autoproteolytic activation of UmuD to the active UmuD' form; (3) and it interacts directly with the UmuD$'_2$C complex at the site of lesion bypass. The region of the RecA structure that interacts with UmuD$'_2$C is located on the surface of the subunit interface that would be at the 3' end of bound DNA, the end of the filament that will be facing the lesion needing to be bypassed by polV.

6.6
Proteins that Interact with Rad51

A number of proteins have been identified that interact with either yeast or human Rad51 using various screens such as the "yeast 2-hybrid assay" or co-immunoprecipitations. This work, and descriptions of the likely biological functions of many of these proteins, has been reviewed recently by Thompson and Schild. A number of ongoing studies are aimed at characterizing the functional consequences of these interactions and providing some molecular detail regarding specific protein domains and residues involved in these interactions. Briefly, one of the best-characterized interactions is between yeast Rad51 and Rad52 proteins. This work led to the identification of Rad52 as a "mediator" of Rad51 function, that is, Rad52 assists with formation of Rad51 filaments at the site of a DNA double-strand break, and also helps Rad51 compete with other DNA binding proteins that may block its function. This particular aspect of Rad52 function is particularly interesting because it appears to have been conserved in all organisms that perform RecA/Rad51-mediated HR. For example, RecFOR assists RecA in a way similar to the Rad52/Rad51 interaction, and, in human cells, the BRCA2 protein displays a "mediator" function regarding its role in regulating assembly of Rad51 nucleoprotein filaments.

Important questions remain regarding how the mechanistic differences seen for bacterial RecA and eukaryotic RadA/Rad51 proteins relate to the various mediators and other proteins that assist each recombinase, as well as varied cellular environments in which these proteins catalyze recombination events.

7
Future Directions

Homologous recombination is one of several major DNA repair pathways dedicated to preserving the integrity of genetic information and is found in all free living organisms. This pathway operates to repair errors that occur frequently during normal metabolic processes such as DNA replication, as well as more catastrophic DNA lesions resulting from exposure to environmental mutagens. In eukaryotic organisms, recombination also plays an important role in proper alignment and segregation of chromosomes during meiotic division. Recent studies show that defects in homologous recombination correlate with the occurrence of several cancer syndromes in humans, thus emphasizing the need for a more detailed mechanistic understanding of recombination proteins. Studies over the past several years have significantly improved our understanding of the molecular design of the bacterial RecA protein and how its many structural domains coordinate their functions to carry out a wide variety of activities that are mechanistically unrelated, such as regulation of gene expression, catalysis of DNA strand exchange, and promotion of DNA synthesis by error-prone polymerases. RecA is a classical allosterically regulated enzyme, and ATP binding triggers a number of conformational changes that activate the protein for its diverse functions. Recent insights from new modeling procedures as well as new structures of archaeal and eukaryotic RecA-like proteins (RadA and Rad51, respectively) have significantly improved our understanding of the catalytic organization of RecA. RecA-like proteins in higher organisms must catalyze the same DNA strand exchange reaction as in bacteria, but the greater complexity of the cellular environment as well as the DNA substrate itself (chromatin) demands a more sophisticated interplay between Rad51 and various other proteins involved directly or indirectly in catalyzing homologous recombination. A major challenge that is currently driving many studies in this area is to fully understand the complexities of homologous recombination in higher organisms, not only with regard to the function of Rad51 and its associated proteins but also to further our understanding of how the cell coordinates various pathways involved in the detection of DNA damage, slowing of cell growth to facilitate DNA repair, and the activation of DNA repair pathways.

Acknowledgments

The authors wish to acknowledge Dr. Martin Marinus, Dr. Nick Rhind, and Dr. Phoebe Rice for their commentary on the manuscript. Thanks to Dr. Phoebe Rice and Dr. Bill Royer for aid in making structural images. This work was supported by NIH grants GM44772 and GM65851 to K.L.K.

See also Aging and Sex, DNA Repair in; DNA Repair in Yeast; Nucleic Acids (DNA) Damage and Repair; Recombination and Genome Rearrangements; Regulome; Repair and Mutagenesis of DNA.

Bibliography

Books and Reviews

Conway, A.B., Lynch, T.W., Zhang, Y., et al. (2004) Crystal structure of a Rad51 filament, *Nat. Struct. Mol. Biol.* **11**, 791–796.

Cox, M.M. (2003) The bacterial RecA protein as a motor protein, *Annu. Rev. Microbiol.* **57**, 551–577.

Datta, S., Ganesh, N., Chandra, N.R., et al. (2003) Structural studies on MtRecA-nucleotide complexes: insights into DNA and nucleotide binding and the structural signature of NTP recognition, *Proteins* **50**, 474–485.

Kelley, J.A., Knight, K.L. (1997) Allosteric regulation of RecA protein function is mediated by Gln194, *J. Biol. Chem.* **272**, 25778–25782.

Kowalczykowski, S.C. (2000) Initiation of genetic recombination and recombination-dependent replication, *Trends Biochem. Sci.* **25**, 156–165.

McGrew, D.A., Knight, K.L. (2003) Molecular design and functional organization of the RecA protein, *Crit. Rev. Biochem. Mol. Biol.* **38**, 385–432.

Pellegrini, L., Yu, D.S., Lo, T., et al. (2002) Insights into DNA recombination from the structure of a RAD51-BRCA2 complex, *Nature* **420**, 287–293.

Shin, D.S., Pellegrini, L., Daniels, D.S., et al. (2003) Full-length archaeal Rad51 structure and mutants: mechanisms for RAD51 assembly and control by BRCA2, *EMBO J.* **22**, 4566–4576.

Story, R.M., Weber, I.T., Steitz, T.A. (1992) The structure of the E. coli recA protein monomer and polymer, *Nature* **355**, 318–325.

VanLoock, M.S., Yu, X., Yang, S., et al. (2003) ATP-Mediated conformational changes in the RecA filament, *Structure (Camb.)* **11**, 187–196.

Wu, Y., Qian, X., He, Y., et al. (2004) Crystal structure of an ATPase-active form of Rad51 homolog from methanococcus voltae: insights into potassium-dependence, *J. Biol. Chem.* **280**, 722–728.

Primary Literature

Aihara, H., Ito, Y., Kurumizaka, H., et al. (1999) The N-terminal domain of the human Rad51 protein binds DNA: structure and a DNA binding surface as revealed by NMR, *J. Mol. Biol.* **290**, 495–504.

Baumann, P., Benson, F.E., West, S.C. (1996) Human Rad51 protein promotes ATP-dependent homologous pairing and strand transfer reactions in vitro, *Cell* **87**, 757–766.

Bork, J.M., Cox, M.M., Inman, R.B. (2001) The RecOR proteins modulate RecA protein function at 5′ ends of single-stranded DNA, *EMBO J.* **20**, 7313–7322.

Campbell, M.J., Davis, R.W. (1999) On the in vivo function of the RecA ATPase, *J. Mol. Biol.* **286**, 437–445.

Churchill, J.J., Kowalczykowski, S.C. (2000) Identification of the RecA protein-loading domain of RecBCD enzyme, *J. Mol. Biol.* **297**, 537–542.

Clark, A.J., Margulies, A.D. (1965) Isolation and characterization of recombination-deficient mutants of Escherichia coli K12, *Proc. Natl. Acad. Sci. U.S.A.* **53**, 451–459.

De Zutter, J.K., Knight, K.L. (1999) The hRad51 and RecA proteins show significant differences in cooperative binding to single-stranded DNA, *J. Mol. Biol.* **293**, 769–780.

Drees, J.C., Lusetti, S.L., Cox, M.M. (2004) Inhibition of RecA protein by the Escherichia coli RecX protein: modulation by the RecA C-terminus and filament functional state, *J. Biol. Chem.* **279**, 52991–52997.

Drees, J.C., Lusetti, S.L., Chitteni-Pattu, S., et al. (2004) A RecA filament capping mechanism for RecX protein, *Mol. Cells* **15**, 789–798.

Egelman, E.H. (1993) What do X-ray crystallographic and electron microscopic structural studies of the RecA protein tell us about recombination?, *Curr. Opin. Struct. Biol.* **3**, 189–197.

Egelman, E.H. (2003) A tale of two polymers: new insights into helical filaments, *Nat. Rev. Mol. Cell Biol.* **4**, 621–630.

Egelman, E.H., Stasiak, A. (1993) Electron microscopy of RecA-DNA complexes: two different states, their functional significance and relation to the solved crystal structure, *Micron* **24**, 309–324.

Eldin, S., Forget, A.L., Lindenmuth, D.M., et al. (2000) Mutations in the N-terminal region of RecA that disrupt the stability of free protein oligomers but not RecA-DNA complexes, *J. Mol. Biol.* **299**, 91–101.

Goodman, M.F. (2002) Error-prone repair DNA polymerases in prokaryotes and eukaryotes, *Annu. Rev. Biochem.* **71**, 17–50.

Hortnagel, K., Voloshin, O.N., Kinal, H.H., et al. (1999) Saturation mutagenesis of the E. coli RecA loop L2 homologous DNA pairing region reveals residues essential for recombination and recombinational repair, *J. Mol. Biol.* **286**, 1097–1106.

Karlin, S., Brocchieri, L. (1996) Evolutionary conservation of RecA genes in relation to protein structure and function, *J. Bacteriol.* **178**, 1881–1894.

Kelley De Zutter, J., Forget, A.L., Logan, K.M., et al. (2001) Phe217 regulates the transfer of allosteric information across the subunit interface of the RecA protein filament, *Structure (Camb.)* **9**, 47–55.

Kiselev, V.I., Glukhov, A.I., Tarasova, I.M., et al. (1988) [Accumulation of N-terminal fragment of recA protein in the htpR- mutant impairs the SOS-function of Escherichia coli cells], *Mol. Biol. (Mosk.)* **22**, 1198–1203.

Koller, T., Dicapua, E., Stasiak, A. (1983) Complexes of RecA with single stranded DNA, in: Cozzarelli N. (Ed.) Alan R. Liss, *Mechanisms of DNA Replication and Recombination*, NY, 723–729.

Konola, J.T., Logan, K.M., Knight, K.L. (1994) Functional characterization of residues in the P-loop motif of the RecA protein ATP binding site, *J. Mol. Biol.* **237**, 20–34.

Konola, J.T., Nastri, H.G., Logan, K.M., et al. (1995) Mutations at Pro67 in the RecA protein P-loop motif differentially modify coprotease function and separate coprotease from recombination activities, *J. Biol. Chem.* **270**, 8411–8419.

Krejci, L., Song, B., Bussen, W., et al. (2002) Interaction with Rad51 is indispensable for recombination mediator function of Rad52, *J. Biol. Chem.* **277**, 40132–40141.

Kurumizaka, H., Aihara, H., Ikawa, S., et al. (1996) A possible role of the C-terminal domain of the RecA protein. A gateway model for double-stranded DNA binding, *J. Biol. Chem.* **271**, 33515–33524.

Kurumizaka, H., Aihara, H., Kagawa, W., et al. (1999) Human Rad51 amino acid residues required for Rad52 binding, *J. Mol. Biol.* **291**, 537–548.

Little, J.W. (1984) Autodigestion of lexA and phage lambda repressors, *Proc. Natl. Acad. Sci. U.S.A.* **81**, 1375–1379.

Little, J.W. (1993) LexA cleavage and other self-processing reactions, *J. Bacteriol.* **175**, 4943–4950.

Little, J.W., Mount, D.W. (1982) The SOS regulatory system of Escherichia coli, *Cell* **29**, 11–22.

Liu, Y., West, S.C. (2004) Happy hollidays: 40th anniversary of the holliday junction, *Nat. Rev. Mol. Cell Biol.* **5**, 937–944.

Liu, Y., Stasiak, A.Z., Masson, J.Y., et al. (2004) Conformational changes modulate the activity of human RAD51 protein, *J. Mol. Biol.* **337**, 817–827.

Logan, K.M., Knight, K.L. (1993) Mutagenesis of the P-loop motif in the ATP binding site of the RecA protein from Escherichia coli, *J. Mol. Biol.* **232**, 1048–1059.

Luo, Y., Pfuetzner, R.A., Mosimann, S., et al. (2001) Crystal structure of LexA: a conformational switch for regulation of self-cleavage, *Cell* **106**, 585–594.

Lusetti, S.L., Shaw, J.J., Cox, M.M. (2003) Magnesium ion-dependent activation of the RecA protein involves the C terminus, *J. Biol. Chem.* **278**, 16381–16388.

Lusetti, S.L., Drees, J.C., Stohl, E.A., et al. (2004) The DinI and RecX proteins are competing modulators of RecA function, *J. Biol. Chem.* **279**, 55073–55079.

Menetski, J.P., Kowalczykowski, S.C. (1985) Interaction of recA protein with single-stranded DNA. Quantitative aspects of binding affinity modulation by nucleotide cofactors, *J. Mol. Biol.* **181**, 281–295.

Menetski, J.P., Kowalczykowski, S.C. (1990) Biochemical properties of the Escherichia coli recA430 protein. Analysis of a mutation that affects the interaction of the ATP-recA protein complex with single-stranded DNA, *J. Mol. Biol.* **211**, 845–855.

Morimatsu, K., Kowalczykowski, S.C. (2003) RecFOR proteins load RecA protein onto gapped DNA to accelerate DNA strand exchange. A universal step of recombinational repair, *Mol. Cells* **11**, 1337–1347.

Nastri, H.G., Guzzo, A., Lange, C.S., et al. (1997) Mutational analysis of the RecA protein L1 region identifies this area as a probable part of the co-protease substrate binding site, *Mol. Microbiol.* **25**, 967–978.

Nguyen, T.T., Muench, K.A., Bryant, F.R. (1993) Inactivation of the recA protein by mutation of histidine 97 or lysine 248 at the subunit interface, *J. Biol. Chem.* **268**, 3107–3113.

Ogawa, T., Wabiko, H., Tsurimoto, T., et al. (1979) Characteristics of purified recA protein and the regulation of its synthesis in vivo, *Cold Spring Harb Symp. Quant. Biol.* **43**(Pt 2), 909–915.

Papavinasasundaram, K.G., Colston, M.J., Davis, E.O. (1998) Construction and complementation of a recA deletion mutant of Mycobacterium smegmatis reveals that the intein in Mycobacterium tuberculosis recA does not affect RecA function, *Mol. Microbiol.* **30**, 525–534.

Pellegrini, L., Venkitaraman, A. (2004) Emerging functions of BRCA2 in DNA recombination, *Trends Biochem. Sci.* **29**, 310–316.

Rehrauer, W.M., Kowalczykowski, S.C. (1993) Alteration of the nucleoside triphosphate (NTP) catalytic domain within Escherichia coli recA protein attenuates NTP hydrolysis but not joint molecule formation, *J. Biol. Chem.* **268**, 1292–1297.

Roca, A.I., Cox, M.M. (1997) RecA protein: structure, function, and role in recombinational DNA repair, *Prog. Nucleic Acid Res. Mol. Biol.* **56**, 129–223.

Sano, Y. (1993) Role of the recA-related gene adjacent to the recA gene in Pseudomonas aeruginosa, *J. Bacteriol.* **175**, 2451–2454.

Saraste, M., Sibbald, P.R., Wittinghofer, A. (1990) The P-loop–a common motif in ATP- and GTP-binding proteins, *Trends Biochem. Sci.* **15**, 430–434.

Sedgwick, S.G., Yarranton, G.T. (1982) Cloned truncated recA genes in E. coli. I. Effect on radiosensitivity and recA+ dependent processes, *Mol. Gen. Genet.* **185**, 93–98.

Selmane, T., Camadro, J.M., Conilleau, S., et al. (2004) Identification of the subunit-subunit interface of Xenopus Rad51.1 protein: similarity to RecA, *J. Mol. Biol.* **335**, 895–904.

Silver, M.S., Fersht, A.R. (1982) Direct observation of complexes formed between recA protein and a fluorescent single-stranded deoxyribonucleic acid derivative, *Biochemistry* **21**, 6066–6072.

Singleton, M.R., Dillingham, M.S., Gaudier, M., et al. (2004) Crystal structure of RecBCD enzyme reveals a machine for processing DNA breaks, *Nature* **432**, 187–193.

Skiba, M.C., Knight, K.L. (1994) Functionally important residues at a subunit interface site in the RecA protein from Escherichia coli, *J. Biol. Chem.* **269**, 3823–3828.

Sonoda, E., Takata, M., Yamashita, Y.M., et al. (2001) Homologous DNA recombination in vertebrate cells, *Proc. Natl. Acad. Sci. U.S.A.* **98**, 8388–8394.

Sprang, S.R. (1997a) G protein mechanisms: insights from structural analysis, *Annu. Rev. Biochem.* **66**, 639–678.

Sprang, S.R. (1997b) G proteins, effectors and GAPs: structure and mechanism, *Curr. Opin. Struct. Biol.* **7**, 849–856.

Stark, J.M., Hu, P., Pierce, A.J., et al. (2002) ATP hydrolysis by mammalian RAD51 has a key role during homology-directed DNA repair, *J. Biol. Chem.* **277**, 20185–20194.

Stasiak, A., Di Capua, E., Koller, T. (1981) Elongation of duplex DNA by recA protein, *J. Mol. Biol.* **151**, 557–564.

Stohl, E.A., Seifert, H.S. (2001) The recX gene potentiates homologous recombination in Neisseria gonorrhoeae, *Mol. Microbiol.* **40**, 1301–1310.

Stohl, E.A., Brockman, J.P., Burkle, K.L., et al. (2003) Escherichia coli RecX inhibits RecA recombinase and coprotease activities in vitro and in vivo, *J. Biol. Chem.* **278**, 2278–2285.

Story, R.M., Steitz, T.A. (1992) Structure of the recA protein-ADP complex, *Nature* **355**, 374–376.

Sugiyama, T., Kowalczykowski, S.C. (2002) Rad52 protein associates with replication protein A (RPA)-single-stranded DNA to accelerate Rad51-mediated displacement of RPA and presynaptic complex formation, *J. Biol. Chem.* **277**, 31663–31672.

Sukchawalit, R., Vattanaviboon, P., Utamapongchai, S., et al. (2001) Characterization of Xanthomonas oryzae pv. oryzae recX, a gene that is required for high-level expression of recA, *FEMS Microbiol. Lett.* **205**, 83–89.

Sutton, M.D., Smith, B.T., Godoy, V.G., et al. (2000) The SOS response: recent insights into umuDC-dependent mutagenesis and DNA damage tolerance, *Annu. Rev. Genet.* **34**, 479–497.

Thompson, L.H., Schild, D. (2001) Homologous recombinational repair of DNA ensures mammalian chromosome stability, *Mutat. Res.* **477**, 131–153.

Thompson, L.H., Schild, D. (2002) Recombinational DNA repair and human disease, *Mutat. Res.* **509**, 49–78.

VanLoock, M.S., Yu, X., Yang, S., et al. (2003a) Complexes of RecA with LexA and RecX differentiate between active and inactive RecA nucleoprotein filaments, *J. Mol. Biol.* **333**, 345–354.

Venkatesh, R., Ganesh, N., Guhan, N., et al. (2002) RecX protein abrogates ATP hydrolysis and strand exchange promoted by RecA: insights into negative regulation of homologous recombination, *Proc. Natl. Acad. Sci. U.S.A.* **99**, 12091–12096.

Vierling, S., Weber, T., Wohlleben, W., et al. (2000) Transcriptional and mutational analyses of the Streptomyces lividans recX gene and its interference with RecA activity, *J. Bacteriol.* **182**, 4005–4011.

Voloshin, O.N., Ramirez, B.E., Bax, A., et al. (2001) A model for the abrogation of the SOS response by an SOS protein: a negatively charged helix in DinI mimics DNA in its interaction with RecA, *Genes. Dev.* **15**, 415–427.

Voloshin, O.N., Wang, L., Camerini-Otero, R.D. (1996) Homologous DNA pairing promoted by a 20-amino acid peptide derived from RecA, *Science* **272**, 868–872.

Voloshin, O.N., Wang, L., Camerini-Otero, R.D. (2000) The homologous pairing domain of RecA also mediates the allosteric regulation of DNA binding and ATP hydrolysis: a remarkable concentration of functional residues, *J. Mol. Biol.* **303**, 709–720.

Walker, G.C. (1984) Mutagenesis of inducible responses to deoxyribonucleic acid damage in Escherischia coli, *Microbiol. Rev.* **48**, 60–93.

Walker, J.E., Saraste, M., Runswick, M.J., et al. (1982) Distantly related sequences in the alpha- and beta-subunits of ATP synthase, myosin, kinases and other ATP-requiring enzymes and a common nucleotide binding fold. *EMBO J.* **1**, 945–951.

Webb, B.L., Cox, M.M., Inman, R.B. (1999) ATP hydrolysis and DNA binding by the Escherichia coli RecF protein, *J. Biol. Chem.* **274**, 15367–15374.

Witkin, E.M. (1976) Ultraviolet mutagenesis and inducible DNA repair in Escherichia coli, *Bacteriol. Rev.* **40**, 869–907.

Wu, Y., He, Y., Moya, I.A., et al. (2004) Crystal structure of archaeal recombinase RADA: a snapshot of its extended conformation, *Mol. Cells* **15**, 423–435.

Xu, L., Marians, K.J. (2002) A dynamic RecA filament permits DNA polymerase-catalyzed extension of the invading strand in recombination intermediates, *J. Biol. Chem.* **277**, 14321–14328.

Xu, L., Marians, K.J. (2003) PriA mediates DNA replication pathway choice at recombination intermediates, *Mol. Cells* **11**, 817–826.

Yasuda, T., Morimatsu, K., Horii, T., et al. (1998) Inhibition of Escherichia coli RecA coprotease activities by DinI, *EMBO J.* **17**, 3207–3216.

Yoon-Robarts, M., Blouin, A.G., Bleker, S., et al. (2004) Residues within the B' motif are critical for DNA binding by the superfamily 3 helicase Rep40 of adeno-associated virus type 2, *J. Biol. Chem.* **279**, 50472–50481.

Yoshimasu, M., Aihara, H., Ito, Y., et al. (2003) An NMR study on the interaction of Escherichia coli DinI with RecA-ssDNA complexes, *Nucleic Acids Res.* **31**, 1735–1743.

Receptor Biochemistry

Tatsuya Haga[1] and Kimihiko Kameyama[2]
[1]*Gakushuin University, Tokyo, Japan*
[2]*National Institute of Advanced Industrial Science and Technology, Tsukuta, Japan*

1	**Classification of Receptors**	554
2	**Ion Channel-Coupled Receptors (ICCR)**	555
2.1	Nicotinic Acetylcholine Receptors	557
2.2	Glutamate Receptors	559
2.3	GABA$_A$ Receptors	564
2.4	Other Ion Channel-Coupled Receptors	564
3	**G Protein-Coupled Receptors (GPCR)**	565
3.1	Structure, Function, and Species of GPCRs	567
3.2	GTP-Binding Regulatory Protein (G-Protein)	572
3.3	Classification of GPCRs and Specificity of GPCR-G Protein Interaction	574
3.4	Regulator of G-Protein Signaling (RGS) and G Protein-coupled Receptor Kinase (GRK)	575
3.5	Neurotransmitter Receptors: Roles of ICCRs and GPCRs	577
4	**Protein Kinase-Coupled Receptors (PKCR)**	578
4.1	Growth Factor Receptors with Tyrosine Kinase Activity	579
4.1.1	Growth Factor Receptors	579
4.1.2	Neurotrophin Receptors	582
4.1.3	Activation of the Ras-MAPK Cascade	583
4.2	Cytokine Receptors Linked with Tyrosine Kinases	585
4.2.1	Cytokine Receptors	585
4.2.2	Activation of the JAK-STAT Pathway	585
4.3	TGF-β Receptors with Serine/Threonine Kinase Activity	588
4.4	Intracellular Signaling: Common Pathways Triggered by GPCRs and PKCRs	589

Encyclopedia of Molecular Cell Biology and Molecular Medicine, 2nd Edition. Volume 11
Edited by Robert A. Meyers.
Copyright © 2005 Wiley-VCH Verlag GmbH & Co. KGaA, Weinheim
ISBN: 3-527-30648-X

| 5 | Other Receptors 589 |

Acknowledgments 590

Bibliography 590
Books and Reviews 590
Primary Literature 591

Keywords

Agonists
Compounds that bind to and activate receptors and include endogenous ligAnds such as hormones and neurotransmitters, chemically synthesized compounds, and natural products like alkaloids.

Antagonists
Compounds that bind to but do not activate receptors, and hence inhibit the action of agonists.

Cytokines and Growth Factors
Proteins that influence proliferation, differentiation, maturation, and survival of cells. Cytokines include interleukins (IL), which are produced by leukocytes and control leukocyte function, and interferons (IFL) which are produced in response to virus infection. Growth factors are produced constitutively and act on nonhematopoietic cells, whereas cytokines are produced transiently and are mostly involved in hematopoiesis and immune or inflammatory responses. In a broader sense, cytokines include growth factors.

GRKs (G Protein–Coupled Receptor Kinases)
Protein serine/threonine kinases that phosphorylate activated forms of GPCRs (e.g. agonist-bound GPCRs and light-activated rhodopsin) specifically.

GTP-binding Regulatory Proteins (G-Proteins or Heterotrimeric G-Proteins)
Proteins that are activated by hormone- or neurotransmitter-bound receptors and activates effectors such as adenylate cyclase, phospholipase C, and ion channels. G-proteins are composed of an α-subunit with GTP-binding activity and $\beta\gamma$ subunits.

Hormones
Chemical substances that are released from one cell and modulate the function of other cells, which are usually located in remote organs.

MAP Kinase (MAPK)
Mitogen-activated protein kinase, a serine/threonine kinase in the signal transduction pathway that is triggered by activation of growth-factor receptors. Extracellular signal-regulated kinase (ERK) is another name. MAP kinase kinase (MAPKK or MEK (MAPK/ERK kinase)) phosphorylates both tyrosine and threonine residues in MAP kinase and activate MAP kinase. MAPKK kinase (MAP kinase kinase kinase, MAPKKK, or MEK kinase) phosphorylates serine residues in MAPKK and activates MAPKK.

Neurotransmitters or Transmitters
Chemical substances that are released from presynaptic terminals of an excited neuron and regulate the excitability of postsynaptic membranes of the other neuron.

Protein Serine/Threonine Kinases
Enzymes that phosphorylate serine or threonine residues in proteins, and include second messenger-activated protein kinases, such as cAMP-dependent protein kinase, protein kinase C, or Ca^{2+}/calmodulin-dependent protein kinase, and protein kinases in the signal transduction pathways such as MAP kinase.

Protein Tyrosine Kinases
Enzymes that phosphorylate tyrosine residues in proteins. Some tyrosine kinases are receptors for growth factors and are activated by binding of growth factors, whereas others are not receptors and are activated by cytokine receptors or other factors.

RGS Proteins (Regulator of G Protein–Signaling)
Proteins that accelerate GTPase activity of G-protein α-subunits and then terminate G-protein signaling.

Second Messengers
Compounds that are formed in the cells in response to the stimulation by hormones or neurotransmitters on the surface of cells (e.g. cyclic AMP, cyclic GMP, diacylglycerol, and inositol triphosphate).

SH2 Domains
The src homology domain 2 that is present in many proteins in the signal transduction pathways and interacts with domains containing phosphorylated tyrosine residues in growth-factor receptors or other proteins. A product of the src gene (sarcoma gene), src protein, contains an SH2, an SH3 (another src homology domain present in many other proteins), and a tyrosine kinase catalytic site.

Communications between cells are performed by cell–cell contact or by chemical substances like neurotransmitters, hormones, cytokines, or growth factors. These chemical substances are secreted by a cell and recognized by receptors in target cells. The binding of these ligands to their receptors initiates a series of chemical reactions. Some hormones such as steroids and thyroxin pass through cell membranes and interact with their receptors inside the cell. Most other hormones and all known neurotransmitters, cytokines, and growth factors cannot penetrate cell membranes and bind to their receptors, which are transmembrane glycoproteins, on the surface of cells. Membrane-bound receptors are classified into three major groups, ion channel-coupled receptors (ICCR), G protein–coupled receptors (GPCR), and protein kinase-coupled receptors (PKCR). In this section, we describe the molecular properties of membrane-bound receptors and intracellular signals triggered by activation of these receptors.

1
Classification of Receptors

Receptors have been named and classified by their endogenous ligands and subclassified by specific agonists and antagonists since the introduction of the receptor concept in the early 1900s by Langley. Molecular entities of receptors have been identified in the last quarter-century, which enables us to classify receptors on the basis of their structure and function. For example, receptors for acetylcholine have been classified into two types, nicotinic acetylcholine receptors and muscarinic acetylcholine receptors, on the basis of their interactions with specific agonists, nicotine and muscarine, respectively. Nicotinic and muscarinic acetylcholine receptors are further characterized by their interaction with specific antagonists, tubocurarine and atropine, respectively. It has been shown by means of purification of nicotinic and muscarinic receptors and their functional reconstitution into artificial membranes that these two receptors have completely distinct structural and functional characteristics, and that nicotinic and muscarinic receptors are, respectively, acetylcholine-gated ion channels and activators of G-proteins.

Cloning of complementary DNAs (cDNAs) encoding receptor proteins and deduction of their amino acid sequences have revealed that a number of receptors have structural and functional characteristics similar to the nicotinic receptor, constituting a superfamily of ion channel-coupled receptors (ICCRs), and a greater number of receptors have structural and functional similarity to the muscarinic receptor, constituting a superfamily of G protein–coupled receptors (GPCRs). Endogenous ligands for ICCRs are restricted to a small number of neurotransmitters. In contrast, endogenous ligands for GPCRs include almost all kinds of neurotransmitters, various hormones, autacoids, chemotactants, pheromones, and external stimulants (e.g. odorants, taste substances, and light). The structural characteristics common to all members of the two superfamilies are several hydrophobic regions, which are assumed to be transmembrane segments: GPCRs have seven such regions, and each subunit of ICCRs has two to four. Thus, GPCRs are also referred to as seven transmembrane receptors or serpentine receptors.

Several groups of receptors for cytokines and growth factors, on the other hand, are characterized by a single transmembrane segment per subunit. Some receptors with a single transmembrane segment are tyrosine kinases with the catalytic activity to phosphorylate tyrosine residues in the receptors themselves and/or related proteins, and others do not have enzyme activity but function as activators of related tyrosine kinases. Members of another group of receptors with a single transmembrane segment have serine/threonine kinase activity. Thus, these receptors with the single transmembrane segment are somehow coupled with protein kinases and are referred to here as protein kinase–coupled receptors (PKCRs).

Table 1 summarizes the characteristics of the three major groups of membrane receptors, ICCR, GPCR, and PKCR. ICCR, GPCR, and PKCR, respectively, can be discriminated from each other by endogenous ligands (neurotransmitters, hormones/neurotransmitters, cytokines/growth factors), coupled function (ion channels, G-protein activators, protein kinases/protein kinase activators), and by the order of response rates (milliseconds, seconds, hours).

Cloning of cDNAs encoding receptor proteins and determination of the human genome sequence has revealed that multiple kinds of receptors or receptor subtypes exist for a given endogenous ligand including most neurotransmitters, hormones, and cytokines. Furthermore, a number of putative receptors are called *orphan receptors* because their cDNAs or genes have been identified on the basis of their sequence homology to those of known receptors but their endogenous ligands remain unidentified. An extreme case is the odorant receptors: approximately five hundred human receptors are believed to exist but their ligands remain to be identified except for a few receptors. One of the future challenges will be to develop specific ligands for a given receptor with a known amino acid sequence.

2
Ion Channel-Coupled Receptors (ICCR)

Endogenous ligands for ICCRs are limited to several neurotransmitters, which are acetylcholine, glutamate (and possibly aspartate), γ-aminobutyric acid (GABA), glycine, serotonin (or 5-hydroxytryptamine, 5HT), and ATP (adenosine triphosphate). ICCRs are not known for catecholamines like adrenaline and dopamine, or for peptides.

ICCRs are oligomers composed of heterogeneous subunits, although some might be homooligomers. They incorporate ion-channel function in the oligomeric structure with a transmitter-gated pore in the center of the oligomers. Receptors for acetylcholine, glutamate, serotonin, and ATP are cation channels, which are permeable to both Na^+ and K^+ (and Ca^{2+} in some cases), and receptors for GABA and glycine are anion (chloride ion) channels. Neurotransmitters are released from the presynaptic terminal and bind to ICCRs in the postsynaptic membranes. The binding leads to the opening of ion channels, followed by depolarization and excitation of the postsynaptic membranes in the case of cation channels and by hyperpolarization and inhibition of excitation in the case of anion channels. In other words, the excitatory postsynaptic potential (EPSP) and inhibitory postsynaptic potential (IPSP)

Tab. 1 Characteristics of three major groups of membrane receptors.

	Ion channel-coupled receptors (ICCR)	G protein-coupled receptors (GPCR)	Protein kinase-coupled receptors (PKCR)
Endogenous ligands	Some neurotransmitters	Neurotransmitters, hormones, autacoids, and external stimulants (odorants, taste substances, and light)	Cytokines, including growth factors, neurotrophins, and TGF-β
Structure and number of transmembrane (TM) segments	Oligomers with a pore in the center Two to four TM segments per subunit	Monomers and some dimers Seven TM segments	Dimers or oligomers Single TM segment per subunit
Immediate function of receptors induced by agonist binding	Permeation of cations or anions through pores causing depolarization or hyperpolarization of membranes	Activation of G-proteins by facilitating GTP–GDP exchange; activation of G protein-coupled receptor kinases (GRKs)	Phosphorylation of tyrosine or serine/threonine residues in receptors or activation of tyrosine kinases
Cellular responses induced by activation of receptors	Excitation of neurons and other excitable cells or inhibition of excitation	Changes in excitability of neurons or other cells: regulation of protein function and gene expression	Regulation of gene expression, cell proliferation, and differentiation
Timescale of responses	Milliseconds	Seconds	Hours

are caused by opening of transmitter-gated cation and anion channels, respectively. The time course of these responses is rapid (within the order of milliseconds). Thus, ICCRs are specified for the purpose of rapid communication in the nervous system. This does not necessarily mean, however, that the role of ICCRs is restricted to the production of EPSP and IPSP. ICCRs may also take part in the stimulation or inhibition of neurotransmitter release through presynaptic receptors, in the long-term facilitation or depression of EPSP through a subtype of glutamate receptor (NMDA receptor), or in the action on glial cells.

2.1
Nicotinic Acetylcholine Receptors

The nicotinic acetylcholine receptor is the most extensively studied and best characterized among all the receptors. Abundant nicotinic receptors are present in the electric organs of the electric eel or *torpedo*, which are homologous to skeletal muscles. Nicotinic receptors were purified from the electric organs of *torpedo* by taking advantage of their abundance and their specific interaction with snake venoms such as bungarotoxin. Purified nicotinic receptors were found to be pentameric glycoproteins composed of four distinct subunits ($2 \times \alpha$, β, γ, δ) and to show acetylcholine-gated cation channel activity when reconstituted in lipid membranes. In 1982, cDNA encoding the α-subunit of nicotinic receptors in the electric organ of *torpedo* was cloned by using its partial amino acid sequence. Thereafter, cDNAs encoding each subunit of nicotinic receptors in the *torpedo* electric organ and in the calf muscle were cloned by using the sequence homology among subunits. Amino acid sequences of each subunit were deduced from corresponding nucleotide sequences of cDNAs. The finding of homology in amino acid sequences among the four subunits indicates derivation from a common ancestor.

Each subunit is composed of 437–501 amino acid residues and has four hydrophobic domains (M1–M4) with approximately 20 amino acid residues each. Extensive studies of nicotinic receptors have been carried out by using biochemical methods such as crosslinking of receptors with acetylcholine analogs or channel blockers, gene technology (e.g. mutagenesis), and physical methods (e.g. electron microscopy and electron diffraction of receptor microcrystals). Results are summarized in the following list, and an atomic model is shown in Fig. 1(A).

1. The amino-terminal portions of each subunit are located extracellularly and are bound with sugars. Two molecules of acetylcholine bind to the amino-terminal portions of the α-subunits.
2. The four hydrophobic domains (M1–M4) of each subunit form α-helices and transverse the plasma membrane forming a pore (ion channel) in the middle of 20 transmembrane segments.
3. The five M2 segments line the pore. Leucine and valine residues in the middle of the five M2 segments are conserved among all four subunits and thought to constitute the gate of the pore.
4. The binding of acetylcholine causes a small rotation of the extracellular parts of the five subunits, which triggers a conformational change in the M2 segments and moves the hydrophobic side chains away from the ion path. Thus, these hydrophobic residues function as a gate (Fig. 1(B)).

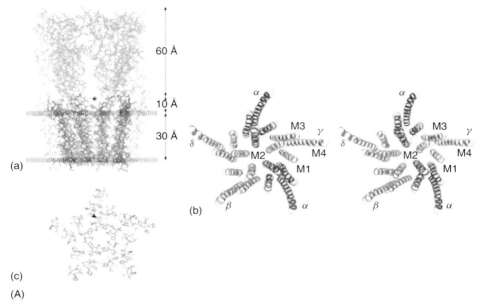

Fig. 1 Structural model of the nicotinic acetylcholine receptor (nAChR: an ion channel-coupled receptor, ICCR). (A) An atomic model with the closed pore was determined from electron images by electron cryo-microscopy of crystalline postsynaptic membranes and their reconstruction by optical diffraction and computation. The side (a) and cross-sectional (c) views show an extracellular ligand-binding domain (green), an inner, and the transmembrane segments composed of pore-facing M2 helices (blue) and lipid-facing M1, M3, M4 helices (red). (b) Stereo view of the pore, as seen from the extracellular face. (c) Cross-sectional view through the pentamer at the middle of the membrane. The pore is shaped by an inner ring of five α helices (M2), which curve radially to create a tapering path for ions, and an outer ring of 15 α-helices (M1, M3, M4), which coil around each other and shield the inner ring from the lipid layer. (B) A model for the gating mechanism. With binding of acetylcholine (ACh), two M2 α-helices of α-subunits rotate to opposite directions. This movement triggers the concerted movement of M2 α-helices of the other subunits resulting in the opening of the gate that consists of hydrophobic residues in the central part of five M2 segments. (From Miyazawa, A. et al. (2003) Structure and gating mechanism of the acetylcholine receptor pore, *Nature* **424**, 949–955) (see color plate p. xli).

5. The rate of channel opening is fast with a time constant of approximately 20 µs. The conductance of a single channel is 30 ps ($= 30 \times 10^{-12}$ A V^{-1}). This means that 2×10^4 cation molecules pass through the channel in an open time of 1 ms when the cation is derived by a potential of 100 mV: 30×10^{-12} (A V^{-1}) $\times$ 0.1 (V) $\times 1 \times 10^{-3}$ (s) $\times 1/96\,500$ (mol C^{-1}) $\times 6.2 \times 10^{23}$ (molecules mol^{-1}) $= 1.9 \times 10^4$ molecules.

6. The channel is permeable to both Na$^+$ and K$^+$ but not to Ca^{2+} or to anions. Anionic charges on both sides of the M2 segments contribute to the permeation of cations but not of anions.

7. In the continued presence of agonists, the channel closes with time constant of 50 to 100 ms at 20 °C (desensitization).

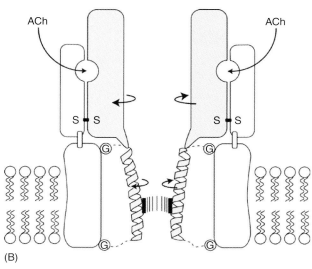

Fig. 1 (Continued)

Phosphorylation of nicotinic receptors by cAMP-dependent protein kinase, protein kinase C, or tyrosine kinase facilitates desensitization.

There are two kinds of nicotinic receptors in the skeletal muscle: the embryonic type ($\alpha_2\beta\gamma\delta$) and the adult type ($\alpha_2\beta\varepsilon\delta$). The denervation of skeletal muscle induces the expression of the γ-subunit and decreases the expression of the ε-subunit, which results in the conversion from the adult type to the embryonic type. The embryonic-type receptor is distributed widely in the whole area of skeletal muscle membranes, whereas the adult-type receptor is localized in the postsynaptic membranes. This specific distribution is regulated by distinct expression of *nAChR* genes in the nucleus adjacent to the synapse.

Different kinds of nicotinic receptor subtypes are present in the autonomous ganglia and the brain. Seven kinds of α-subunits (α_2–α_9, no α_8 in human), which are homologous to the skeletal α-subunit (α_1), and three kinds of β-subunits (β_2–β_4) have been identified (see Table 2). Functional receptors are thought to be pentamers like (α_4)$_2$(β_3)$_3$. The effect of bungarotoxin on brain nicotinic receptors is weak or absent, except that α_7 and α_9 bind the toxin. Some brain nicotinic receptors are present in presynaptic membranes and regulate the release of glutamate. Ca^{2+} is reported to permeate through the α_7 homooligomer, while Ca^{2+} does not permeate through other nicotinic receptors including the skeletal muscle type.

2.2
Glutamate Receptors

Glutamate receptors are classified into ionotropic receptors (ion channel-coupled receptors, ICCRs) and metabotropic receptors (G protein–coupled receptors, GPCRs). Ionotropic receptors are divided by the use of a specific agonist, N-methyl D-aspartic acid (NMDA), into NMDA and non-NMDA types. Non-NMDA types are

Tab. 2 Subunits, genes, and gene loci of human ion channel-coupled receptors (ICCRs).

Subunit	Gene	Gene locus	Subunit	Gene	Gene locus	Subunit	Gene	Gene locus
Nicotinic acetylcholine receptor			GABA$_A$ receptor			Glutamate receptor (AMPA Type)		
nAChRα1	CHRNA1	2q24-q32	GABA$_A$Rα1	GABRA1	5q34–q35	GluR1	GRIA1	5q31.1
nAChRα2	CHRNA2	8q21	GABA$_A$Rα2	GABRA2	4p12	GluR2	GRIA2	4q32-q33
nAChRα3	CHRNA3	15q24	GABA$_A$Rα3	GABRA3	Xq28	GluR3	GRIA3	Xq25-q26
nAChRα4	CHRNA4	20q13.2-q13.3	GABA$_A$Rα4	GABRA4	4p12	GluR4	GRIA4	11q22
nAChRα5	CHRNA5	15q24	GABA$_A$Rα5	GABRA5	15q11.2-q12	(Kainate type)		
nAChRα6	CHRNA6	8p11.21	GABA$_A$Rα6	GABRA6	5q34	GluR5	GRIK1	21q22.11
nAChRα7	CHRNA7	15q14	GABA$_A$Rβ1	GABRB1	4p12	GluR6	GRIK2	6q16.3-q21
nAChRα9	CHRNA9	4p14	GABA$_A$Rβ2	GABRB2	5q34	GluR7	GRIK3	1p34-p33
nAChRα10	CHRNA10	11p15.5	GABA$_A$Rβ3	GABRB3	15q11.2-q12	KA1	GRIK4	11q22.3
nAChRβ1	CHRNB1	17p13.1	GABA$_A$Rδ	GABRD	1p36.3	KA2	GRIK5	19q13.2
nAChRβ2	CHRNB2	1q21.3	GABA$_A$Rε	GABRE	Xq28	(NMDA type)		
nAChRβ3	CHRNB3	8p11.2	GABA$_A$Rγ1	GABRG1	4p12	NR1	GRIN1	9q34.3
nAChRβ4	CHRNB4	15q24	GABA$_A$Rγ2	GABRG2	5q31.1-q33.1	NR2A	GRIN2A	16p13.2
nAChRδ	CHRND	2q33-q34	GABA$_A$Rγ3	GABRG3	15q11-q13	NR2B	GRIN2B	12p12
nAChRε	CHRNE	17p13-p12	GABA$_A$Rπ	GABRP	5q33-q34	NR2C	GRIN2C	17q25

Receptor	Gene	Locus	Receptor	Gene	Locus	Receptor	Gene	Locus
nAChRγ	CHRNG	2q33–34	GABA$_A$Rθ	GABRQ	Xq28	NR2D	GRIN2D	19q13.1-qter
			GABA$_A$Rρ1	GABRR1	6q13-q16.3	NR3A	GRIN3A	9q31.1
			GABA$_A$Rρ2	GABRR2	6q13-q16.3	NR3B	GRIN3B	19p13.3
Glutamate receptor (δ type)			**Glycine receptor**			**Purinergic receptor (P2X type)**		
GluRδ1	GRID1	10q22	GlyRα1	GLRA1	5q32	P2X1	P2RX1	17p13.3
GluRδ2	GRID2	4q22	GlyRα2	GLRA2	Xp22.1-p21.3	P2X2	P2RX2	12
			GlyRα3	GLRA3	4q33-q34	P2X3	P2RX3	4q33-q34
			GlyRβ	GLRB	4q31.3	P2X4	P2RX4	4q31.3
Serotonin receptor (5-HT$_3$ type)						P2X5	P2RX5	17p13.3
5HTR3A	HTR3A	11q23.1-q23.2				P2X7	P2RX7	12q24
5HTR3B	HTR3B	11q23.1				P2X1like	P2RXL1	22q11.21

Receptors are classified by their neurotransmitters and types, which are written in **bold** face. Most commonly used names are listed for receptors and receptor subunits. Genes for receptor subunits are expressed by names approved by the gene nomenclature committee in HUGO (Human Genome Organization). Gene loci are mapped on chromosomes of the human genome.

further subdivided into AMPA (α-amino-3-hydroxy-5-methyl-4-isoxalone propionic acid) and kainic acid (KA) types, where AMPA and KA are specific agonists for the respective subtypes. A number of different genes have been identified for subunits of each glutamate receptor (seven genes for NMDA receptor subunits, four for AMPA receptors, five for KA receptors)(see Table 2). There are also two genes encoding δ-type glutamate receptors, which have significant homology to other ionotropic glutamate receptors but whose functions remain to be identified. Some of the genes encoding glutamate receptor subunits are reported to have splicing variants: for example, eight kinds of polypeptides derived from the NMDA receptor 1 (*NMDAR1*) gene are expressed in the brain. Each glutamate receptor is believed to be a heterotetramer composed of the same type of receptor subunits.

The molecular properties of glutamate receptors are not as well understood as those of the nicotinic receptors, although the fundamental properties are assumed to be similar. Each subunit of AMPA, KA, and NMDA receptors has four hydrophobic domains (M1–M4), like the nicotinic receptors and other ICCRs, and there are some similarities in these domains between glutamate receptors and other ICCRs. Each glutamate receptor subunit, however, is thought to have three, rather than four, transmembrane domains, where M1, M3, and M4 but not M2 are the transmembrane segments. In addition, glutamate receptors are thought to be composed of four, but not five, subunits in contrast with nicotinic receptors. Thus, the extracellular part of glutamate receptors is thought to be composed of four amino-terminal parts and four loops between the M3 and M4 transmembrane segments, and to constitute the glutamate-binding site. Actually the protein complex corresponding to the extracellular domains of an AMPA receptor has been expressed, purified, crystallized, and used for X-ray analysis. The X-ray analysis showed an atomic structure composed of two domains with a clam-shaped structure and with a glutamate-binding site in the cleft of the clam shell. Four clams with eight halves of shells are assembled, and the binding of multiple glutamate molecules is supposed to cause a conformational change so as to close the shells, which should be somehow transmitted to the transmembrane domains. Recently, an atomic structure of the extracellular domain of an NMDA receptor subunit, NR1, was determined and shown to be clam shaped, as seen with the AMPA receptor. NMDA receptors have to bind both glutamate and glycine (or D-serine) to be activated, where the NR1 subunit binds to glycine (or D-serine) and the NR2 subunit to glutamate. The cleft of the clamshells in NR1 was shown to be open in the presence of antagonist, and closed in the presence of glycine or D-serine.

The M2 segment is thought to be inserted into the membrane from the intracellular side forming a loop. A certain amino acid residue in the M2 segment is critical for the selectivity of permeable ions. The genetic code for the amino acid is asparagine for NMDA receptors and glutamine for AMPA receptors. This site, however, is converted to arginine by RNA editing for the GluR2 subunit of AMPA receptors. NMDA receptors are permeable to Na^+, K^+, and Ca^{2+} ions. AMPA receptors comprised of the subunits with glutamine in this site are permeable to Na^+, K^+, and also Ca^{2+} ions with inwardly rectified voltage dependency in the current. On the other hand, AMPA receptors including the edited GluR2 subunit with the arginine residue are not permeable to Ca^{2+}

ions, and are permeable only to Na^+ and K^+ ions with linear voltage dependency in the current. The ratio between edited and nonedited GluR2 mRNA increases after birth, and almost 100% of GluR2 mRNA is edited in the adult brain.

Non-NMDA glutamate receptors play a major role in the excitatory synaptic transmission in the central nervous system. NMDA receptors are different from non-NMDA receptors in that NMDA receptors are usually not active, as the ion permeability of NMDA receptors is inhibited by Mg^{2+} ions. The inhibitory effect of Mg^{2+} is suppressed by depolarization. The NMDA receptor channel opening is also regulated by glycine or D-serine. Thus, the NMDA receptor opens only when both glutamate and glycine (or D-serine) are present and the cell membrane containing NMDA receptors has been depolarized.

Phenomena referred to as long-term potentiation (LTP) and long-term depression (LTD) have been studied in detail, particularly in the hippocampus, and are considered to represent cellular models for memory formation. LTP is defined as a sustained increase in the efficiency of synaptic transmission, and is caused by brief trains of high-frequency stimulation (e.g. 30–100 stimuli/s with duration of 5–10 s, that is, consecutive 150–1000 stimuli). On the other hand, LTD is defined as a sustained decrease in the efficiency of synaptic transmission, and is caused by low-frequency stimulation (1–3 stimuli/s with duration of 5–15 min, that is, consecutive 300–1000 stimuli). NMDA receptors are involved in both LTP and LTD in the hippocampal CA1 area. High-frequency stimuli induce depolarization of postsynaptic neurons through the AMPA receptor, which is followed by opening of NMDA receptors, Ca^{2+} entry, and activation of Ca^{2+}-calmodulin-dependent protein kinase II (CaMKII). Autophosphorylation of CaMKII makes it constitutively active, and the persistent active form of CaMKII binds to the NMDA receptor NR2B subunit in postsynaptic membranes. High-frequency stimulation causes the translocation of the AMPA receptor containing the GluR1 subunit from intracellular organelles into synaptic membranes, which results in increases of the excitatory postsynaptic current. CaMKII activation and AMPA receptor translocation are thought to be somehow related to the formation of LTP. On the other hand, low-frequency stimuli induce mild activation of NMDA receptors followed by low-level elevation of Ca^{2+} concentrations, which causes the activation of protein phosphatases. Phosphatase activity induces the internalization of AMPA receptors into intracellular vesicle pools, which decreases the excitatory postsynaptic current. Thus, LTP and LTD, which are measured as the increase and decrease in the excitatory postsynaptic current, respectively, are linked with recruiting and internalization of the AMPA receptor on the synaptic surface and also with the activation of CaMKII and phosphatases, but the exact molecular mechanisms are yet to be identified.

Glutamate receptors are highly enriched in postsynaptic membranes and are anchored by so-called scaffolding proteins. Excitatory synapses in the central nervous system are characterized by the electron dense regions in postsynaptic sites, which are known as postsynaptic density (PSD). One major component in postsynaptic densities is PSD-95, which binds to NMDA receptor NR2 subunits and some voltage-gated potassium channels. The carboxy termini of the latter proteins have the consensus sequence of Thr/Ser-X-Val (X is any amino acid residue), to which the so-called PDZ domain in PSD-95 binds. As there are three PDZ domains in a single PSD-95 molecule, coexpression of PSD-95 with NMDA receptors in cultured cells induces their clustering. Proteins referred to as GRIP (glutamate receptor interacting protein) also have PDZ domains and bind to AMPA receptors. These molecules function as scaffolding proteins contributing to assembling receptors, ion channels, and enzymes in postsynaptic sites. These molecules bind directly or indirectly to motor proteins of the kinesin family, which are supposed to transport receptor molecules from the cell body to the dendrite.

2.3
GABA$_A$ Receptors

There are two kinds of GABA receptors, GABA$_A$ and GABA$_B$, which are ion channel-coupled receptors (ICCR) and G protein–coupled receptors (GPCR), respectively. Eighteen kinds of GABA$_A$ receptor subunits have been identified in the human genome (six α, three β, three γ, one ε, one δ, one π, one θ, two ρ). Various oligomers with different combinations of these subunits are thought to exist *in situ*. The fundamental structure of GABA$_A$ receptors is thought to be similar to that of nicotinic receptors, that is, a pentamer of subunits with four transmembrane segments. GABA$_A$ receptors, however, are anion channels, in contrast with nicotinic receptors, and there are more positive amino acid residues on both sides of the M2 segments of GABA$_A$ receptor subunits compared with the M2 segments of nicotinic receptors. These positive charges are considered to help attract chloride ions before their permeation through the pores. Binding of GABA to GABA$_A$ receptors and the subsequent opening of chloride channels causes the hyperpolarization of postsynaptic membranes and thereby the suppression of excitation.

GABA$_A$ receptors take a major role in the inhibitory synaptic transmission in the brain. Different kinds of drugs are known to interact with GABA$_A$ receptors. Benzodiazepines and related compounds are tranquilizers and facilitate the action of GABA by increasing the affinity of the receptor for GABA. Barbiturates, which induce sleep, also increase the action of GABA. Picrotoxin, on the other hand, blocks the channel and then acts as an inducer of seizures. The effect of drugs appears to be subtype-specific. GABA$_A$ receptors that do not contain γ-subunits are not affected by benzodiazepines. This suggests the possibility that various drugs with different subtype specificity might be developed in the future and be used for different kinds of neurological or psychiatric diseases.

2.4
Other Ion Channel-Coupled Receptors

Glycine receptors are chloride channels and play a role in the inhibitory synaptic transmission mostly in the spinal cord. There are four subunit genes in human (three α, one β). Glycine receptors are thought to have a structure similar to nicotinic receptors and GABA$_A$ receptors.

A number of receptors for serotonin (or 5-hydroxytryptamine, 5HT) are GPCRs, and only serotonin 5HT$_3$ receptors are cation channels. The 5HT$_3$ receptors are expressed in forebrain, hippocampus, brainstem, and spinal cord, although the expression is restricted to some populations of neurons. At first only one gene corresponding to 5HT$_{3A}$ was identified, and the channel properties of the 5HT$_{3A}$ homopentamer were found to be different from those of native brain 5HT$_3$ receptors. After homology search of the human genome sequence, another gene corresponding to 5HT$_{3B}$ was found, which had significant homology to the *5HT$_{3A}$* gene. When both 5HT$_{3A}$ and 5HT$_{3B}$ cDNAs were coexpressed, the expressed receptor showed channel properties similar to those of native brain receptors.

Purinergic receptors are divided into two groups, P2X receptors and P2Y receptors, which are ion channel-coupled receptors (ICCR) and G protein–coupled receptors (GPCR), respectively. There are at least seven different subunit genes for P2X receptors in the human genome. These subunits are clearly distinct from subunits of either nicotinic or glutamate receptors in that P2X subunits have only two transmembrane domains. The loop between the two transmembrane domains is located extracellularly and is assumed to be the ATP-binding domain because it has an ATP-binding motif and ATP acts on the channel from the extracellular phase. P2X receptors appear to form either homo- or heterooligomers, and some receptors are suggested to be trimers although this remains to be proven in a more direct manner. P2X receptors are ATP-gated cation channels with equal permeability to Na$^+$ and K$^+$ and a significant permeability to Ca^{2+}. P2X receptors are expressed in post- and presynaptic membranes in the central nervous system, and may contribute to ATP-driven excitatory postsynaptic current (EPSC) although the ATP-driven EPSC is relatively small and less abundant as compared with that driven by glutamate receptors. P2X3 is expressed in some sensory neurons, and is thought to serve as a pain sensor. Recently, P2X4 in microglia is reported to play an important role in evoking and maintaining a kind of neuropathic pain.

3
G Protein-Coupled Receptors (GPCR)

Endogenous ligands of GPCRs include all neurotransmitters except glycine, most hormones and autacoids, several chemotactic factors, and external stimulants such as pheromones, odorants, and taste substances. Chemically, these ligands are diverse and include amino acids, peptides, proteins, amines, nucleosides, nucleotides, lipids such as prostaglandins and leukotrienes, Ca^{2+} ion, H$^+$ ion, and various organic compounds acting as odorants. Rhodopsin and optic pigments, which are receptors for light, are also GPCRs (Tables 3 and 4). The number of GPCR gene species in the human genome is estimated to be 800 to 900, approximately half of which are odorant receptors. Three to four hundred GPCRs are known to have endogenous ligands, but more than one hundred GPCRs are considered to be orphan receptors for which the endogenous ligands remain to be identified.

Each GPCR was thought to be a homogeneous protein and to exist as a monomer. However, recently GABA$_B$ and metabotropic glutamate receptors (mGluRs) have been shown to exist as dimers. Some GPCRs in the rhodopsin family, which is the major group of GPCRs, are also reported to form homo- or heterooligomers with intact functions. However, it is not clear yet whether oligomerization is a prerequisite for the function of most GPCRs or whether it is required for only a small number of GPCRs.

The function of GPCRs is to activate heterotrimeric G-proteins ($\alpha_{GDP}\beta\gamma$), that is, to stimulate the dissociation of GDP from G-proteins and thereby facilitate the binding of GTP to G-proteins followed by dissociation into α_{GTP} and $\beta\gamma$ subunits (see Fig. 2). Activated G-proteins, that is α_{GTP} and $\beta\gamma$ subunits, stimulate or inhibit enzymes that synthesize or break down second messengers such as cAMP, diacylglycerol and inositol 1,4,5 triphosphate (IP_3). Thus, the information of the presence of hormones, neurotransmitters, odorants or other molecules outside of cells is converted to changes in concentrations of second messengers, which affect activities of protein serine/threonine kinases; for example, cAMP and diacylglycerol activate cAMP-dependent protein kinase (A kinase) and protein kinase C (C kinase), respectively. In addition, IP_3 interacts with and opens the IP_3-gated Ca^{2+}-channel in intracellular vesicles causing an increase in intracellular Ca^{2+} concentrations, which may activate Ca^{2+}-calmodulin-dependent protein kinases. These protein kinases affect activities of different kinds of proteins including enzymes, ion channels, and so on. In addition, expression of some genes is regulated through these protein kinases; for example, CREB (cAMP responsive element (CRE) binding protein) is phosphorylated by cAMP-dependent protein kinase. Phosphorylated CREB, together with CREB binding protein, stimulates the transcription of genes that reside downstream of CRE with a sequence of TGACGTCA.

As for the relation between GPCR and ion channels, there are different types depending on the species of ion channels; some ion channels like the inward-rectified K^+ channel in the heart are directly regulated by G-protein $\beta\gamma$ subunits, cation channels in visual or olfactory cells are regulated by cGMP or cAMP, and many other ion channels are thought to be directly or indirectly regulated by phosphorylation by second messenger-dependent protein kinases. In any case, the time courses of their opening or closing after ligand binding to the receptor are much

---

Fig. 2 Structures of rhodopsin and a heterotrimeric G-protein. (A) An atomic model for rhodopsin was obtained by X-ray analysis of the rhodopsin crystal. (a) A stereoview in parallel to the plane of the membrane. The upper and lower parts of this model correspond to the intra- and extracellular domains respectively. There are seven transmembrane α-helixes (I-VII), another α-helix in the C-terminal region (VIII), 11-cis-retinal (yellow) in the transmembrane region, and two β-sheets (1–2 and 3–4) and sugars in the N-terminal region. View of the receptor from the cytoplasmic (b) and intradiscal (extracellular) side (c) of the membrane. (From Palczewski, K. et al. (2000) Crystal structure of rhodopsin: A G protein-coupled receptor, *Science* **289**, 739–745). (B) Receptor-G protein interaction. (a) A structural model for the cytoplasmic surface of the M_1 muscarinic receptor. Residues important for interaction with G protein are shown in red, magenta (i3 loop), and yellow. The arrows indicates the outward movement of TMb and increased mobility of TM7 and helix 8, which are thought to accompany receptor activation. (b) A structural model for the interaction between G protein-coupled receptor (GPCR) and G-protein $\alpha\beta\gamma$ trimer. Structures are taken from crystal structures of rhodopsin and a chimera of transducin (Gt) and Gi. The N- and C-termini of the transducin α-subunit and the $\alpha4$-$\beta6$ loop sequences, which are thought to be involved in the interaction with GPCR, are shown in red. The switch-II helix in the G-protein α-subunit is bound to the β-subunit when the α-subunit is bound to GDP, and is subject to conformational change and is released from the β-subunit when the α-subunit is bound with GTP. (From Lu, Z.L. et al. (2002) Seven-transmembrane receptors: crystals clarify, *Trends Pharmacol. Sci.* **23**, 140–146) (see color plate p. xlii).

slower (10 ms–1 min) as compared to IC-CRs (ms).

3.1
Structure, Function, and Species of GPCRs

The amino-terminal domain of GPCRs is considered to be located extracellularly, and the carboxy-terminal tail in the cytoplasm. This assumption is consistent with the facts that the consensus sequence for attachment of sugars (Asn-X-Ser/Thr) is present in the amino-terminal tail for almost all members of this family and that phosphorylation sites are in the carboxy-terminal tail for some GPCRs. The amino and carboxy-terminal tails have a variable length ranging from several tens to several hundreds of amino acid residues. Seven hydrophobic regions (20–25 amino acid residues each) are assumed to transverse the plasma membrane forming α-helices. The three intracellular and extracellular loops are generally short (10–30 amino acid residues) except that some receptors have a long third intracellular loop. Recently, the crystal structure of rhodopsin was determined at atomic resolution (see Fig. 2). This structure confirmed the above assumption

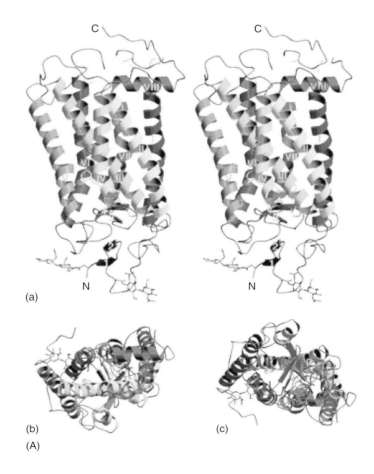

(a)

(b) (c)

(A)

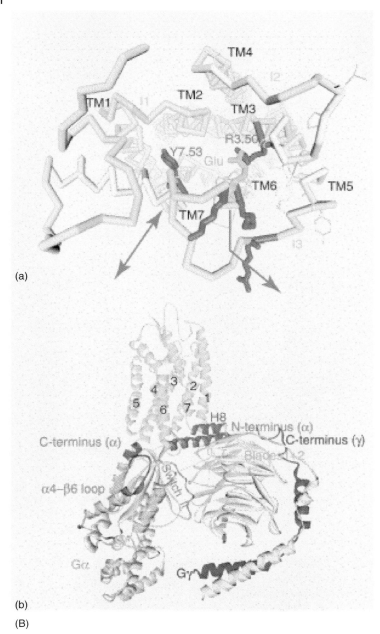

Fig. 2 (Continued)

Tab. 3 Neurotransmitters and receptors.

Neurotransmitters	Ion channel-coupled receptors (ICCR) (Species of subunits)	G protein–coupled receptors (GPCR) (Species of coupled G proteins)
Acetylcholine	Nicotinic Muscle type($\alpha 1_2 \beta 1 \gamma \delta$ or $\alpha 1_2 \beta 1 \varepsilon \delta$) Neuronal type (pentamer of α_{2-9} and β_{2-4})	Muscarinic $M_{1,3,5}$ (G_q) $M_{2,4}$ (G_i/G_o)
Glutamate	Ionotropic AMPA (GluR1-4) Kainate (GluR5-7, KA1,2) NMDA (NMDAR1, 2A-2D, 3A-3B)	Metabotropic mGluR1,5 (G_q) mGluR2-4, mGluR6-8 (G_i/G_o)
Serotonin (5HT)	5HT$_3$ (5HT$_{3A,3B}$)	5HT$_{2A-2C}$ (G_q) 5HT$_{1A,1B,1D-1F}$ (G_i/G_o) 5HT$_{4,6,7}$ (G_s)
ATP	P2X$_{1-5,7}$	P2Y$_1$ (G_i/G_o) P2Y$_2$ (G_s)
GABA	GABA$_A$ (six α, four β, three γ, one δ, one ε, one π, one θ, two ρ)	GABA$_B$ (GABA$_B$R$_1$, GABA$_B$R$_2$) (G_i/G_o)
Glycine	Glycine (three α, one β)	
Noradrenaline (Adrenaline)		$\alpha_{1A,1B,1D}$ (G_q) $\alpha_{2A,2B,2C}$ (G_i/G_o) $\beta_{1,2,3}$ (G_s)
Dopamine		D_{2-4} (G_i/G_o) D_1, D_5 (G_s)
Histamine		H_1 (G_q) H_2 (G_s) H_3 (G_i/G_o)
Opioid		μ, δ, κ, (G_i/G_o)
Tachykinin		NK1-3 (G_q)
Adenosine		A_{1-3} (G_i/G_o) $A_{2A,2B}$ (G_s)

and demonstrated that rhodopsin actually has seven transmembrane α-helices with amino- and carboxy-terminal tails in the outside and inside of the cells respectively. In addition, it was revealed that rhodopsin has an eighth α-helix between the seventh transmembrane α-helix and the palmitoylated cysteine residues in the carboxy terminus, and a disulfide bond between cysteine residues in the third transmembrane segment and in the loop between the fourth and fifth transmembrane segments.

On the basis of sequence similarities, mammalian GPCRs are divided into three major groups. Group A (rhodopsin family) is the largest one and includes rhodopsin and odorant receptors, receptors for neurotransmitters like acetylcholine and other amines, and receptors for peptides and other ligands such as nucleosides, nucleotides, and lipids. Approximately 20 amino acid residues in the transmembrane segments are highly conserved among members of this group, and most members have cysteine residue(s) at conserved

Tab. 4 G protein-coupled receptors (GPCRs) for hormones, exogenous stimulants, and other ligands except typical neurotransmitters.[a]

Ligands	Receptor subtypes	G-proteins	Effectors	Second messengers and their downstreams
Group 1[b]				
CGRP (calcitonin gene-related peptide)	CGRP	G_s	Adenylate cyclase activation	cAMP increase Activation of cAMP-dependent protein kinase
Vasopressin/oxytocin	V2			
Prostanoid	DP, IP, EP2			
Glucagon				
Thyrotropin				
Lutropin-gonadotropin				
Odorants	300–500 species	G_{olf}		
Group 2				
Cannabinoid	CB1	G_{i1}, G_{i2}, G_{i3}	Adenylate cyclase inhibition	cAMP decrease
Somatostatin	SSTR1,2	G_o		
VIP (Vasoactive intestinal peptide)	VIP1,2,3		K^+ channel open Ca^{2+} channel close	Hyperpolarization Ca^{2+} decrease
Chemokines	MIP1a, MCP1, IL8A,B			
Neuropeptide Y	Y1			
Light	Rhodopsin cone pigments (blue, red, green)	G_{t1} G_{t2}	cGMP – phosphodiesterase activation	cGMP decrease Hyperpolarization
Taste	20–30 species	G_g	cAMP – phosphodiesterase activation	cAMP Decrease Depolarization

Group 3			
PAF (platelet-activating factor)		G_q	Phosphatidylinositol-specific phospholipase C β types	Increase in diacylglycerol (DG) and IP_3
Thromboxane		G_{11}		Activation of protein kinase C
Prostanoid	FP, TP, EP1,3	G_{14}		
Bombesin	B1,2	G_{15}		Increase of Ca^{2+}, and activation of Ca^{2+}-calmodulin kinase
Endothelin	ET(A), (B)			
Angiotensin	AT1			
Vasopressin/oxytocin	V1A,1B, OT			
Cholecystokinin/gastrin	CCK(A), (B)			
Bradykinin	BB1,2			
Group 4				
Thrombin		G_{12}	ρ GEF activation	Activation of ρ and then ρ-kinase
Thromboxan		G_{13}		Reorganization of cytoskeleton and change in cell motility

[a] GPCRs for typical neurotransmitters are shown in Table 3. It is difficult in some cases to distinguish between neurotransmitters and nonneurotransmitters (hormones, autacoids, or neuromodulators), and the distribution of some compounds between Table 3 and Table 4 is more or less arbitrary. Noradrenaline and adrenaline may be taken as both neurotransmitters and hormones, and serotonin; histamine, and adenosine are both neurotransmitters and autacoids. Somatostatin, VIP, and other peptides are known to be released from neurons and to act on neurons although they are listed in Table 4.
[b] Receptors are grouped by their G-protein coupling properties.

positions in the carboxy termini, homologous to the ones that are palmitoylated in rhodopsin. Group B (secretin receptor family) consists of receptors for rather large peptides like secretin, glucagon, and so on, and these receptors have a large amino-terminal domain with conserved disulfide bonds. Group C (glutamate receptor family) consists of metabotropic glutamate receptors (mGluR), $GABA_B$ receptors, and putative taste receptors. These receptors have large amino- and carboxy-terminal domains and short and highly conserved third intracellular loops. Cysteine residues in the extracellular loops two and three, which are known to form a disulfide bond in rhodopsin, are conserved in most GPCRs. The ligand-binding pocket is formed by transmembrane domains in the case of Group A receptors. On the other hand, Group B and C receptors contain the ligand-binding domain in the large amino-terminal segment. The glutamate-binding domain of mGluRs is homologous to the amino acid binding domain of bacterial periplasmic binding proteins.

GPCRs were believed to exist as monomers, but various lines of evidence have indicated that class C receptors form dimers and function as dimers. An atomic structure of the ligand-binding domain of the mGluR1 indicated that the ligand-binding domain consists of a homodimer, each subunit of which has two globular domains with a cleft for the ligand. In the case of the $GABA_B$ receptor, two kinds of $GABA_B$ receptors, $GABA_BR1$ and $GABA_BR2$, were shown to have a higher affinity for GABA when they were coexpressed in cultured cells than when each receptor was expressed separately, indicating the formation of a heterooligomer. Native $GABA_B$ receptors are thought to be heterooligomers because they have high affinity for GABA, and $GABA_BR1$ and $GABA_BR2$ are coimmunoprecipitated from extracts solubilized from brain membranes. Taste receptors with a long amino terminus were also shown to have a high affinity for amino acids or nucleosides when heterooligomers were formed. A number of examples were reported indicating that class A receptors also may form homo- or heterooligomers and that the formation of oligomers may affect the affinity and specificity for ligands and receptor trafficking. Oligomerization with components other than GPCRs was also reported: for example, calcitonin receptor-like receptor interacts with a protein with a single transmembrane segment called *RAMP* (receptor activity–modifying protein), and the dopamine D5 receptor interacts with the $GABA_A$ receptor. It is clear that some class A receptors may form and function as oligomers, but it remains to be clarified whether their oligomerization is a general phenomenon under physiological conditions.

The binding sites in GPCRs for G-proteins are thought to be the second and third intracellular loops and the carboxy-terminal tail. Following ligand binding, GPCRs are supposed to undergo conformational changes in the ligand-binding site at first and then, secondarily, in the G protein–binding sites. The molecular details involved in these conformational changes, however, remain to be clarified although extensive modeling studies have been carried out with the rhodopsin system.

3.2
GTP-Binding Regulatory Protein (G-Protein)

G-proteins are composed of three subunits, $\alpha\beta\gamma$, the molecular weights of which are approximately 40, 35, and 5–10 kDa, respectively. The crystal structure of the $\alpha\beta\gamma$ trimer was determined (Fig. 2). G-proteins are not transmembrane proteins but are bound to cell membranes through prenyl groups and fatty acids attached to the carboxy terminus of the γ-subunit and the amino terminus of the α-subunit, respectively. The α-subunit has a binding site for GTP or GDP and harbors GTPase activity. The β-subunit has seven repeated domains forming a tertiary structure like a seven-bladed propeller, and the γ-subunit is tightly bound with the amino-terminal part of the β-subunit. G-proteins bound with GTP dissociate into α_{GTP} and $\beta\gamma$ subunits, and the α_{GTP} and $\beta\gamma$ subunits activate or inhibit effectors. The activity of the α_{GTP} and $\beta\gamma$ subunits is terminated by hydrolysis of GTP and the subsequent formation of $\alpha_{GDP}\beta\gamma$ trimer. The $\alpha_{GDP}\beta\gamma$ trimer is converted into the α_{GTP} and $\beta\gamma$ subunits by action of the agonist-bound receptor. The agonist-bound receptor (aR) facilitates release of GDP from G-proteins and forms a complex with guanine nucleotide-free G-proteins (aRG complex), as evidenced by reconstituting purified receptors and G-proteins in artificial membranes. The aRG complex is thought to be formed as a transient intermediate *in vivo*, forming the transition state. The activation energy of the reaction path G_{GDP}–G–G_{GTP} is considered to be reduced by taking the route of G_{GDP}–aRG–G_{GTP}, as is the case for the formation of enzyme–substrate complexes. Actually, aR activates multiple G-proteins like a catalyst.

Seventeen different α-subunit genes have been identified, and they are classified into four groups according to the homology of their amino acid sequences: G_s, G_i/G_o, G_q, and G_{12} (see Table 3). Multiple species of β (5) and γ (11) subunits have been identified, although their functional differences remain to be clarified. Several kinds of $\beta\gamma$ subunits are shown to interact equally well with both G_s-α and G_i-α, and several combinations of β- and γ-subunits have been expressed in cultured cells and shown to form a complex with each other and with G_i-α subunits. The specificity of interaction of the three subunits, however, has not been fully elucidated.

G_s-α (or α_s) that is bound with GTP activates adenylate cyclase. G_{olf}-α is an olfactory bulb-specific α-subunit and activates adenylate cyclase as G_s-α does. At least nine different adenylate cyclases have been cloned and all of them are activated by G_s-α. The addition of $\beta\gamma$ subunits in the presence of G_s-α further stimulates adenylate cyclase II but inhibits adenylate cyclase I. The $\beta\gamma$ subunit alone neither stimulates nor inhibits adenylate cyclase I or II. Ca^{2+}-calmodulin further stimulates adenylate cyclase I in the presence of G_s-α, whereas Ca^{2+} inhibits adenylate cyclase V and VI. These synergistic mechanisms of stimulation or inhibition may be utilized for detection of coincident stimulation of a single neuron by one receptor linked to G_s and the other receptor linked to elevation of $\beta\gamma$ subunits or Ca^{2+}.

Most α-subunits in the second group of G-proteins (G_{i1}-α, G_{i2}-α, G_{i3}-α, G_o-α, G_{t1}-α, G_{t2}-α, and G_g-α) are ADP-ribosylated by pertussis toxin. G-proteins in the other groups and G_z in this group are not ADP-ribosylated by pertussis toxin. The ADP-ribosylated G-proteins lose the ability to interact with receptors, and hence the inhibition by pertussis toxin of receptor-mediated reactions indicates the involvement of these G-proteins. G_i-α and G_z-α inhibit adenylate cyclase I, V, and VI, and G_o-α inhibits adenylate cyclase I. The amounts of G-proteins present in the mammalian brain show the following order: $G_o > G_{i1} > G_{i2}, G_s >$

G_q. This indicates that the major source of G-protein $\beta\gamma$ subunits in the brain is from members of this second group. Thus, activation of G-proteins in this group causes the stimulation of adenylate cyclase II through $\beta\gamma$ subunits and inhibition of adenylate cyclase I through both α- and $\beta\gamma$ subunits. In addition, the $\beta\gamma$ subunit directly interacts with and opens an inward-rectified K^+ channel (GIRK or Kir3). The K^+ channel in the heart is regulated by acetylcholine released from the vagus nerve through muscarinic M_2 receptors and G_i, probably G_{i2}. Similar inward-rectified K^+ channels are known to be present in the brain. A certain kind of voltage-gated Ca^{2+} channel is known to be closed by G-proteins belonging to this group through direct interaction between channels and G-protein α-subunits.

G_t (or transducin) is a retina-specific G-protein, and G_{t1}-α and G_{t2}-α are present in rod and cone cells, respectively. G_{t1} and G_{t2} are activated by light-absorbed rhodopsin and cone pigments, respectively. G_t-α bound with GTP activates cGMP-phosphodiesterase resulting in a decrease of cGMP concentrations. As cGMP interacts with and opens cation channels and depolarizes visual cells, light stimulation causes hyperpolarization of visual cells. G_g (or gustducin) is a taste bud–specific and transducin-like G-protein, and is activated by receptors for taste substances and activates cAMP phosphodiesterase.

G_q-α and related α-subunits (α_{11}, α_{14}, α_{15} (or α_{16})) activate phosphatidylinositol-specific phospholipase C β types (PLCβ), which catalyze the formation of inositol 1,4,5 triphosphate (IP_3) and diacylglycerol (DG) from phosphatidyl inositol 4,5-bisphosphate. The GTPase activity of G_q-α, G_{11}-α, or G_{14}-α is stimulated by PLCβ, which shortens the lifetime of the active α_{GTP} forms. There are several species of PLCβ; PLCβ2 and PLCβ3 are stimulated by G-protein $\beta\gamma$ subunits, whereas PLCβ1 is either inhibited or stimulated by $\beta\gamma$ subunits. Stimulatory effects of $\beta\gamma$ subunits are observed even in the absence of G_q-α, in contrast with the fact that the stimulation by $\beta\gamma$ subunits of adenylate cyclase II requires the presence of G_s-α. Stimulation of phosphatidyl inositol turnover by G_i- or G_o-coupled receptors may be due to activation of PLCβ2 or PLCβ3 by G-protein $\beta\gamma$ subunits.

G_{12} and G_{13} constitute a different group and are present ubiquitously. Receptors for thrombin and thromboxan A2 are coupled with $G_{12/13}$, and the GTP-bound form of $G_{12/13}$-α activates ρ GEF (a guanine nucleotide exchange factor of the small GTP-binding protein ρ). The activated GTP-bound form of ρ activates ρ-kinase resulting in the rearrangement of the actin cytoskeleton. Thus, G_{12} appears to be involved in cell motility.

3.3
Classification of GPCRs and Specificity of GPCR-G Protein Interaction

GPCRs may be classified according to the homology of amino acid sequences, the kind of endogenous ligands, or the species of G-protein α-subunits with which they interact. Neurotransmitter receptors are summarized in Table 3, where GPCRs are classified by their ligands. Other GPCRs are classified by G-protein interactions and are shown in Table 4. The best-characterized receptors, which interact with G_s-, G_i/G_o-, and G_q-type G-proteins, are β-adrenergic receptors, M_2 muscarinic and α_2 adrenergic receptors, and M_1 muscarinic and α_1 adrenergic receptors, respectively. There are at least three subtypes for each of the β, α_2, and α_1 adrenergic receptors (Table 3). Thus, adrenaline or noradrenaline may activate at least three different G-proteins through nine different receptors. The function of adrenaline (noradrenaline) on a certain cell will depend on species of receptors expressed in the cell. In the case of acetylcholine, there are five different muscarinic acetylcholine receptors; M_2 and M_4 muscarinic receptors are coupled with G_i/G_o family G-proteins, and M_1, M_3, and M_5 receptors with G_q family G-proteins.

The specificity of interaction between receptor subtypes and distinct G-proteins is different depending on receptor and G-protein species. In reconstituted systems of purified receptors and G-proteins, the M_2 muscarinic receptor interacts equally well with G_{i1}, G_{i2}, and G_o and the M_1 muscarinic receptor interacts equally well with G_q, G_{11}, G_{14}, but the M_2 receptor does not interact with G_q family G-proteins nor does the M_1 receptor couple to G_i/G_o family G-proteins. Some receptors show a stricter specificity for members of the G_i/G_o family than the M_2 receptor; the α_{2A} adrenergic and A_1 adenosine receptors show higher affinity for G_{i3} than for G_{i1}, G_{i2}, and G_o, and the D_{2L} dopamine receptor shows a higher affinity for G_{i2} than for other G_i/G_o proteins. On the other hand, some receptors including the turkey β-adrenergic and the substance P (NK1 tachykinin) receptors are known to interact with both G_s and G_q family G-proteins. Among 146 GPCRs examined, 36 (25%), 76 (52%), and 51 (35%) interact with G_s, G_i/G_o, and G_q respectively, and 14 GPCRs interact with more than two kinds of G-proteins: 1, 3, 7, and 3 interact with G_s/G_i, G_s/G_q, G_i/G_q, $G_s/G_i/G_q$, respectively (cf. The IUPHAR compendium of receptor characterization and classification, IUPHAR Media, 1998). In the interactions of muscarinic receptors and G-proteins, amino acid sequences (approximately 20 amino acids) in the N- and C-terminal portions of the third intracellular loop in muscarinic receptors are critical for determining the specificity of their interaction (M_1, M_3, M_5-G_q and M_2, M_4-G_i/G_o). On the other hand, alternatively spliced forms of prostaglandin EP_3 receptors, which have different carboxy-terminal tails with distinct amino acid sequences, interact with different G-proteins including G_i/G_o, G_s, and G_q. As for G-proteins, the last 10 amino acids of the α-subunits are reported to be most critical for determining the specificity for receptors. $G_{q/i}$-α, which is a chimera of G_q-α and the carboxy-terminal tail of G_i-α, interacts with G_i-coupled receptors as G_i-α does, and then activates PLCβ1 as G_q-α does. G_{16} belongs to the G_q family but interacts promiscuously with G_i-, G_s-, and G_q-coupled receptors. The specificity or promiscuity of $G_{q/i}$ or G_{16} is not general, since each of them does not interact with some G_i-coupled receptors.

3.4
Regulator of G-Protein Signaling (RGS) and G Protein-coupled Receptor Kinase (GRK)

Signaling from GPCRs to G-proteins is regulated by regulators of G-protein signaling (RGS) and G protein-coupled receptor kinases (GRK). RGS proteins are GTPase-activators of G-protein α-subunits and accelerate the inactivation of G-protein. GRKs are serine/threonine kinases, which phosphorylate activated forms of GPCRs specifically, and initiate the attenuation of GPCR-mediated responses, which is called *desensitization*.

At least 19 different RGS members have been identified, which are subdivided into several groups such as R4 (RGS1, 2, 4, 5, 8, 13, 16), R3 (RGS3), RGSZ (RGS17, 19, 20), R7 (RGS6, 7, 9, 11), and R12 (RGS10, 12, 14). All members have RGS domains, which are responsible for the activation of GTPase of G-protein α-subunits. The specificity for G-proteins depends on the species: for example, R4 and R3 RGS act on both G_i-α and G_q-α: RGS20 on G_z-α and RGS17, 19 on G_i-α; RGS12, 14 on G_i-α. R7 RGS proteins have a G-protein γ-subunit-like domain, besides an RGS domain, and bind to the G-protein β_5 subunit. The RGS–Gβ_5 complex acts on $G_o\alpha$. RGS9 is expressed in the retina and the RGS9–Gβ_5 complex acts on the transducin α-subunit (G_t-α). The specificity for G_t-α appears to be due to the binding of Gβ_5, because the RGS domain of RGS9 can bind to any α_{GTP} subunits of G_t, G_i, or G_q and activate their GTPase activity. RGS9 is anchored to photoreceptor membranes by the transmembrane protein R9AP. Patients with recessive mutations in the RGS9 or R9AP genes are reported to have difficulty in adapting to sudden changes in luminance levels. RGS proteins for G_s were missing until recently, but RGS-PX1 was found to stimulate the GTPase activity of α-subunit of G_s but not other G-proteins. Some RhoGEFs have an RGS-like domain, which functions as GTPase activator of α-subunits of G_{12} or G_{13}. In the case of p115-Rho-GEF, not only RGS-like domains but also other regions are required for full GTPase activation, although most other RGS proteins have essentially the same GTPase-activating activity as the RGS domain alone.

There are other subfamilies called *RGS-like proteins* containing RGS-like domains, which share significant similarities with RGS domains, although some of the RGS-like domains do not have GTPase-activating activity. RGS and RGS-like proteins have other domain(s) that interact with other proteins including GPCRs. Some RGS proteins of the R4 family directly interact with specific receptors via their amino-terminal domain. RGS1, 4, and 16 attenuate the IP_3 response mediated by M_3 muscarinic receptors in pancreatic acinar cells more effectively than that mediated by cholecystokinin receptor, whereas RGS2 attenuates both signals equally. RGS3 has several splicing variants, and the difference in their amino-terminal sequences specifies their target proteins. RGSL interacts with $\beta\gamma$ subunits suppressing their stimulating effect on PLCβ. RGS3T binds directly to adenylate cyclase inhibiting its activity, and has a nuclear localization signal, which enables it to translocate into the nucleus and induce apoptosis of the cell. PDZ-RGS3 has PDZ domains, which bind to the ephrin-B receptor and attenuate signaling through GTPase activation. One RGS12 splicing variant also has PDZ domain, which binds to the interleukin 8 (IL8) receptor and terminates the receptor signal specifically.

GPCRs such as rhodopsin, β-adrenergic, α_2 adrenergic, M_2 muscarinic, and other receptors are known to be phosphorylated by G protein–coupled receptor kinases (GRKs) in a light- or agonist-dependent manner. There are six kinds of GRKs. GRK1 is rhodopsin kinase and phosphorylates light-absorbed rhodopsin. Phosphorylated rhodopsin interacts with a protein called *arrestin*, and this interaction prevents rhodopsin from interacting with and activating G-proteins. The phosphorylation of rhodopsin by rhodopsin kinase is inhibited by a calmodulin-like protein called *s-modulin* or *recoverin* in a Ca^{2+}-dependent manner. The phosphorylation of rhodopsin by rhodopsin kinase and its inhibition by s-modulin are thought to be linked to desensitization and light adaptation, respectively.

GRK2 and GRK3 are originally known as β-adrenergic receptor kinases or βARK1 and βARK2. There are $\beta\gamma$-binding sites in the carboxy terminus of GRK2 and GRK3 but not in other GRKs. The binding of $\beta\gamma$ subunits to GRK2 or 3 facilitates their translocation from the cytoplasm to the plasma membrane. GRK2 is synergistically stimulated by G-protein $\beta\gamma$ subunits and agonist-bound receptors, providing an explanation why GRK2 phosphorylates agonist-bound receptors specifically. The phosphorylation of β-adrenergic receptors by GRK2 causes uncoupling of β-adrenergic receptors from G-proteins through action of β-arrestin, which is a protein similar to arrestin and binds to phosphorylated receptors. The agonist-dependent phosphorylation of muscarinic receptors and other GPCRs by GRK2 and other GRKs facilitates their internalization into intracellular vesicles. GRK2 phosphorylation sites and sites responsible for receptor internalization reside in the carboxy terminus for β-adrenergic receptors and in the central part of the third intracellular loop for muscarinic receptors. Agonist-dependent internalization may occur through either dynamin-dependent or -independent processes depending on the species of receptors or receptor subtypes involved.

GRK2 and 3 have an RGS-like domain in their amino termini, which binds to G_q-α and suppresses its activity. The RGS-like domain has GTPase-activating activity at G_q-α although the activity is much less compared with other RGS proteins. Thus, signaling by G_q-coupled receptors may be attenuated by GRK2 in two parallel ways, that is, the phosphorylation of the receptor followed by uncoupling from G-protein and receptor internalization, and the suppression of G_q-α activity by GRK2 binding. The activity of some G_q-coupled receptors is attenuated by GRK2 although they are not substrates of GRK2.

3.5
Neurotransmitter Receptors: Roles of ICCRs and GPCRs

Neurotransmitters may be classified based on the type of receptors they interact with (cf. Table 3). Glutamate, acetylcholine, serotonin, and ATP constitute the first family, and there are both IC-CRs with cation channels and GPCRs for these neurotransmitters. The fast excitatory synaptic transmission is mostly mediated by nicotinic acetylcholine receptors in the peripheral nervous system and by AMPA-type glutamate receptors in the central nervous system. GABA and glycine form the second group of neurotransmitters and participate in fast inhibitory synaptic transmission through their IC-CRs with anion channels. The GPCRs for GABA are pharmacologically identified as $GABA_B$ receptors. $GABA_B$ receptors have characteristics similar to metabotropic glutamate receptors. Glycine is a unique neurotransmitter that does not seem to act on GPCRs. Amines like dopamine, noradrenaline, and histamine, peptides like opioids and tachykinins, and adenosine constitute the third group and only act on GPCRs. Some peptides like VIP and neuropeptide Y, which are listed in Table 4, are also released from neurons and act on GPCRs on the post- or presynaptic membranes. In that sense, these substances could be considered neurotransmitters, although they are usually not classified as typical neurotransmitters.

ICCRs are specified for rapid information transfer in the nervous system. GPCRs, on the other hand, are present ubiquitously in all tissues and in different species from yeasts to human beings, and the rate of signal transduction is not so fast as observed with ICCRs (Table 1). Thus, neurotransmitters that only act on GPCRs should participate only in slow synaptic transmission. The question is raised what are the functions of GPCRs for neurotransmitters that act on both ICCRs and GPCRs. Neurotransmitters released at time zero may depolarize or hyperpolarize postsynaptic membranes through ICCRs in less than a millisecond and may also activate various events through GPCRs more than 10 to 100 ms later. The fast signal transmission mediated through activation of ICCRs by neurotransmitters cannot be affected by activation of GPCRs by the same neurotransmitters released at the same time. GPCRs can be considered monitors of synaptic activities at time zero and affect synaptic activities later; for example, modification of subsequent generation of action potentials through slow excitatory or inhibitory postsynaptic potentials in a timescale of around 100 ms, modification of subsequent release of neurotransmitters through presynaptic receptors, or modification of excitability in a longer timescale by phosphorylation of ion channels or other proteins or by modification of gene expression. It will be reasonable to assume that GPCR and subsequent signal transduction are involved in LTP and LTD as well as in learning and memory formation. Actually, various lines of experimental evidence support the idea that GPCR-mediated processes are involved in the long-term regulation of brain function.

All receptors for dopamine, noradrenaline, and orexin are GPCRs, and those for acetylcholine and serotonin are also mostly GPCRs in the brain. Neurons containing these neurotransmitters are localized in specific regions of the brainstem or basal forebrain, and send axons to many regions of the brain. These neurons are thought to regulate various general activities that do not need fast information transfer, such as emotion, sleep/wakefulness, depression/excitation, or attention.

GPCRs are also involved in the recognition of external stimuli such as light, odor, or taste, although it will take at least 10 ms or may be even 100 ms or 1 s. Speed of information transfer may be critical in some cases such as a batter who is going to hit a ball thrown at a speed of 150 km h^{-1} (= 40 m s^{-1} = 4 cm m s^{-1}) or an antelope who smells a panther dashing up at a speed of 150 km h^{-1}. The delay of 10 or 100 ms corresponds to 40 cm to 4 m of movement of a ball or a panther. It is interesting to note that GPCRs have survived as the cell sensor in spite of their slow speed and ICCR-type of sensors have not appeared in the process of evolution.

4
Protein Kinase-Coupled Receptors (PKCR)

The proteins that control the proliferation, survival, maturation, differentiation, and function of cells in specific tissues or organs are grouped together as cytokines and growth factors. Cytokines include interleukin (IL) and interferons (IFN), and in the broader definition, growth factors are also considered cytokines. Different tissues produce distinct growth factors such as epidermal growth factor (EGF), platelet-derived growth factor (PDGF), hepatocyte growth factor (HGF), nerve growth factor (NGF), and so on. Both cytokines/growth factors and hormones are defined as molecular signals that are released from one cell and regulate the function of other cells. However, they are distinct in several respects:

1. Hormones generally act on cells in tissues remote from their tissue of origin, whereas cytokines/growth factors exert major actions within the tissues that produce them.
2. Hormones are amines, lipids, peptides, and also proteins, whereas cytokines/growth factors are proteins.
3. Receptors for most hormones, which do not penetrate cell membranes, are GPCRs, whereas receptors for cytokines/growth factors are directly or indirectly linked to activation of tyrosine kinases.
4. In general, activation of GPCRs causes metabolic changes in target cells through activation or inhibition of enzymes by protein serine/threonine phosphorylation, whereas activation of tyrosine kinases through cytokine/growth-factor receptors leads to expression of specific genes and proliferation or differentiation of target cells. However, there are cases in which gene expression and metabolic functions are regulated by hormones and cytokine/growth factors, respectively.

Insulin is a hormone in the sense of (1) but a growth factor in the sense of (3). Receptors for insulin and insulin-like growth factor (IGF-1) are tyrosine kinases. Insulin itself has an activity to stimulate proliferation of cells, and insulin-like growth factor (IGF-1) has an even stronger activity. On the other hand, the receptor for interleukin 8 is a GPCR despite "rule" (3).

Receptors for growth factors/cytokines share the following characteristics:

1. With some exceptions (e.g. the α-subunit of the insulin receptor), they have a single transmembrane segment per subunit (Fig. 3).
2. The binding of agonist (protein ligand) to the extracellular domain of receptors induces the dimerization or oligomerization of receptors.
3. Dimerization or oligomerization results in receptor activation.
4. Receptor activation is directly or indirectly linked to activation of tyrosine kinases.
5. The activation of tyrosine kinases finally leads to the activation or inhibition of cell-specific transcription factors and the expression of specific genes.

Receptors for growth factors/cytokines are divided into three subgroups according to their function (Table 5). Group 1 receptors, which include receptors for growth factors like insulin, EGF, PDGF, FGF, and neurotrophins have a catalytic domain of tyrosine kinase in their intracellular domain. Group 2 receptors, which include receptors for most cytokines except growth factors and transforming growth-factor β (TGFβ), do not have catalytic domains but interact with and activate tyrosine kinases. Group 3 receptors, which include receptors for TGFβ, have a catalytic domain of serine/threonine kinase in their intracellular domain.

4.1
Growth Factor Receptors with Tyrosine Kinase Activity

4.1.1 Growth Factor Receptors

Receptors for insulin, IGF-1 and HGF are composed of two subunits, α and β. After removal of the precursor signal peptide, the insulin and HGF receptors (c-met protooncogene) are posttranslationally cleaved into two chains (α and β) that

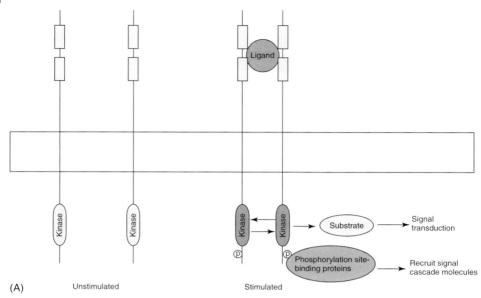

Fig. 3 Interaction of growth factors and growth-factor receptors. (A) A general model for the growth-factor receptor and other PKCRs with a single transmembrane segment. Following ligand binding, most receptors of this family form dimers (or trimers or tetramers in some cases). In many cases, ligands are dimers and then ligand binding automatically induces the oligomerization of receptors, as shown here. Dimerization of receptors triggers the signal transduction cascade, starting with autophosphorylation of a receptor subtype by the other and *vice versa*, followed by binding to phosphorylated tyrosine residues of signaling molecules, which are bound by other molecules. (B) Structure of the complex of epidermal growth factor (EGF) with the extracellular domain of EGF receptor. Binding of two EGF molecules induces the direct interaction of two EGF receptor molecules without direct interaction of EGF molecules. Two EGF molecules are observed on opposite sites. This is in contrast with the model shown in A regarding the mode of dimerization, although the outcome (receptor dimerization) is the same in both cases. (a) the ribbon diagram, (b) the top view of (a), (c) a surface model corresponding to (a). (From Ogiso, H. et al. (2002) Crystal structure of the complex of human epidermal growth factor and receptor extracellular domains, *Cell* **110**, 775–787).

are covalently linked. The α-subunit is in the extracellular space and has a ligand-binding domain, whereas the β-subunit has a single transmembrane domain and a tyrosine kinase catalytic domain in the intracellular space. Receptors for insulin and IGF-1 are composed of two α- and two β-subunits, which are linked to each other by S−S bonds between the α- and β-subunits and between the two β-subunits. Thus, the receptors for insulin and IGF-1 are tetramers even in the absence of ligands. Hence, ligand binding does not induce the dimerization of receptors, in contrast with other growth-factor receptors. The binding of insulin or IGF-1 is thought to cause conformational changes in their receptors, which lead to their activation. A major substrate for tyrosine kinase of activated receptors is insulin receptor substrate (IRS). There are four kinds of IRSs in the human genome. The expression levels of these molecules differ

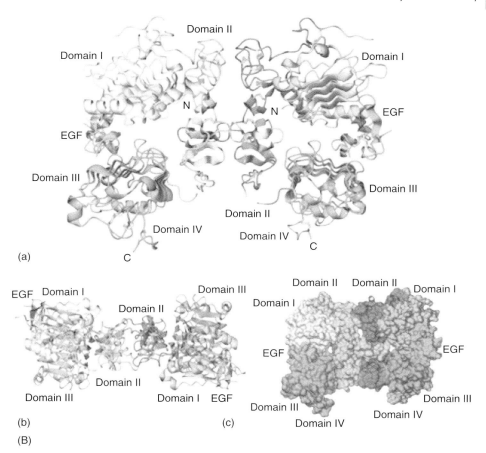

Fig. 3 (Continued)

from one tissue to the other, explaining the different effects of insulin in different tissues. IRS proteins have PH and PTB domains, and function as adaptor molecules. Following phosphorylation of specific tyrosine residues, IRS proteins recruit SH2 domain-containing proteins to the membrane. One of the molecules that binds to IRSs is phosphatidyl inositol 3 kinase, which is one of the key mediators in intracellular signal cascades.

Each receptor for PDGF, EGF, and FGF is a membrane protein with an extracellular ligand-binding domain, a transmembrane domain, and an intracellular tyrosine kinase domain. The molecular weights of the EGF and PDGF receptors are 120 to 130 kDa for the protein portions and 170 to 180 kDa when sugar is added. There are several characteristic domains in the extracellular part, such as cysteine-rich and immunoglobulin (Ig)-like domains. Receptors for EGF, PDGF, FGF have two cysteine-rich domains, five Ig-like domains, and two to three Ig-like domains, respectively. The cysteine-rich domains in EGF receptors and the Ig-like domains in PDGF and FGH receptors are known to be involved in the binding of their respective growth factors.

PDGF is a dimer composed of A- and B-subunits and there are three kinds of dimers, AA, AB, and BB. There are two kinds of PDGF receptors, α- and β-receptors. The A-subunit binds only with the α-receptor, while the B-subunit binds with either α- or β-receptor. Thus, PDGF-AA induces $\alpha 2$ receptor homodimerization, PDGF-AB leads to $\alpha 2$ and $\alpha\beta$ homo- or heterodimerization, and PDGF-BB induces the dimerization of three different forms $\alpha 2$, $\beta 2$ or $\alpha\beta$. Receptor dimerization results in cross phosphorylation of tyrosine residues in one receptor subunit by the other receptor subunit and *vice versa*. Three different dimers may allow the binding of different kinds of proteins to their phosphorylated regions, which may lead to different kinds of cellular responses. Recently, other kinds of PDGF named PDGF C and D have been reported. They induce homodimers, and PDGF-CC interacts with $\alpha 2$ receptors, and PDGF-DD with $\beta 2$ (and possibly $\alpha\beta$) receptors.

EGF is a monomer, and two molecules of EGF bind to two molecules of EGF receptors (EGFR) causing formation of an EGF–EGFR–EGFR–EGF complex, where the direct interaction between two EGFRs is induced by binding of EGF but there is no direct interaction between the two EGF molecules. This is in contrast with the dimerization of NGF receptors, which is based on the direct interaction of two NGF molecules within the dimer. The EGF receptor also forms heterodimers with ErbB family receptors.

There are 23 members of the FGF family in the human genome and some of them have alternative splicing variants. There are four distinct genes for FGF receptors, *FGFR1–4*, and alternative splicing variants. These receptors form homo- and heterodimers when bound to FGFs, and intermolecular transphosphorylation occurs. FGF mediates either proliferation and/or differentiation; for example, FGF causes proliferation of NIH3T3 cells, neurite outgrowth of pheochromocytoma-derived cells (PC12 cells), and mesoderm induction in *Xenopus* oocytes. These responses appear to be mediated by a common pathway involving activation of a small GTP-binding protein, ras, but it remains unknown how these responses are differentially induced in different cells.

4.1.2 Neurotrophin Receptors

Neurotrophins are survival and differentiation factors of neurons. NGF was characterized as a survival factor for sympathetic and sensory neurons. Effects of NGF in the central nervous system are restricted to cholinergic neurons (neurons containing acetylcholine as a transmitter) in the striatum and basal forebrain. Four other neurotrophins, BDNF (brain-derived neurotrophic factor), NT-3, and NT-4/5, have been identified. All these neurotrophins are proteins with approximately 120 amino acids and form homodimers with a common three-dimensional structure.

There are two kinds of neurotrophin receptors, $p75^{NTR}$ (p75) and $p140^{Trk}$ (Trk; three subtypes TrkA, TrkB, and TrkC). Each Trk is a membrane protein with an extracellular ligand-binding domain, a transmembrane domain, and an intracellular tyrosine kinase domain. The binding of neurotrophins is thought to induce the dimerization of Trks and the autophosphorylation of dimerized receptors. p75 is also a transmembrane protein with a ligand-binding domain but does not have a tyrosine kinase domain (see Fig. 3). p75 receptors also undergo ligand-induced dimerization. Neurotrophin-receptor complexes are internalized and transported from the axon to the soma in a retrograde fashion. The complex of phosphorylated Trk and neurotrophin activates Erk5 (a kinase of the MAP kinase family), which is followed by phosphorylation of CREB (Cyclic AMP-responsive element binding protein) and expression of specific genes for cell survival.

TrkA and TrkC specifically interact with NGF and NT-3, respectively, whereas TrkB interacts with either BDNF or NT-4/5. Expression of TrkA is localized to only a few neuronal types in the central and peripheral nervous systems, whereas both TrkB and TrkC are widely distributed throughout the brain and exist in the majority of neurons in the central nervous system, but not in nonneuronal cells such as glia. On the other hand, p75 is broadly expressed in neuronal and nonneuronal tissues. p75 is able to interact with each neurotrophin with equal affinity, but is reported to have a higher affinity for neurotrophin precursors rather than for matured neurotrophins cleaved by proteases. p75 has also been shown to bind to aggregated forms of β-amyloid and prion proteins, and to have a death domain in its intracellular region, which is somehow related to apoptotic signals.

4.1.3 Activation of the Ras-MAPK Cascade

Ligand binding to receptors results in autophosphorylation of tyrosine residues in receptors (or in IRSs in the case of insulin receptors). It is not clear and may be different among receptors if the phosphorylation of tyrosine residues further activates the catalytic activity or not. Phosphorylated tyrosine residues become targets of several proteins, which contain specific domains called SH2 (src homology domain 2). SH2 domains consist of approximately 100 amino acid residues, and recognize phosphorylated tyrosine (Y*)-containing motifs such as Y*DNV, Y*ENP, and Y*MXM. SH2 domains are included in different proteins such as nonreceptor type tyrosine kinases like src, phospholipase C γ (PLCγ), phosphatidyl inositol 3 kinase (PI3 kinase), phosphotyrosine phosphatase (PTP1C), GTPase-activating proteins (GAP), and adaptor proteins like Grb2 (growth-factor receptor bound protein 2). Amino acid sequences of various SH domains are similar but distinct from each other, and the amino acid sequences around the phosphorylated tyrosine-residue are also distinct. Thus, the interaction between SH2 domains and phosphorylated tyrosine-containing motifs is specific. For example, nine different tyrosine residues in the PDGF-β receptor are phosphorylated when activated by PDGF, and motifs containing residues 716, 771, 1009 interact with Grb2, GAP, and PTP-1D, respectively, motifs containing residues 579 and 581, 740 and 751, and 1009 and 1021 with src, PI3-kinase, and PLC-γ, respectively. Thus, a single kind of receptor may interact with and activate distinct proteins that have different kinds of SH2 domains.

A major downstream effector pathway of tyrosine kinase receptors is the MAPK (mitogen-activated protein kinase) cascade. Tyrosine kinase receptors recruit the adaptor protein Grb2. Adaptor proteins bind two different proteins and convert the binding of one protein into activation of the other protein; Grb2 has one SH2 and two SH3 domains, which are another kind of src homology domain, and binds with tyrosine kinase receptors through the SH2 domain and activates a protein called *son of sevenless* (sos) by interacting with it through SH3 domains. Sos is a GDP–GTP exchange protein, and facilitates the release of GDP from ras, a small GTP-binding protein, and thereby facilitates the binding of GTP to ras. GTP-bound ras is the activated form of ras and interacts with and activates a serine/threonine kinase, raf kinase (or MAPKKK, MAP kinase kinase kinase). Raf kinase phosphorylates and activates MAPKK (or MEK (MAPK/ERK kinase)), which phosphorylates both tyrosine and threonine residues in MAPK (or ERK, extracellular signal-regulated kinase). Phosphorylated- and activated-MAPK is translocated into the nucleus and phosphorylates several proteins such as CREB. Finally, specific transcription factors are activated and specific genes are expressed. MAPK in PC12 chromaffine cells is activated through activation of either NGF or EGF receptors, but the final outcomes of activation of NGF and EGF receptors are opposite; activation of NGF receptors results in differentiation and arrest of proliferation, whereas activation of EGF receptors results in proliferation of PC12 cells. It seems that the duration of activation of MAPK is a critical factor and that differentiation is caused by a longer duration of MAPK activation. Activation of MAPK is also mediated through activation of other tyrosine kinase receptors including those for PDGF, insulin, and so on. There appear to be several routes from receptors to activation of ras, and also several routes for ras-mediated and ras-independent activation of MAPK.

PLC-γ binds to PDGF or FGF receptors and is translocated to membranes by interaction between phosphorylated tyrosine-containing motifs and SH2 domains, and is phosphorylated by receptors. PLC-γ is activated through translocation to membranes. Phosphorylation of PLC-γ by receptors does not affect its activity *in vitro* but is necessary to activate it in intact cells.

PLC-γ catalyzes the formation of diacylglycerol and IP$_3$ from phosphatidylinositol 4,5-bisphosphate (PIP$_2$), as does PLC-β activated by the G-protein G$_q$. PIP$_2$ is known to bind to actin, α-actinin and other cytoskeletal proteins. PIP$_2$ bound to these proteins is not a substrate of PLC-γ but becomes a substrate of PLC-γ phosphorylated by tyrosine kinase receptors. There is a possibility that the activation of PLC-γ affects the organization of the cytoskeleton by allowing the interaction between actin and actin binding proteins through degradation of PIP$_2$.

PI$_3$ kinase catalyzes the phosphorylation of the 3′-OH of phosphatidyl inositol (PI), PI 4-phosphate, and PI(4,5)P$_2$, which results in the production of PI3-phosphate (PI3P), PI3,4-bis phosphate (PI(3,4)P$_2$), and PI3,4,5-triphosphate (PI(3,4,5)P$_3$). PI(3,4)P$_2$ and PI(3,4,5)P$_3$ bind to the Pleckstrin homology (PH) domain of intracellular signaling proteins and anchor them to cell membranes. PI3 kinase is composed of two subunits, p110 and p85, representing the catalytic subunit and an SH2-containing protein, respectively. The catalytic subunit is thought to be activated through interaction of p85 with IRSs phosphorylated by insulin receptors or with other phosphorylated proteins including receptor tyrosine kinases. Activation of PI3 kinase is necessary for insulin-mediated increase of glucose uptake, which is due to the translocation of the glucose transporter GLUT4 into cell membranes, in adipocyte, liver, and skeletal muscles. PI3 kinase increases the concentration of 3-phosphoinositide, which activates 3-phophoinositide-dependent protein kinase (PDK-1). PDK-1 phosphorylates the atypical protein kinase C and protein kinase B (PKB): PDK-1 and PKB have PH domains.

These kinases phosphorylate critical serine residues of GLUT4 resulting in its translocation to the plasma membrane. It has been suggested that PI3 kinase is involved in the regulation of cytoskeletal proteins and cell motility through activation of a subfamily of small GTP-binding proteins, rac.

4.2 Cytokine Receptors Linked with Tyrosine Kinases

4.2.1 Cytokine Receptors

Receptors for cytokines including interleukins (ILs) and interferons (IFNs) do not have tyrosine kinase activity, in contrast with receptors for growth factors. There are 30 IL genes and 25 IFN genes in the human genome. ILs are largely synthesized in lymphocytes and affect proliferation, differentiation, and functions of lymphocytes, which are involved in immune reactions. IFNs are secreted in response to virus infection, and suppress the proliferation of virus and also affect cell growth or immune responses. Receptors for ciliary neurotrophic factor (CNTF) and leukemia inhibitory factor (LIF) also lack tyrosine kinase activity, in contrast with NGF or other neurotrophin receptors. CNTF exerts potent survival effects on many peripheral neurons, in common with neurotrophins and FGF. LIF was identified also as a cholinergic differentiation factor, and has the ability to convert sympathetic adrenergic neurons into cholinergic neurons in *in vitro* experiments.

Most cytokine receptors are composed of an extracellular part, a single transmembrane domain, and an intracellular part. The extracellular parts of cytokine receptors contain cysteine-rich, Ig-like, or fibronectin type III-like domains, depending on the receptor species. The intracellular parts of the receptors do not have any known catalytic domains but have conserved motifs called *box 1 and box 2*, which are located adjacent to the transmembrane segments.

Table 5 shows some examples of cytokines and their receptors. Each of the receptors for growth hormones, erythropoietin, and prolactin is homogeneous and forms a homodimer when activated by the respective ligand. On the other hand, most cytokine receptors consist of more than two subunits, and one subunit is often shared by other cytokines. LIF, CNTF, and interleukin 6, 11 share a common receptor called gp130. LIF binds to a heteromeric complex of LIF receptor and gp130. CNTF binds to the CNTF receptor, which lacks a cytoplasmic domain, and the ligand–receptor complex then induces a heteromeric complex with the LIF receptor and gp130. Similarly, the complex of IL6 receptor with IL6 induces a complex containing a homodimer of gp130. The gp130 homodimer or gp130-LIF receptor heterodimer are thought to trigger signal transduction, while the CNTF and IL6 receptors facilitate the formation of dimers containing gp130. Receptors for IL3, GM-CSF (granulocyte macrophage-colony stimulating factor) are composed of a ligand-specific α-subunit and a common β_c-subunit, where the subscript c stands for "common." Receptors for IL2, 4, 5, 7, 9, 13, 15, and 21 are composed of a specific β-subunit and a common γ-subunit. An α-subunit contributes to an increase in the affinity of ligands to receptors but its significance in signal transduction is not known.

4.2.2 Activation of the JAK-STAT Pathway

Cytokine receptors do not have protein kinase activity but have the ability to activate nonreceptor tyrosine kinases such

Tab. 5 Protein kinase-coupled receptors (PKCRs) for growth factors, cytokines, and TGF-β, and TNF receptors.

Endogenous ligands	Receptors	Events induced by ligand binding
Group 1: Growth factor receptors with tyrosine kinase activity		
(a) Insulin, insulin-like growth factor 1 (IGF-1), hepatocyte growth factor (HGF)	$\alpha_2\beta_2$ tetramer; α-subunit has ligand binding activity and β-subunit has a tyrosine kinase catalytic domain	(1) Homo- or heterodimerization of receptors (2) Transphosphorylation of tyrosine residues (3) Binding of phosphorylated tyrosine-containing motifs and SH2 domains in Grb2, PLCγ and so on. (4) Activation of SH2-containing proteins and a signal transduction cascade, Grb2-sos-ras-raf1-MAPKK-MAPK, and other cascades (5) Translocation of MAPK into nucleus and expression of specific genes
(b) Epidermal growth factor (EGF), Transforming growth factor α (TGF-α)	Homo- or heterodimer Two cysteine-rich domains	
(c) Platelet-derived growth factor (PDGF) (AA, AB, and BB dimers)	$\alpha\alpha$, $\alpha\beta$, $\beta\beta$ dimers Five Ig-like domains	
(d) Fibroblast growth factor, 23 members including acidic FGF (aFGF) and basic FGF (bFGF)	Four receptors FGFR1–4 Three Ig-like domains	
(e) Neurotrophin (NGF, brain-derived neurotrophin (BDNF), NT-3, NT-4/5)	trkA, trkB, or trkC with tyrosine kinase activity and p75 without tyrosine kinase activity	
Group 2: Cytokine receptors linked with tyrosine kinases		
(a) Growth hormone, erythropoietin (EPO), prolactin, granulocyte colony stimulating factor (G-CSF),	Homodimers	(1) Homo- or heterodimerization or oligomerization of receptors (2) Binding of Janus kinases (JAK) to receptors and activation of JAK (3) Phosphorylation by JAK of STAT (Signal transducer and activator of transcription) (4) Dimerization and translocation into nucleus of phosphorylated STATs (5) Stimulation of transcription of specific genes by dimerized STATs with or without other proteins
(b) Leukemia inhibitory factor (LIF), ciliary neurotrophic factor (CNTF), interleukin 6 (IL-6) IL-11,	Common subunit gp130	
(c) IL-3, IL-5, IL-21, granulocyte macrophage-colony stimulating factor (GM-CSF)	Common β_c subunit and specific α-subunits	
(d) IL-2, 4, 7, 9, 15	Common γ subunit and specific β subunits	
IL-10, 22, 28, 29	Common IL-10β-subunit and specific α-subunit	
IL-12, 23	Common IL–12β subunit and specific α subunit	

(e) Type 1 interferon (IFN)
IFNα1, 2, 4, 5, 6, 7, 8, 10, 14, 16, 17, 21
IFN-β1, 3, IFN-τ, IFN-ω, FN-κ

Heterodimer of IFNαR1 and IFNαR2

(f) Type 2 interferon IFN-γ

Tetramer of two IFNγR1 and two IFNγR2

Group 3: TGFβ receptor with serine/threonine kinase activity

Transforming growth factor β1 (TGF-β1), TGF-β2, TGF-β3, activin
BMPs (bone morphogenetic proteins) (2–7)

Tetramers composed of two type I and two type II receptors
Type II receptors exist as dimers and are constitutively active even in the absence of ligands

(1) Binding of TGFβ to type II receptor dimers
(2) Binding of type I receptors to the TGFβ- type II receptor complex
(3) Phosphorylation of type I receptor and its activation
(4) Phosphorylation of SMAD
(5) Heterodimerization of phosphorylated SMAD with SMAD 4
(6) Translocation of SMAD heterodimer into nucleus and stimulation of specific gene expression

Group 4: TNF receptor family

TNF,
FasL
NGF

TNFR Homo-trimers
FAS homotrimer
NGFRp75 homodimer

(1) Binding of TNF trimer to TNFR-1, release of the inhibitory protein silencer of death domains from TNF-R1 intracellular domain (ICD)
(2) Binding of the adaptor protein TRDD (TNF receptor-associated death domain) to aggregated ICD
(3) TRDD recruits various adaptor proteins including caspase 8, IκB kinase, and so on.
(4) Caspase 8 *triggers* a protease cascade, IκBK phosphorylates IkB that results in its dissociation from NFκB, and JNK is activated leading to phosphorylation of c-Jun
(5) Initiation of apoptosis, and stimulation of specific gene transcription by NFκB and c-Jun

as JAK. JAK was named after the Roman god with two faces, Janus, because they have two kinase-like domains, although only one of them turned out to be active. Agonist binding to cytokine receptors induces the dimerization or oligomerization of receptors, which leads to activation of JAKs. JAKs bind to box 1 and box 2 of cytokine receptors. Four members of the JAK family are known; JAK1, 2, 3, and TYK2. The common receptors, gp 130, β_c, and γ, appear to interact with and activate JAK1/JAK2/TYK2, JAK2, and JAK1/JAK3, respectively. Two molecules of JAK kinases are thought to interact with ligand-bound receptor dimers and cross-phosphorylate and activate each other. Major targets of activated JAKs are STATs (signal transducer and activator of transcription). Seven members of the STAT family have been identified in the human genome (STAT1, 2, 3, 4, 5A, 5B, and 6), and each cytokine is thought to activate a specific set of STATs. STATs have SH2 domain, and the interaction between SH2 domain and phosphorylated tyrosine-containing motifs may contribute to the formation of homo- or heterodimers of phosphorylated STATs. Phosphorylated and dimerized STATs are translocated from the cytoplasm into the nucleus. STAT dimers alone or with another protein (e.g. p48) interact with DNA and stimulate transcription of specific genes.

4.3
TGF-β Receptors with Serine/Threonine Kinase Activity

Transforming growth factor-β (TGF-β) is a prototype for a large family of factors that regulate cell growth and differentiation, including activin/inhibin, bone morphogenic protein (BMP), Müllerian inhibition substances (MIS), and glial cell line-derived neurotrophic factor (GDNF). These factors have homology in amino acid sequence and tertiary structure, and all factors except MIS are dimers connected with an S–S bond. TGF-β was originally identified as a factor that causes transformation of fibroblast, and is now known to suppress proliferation of cells. Activin and inhibin were isolated as factors that stimulate or inhibit the release of follicle-stimulating hormone from cells in the pituitary, respectively. Activin is now known to function as an inducer of mesoderm. There are two kinds of inhibin dimers $\alpha\beta_A$ and $\alpha\beta_B$, and three kinds of activin homo- or heterodimers, $\beta_A\beta_A$, $\beta_A\beta_B$, $\beta_B\beta_B$. BMPs are known to induce the formation of bone and cartilage. GDNF was found as a protein that functions as a trophic factor for dopaminergic cells in the midbrain and was recently shown to support survival of motor neurons.

Receptors for TGF-β contain serine/threonine kinase activity in the intracellular part. Ligand binding to the extracellular surface of TGF receptors induces the oligomerization of type I and type II subtypes, most likely resulting in the tetramerization of two type I and two type II receptors. Type II receptors exist as dimers and are constitutively active even in the absence of TGF-β. The binding of TGF-β to type II receptor dimers causes the binding of type I receptor to the TGF-β-type II receptor complex and the phosphorylation of type I receptors. The phosphorylation of type I receptors activates its own kinase and induces the phosphorylation of downstream signal molecules named SMAD. There are eight kinds of SMAD genes in the human genome. SMAD1, 2, 3, 5, and 8 are directly phosphorylated by the type I receptor on conserved serine residues in the carboxy terminus. Phosphorylated SMAD forms a heterodimer with SMAD4, and the heterodimer is translocated to the nucleus where it binds to transcription regulatory proteins. On the other hand, SMAD 6 and 7 inhibit the signals induced by phosphorylated SMAD. Disruption of the activity of TGF-β and SMAD family members may cause a variety of human diseases including cancer.

4.4
Intracellular Signaling: Common Pathways Triggered by GPCRs and PKCRs

Major downstream effector cascades of growth-factor receptors and cytokine receptors are the ras-MAPK and JAK-STAT pathways, respectively. There are also crossing pathways. Most cytokines also activate the ras-MAPK cascade. The src family tyrosine kinase, the adaptor protein SHC, and the carboxy-terminal domain of cytokine receptors are suggested to be involved in this response. At the same time, tyrosine kinase receptors activate STATs; for example, EGF and PDGF induce the phosphorylation of STAT1a. In addition, a pathway similar to the MAPK cascade is suggested to be present downstream of TGF-β receptors. Crosstalk between pathways triggered by GPCRs and PKCRs is also known. MAPK may be activated through GPCRs. One pathway is likely to be mediated through activation of G_q followed by activation of protein kinase C and/or increase in intracellular Ca^{2+} concentrations and another through G-protein $\beta\gamma$ subunits. In addition, activation of JAKs and STATs is reported to be caused by activation of the angiotensin II receptor (a GPCR).

Activation of PLC is mediated through different receptors. Receptors linked to G_q, G_i/G_o and growth-factor receptors may activate PLC-β1, PLC-β2/β3, and PLC-γ, respectively, and therefore any agonists interacting with these receptors may cause the increase of intracellular Ca^{2+}, and activation of Ca^{2+}-calmodulin-dependent protein kinases and protein kinase C. In addition, it has been reported that there are two kinds of PI3 kinases, which are activated by growth-factor receptors and G-protein $\beta\gamma$ subunits, respectively. PLC-γ and PI3 kinase are also known to be activated by some cytokines.

These findings indicate that the relation between receptors and intracellular signaling pathways is not straightforward and that the same outcome may be induced by activation of either GPCRs or PKCRs. Major consequences of simulation of PKCRs and GPCRs are expression of specific genes and regulation of specific protein functions, respectively. It is also possible, however, that PKCRs and GPCRs may regulate protein functions and gene expression, respectively.

5
Other Receptors

Receptors for natriuretic peptides are composed of three segments; an extracellular ligand-binding domain, a single transmembrane domain, and a cytoplasmic domain with guanylate cyclase activity and a protein kinase-like domain. There are three kinds of natriuretic peptides, ANP (atrial type), BNP (brain type) and CNP (C type), and two kinds of receptors, GC-A and GC-B. GC-A interacts with both ANP and BNP, and GC-B with CNP. Another kind of receptor for all three natriuretic peptides lacks catalytic domains, serves to internalize these peptides, and is called a *clearance receptor*. The other type of guanylate cyclase is a soluble protein, lacks the transmembrane segment, and is activated by nitric oxide through its interaction with a heme group in the enzyme.

There are approximately 20 members in the superfamily of tumor necrosis factor receptors (TNFRs), including TNFR1, TNFR2, FAS, and CD40. Tumor necrosis factor (TNF) is a major mediator of apoptosis, inflammation, and immunity, and is thought to be involved in various diseases. TNF is a homotrimer of 157 amino acid subunits produced by activated macrophages. Most other ligands also act as a trimer. They share the homologous sequence in the domains responsible for trimerization, whereas the other regions that recognize their receptors are divergent. The receptor also forms a trimer after ligand binding. These receptors have several cysteine-rich domains in the extracellular domain, a membrane-spanning domain, and a cytoplasmic tail where some members have a death domain and a TRAF (TNF receptor-associated factor) binding site. Upon ligand binding, inhibitory proteins are released from the aggregated form of intracellular domains in the receptor and various proteins are bound there. There are six TRAFs in the human genome: TRAF1 and TRAF2 form homo- or heterodimers, and recruit IκB kinase (IKKα and IKKβ) and upstream kinases for c-jun NH$_2$-terminal kinase (JNK). IκB (inhibitor of nuclear factor κB (NF-κB) retains NF-κB in the cytoplasm in an inactive form, and phosphorylation of IκB results in its dissociation from NF-κB and translocation of NF-κB into the nucleus. NF-κB and c-jun stimulate transcription of specific genes in the nucleus. FAS-associated death domain (FADD) proteins are bound to the death domain, and caspase 8 is recruited to FADD. Caspase 8 initiates a proteolysis cascade leading to apoptosis. The molecular events involved in these pathways are complex and many details remain to be elucidated.

Acknowledgments

We are deeply indebted to Dr. J. Wess in N.I.H. for detailed comments and corrections throughout the whole manuscript. We also thank Dr. K. Inoue in National Institute of Health Sciences, Dr. Hiroyasu Furukawa in Columbia University, Dr. U. Langel in Stockholm University, Dr. G. Berstein in Wyeth Pharmaceuticals for reading a part or all of this manuscript and giving us valuable comments.

See also Endocrinology, Molecular; Membrane Traffic: Vesicle Budding and Fusion; Receptor Targets in Drug Discovery; Receptor, Transporter and Ion Channel Diseases; Signal Transduction Mediated by Heptahelical Receptors and Heterotrimeric G Proteins; Vitamin Receptors.

Bibliography

Books and Reviews

Aaronson, D.S., Horvath, C.M. (2002) A road map for those who do not know JAK-STAT, *Science* **296**, 1653–1655.

Attisano, L., Wrana, J.L. (2002) Signal transduction by the TGF-beta superfamily, *Science* **296**, 1646–1647.

Chen, G., Goeddel, D.V. (2002) TNF-R1 signaling: a beautiful pathway, *Science* **296**, 1634–1635.

Dijke, P.T., Hill, C.S. (2004) New insights into TGF-β-Smad signalling, *Trends Biochem. Sci.* **28**, 265–273.

Grotzinger, J. (2002) Molecular mechanisms of cytokine receptor activation, *Biochim. Biophys. Acta* **1592**(3), 215–223.

Ihle, J.N. (2001) The Stat family in cytokine signaling, *Curr. Opin. Cell Biol.* **13**, 211–217.

Khakh, B.S. (2001) Molecular physiology of P2X receptors and ATP signaling at synapses, *Nat. Rev. Neurosci.* **2**, 165–174.

Locksley, R.M., Killeen, N., Lenardo, M.J. (2001) The TNF and TNF receptor superfamilies: integrating mammalian biology, *Cell* **104**, 487–501.

Lu, Z.L., Saldanha, J.W., Hulme, E. C. (2002) Seven-transmembrane receptors: crystals clarify, *Trends Pharmacol. Sci.* **23**, 140–146.

Malenka, R.C., Nicoll, R.A. (1999) Long-term potentiation–a decade of progress?, *Science* **285**, 1870–1874.

Malinow, R., Malenka, R.C. (2002) AMPA receptor trafficking and synaptic plasticity, *Annu. Rev. Neurosci.* **25**, 103–126.

Okada, T., et al. (2001) Activation of rhodopsin: new insights from structural and biochemical studies, *Trends Biochem. Sci.* **26**, 318–324.

Pestka, S. (2000) The human interferon alpha species and receptors, *Biopolymers* **55**, 254–287.

Pierce, K.L., Premont, R.T., Lefkowitz, R.J. (2002) Seven-transmembrane receptors, *Nat. Rev. Mol. Cell Biol.* **3**, 639–650.

Rios, C.D., Gomes, J.I., Devi, L.A. (2001) G-protein-coupled receptor dimerization: modulation of receptor function, *Pharmacol. Ther.* **92**, 71–87.

Sierra, D.A., Popov, S., Wilkie, T.M. (2000) Regulators of G-protein signaling in receptor complexes, *Trends Cardiovasc. Med.* **10**, 263–268.

Song, I., Huganir, R.L. (2002) Regulation of AMPA receptors during synaptic plasticity, *Trends Neurosci.* **25**, 578–588.

Wieland, T., Mittmann, C. (2003) Regulators of G-protein signaling: multifunctional proteins with impact on signaling in the cardiovascular system, *Pharmacol. Ther.* **97**, 95–115.

Primary Literature

Furukawa, H., Gouaux, E. (2003) Mechanisms of activation, inhibition and specificity: crystal structure of the NMDA receptor NR1 ligand-binding core, *EMBO J.* **22**: 2873–2885.

Kunishima, N., Shimada, Y., Tsuji, Y., Sato, T., Yamamoto, M., Kumasaka, T., Nakanishi, S., Jingami, H., Morikawa, K. (2000) Structural basis of glutamate recognition by a dimeric metabotropic glutamate receptor, *Nature* **407**, 971–977.

Miyazawa, A., Fujiyoshi, Y., Unwin, N. (2003) Structure and gating mechanism of the acetylcholine receptor pore, *Nature* **424**, 949–955.

Ogiso, H., et al. (2002) Crystal structure of the complex of human epidermal growth factor and receptor extracellular domains, *Cell* **110**, 775–787.

Palczewski, K., Kumasaka, T., Hori, T., Behnke, C., Motoshima, H., Fox, B.A., Trong, I.Le, Teller, D.C., Okada, T., Stenkamp, R.E., Yamamoto, M., Miyano, M. (2000) Crystal structure of rhodopsin: A G protein-coupled receptor, *Science* **289**, 739–745.

Schlessinger, J. (2002) Ligand-induced, receptor-mediated dimerization and activation of EGF receptor, *Cell* **110**, 669–672.

Unwin, N., Miyazawa, A., Li, J., Fujiyoshi, Y. (2002) Activation of the nicotinic acetylcholine receptor involves a switch in conformation of the alpha subunits, *J. Mol. Biol.* **319**, 1165–1176.

Receptor Targets in Drug Discovery

Michael Williams[1], Christopher Mehlin[2], Rita Raddatz[1], and David J. Triggle[3]
[1] *Cephalon Incorporated, West Chester, PA*
[2] *University Of Washington, Seattle, WA*
[3] *State University of New York, Buffalo, NY*

1   Drug Discovery   596

2   Cellular Communication   597

3   Receptor Concepts   600
3.1   Occupancy Theory   600
3.2   Rate Theory   604

4   Receptor Complexes and Allosteric Modulators   604

5   Ternary Complex Models   606

6   Constitutive Receptor Activity   606

7   Efficacy Considerations   607

8   Receptor Dynamics   608

9   Receptor Nomenclature   609

10   Receptor Classes   609
10.1   G-protein–Coupled Receptors   610
10.1.1   G-protein–Associated or – Modulating Proteins   611
10.2   Ligand-gated Ion Channels   611
10.2.1 $GABA_A$/Benzodiazepine (BZ) Receptor   611
10.2.2   The NMDA Receptor   612
10.2.3   The Nicotinic Cholinergic Receptor, nAChR   612
10.2.4   The P2X Receptor   612

Encyclopedia of Molecular Cell Biology and Molecular Medicine, 2nd Edition. Volume 11
Edited by Robert A. Meyers.
Copyright © 2005 Wiley-VCH Verlag GmbH & Co. KGaA, Weinheim
ISBN: 3-527-30648-X

10.3	Steroid Receptor Superfamily	613
10.4	Intracellular Receptors	613
10.5	Non-GPCR-linked Cytokine Receptors	613
10.6	Orphan Receptors	613
10.7	Neurotransmitter Transporters	614
10.8	Neurotransmitter Binding Proteins	614
10.9	Drug Receptors	614

11 Molecular Biology of Receptors   615

12 Functional Genomics: Target Validation/Confidence Building   615

13 Compound Properties   618

13.1	Structure-activity Relationships	618
13.2	Defining the Receptor–Ligand Interaction	619
13.2.1	Receptor-binding Assays	620
13.2.2	Functional Assays	622
13.3	Receptor Sources	623
13.4	ADME	623
13.5	Compound Databases	624

14 Lead Compound Discovery   625

14.1	High-throughput Screening	625
14.2	Compound Sources	626
14.2.1	Natural Product Sources	626
14.2.2	Pharmacophore-based Ligand Libraries	628
14.2.3	Diversity-based Ligand Libraries	631
14.3	Biologicals and Antisense	632

15 Future Directions   633

Acknowledgment   633

Bibliography   633
Books and Reviews   633
Primary Literature   635

Keywords

Agonist
A ligand that interacts with a receptor to activate and produce a defined response.

Antagonist
A ligand that interacts with a receptor to block the effects of an agonist.

Drug
A compound that, via its interaction with a defined receptor target, reverses a disease phenotype. Drugs are characterized by their selectivity, safety, and pharmacokinetic properties, which are consistent with their reaching their target site of action within the body.

G-protein–Coupled Receptors
The best-known receptor family through which the largest number of known drugs act to produce their pharmacological effects, for example, cimetidine (histamine H_2 receptor), and propranolol (β-adrenoceptor). These receptors are defined by a common structure of seven transmembrane domains and their interactions with members of the heterotrimeric G-protein family.

Heterotrimeric G-protein
A family of oligomeric proteins consisting of three subunits (α, β, and γ). These protein complexes transduce signals from activated GPCRs to a variety of intracellular signaling cascades.

High-throughput Screening
Testing of large sets of compounds (typically 100 000–2 million) for interaction with a receptor target.

Ion Channels
Typically composed of protein multimers that form channels to conduct anions and/or cations through the membrane (e.g. the neuronal nicotinic receptor). Activation of ion channels can occur by ligand activation (e.g. nicotine, glutamate), by changes in membrane voltage, pH, mechanical distension, and temperature.

Ligand
A defined chemical entity that is natural (e.g. histamine) or synthetic and (e.g. propranolol) that binds to a receptor.

New Chemical Entity (NCE)
A compound with potential druglike properties.

Orphan Receptor
A receptor, identified on the basis of its structural motif for which the endogenous ligand and physiological function have not yet been identified.

Receptor

1. A membrane associated protein that selectively responds to extracellular messenger molecules to alter cellular function to maintain homeostasis and normal cell/tissue function. Receptors comprise two major classes, G-protein–coupled receptors (GPCRs) and ion channels.

2. A drug target – a concept that expands the definition of a receptor to intracellular sites that include enzymes, transporters, and signal transduction elements.

SAR
Dose/concentration dependent differences in the efficacy (agonist or antagonist) of compounds in a ligand series, which can be associated with changes in chemical structure.

Therapeutic Index
The ratio between the dose at which a drug or NCE produces efficacy in a disease state or disease model, and the dose at which it produces side effects.

> Receptors, located on both the cell surface and within the cell are the defined molecular targets through which drugs produce their beneficial effects in various disease states. Initially conceptualized over a century ago by Ehrlich and Langley, receptor concepts and receptor theory have undergone continuous modification as their behavior in normal and disease states/tissues have been more clearly characterized. Since the isolation of the nicotinic acetylcholine receptor (nAChR) from the *Torpedo*, some 50 years ago, new techniques of molecular biology have made it relatively routine to isolate receptors, including orphan receptors. Once these have been validated, they can be used in conjunction with high-throughput screening approaches to identify "hits," molecules that bind with relatively high affinity to these targets. Such hits can then be optimized to druglike entities using combinatorial/parallel synthesis technology platforms to yield clinical candidates that are potent, efficacious, and bioavailable entities with appropriate safety profiles.

1
Drug Discovery

Drug discovery, the process of identifying and developing novel chemical entities (NCEs) to treat human disease states, has entered the new century with a wealth of sophisticated technologies and information generation platforms, which have the potential to allow more rapid development of new drugs with improved selectivity and safety profiles.

Successful drugs from the pharmaceutical industry that include small-molecule drugs (NCEs), biologicals, vaccines, and antibodies, have resulted in an increase in human life expectancy, with the concomitant expectation that new drugs will be available for the aging population, which will improve the quality of life while adding to its span. The draft map of the human genome has led to the possibility of understanding diseases at a level of precision that was never before possible. Thus, in the last decade, there has been an exponential increase in understanding the potential pathophysiologies of disorders like pain, diabetes, some forms of cancer, and neurodegenerative processes (Alzheimer's and Parkinson's diseases).

The challenge in effectively using the copious information now available for disease-related gene identification and the molecular targets involved in the cause(s) of various diseases is how to handle and productively focus the bewildering flow of data, much of it archival, to cost-effectively discover and develop new drugs – a challenge that has given rise to key disciplines like bio- and chemo-informatics.

Irrespective of compound or drug target sources, more productive high-throughput screening (HTS) processes, and more detailed structural information on putative drug targets, it is clear that continued success in the drug discovery process will depend on the iteration and integration of lead compound identification and optimization through the hierarchical complexity of *in vitro* and *in vivo* assays that measure efficacy, selectivity, side effect liability, absorption, distribution, metabolism and excretion (ADME), and potential toxicity. To do this in an effective manner, it is critical to understand how existing knowledge of drug targets has evolved and what the realistic potential is for identifying, validating, and prioritizing new ones.

2
Cellular Communication

The transfer of information between cells and the subsequent integration of multiple inputs that are necessary to maintain cellular homeostasis and tissue viability – under both normal and adverse disease-related conditions – involves a variety of different external signaling modalities. These include: temperature, membrane potential, mechanical distension and stress, alterations in ion (H^+/K^+) concentrations, pheromones, oderants, as well as more traditional neurotransmitters, neuromodulators, and hormones. These physical stimuli and endogenous chemicals elicit their effects through interactions with cell surface targets, usually proteins, which are termed *receptors* (Table 1). Once receptors are activated by an *agonist* ligand, a compound capable of producing a cellular response, they transduce or couple the energy associated with the binding event to a cellular effect via a signal transduction process(es) involving protein–protein interactions or second messenger signaling systems, for example, G-proteins, protein kinase or phosphatase modulation, alterations in lipid metabolism, calcium translocation, and so on, to produce acute or more long-term effects on cell and tissue function. Thus, neurotransmitters, neuromodulators, and hormones can produce transient increases in second messengers like cyclic AMP or inositol triphosphate (IP_3) or more long-term changes that involve changes in gene expression through activation of transcription factors.

Alterations in receptor function occur by: (1) a functional overstimulation of receptors with their consequent desensitization, a phenomenon resulting from excess ligand availability or an enhanced coupling of the ligand-activated receptor to second messenger systems, or (2) a reduction in stimulation resulting from decreased ligand availability or dysfunctional receptor coupling processes.

Drugs that effectively treat human disease states by restoring disease-associated defects in signal transduction can act either by replacing endogenous transmitters, for example, L-dopa treatment to replace the dopamine (DA) lost in Parkinson's disease, or blockade of excess agonist stimulation,

Tab. 1 Drug recognition sites classified as receptors.

Type	Class	Examples
Receptor	GPCR	Dopaminergic, adrenergic, GABA$_B$, mGluR
	Ion channel	
	LGIC	NMDA, nicotinic, GABA$_A$
	VGIC	Na$_V$, Ca$_V$
	Others	ASIC, TRPV
	Enzyme Assoc. R	
	Tyrosine kinase	PDGF, EGF, Trk
	Others	CNTF, cytokine
	Nuclear hormone R	
	Steroid R	GR, PR
	Others	PPAR, RAR
Neurotransmitter Transporter	Plasma membrane neurotransmitter transporters	SERT, NET, GAT, DAT
Enzyme	Cell surface/ extracellular	ACE, NOS, HMG CoA reductase, acetylcholinesterase
	Intracellular	Caspase, CDK, ROCK DNA polymerase

Note: GABA: γ-aminobutyric acid; mGluR: metabotropic glutamate receptor; LGIC: ligand-gated ion channel; VGIC: voltage-gated ion channel; NMDA: N-methyl-D-aspartic acid; ASIC: acid sensing ion channel; TRPV: transient receptor potential vanilloid; PDGF: platelet-derived growth factor; EGF: epidermal growth factor; CNTF: ciliary neurotrophic factor; GR: glucocorticoid receptor; PR: progesterone receptor; PPAR: peroxisome proliferator-activated receptor; RAR: retinoic acid receptor; SERT: serotonin transporter; NET: norepinephrine transporter; GAT: GABA transporter; DAT: dopamine transporter; ACE: angiotensin-converting enzyme; NOS: nitric oxide synthase; CDK: cyclin-dependent kinase; ROCK: rho-dependent kinase.

for example, the histamine H$_2$ receptor antagonist, cimetidine, which blocks histamine-induced gastric acid secretion and thus reduces ulcer formation. Of the approximately 450 targets through which known drugs act, 71% are receptors in the G-protein–coupled receptor (GPCR), ion channels, or neurotransmitter transporter (SCDNT) families. Of these drugs, approximately 90% produce their effects by antagonizing the actions of endogenous agonists

Enzymes, by producing products that regulate second messenger availability, for example, adenylyl cyclase-catalyzed production of cyclic AMP or phosphodiesterase-mediated degradation of cAMP, or by modifying protein targets, for example, by adding (protein kinases) or removing (protein phosphatases) phosphates on serine or threonine residues, serve a similar role in cellular homeostasis, and as such, also represent key drug targets that can be conceptually included in the category of receptors.

The seminal concept that therapeutic agents produce their effects by acting as "magic bullets" at discrete molecular targets on tissues within the body is attributed to Ehrlich and Langley, who at the beginning of the twentieth century independently generated experimental data that led to the evolution of the "lock and key" hypothesis. A ligand (L) thus acted as the unique "key" to selectively modulate receptor (R) activity, the latter functioning as the "lock" for the "entry" of external signals into the cell. In this model, an agonist ligand forms an RL complex and has the ability to "turn" the lock (Equation 1), whereas a receptor antagonist would occupy the lock and prevent agonist access.

$$\text{Receptor} + \text{ligand} \underset{k_{-1}}{\overset{k_{+1}}{\rightleftharpoons}} \text{RL} \longrightarrow \text{Signaling event} \quad (1)$$

Despite quantal advances in the technology used to study receptor interactions that have occurred over the past century, the receptor–ligand (RL) concept and the similar enzyme–substrate (ES) complex have remained the conceptual foundations for understanding receptor and enzyme function, disease pathophysiology, and medicinal chemistry-driven approaches to drug discovery. Over the past decade, however, with the explosion in the number and diversity of receptors driven by receptor cloning approaches and the mapping of the human genome, there has been an increased appreciation of the inherent complexity of receptor function.

At one time, it was thought that a receptor-mediated response was a predictable, linear process that involved ligand-induced activation of a protein monomer and its signal transduction pathway independently, or with very little influence, from other membrane proteins. It has become increasingly evident, however, that receptors can physically interact both with one another and with other membrane proteins. Numerous examples exist of receptor coexpression and interactions, for example, $GABA_BR1$ (γ-aminobutyric acid) and $GABA_BR2$, dopamine with somatostatin, and $GABA_A$ receptors and opioid receptors with α_2-adrenoceptors.

These interactions are often necessary to obtain functional cell surface receptors and allow interactions with one protein partner to modulate the function of the entire signaling complex. Effects of receptors on protein subunits can be mediated through orthosteric sites, for example, the site that the endogenous ligand(s) interact with, or through allosteric sites (e.g. the benzodiazepine (BZ) site in the $GABA_A$ receptor, and the glycine receptor on the N-methyl-D-aspartate (NMDA) receptor complex). More recently, allosteric interactions with GPCRs (muscarinic adenosine A_1) and neurotransmitter transporters (NET) have been reported. Integration of the functional effects of signal transduction pathways also affects receptor functions in a cell-specific manner (receptor cross talk). There also exist several discrete classes of receptor-associated proteins including receptor-activity-modifying proteins (RAMPs) and trafficking chaperones, which currently number in the many hundreds and play key roles in cellular events like receptor trafficking from the endoplasmic reticulum to the cell surface, and modulation of cell surface responses. Examples of the complexity of receptor signaling at the postsynaptic level include the NMDA receptor, where proteomic analysis demonstrated more than 70 proteins other than the receptor potentially involved in the function of the receptor complex; and the ATP-sensitive $P2X_7$ receptor, where a signaling complex comprised of 11 proteins including laminin-3, integrin 2, actin, supervillin, MAGuK, three heat shock proteins, and phosphatidylinositol 4-kinase; and the receptor protein tyrosine phosphatase (RPTP) was identified, which appears to modulate $P2X_7$ receptor function through control of its phosphorylation state.

3
Receptor Concepts

Receptor theory is based on the classical Law of Mass Action as developed by Michaelis and Menten for the study of enzyme catalysis. The extrapolation of classical enzyme theory to receptors is, however, an approximation. In an enzyme–substrate (ES) interaction, the substrate S undergoes an enzyme-catalyzed conversion to a product or products. Because of the equilibrium established, product accumulation has the ability to reverse the reaction process. Alternatively, the latter can be used in other cellular pathways and is thus removed from the equilibrium situation or can act as a feedback modulator to alter the ES reaction either positively or negatively (Equation 2).

$$\text{Enzyme} + \text{Substrate} \rightleftharpoons \text{ES} \rightleftharpoons \text{E} + \text{Product} \quad (2)$$

For the receptor–ligand interaction, binding of the ligand to the receptor to form the RL complex results in a response driven by the thermodynamics of the binding reaction that leads to functional changes in the target cell (Equation 1). Whereas conformational changes occur in the ligand, the receptor, or both, there is no chemical change in the ligand resulting from the RL interaction such that there is no chemical product derived from the ligand that results from the RL interaction. Despite events such as receptor internalization, receptor phosphorylation, second messenger system activation, and so on, the bound ligand is chemically unchanged by the binding event, and thus, there is no equilibrium established between the RL complex and the consequences of receptor activation.

After the formation of an RL complex, a functional response to receptor activation can be related to the concentration of the ligand that is present (Fig. 1a). *Occupancy theory*, developed by Clark in 1926, is based on a dose/concentration-response relationship. This theory has undergone continuous refinement on the basis of experimental data to aid in further delineating the increasingly complex concept of ligand efficacy and to accommodate allosteric site modulation of receptor function, ternary complex models (TCM), and their extension to constitutively active receptor systems – those that are functional in the absence of an identified ligand.

3.1
Occupancy Theory

The basic premise of Clark's occupancy theory, based on Michaelis–Menten theory, was that the effect produced by an agonist was dependent on the number of receptors occupied by that agonist, a reflection of the agonist concentration present. This theory was developed from Clark's observations that acetylcholine (ACh) receptor antagonists such as atropine caused a rightward shift in the ACh dose-response (DR) curve in muscle preparations when plotted logarithmically. The basic tenets of occupancy theory were as follows: (1) the RL complex was assumed to be reversible; (2) the association of the receptor with the ligand to form the RL complex was defined as a bimolecular process with dissociation being a monomolecular process; (3) all receptors in a given system were assumed to be equivalent to one another and able to bind ligand independently of one another; (4) formation of the RL complex did not alter the free (F) concentration of the ligand or the affinity of the receptor for the ligand; (5) the

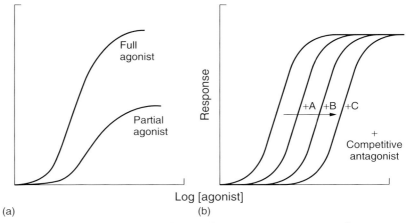

Fig. 1 Dose-response curve. The addition of increasing concentrations of an agonist ligand causes an increase in a biological response. Plotted on a logarithmic scale, a sigmoidal curve is obtained. (a) A full agonist produced a maximal response, whereas a partial agonist reaches a plateau that is only part of the response seen with a full agonist. (b) In the presence of antagonist concentrations A, B, C, the dose-response curve is progressively moved to the right. Increasing agonist concentrations overcome the effects of the antagonist.

response elicited by receptor occupancy was directly proportional to the number of receptors occupied; and (6) the biological response was dependent on attaining an equilibrium between R and L according to Equation 1.

Although it is not always possible to determine the concentration of free ligand (F) or that of the RL complex, rearranging the latter, the equilibrium dissociation constant, K_d, equaling $k-1/k+1$, can be derived from the equation:

$$K_d = \left(\frac{[R][L]}{[RL]} \right) \qquad (3)$$

and is equal to the concentration of L that occupies 50% of the available receptors.

Antagonist interactions with the receptor were defined by Gaddum as being the result of receptor occupancy with the antagonist ligand being unable to elicit a functional response. Antagonists thus block agonist actions. Agonists overcome the effects of a competitive (e.g. reversible/surmountable) antagonist when their concentration is progressively increased (Fig. 1b) such that in the presence of increasing fixed concentrations of a competitive antagonist, a series of parallel agonist DR curves can be generated that shift progressively to the right (Fig. 1b). A Schild regression relationship, a plot of log (DR-1) versus the log antagonist concentration (Fig. 2), can be used to derive the pA_2 value for an antagonist from the intercept of the abscissa. The pA_2 value is the negative logarithm of the affinity of an antagonist for a given receptor in a defined biological system and is equal to $-\log_{10} K_B$ where K_B is the dissociation constant for a competitive antagonist with a slope of near unity. Not all antagonists are competitive. Non- or uncompetitive antagonists that act at allosteric sites or that bind irreversibly to the agonist site have slopes that are

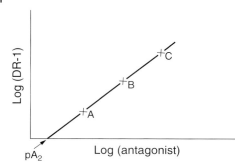

Fig. 2 Schild plot regression. Data from Fig. 1(b) for antagonist concentrations A, B, and C can be plotted by the method of Schild (29) to yield a pA_2 value, a measure of antagonist activity. A slope of unity indicates a competitive antagonist.

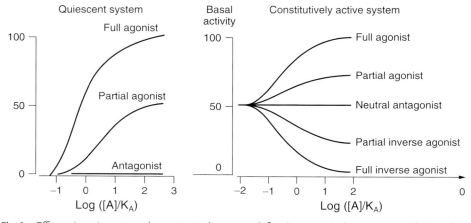

Fig. 3 Efficacy in quiescent and constitutively active systems. In a quiescent system, three types of ligands can be defined, full agonist, partial agonist, and antagonist, depending on the response elicited. In a constitutively active system, an antagonist from a quiescent system is defined as a neutral antagonist with ligands that inhibit the activity of the constitutively active system and are defined as full and partial inverse agonists depending on the degree of inhibition.

significantly less than unity. The Schild plot can thus be used to determine the mechanism by which an antagonist produces its effects.

Occupancy theory was modified by Ariens on the basis of data showing that not all cholinergic agonists were able to elicit a maximal response in a skeletal muscle preparation, even when administered at supramaximal concentrations. This led to the introduction of the concept of the *intrinsic activity* of a ligand. A full agonist was defined as having a value of 1.0 with the value for an antagonist being zero. However, many compounds were subsequently identified that bound to the receptor but were able to only produce a portion of the response seen with a full agonist. These were defined as partial agonists (Fig. 1a). By definition, these compounds were also partial antagonists.

Other agonists have also been identified that produce a response greater than the previously defined full agonist. These have been termed *super agonists*. One example is the muscarinic cholinergic receptor agonist, L 670,207 (Fig. 3), an arecoline bioisostere that was 70% more active than arecoline and thus had an intrinsic activity of 1.7. From the activity seen in response to L 670,207, the system that is used to characterize these muscarinic agonists was obviously capable of a greater response than that seen with arecoline, making the latter compound a partial agonist.

The partial agonist concept was additionally refined by Stephenson in introducing the concept of *efficacy*, ε, which differed from intrinsic activity in that the latter was defined as a proportion of the maximal response [effect = (RL)]. This concept was extended to situations in which a maximal response to an agonist could occur when only a small proportion of the total number of receptors on a tissue were occupied (a condition termed *receptor reserve*), as in the situations when receptors were inactivated by alkylating agents. This resulted in a nonlinear occupancy relationship with the response then being defined as the stimulus, S, a product of the fraction of receptors occupied and the ligand efficacy. A nonlinear functional response clearly complicates data interpretation, especially when spare receptor or receptor reserve concepts are introduced to rationalize individual data sets. An additional issue in defining efficacy was the degree to which the receptor activation event and its blockade by antagonists was measured through events that were spatially and temporally removed from the receptor activation event, and also the degree to which the response could be amplified through cofactor and signal transduction cascades.

Kenakin described ligand-mediated responses in a given tissue in terms of four parameters: (1) receptor density; (2) the efficiency of the transductional process; (3) the equilibrium dissociation constant of the RL complex; and (4) the intrinsic efficacy of the ligand at the receptor. *In vivo*, receptor occupancy for exogenous ligands is primarily dependent on pharmacokinetic parameters, whereas that for native endogenous ligands is most probably under intrinsic homeostatic controls including rates of production, release, metabolism, and reuptake of ligands.

Classical receptor theory assumed that affinity and efficacy were independent parameters, suggesting that no consistent relationship existed between the affinity of a ligand and its ability to elicit a full response. A ligand with relatively low affinity ($<10^{-6}$ M) could still be a full agonist when a sufficient concentration interacted with the receptor. With the discovery of constitutive receptor activity, it seems that this lack of a consistent relationship is more reflective of an inability to measure receptor-mediated activity than a potency disconnect. Thus, one may conclude that all ligands have efficacy if the appropriate system is used to measure this parameter.

The relationship between the receptor and ligand in the classical lock and key model with the RL interaction resulting in no change in the receptor conformation essentially described a static situation. In 1958, however, Koshland described the induced fit model of the ES complex, where substrate binding to the enzyme caused a change in the three-dimensional structure of the protein, leading to a change in activity. Together with the pioneering work of Hill on hemoglobin allosterism, the concept of allosteric modulation of receptor function was proposed.

3.2
Rate Theory

On the basis of experimental data showing the persistence of antagonist-mediated responses and agonist "fade" where maximal responses occur transiently, to be followed by lesser responses of longer duration and agonist-mediated blockade of agonist effects, Paton modified the concept of occupancy to include a chemically based rate term. According to rate theory, it was not only the number of receptors occupied by a ligand that determined the tissue response but also the *rate* of RL formation. The resultant effect, E, was considered equal to a proportionality factor, ϕ, that included an efficacy component and the velocity of the RL interaction, V. Thus,

$$E = \omega V eq \qquad (4)$$

The rate of RL formation, like that of neurotransmitter release, was measured in quantal terms with discrete "all or none" changes in receptor-mediated events. Pharmacokinetic considerations also play a major role determining the rate of RL formation.

The primary factor delineating occupancy and rate theory seemed to be the dissociation rate constant. Thus, if this factor was large, the ligand was an agonist; if the factor was small, reducing the quantal response to receptor occupancy, the ligand functioned as an antagonist. The kinetic aspects of rate theory did not, however, take into account the efficacy of transductional coupling and the potential for amplification after the initial binding event leading to its description as "a provocative conceptualization ... with limited applicability," a phrase that can be applied to many of the newer aspects of receptor theory.

4
Receptor Complexes and Allosteric Modulators

The studies of A.V. Hill on the binding of oxygen by hemoglobin demonstrated that identical binding sites on protein oligomers could influence one another such that the binding of the first ligand (in the case of hemoglobin in oxygen) facilitated the binding of a second, identical ligand, and so on for sequentially bound ligands such that the saturation curve describing the interaction of the ligand with its recognition site is steeper than that which would be predicted from classical Michaelis–Menten kinetics. The process of one ligand, homologous or heterologous, interacting with the binding of another is thought to occur by a cooperative, conformational change in the binding protein for the second ligand from a site adjacent to the ligand recognition site. Koshland's induced fit model extended these findings, leading to the development of two models of cooperativity or allosterism: the *sequential or induced fit* model described by Koshland; and the *concerted model* of Monod, Wyman, and Changeux, the key elements of which are outlined in Table 2.

Both models assume the existence of oligomeric protein units existing in two states that are in equilibrium with one another in the absence of ligand. Ligand binding induces a conformational change in the protein(s), moving the equilibrium of the two states to favor that with the higher affinity for the ligand. This in turn alters the kinetic and functional properties of the oligomeric complex. This model has been further refined in terms of *ligand-stabilizing conformational ensembles*.

The site on a receptor that defines its pharmacology and membership of

Tab. 2 Allosteric receptor models.

Monod, Wyman, Changeux Concerted Model
Receptor complex is a multicomponent oligomer comprised of a finite number of identical binding sites.
Subunits are symmetrically arranged each having a single ligand binding site.
Receptor complex exists in two conformational states, one of which has a preference for ligand binding.
Conformational transition state involves a simultaneous shift in the state of all subunits.
No hybrid states exist implying cooperativity.

Koshland Nemethy Filmer Sequential Model
Receptor complex is a multicomponent oligomer with symmetrically arranged protomers each with a single ligand binding site.
Protomers exist in two conformational states with transition induced by ligand binding.
Receptor symmetry is lost on ligand binding.
Hybrid states of the receptor complex can be stabilized by protomers.
Stabilization is equivalent to negative cooperativity.

---

a particular receptor superfamily, for example, 5HT, nicotine, and so on, is termed the *orthosteric site*. Ligands that bind to this receptor have a spatial overlap for the binding site such that their binding is mutually exclusive (unless an antagonist covalently binds to the orthosteric site). In contrast, the *allosteric site* (of which there may be more than one associated with a single orthosteric site and which can affect that site) is distinct from the latter in that ligands that bind to the allosteric site(s) can produce effects on ligand binding and efficacy to the orthosteric site through an indirect, conformational modulation of this site, which probably involves alterations in either the association or dissociation rates of the orthosteric ligand. While much of the early work on allosterism derived from studies on enzymes and ligand-gated ion channels (LGICs), it is now clear that GPCRs that were once considered as monomeric proteins with only an orthosteric site are now known to contain allosteric sites and can form oligomeric complexes. The concept of allosterism becomes more complex when considering multiple ligand sites on a receptor that have different pharmacological profiles, for example, when the ligand recognition sites are totally heterogeneous.

The identification of allosteric ligands that can have both positive and negative effects on the function of the orthosteric site occurred in a largely serendipitous manner. The first drug identified as an allosteric modulator was the BZ, diazepam, which has anxiolytic, hypnotic, and muscle relaxant activities and produces its effects by facilitating the actions of the $GABA_A$ receptor. Unlike the majority of directly acting $GABA_A$ receptor agonists, diazepam has a relatively safe side effect profile.

Allosteric modulators have three apparent advantages over drugs acting via orthosteric sites: their effects are saturable such that there is a ceiling effect to their activity, which results in a good margin of safety in human use; their effects are selective, and as mentioned earlier, frequently "use-dependent." Thus, the actions of an allosteric modulator occur only when the endogenous orthosteric ligand is present. In the absence of the latter, an allosteric modulator is theoretically quiescent and may thus represent an ideal prophylactic treatment for disease states associated with sporadic or chronotropic occurrence. Finally, allosteric modulators are considered to be more specific in their effects, partly because of the nature of their binding sites that are distinct from the orthosteric site and partly because of the extent to which their effects depend on the degree of cooperativity between the allosteric and orthosteric sites.

The "cys-loop" family of LGICs including the $GABA_A$, glycine, $5HT_3$, and nicotinic cholinergic receptors are the best characterized of the allosterically regulated receptors. GPCRs are also subject to allosteric modulation. The families identified to date demonstrating this property include adenosine (P1) and α-adrenergic, dopamine, chemokine, $GABA_B$, endothelin, metabotropic glutamate, neurokinin-1, P2Y, and muscarinic cholinergic, as well as some members of the 5HT superfamily. The cone snail conotoxin, ρ-TIA is an allosteric modulator of the α_{1B}-adrenoceptor. Changes in GPCR function resulting from alterations in the ionic milieu also reflect the potential for allosteric modulation of receptor function.

5
Ternary Complex Models

The ternary complex model or TCM describes allosteric interactions between orthosteric and allosteric sites present on a single protein monomer. It can also be extended to reflect the other two-state interactions involving sites on adjacent proteins and the effect of signaling proteins, for example, G-proteins on the function of the receptor. Christopoulos notes allosteric interactions as being reciprocal such that the effects of a ligand A on the binding properties of ligand B also imply an effect on the binding of ligand B on the properties of ligand A. Similarly, because GPCRs alter the conformation of G-proteins to elicit transductional events and an alteration in cell function, changes in G-protein conformation and interactions alter receptor function, and this may be reflected in desensitization.

6
Constitutive Receptor Activity

A basic principle of receptor theory is that when a receptor is activated by a ligand, the effect produced by the ligand is proportional to the concentration of the ligand; for example, it follows the Law of Mass Action. It is now becoming apparent that receptors spontaneously form active complexes as a result of interactions with other proteins. This is especially true when receptor cDNA is *overexpressed* in cell systems such that the relative abundance of a receptor is in excess of that normally occurring in the native state, or associates with proteins that reflect the host cell milieu in which the receptor is expressed, rather than an intrinsic property of the receptor in its natural environment. This is a major issue in the characterization of ligand efficacy. A spontaneous interaction between receptor and effector can occur more frequently in a system where the proteins are in excess and where the normally present factors that control such interactions are absent. This is shown graphically in Fig. 3, where constitutive activity is shown in the range of 0–50 and where the theoretical effects of inverse agonists, full and partial, are shown. A quiescent system that is more reflective of classical receptor theory shows a full agonist, partial agonist, and what is now defined as a neutral antagonist. Constitutive receptor activation has been described in terms of *protein ensemble* theory and in terms of *allosteric transition*, where changes in receptor conformation can occur through random thermal events.

7
Efficacy Considerations

Historical receptor theory describes a ligand efficacy continuum, with full agonism at one end and full antagonism at the other. Between the two ends of this continuum lie partial agonists that, as already noted, imply that ligands can also be partial antagonists. Antagonism *per se* implied that a ligand could bind to a receptor without producing any effect and limiting access to the native agonist – block receptor activation. Such a compound is now called a *neutral antagonist*.

With the ability to measure constitutive receptor activity, some compounds, for example, the $\beta 2$-adrenoceptor antagonist, ICI 118551, inhibited constitutive activity, thus functioning as an *inverse agonist* or *negative antagonist*.

The actions of a competitive antagonism can be surmounted by the addition of increasing concentrations of the agonist ligand, resulting in a functional DR curve that undergoes a rightward shift with approximately the same shape and maximal effect (Fig. 1). Noncompetitive or uncompetitive antagonists interact at sites distinct from the agonist recognition site, and can modulate agonist binding either by proximal interactions with this site from a site adjacent to the recognition site or by allosteric modulation. The effects of noncompetitive antagonists are usually not reversible by the addition of excess agonist. This type of antagonism, whether competitive or noncompetitive, occurring at a distinct molecular target is known as *pharmacological antagonism* and involves the interactions between ligands and the receptor site (Fig. 4). In contrast, *functional antagonism* refers to a situation in which an antagonist that does not interact with a given receptor can still block the actions of an agonist of that receptor, and is typically measured in intact tissue preparations or whole animal models.

In the cartoon outlined in Fig. 4, neurotransmitter A released from neuron A interacts with A-type receptors on neuron B. Antagonist α can block the effects of A on cell B by interacting with A receptors. Antagonist α is thus a pharmacological

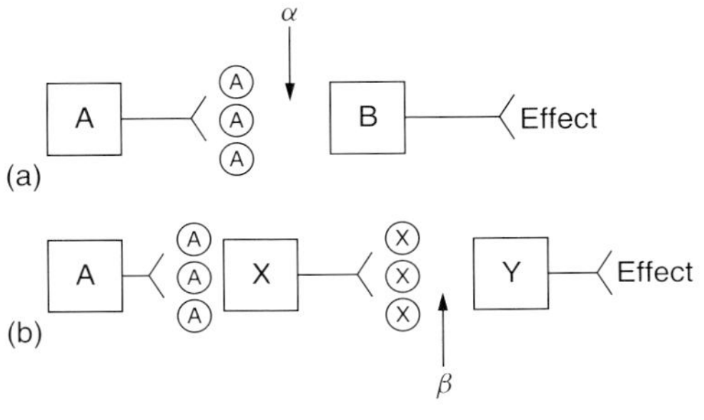

Fig. 4 Pharmacological versus functional antagonism. In panel (a), neurotransmitter A is released from neuron A, directly interacting with neuron B to produce a functional response. Antagonist blocks the effects of A, a direct pharmacological antagonism of the effects of A. In panel (b), neurotransmitter A is released from neuron A, directly interacting with neuron X, which in turn releases neurotransmitter X, which acts on cell Y to produce a functional response. Antagonist blocks the effects of X on cell Y, but in the absence of other data on the actions of antagonist, seems to block the functional effects of A because of the circuitry involved. Antagonist thus acts as a functional antagonist.

antagonist of A receptors. In the second example in Fig. 4, neuron A releases neurotransmitter A, which interacts with A-type receptors located on neuron X. In turn, neuron X releases neurotransmitter X that interacts with X-type receptors on neuron Y. Antagonist β is a competitive antagonist that interacts with X receptors to block the effect of neurotransmitter X, and in doing so, indirectly blocks the actions of neurotransmitter A. Antagonist β is thus a pharmacological antagonist of receptors for the neurotransmitter X, but a functional antagonist for neurotransmitter A.

In interpreting functional data in complex systems, it is always important to consider the possibility that a ligand has more than one effect mediated through a single class of receptor. For this reason, in advancing new ligands from *in vitro* evaluation to more complex tissue systems or animal models, it is extremely helpful to have a ligand-binding profile, for example, the activity of a compound at a battery of 90 or more receptors and enzymes (a Cerep profile), to fully understand any new findings. For instance, when a ligand for a new receptor is advanced to animal models and found to elicit changes in blood pressure, it is extremely helpful to know whether in addition to its defined activity at the new receptor, it has some other properties that relate to the blood pressure effects, rather than assuming that some unknown mechanism related to activation of the new receptor has cardiovascular-related liabilities.

8
Receptor Dynamics

Receptors, present in their active form at the cell surface are not static entities.

Ligand binding, alterations in gene function and tissue dysfunction due to disease, trauma or aging, can alter the number, half-life, and responsiveness of receptors, channels, and transporters. Receptors turnover as a normal consequence of cell growth, with half-lives that vary between hours and days. Ligand binding can result in receptor internalization through phosphorylation-dependent events often initiated as a result of the ligand-binding process, exposing serine and threonine residues in the receptor protein. Histamine H_2 antagonists acting as inverse agonists in constitutively active systems can upregulate their cognate receptors, potentially increasing cell sensitivity. It is reasonable to assume that the target cells for endogenous effector agents, neurotransmitters, neuromodulators, and neurohormones operate under tonic control, one example being chronotropic, varying with the circadian rhythm of the organism. In contrast, the effects of exogenously administered ligands, for example, drugs, are rarely under normal homeostatic control, and as a result, their effects frequently become blunted on repeated administration or when they are administered in controlled release forms. Therefore, it should not be a surprise that the majority of effective therapeutic agents are antagonists of receptor function.

In Parkinson's disease, the presynaptic nerve cells in the *substantia nigra* that normally produce dopamine die as the result of an as yet unknown disease etiology. This defect results in a decrease in dopamine levels and a consequent hypersensitivity of postsynaptic responses as the cellular homeostatic events attempt to compensate for a lack of endogenous ligand. Agonists that rapidly desensitize a receptor, for example, ATP at the $P2X_3$ receptor and nicotine at the $\alpha 4\beta 2$ receptor, appear to be antagonists because their net effect is to attenuate the normal receptor response.

9 Receptor Nomenclature

Even before the publication of the draft maps of the human genome, the techniques of molecular biology had already resulted in an explosion in the number of putative receptor families and subtypes within families, as well as classes of receptors known as *orphan receptors*. These were structurally related members of receptor classes for which the endogenous ligand and associated function was unknown.

New receptors have been (and continue to be) identified in different laboratories often simultaneously, leading to different names causing considerable confusion in the literature. The *International Union of Pharmacology* (IUPHAR) has undertaken the development of a systematic nomenclature system, based in part on naming families of receptors for their cognate endogenous ligands. The deliberations of the various committees enlisted to devise a systematic nomenclature are published periodically in *Pharmacological Reviews*. Compendiums are also regularly published, the most current and comprehensive being *The Sigma-RBI Handbook*.

10 Receptor Classes

Receptors can be divided into four major classes (Table 1): heptahelical, 7-transmembrane (7-TM) GPCRs, ion channels, transcription factor receptors and enzyme-associated receptors. Of these, the 7-TM receptors have historically represented the most fertile class for drug discovery, as these have been the most studied. This receptor superfamily includes receptors for the bioamine neurotransmitters (5-HT, dopamine, etc), peptide hormones, and lipid signaling molecules. Ion channels are further subdivided into LGIC, voltage-sensitive calcium channel (VSCC) and potassium (Kir), and ion/pH-modulated (ASIC) subtypes, all of which have similar but distinct, multimeric structural motifs. TRPV (transient receptor potential vanilloid) ion channels can be modulated by temperature, for example, vanilloid receptors. The largest group within the transcription factor receptor superfamily is the nuclear hormone receptors that include receptors for steroids (glucocorticoid (GR), progesterone (PR), mineralocorticoid (MR), androgen (AR)), and nonsteroids (thyroid hormones (TRs), PPARs, the retinoic acid receptors (RXR, RAR), and vitamin D_3 (VDR)). In addition, many orphan receptors have been identified with structural similarity to the known transcription factor receptors. The enzyme-associated receptor superfamily is a family of single- or multi-subunit proteins, which contain a subunit with a single transmembrane domain. The largest groups within this superfamily are the single-subunit–tyrosine kinase receptors, such as the PDGF (platelet-derived growth factor) and EGF (epidermal growth factor) receptors, and the multimeric complexes that utilize kinases such as the JAK-type kinases for their signal transduction (e.g. cytokine and growth hormone receptor families). Also included are receptors with serine/threonine kinase or guanylyl cyclase activity.

In addition to this multitude of receptor classes, further complexity in conceptualizing receptors as distinct, classifiable entities is exemplified by recent findings related to GPCRs. By implicit definition, these receptors produce their physiological effects by coupling through the G-protein family. However, there are instances in which ion channels produce their effects by coupling through G-proteins. Receptor classification thus has few absolutes.

10.1
G-protein–Coupled Receptors

The 7-TM motif of GPCRs based on the X-ray structure of rhodopsin is a single protein comprised of approximately 300–500 amino acids with discrete amino acid motifs in the transmembrane (TM) regions and on the C-terminal extracellular loop, determining the ligand specificity of the receptor and those on the amino terminal and intracellular loops designating G-protein interactions. For example, RL interactions for many of the bioamine neurotransmitter receptors are thought to occur within a pocket in the 7-TM motif that is generically designated to lie between TMs III, IV, and VI. Posttranscriptional alternative splicing can alter the GPCR to create isoforms that may be species, tissue, and disease-state dependent.

The number of GPCRs present in the human genome, including orphan receptors, has been estimated to be 1000–2000 with over 1000 of these coding for odorant and pheromone receptors. Currently of greatest interest in drug discovery efforts are the 367 nonchemosensory GPCRs, over 100 of which remain "orphan receptors."

GPCRs are organized into four main families. *Family 1* includes the majority of GPCRs for peptide hormones, neurotransmitters, odorants, and a large group of orphans. Family 1 GPCRs can be divided by structural similarity into subfamilies: 1a that includes rhodopsin, adrenoceptors, thrombin, and the adenosine A_{2A} receptor, with a binding site localized within the 7-TM motif; 1b that includes receptors for peptides with the ligand-binding site in the extracellular loops, the N terminal, and the superior regions of the TM motifs; and 1c that comprises receptors for glycoproteins. Ligand binding to this receptor class is mostly extracellular. *Family 2* is morphologically related, but not sequence related to the 1c family and consists of four GPCRs activated by the hormones glucagon, secretin, and VIP-PACAP. *Family 3* contains four metabotropic glutamate receptors (mGluR) and three $GABA_B$ receptors. *Family 4* is a group of GPCRs that includes "frizzled" and "smoothened," both involved in embryonic development. Additional GPCRs have been described in invertebrate species that may define additional receptor families.

Interactions with heterotrimeric G-proteins and other associated signaling molecules have the potential for considerable complexity. Heterotrimeric G-proteins exist as complexes formed from α, β, and γ-type subunits, for each of which there are multiple subtypes, thereby offering a considerable variety of potential functional G-proteins. The α subunits are best characterized and often determine the signaling pathway activated There are four major G-protein α-subunit families that interact with GPCRs: (1) G_s that activates adenylyl cyclase; (2) $G_{i/o}$ that inhibits adenylyl cyclase and can also regulate ion channels and activation of cGMP PDE; (3) G_q that activates phospholipase C; and (4) G_{12} that regulates Na^+/H^+ exchange. The β and γ subunits are less well studied, but are also involved in signaling events such as interactions with ion channels and certain isoforms of phospholipase C, as well as in membrane trafficking and receptor interaction events.

10.1.1 G-protein–Associated or – Modulating Proteins

The cyclic nucleotide phosphodiesterases (PDEs) that are responsible for hydrolytic degradation of the cyclic nucleotides, cAMP and cGMP, exist in more than 15 isoforms. While the protein kinases, PKA and PKC, are responsible for protein phosphorylation, the GPCR kinases, GRK 1–6, are responsible for GPCR phosphorylation, and the protein phosphatases are responsible for dephosphorylation, the latter potentially numbering in excess of 300, significantly increasing the complexity of GPCR-associated signaling processes.

Superimposed on these signaling proteins are the calmodulins that mediate calcium modulation of receptor function, the β-arrestins, involved in inactivation of phosphorylated receptors (6 members), a group of proteins termed RGS (regulators of G-protein signaling), RAMPs, and a protein known as Sst2p that is involved in receptor desensitization.

Given the number of GPCRs present in the human genome, including orphan receptors, multiple G-protein subtypes, phosphatases, and the various GPCR-associated signaling proteins described earlier, there is obviously considerable scope for complexity in cell signaling associated with the GPCR family. GPCRs can also form homo- and heteromeric forms (e.g. $GABA_B$, adenosine A_1 and A_{2A}, angiotensin, bradykinin, chemokine, dopamine, metabotropic glutamate, muscarinic, opioid, serotonin, and somatostatin), further increasing the potential complexity of ligand-driven GPCR signaling processes and offering an opportunity to explore new targets in medicinal chemistry.

Applying an evolutionary trace method (ETM) to assess potential protein–protein interactions, Dean et al., identified functionally important residue clusters on TM helices 5 and 6 in over 700 aligned GPCR sequences. Similar clusters were found on TMs 2 and 3. TM 5 and 6 clusters were consistent with 5,6-contact and 5,6-domain swapped dimer formation. Additional application of ETM to 113 aligned G-protein sequences identified two functional sites: one associated with adenylyl cyclase, and regulator of G-protein signaling (RGS) binding, and the other extending from the ras-like to helical domain that seems to be associated with GPCR dimer binding. From such findings, it was concluded that GPCR dimerization and heterodimerization occur in all members of the GPCR superfamily and its subfamilies.

From these findings, potential new approaches to ligand design and by extrapolation, drug discovery, include: (1) antagonists that can act by inhibiting dimer formation, for example, transmembrane peptide mimics; (2) bivalent compounds/binary conjugates; and (3) compounds targeting the GPCR–G-protein interface.

10.2 Ligand-gated Ion Channels

Ion channels consist of homo- or heteromeric complexes numbering between three (P2X) and eight (Kirs) subunits. Examples of these are the $GABA_A$/benzodiazepine and NMDA/glycine receptor, neuronal nicotinic receptors (nAChR), and P2X receptors.

10.2.1 $GABA_A$/Benzodiazepine (BZ) Receptor

$GABA_A$/benzodiazepine receptor is an LGIC that is the target site for numerous clinically effective anxiolytic, anticonvulsant, muscle relaxant, and hypnotic drugs, which produce their therapeutic effects by enhancing the actions of the inhibitory neurotransmitter, GABA. It is a pentameric LGIC, the constituent subunits of which are formed from a family of six, four, one, and two subunits, leading to the potential existence of several thousand different pentamers. The functional receptor complex contains a $GABA_A$ receptor, a BZ recognition site, and by virtue of its pentameric structure, a central chloride channel. Allosteric recognition sites on this complex include those for ethanol, avermectin, barbiturates, picrotoxin, and neurosteroids like allopregnanolone. The pharmacology and function of the allosteric sites depends on the actual subunit composition. The 1 subunit, which is present in nearly 60% of $GABA_A$ receptors, in mouse brain mediates the sedative, anticonvulsant, and amnestic effects of diazepam; the 2 subunit, the antianxiety effects of diazepam; and the 5 subunit, associative temporal memory. With this knowledge, rather than screening new ligands for the $GABA_A$ receptor in various animal models to derive a profile for a nonsedating anxiolytic, by understanding the structural determinants of BZ interactions with 1 and 2 subunits and designing compounds that preferentially interact with the latter, the process of designing novel ligands can be considerably enhanced.

10.2.2 The NMDA Receptor

The NMDA receptor is a member of the glutamate receptor superfamily that mediates the effects of the major excitatory transmitter, glutamate. It is composed of an NMDA receptor, a central ion channel that binds magnesium, the dissociative anesthetics, ketamine, and phencyclidine (PCP), the noncompetitive NMDA antagonist, dizocilpine (MK 801), glycine and polyamine binding sites, activation of which can markedly alter NMDA receptor function, and some 70 other ancillary proteins, the physiological function of which remains to be determined. The activation state of the receptor can define the effects of the allosteric modulators. Thus, some are termed *use-dependent*, reflecting modulatory actions only when the channel is opened by glutamate.

10.2.3 The Nicotinic Cholinergic Receptor, nAChR

The nicotinic cholinergic receptor is another pentameric LGIC composed of distinct functional regions and subunits. The subunit composition of the receptor varies, imparting different functionality when the channel is activated by nicotine, or by the endogenous ligand, ACh. Allosteric modulators of neuronal nAChRs include dizocilpine, avermectin, steroids, barbiturates, and ancillary proteins.

10.2.4 The P2X Receptor

The P2X receptor is an LGIC responsive to ATP that functions as a trimer formed from a family of seven subunits that can form both homo- and heteromers leading to the existence of at least seven distinct receptor subtypes.

Little is known regarding the structural elements involved in the ligand pharmacology and function of LGICs, such that signaling transduction pathways have not been extensively characterized. For nAChRs, it is known that the recognition site for ACh is formed between two subunits. Thus multiple orthosteric sites for ACh are possible, depending on the types of subunit forming the receptor. Many new classes of LGIC are still being discovered.

10.3
Steroid Receptor Superfamily

The steroid receptor superfamily includes the glucocorticoid (GR), progesterone (PR), mineralocorticoid (MR), androgen (AR), thyroid hormone (TR), and vitamin D_3 (VDR) receptors. These receptors bind steroid hormones and are then translocated to the nucleus, where they bind to hormone responsive elements on DNA promotor regions to alter gene expression. While steroids are very effective anti-inflammatory agents, they have a multiplicity of serious side effects that limit their full use.

The antiestrogen, tamoxifen, is the most commonly used hormonal therapy for breast cancer and has demonstrated positive effects on the cardiovascular and skeletal systems of postmenopausal women but is associated with an increased risk of uterine cancer. Tamoxifen is described as a selective estrogen receptor modulator (SERM) with a tissue selective profile dependent on the different distribution of the α- and β-subtypes of the estrogen receptor (ER-α and ER-β) that activate and inhibit transcription respectively. These selective effects have been ascribed to differential interactions with gene promotor elements and coregulatory proteins depending on whether the ER interacts directly, or in a tethered manner with DNA.

10.4
Intracellular Receptors

Members of the intracellular receptor family include the cytochrome P450 (CYP) family, the SMAD family of tumor suppressors, intracellular kinases and phosphatases, nitric oxide synthases (NOS), caspases, the RXR, RAR superfamilies, receptor activated transcription factors (RAFTs), and signal transducers and activators of transcription (STATs) such as AP-1, NFκB, NF-AT, STAT-1, PPARs, various hormone responsive elements on DNA and RNA promoters, and ribozymes. These represent a bewildering number of potential drug targets especially in the metabolic disease and cancer areas. Given their roles in normal cell function, it will be a challenge to ensure specificity in ligands interacting with these targets.

10.5
Non-GPCR-linked Cytokine Receptors

Cytokines are polypeptide mediators and are involved in the inflammatory/immune response. There are three cytokine families: *hematopoietin*, which includes IL-2–IL7, L-9–13; IL-15–IL-17; IL-19, 1L-21, IL-22, GMCSF, GCSF, EPO, LIF, OSM, and CNTF, with primary signal transduction through the Jak/STAT pathway; *tumor necrosis factor*, comprising the receptors, TNFRSF1–18 that signal through NFB, TRAF, and caspases; and the *interleukin 1/TIR* family that includes IL-1RI and IL-1RII, IL-1Rrp2, and IL1RAPL. TIGGR-1, ST2, IL-18, and Toll 1–9 also signal through NFB and TRAF.

10.6
Orphan Receptors

Orphan receptors have been generally defined as proteins with a receptor motif that lack both a ligand and function. Approximately 160–300 orphan GPCRs are thought to be in the draft sequence of the human genome and intense efforts are currently ongoing to identify the ligands for these and their function as novel intellectual property for the drug discovery process. Orphan receptor validation can be done using expression profiling to identify tissues rich in the expression of the receptor of interest, and a technique known as *reverse pharmacology* that can be used to identify a ligand for the orphan receptor. In the latter, the orphan receptor is used as "bait" to bind selective ligands. These can then be used to further characterize receptor function. Nearly 30 orphan GPCRs have been validated in this manner. While most of the current interest on orphan receptors is focused on GPCRs because of the considerable body of existing knowledge about this receptor class, it is anticipated that orphan receptors will also be discovered for other receptor classes, for example, orphan nuclear receptors.

The orphan receptor approach to drug discovery is exemplified by the orphanin/FQ receptor, ORL1, a structural homolog of the opioid receptor. Identified in 1995 using a homology-based screening strategy, ORL1 had low affinity for known opioid ligands. A novel heptadecapeptide ligand for the ORL1 receptor, orphanin/FQ, was subsequently isolated from brain regions rich in ORL1 providing the key tool to validate the target and identify a functional role for the receptor in stress-related situations in animal models. From an intensive screening program, an antagonist of this receptor was identified, Ro 64–6198, which represented a novel anxiolytic/antidepressant drug candidate.

10.7
Neurotransmitter Transporters

Neurotransmitter transporters (NTs) are responsible for terminating the effects of neurotransmitters by removing them from the extracellular space. NTs are integral membrane proteins present on both the plasma membrane and on intracellular vesicles where they effect vesicular packaging of neurotransmitters. Their activity is dependent on the Na+ intracellular/extracellular gradient. Most NTs (e.g. dopamine transporters (DAT), norepinephrine transporters (NET), serotonin transporters (SERT) also require Cl^- and are termed Na^+/Cl^- *dependent NTs*, (SCDNTs) Transporters for glutamate also require K^+ (SKDGTs). These proteins bind their cognate neurotransmitters with high affinity and transport them across the plasma membrane into the cell. GABA transporters comprise the GAT family.

NT inhibitors represent a major class of drugs producing their effects by blocking neurotransmitter uptake, thus potentiating the actions of endogenous neurotransmitters. As such, NTs can be considered as indirect receptor agonists. Inhibitors of SERT (also known as *selective serotonin reuptake inhibitors*, SSRIs) are major antidepressants, for example, fluoxetine. Mixed SERT/NET inhibitors (SNRIs) are also antidepressants and analgesics while DAT inhibitors are well-known stimulants, e.g. cocaine, amphetamine and so on. Inhibitors of GABA transporter (GAT) are anticonvulsants (e.g. gabatril).

10.8
Neurotransmitter Binding Proteins

Binding proteins for corticotrophin releasing factor (CRF) and acetylcholine (AChBP) have been identified and exploited as potential drug discovery targets. These proteins can act as reservoirs or "decoy receptors" for their respective neurotransmitters, playing a general role in buffering synaptic message transfer.

10.9
Drug Receptors

Most drugs interact with receptors (or enzymes) for which the natural ligand (or substrate) is known. There are, however, a number of receptors, distinct from the evolving class of orphan receptors, for which the synthesized drug is the only known ligand. The best example of this is the central benzodiazepine receptor present on the $GABA_A$ ion channel complex that was originally identified using radiolabeled diazepam. Because this is the site of action of the widely used BZ anxiolytic drug class, this is a *bona fide* receptor with clinical relevance. However, despite considerable efforts and a number of interesting candidate compounds, no endogenous ligand has yet been unambiguously identified which would represent an endogenous modulator of anxiety acting through this site. Other examples of drug receptors for which synthetic ligands were identified before the endogenous agonists, are the cannabinoid receptors, CB_1 and CB_2, through which the following compounds act: 9-Δ-tetrahydrocannibol, the active ingredient of the psychoactive recreational drug/analgesic/antiemetic drug, marijuana; the vanilloid receptor, VR-1, the known ligand for which is capsaicin, the ingredient in red pepper that evokes heat sensation; and the opioid receptor family, the site of action of morphine and other derivatives of the poppy. For each of these drug receptors, endogenous mammalian ligands, anandamide, the endovanilloid, *N*-arachidonyl-dopamine (NADA), enkephalins, and orphanin/FQ have been identified.

11
Molecular Biology of Receptors

Recombinant DNA methods have been extensively used to isolate and analyze the sequence of numerous receptors from various mammalian complementary DNA (cDNA) libraries using the polymerase chain reaction (PCR) technique to clone receptors that can then be expressed in various prokaryotic and eukaryotic cell lines and *Xenopus* oocyte. Conserved regions in receptors that are involved in ligand binding, coupling to transductional systems and ion channel formation, have thus been identified. Receptor cloning, sequencing, and expression are now an intimate part of the drug discovery process.

From the receptor/ligand modeling standpoint, the ability to specifically alter the structure of cloned receptors through the modification of a small number of nucleotides, the process known as *site-directed mutagenesis*, the removal of specific regions of the receptor gene "deletion mutagenesis," and the construction of hybrid (chimeric) receptor proteins can provide additional clues on the relative importance of individual amino acids in the process of ligand recognition and receptor function. These techniques are also used to manipulate receptors to produce recombinant expression systems with better cell surface expression, more convenient signal transduction coupling, and other qualities important for efficient drug discovery.

12
Functional Genomics: Target Validation/Confidence Building

The identification of novel receptor targets from the human genome and their subsequent use in the drug discovery process requires that the target be validated. Their validation requires that the ligand and function of the receptor be known, and this topic has been addressed in relation to orphan receptors mentioned earlier.

Target validation, especially that related to novel, genomically derived targets, is a highly complex and resource intense process, adding significantly to the cost of drug discovery. In addition to orphan receptors, identified on the basis of their homology to members of known receptor classes, novel genomic targets can be identified by drug-related differential gene expression analysis, population genetic approaches, or a combination of both. Once a gene is associated with a disease, its protein product needs to be characterized and a biochemical function, for example, receptor, enzyme, or transporter, ascribed to it. The probable function of the protein can then be assessed and its cognate ligand or substrate identified. Not all proteins identified in this manner are obviously involved in mammalian cell function. The protein product of a novel lithium-related (NLR) gene identified in mice exposed to lithium has sequence homology to a bacterial nitrogen permease. The potential role of this protein in the etiology of bipolar affective disorder remains unclear. Patients can also be genotyped for susceptibility related to a drug target or drug side effects, to identify "responsive" patients for more efficient and safer drug use.

To identify what are termed *druggable* gene products, the still somewhat mystical process of functional genomics can be used to construct protein–protein interaction maps, to identify other proteins in a pathway/network that can represent more facile entry points to the protein cluster associated with the genetically identified target and with the disease state. Using techniques like yeast two-hybrid and *gal*-pull down, the function of an uncharacterized protein can be assigned on the basis of the known function of its interacting partners, involving the extensive use of bioinformatics tools.

Targeted gene disruption, antisense, and ribozyme inhibition (RNAi) are other techniques for assessing the phenotype of a given gene. Antisense to the rat $P2X_3$ receptor had marked hyperalgesic activity showing an unambiguous role for this receptor in chronic inflammatory and neuropathic pain states. Transgenic animals, in which the function of a gene is knocked out by genetic techniques, are less useful in understanding the function of a protein. In addition to being restricted to the mouse, this approach is high in cost and time (taking 1–2 years to generate sufficient animals for evaluation), and in many instances, the absence of a given protein has no overt effect on the phenotype of an animal because of compensatory changes during the developmental phase. Alternatively, the gene knockout may lead to an animal that has a limited, if any, life span or reproductive capability.

Another target validation approach involves the use of ligands to define the function of a new target. Already discussed in regard to orphan receptors, other ligand-directed target validation models fall under the rubric of *chemical genomics*. A key to identifying useful ligands that selectively interact with proteins of unknown function to define their function is a sufficient level of pharmacophore diversity to ensure success.

Assuming that a novel protein target is amenable to HTS approaches to identify its cognate ligand, that the ligand is identified, and that other approaches are able to define a putative function for the target, the next step in target validation is to identify a "druggable" molecule that can be advanced to the clinic. For ORL1, Ro 64–6198 is an excellent example of this approach.

This raises the key test of *target validation* or *target confidence building*. Does a compound that has the appropriate potency, selectivity, and ADME properties, that is active in "predictive" animal models, and is free of other systems toxicology work in the targeted human disease state? This is the ultimate, and only, validation of a target-based drug discovery program. All events leading up to this point are more accurately defined as target confidence building. With the considerable compound attrition rates in moving from animals to diseased humans, this has not proven to be a predictable transition. A case in point is that of the NK1 receptor activated by the peptide, substance P.

Over the past decade, data from both animals and humans has implicated this receptor in a variety of human pain states including migraine and neuropathic pain. As the result of several successful HTS campaigns run in parallel at Pfizer, Lilly, Merck, and Sanofi Aventis, a number of highly efficacious and selective antagonists of the NK1 receptor were identified which were then chemically optimized for use as drug candidates. These compounds, with varying degrees of potency, were active in animal models of pain and were free of overt side effects in phase I human clinical trials. However, they uniformly failed in phase II studies as novel analgesic agents in patients with various pain conditions.

The reasons for this remain unknown and have focused on the relevance of animal pain models to the human condition, for example, a lack of understanding of the true human disease condition and various nuances of substance P signaling pathways. However, before the results with NK1 antagonists, drug discovery in the analgesia area was considered one of the most robust. All known analgesics, for example, aspirin and morphine, were active in one or more of the animal pain models; new receptors could be mapped in pain pathways and their function assessed using knockout and antisense procedures, and the occupancy of receptors in human brain and pain-sensing pathways could be noninvasively imaged. The only limitation was whether the side effect profile of a putative analgesic agent acting on a novel target or suboptimal ADME characteristics would limit human exposure.

On the basis of the available data, the NK1 receptor has not been validated as a target for pain treatment despite an overwhelming body of robust preclinical data. This example thus serves to underline the many significant challenges of validating targets in the drug discovery process at the present time.

A second example, more directly related to the human genome, is the search for genes that are associated, and by inference may be causative, in producing the psychiatric disorder schizophrenia. This disease affects nearly 1% of the population and has a strong genetic association. From studies comparing schizophrenic patients with individuals lacking symptoms of the disorder, LOD scores of 2.4–6.5 on markers between or close to markers D1S1653 and D1S1679 on chromosome 1q have been reported. An LOD score of above a numerical value of 3 is considered to be indicative of a relevant association akin to a significant P value ($P < 0.05$). With this information, a search for the gene on chromosome 1q and the delineation of its function in humans would be a logical approach to finding novel genomic targets that could lead to new drugs that would more effectively and safely treat schizophrenia. However, a subsequent study using eight individually collected schizophrenia populations and a sample set that was "100% powered to detect a large genetic effect under the reported recessive model," showed no evidence for a linkage between chromosome 1q and schizophrenia; a finding that could not be ascribed to ethnicity, statistical approach, or population size. Another group using two prefrontal cortex tissues from two separate schizophrenic populations showed an upregulation of apolipoprotein L1 gene expression that was not seen in tissue from patients with bipolar disorder or depression. The genes related to apolipoprotein L1 gene expression are clustered on the chromosome locus 22q12, providing another target for the functional genomic approach to target discovery and validation. Another locus at chromosome 3q has been reported.

To the researcher embarking on a well-funded, science-driven approach to new targets for schizophrenia in 1997, these new findings in 2002, as well as others showing gene association with schizophrenia on chromosomes 6 and 13, would give pause to wonder how best to proceed. Do all five locations represent *bona fide* targets for functional genomics? Should one continue work on the original chromosome 1 findings in light of the failure to replicate? Is the only real validation, in light of conflicting findings, to proceed to the identification of multiple compounds that can be tested in schizophrenic populations – to validate the genome-based approach? Given that the cost of initiating a project, finding leads, optimizing these, and running a single clinical candidate to phase IIa clinical proof of principle is in the range of $24–28 million, how many organizations can afford the $120 million+ cost to undertake scientifically logical yet financially prohibitive strategies? These are difficult questions that can be applied to many other psychiatric and neurological diseases, as well as to any other disease state with a potential genetic causality. One answer to these questions obviously points to multifactorial genetic causes in the genesis of many diseases and thus negates the overly simplistic "one gene, one disease" mantra that heralded the age of drug discovery based on the human genome map. Strategies resulting from the answers to these questions also make the process of genome-based target and drug discovery a much more costly endeavor, with good evidence for two to three or more potential targets for each disease that need to be examined in parallel. Hopefully, the quality of the target finally validated using compounds in the human disease state will be a magnitude of order superior to currently existing drugs and will thus justify the cost of this approach.

13
Compound Properties

Once a target has been identified, the challenge then becomes the identification of compounds that selectively interact with the target and can be used as the basis of a lead optimization program to identify potential drug candidates.

An ideal compound should have the appropriately unique recognition characteristics to impart affinity and selectivity for its target, have the necessary efficacy to alter the assumed deficit in cell function associated with the targeted disease state, be bioavailable, metabolically, and chemically stable (drug like, be chirally pure, and be easy and cost-effective to synthesize).

13.1
Structure-activity Relationships

The structure-activity relationship (SAR) of a compound series is a means to relate changes in chemical diversity to the biological activity of the compound *in vitro* and *in vivo*, as well as the pharmacokinetic (gut/blood brain barrier transit, liver metabolic stability, plasma protein binding, etc.) and toxicological properties of the molecule. These SARs, when known, are frequently distinct such that changes that improve the bioavailability of a compound often decrease its activity and/or selectivity. Compound optimization is thus a highly iterative and dynamic process.

For the purpose of the present review, SAR is used to describe classical compound efficacy unless otherwise stated. Quantitative SAR (QSAR) involves a more mathematical approach involving neural networks and computer-assisted design.

Following the characterization of the SAR and the documentation of the effects of different pharmacophores and various substituents on biological activity, it is possible to theoretically model the way in which the ligand interacts with its target and thus derive a two- or three-dimensional approximation of the active site of the receptor or enzyme. Computer-assisted molecular design (CAMD) techniques can then be used to predict – sometimes successfully – the key structural requirements for ligand binding, thus defining those regions of the receptor target that are necessary for ligand recognition and/or functional coupling to second messenger system to permit the design of new pharmacophores.

The SAR can be further delineated in terms of the type of activity measured as the readout. *In vitro*, this can be displacement of a radiolabeled ligand from a receptor, receptor activation as measured in a functional assay, blockade of receptor function by an antagonist ligand, and so on. An additional ligand property is that of selectivity, the degree to which a ligand interacts with the target of choice compared with related structural targets. The degree of selectivity typically determines the side effect profile of a new compound, given that the targeted mechanism itself does not produce untoward effects when stimulated beyond the therapeutic range. As noted, the development of a ligand binding profile for a ligand active at a new target is very useful in assessing its effects in more complex tissue systems.

13.2
Defining the Receptor–Ligand Interaction

The complex physiochemical interactions that describe the interaction of a small molecule with a protein, despite the many sophisticated technologies used to study this interaction, are still highly empirical, being implicitly defined by the SAR for a series of active and less active compounds. In an increasing number of instances, however, the ability to clone, express, and readily derive crystals of a receptor or enzyme and analyze the interaction of a ligand or substrate with these using X-ray crystallographic, NMR, and amino acid point mutation approaches, has provided information on the actual topography of the selected drug target that can then be used in *de novo* ligand design.

There are a variety of approaches to derive information on the RL interaction for use in understanding the key features required in ligand necessary to dictate a potent and selective interaction with its target. Analysis of the interactions of a series of structurally distinct pharmacophores, agonists, and antagonists with the target can be combined with point mutation changes in the target to elucidate key amino acids involved in compound recognition. This can be complemented by X-ray crystallographic and

NMR-derived data to design novel pharmacophores. However, many proteins of interest, especially membrane-bound receptors and ion channels, are not available in the soluble form amenable to the use of these techniques such that the design of new compounds is based on the conceptualization of the target, for example, virtual receptors. There are also limitations to structure-based design approaches, for example, approximations of the hydration state of the isolated protein, the impact of removal from its native environment, and three-dimensional structural issues with recombinantly derived proteins. Nonetheless, these technologies have been useful in compound design. More recently, automated, high-throughput approaches have been applied to X-ray crystallography, providing a more rapid means to generate information on multiple ligand interactions with a given crystal.

13.2.1 Receptor-binding Assays

Until the 1960s, newly synthesized compounds and compounds isolated from natural sources were assessed by a mixture of *in vivo*, whole animal screens, and classical tissue assays. While many useful therapeutic agents were identified by this approach, the cost, in terms of compound quantity as well as time and animal use was considerable. *In vivo* test paradigms also suffered from the possible elimination of interesting compounds on the basis of unknown pharmacokinetic properties as test paradigms were usually routine in terms of timing, and as a result, many potentially interesting compounds that exhibited short plasma half-lives were considered "inactive" because data on their actions was sought after their peak plasma concentration. This type of screening approach also provided little useful chemical information about discrete interaction of the drug with its target, limiting the design of analogs. The specificity of the response, ignoring caveats related to pharmacokinetics, was not ideal because the mechanism inducing the overt response and potential points of intervention to block the response were unknown.

This empirical approach would predict, in the absence of any data related to specific interactions with a molecular target, that because β-adrenoceptor antagonists lower blood pressure, then any compound that lowers blood pressure is by definition a β-adrenoceptor antagonist. The 1970s saw the development of a number of *in vitro* biochemical screens that moved the measurement of the RL interaction a little closer to the molecular level. Nonetheless, the major challenge was to develop assays that measured the RL interaction independently of "downstream" events such as enzyme activation and second and third messenger systems. By such means, the ability of a compound to bind to a receptor could be determined on the basis of the SAR and thus provide the chemist with a more direct means to model the RL interaction.

Snyder, building on pioneering work by Roth, Cuatrecasas, and Rang developed radioligand binding as a valuable tool in the drug discovery process in 1973. This led to an explosion in the identification and characterization of new receptors and their subtypes, which was enhanced by the application of tools of recombinant DNA technology to the process.

While the technique of radioligand binding has been largely supplanted by activity-based assays in many high-throughput settings, it remains the gold standard for compound characterization and SAR development. It is done by measuring the RL interaction (Equation 1) *in vitro* using a radioactive ligand, R^*, to bind with high affinity and selectivity to receptor sites. The interaction of unlabeled ("cold") ligands with the receptor competes with the radioligand, decreasing its binding. This simple technique revolutionized compound evaluation in the 1970s, allowing SARs to be determined with milligram amounts of compound in a highly cost- and resource-effective manner. At steady state, the RL* complex can either be separated out from free radioligand using filtration or assayed using a scintillation proximity assay.

The parameters measured in a binding assay are the dissociation constant, K_d, the reciprocal of the affinity constant, K_a. The K_d is a measure of the affinity of a radioligand for the target site: the B_{max}, usually measured in moles per milligram protein, a measure of the concentration of binding sites in a given tissue source and the IC_{50} value. The K_d and B_{max} values can be determined using a saturation curve where the concentration of radioligand is increased until all the ligand recognition sites are occupied, or by measuring radioligand association and dissociation kinetics; K_d is the ratio of the dissociation and association rate constants.

The IC_{50} value is the concentration of unlabeled ligand required to inhibit 50% of the specific binding of the radioligand. This value is determined by running a competition curve (Fig. 5) with a fixed concentration of radioligand and tissue, and varying concentrations of the unlabeled ligand. To accurately determine the IC_{50}, it is essential that sufficient data points be included. As shown in Fig. 5, if the data used to derive the IC_{50} value are clustered over a range that reflects 40 to 60% of the competition curve, much useful information is lost. Ideally, the competition curve should encompass the range of 10 to 90% of the competition curve and include a minimum of 20 data points. On the basis of Michaelis–Menten kinetics, when binding is the result of the interaction of the displacer with one recognition site, 10 to 90% of the radioligand is inhibited over an 82-fold concentration range of the displacer. The slope of a competition curve can then be analyzed to assess the potential cooperation of the RL interactions. When binding is complex resulting in the interaction of the displacer with more than one recognition site, a greater than 82-fold concentration of displacer is required to inhibit the same 10 to 90% of specific radioligand binding.

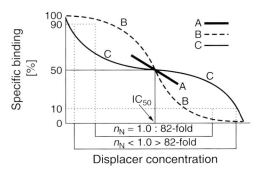

Fig. 5 Measuring ligand interactions in a receptor-binding assay. Binding of the radioligand is 100%. In curve A, close to the 50% point, an IC_{50} value can be obtained but ignores complexities of the displacement curve. More complete displacement, as in curves B and C, provides more information on the ligand. Displacement curve B has a Hill coefficient of unity requiring an 82-fold difference in displacer concentration to displace 10 to 90% of binding. Displacement curve C has a Hill coefficient of less than one, requiring a greater than 82-fold difference in displacer concentration to displace 10 to 90% of binding, and indicating the presence of more than one site. By extending the range of concentrations used, curves B and C can assess the possibility of multiple sites being present.

The IC_{50} value for a given compound is dependent on the assay conditions: the concentration of the radioligand used, the receptor density, and the affinity of the receptor, the K_d, for the radioligand. To compare the activity of a "cold" ligand across different radioligand binding assays, the Cheng–Prusoff equation is used to compensate for differences in K_d and the radioligand concentration to obtain a K_i value derived by the relationship:

$$K_i = \frac{IC_{50}}{1 + \frac{[L]}{K_d}} \quad (5)$$

where [L] is the concentration of radioligand used and K_d is the dissociation constant for the radioligand at the receptor. This relationship thus corrects for inherent differences in assay conditions.

Binding assays can be rapidly used to assess compound recognition characteristics but are generally limited in their ability to delineate agonists from antagonists, especially in a high-throughput setting.

13.2.2 Functional Assays

Biochemical assays involving the measurement of cAMP production or phosphatidylinositol turnover have given way in HTS scenarios to reporter systems in which receptor activation or inhibition can be measured using a fluorescence-based readout. This depends on the use of calcium sensing dyes coupled with real-time measurement using charge coupled device (CCD) cameras and data capture. While many types of plate reader are available, a widely used system is the fluorescence imaging plate reader (FLIPR). Using a 96- or 384-well microtiter plate format, the throughput on an FLIPR is such that compound libraries of 0.5 million distinct compounds can be assayed in weeks.

Other approaches to functional assay include various reporter gene constructs in which formation of the RL complex leads to the expression of a gene that produces a response that can be read immediately. These include various luciferase reporter gene or aequorin-based assays that produce light that can be measured by multiple photodiode arrays, and α-galactosidase reporter gene that leads to a colorimetric readout. Frog melanocytes transfected with GPCR cDNA represent yet another approach to determining whether a ligand is an agonist or antagonist that is independent of a reporter gene construct. Addition of melatonin to a transfected oocyte reduces intracellular cAMP concentrations resulting in the aggregation of the pigment in the cells. Agonist stimulation of the transfected cell increases cAMP resulting in pigment dispersal, a reaction that can not only be immediately determined visually on a plus/minus basis but can also be quantified in terms of light transmission.

The ability to measure colocalization of two cellular components in living cells is an increasingly powerful technique in understanding ligand–receptor and protein–protein interactions. Two such approaches are fluorescence resonance energy transfer (FRET) that measures the proximity of two proteins through the use of a luciferase tag on one partner and green fluorescent protein (GFP) on the other, and fluorescence polarization (FP) that relies on a polarized excitation light source to illuminate a binding reaction mixture. The smaller partner (e.g. the ligand) must be fluorescently labeled, and if this is unbound then it will, by tumbling in solution, emit depolarized fluorescence. This is thus a means of looking at the amount of unbound ligand in the binding mixture and has the benefit of not requiring any washing steps. Current systems do not work well with turbid solutions or fluorescent compounds, although the latter has largely been dealt with through the use of red fluorescent dyes.

13.3
Receptor Sources

The choice of a tissue or cell line as a receptor source has a significant impact on the data generated. The natural receptor concentration in most tissues is in the femtomole to picomole range; brain tissue has a much higher density of receptors because of more extensive nerve innervation. Expression of the drug target in a cell line can, however, lead to differences in the number of receptors expressed per clone, a factor dependent on the relative proportion of transient to stable expressed cells and the passage number of the transfectants. Thus, the number of receptors can vary, affecting the apparent activity of unknown ligands that compete for binding with the radioligand.

As the drug targets of greatest interest are those in the human, the use of a human receptor or enzyme would seem ideal in defining the SAR of a potential drug series. The drawback, however, is that nearly all the toxicology and safety studies done in preparing a compound for clinical trials are conducted in rodents, dogs, and nonhuman primates. If there are no species differences between rat and human, this testing becomes a moot point. If on the other hand, the human target is substantially different from that in rat or dog or monkey, and there are many examples of this, the safety and toxicology studies may be conducted on a compound that has limited interactions with the drug target in species other than human. One approach is to incorporate human receptor orthologs into mice.

The use of transfected cell lines in compound evaluation can lead to a number of potential artifacts. Activation of transfected receptor may result in an increase in cAMP, a second messenger effect that may already be known to be a consequence of ligand activation of the receptor in its natural state. It is also possible that the transfected receptor may activate a cell-signaling pathway in the transfected cell that is not linked to the receptor in its normal tissue environment. In this instance, the second messenger readout actually functions as a "reporter," a G-protein–linked phenomenon that results from the introduction of the cDNA for a GPCR and the generic or promiscuous interaction of the receptor with the G-protein systems. The introduction of the cDNA for any GPCR may then act to elicit a similar response. It is then advisable to use caution in extrapolating events occurring in the transfected cell to the physiological milieu of the intact tissue, and there is at least one case in which a compound identified as an antagonist in a cell system overexpressing the receptor of interest was subsequently found to be a partial agonist after it exacerbated disease symptomatology in phase II clinical trials.

13.4
ADME

In evaluating new compounds, it is only in the past 15 years that the ability of a compound to reach its putative site of action has been a priority in the discovery phase of compound identification. For many years, it was naively assumed that there was a generic approach that could be used on compounds with poor bioavailability that would turn them into drug candidates. With attrition rates of lead compounds in the clinical development process of 50 to 60% range, with one estimate of greater than 90%, this was clearly not the case. Hodgson has noted, "a chemical cannot be a drug, no matter how active nor how specific its action, unless it is also taken appropriately into the body (absorption), distributed to the right parts of the body, metabolized in a way that does not instantly remove its activity and eliminated in a suitable manner – a drug must get in, move about, hang around and then get out." The factors involved in defining in vivo activity form the basis of Lipinski's "Rule of 5." In this widely cited, retrospective study, a number of compounds have been assessed and used to design new molecules. The physical properties that were determined as limiting bioavailability were as follows: molecular weight greater than 500; more than 5 hydrogen bond donors; more than 10 hydrogen bond acceptors; and a $C \log P$ value less than 5.

A retrospective evaluation of the oral bioavailability of 1100 novel drug candidates with an average molecular weight of 480 from SmithKline Beecham's drug discovery efforts in rats by Veber established that reduced molecular flexibility, measured by the number of rotatable bonds and low polar surface area or total hydrogen bond count (sum of acceptors and donors), were important predictors of good (>20–40%) oral bioavailability, independent of molecular weight. A molecular weight cutoff of 500 did not significantly separate compounds with acceptable oral bioavailability from those with poor oral bioavailability, the predictive value of molecular weight was more correlated with molecular flexibility than molecular weight per se. From this retrospective analysis, Veber et al. suggested that compounds with 10 or fewer rotatable bonds and a polar surface area equal to or less than 140 A^2 (representing 12 or fewer H-bond acceptors and donors) have a high probability of having good oral bioavailability in rats. Reduced polar surface area was a better predictor of artificial membrane permeation than lipophilicity ($C \log P$), with increased rotatable bond count having a negative effect on permeation and having no correlation with in vivo clearance.

ADME is now a routine part of the efforts of a drug discovery team with the use of a number of in vitro approaches, for example, Caco 2 intestinal cell lines, human liver slices, or homogenates to assess potential metabolic pathways, in addition to classical rat, dog, and nonhuman primate in vivo studies, none of which has yet shown reliable predictability for the human situation. More recently, proteomic approaches have been used to derive potential toxicological profiles.

13.5
Compound Databases

With the exponential increase in information flow resulting from the ability to make many more compounds and test these in multiple assays, the capture, analysis, and management of data is a critical success factor in drug discovery.

Many databases used in drug discovery are ISIS/Oracle-based using MDL structural software and a variety of data entry and analysis systems, some are PC/Mac-based, and some are on a server. The ability to capture data and then reassess its value through in silico approaches has the potential to be a vast improvement over the "individual memory" systems that many drug companies used for the better part of the last century. Thus, with the retirement of a key scientist, the whole history of a project or even a department disappeared, and whatever folklore existed regarding unexplained findings with compounds 10 or 20 years before was lost with the individual.

14
Lead Compound Discovery

As evidenced by the compound code numbers used by pharmaceutical companies, many thousands of new chemical entities have been made since the industry began in the late nineteenth century. Until the advent of combinatorial chemistry, the chemical libraries at most of the major pharmaceutical companies numbered from 50 000 to 800 000 compounds and comprised of newly synthesized compounds as well as those from fermentation and natural product sources. Approximately 2 to 5 million compounds were identified in the search for new drugs to treat human disease states over the past century. This number has obviously leaped to the billions with combinatorial approaches. Given decomposition and/or depletion of compounds, the 2 to 5 million compounds could be rounded down to 1 million.

The 2001 edition of the *Merck Index*, the compendium of drugs and research tools, lists a total of 10 250 compounds. Thus, from a hypothetical million compounds, only 1% has proven to be of sustained interest as either therapeutic agents or research tools.

In the early 1980s, with the promise that compounds could be created and tested on a computer screen (a promise as yet unrealized), the screening of large numbers of compounds against selected targets in biological systems was viewed as irrational. Indeed, the head of research at one of the top 20 pharmaceutical companies told the medicinal chemists at that company in the early 1980s that the demand for their skills was becoming less and would cease by the 1990s, a viewpoint somewhat akin to the apocryphal story of the head of the US Patent Office in the early 1900s who recommended its closure because "everything that could be discovered had been discovered."

The breakthrough for screening came in 1984 with the identification by Chang and colleagues at Merck of the CCK antagonist, asperlicin, which led to the clinical candidate, MK 329. Considerable effort has been expended in the pharmaceutical industry worldwide to capitalize on this approach, with significant successes, which have enhanced the search for novel chemical entities as well as providing research tools to better understand receptor and enzyme function.

14.1
High-throughput Screening

To identify compounds, it is imperative that a rapid, economical, and information-rich evaluation of biological activities be available. The term high-throughput screening describes a set of techniques designed to permit rapid and automated (robotic) analysis of a library of compounds in a battery of assays that generate specific receptor- or enzyme-based signals. These signals may be membrane-based (radioligand binding, enzyme catalysis) or cell-based (flux, fluorescence). The primary purpose of HTS is not to identify candidate drugs, but rather to identify lead structures, preferably containing novel chemical features, which may serve as a guide for more tailored iterative optimization. HTS should generate as few false-positive leads as possible because exploitation of leads is an expensive component of the drug discovery process. HTS is designed to give information principally about potency, and a combination of follow-up screens may provide information about selectivity and specificity. The increasing use of cell-based assays provides additional information, including agonist/antagonist characteristics, and a biological "readout" under physiological or nearly physiological conditions. Additionally, cell-based assays can provide information about the cytotoxicity and bioavailability of molecules. The increased use of "designer cells" with visual and fluorescent signal readouts will continue to facilitate the screening process, and in the future, will doubtlessly include measures of metabolism, toxicity, bioavailability, and other important pharmacokinetic parameters.

Using FLIPR-like instrumentation, 100 000 or more compounds can be screened through 10 to 20 targeted assays in less than a week, and complete libraries numbering in the millions can be assessed at single concentration points in about a month. The issue is the success rate, typically in the 0.1% range, and the ability to capture and store the data for posterity. Clearly, the quality and diversity of the compounds and the robustness of the assays play a key role in a successful screening program.

14.2
Compound Sources

New chemical entities (NCEs) are discovered or developed/optimized from the following: (1) natural products and biodiversity screening; (2) exploitation of known pharmacophores; (3) rationally planned approaches, for example, computer-assisted molecular design; (4) combinatorial or focused library chemistry approaches; and (5) evolutionary chemistry.

14.2.1 Natural Product Sources

Approximately 70% of the drugs currently in human use originate from natural sources including morphine, pilocarpine, physostigmine, theophylline, cocaine, digoxin, salicylic acid, reserpine, and many antibiotics. Medicinal and herbal extracts form the basis for the health care of approximately 80% of the world's population; some 21 000 plant species are used worldwide. Screening of natural products led to the discovery of the immunosuppressants, cyclosporin, rapamycin, and FK 506, and there is a continued search for new compounds, even in relatively well-explored areas such as China.

The continued destruction of habitat with the accompanying loss of animal and plant species may impede further natural product-based drug discovery. Many interesting drugs have vanished. For example, a plant called *silphion* by the Greeks and sylphium by the Romans grew around Cyrene in North Africa. It may have been an extremely effective antifertility drug in the ancient world but was harvested to extinction.

While 100% of the world's mammals are known, fewer than 1 to 5% of other species, bacteria, viruses, fungi, and most invertebrates, are well characterized. It has been estimated that only 0.00002 to 0.003% of the world's estimated 3 to 500×10^6 species are used as a source of modern drugs. Exploration of environments previously assumed to be hostile to life has revealed bacterial species living at extreme depths, at extraordinary temperatures, and in the presence of high concentrations of heavy metals.

The sea covers almost three-quarters of the earth's surface and contains a broader genetic variation among species relative to the terrestrial environment. Although a number of important molecules have been derived from marine sources, including arabinosyl nucleotides, didemnin B, and bryostatin 1, there has been an inadequate focus on this potentially chemically productive biosphere. Sea snails, termed *nature's combinatorial chemistry factories*, produce a bewildering array of novel conotoxins active as mammalian drug targets. These toxins are typically 10–30 amino acid residues in length, contain several disulfide bridges, and are rigid in structure. There are several hundred varieties of cone snails, and each may secrete more than 100 toxins. Therefore, there are likely to be several tens of thousands of these toxins, representing a library of substantial structural and functional diversity. These peptides are first synthesized as larger precursors from which the mature peptide is cleaved. In the mature peptide, there is a constant N terminus region and a hypervariable C terminus region from which the biological diversity is derived. Conus toxins have proven to be invaluable as molecular probes for a variety of ion channels and neuronal receptors and as templates for drug design.

Poison frogs of the *Dendrobatidae* family contain a wide variety of skin-localized poisonous alkaloids that are presumably secreted for defensive purposes. Among the chemical structures present are the batrachotoxins, pumiliotoxins, histrionicotoxins, gephyrotoxins, and decahydroquinolines (Fig. 6) that target both voltage- and ligand-gated ion channels. The alkaloid epibatidine (Fig. 6), present in trace amounts in *Epipedobates tricolor*, is of particular interest because it has powerful analgesic activities, being 200 times more potent than morphine. The alkaloid is a potent neuronal nicotinic channel agonist selective for the $\alpha 4\beta 2$ subtype. Isolated by Daly and Myers in 1974, the structure of epibatidine was not determined until 1992. The discovery of epibatidine as a novel and potent analgesic led to the identification of ABT-594, which had equivalent analgesic efficacy to epibatidine but with reduced side effect liabilities. Despite continued successes in isolating new compounds with pharmaceutical potential from natural sources, the pharmaceutical

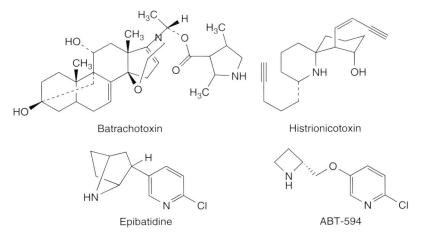

Fig. 6 Alkaloids and related structures.

industry has tended to loose interest in this approach limiting an important aspect of chemical diversity.

14.2.2 Pharmacophore-based Ligand Libraries

The majority of pharmaceutical companies have relatively large chemical libraries representing the cumulative synthetic efforts of the medicinal chemists within the company. Typically, the chemical diversity in these libraries is limited as the synthetic approach to drug design revolves around defined pharmacophores in lead series and the rational and systematic development of the SAR. Companies will thus have a large number of similar compounds based on the approaches and successes attendant to a therapeutic area.

The systematic modification of existing structures, both natural and synthetic, is an approach to compound optimization with improvements in potency, selectivity, efficacy, and pharmacokinetics being linked to discrete changes in molecular constituents on the basic pharmacophore.

Angiotensin-converting enzyme (ACE) inhibitors are important cardiovascular drugs that block the conversion of angiotensin I, formed by the action of renin on substrate angiotensinogen, to angiotensin II (a powerful pressor and growth factor agent). Until 1973, peptide inhibitors from the venom of the Brazilian viper were the only known inhibitors of this enzyme. The nonapeptide, teprotide, was an orally active and competitive inhibitor of ACE. Benzylsuccinic acid, a potent inhibitor of carboxypeptidase A, an enzyme with structural and mechanistic similarities to ACE, led Ondetti to select N-succinyl-l-proline as a lead ($IC_{50} = 330$ pM). This was subsequently optimized on the basis of the presence of a Zn^{2+} in the active site of both ACE and carboxypeptidase and the likely presence of hydrophobic pockets. An SH group to coordinate Zn^{2+} was incorporated into the stereoselectively active methyl analog, which eventually led to captopril and the analogs, enalapril, cilazapril, and lisinopril (Fig. 7).

14.2.2.1 Molecular modeling.
Rationally planned approaches to structure design are an increasingly important part of the drug discovery process, whether planned *ab initio*, derived from structural knowledge of a putative ligand-binding site on a biological target in the absence of ligand information, or derived by rational exploitation of an existing chemical lead.

14.2.2.2 Privileged pharmacophores.
It is increasingly apparent that a small number of common structures – "basic pharmacophores," "templates," or "scaffolds" – are associated with a multiplicity of diverse biological activities. These structures represent facile starting points for combinatorial chemical approaches to ligand diversity.

Benzodiazepines The benzodiazepines are well established as anxiolytics, hypnotics, and muscle relaxants as represented by diazepam, clonazepam, midazolam, and triazolam (Fig. 8). The BZ nucleus also occurs in natural products. Asperlicin is a naturally occurring ligand that is a weak, albeit selective, antagonist at cholecystokinin receptors and contains a benzodiazepine nucleus. From this lead was derived a series of potent and selective benzodiazepine ligands, active at CCK_1 and CCK_2 receptors, for example, L-364 718. Other BZs (Fig. 8) with activity at receptors distinct from the BZ receptor are as follows: tifluadom (opioid receptor);

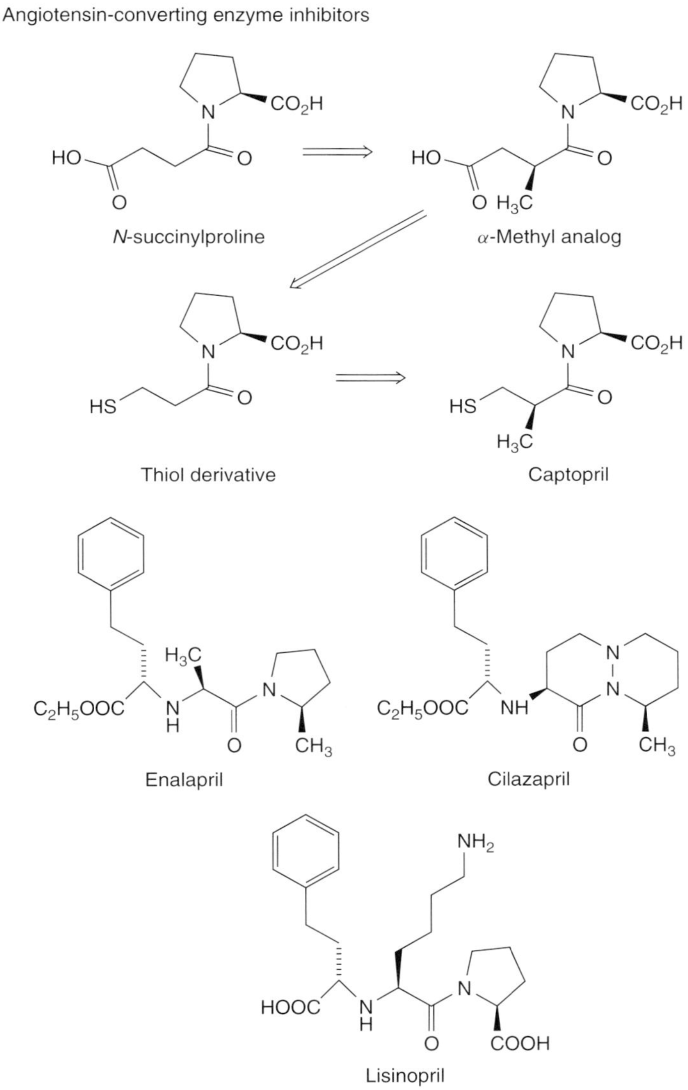

Fig. 7 ACE Inhibitors.

somatostatin; GYKI 52466 (glutamate receptor); and inward-rectifying potassium channels.

1,4-Dihydropyridines(DHP) The DHP pharmacophore is a well-established chemical entity. Nifedipine (Fig. 9) and several related DHPs, including amlodipine, felodipine, nicardipine, nimodipine, and nisoldipine, are well-established antihypertensive and vasodilating agents that act through voltage-gated L-type Ca^{2+} channels in the vasculature. However, DHPs also interact with lower activity at

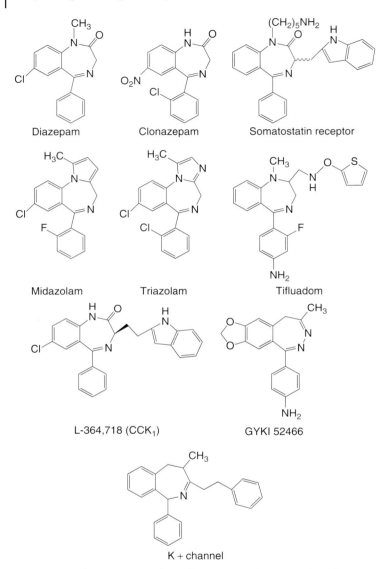

Fig. 8 Benzodiazepines: anxiolytics, hypnotics, and compounds with activity at other targets.

other classes of Ca^{2+} channels, including N- and T-type channels, and "leak" channels. DHPs can block delayed rectifier K^+ channels and cardiac Na^+ channels. Other DHP analogs (Fig. 9) are active at PAF receptors (UK 74,505), adenosine A_3 receptors, K^+ ATP channels (ZM 24 405 and A-278 367), capacitative SOC channels (MRS 1845), and α_{1A}-adrenoceptors (SNAP 5089).

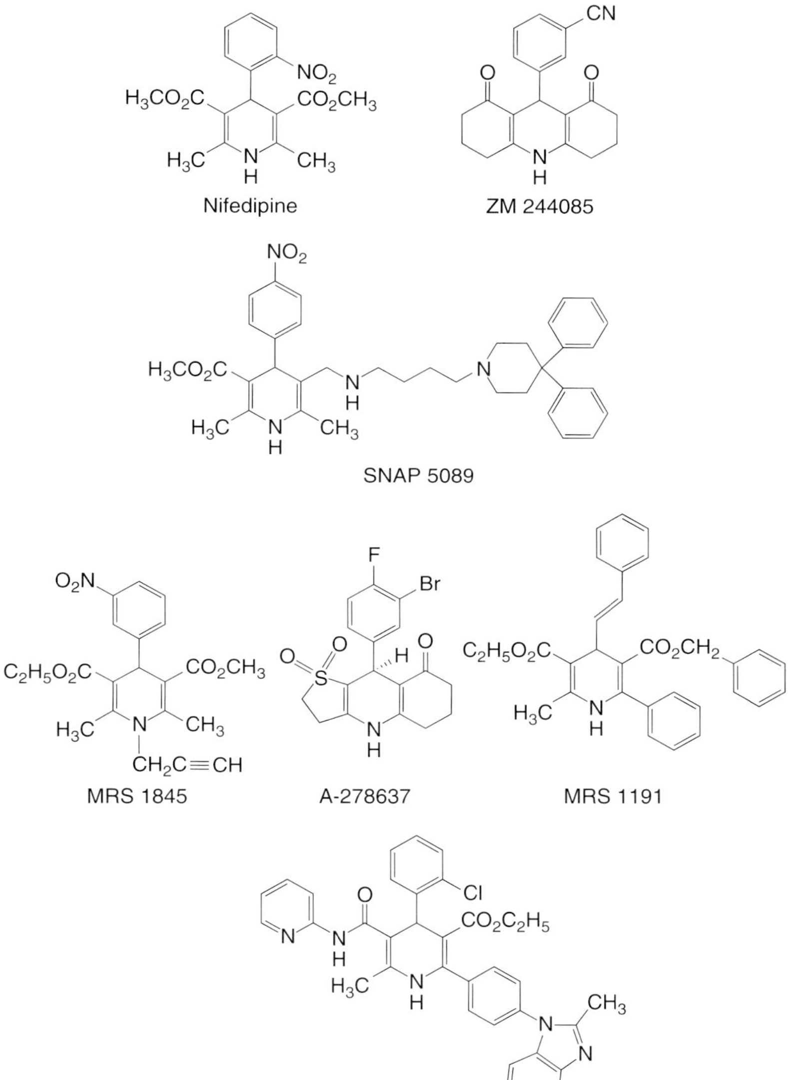

Fig. 9 Dihydropyridines; calcium channel, and other receptor antagonists.

14.2.3 Diversity-based Ligand Libraries

The issue of diversity reflects the need to enhance the scope of the library beyond those available within a company. This can be done by compound exchange with other companies, by acquiring compounds from university departments and commercial sources (Sigma, Asinex, Bader, Mayhew, Cookson, etc.). Additional sources of novel synthetic compounds include the libraries of major present or former chemical/agricultural companies and their successors like Eastman Kodak, Stauffer, FMC, and Shell, and chemicals made in the former Eastern Block countries that are now being brokered to pharmaceutical companies. In conjunction with the use of computerized cluster programs, the selection of compounds based on diverse structures can be considerably enhanced to provide maximum coverage of molecular space. This can be done by generating libraries of approximately 2000 compounds that is used to rapidly identify potential leads for a new drug target using the SAR generated in an HTS assay. Such libraries are assembled from compounds that are available in relatively large supply and may not necessarily be proprietary to the company. Their value is in rapidly eliminating unlikely structures in a systematic manner for each new target.

Combinatorial chemistry has provided the means to synthesize literally billions of molecules, a major step forward in the exploration of "molecular space." This random or "brute-force" approach to the search for new leads has been a major disappointment in its contribution to product pipelines and the sacrifice of quality for quantity. However, combinatorial exploration around lead compounds or pharmacophores to generate dedicated libraries, when coupled with HTS, clearly provides an economical and efficient way to rapidly generate lead compounds.

Combinatorial diversity can be exploited in nature in the repertoire of antibodies with their remarkable combination of high affinity and selectivity and is also reflected in the conotoxins.

The polyketides are a family of natural products containing many important pharmaceutical agents that are synthesized through the multienzyme complex, polyketide synthase, which can display substantial molecular diversity with respect to chain length, monomer incorporation, reduction of keto groups, and stereochemistry at chiral centers. This variability, together with the existence of several discrete forms of polyketide synthase, allows the generation of diverse structures like erythromycin, avermectin, and rapamycin. This biochemical diversity has been considerably expanded by the introduction of new substrate species that were used by the enzymes to produce new or unnatural polyketides.

These methods are likely to be more extensively used in the future to provide a highly focused combinatorial approach to generating molecular diversity. The multienzyme pathway responsible for converting simple linear unsaturated allylic alcohols to sterols, carotenoids, and terpenes is incompletely characterized, but offers excellent potential for genetic manipulation to provide directed biocombinatorial chemistry.

14.3
Biologicals and Antisense

The use of the naturally occurring hormones as drugs is not new as evidenced by the use of insulin and epinephrine. The techniques of molecular biology allow the production of an increasing number of native and modified hormones and their soluble forms. Erythropoietin (EPO) is a classic example of a successful drug taken from the human body. Soluble receptors like Enbrel and humanized antibodies, for example, herceptin and D2E7 (Humira), are additional examples of how cloning techniques have altered the concept of rational drug design and what can be considered as a drug. Anticytokine therapies are particularly amenable to this approach.

Antisense oligonucleotides and RNA interference (RNAi) approaches may similarly provide novel, specific to disease treatment by the suppression/modification of a gene product. The fundamental concept is that an oligonucleotide, complementary to a disease-causing gene will anneal to that gene and prevent its transcription. In a similar manner, RNAi approaches use the fact that double-stranded RNA is rapidly degraded, and double-stranded RNA oligomers can trigger degradation of the genes encoded by them. While delivery issues remain a concern, one topically active antisense drug, Vitravene (ISIS) is approved for the treatment of CMV-induced retinitis.

15
Future Directions

The gradual evolution of the receptor concept as the basis for drug discovery over the past century has led to major advances in the understanding of biological systems and human disease states.

While the development of ever more sophisticated structure-based technologies and ultra, micro-HTS that encompass activity, ADME, toxicity, and structural data generation has resulted in an ever-increasing body of knowledge, it does not seem to have added significantly to success as measured by increased numbers of quality INDs (investigative new drug applications), but rather, together with genomics and proteomics, significantly added to the cost of the search for new medicines. The medicinal chemist and the pharmacologist play key roles in integrating and interpreting the data flow that constitutes much of the day-to-day workings of the drug discovery environment. Their challenge is to provide the intellectual framework to use this data to find drugs rather than play a technology-driven numbers game.

Acknowledgment

The present text is a modified and updated version of the chapter "Receptor-Targets in Drug Discovery and Development" published in "Burger's Medicinal Chemistry and Drug Discovery" (ISBN 0-471-37032-0) edited by D.J. Abraham in 2003.

See also Biotransformations of Drugs and Chemicals; Medicinal Chemistry; Pharmacogenomics and Drug Design; Pharmacokinetics and Pharmacodynamics of Biotech Drugs; Receptor Biochemistry; Structure-based Drug Design and NMR-based Screening; Synthetic Peptides: Chemistry, Biology, and Drug Design; Targeting and Intracellular Delivery of Drugs.

Bibliography

Books and Reviews

Ariens, E.J. (1954) Affinity and intrinsic activity in the theory of competitive inhibition. I. Problems and theory, *Arch. Int. Pharmacodyn.* **99**, 32–49.

Blundell, T.L., Jhoti, H., Abell, V. (2002) High-throughput crystallography for lead discovery in drug design, *Nat. Rev. Drug Discov.* **1**, 45–54.

Bockaert, J., Pin, J.P. (1999) Molecular tinkering of G protein-coupled receptors: an evolutionary success, *EMBO J.* **18**, 1723–1729.

Bredt, D.S., Snyder, S.H. (1994) Nitric oxide: a physiologic messenger molecule, *Annu. Rev. Biochem.* **63**, 175–195.

Christopoulos, A. (2002) Allosteric binding sites on cell-surface receptors: novel targets for drug discovery, *Nat. Rev. Drug Discov.* **1**, 198–210.

Copeland, R.E. (2005) *Evaluation of Enzyme Inhibitors in Drug Discovery*, Wiley Interscience, Hoboken, NJ.

Daly, J.W., Garraffo, H.M., Spander, T.F., Decker, M.W., Sullivan, J.P., Williams, M. (2000) Alkaloids from frog skin: the discovery of epibatidine and the potential for developing novel non-opioid analgesics, *Nat. Prod. Rep.* **17**, 131–135.

De Souza, E., Grigoriadis, D.E. (2002) in: Davis, K.L., Charney, D., Coyle, J.T., Nemeroff, C. (Eds.) *Psychopharmacology: Fifth Generation of Progress*, Lippincott Williams and Wilkins, Philadelphia, PA, pp. 91–107. Eds.),

Dower, S.K. (2000) Cytokines, virokines and the evolution of immunity, *Nat. Immunol.* **1**, 367–368.

Drews, J. (2000) Drug discovery: a historical perspective, *Science* **287**, 1960–1964.

Edwards, S.W., Tan, C.M., Limbird, L.L. (2000) Localization of G-protein-coupled receptors in health and disease, *Trends Pharmacol. Sci.* **21**, 304–308.

Ellis, C. (2004) The state of GPCR research in 2004, *Nat. Rev. Drug Discov.* **3**, 577–626.

Estibeiro, P., Godfray, J. (2001) Antisense as a neuroscience tool and therapeutic agent, *Trends Neurosci.* **24**, S56–S62.

Flower, R. (2002) Drug receptors: a long engagement, *Nature* **415**, 587.

Furchgott, R.F. (1964) Receptor mechanisms, *Annu. Rev. Pharmacol.* **4**, 21–50.

Hill, R. (2000) NK1 (substance P) receptor antagonists–why are they not analgesic in humans? *Trends Pharmacol. Sci.* **21**, 244–246.

Hill, S.J., Baker, J.G., Rees, S. (2001) Reporter-gene systems for the study of G-protein-coupled receptors, *Curr. Opin. Pharmacol.* **1**, 526–532.

Hodgson, J. (2001) ADMET? turning chemicals into drugs, *Nat. Biotechnol.* **19**, 722–726.

Horrobin, D. (2001) Realism in drug discovery-could Cassandra be right? *Nat. Biotechnol.* **19**, 1099–1100.

Jacobson, K.A., Jarvis, M.F., Williams, M. (2002) Purine and pyrimidine (P2) receptors as drug targets, *J. Med. Chem.* **45**, 4057–4093.

Kenakin, T.P. (2001) Inverse, protean, and ligand-selective agonism: matters of receptor conformation, *FASEB J.* **15**, 598–611.

Kenakin, T. (2002) Efficacy at G-protein-coupled receptors, *Nat. Rev. Drug Discov.* **1**, 103–110.

Kenakin, T. (2003) Predicting therapeutic value in the lead optimization phase of drug discovery, *Nat. Rev. Drug Discov.* **2**, 429–438.

Kenakin, T.P., Bond, R.A., Bonner, T.I. (1992) Definition of pharmacological receptors, *Pharmacol. Rev.* **44**, 351–362.

Khosla, J.R., Zawada, R.J.X. (1996) Generation of polyketide libraries via combinatorial biosynthesis, *Trends Biotechnol.* **14**, 137–142.

Kopec, K., Bozyczko-Coyne, D.B., Williams, M. (2005) Commentary: target identification and validation in drug discovery: the role of proteomics, *Biochem. Pharmacol.* **69**, 1133–1139.

Legrain, P., Wojcik, J., Gauthier, J.-M. (2001) Protein–protein interaction maps: a lead towards cellular functions, *Trends Genet.* **17**, 346–352.

Le Novere, N., Changeux, J.P. (1995) Molecular evolution of the nicotinic acetylcholine receptor: an example of multigene family in excitable cells, *J. Mol. Evol.* **40**, 155–172.

Liebmann, C. (2004) G protein-coupled receptors and their signaling pathways: classical therapeutical targets susceptible to novel therapeutic concepts, *Curr. Pharm. Des.* **10**, 1937–1958.

Lipinski, C.A., Lombardo, F., Dominy, B.W., Feeney, P.J. (1997) Experimental and computational approaches to estimate solubility and permeability in drug discovery and development settings, *Adv. Drug. Deliv. Rev.* **46**, 3–26.

Lloyd, G.K., Williams, M. (2000) Neuronal nicotinic acetylcholine receptors as novel drug targets, *J. Pharmacol. Exp. Ther.* **292**, 461–467.

Lutz, M., Kenakin, T.P. (1999) *Quantitative Molecular Pharmacology and Informatics In Drug Discovery*, Wiley, Chichester, England.

Mangelsdorf, D.M., Evans, R.M. (1995) The RXR heterodimers and orphan receptors, *Cell* **83**, 841–850.

Masson, J., Sagné, C., Hamon, M., El Mestikawy, S. (1999) Neurotransmitter transporters in the central nervous system, *Pharmacol. Rev.* **51**, 439–464.

Milligan, G. (2001) Oligomerisation of G-protein-coupled receptors, *J. Cell. Sci.* **114**, 1265–1271.

Milligan, G. (2003) Constitutive activity and inverse agonists of G protein coupled receptors: a current perspective, *Mol. Pharm.* **64**, 1271–1276.

Monod, J., Wyman, J., Changeux, J.-P. (1965) On the nature of allosteric transitions: a plausible model, *J. Mol. Biol.* **12**, 88–118.

Neubig, R.P. (2002) Regulators of G-protein signaling as new central nervous system drug targets, *Nat. Rev. Drug Discov.* **1**, 187–197.

Neubig, R.R., Spedding, M., Kenakin, T., Christopoulos, A. (2003) International Union of Pharmacology Committee on Receptor Nomenclature and Drug Classification. XXXVIII, update on terms and symbols in quantitative pharmacology, *Pharmacol. Rev.* **55**, 597–606.

Onaran, H.O., Scheer, A., Cotecchia, S., Costa, T. (2000) *Handbook of Experimental Pharmacology*, Vol. 148, Springer, New York, 217–280.

Ondetti, M.A. (1994) From peptides to peptidases: a chronicle of drug discovery, *Annu. Rev. Pharmacol. Toxicol.* **34**, 1–16.

O'Neil, M.J., Smith, A., Heckelman, P.E., Obenchain, J.R., Gallipeau, J.A.R., D'Arecca, M.A. (2001) *The Merck Index*, 13th edition, Merck, Rahway, NJ.

Schreiber, S.L. (2000) Target-oriented and diversity-oriented organic synthesis in drug discovery, *Science* **287**, 1964–1969.

Triggle, D.J. (1992) Calcium-channel antagonists: mechanisms of action, vascular selectivities, and clinical relevance, *Cleve. Clin. J. Med.* **59**, 617–627.

Walters, W.P., Stahl, M.T., Murko, M.A. (1998) Virtual screening – an overview, *Drug Discov. Today* **3**, 160–178.

Watling, K.J. (2001) *The RBI Handbook*, 4th edition, Sigma-RBI, Natick, MA.

Williams, M., Coyle, J.T., Shaikh, S., Decker, M.W. (2001) Same brain, new century. Challenges in CNS drug discovery in the postgenomic, proteomic era, *Annu. Rep. Med. Chem.* **36**, 1–10.

Williams, M., Giordano, T., Elder, R.A., Reiser, H.J., Neil, G.L. (1993) Biotechnology in the drug discovery process: strategic and management issues, *Med. Res. Rev.* **13**, 399–448.

Wise, A., Gearing, K., Rees, S. (2002) Target validation of G-protein coupled receptors, *Drug Discov. Today* **7**, 235–246.

Primary Literature

Chang, R.S., Lotti, V.J., Monaghan, R.L., Birnbaum, J., Stapley, E.O., Goetz, M.A., Albers-Schonberg, G., Patchett, A.A., Liesch, J.M., Hensens, O.D., Springer, J.P. (1985) A potent nonpeptide cholecystokinin antagonist selective for peripheral tissues isolated from Aspergillus alliaceus, *Science* **230**, 177–179.

Davis, R.J. (2000) Signal transduction by the JNK Group of MAP kinases, *Cell* **103**, 239–252.

Dean, M.K., Higgs, C., Smith, R.E., Bywater, R.P., Snell, C.R., Scott, P.D., Upton, G.J., Howe, T.J., Reynolds, C.A. (2001) Dimerization of G-protein-coupled receptors, *J. Med. Chem.* **44**, 4595–4614.

Enmark, E., Pelto-Huikko, M., Grandien, K., Lagercrantz, S., Lagercrantz, J., Fried, G., Nordenskjold, M., Gustafsson, J.A. (1997) Human estrogen receptor beta-gene structure, chromosomal localization, and expression pattern, *J. Clin. Endocrinol. Metab.* **82**, 4258–4265.

Honore, P., Kage, K., Mikusa, J., Watt, A.T., Johnston, J.F., Wyatt, J.R., Faltynek, C.R., Jarvis, M.F., Lynch, K. (2002) Analgesic profile of intrathecal $P2X_3$ antisense oligonucleotide treatment in chronic inflammatory and neuropathic pain states in rats, *Pain* **99**, 19–27.

Husi, H., Ward, M.A., Choudhary, J.S., Blackstock, W.P., Grant, S.G. (2000) Proteomic analysis of NMDA receptor-adhesion protein signaling complexes, *Nat. Neurosci.* **3**, 661–669.

Jacobsen, J.R., Hutchinson, C.R., Cane, D.E., Khosla, C. (1997) Precursor-directed biosynthesis of erythromycin analogs by an engineered polyketide synthase, *Science* **277**, 367–369.

Jenck, F., Wichmann, J., Dautzenberg, F.M., Moreau, J.L., Ouagazzal, A.M., Martin, J.R., Lundstrom, K., Cesura, A.M., Poli, S.M., Roever, S., Kolczewski, S., Adam, G., Kilpatrick, G. (2000) A synthetic agonist at the orphanin FQ/nociceptin receptor ORL1: anxiolytic profile in the rat, *Proc. Natl. Acad. Sci. U.S.A.* **97**, 4938–4943.

Jones, K.A., Borowsky, B., Tamm, J.A., Craig, D.A., Durkin, M.M., Dai, M., Yao, W.J., Johnson, M., Gunwaldsen, C., Huang, L.Y., Tang, C., Shen, Q., Salon, J.A., Morse, K., Laz, T., Smith, K.E., Nagarathnam, D., Noble, S.A., Branchek, T.A., Gerald, C. (1998) GABA(B) receptors function as a heteromeric assembly of the subunits GABA(B)R1 and GABA(B)R2, *Nature* **396**, 674–679.

Kim, M., Jiang, L.H., Wilson, H.L., North, R.A., Surprenant, A. (2001) Proteomic and functional evidence for a P2X$_7$ receptor signalling complex, *EMBO J.* **20**, 6347–6358.

Koshland, D.E., Nemethy, G., Filmer, D. (1966) Comparison of experimental binding data and theoretical models in proteins containing subunits, *Biochemistry* **5**, 365–385.

Morisset, S., Rouleau, A., Ligneau, X., Gbahou, F., Tardivel-Lacombe, J., Stark, H., Schunack, W., Ganellin, C.R., Schwartz, J.C., Arrang, J.M. (2000) High constitutive activity of native H$_3$ receptors regulates histamine neurons in brain, *Nature* **408**, 860–864.

Schoonbroodt, S., Piette, J. (2000) Oxidative stress interference with the nuclear factor-kappa B activation pathways, *Biochem. Pharmacol.* **60**, 1075–1083.

Shang, Y., Brown, M. (2002) Molecular determinants for the tissue specificity of SERMs, *Science* **295**, 2465–2468.

Sharpe, I.A., Gehrmann, J., Loughnan, M.L., Thomas, L., Adams, D.A., Atkins, A., Palant, E., Craik, D.J., Adams, D.J., Alewood, P.F., Lewis, R.J. (2001) Two new classes of conopeptides inhibit the alpha1-adrenoceptor and noradrenaline transporter, *Nat. Neurosci.* **4**, 902–907.

Smit, M.J., Leurs, R., Alewijnse, A.E., Blauw, J., Van Nieuw Amerongen, G.P., Van De Vrede, Y., Roovers, E., Timmerman, H. (1996) Inverse agonism of histamine H2 antagonist accounts for upregulation of spontaneously active histamine H2 receptors, *Proc. Natl. Acad. Sci. U.S.A.* **93**, 6802–6807.

Smit, A.B., Syed, N.I., Schaap, D., van Minnen, J., Klumperman, J., Kits, K.S., Lodder, H., van der Schors, R.C., van Elk, R., Sorgedrager, B., Brejc, K., Sixma, T.K., Geraerts, W.P. (2001) A glia-derived acetylcholine-binding protein that modulates synaptic transmission, *Nature* **411**, 261–268.

Stephenson, R.P. (1956) A modification of receptor theory, *Br. J. Pharmacol.* **11**, 379–392.

Thoma, J.A., Koshland, D.E. (1960) Competitive inhibition by substrate during enzyme action – evidence for the inducedfit theory, *J. Am. Chem. Soc.* **82**, 3329–3333.

Vassilatis, D.K., Hohmann, J.G., Zeng, H., Li, F., Ranchalis, J.E., Mortud, M.T., Brown, A., Rodriguez, S.S., Weller, J.R., Wright, A.C., Bergamnn, J.E., Gaitanaris, G.A. (2003) The G-protein-coupled receptor repertoires of human and mouse, *Proc. Natl. Acad. Sci. U.S.A.* **100**, 4903–4908.

Veber, D.F., Johnson, S.R., Cheng, H.Y., Smith, B.R., Ward, K.W., Kopple, K.D. (2002) Molecular properties that influence the oral bioavailability of drug candidates, *J. Med. Chem.* **45**, 2615–2623.

Wolf, B., Green, D.R. (1999) Suicidal tendencies: apoptotic cell death by caspase family proteinases, *J. Biol. Chem.* **274**, 20049–20052.

Receptor, Transporter and Ion Channel Diseases

J. Jay Gargus
University of California, Irvine, CA, USA

1	Introduction	645
2	Physiological Function of Ion Channels	646
3	Biochemical Structure of Ion Channels	650
4	**Ion Channels as a Mechanism of Disease**	**656**
4.1	Contribution of Genes and Environment to Disease	656
4.2	Monogenic Ion Channel Disease	656
4.3	Polygenic Ion Channel Disease	658
4.4	Pharmacogenetic Ion Channel Disease	658
4.5	Ion Channel Toxins	659
4.6	Mechanism of Ion Channel Disease Pathogenesis	659
5	**Ion Channel Diseases of the Heart**	**660**
5.1	Function of Ion Channels in the Heart	660
5.2	Ion Channel Dysfunction in Arrhythmia	661
5.3	Long QT Syndrome	662
5.4	Nature of LQT Genetic Loci and Alleles	663
5.5	The Dominant LQT Alleles	663
5.6	The Recessive LQT Alleles	665
5.7	Acquired LQT Channelopathies	666
5.8	Pharmacogenetic and Polygenic Arrhythmia Syndromes	666
6	**Ion Channel Diseases of Muscle**	**667**
6.1	Function of Ion Channels in Muscle	667
6.2	Ion Channel Dysfunction in Paralysis and Myotonia	670
6.3	Periodic Paralysis Syndromes	672
6.4	Malignant Hyperthermia	673

Encyclopedia of Molecular Cell Biology and Molecular Medicine, 2nd Edition. Volume 11
Edited by Robert A. Meyers.
Copyright © 2005 Wiley-VCH Verlag GmbH & Co. KGaA, Weinheim
ISBN: 3-527-30648-X

6.5	Other Monogenic Muscle Ion Channel Syndromes	674
6.6	Autoimmune Channelopathies in Muscle	674

7   Ion Channel Diseases of Nerve   675

7.1	Function of Ion Channels in Neurons	675
7.2	Ion Channel Dysfunction in Seizures	676
7.3	Hereditary Epilepsy Syndromes	677
7.4	LQT-like Loci and Alleles in Neuronal Syndromes	678
7.5	Other Monogenic Ion Channel Epilepsy Syndromes	678
7.6	Monogenic Ataxia and Migraine Syndromes	681
7.7	Other Monogenic Neuronal Ion Channel Syndromes	682
7.8	Acquired CNS Ion Channel Syndromes	683
7.9	Polygenic Syndromes of Altered CNS Excitability	684

8   Ion Channel Diseases of Epithelia   684

8.1	Structure and Function of Epithelial Tissues	684
8.2	Function of Ion Channels in Epithelia	685
8.3	Ion Channel Dysfunction and Diseases in Epithelia	686
8.4	Cystic Fibrosis	686
8.5	Renal Channelopathies	688
8.6	Pseudohypoaldosteronism and Liddle Syndrome	690
8.7	Bartter and Gitelman Syndrome	691
8.8	Hypomagnesemia	694
8.9	Polycystic Kidney Disease	695
8.10	Glucosuria	695
8.11	Nephrogenic Diabetes Insipidus	696
8.12	Hereditary Kidney Stone	697

9   Ion Channel Disease of Other Nonexcitable Tissue   698

10   Ion Channels as Targets for Drug Therapy in Common Diseases   699

Bibliography   700
Books and Reviews   700
Primary Literature   700

Keywords

Action potential
An all-or-none transient electrical depolarization propagated in excitable tissues.

ADH
Antidiuretic hormone or vasopressin; peptide hormone from the pituitary that controls plasma osmolarity by regulating aquaporins in the apical membrane of renal Principle cells.

Aldosterone
Adrenal steroid hormone that controls plasma volume and salt balance by regulating EnaC channels in the apical membrane of renal Principle cells.

Allele
The specific sequence of a given gene. Each possible variation of that sequence, functional or not, reflects a different allele.

Apical membrane
The specialized face of an epithelial cell that generally faces into the lumen of the gland and often is folded into a brush border with microvilli.

Aquaporin
A transmembrane protein that allows the transport of water free of solvated ions.

Ataxia
Inability to coordinate voluntary muscle activity, generally due to defect in cerebellum or spinal cord.

Channel
Transmembrane protein that allows the conductive transmembrane transport of small molecules, most commonly ions and water.

Channelopathy
A disease originating from dysfunction of a channel, or channel-like membrane protein.

Depolarization
A change in membrane potential tending to make the cell interior more positive.

Diuretic
A drug that serves to enhance urine flow, generally by acting to inhibit a renal sodium transport mechanism.

Dominant Allele
One copy of a mutation is sufficient to produce the mutant phenotype, such that the heterozygote expresses the phenotype.

Dominant Negative Mutation
A dominant mutation that results in less activity of the gene product than would be produced in a heterozygote with one wild-type and one null allele. Most commonly occurs because gene product functions as an inhibitory subunit in protein multimer.

EKG
Electrocardiogram; standardized clinical recording of extracellular electrical activity generated by the heart.

EMG
Electromyogram; standardized clinical recording of electrical activity generated by muscle contraction.

Environmental Disease
A disease caused by an agent external to the host.

Epilepsy
A chronic condition of seizure susceptibility, characterized by paroxysmal brain dysfunction and synchronized neuronal depolarizations.

Epithelium
A coupled sheet of polarized cells.

Excitation-contraction Coupling
The process by which electrical depolarization of a muscle membrane leads to activation of contractile fibers and cell contraction.

Gain-of-function Mutation
A mutation that increases the activity of the gene product to any degree. Syn = hypermorphic.

Gate
Portions of channel protein that close access to permeability pore and is operated in response to ligand binding, voltage, or other stimulus.

Gene Family
An assembly of genes related by sequence homology, likely reflecting decent from a common ancestor followed by divergence.

G-protein Coupled Receptor
A large family of membrane proteins having seven transmembrane spans, the prototype being rhodopsin, which serve in sensory transduction by coupling to a signaling cascade through trimeric GTP binding proteins.

Haploinsufficiency
A mechanism of dominant inheritance whereby a heterozygote with one wild-type and one null mutant allele manifests the mutant phenotype. This is an atypical situation since rarely is 50% of wild-type activity insufficient for physiological function.

Heteromultimer
A combination of protein product subunits encoded by distinct genes uniting to produce a functional protein assembly.

Heterozygote
An individual having two different alleles of a gene. The default use implies that one copy is wild-type and the other mutant; however, this can also be an individual having two different mutant alleles, in which case they are specified.

Homozygote
An individual having identical copies for both alleles of a gene. These can both be wild-type or mutant.

Hypermorphic
A mutation that increases the activity of the gene product to any degree. Syn = gain of function.

Hyperpolarization
A change in membrane potential tending to make the cell interior more negative.

Hypomorphic Mutation
A mutation that inactivates the gene product to any degree. Syn = loss of function.

Inactivation
A nonconducting conformation of an ion channel from which it is incapable of carrying out an activating transition into a conducting conformation.

Ion channel
A transmembrane protein that allows the selective transport of ions freed from solvating water.

Ligand-gated Ion Channel
Binding a chemical compound to a site on the channel protein operates the gating mechanism to open or close substrate access to the permeability pore.

Loss-of-function Mutation
A mutation that inactivates the gene product to any degree. Syn = hypomorphic.

Loss of Heterozygosity
Regional loss of the chromosome carrying a normal copy of a gene, exposing clinical manifestations of a recessive mutation that is otherwise carried silently on the other copy of the chromosome.

Mendelian Disease
The same as a monogenic disease. Its inheritance follows Mendel's laws and can be either dominant or recessive.

Migraine
A condition characterized by paroxysmal onset of severe headache, generally unilateral, often associated with vertigo, nausea and photobia and proceeded by a sensory aura.

Miniature Endplate Potentials (MEPPs)
Spontaneous depolarizations recorded at the motor endplate, which represent postsynaptic neuroreceptor response to spontaneous quantal release of neurotransmitter vesicles into the synapse.

Monogenic Disease
A disease caused by a major-effect gene mutation, such that the presence of this one mutation alone is highly predictive of the disease phenotype.

Motor Endplate
Postsynaptic membrane of specialized synapse of the motor neuron onto the skeletal muscle sarcolemma.

Myotonia
Delayed relaxation of skeletal muscle.

Neomorphic Mutation
A mutation that conveys to the gene product a new activity not present in wild-type.

Nernst Equilibrium Potential
The electrical potential difference at which no net ionic flux is observed across a permeable membrane despite the presence of a concentration difference between the sides. $E_K = [RT/nF]\ln[K_o/K_i]$ gives the Nernst potential for potassium, E_K, where R is the gas constant, T the absolute temperature Kelvin, n the valance, F the Faraday constant, and K_o and K_i the extracellular and intracellular potassium concentration, respectively.

Neuromuscular Junction
Specialized synapse of the motor neuron onto the skeletal muscle.

Null Mutation
A total loss of gene function, the extreme limit of a hypomorphic mutation, can include deletion or disruption (as in a knockout allele) of the gene itself.

Pacemaker
Endogenous source of rhythmic activity. In the heart, this is the sinoatrial node and the rhythm arises via spontaneous depolarization of the membrane.

Paracellular
Pathways across an epithelial sheet that pass between cells.

Paralysis
Inability to contract muscles.

Patch Electrode
Fire-polished glass electrode with 1-µm tip diameter that forms a molecularly-tight high resistance seal when applied to cell membrane, allowing recording from single channel molecules.

Pathophysiological Mechanism
Alteration in biological function leading to a disease process. This can be resolved from the molecular to the organismal and ecological levels.

Permeability Pore
Pathway though the channel protein taken by the conducted substrate.

Pharmacogenetic Disease
A disease caused by the interaction of an inciting "triggering" drug and an individual with a susceptible "at risk" genotype.

Plateau Phase
Segment of the cardiac myocyte action potential that coincides with the QT interval as observed on the EKG.

Polygenic Disease
A disease caused by an assembly of minor-effect gene mutations, such that the presence of any one of the mutations contributes only a small additional susceptibility to the disease.

Primary Active Transport
A transmembrane transport process that is capable of moving a substrate against its electrochemical potential gradient by directly utilizing the chemical energy of ATP hydrolysis.

QT Interval
The phase of cardiac ventricular repolarization as observed on the EKG. This coincides with the plateau phase of the cardiac myocytes action potential.

Recessive Allele
Carrying one copy of the mutation, the heterozygote appears like wild-type, not mutant. Only when both gene copies are mutant is the mutant phenotype observed.

Resorptive Epithelium
An epithelial sheet designed to create a flow of salt and water out of the lumen and into the blood.

Sarcoplasmic Reticulum
The specialized endoplasmic reticulum of skeletal muscle cells.

Secondary Active Transport
A transmembrane transport process that is capable of moving a substrate against its electrochemical potential gradient by coupling to the electrochemical potential of another chemical species, often sodium.

Secretory Epithelium
An epithelial sheet designed to create a flow of salt and water into the lumen and out of the blood.

Segregation
The disjunction of homologous chromosomes at meiosis such that offspring receive only one copy of each parental chromosome. This process underlies the characteristic Mendelian patterns of inheritance.

Seizure
Paroxysmal brain dysfunction associated with synchronized neuronal depolarizations.

Selective Advantage
An attribute of an allele that evolution favors, reflected by the survival of an excess of its descendants in subsequent generations.

Selectivity Filter
Portion of channel protein that restricts conductance to only specific substrates.

Splice Variant
Gene products that result from the use of alternative exons, often by exon skipping.

Subcellular Compartment
Membrane enclosed regions within eukaryotic cells. Examples include the endoplasmic reticulum, Golgi, mitochondria, and lysosomes.

Toxin
A natural biomolecule with potent biological activity. Ion channel toxins typically act to block or open ion channels at picomolar concentrations.

Transcellular
Pathways across an epithelial sheet that go through the cytosol.

T-tubules
Specialized invaginations of sarcolemma that play specialized role in excitation-contraction coupling of muscle fiber contraction.

Voltage Clamp
A feedback amplifier that adjusts externally applied current across a membrane such that the membrane potential is kept at a constant value even though its permeability character is varying.

Voltage-gated Ion Channel
Ion channel that is gated in response to changes in the membrane potential.

Wild-type
The allele of a gene most abundantly found in nature, for simplicity this can be considered the normal, nonmutant version of the gene.

Ion channels reflect an important but still very new mechanism of disease, first formalized in 1989 with the discovery of mutations in the *CFTR* ion channel gene causing cystic fibrosis, thereby carving out a new category of genetic inborn error, the channelopathy. Now, however, ion channel disorders are recognized to cover the gamut of medical disciplines and to cause significant pathology in virtually every organ system. The opening of this era was driven largely by the success of positional cloning strategies in isolating the genes underlying the rare monogenic disorders of ion channels and the explosive growth in basic ion channel physiology and biophysics that rendered these pathogenic mutant alleles interpretable, but in addition crystal structures have recently become available. Ion channels are a large family of over 400 related proteins representing over 1% of our genetic endowment, and many of the features shared among the rare monogenic ion channel diseases provide a solid foundation to begin to target these mechanisms for the development of novel therapeutics. They also render members of this gene family important "functional candidates" in the much more common complex polygenic diseases.

1 Introduction

Ion channel diseases reflect a relatively new category of genetic disease, or inborn error. Monogenic Mendelian ion channel diseases, or *channelopathies*, were first recognized in 1989 with the isolation of pathogenic alleles causing cystic fibrosis. The field explosively advanced as positional cloning revealed new pathogenic loci and alleles, and patch electrode electrophysiology allowed detailed characterization of the functional effects of mutations. This work most powerfully converged in the dissection of components of the action potential of excitable tissues – nerve, muscle and heart and while it remains true that diseases of excitable tissue still most clearly illustrate this family of diseases, ion channel disorders now cover the gamut of medical disciplines, causing significant pathology in virtually every organ system, producing a surprising range of often unanticipated symptoms, and providing valuable targets for pharmacological intervention.

2
Physiological Function of Ion Channels

Ion channels are a large family of related transmembrane proteins that provide ions, predominantly K^+, Na^+, Ca^{++}, and Cl^-, a passive pathway through which they can rapidly diffuse down their electrochemical gradient across the hydrophobic barrier of the plasma membrane. The standing electrochemical gradients that drive ion movements though channels are established by ion pumps – such as the Na/K-ATPase, and ion carriers – such as the Na-K-Cl cotransporter and the Na-H antiporter. While classical kinetic studies portray channels, pumps, and carriers as radically different transport mechanisms, with channels conducting ions 4 orders of magnitude faster and seemingly quite differently than the one-at-a-time transmembrane turnover of the pumps and carriers, we now recognize much similarity in their molecular mechanisms. Most cells maintain a high intracellular K^+, and a low intracellular Na^+ and Ca^{++}, compared with the surroundings, while Cl^- is often passively distributed at its *Nernst equilibrium*, which in a cell with a typical -60 mV membrane potential (inside negative) would be 10-fold lower on the inside (i.e. $E_{Cl^-} = (RT/nF) \log [Cl^-]_i/[Cl^-]_o \sim -59$ mV). The K^+ gradient, created at the cost of ATP hydrolysis by the ion pumps, is typically 10-fold in mammalian cells, and since the K^+ permeability, P_{K^+}, of most membranes overwhelms that of other ionic species, the K^+ diffusion potential predominates, setting the cell membrane potential, and giving rise to the typical -60 mV resting membrane potential near E_{K^+}. *Note that the physiological ions both create and respond to electrical phenomena across the cell membrane.*

In many ways, channels act like highly selective water filled pores that can be opened and closed in a controlled fashion to allow a specific ion species to flow, a flow that in turn can be perceived as both a miniscule chemical flux, with the use of isotopic tracer, or, more commonly, as an electrical current that changes the membrane potential toward the Nernst potential of the conducted ion, an interior positive *depolarization* created by opening Na^+ and Ca^{++} channels, an increasingly negative *hyperpolarization* created by opening K^+ channels, and a stabilization of the membrane potential by opening Cl^- channels. Note that unlike enzymes, channels carry out no biochemical transformations; the product and substrate ions differ only in regard to the side of the membrane on which they are found. Ion channels are further distinguished from enzymes in that rarely is the transported ion itself of any physiological consequences, since the flux through channels minimally changes the ion's concentration across the cell membrane; it is predominantly the electrical consequences of the ion current flow that underlies the physiology of the ion channels. Calcium is often an important exception to this rule, since it plays a critical role in coupling electrical activity to biochemical pathways. A second exception is that the chemical consequences of ion channel flux often predominates their function in epithelial cells.

We now recognize that ion channel proteins have structures that impose the three cardinal functions of a channel mentioned above: a *permeability pore* that provides ions but not water a passageway across the bilayer, a *selectivity filter* that allows some ions access while excluding others, and one or more *gates* that opens and closes the passageway. Three eras of Nobel Prize-winning technological and conceptual breakthroughs have led to our current level of understanding channel function (Fig. 1). All three key features of channel function were first defined, as abstract entities modeled with mathematical exactitude, in the era of the pioneers in membrane biophysics, an era that was opened up by *voltage clamp* technology fifty years ago and crowned by the Nobel Prize-winning classic squid axon studies of Hodgkin and Huxley (Fig. 1a). However, recent Nobel Prize-winning crystallographic work from the lab of Rod MacKinnon has now defined all three aspects in exquisite molecular detail for at least a few prototype channels (Fig. 1c). An essential era linking the mathematical and structural eras was that brought about by the breakthrough of *patch electrode* electrophysiology and crowned by the Nobel Prize-winning work of Neher and Sakmann. This era was critical in defining channels as biochemical mechanisms, and worked hand-in-glove with the molecular genetics of gene isolation and *in vitro* expression studies of cloned wild type and mutant channels, a breakthrough enabling paradigm pioneered in the lab of Shosaku Numa. Since the patch electrode would seal onto the plasma membrane so tightly that ions, and therefore ionic currents, could not leak between the two structures, magically tight gigaohm seals could be achieved that allowed electronic amplifiers the sensitivity to monitor in real time a single ion channel molecule in its millisecond stochastic dance between conformations: "closed," where it is nonconducting, and "open," where it passes on the order of 10^7 ions s^{-1}, producing a current of 10^{-12} amps, a picoamp, with stochastic transitions between the two states giving rise to the beautiful square wave traces of single-channel current records that are emblematic of the era (Fig. 1b).

One can separate the greater family of ion channels into various large classes based upon the cardinal channel features of selectivity and gating. One major distinction is based upon the nature of the selectivity filter and the predominant permeant ion species: hence, since the era of Hodgkin and Huxley, there have been K^+ channels, Na^+ channels and so on (Fig. 1a). Another major classification is based upon the gating mechanism. One large class of channels gate in response to their direct binding of ligand – the *ligand-gated ion channels*. Many of these ligands are classical neurotransmitters: hence, there are acetylcholine-gated ion channels, dopamine-gated ion channels and so on. But a wide range of other types of extracellular and intracellular ligands are also able to directly gate ion channels through binding. In addition, a large family of ion channels is indirectly gated by ligands, many by the same neurotransmitters mentioned above, but in this case neurotransmitter binding occurs to a heptahelical *G-protein coupled receptor* (GPCR) and the channel is activated by a second messenger ligand or a covalent modification, such as a protein

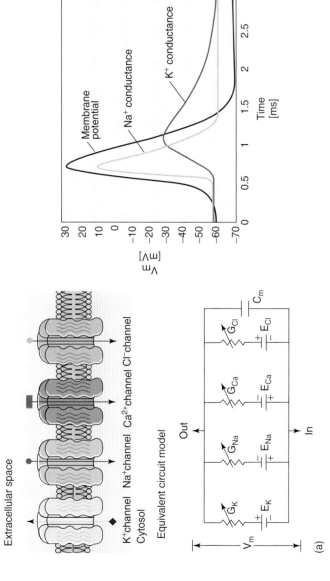

Fig. 1 3 ERAS of the ion channel. (a) Hodgkin–Huxley era, (b) Patch-clamp era, (c) Crystallography era.

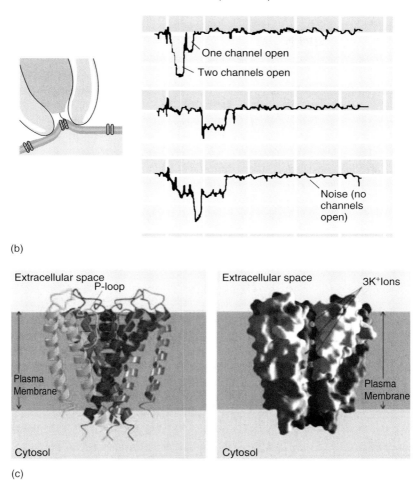

(b)

(c)

Fig. 1 (Continued)

phosphorylation or dephosphorylation reaction. Finally, a very large class of ion channels gate in response to changes in the electrical potential across the membrane, the *voltage-gated ion channels*. You can imagine that these are particularly tricky to study since the channel's own ionic current that is being controlled by the membrane potential *itself* alters the membrane potential, a vicious cycle that could only be broken by the voltage clamp amplifier. This experimentally vexing behavior of voltage-gated ion channels is, however, critical to their function in the perpetuation of a *propagating action potential* (AP). As a patch of membrane begins to depolarize, often because of the activation of a ligand-gated channel, voltage-gated Na^+ and Ca^{++} channels begin to respond to the voltage change by undergoing a conformational change from the "closed" into the "open" conformation (Fig. 2), increasing the membrane's permeability to sodium and calcium (P_{Na^+} and $P_{Ca^{++}}$) and hence

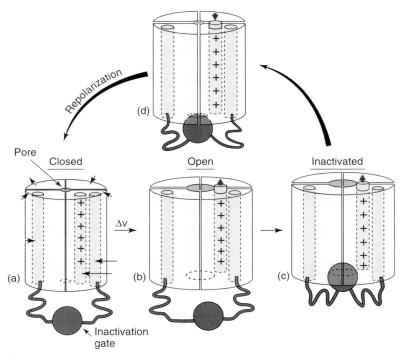

Fig. 2 Three states of the channel.

driving the membrane potential toward *their* inside-positive Nernst potential, a depolarization that serves to further activate more channels in surrounding patches of membrane in an explosive self-reinforcing cycle, recognized as a propagating AP, a depolarizing wave that rapidly spreads through an excitable tissue until all of its channels have been activated. Those channels next spontaneously enter an *inactive*, nonconducting conformation that is distinctly different from the "closed" state, since while "inactive," a channel cannot be opened and the tissue is rendered nonexcitable. Only by restoring the resting membrane potential does the channel's conformation become reset to the "closed" state that has the potential to open, and excitability is restored to the tissue. This resetting and restoring job is carried out by the more slowly activating voltage-activated K^+ channels. Since cells have but one membrane potential, it is always integrating inputs from all channel mechanisms in the membrane, and for this reason it is the great integrator in signaling pathways. Also, unlike biochemical pathways whose overall flux is predominated by a single rate-limiting step, membrane potential is intrinsically a continuous variable fully reflecting subtle changes in the entire host of contributing channels.

3
Biochemical Structure of Ion Channels

Ion channel proteins share many structural features in common that today allow

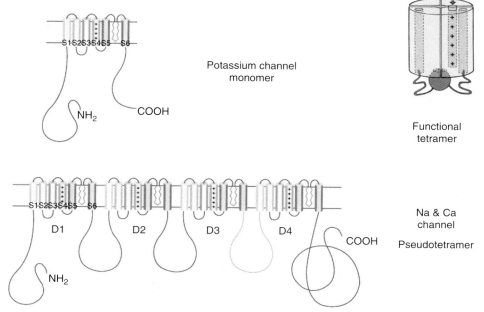

Fig. 3 Voltage-gated channels.

us to model their consensus structures and functions based upon the few prototype channel structures that have emerged from crystallography (Fig. 1c). For example, the large superfamily of voltage-gated potassium channels share a monomer structure of 6 transmembrane helices, each helix characteristically 20 to 25 amino acids in length, and both the N- and C-termini facing the cytosol (Fig. 3). The N-terminus forms a tethered "ball and chain" motif that gates the channel under the influence of electrostatics and the altered membrane potential its activation has brought about. It does this by swinging into the inner mouth of the pore and thereby effects the "opened" to "inactive" transition of the channel. The forth helix characteristically contains a cluster of 4 to 8 positively charged residues, converting that helix into a dipole that allows it to move under the influence of the membrane potential. This allows it to serve as the voltage sensor that gates the channel from the "closed" to the "opened" conformation. Between the fifth and sixth helix is found the heart of channel function, a signature pore domain characterized by a hairpin motif that dips partially through the membrane to form the conductive pathway used by the ion (Fig. 4). Some families of channels, such as the inward rectifiers, contain only this pore and the two flanking helices (Fig. 5). This pore domain contains the K^+ selectivity filter, a nonhelical motif conveyed by the amino acid sequence, TVGYG. It is, however, the strongly dipolar carbonyl groups of the α-carbons in the peptide backbone of this amino acid sequence, not the specialized side chains, that line the pore and create, through electrostatics and geometry, the selectivity filter that allows potassium ions

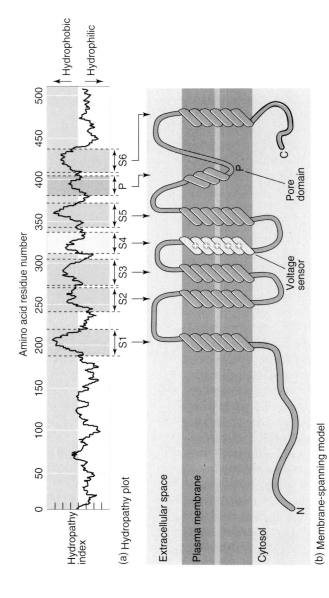

Fig. 4 Hydropathy plot and secondary structure prediction of amino acid sequence of potassium channel monomer.

Receptor, Transporter and Ion Channel Diseases | 653

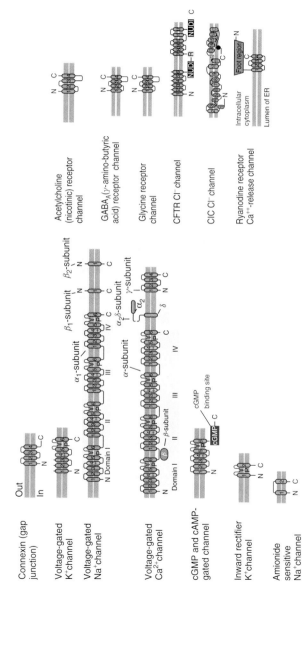

Fig. 5 Secondary structure of channel families.

entrance, but excludes sodium and other ions. The functional potassium channel that creates the transmembrane pore is a tetramer and the overall structure of the tetrameric assembly is that of an inverted teepee created by the tertiary structure of the fifth through sixth helices of the four interacting subunits, with the large vestibule of the teepee facing outward and lined by the selectivity filter residues from each monomer (Fig. 1c). Helices 1 to 4 of the monomers surround this core structure and present a surface that interacts with the lipid bilayer. Upon channel opening, the potassium ions move in a single file fashion along the pore, rapidly hopping between interactions with each of the critical residues of the filter motif, thousands of ions flowing during each channel's opening.

Conserved features that families of ion channels share in common have also proven useful in the identification of their genes in the genome. The human genome contains over 400 channel genes representing between 1 to 2% of our genetic endowment, although the substrate transported by most remains to be identified. While there are already known several large families of ion channels that apparently carry out extremely similar biochemical functions (Figs. 5 and 6), for instance, over 100 voltage-gated channels, the large number of channel genes almost certainly does *not* reflect redundancy. Since their job is not primarily contingent on their ability to transport ions from one side of the membrane to the other, but rather on the shape of the electrical signal they produce during conductive ion flow, it is apparently this that is under evolutionary selective pressure and it suggests that physiology requires the diverse vocabulary provided by multiple members of a gene family, which carry out nearly indistinguishable transport reactions with subtly differing kinetics. Furthermore, it is becoming increasingly clear that gene-family diversity is additionally greatly expanded through alternative exon splicing, allowing one channel gene locus to produce multiple splice isoforms and hence multiple functionally distinct protein products. Finally, for at least a handful of channels, and best characterized in the glutamate receptor channels, the exotic complex process of RNA editing is carried out to change single amino acid residues in the permeability pathway that critically tune ion selectivity and other aspects of channel function. Clearly, very subtle differences in channel function are important since evolution seems to have gone to very great lengths to preserve and enhance channel gene diversity.

For the potassium channels, diversity is further tremendously amplified through combinatorial monomer assembly into the functional tetramer (Fig. 3). This predominantly arises because channel protein subunits encoded by all members of a channel gene *family*, not just the allelic splice-variant protein products, coassemble to create a functional channel that differs in biophysical behavior as a function of the stoichiometry of the constituent subunits in the assembly. The panoply of combinatorial heteromultimers formed as monomers from family members coassemble, creates tremendous functional diversity in the potassium currents. This finely tunes the behavior of the population of expressed functional channels. This diversity conveys the essence of potassium channel function, allowing them to be tuned to modulate signal processing as they interact and jointly shape the membrane potential, thereby "deciding" the cell's aggregate signaling output.

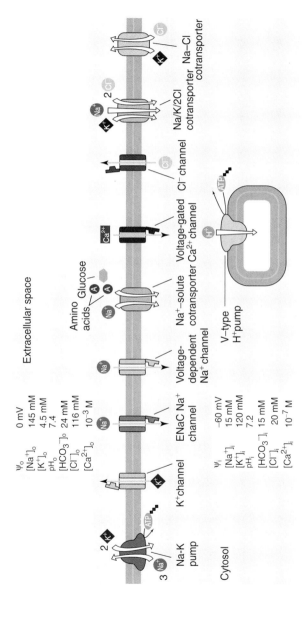

Fig. 6 Ion pumps, carriers, and channels.

In contrast to the broad heterogeneity of composition and function of the "contemplative" potassium channels, the voltage-gated sodium and calcium channels that are the "do-ers," able to execute the result of signal processing, are extremely homogeneous in their structure and function in a given tissue. While they can be perceived to share the structure of the potassium channel monomer, in their case, where a crisp uniformity of channel function is required within a tissue, a fixed pseudo-tetramer structure is hard-wired into the large endoduplicated genes that encode a pseudotetrameric functional monomer (Figs. 3 and 5). Tissue-specific diversity in the function of these channels is conveyed by tissue-specific isoforms of the gene family of voltage-gated sodium and calcium channels.

In addition to the major, α, channel subunits that directly contribute to the conducting pore of ion channels, discussed above, there are also a host of auxiliary subunits, each represented by gene families often labeled $\beta, \gamma, \delta, \varepsilon$, that modify channel function (Fig. 5). This obviously adds another enormous combinatorial explosion of channel diversity. Much of this diversity allows channel function to be tuned for its specific use in a specific tissue, since many channel genes are expressed in a remarkably tissue-selective fashion. On the other hand other channels may be found to be broadly expressed, but to predominate in the physiology of only a few tissues. As will become clear below, both of these features allow channels to produce tissue-selective disease. This tissue selectivity also points out that ion channels should offer highly focused targets for therapeutic intervention, perhaps devoid of the cross-tissue "side effects" that complicate many current receptor-based drug targets.

4
Ion Channels as a Mechanism of Disease

4.1
Contribution of Genes and Environment to Disease

While classically diseases are recognized to arise from a host of extrinsic environmental causes, such as infectious agents, or from intrinsic causes, most simply a reflection of our genes, we now recognize that genes and the environment are inextricably connected in all disease, and in most we see various contributions from a small number of "major-effect" genes, smaller contributions from a large number of "minor-effect" genes, and variable contributions from a host of factors in the environment. While recognizing this interconnectedness of genes and environment, we still tend to refer to diseases as being "caused" by the most salient of these three factors, the *monogenic* diseases, the *polygenic* diseases, and the environmental diseases. *Most parsimoniously, disease reflects a mismatch between genotype and environment.*

4.2
Monogenic Ion Channel Disease

Rare mutations arise *de novo* in the population each generation, breaking the molecular mechanism encoded by every gene in the genome in every possible way, most mutations being either neutral or deleterious to the gene's function; should the effect disrupt a critical physiological pathway, they are "disease-causing." *Dominant* mutations are expressed in a phenotype even in the presence of one normal allele, so the phenotype occurs in the individual in whom they first arise; those producing a phenotype incompatible with reproduction being immediately eliminated from the population, but being constantly replenished though new mutations. Thus, dominant mutations that are maintained and transmitted in the population tend to produce a later onset or milder course of disease. Recessive mutations that arise *de novo* are carried silently in a *heterozygote*, only rarely producing a disease on those rare occasions when two copies of the mutant gene are contributed to a *homozygous* offspring of two rare heterozygotes, this likelihood tremendously enhanced in a consanguineous mating. As most recessive alleles in the population are carried in heterozygotes, even lethal homozygote phenotypes have little effect on the disease-allele's frequency. On the other hand some "disease-causing" alleles are now very abundant in a given population, exceeding 1% of the population's alleles, frequencies typical of *polymorphic variants* that are commonly considered to be of "neutral" functional significance. An example of such a common polymorphism, carried by approximately 1 in 20 Caucasians, is the mutation that causes the most common genetic disease in Caucasians and the first recognized genetic ion channel disease, cystic fibrosis (CF). To understand the apparent anomaly of a lethal mutation rising to such abundance, a historical perspective is required. Often such high "deleterious" allele frequencies arise because of some *selective advantage* conveyed to the heterozygote carrier. In other instances, it simply appears that by chance heterozygotes for a rare mutation were among the founders who established an isolated population; with inbreeding, this allele became abundant and the disease frequency in this isolate became appreciable. Many common disease-causing mutations seem to have served in our evolutionary past as primitive "inoculations" protecting heterozygous carriers, who possess most of the "disease-alleles" in the population, against specific disease agents. The best-known example is the assent of the globin sickle cell and thalassemia hemolytic anemia alleles in the malaria belt, but CF alleles appear to be particularly protective against the infectious secretory diarrheas that imposed a high infant mortality in our recent past. While carrying two copies of the common CF allele is sufficient to produce a disease phenotype, independent of the contribution of other genes or environmental agents, and hence is recognized as a monogenic disease, the rare, non-polymorphic mutant alleles of genes more commonly cause the rare monogenic diseases, such as the Long QT syndrome ion channel disorder, or channelopathy, to be discussed below.

Diseases such as those discussed above, in which there is a major single gene contribution to the disease phenotype are classically recognized as monogenic or *Mendelian* genetic diseases. The classical Mendelian pattern of recessive disease arises most commonly from *loss of function, null,* or *hypomorphic*, mutations, predominantly in a situation where sufficient gene product is available from the sole remaining copy of an essential gene for homeostasis. The dominant pattern of inheritance commonly requires a *gain of function*, such that a phenotype is created even in the presence of a normal copy of the gene; this can be an enhanced normal *hypermorphic* activity, a *neomorphic* activity not present in the native protein, or an inhibitory activity, *negative-dominance*, generally arising from an inactivating multimerization with the native protein. An alternative mechanism of dominance is *haploinsufficiency*, the less common situation where 50% of the physiological level of the gene product is *not* enough for homeostasis and in such cases a loss-of-function allele can be dominant. Even for these "monogenic" diseases we recognize that other genes in the genome and the environment still contribute to the disease phenotype, and these effects are acknowledged under the classical terms of *penetrance* and *expressivity*; two individuals carrying exactly the same two alleles of a major-effect gene may have vastly different phenotypes with one showing the classical disease and the other ranging from having such minor manifestations that no aspect of the phenotype reveals the presence of the "disease genotype," in which case he is called *nonpenetrant*, to perhaps manifesting only a minor, yet still diagnostic, feature of the disease, in which case he would demonstrate the variable expressivity of the disease.

4.3
Polygenic Ion Channel Disease

The situation described above is magnified in the polygenic diseases, common diseases of ill-defined pathogenesis, such as hypertension, diabetes, and epilepsy, where typically one mutant gene is insufficient to create the disease phenotype, most of the genes still remain to be defined, and the pattern of inheritance observed is simply familial clustering, not the clear segregation patterns of recessive or dominant inheritance seen in the Mendelian disorders. While many gene loci contribute together to produce the common polygenic diseases, both common polymorphisms and rare mutant alleles seem to participate, and while major-effect genes play a very minor role in the population, not uncommonly rare Mendelian forms of the disease phenotype do occur, their etiological gene providing a peak into the pathogenic mechanisms likely underlying the common form of the disease, such as the monogenic channelopathies recognized to cause epilepsy in benign familial neonatal convulsions (BFNC), hypertension in Liddle syndrome, and diabetes in permanent neonatal diabetes (all to be discussed in detail below). Most polygenic diseases additionally have a significant environmental component, hence their more comprehensive title of *complex polygenic diseases*. The environmental component of these diseases can be recognized since even identical monozygotic twins, sharing all of their genes in common, are discordant for the disease phenotype a high percentage of the time.

4.4
Pharmacogenetic Ion Channel Disease

Finally, some diseases appear to be caused by environmental agents. However, we now recognize that only in the face of an appropriate "at risk" genotype is an appropriate physiological target available such that the agent can be a toxin. Should the toxin be fairly rare in the environment, such as a general anesthetic, it may appear initially to cause an idiosyncratic environmental disease, such as the lethal rare adverse reaction to anesthetics referred to as *malignant hyperthermia* (MHS). Much later it was recognized that specific heritable ion channel mutations cause one to be vulnerable to this syndrome, an individual inheriting an MHS allele being perfectly healthy, just at risk for malignant hyperthermia, and not developing the lethal disease phenotype until exposed to the inducing anesthetic. The recognition of major genetic and environmental components in the disease classifies it as a *pharmacogenetic* disease, and likely many of the idiosyncratic drug reactions we observe, either toxic or therapeutic, reflect this same phenomenon. On the other hand, should the environmental agent be ubiquitous, such as phenylalanine, an amino acid found in all foods, a disease caused by an environmental agent may appear to be a genetic disease, in this case phenylketonuria (PKU), the archetypical "genetic" disease for which every newborn is screened, but which produces no disease phenotype if phenylalanine exposure is controlled.

4.5
Ion Channel Toxins

That ion channels are exquisitely sensitive, finely tuned, evolutionarily highly conserved mechanisms essential to an organism's physiology has not been lost to the combatants on Evolution's fierce battleground. Nearly all of the most potent toxins in Nature target ion channels, most with nanomolar to picomolar affinity. Tetrodotoxin from the puffer fish, saxitoxin from red tide dinoflagellates, and a huge number of scorpion and coelenterate peptide toxins target the voltage-gated sodium channels; ω-conotoxin from the cone snails and a host of spider venom toxins target the voltage-gated calcium channels; a number of scorpion, insect, and coelenterate toxins target families of potassium channels; α-bungarotoxin, from snake venom, and the Amazon blow-dart alkaloid toxin curare target the nicotinic acetylcholine receptor and the plant alkaloid picrotoxin targets the GABA receptor channel. Most toxins act by relying upon intimate, high-affinity contacts within the channel molecule, and for this reason many have proven to be essential tools in defining classes of ion channels and their mechanisms of action. Since most of these toxins are quite rare in the environment and none seem to have participated in major human evolutionary bottlenecks, nearly all human genotypes are vulnerable to these toxins, despite the fact that in the laboratory mutations can be selected to convey toxin resistance. For this reason, a huge array of toxins can be recognized to produce quite pure forms of environmental or toxic channelopathies.

4.6
Mechanism of Ion Channel Disease Pathogenesis

Ion channel diseases span the gamut of pathogenic mechanisms discussed above. Furthermore, the pathophysiological mechanisms that lead from a defective ion channel protein to a disease phenotype are complex and multifaceted. That Nature requires a genetic investment in the huge array of subtly differing channels leads one to expect that subtle changes in ion channels matter, foreshadowing the wide range of phenotypic abnormalities different mutants of the same gene might produce. It is therefore not surprising that ion channel diseases as apparently distinct as migraine and ataxia, paralysis, and myotonia or even as having dominant versus recessive inheritance, are found to be allelic, caused by different mutations in the same gene.

A *periodic* disturbance of rhythmic function is the cardinal feature of ion channel disease of excitable tissues. In the heart, this produces a fatal *arrhythmia*, observed as a loss of the normal synchronous contraction of the muscle composing the pumping chambers secondary to a disruption in the rhythmic action potentials arising in this electrically coupled tissue. In skeletal muscle, abnormal ion channel function produces periodic alterations in contractility, ranging from the inability to contract – *paralysis*, to the inability to relax – *myotonia*. And in the central nervous system (CNS), the rhythmic disturbance is observed as an abnormally synchronous electrical discharge, a seizure. The remarkable finding in all of these syndromes is the absence of an overt functional abnormality the vast majority of the time. In the extreme, this results in the tragic situation where no phenotype is obvious until the moment of premature death. While many insights have been garnered as to how a variety of stresses serve to create the decompensations that allow these periodic phenotypes to become manifest, the mechanisms that serve to compensate for a constitutionally defective channel such that healthful homeostatic function is maintained the majority of the time, remain to be defined.

Ion channel phenotypes in nonexcitable tissues are more diverse because of the very different roles channels play in such tissues, participating in signaling involved in endocrine secretion, the function of cytosolic compartments and complex epithelial secretory and resorptive function, which in the kidney, produce secondary changes in systemic electrolyte balance and blood pressure.

It is simplest to develop the pathophysiology of ion channel disease by starting with the cardiac phenotype of arrhythmia and the Long QT (LQT) syndromes, demonstrating how the ion channels in the membrane work together in tissue function and how component parts of the membrane potential are dissected by the genetic lesions. It further provides specific paradigms for how subunit interactions can be manifest, how loss-of-function and gain-of-function phenotypes occur and how dominance can arise through either haploinsufficiency, gain-of-function, or in a "dominant negative" fashion. The muscle and CNS phenotypes, while not lending themselves to as simple a dissection of function are, however, easily built by extrapolation from this paradigm. They further illustrate the power of gene families in identifying candidate disease genes and reinforce the intrinsic polygenic nature of ion channel phenotypes. Finally, the nonexcitable tissue phenotypes can be used to show the range of functions ion channels carry out.

5
Ion Channel Diseases of the Heart

5.1
Function of Ion Channels in the Heart

Cardiac cells possess an intrinsic instability of their membrane potential, such that spontaneous depolarizing action potentials arise (as described at the end of Sect. 2 above), that can be observed as contractions in dissociated cardiac cells in culture. Muscle contraction is directly coupled to the propagation of an action potential, a phenomenon referred to as *excitation-contraction* (EC) *coupling*, since the calcium ions that enter the cell, amplified by other calcium sources inside the cell, produce the biochemical activation of the contractile proteins. Briefly, this entails calcium binding to troponin and releasing tropomyosin's block on myosin's cycling cross-bridge formation with actin. The electrical *pacemaker* function in a healthy heart only has the opportunity to manifest itself at the sinoatrial (SA) node, the physiological pacemaker of the heart. Since all of the other cardiac cells have a slower endogenous rate of spontaneous depolarization (pacing), and since all cells in the heart are electrically coupled, the fastest spontaneously depolarizing pacemaker, the SA node, is the origin of a propagated action potential that drives though the remaining cells of the organ,

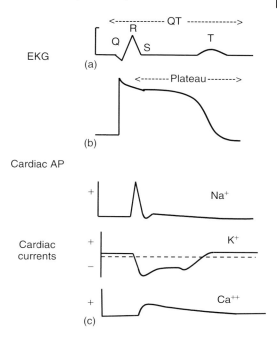

Fig. 7 Synchronized EKG, cardiac action potential and component currents. (a) EKG trace, (b) cardiac ventricle myocyte action potential, (c) resolved ionic currents underlying ventricular myocyte action potential. From Gargus, J.J. (2003) Unraveling monogenic channelopathies and their implications for complex polygenic disease.

forcing them to depolarize before they would do so spontaneously. Should the fastest pacemaker cells be lost, the next fastest pacemakers progressively take over, a very reassuring redundancy in a complicated process that must be repeated several billion times, failure rapidly resulting in death. The directional flow of the propagated action potential through the heart muscle creates the vector of the surface current revealed by the EKG traces, and produces a synchronous rhythmic coordinated contraction of the muscle cells composing the pumping chambers of the heart. If the cardiac cells of a pumping chamber all contract synchronously, the chamber volume rapidly shrinks, forcefully pumping blood that is directed by the cardiac valves. Without synchrony there is no pumping, only ineffectual fibrillation of individual cardiocytes and death to the individual. The cardiac pump cycle thus can be followed with the EKG. The QT interval measured on the EKG is the long interval of the cardiac cycle during which the ventricles, the major pumping chambers, repolarize, their conducted depolarizing action potential being reflected by the QRS waves (Fig. 7).

5.2
Ion Channel Dysfunction in Arrhythmia

A design feature engineered into the ventricular action potential by the mix of ion channels in its membrane is a prolonged plateau phase of depolarization, a phase *not* present in the classic rapid action potential of nerve (Fig. 1a). LQT syndrome is caused by mutations in the cardiac ion channel genes contributing to this component of the cardiac action potential. The plateau phase is a time during which calcium enters ventricular myocytes to produce contraction, but as importantly, the depolarized plateau phase serves to hold sodium channels in the "inactive" nonexcitable state until the inciting depolarizing wave has spread to all of the electrically coupled myocardium. In this manner, it can be assured that all ventricular myocytes will repolarize and regain the ability to again depolarize in synchrony. If this mechanism fails, and one group of cells fires an action potential that cannot be conducted through a still-recovering inactive area, those inactive cells may subsequently have sufficient time to develop their endogenous pacemaker-like depolarization and asynchrony in the tissue is initiated as an endlessly looping futile depolarization wave spreads through any newly excitable domains of the tissue. Such asynchrony is lethal since unless the entire ventricular muscle depolarizes and contracts together, it does not pump blood, but only fibrillates. During the plateau phase, the membrane potential struggles between the depolarizing influence of sodium and calcium currents and the repolarizing influence of potassium currents, with the slowly developing potassium currents finally overcoming the depolarizing currents and restoring the resting membrane potential (Fig. 7). Mutations that serve to either enhance the sodium current or to reduce the potassium currents predictably prolong the plateau phase and are associated with LQT syndrome. A long plateau phase is a dangerous situation since prolonging it opens the opportunity for asynchrony between the cells and susceptibility to arrhythmia, as described above.

5.3
Long QT Syndrome

Classically the LQT syndromes are divided into those with a dominant pattern of inheritance and a phenotype limited to the heart, the Romano–Ward syndrome (RW), and those having recessive inheritance and a phenotype that includes sensorineural hearing loss, the Jervell and Lange-Nielsen syndrome (JLN). In addition, there are now recognized acquired forms of the disease, primarily developing in a failing heart, and pharmacogenetic forms, a serious complicating side effect for a wide spectrum of drugs. As the name suggests, LQT prolongs the QT interval measured on the EKG and predisposes to a fatal arrhythmia; however, typically the pedigree may only reveal early sudden death of no known etiology. There are no diagnostic pathological findings and commonly no symptoms before death such that the disease is commonly not diagnosed until repeated deaths in a family have occurred. Diagnosis of the syndrome, on the other hand, does not depend upon a clinically recognized phenotype; rather, it depends only upon the EKG findings. There is a characteristic context of death observed in the syndrome that further illuminates the critical balancing act performed by multiple cardiac channels in normal physiology. The context is that of excess adrenergic outflow, as occurs in high emotion or exertion. This is a risk since the heart rate dramatically increases during the "fight or flight" responses these transmitters mediate. While we may initially focus on this response in terms of the β-adrenergic receptor increasing the rate at which the cardiac pacemaker depolarizes and the heart contracts, it is clear that to depolarize more rapidly, it must also *repolarize* more rapidly, and in fact specific cAMP-dependent channel regulatory mechanisms downstream of the β-adrenergic receptor assure this coordination, effectively shortening the QT as the heart rate rises. This explains why the rational way to evaluate a QT is in the context of the underlying heart rate, the QTc. In a patient with an intrinsic LQT defect that lengthens the QT, this adjustment of the rate of repolarization can not keep pace, the QT can not sufficiently shorten at high heart rates, and an arrhythmia ensues.

5.4
Nature of LQT Genetic Loci and Alleles

LQT mutations have been found in six cardiac ion channel genes. One LQT gene encodes the cardiac sodium channel and five encode the primary and auxiliary subunits of three different potassium channels (Fig. 8). While this provides a nearly complete dissection of the major channels involved during the critical plateau phase, one major actor, the cardiac calcium channel, encoded by *CACNA1C*, has yet to yield a pathogenic allele, perhaps being an early embryonic lethal. However, the most recently discovered LQT locus, LQT4, encoding ankyrin-B, while being unique in *not* encoding an ion channel protein, regulates a complex of transporters in the cardiac membrane mediating calcium homeostasis. *LQT1*, the first LQT gene identified, was found using a strategy involving linkage and positional cloning. All of the other LQT genes were identified as functional candidates within a mapped chromosomal interval.

5.5
The Dominant LQT Alleles

LQT1 was located on chromosome 11p 15.5, and it was ultimately demonstrated that mutations in a potassium channel α-subunit gene, *KCNQ1*, were causal. When expressed *in vitro*, *KCNQ1* produces a current that is not found in the heart; however, when expressed together with its β-subunit, encoded by *KCNE1*, they can be recognized to produce the I_{Ks} channel underlying the slow delayed rectifier K$^+$ current that participates in repolarization (Fig. 8). Over 35 different pathogenic alleles have been reported in this gene and they are the most common cause of RW syndrome. They primarily confer a dominant-negative phenotype when studied *in vitro*. Defective subunits coassemble with wild type copies producing, through combinatorials, a supermajority of defective channel tetramers, and hence reduced repolarizing current.

LQT2 was the first LQT gene cloned, taking advantage of a candidate gene approach within the region of chromosome 7q35–36 where a second locus was identified in families not mapping to chromosome 11. The gene, *KCNH2*, was a strong functional candidate based upon its homology to a fly gene with a proven ability to create a rhythm disorder phenotype, ether-a-go-go. Like *KCNQ1*, coexpression of *KCNH2* with its β-subunit, *KCNE2*, was required to produce a current that can be recognized in the heart. The channel they form is I_{Kr}, the rapidly activating delayed rectifier K$^+$ current (Fig. 8). Over 16 alleles of *KCNH2* have been reported, and like *KCNQ1*, many confer a "dominant-negative" phenotype when expressed *in vitro*. However, other dominant alleles appear to be simple loss-of-function alleles, suggesting that the membrane current conducted by this channel is so finely tuned that a haploinsufficiency mechanism is adequate to produce dominance.

LQT3 is the only sodium channel locus involved in the disease, and it is caused by mutations in the *SCN5A* gene found at chromosome 3p21–24. It encodes the cardiac-specific voltage-gated sodium channel that underlies the rapid depolarization phase of the ventricular action potential that produces the QRS complex and ventricular contraction (Fig. 8). Over 30 alleles of this gene have been

Receptor, Transporter and Ion Channel Diseases

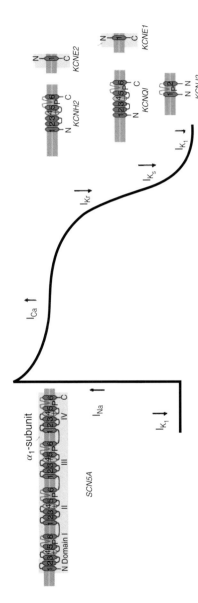

Fig. 8 Cardiac ion channel LQT genes. Each channel subunit gene product shown carries known LQT mutations. ↑ signifies depolarizing current, excess leading to LQT; ↓ signifies hyperpolarizing current, deficiency leading to LQT.

reported. Channels encoded by pathogenic alleles have delayed or decreased inactivation after opening, leaving excess inward depolarizing current during the plateau phase, delaying the time at which potassium currents can bring about repolarization. They are thus dominant "gain-of-function" mutations.

A distinct set of *SCN5A* alleles produce the *opposite* effect on inactivation from the *LQT3* alleles; they produce a rapid recovery from inactivation. They produce a different dominant arrhythmia syndrome, Brugada syndrome. The myocardium of these individuals contains a mixed population of sodium channels that are no longer locked in synchrony by a common period of inactivation, forming an ideal substrate for arrhythmia. Some alleles, surprisingly, are capable of producing either sodium channel phenotype within a family, demonstrating the delicate balancing act between excitation and inactivation carried out by the sodium channel that will be more clearly illustrated in the skeletal muscle disorders, and hinting at modifier genes that alter this balance, and therefore polygenic disorders.

LQT4 is the exception that proves the rule that LQT is an ion channel disease. The *LQT4* locus, found at 4q25 encodes ankyrin-b, a scaffold protein that colocalizes and regulates the Na/K-ATPase, the Na/Ca exchanger and the IP3 receptor. It is not a channel, but pathogenic loss-of-function alleles disturb calcium homeostasis through the functional loss of these critical mechanisms. It thus creates a dominant arrhythmia syndrome that includes LQT, but is not restricted to LQT.

LQT5 and *LQT6* are two adjacent loci on chromosome 21q22.1, and their genes, *KCNE1* and *KCNE2*, respectively, encode the two highly homologous potassium channel β-subunits discussed above as the accessory subunits that interact with their α-subunit partner to form I_{Ks} and I_{Kr}, respectively (Fig. 8). Both are proteins with only a single transmembrane α-helix and both function only to modify the behavior of the α-subunit with which they multimerize. Dominant pathogenic alleles are missense and effectively achieve dominance by altering channel gating to produce less current.

LQT7 is an alternative name for Andersen syndrome, a dominant multisystem disorder that includes long QT, but also extracardiac findings such as periodic paralysis and, more surprisingly, dysmorphology. Over nine different dominant-negative alleles of the potassium channel gene *KCNJ2* are responsible for this syndrome. The *KCNJ2* channel subunits multimerize to form the inwardly rectifying potassium channel that governs the resting membrane potential of the cardiac myocyte (Fig. 8). Current through this channel participates at the very end of repolarization, and hence dominant-negative loss of its function prolongs the process. The extracardiac findings imply that it also contributes in a significant way to membrane signaling in muscle and perhaps to the developmental processes underlying the dysmorphology.

5.6
The Recessive LQT Alleles

The recessive JLN syndrome differs from the dominant RW syndrome not only in its pattern of inheritance but also in its extracardiac manifestation of a constitutive sensorineural hearing loss. While in both syndromes the cardiac arrhythmia phenotype is lethal and intermittent, the striking differences between the two syndromes would never lead you to guess they might be allelic. In fact, the JLN alleles are strong loss-of-function or null alleles in the same genes that cause dominant *LQT1* and *LQT5*, *KCNQ1* and *KCNE1*, the two channel genes encoding the I_{Ks} channel (Fig. 8). Surprisingly the same channel is expressed selectively in the heart and the inner ear, where its abundant expression is involved in the production of the potassium-rich endolymph fluid that fills this chamber and participates in the transduction of auditory sensation. Only complete loss of the channel significantly disturbs this secretory process. It is of note that another inner ear *LQT1* homolog, encoded by *KCNQ4*, has dominant alleles that produce only deafness without LQT (*DFNA2*).

5.7
Acquired LQT Channelopathies

While heart failure is the very common end-stage phenotype of a dilated and poorly contractile myocardial pump, and is produced by a variety of common insults to the heart, most deaths in this disorder arise from a fatal arrhythmia, and it likely reflects one of the most common forms of the LQT syndrome. The failing myocytes downregulate expression of potassium currents, perhaps in an attempt to enhance contractility by prolonging the period of calcium entry, the QT interval, but doing so predisposes them to the arrhythmia discussed above.

5.8
Pharmacogenetic and Polygenic Arrhythmia Syndromes

Initial studies suggested that LQT was a highly penetrant Mendelian phenotype with "strong" alleles that, while clinically silent, would dependably be revealed by EKG. Not surprisingly, it has become clear that there are both "weak" and "subclinical" alleles of these genes as well and likely polygenic interactions between them. Additionally, there is clearly an environmental component to the disease, seen as a pharmacogenetic syndrome, which greatly broadens the scope of the diseases involving these cardiac ion channel genes. These more subtle features of the disease became most apparent in studies on families initially considered to have a case of "sporadic" LQT. While half of the "sporadic" probands in the study had the predicted new dominant mutation in one of the LQT genes, the other half were found *not* to have new mutations, but rather to have families segregating a "weak" pathogenic LQT allele. This allele in these families had many silent carriers with no EKG abnormality, or they had carriers who only expressed the abnormality once taking a medication with potassium channel blocking activity. The silent carriers presumably lacked a phenotype because they lacked other unidentified susceptibility alleles at other loci carried by the proband. It is easy to imagine, based upon the descriptions above of the interactions between the multiple normally and abnormally functioning cardiac ion channels in maintaining and terminating the plateau phase of the cardiac action potential, how minor functional variants in these same genes could sum with one another to produce a major alteration in the plateau phase and hence produce a polygenic LQT, arrhythmia or sudden death phenotype. It is likewise easy to imagine how they might sum with a drug effect that acts like one of these mutations by blocking a potassium channel.

The pharmacological induction of the phenotype in an individual such as those in the Priori study suggests that environmental or genetic liability can sum with that contributed by the weak LQT alleles. Since the drugs that unmask the phenotype produce potassium channel block, similar to known LQT mutation, so presumably do the hidden unmasking liability alleles. The pharmacogenetic LQT syndrome is clinically quite important; causing the recall of several otherwise valuable drugs, and is now recognized as a potentially lethal side effect of many common medications. These include antihistamines, such as Seldane, antibiotics and cisapride, as well as the more predictable antiarrhythmics. Most of the implicated drugs share a common mechanism of action that involves the block of I_{Kr} potassium channels, and a number of cases now explicitly demonstrate that weak or subclinical alleles in the LQT genes, encoding both Na^+ and K^+ channels, are contributory to the pharmacogenetic disease.

A special consideration of pathogenic weak LQT alleles is raised by the contribution of these genes to sudden infant death syndrome (SIDS). In a landmark prospective study spanning an 18-year period, Schwartz and coworkers from nine large maternity hospitals performed EKGs on all healthy newborns between the third to fourth day of life, studying a total of over 34 000 neonates. At the 1-year follow-up evaluation, it could be observed that infants who had died of SIDS had a longer QTc interval *at birth* than did survivors or infants dying of other causes, even though none had a positive family history for LQT. Half of the infants dying of SIDS (12 of 24) had a QTc at birth greater than two standard deviations above the mean for the cohort, a finding that alone predicted greater than a 40-fold increase in the odds of dying of SIDS. However, because SIDS is so rare, despite the finding's dramatic odds ratio, only 2% of the neonates found to have a long QT die of SIDS. More recent studies have begun to add a molecular dimension to these cases, and for a few cases, a new mutation in one of the LQT genes, producing an allele previously seen in a family with the classical syndrome, can be identified. It remains to be determined whether weak inherited mutations, or even polymorphisms, in two or more of the LQT genes contribute to the rest.

6
Ion Channel Diseases of Muscle

6.1
Function of Ion Channels in Muscle

In many ways, the skeletal muscle AP is similar to the initial and final portions of that in the heart, but lacks the long plateau phase (Fig. 7). It is a fast depolarization, driven by sodium and calcium currents, followed immediately by a rapid repolarization driven by potassium currents, and in this regard is much like that of nerve (Fig. 1a). Instead of arising endogenously within the tissue, as at the cardiac sinoatrial node, the inciting muscle depolarization arrives *synaptically* via the quantal release of a chemical neurotransmitter, acetylcholine (ACh), from a motor nerve terminal bouton that forms a specialized neuromuscular junction with the *motor endplate* of the muscle plasma membrane, or *sarcolemma*. As the motor nerve's propagating AP reaches its terminus, cytosolic calcium rises, triggering a few hundred *synaptic vesicles*, each holding several thousand ACh molecules, to fuse into the presynaptic neuronal membrane (Fig. 9). As they fuse, each vesicle releases, in a quantal fashion, all its contents. This released ACh diffuses, in a matter of milliseconds, across the *synaptic cleft* to the muscle, where it is degraded by acetyl

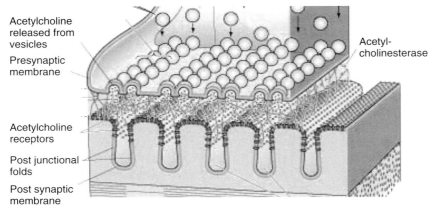

Fig. 9 Neuromuscular junction.

cholinesterase. A ligand-gated cation channel found on the postjunctional folds of the sarcolemmal endplate membrane, the heteropentameric nicotinic ACh receptor (nAChR), is activated by ACh binding (Figs. 5 and 9). Each receptor activated progressively brings about an initial incremental muscle membrane depolarization, and these events, referred to as *miniature endplate potentials* (MEPPs), were critical in uncovering the quantal nature of synaptic transmission. This receptor is the target of autoimmune and genetic myasthenia syndromes, and these quantal synaptic events are disturbed by these diseases, which are discussed below. In the physiological condition, a large number of tiny receptor-mediated MEPPs summate, producing an endplate potential that brings the membrane potential to the *threshold* at which a few sodium channels begin to open (Fig. 10). The sodium channels in turn further depolarize the membrane, causing more adjacent sodium channels to open in a reinforcing cycle until all sodium channels in the muscle have explosively opened while rapidly spreading a depolarizing wave across the surface of the muscle. At this point, all of the sodium channels are rendered "inactive" and nonexcitable, beginning the process of repolarization, a process completed when more slowly activating potassium channels restore the resting potential and return the sodium channels to the "closed" state. Unlike the situation in cardiac muscle, the rapid recovery of skeletal muscle excitability allows rapid repetitive contractions, an advantage in performing fine precision movements. The major subunit of muscle sodium channels is the muscle-specific isoform α_4, encoded by *SCN4A*, a relative of the gene involved in *LQT3*. Alleles of this one gene produce the diverse array of muscle phenotypes that will be discussed below.

As the depolarization wave of the muscle AP spreads across the sarcolemmal plasma membrane, it extends into specialized membrane invaginations, called *T-tubules*, which dive deep into the cytoplasm and ramify near the contractile proteins (Fig. 9). Their function is to facilitate calcium-dependent activation of contraction throughout the large muscle fiber. In the muscle T-tubule, as in the motor nerve bouton, the depolarization wave activates the voltage-gated calcium channel, which is similar in structure to

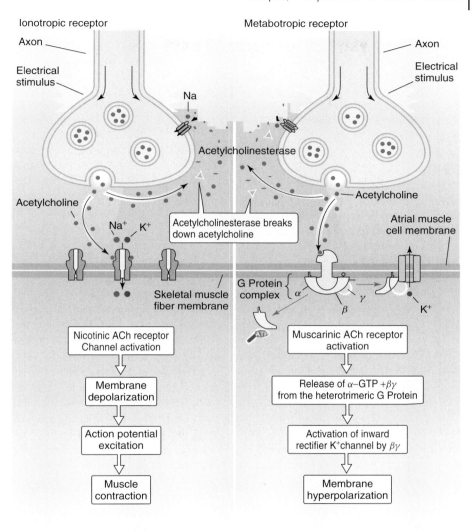

(a)　　　　　　　　　　　　(b)

Fig. 10 Mechanisms of acetylcholine receptor activation.

the sodium channel, discussed above. The major, α_1, subunit is pseudotetrameric and contains the ion-conducting pore. Its function is modified by auxiliary β, γ and α_2/δ subunits, the skeletal muscle calcium channel's isoforms being encoded by *CACNA1S, CACNB1, CACNG1,* and *CACNA2D1*, respectively (Fig. 5). This channel is responsible for rapidly immersing the contractile proteins throughout the large muscle cell in elevated concentrations of ionized calcium, thereby

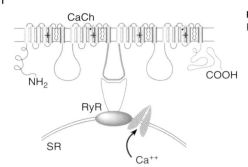

Fig. 11 Ca channel and Ryanodine receptor.

bringing about concerted contraction, much as was described for the cardiac contractile machinery. It does this in part by allowing the passage of extracellular calcium through its pore, down its electrochemical gradient into the cytoplasm. But additionally a large cytosolic loop in the skeletal muscle α_1-subunit contacts a *different* calcium channel located in a closely opposed *different* membrane compartment, the sarcoplasmic reticulum (SR) (Figs. 5 and 11). The SR membrane contains calcium pumps that rapidly sequester intracellular calcium after each AP, creating a releasable calcium store and restoring muscle relaxation. Through the physical interaction with its cytosolic loop, the voltage-gated calcium channel directly gates the SR calcium release channel. The release channel is called the *ryanodine receptor*, encoded by *RYR1*, and opening it spills the SR intracellular calcium stores into the cytoplasm. After activation, both the sodium and calcium voltage-gated channels spontaneously enter the nonconducting "inactive" conformation, beginning the process of membrane repolarization, a process completed by potassium channels that hyperpolarize the membrane, resetting the sodium and calcium channels into the "closed" conformation. The sodium and calcium channels then are kept closed by the large stabilizing current of the chloride channels, encoded by *CLCN1* (Fig. 5).

6.2
Ion Channel Dysfunction in Paralysis and Myotonia

The two major skeletal muscle phenotypes produced by ion channel disorders reflect either an inability to contract the muscle, ranging from weakness to *paralysis*, or an inability to relax the muscle, *myotonia*. While a third muscle phenotype, malignant hyperthermia (MHS), appears quite different, it is surprisingly closely related to the paralysis syndromes, in fact, being allelic. Two major themes in the ion channel diseases begin to develop in discussing the skeletal muscle diseases. First, it becomes clear that the LQT syndrome provides a powerful paradigm for understanding the types of alleles that produce disease and the mechanisms by which they act. Second, the proclivity of certain ion channel gene families to produce disease begins to emerge as does their ability to produce distinctly different clinical phenotypes with subtly differing alleles (Table 1).

While the pathophysiology of paralysis and hyperthermia is complex, that of myotonia is fairly straightforward. Since the high chloride conductance of *CLCN1* in the muscle membrane serves to stabilize

Tab. 1 Channelopathy gene families.

Gene family	Tissue	Disease phenotype	Pathogenic mutation
KCNE1	Heart/ear	LQT5/JLN	Dominant negative/recessive loss-of-function
KCNE2 (100%)[a]	Heart	LQT6	Dominant negative
KCNE3	Muscle	HOKPP	Dominant negative
KCNQ1	Heart/ear	LQT1/JLN	Dominant negative/recessive loss-of-function
KCNQ2	CNS	BFNC1	Dominant loss-of-function
KCNQ3 (80%)[a]	CNS	BFNC2	Dominant loss-of-function
KCNQ4	Ear	DFNA2	Dominant missense
KCNQ5	CNS	6q14	
RYR1	Muscle	MHS1	Missense
RYR2 (66%)[a]	Heart	Ventricular tachycardia	Missense
RYR3	CNS	15q14	
SCN1A	CNS	GEFS+/SMEI	Gain-of-function/loss-of-function
SCN2A	CNS	GEFS+	Gain-of-function
SCN3A	CNS	2q24	
SCN4A (40%)[a]	Muscle	HYPP/HOKPP/MHS/etc.	Gain-of-function
SCN5A	Heart	LQT3/Brugada S	Gain-of-function
SCN6A and SCN7A	Neuronal	2q23	
SCN8A	Motor endplate	12q13	
SCN9A	Neuroendo	2q24	
SCN10A	Nerve/muscle	3p22	
SCN11A and SCN12A	Sensory neurons	3p24	
SCN1B	CNS	GEFS+	Gain-of-function
SCN2B	Neuronal	11q23	
CACNA1S	Muscle	HOKPP/MHSS	Missense
CACNA1A	CNS	FHM1/EA2/SCA6/seizures	Many types
CACNA1F	Retina	Night blindness	Hemizygous loss-of-function
CHRNA1	Muscle	SCCMS	Gain-of-function

Sources: Gargus, J.J. (2003) Unraveling monogenic channelopathies and their implications for complex polygenic disease, Am. J. Hum. Genet. 72(4), 785–803.
[a] Percent of gene family members already associated with disease.

the membrane potential, thereby preventing "echo" contractions after an induced contraction, its functional loss is associated with the inability to normally relax the muscle, and hence the slow muscle relaxation and muscle stiffness of myotonia. Both dominant and recessive mutations of this gene, which decrease channel current, produce myotonia and a characteristic "dive bomber" EMG record of repetitive contractions. The repetitive volleys of contraction can arise in skeletal muscle because unlike in the heart, where sodium channels are held in the inactive state for a prolonged period to prevent arrhythmia, in skeletal muscle, sodium channels must be ready to accept a rapid train of controlled activations from the motor nerve. Problems arise, however, in the absence of a strong chloride stabilizing current, since instability of the membrane potential during recovery from a depolarization leaves a high probability that a secondary "echo" depolarization will reach threshold and trigger a repeated activation of the sodium channels.

The recessive loss-of-function alleles of *CLCN1* produce Becker myotonia, with a phenotype characterized by stiffness that warm-up exercise improves, and paradoxically weak hypertrophied muscles. The dominant-negative alleles leave more residual current and produce Thomsen myotonia, having a similar but milder phenotype without weakness. The hypertrophy arises because the muscles are constantly "exercising" even at rest secondary to their electrical instability, and the weakness arises because of incomplete recovery from sodium channel inactivation, as will be described for the periodic paralysis syndromes below.

6.3
Periodic Paralysis Syndromes

To understand how the array of periodic paralysis syndromes arise, it is essential to understand how very subtle "tweaks" in sodium channel function give rise to macroscopic phenotypes that are quite distinct – superficially the antithesis of one another – yet are so closely related that a given patient can transition from one state to the other. All of these pathogenic muscle sodium channel alleles appear to share with the cardiac sodium channel *LQT3* alleles a dominant gain-of-function mechanism of pathogenesis – *delayed and incomplete inactivation*. The essential biophysical feature of sodium channel function that makes this interpretable is the fact that while a depolarization will make it more probable that the voltage-sensitive gate of the closed channel will open, when sustained, depolarization also leaves behind more of the channels in an "inactive" nonexcitable conformation. As the sodium channel is the major mechanism for propagating APs in muscle (as it is in most tissues), this delicate balancing act has the potential to move these tissues from a state of hyperexcitability to a state of inexcitability, manifested as weakness and paralysis. Plasma potassium levels are commonly altered during these attacks of paralysis, and, while incompletely understood, probably has to do with the fact that the vast majority of the potassium in the body, >95%, is found in the muscle. The anxiety that surrounds clinical management of such changes is that very small amounts of potassium dramatically change the plasma potassium concentration and this is a critical determinant of the potassium Nernst potential (hence the membrane potential of all cells) and such changes in the heart pose the risk of fatal arrhythmia.

The ranges of phenotypes gain-of-function *LQT3*-like incompletely inactivating *SCN4A* alleles produce include: (1) hyperkalemic periodic paralysis (HYPP), (2) hypokalemic periodic paralysis (HOKPP), (3) paramyotonia congenita, a cold exacerbated myotonia and (4) potassium-activated myotonia.

HYPP alleles are the most common, and the disorder is characterized by short, mild frequent attacks of profound weakness beginning in infancy and provoked by rest just after exercise or stress. Commonly, the individual performs significant physical exertion, then sits to rest and finds he is unable to move for an hour or more (although respiration is not compromised). There is often myotonia between attacks. During an attack, plasma potassium levels rise to pathological levels, probably via release from the depolarized muscle. The alleles are characteristically missense mutations altering amino acids in the transmembrane spans of the channel protein. Andersen syndrome, discussed above as caused by *LQT7* alleles, additionally has the muscle phenotypes of HYPP and that of MHS, discussed below, suggesting an intrinsic relation between the two. Furthermore, some HYPP alleles of *SCN4A* additionally produce MHS, and the gene is probably also the *MSH2* locus, a rare cause of the phenotype.

HOKPP is characterized by an onset in the second decade of infrequent, profound painless episodes of weakness that can last for days. During an attack, the plasma potassium can fall to dangerous levels, presumably being driven into the muscles. The factors known to incite the attacks are known to physiologically have this activity. They include insulin or glucose intake (which stimulates insulin release) and the hormonal changes that occur upon awakening. Two other loci, *CACNA1S* and *KCNE3*, also have alleles that produce HOKPP. Missense dominant alleles of *CACNA1S*, encoding the major α_1, subunit of the muscle voltage-gated calcium channel, were the first recognized and remain the most common cause of hypokalemic periodic paralysis. Most recently mutations in one of the skeletal muscle potassium channels involved in the rapid repolarization of the membrane, encoded by *KCNE3*, a relative of the *LQT5* and *LQT6* genes also have been shown to cause HOKPP. With this discovery, all of the members of the fully ascertained *KCNE* gene family are proven pathogenic loci, suggesting such familial relationships may be useful in promoting candidate genes for evaluation in disease association (Table 1).

6.4
Malignant Hyperthermia

The pathophysiology of malignant hyperthermia is typical of a pharmacogenetic syndrome; it involves an inciting agent, typically an inhalation anesthetic such as halothane, and an individual with a predisposing genotype. The pathogenic mechanism arises from a dramatic rise in muscle cytoplasmic calcium as the hypersensitized SR stores are discharged. This then secondarily produces contracture of the muscles and a hyper-metabolic state of glycogenolysis, ATP depletion, and lactic acidosis driven by the attempt to resequester calcium. This process evolves to produce hyperthermia (not fever) from activity of the contractile machinery and mitochondrial respiration, and disruption of the muscle membranes, releasing myoglobin, creatine kinase, and potassium, and leading to renal failure, disseminated intravascular coagulation and death The predisposing genotypes enhance the sensitivity of the calcium release process, which primarily involves the sarcolemmal calcium channel and the SR calcium release channel. Missense alleles altering either the cytosolic loop of the *CACNA1S*-encoded α_1 calcium channel subunit or its contact domains on the ryanodine receptor render this complex hypersensitive and capable of triggering massive calcium release, muscle activation, and malignant hyperthermia (Fig. 11). Over 50% of the cases are caused by dominant *MHS1* alleles of *RYR1* and the next most common cause is dominant *MHS5* alleles of *CACNA1S*. Some of the HYPP alleles of *SCN4A* additionally seem to produce this syndrome, and it probably reflects *MHS2*. Three other *MHS* loci have been mapped but susceptibility alleles have yet to be identified. Additionally, this phenotype is also a component of Andersen syndrome, discussed above, affecting the potassium channel that controls the resting membrane potential in muscle and heart, *KCNJ2* (Fig. 8). Finally, alleles of *RYR1* also cause central core disease (CCD), a congenital myopathy characterized by nonprogressive proximal muscle weakness and distinctive muscle fiber histology. While individuals with CCD do have a susceptibility to anesthetic-induced MHS, the remainder of the disease phenotype is constitutive and does not require pharmacological induction, and therefore the disorder is illustrative of the continuum between genetic and environmental disease.

6.5
Other Monogenic Muscle Ion Channel Syndromes

Dominant gain-of-function mutations in four of the five subunits that compose the nAChR found at the neuromuscular junction and encoded by *CHRNA1*, *CHRNB1*, *CHRND* and *CHRNE* produce slow-channel congenital myasthenic syndrome (SCCMS), a phenotype characterized by muscle weakness. Acetylcholine dissociates slowly from these mutant receptors, leaving them persistently activated, depolarizing the membrane. One might guess that the prolonged receptor activation the mutant alleles produce will give a hyperexcitable muscle; however, as was discussed above in the periodic paralysis syndromes, the phenotype seen is produced by the loss of sodium channels into the inactive state in this chronically depolarized tissue.

6.6
Autoimmune Channelopathies in Muscle

Myasthenia gravis (MG) is one of the classic autoimmune diseases, first placed on a molecular footing in the early 1970s, well before the era of the monogenic channelopathies. It develops in mid-life, produces facial, respiratory, and limb muscle weakness and is associated with tumors of the thymus and antibodies to the nAChR. It is clear that the antibodies produce the disease, since they can passively transfer the symptoms and their removal is therapeutic. The autoantibodies accelerate the destruction of the nAChR, reducing their abundance in the synapse and hence the efficacy of synaptic transmission. This was first observed as a decrease in the size, but not number, of MEPPs, suggesting that the ACh-filled vesicles were normally releasing quanta of neurotransmitter, but that they had a smaller postsynaptic effect, causing weakness. Enhancing the synaptic concentration of ACh by inhibiting acetyl cholinesterase rapidly improves strength, and is used diagnostically and therapeutically (Fig. 9). Note that the muscle phenotype is very similar to that created by SCCMS alleles of the nAChR subunit genes, discussed above, but that those genetic syndromes do not have thymus tumors or autoantibodies, confirming that receptor dysfunction does not cause those findings in MG, the tumor appearing to be primary and causal in autoimmunity.

Lambert-Eaton myasthenia syndrome (LEMS) is another autoimmune disorder that produces a clinical picture very similar to that of MG, but preferentially targeting proximal limb muscles and being associated with small-cell carcinoma of the lung. It additionally produces symptoms referable to the autonomic nervous system, such as dry mouth, so it is clear that the epitope targeted by the antibodies is not strictly limited to muscle. In fact, the antibodies are directed at the P/Q-type voltage-gated calcium channels of the presynaptic neuron. This channel is encoded by *CACNA1A* and is similar to its family members expressed in heart and skeletal muscle, but it participates in excitation-*secretion* coupling (not EC coupling), serving as a calcium entry pathway that couples the propagated AP to the biochemical pathways underlying neurotransmitter release at the presynaptic bouton, to be discussed below. It is distinguished from its family members by its high affinity for ω-conotoxin. Antibodies do not cross the blood-brain barrier, so symptoms in the autoimmune disorders are confined to effects in the periphery. As will be seen below, very different symptoms arise from genetic lesions in *CACNA1A* where CNS symptoms predominate.

7
Ion Channel Diseases of Nerve

7.1
Function of Ion Channels in Neurons

The neuronal AP is very similar to that of muscle: typically, activation is initiated across a *synapse*, neuroreceptors and their endogenous or coupled channels are activated, the supra-threshold summated postsynaptic currents in turn trigger a rapidly spreading depolarization wave by voltage-gated sodium and calcium channel currents, this is rapidly followed by a repolarization driven by potassium currents, and the AP couples to biochemical effector pathways through a rise in the cytosolic calcium concentration. Despite this gross oversimplification of all of the diversity of neuronal structure and function, and their vastly more complex geometry, the major distinctions from muscle lie at the afferent and efferent ends of the neuronal transduction pathway, not in the propagated AP in between.

First, neurons have a vastly more complex variety of activating, as well as inhibitory, neuroreceptors and synapses than muscle, and they are primarily found at the branching *dendritic*, or afferent, ends of the prototypical neuron (Fig. 12). The receptors produce both depolarizing, excitatory, as well as hyperpolarizing, inhibitory, *postsynaptic potentials* (EPSPs and IPSPs) that, like the MEPPs of muscle, summate over time, but also through the complex spacial geometry of the arborized dendritic tree. This summation serves the "computational" function of the neuron; some combinations of dendritic inputs will drive the soma, or neuronal cell body, to initiate an AP that is propagated down the axon, while others that differ slightly in timing, spacial geometry of the activated synapses, or the mix of incoming neurotransmitters will not.

Second, when the neuron does "decide" to produce a propagated AP, the elevated cytosolic calcium wave does not participate in EC coupling, but rather in excitation-secretion (ES) coupling, the controlled release of neurotransmitter into another synapse (Fig. 10), as was briefly discussed for LEMS, above. In the neuron, the calcium-coupled neurosecretion events occur after the AP has been conducted down the *axon*, the long cellular projection that serves as the efferent and effector side of the neuron. There a wide variety of small-molecule neurotransmitters can be released into a synapse to initiate in the downstream neuron the process just described above (Fig. 12).

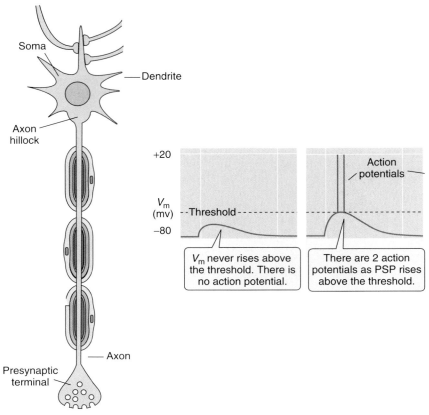

Fig. 12 Neuron structure and initiation of neuronal AP.

7.2
Ion Channel Dysfunction in Seizures

As opposed to the situation in heart and muscle, synchrony of electrical activity in the CNS is actively prevented. An abnormally synchronous discharge of neurons in the CNS produces a *seizure*, a pathological state characterized by a stereotyped alteration in behavior. Seizures are the most common monogenic ion channel phenotype in the CNS and in many ways are similar to the arrhythmia phenotype that arises in the hyperexcitable LQT heart. Both are periodic disorders in constitutionally hyperexcitable tissues in which the normal rhythmic electrical activity of the tissue is temporarily lost. However, whereas the heart is typified by its extremely homogeneous set of responding cells, the CNS displays maximal tissue complexity both in terms of the number of different cell types present and in the number of different ways they connect with and influence one another in stimulatory or inhibitory fashions. Therefore, while one can visualize the very specific and individual contribution each channel type makes to components of an action potential perturbed in LQT in the heart or in the skeletal muscle syndromes, one can much less specifically interpret how

Tab. 2 Unlike heart, must still guess WHERE neuronal channels act.

Seizures:		
BFNC1	KCNQ2	Haploinsufficiency, decrease I_K
BFNC2	KCNQ3	Haploinsufficiency, decrease I_K
ADNFLE	CHRNA4	GOF increase activation, increase I_{Na}
ADNFLE	CHRNB2	GOF increase activation, increase I_{Na}
GEFS+	SCN1B	GOF decrease inactivation, increase I_{Na}
GEFS+	SCN1A	GOF decrease inactivation, increase I_{Na}
GEFS+	SCN2A	GOF decrease inactivation, increase I_{Na}
GEFS+	GABRG2	Dominant missense, decrease I_{Cl}
JME	GABRA1	Dominant missense, decrease I_{Cl}
JME	CACNB4	Dominant missense, I_{Ca}
SMEI	SCN1A	Haploinsufficiency, decrease I_{Na}
...f	KCNA1	Haploinsufficiency, decrease I_K
...f	* CACNA1A	Dominant negative, decrease I_{Ca}
Ataxia		
...f	* CACNA1A	Dominant negative, decrease I_{Ca}
SCA6	* CACNA1A	Dominant CAG expansion, decrease I_{Ca}
EA2	* CACNA1A	Haploinsufficiency, decrease I_{Ca}
EA1	KCNA1	Haploinsufficiency, decrease I_K
Migraine:		
FHM1	* CACNA1A	Dominant missense, decrease I_{Ca}
FHM2	ATP1A2	Dominant loss-of-function

Sources: Gargus, J.J. (2003) Unraveling monogenic channelopathies and their implications for complex polygenic disease, *Am. J. Hum. Genet.* **72**(4), 785–803.
* Different manifestations of P/Q channel dysfunction;
f – case reports, discussed in text.

or even where the channel dysfunction underlying seizures occurs. Also, rhythm disorders in the heart are often a once in a lifetime event; in the CNS they are usually a lifelong chronic condition, *epilepsy*. Epilepsy is also a useful phenotype to serve as a springboard to understanding the other major monogenic CNS channelopathy phenotypes, *ataxia* and *migraine*, as they can be allelic.

While the complexity of neuroanatomy obscures functional analysis of the mutations causing monogenic epilepsy syndromes, extrapolation from the cardiac and muscle syndromes provides a simplified predictive framework of molecular pathology: K^+ and Cl^- channels, which physiologically stabilize excitable tissue, will likely have pathological lesions that diminish their current, and Na^+ and Ca^{++} channels, which physiologically excite the tissue, will likely have gain-of-function lesions (Table 2).

7.3
Hereditary Epilepsy Syndromes

Nearly 40% of all epilepsy is associated with no macroscopic pathological lesion, and is therefore classically referred to as *idiopathic epilepsy*, a common disorder that affects ~ 1% of the population. It is overwhelmingly a genetic disease, with monozygotic twins greater than 95% concordant for the phenotype; however, nearly all cases are polygenic and most of those genes still remain to be identified. Obviously, since ion channels control electrical activity in the CNS, as in the heart and muscle, they are strong functional candidates for this disorder. In man, they were the first identified, and remain nearly the exclusive, proven genetic causes of epilepsy. Only a handful of rare monogenic epilepsy syndromes are recognized, but predictably success at pathogenic allele definition has thus far come only from these rare monogenic syndromes.

7.4
LQT-like Loci and Alleles in Neuronal Syndromes

The first rare monogenic seizure syndrome for which the major etiological genes were identified was benign familial neonatal convulsions (BFNC), a rare autosomal dominant disorder characterized by a brief period of seizures in the neonatal period, generally resolving in weeks, but 10% having persistent adult epilepsy. The BFNC loci were mapped to chromosomes 20 and 8. The *BFNC1* gene was positionally localized within chromosome 20q13.3 taking advantage of a family with the syndrome and a microdeletion chromosome. The interrupted locus contained a promising candidate gene, *KCNQ2*, a member of the *LQT1* gene family that was predominantly expressed in neurons. Unlike the *LQT1* alleles that demonstrate dominant-negative interactions *in vitro*, this first *BFNC1* allele was a null, and most subsequent alleles are simply loss-of-function. The relevant neuronal currents mediated by this channel must be so critically tuned that pathogenic alleles can achieve dominance simply via haploinsufficiency. Heterozygous null mice display a milder inducible pharmacogenetic syndrome, showing no basal seizures, but only an increased sensitivity to seizure-inducing drugs. Presumably, the strain's polygenic background raises their seizure threshold making them appear much like silent-carrier family members of the "sporadic" LQT syndrome probands who have disease produced by weak alleles.

Since the *KCNQ* family had the demonstrated ability to produce disease, other family member genes were sought by homology. *KCNQ3* was thus identified, and mapping to chromosome 8q, was a tempting candidate for the second BFNC locus, a hypothesis proven by finding loss-of-function alleles segregating in a family. While both KCNQ subunits produced channels when expressed *in vitro*, neither subunit alone produced a recognizable current. However, KCNQ2 and KCNQ3 were subsequently recognized to heteromultimerize to form the "M current," a long-sought signature potassium current activated by muscarinic acetylcholine receptors (mAChR) (Fig. 10). Therefore, both their gene-family relationship and their functional subunit interaction help explain the common phenotype that mutations in either gene produce. The final member of this *KCNQ* family, *KCNQ5*, mapping to chromosome 6q14, and the only member still to be associated with a disease, is expressed in neurons and can also interact with KCNQ3 to produce M current, making it an extremely strong functional candidate disease gene.

7.5
Other Monogenic Ion Channel Epilepsy Syndromes

The gene *CHRNA4* was the first ion channel gene demonstrated to contribute pathogenic alleles to an epilepsy syndrome. It is found on chromosome 20q13 and encodes the most abundant neuronal isoform of the major subunit, α_4, of the nAChR. Unlike the mAChR, discussed immediately above, which couples to channel gating via G-protein activation (a G-protein coupled receptor, GPCR), the nAChR, like that found in muscle, is a heteropentameric receptor/channel that is *itself* a ligand-gated nonselective cation channel, its activation depolarizing the membrane (Fig. 10). It is closely related to the muscle nAChR, which participates in the phenotype of SCCMS congenital myasthenia. In the nerve, it is composed of two α- and three β-homologous subunits, each with four transmembrane α-helices (Fig. 5). The rare seizure syndrome, autosomal dominant nocturnal frontal lobe epilepsy (ADNFLE), producing a phenotype with brief clusters of frontal seizures that occur at night, was mapped to this region in a family and missense alleles identified. To date, three alleles have been recognized, all altering the channel pore region and receptor function but having no obvious common effect on *in vitro* channel behavior. More recently, mutations in the most abundant neuronal β-subunit, β_2, were also recognized to cause this syndrome, suggesting that the channel isoform relevant to this phenotype is an $\alpha_4\beta_2$ pentamer. These missense alleles in *CHRNB2*, located on chromosome 1q21, produced the common *in vitro* effect of receptor activation, suggesting that some depolarizing gain of function is probably the mechanism of dominant pathogenesis, like that found in the SCCMS muscle pathogenic homologs. The vulnerable cell in the CNS whose function is perturbed by these mutations, however, remains to be defined.

Fever, a common accompaniment of many illnesses, lowers the seizure threshold so it is not surprising that febrile seizures are by far the most common polygenic seizure disorder, affecting 3% of children worldwide. While these susceptibility genes remain to be defined, a rare Mendelian dominant seizure syndrome that includes febrile seizures provides the first lead to their identification. This febrile seizure syndrome additionally evolves to include a variety of afebrile seizures, and is called *generalized epilepsy with febrile seizures plus* (GEFS+). The first locus responsible for GEFS+ was mapped in a large family to chromosome 19q13.1 and a missense mutation was identified in an auxiliary subunit of the voltage-gated sodium channel. This subunit, the β_1 subunit encoded by the gene *SCN1B*, has a single transmembrane helix and functions only to modify the activity of the large α-subunit (Fig. 5). Subsequently, mutations producing GEFS+ or closely related syndromes were identified in two different adjacent neuronal sodium channel α-subunit genes, *SCN1A* and *SCN2A*, on chromosome 2q24. Not until the α_1 subunit alleles were coexpressed with their auxiliary β-subunits did a reproducible picture of defective channel inactivation and sustained sodium current appear. This is a picture similar to that observed with the LQT3 and HYPP and HOKPP alleles of the gene-family members *SCN5A* and *SCN4A*, respectively. As in the heart, most simply, a persistent sodium channel current can be viewed as a gain-of-function dominant lesion that favors a hyperexcitable state; however, the periodic paralysis lesions caution that hyperexcitability can easily transition into inexcitability, through the inactivation mechanism, and this dichotomous nature of sodium channel lesions is reinforced by the recent observations that a rare, very severe dominant seizure syndrome, initially associated with febrile seizures, called *severe myoclonic epilepsy of infancy* (SMEI), proved to be allelic with GEFS+. All of dozens of known alleles were shown to be new mutations in *SCN1A*, and their nature (e.g. frameshift, nonsense) suggests they are predominantly functional nulls. While the physiological interpretation of these findings is clear, together they are puzzling; haploinsufficiency for the major subunit of the sodium channel produces a very severe dominant *loss*-of-function phenotype via some critically tuned cell type; a milder syndrome being caused by the dominant *gain*-of-function GEFS+ alleles.

The remaining GEFS+ locus, *GABRG2*, found within a cluster of GABA receptor genes on chromosome 5q31, encodes the γ_2 subunit of the inhibitory GABA$_A$ receptor. The GABA$_A$ receptor, like the nAChR discussed above, is a pentameric ligand-gated channel, but it differs in that instead of conducting a depolarizing cationic current, it conducts a stabilizing chloride current (Fig. 5). This receptor is the major mechanism through which GABA, the principle *inhibitory* neurotransmitter in the brain, functions. It is at this site that the benzodiazepine drugs act to treat seizures, by potentiating GABA effects, and that toxins, such as picrotoxin, act to induce seizures, by blocking GABA. So physiologically the inhibitory, neuronal silencing, activity of GABA participates in preventing seizures. The GEFS+ allele produces a decrease in GABA-induced chloride current *in vitro*. Therefore, like the seizure-inducing toxins, seizure-inducing mutations alter receptor/channel function to produce *less* GABA-induced stabilizing current.

While some human "seizure" genes have not recapitulated that phenotype in mice (for instance, mice homozygous for null alleles of *CHRN4A* and *CHRNB2* – the two loci thus far recognized with alleles causing ADNFLE in man – have a phenotype that alters the pain response, not seizures), spontaneous mutations producing a seizure phenotype in mice have led to the recognition of an important new category of ion channel seizure disorders in man, those altering voltage-gated calcium channel function. More importantly, those loci have proven to be particularly informative for their ability to tie the various neurological phenotypes together in an interpretable fashion. As discussed above, the voltage-gated calcium channels are similar to the voltage-gated sodium channel, and the major pore-containing α_1 subunit that dictates the channel's subtype is pseudotetrameric with its function modified by the auxiliary β, γ, and α_2/δ subunits (Fig. 5). Alleles in all four different types of neuronal voltage-gated calcium channel subunit genes produce a monogenic absence seizure phenotype in mice. The "lethargic," "stargazer," and "ducky" mice are recessive phenotypes produced by mutations altering neuronal calcium channel auxiliary subunits, *CACNB4* encoding a β-subunit, *CACNG2* encoding a γ-subunit, and *CACNA2D2* encoding a protein cleaved to produce the α_2 and δ-subunit, respectively. Of these three genes, a human phenotype has been reported only for mutations in the first. The human *CACNB4* gene maps to human chromosome 2q22-q23 and two dominant alleles, a truncation and missense allele, were found in families with a dominant seizure syndrome. The human *CACNG2* maps to chromosome 22q13.1 and the human *CACNA2D2* gene maps to 3p21.3, both remaining only promising seizure candidate genes in man. The "tottering" and "leaner" mice are recessive seizure phenotypes produced by alleles of *CACNA1A*, the gene encoding the major α_1 subunit of the neuronal P/Q-type calcium channel. The P/Q-type calcium channel plays a major role in calcium entry underlying synaptic release of the major *excitatory* neurotransmitter, glutamate, and while this process is decreased in the mutants, suggesting they are hypomorphic alleles, it is unclear how this produces a hyperexcitable seizure phenotype.

7.6
Monogenic Ataxia and Migraine Syndromes

While the "tottering" and "leaner" mice are recessive seizure phenotypes produced by alleles of *CACNA1A*, a seizure phenotype is *not* the major phenotype caused by mutations in this gene in man. Human mutations in *CACNA1A*, which maps to chromosome 19p13, have been shown to cause three apparently distinct and different late-onset neurological disease phenotypes: episodic ataxia type 2 (*EA2*), familial hemiplegic migraine type 1 (*FHM1*), and spinocerebellar ataxia type 6 (*SCA6*). The *EA2* alleles are predominantly truncations, with over 22 known alleles, including frameshift and splice site mutations, but also five missense alleles, some altering conserved pore residues producing complete loss of function. It is therefore presumably a haploinsufficiency dominant loss-of-function syndrome. Familial hemiplegic migraine is caused by at least nine known missense alleles producing no obvious uniform functional change in calcium current *in vitro*; therefore, the relevant functional change they share in common is presumably subtle. Finally, SCA6 is a progressive degenerative phenotype primarily caused by trinucleotide expansion alleles. The polymorphic CAG repeat encodes a C-terminal polyglutamine repeat of 5 to 20 residues in unaffected individuals. Pathogenic alleles encode 21 to 30 glutamines. The only point mutant allele recognized to produce the syndrome is G293R, which changes a conserved pore residue. An intermediate-length CAG repeat allele with only 20 repeats produced the milder phenotype of EA2.

Thus, the four allelic *CACNA1A* diseases in man and mouse are clearly closely related at the molecular level. The phenotypes also can blend one into the other, for instance, many individuals with FHM have ataxia as do most of the murine seizure mutants, and recently a dominant-negative truncation allele has been reported to produce a syndrome of progressive ataxia and seizures in a child. In man, the mildest alleles appear to be the missense mutations producing FHM1. On a phenotypic continuum, these appear to be hypomorphic since the truncations, pore mutations, and intermediate repeat expansions produce EA2, apparently a haploinsufficiency syndrome. Expanding the polyglutamine repeat by only a few additional residues yields the potent long-repeat alleles producing the progressive SCA6 syndrome, which can also be produced by one specific missense pore mutation. While the mechanism of action of all pathogenic polyglutamine expansion alleles are controversial, if the three human *CACNA1A* phenotypes are truly the continuum suggested by the EA2 intermediate repeat phenotype, these repeat expansion alleles should be dominant negative to eliminate more of the P/Q current than haploinsufficiency and to keep the progression of allele potency parallel to the severity of the symptoms and pathology. While all of the human alleles produce the phenotype in heterozygotes, the mouse seizure alleles, which are missense, like FHM1, and demonstrated to be hypomorphic, produce no phenotype in heterozygotes, including knockout null alleles; decreased synaptic glutamate neurotransmitter release and seizures being produced only in the homozygotes.

Another strong clue of the relationship between ataxia and seizures comes from the other known episodic ataxia locus, *EA1*. EA1 is produced by dominant loss-of-function alleles of perhaps the most "famous" potassium channel, "shaker", named after its mutant phenotype in fly, and the paradigmatic first K$^+$ channel cloned. It is a voltage-gated potassium channel formally called *KCNA1*. Knockout null alleles in the mouse produce a recessive seizure disorder, with the heterozygotes lacking a phenotype, just like the calcium channel mutants, but like them showing a reduced threshold for induced seizures, hinting at a potential mechanism for pharmacogenetic neurological syndromes in man. While the vast majority of EA1 patients have only ataxia, a few cases have been reported to additionally have seizures.

Just as examining both loci that produce episodic ataxia, *EA1* and *EA2*, informs the underlying pathophysiology of the indistinguishable diseases they produce, so too does examining both loci that cause familial hemiplegic migraine, *FHM1* and *FHM2*. Whereas *FHM1* alleles are subtle missense mutations of the calcium channel-encoding *CACNA1A* gene, recently *FHM2* alleles were discovered to be mutations in *ATP1A2*, the gene encoding the $\alpha 2$ subunit of the Na/K-ATPase, one of the major ion pump mechanisms responsible for establishing the ion gradients tapped by ion channels (Fig. 6). This is the first disease that is established to be caused by mutations in the Na/K pump, a central physiological mechanism. All FHM2 alleles identified to date are missense and alter narrowly conscribed domains of the pump protein. While loss-of-function and haploinsufficiency mechanisms are more suggestive of deletions and other forms of more readily produced null alleles than the observed constrained missense, this has been suggested to be the mechanism of some pathogenic alleles; however, detailed enzyme kinetics have more recently shown others to achieve dominance by gain of function, producing altered kinetics of the pump. It is still not clear what biophysical property of the neuron is perturbed in common by all of the FHM lesions, although a common denominator seems to be membrane depolarization and elevated cytosolic calcium.

7.7
Other Monogenic Neuronal Ion Channel Syndromes

One of the first monogenic CNS ion channel disorders for which the underlying pathogenic alleles were identified is the rare disorder hyperekplexia (HE), more commonly called *Startle syndrome*. The phenotype includes hypertonia and a severe startle reaction that simulates a drop seizure. It is produced by dominant and recessive mutations in *GLRA1*, the gene encoding the $\alpha 1$ subunit of the glycine receptor/channel. Since like the inhibitory GABA – and excitatory nAChR – receptor/channels, described above, the chloride-conducting inhibitory glycine receptor is a ligand-gated ion channel, sharing with other members of this channel family a subunit structure with four membrane-spanning helices and a characteristic heteropentameric composition (Fig. 5), it is not surprising that recessive mutations in the *GLRB* gene, encoding the glycine receptor β-subunit, also cause the same HE phenotype. Further, it is interesting that pathogenic alleles in all three types of ligand-gated channels, causing the discussed syndromes of HE, SCCMS, ADNFLE, and GEFS+, are missense mutations primarily altering residues in the extracellular loop between the second and third transmembrane helix that controls gating, suggesting a related molecular pathogenesis of these macroscopically distinct diseases.

Many of the mutations that underlie hereditary defects in the special senses, such as vision and hearing, are found in ion channel genes. X-linked congenital stationary night blindness type 2 is caused by defects in the retinal-specific close relative of both the neuronal calcium channel "seizure-ataxia-migraine" gene *CACNA1A* and the skeletal muscle "HOKPP periodic paralysis" gene *CACNA1S*, discussed above. The syndrome is caused by several loss-of-function mutations in *CACNA1F*. Color vision requires, in addition to the three cone opsin pigment genes, mutations in which cause loss of color detected by one cone type, only the genes encoding the two subunits of a specific cyclic nucleotide-gated heteromultimeric ion channel found in the photoreceptor membrane (Fig. 5). Defects in these channel subunits cause color blindness due to dysfunction of all cones. Rod monochromacy is produced by at least 10 recessive missense alleles of *CNGA3* and Achromatopsia type 3 is produced by missense loss-of-function alleles in *CNGB3*.

Nonsyndromic deafness (NSD) is produced by mutations in dozens of genes, including those potassium channel genes that contribute alleles to produce Jervell and Lange-Nielsen recessive LQT syndrome, discussed above. Additionally, the genes encoding subunits of a very different type of channel, the connexin gap junctions (Figs. 5 and 13), are the most common source of alleles for NSD, being polymorphic in many populations. Gap junctions serve to connect all types of cells via a large nonselective pore that mediates the transit between cells of not only ions but also small macromolecules and metabolites. They are functionally dimers contributed by each member of a pair of cells, each cell membrane having a hexameric hemichannel that seals together with its partner. Each hemichannel is composed of the family of connexin subunits. Alleles of *GJB2*, encoding connexin 26, are the most common cause of NSD. There are a few dominant alleles, but several dozen recessive alleles, including one specific polymorphic pathogenic allele, 35delG, which accounts for as many as 4% of the European Caucasian alleles. Alleles of *GJB6* and *GJB3* also contribute to this syndrome. The mechanism by which they act appears to be similar to that of the Jervell and Lange-Nielsen LQT alleles; they perturb the formation of the potassium-rich endolymph fluid in the inner ear that is required for signal transduction. Curiously, some alleles of all three connexin genes also cause distinct hyperkeratotic skin syndromes with or without deafness.

7.8
Acquired CNS Ion Channel Syndromes

Just as cardiac failure produces an altered pattern of ion channel gene expression, and thereby creates an acquired channelopathy arrhythmia syndrome (see Sect. 5.7), experimental forms of induced temporal lobe epilepsy, the most common form of adult epilepsy, greatly reduce expression of the *KCND2* potassium channel gene, producing a downregulation of the neuronal A-type potassium current and a reduced threshold for seizure initiation in the hippocampus, a common origin for such seizures. Thus, seizure activity produces a reduced threshold for subsequent seizures.

7.9
Polygenic Syndromes of Altered CNS Excitability

While most common seizure, migraine, and ataxia syndromes behave as complex polygenic traits, _not_ as the simple monogenic Mendelian traits discussed above, and none of the genes involved in the polygenic forms of the disease have yet been convincingly demonstrated, a definition of the shared pathophysiology of these monogenic syndromes is likely to provide the best leads to the nature of those genes and to novel therapeutics for these common diseases. Likewise, while the definition of specific pharmacogenetic syndromes of altered neural excitability is only in its infancy, largely limited at this time to the definition of those individuals with altered rates of metabolism of seizure or antipsychotic medications, one would anticipate a dramatic impact on neuropsychopharmacology just by defining the genotypic basis of the wide variance now observed in patient response to current neuropsychiatric drugs. Additionally, it is likely that polygenic ion channel disorders produce a wider range of phenotypes in the CNS than those already revealed by the monogenic disorders. The salient feature of the CNS ion channel disorders, as well as those in the heart and muscle, was that they arose from a _periodic_ disturbance in normal rhythmic activity. They were intrinsically episodic disorders, although many progressed to become constant. Many of the common complex polygenic neuropsychiatric disorders share this character. Further, several of the drugs used to treat the recognized channel phenotypes are additionally used to treat these disorders (e.g. valproate is used to treat both seizures and bipolar disease). Ion channel genes are thus important "functional candidates" to pursue within the broad chromosomal regions shown through linkage or linkage disequilibrium to be associated with these diseases, and are particularly intriguing should alleles be recognized that are similar in activity to those seen in the monogenic channel disorders.

8
Ion Channel Diseases of Epithelia

8.1
Structure and Function of Epithelial Tissues

Epithelia are sheets of _polarized_ cells (sometimes consisting of many layers but only one cell layer of which is functionally significant) that participate in the transport of ions, water, and metabolites, not merely in and out of the cell, but fully _across_ the entire cell sheet (Fig. 13). The epithelial cells are not excitable, like heart, muscle, and nerve, but they are electrically and metabolically coupled horizontally by gap junctions, described above, and the epithelial sheet therefore functions nearly as though it was two sheets of parallel membrane with cytosol between. One side of the epithelial sheet, called the _basolateral membrane_, faces the blood and is relatively unspecialized; the other, called the _apical membrane_, lines a nonblood compartment, often the lumen of a tubule, and the membrane is often elaborated into a brush border and microvilli (Fig. 13). It is primarily in the apical membranes that are found the many special channels and transporters that distinguish one epithelium from another. The point of demarcation between the two faces of

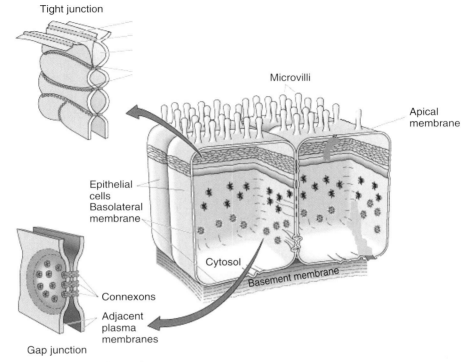

Fig. 13 Structure of epithelia.

an epithelial cell is the *tight junction*, the special barrier-forming connection made between adjacent cells in the epithelium (Fig. 13).

Epithelia serve two essential functions in physiology that can be perturbed in disease: to transport salt, water and other substances from one compartment to another, and to maintain gradients of salt, water, or other substances between compartments. While all epithelia share a fundamental organization and have in common a basic set of transport functions, various epithelia are optimized to perform one or the other of these two principle functions, and this is primarily conferred by the nature of the tight junctions between cells in the sheet. Those optimized for rapid transepithelial transport have "leaky" tight junctions and large *paracellular* pathways between cells providing a shunt parallel to the *transcellular* pathway across the sheet, and those optimized to form large electrical or chemical gradients across the membrane sheet have "tight" tight junctions that do not allow the passage of ions and water except through the cells.

8.2
Function of Ion Channels in Epithelia

Even though various epithelia transport a wide spectrum of compounds, there are a number of basic membrane properties that they all share, and these shared properties even extend to most nonpolarized mammalian cells, as introduced in Sect. 2. This generalization is most particularly true for epithelial basolateral membranes, which are much less specialized and more like those of other cells. First there is a Na/K-ATPase ion pump located in the basolateral membrane (Figs. 6 and 14). This ATPase is oriented so as to extrude Na^+ from the cytoplasm in exchange for the uptake of K^+. Second, the electrical potential difference across the basolateral membrane is electrically negative (cell interior with respect to the blood) and on the order of 50–60 mV, like most mammalian cells. Third, the intracellular ion concentrations reflect the activity of the Na/K pump, with a high internal K^+, a low internal Na^+ and a Cl^- concentration near its Nernst electrochemical equilibrium potential (Fig. 6). Fourth, the passive permeability of the basolateral membrane is predominated by that for K^+, with low permeabilities to Na^+ and Cl^-. Since K^+ is not in electrochemical equilibrium across the K^+-permeable basolateral membrane, the −60 mV resting membrane potential is then a K^+ diffusion potential attenuated by low finite Na^+ and Cl^- permeabilities. Finally, the properties of the apical membrane of epithelia are very diverse, but commonly are predominated by a high Na^+ permeability. The specifics of these apical Na^+ permeabilities and their properties, however, vary such that broad generalizations can not be made. Major breakthroughs in understanding epithelial function occurred as Ussing defined how simply opening a passive Na^+ channel in the apical membrane of an epithelial sheet with the five cardinal characteristics described above allowed transepithelial transport of sodium, and later Curran defined how such vectorial transport of sodium could be coupled to the iso-osmotic vectorial transepithelial flow of water and other solutes. To allow these mechanisms to create a net flux, anion had to be allowed to passively accompany the transported Na^+ since macroscopic electroneutrality had to be maintained on both sides of the epithelium, as it required an enormous amount of energy to separate charge across a membrane. As will be discussed below, if there is no pathway for chloride movement, as occurs in cystic fibrosis, transepithelial salt and water flow are greatly impeded.

8.3
Ion Channel Dysfunction and Diseases in Epithelia

On one hand, the macroscopic phenotypes produced by defects in epithelial channels can be seen to reflect the wide range of specific uses to which Nature has put epithelia: failure of lung, kidney, digestive and other internal organs, deafness, systemic fluid and electrolyte abnormalities, and even abnormalities of morphogenesis. On the other hand, these phenotypes all ultimately can be recognized to arise from defects in either of the two cardinal functions of epithelia, transepithelial transport, or maintenance of transepithelial gradients.

8.4
Cystic Fibrosis

Cystic fibrosis is the most common simple genetic disease in Caucasians, nearly 5% of the members of those populations carrying one pathogenic allele for this lethal autosomal recessive disease, with one common allele, Δ F508, reflecting 70%, but at least four other pathogenic alleles also reaching polymorphic gene frequencies (~1%) and over 1000 alleles now being known. It is also the first genetic disease for which the etiological gene and pathogenic alleles were discovered strictly on the basis of only its chromosomal position, without benefit of a known protein or a pathogenic allele created by a chromosomal rearrangement. It was additionally a landmark since it represented the first genetic channelopathy. The clinical phenotype of this disease – predominated by failure of lung, pancreas, and intestinal function – suggests nothing to bring to mind ion channel processes in the context discussed for the excitable tissues, and only slowly is the full molecular pathophysiology of this disease coming to light.

The gene, *CFTR*, standing for "cystic fibrosis transmembrane regulator," encodes a large transmembrane protein with 12 transmembrane helical spans, two cytosolic ATP-binding domains, and a large cytoplasmic regulatory domain that has multiple phosphorylation sites (Fig. 5). It is itself an ATP-regulated chloride channel, activated by the combination of both direct ATP binding and ATP-dependent phosphorylation via a cAMP-dependent kinase.

How does loss of function of this epithelial chloride channel produce the multifaceted phenotype of CF? The key features of the disease reflect epithelial dysfunction and include very thick dehydrated mucus secretions in the respiratory tree, which become purulent with chronic infections and progress to destruction of the lung parenchyma and therefore chronic respiratory failure. Additionally, a majority of patients have malabsorption caused by pancreatic insufficiency, caused by plugging of the secretory ducts resulting in activation of the digestive enzymes within the ducts and eventually pancreatic fibrosis (hence CF's old names "mucoviscidosis" and "cystic fibrosis of the pancreas"). As medical treatment allowed a longer lifespan, it was recognized that nearly all males at adulthood have infertility caused by plugged atretic vasa deferens and that a few percent of the patients first presented as newborns with meconium ileus, an obstruction of the newborn's intestines caused by cement-like intestinal contents. The physiology of these complex secretory processes was insufficiently understood to lead to an understanding of this disease or its diagnosis. However, the modern scientific underpinnings of CF began during a New York heat wave in 1953, when it was astutely observed that many of the children who were hospitalized for dehydration and salt depletion came to be diagnosed with CF. The dehydration arose because they secreted a salt-rich sweat. This observation led di Sant'Agnese and coworkers to the development of the diagnostic "sweat chloride test" based upon the recognition that the high concentration of chloride in sweat of CF homozygotes, well before they became symptomatic, could be distinguished from unaffected controls or carriers. Not until 30 years later did patch-clamp electrophysiology studies on the sweat gland by Quinton and coworkers establish that a defect in the function of an epithelial chloride channel prevented the physiological resorption of salt from sweat during its transit through the duct. Therefore, the cardinal aspects of transepithelial transport, salt and water movements, were impeded by the CF chloride channel defect. As the sweat gland was a *resorptive epithelium*, salt resorption was prevented, so salt was wasted in sweat. The same molecular lesion in the *secretory epithelium* of the respiratory mucus gland, the pancreatic acinus, and the vas deferens resulted in concretions of dehydrated product that plugged and scarred the ducts, ultimately destroying the tissues.

Despite the complicated name of the CFTR channel it is clear that this protein *itself* functions as an ion channel, not just an indirect regulator of other channels. This point is most simply made by classes of mutant alleles that alter ion selectivity and single-channel conductance, functions intrinsic to a channel and *not* subject to its regulation. There are also informative classes that alter channel gating; however, this *is* a process subject to regulation, so such alleles do not prove intrinsic channel activity. Likewise, the vast majority of the mutant alleles are even less informative in that they merely alter the membrane protein's biosynthesis, maturation, and stability, telling little about the protein's function, but perhaps pointing to vulnerabilities that can be targeted in therapy.

8.5 Renal Channelopathies

The range of the diversity of apical sodium entry pathways available in epithelial designs is reflected in channelopathies of the kidney tubules, but mutations in the subunits of the heterotrimeric EnaC epithelial sodium channels, encoded by *SCNN1A*, *SCNN1B*, and *SCNN1G*, provide the simplest entry to the pathophysiology of these diseases. The α-, β-, and γ-subunits of this channel are homologous to each other and to the degenerins that participate as an olfaction and neurodegeneration channel in *Caenorhabditis elegans*. Each gene encodes a simple ion channel subunit that contains cytoplasmic termini and only two transmembrane helices (Figs. 5 and 6). Expression of the α-subunit alone produced ion channel activity *in vitro*, but it did not reflect a recognized native current. Neither the β- nor γ-subunits alone produced any channel activity. All three subunits expressed together, however, could be recognized to produce the amiloride-sensitive, aldosterone-regulated sodium channel best characterized in the principal cell of the distal nephron (Fig. 14d), but also found in the colon, lung, and exocrine glands, many places CFTR was expressed. Amiloride is a diuretic drug, and drugs we call diuretics are in fact natruretics in that they cause sodium wasting by inhibiting a renal sodium resorption mechanism. Each class of clinically useful diuretics targets a different renal epithelial apical sodium entry pathway, and, remarkably, mutations have been found targeting each of these pathways, simulating the effects of the diuretic drugs furosemide, thiazide, and amiloride.

Just as without a rudimentary understanding of transepithelial transport one would be unlikely to guess the phenotype of CFTR mutations, it takes at least a superficial appreciation of renal physiology to derive the phenotype produced by these mutations: alterations in blood pressure. These days we tend to forget that salt is a precious essential ingredient for life and that loss of body salt (much more than loss of body water, as will be discussed below) results in blood volume depletion and a syndrome of shock similar to that produced by a severe hemorrhage. While the sweat gland is a relatively simple structure that does a fair job of resorbing some salt from the small volume of blood ultrafiltrate that gives rise to the excreted fluid, the kidneys are Olympic performers in this field in that they cope with fully one-fifth of the cardiac output, sending one-fifth of that across the glomerular filter (G) into the epithelial nephron tubules, where in excess of 99.5% of the salt can be reabsorbed by crossing the luminal apical membrane and then the basolateral membrane back into the blood, leaving by default unspecified waste to be lost in the urine (Fig. 14). The epithelial sheet that forms the nephron tubule changes in character along its length, and classical renal tubule

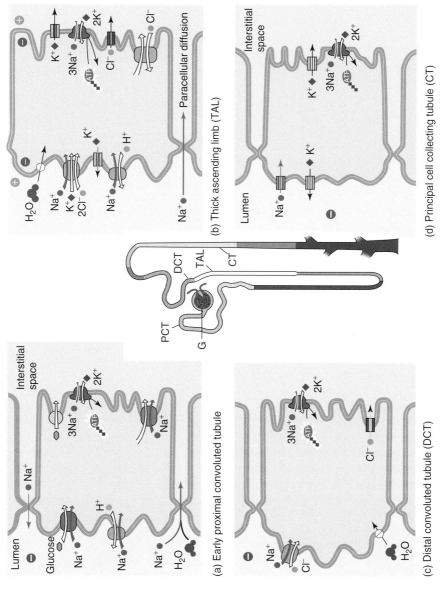

Fig. 14 Renal epithelia.

segment names have now come to be associated with specific molecular transport mechanisms. For instance, 66% of the filtered salt is resorbed in first segment, the proximal tubule (PCT), 25% in the thick ascending limb (TAL) of the loop of Henle via the furosemide-sensitive apical Na-K-Cl cotransporter, 7% in the distal convoluted tubule (DCT) via the thiazide-sensitive apical Na-Cl cotransporter, and the final 2% via the amiloride-sensitive, aldosterone-regulated heterotrimeric EnaC epithelial sodium channels in the collecting tubule (CT) (Fig. 14).

The kidney is capable of regulating sodium resorption over a wide range to compensate for changes in intake, using only the final 2% of the filtered load. This is all achieved by control of the abundance of the EnaC epithelial sodium channels in the principal cells of the CT by the adrenal steroid hormone aldosterone (Fig. 14d). Aldosterone is secreted in response to angiotensin II, which in turn is produced from angiotensin I by angiotensin converting enzyme (ACE), an important pharmacological target for the control of blood pressure. The entire cascade leading to aldosterone is initiated in response to sensors that detect aspects of blood volume, critical ones being the intrarenal and central baroreceptors that trigger specialized cells in the nephron to release an enzyme, called *renin*, which cleaves the inactive angiotensinogen precursor to angiotensin I. Therefore, hemorrhage causes a low blood volume and pressure to be sensed, triggering renin release, leading to the production of angiotensin and aldosterone, the insertion of excess EnaC epithelial sodium channels in the principal cells of the CT, enhanced distal nephron sodium resorption into the blood and a physiological trend to restoring blood volume. On the other hand, artificially producing high blood pressure will homeostatically decrease renin release, leaving angiotensin and aldosterone levels low, reducing the abundance of EnaC epithelial sodium channels and distal nephron sodium resorption, wasting sodium in urine and diminishing blood volume. Because the transepithelial electrical potential across the "tight" CT epithelium is used to drive the secretion of potassium and hydrogen into the nephron lumen, and that electrical potential results from EnaC sodium channel activation, a low abundance of EnaC sodium channels results in reduced K^+ and H^+ wasting in the urine, and hence elevated plasma levels of both. Conversely, high levels of EnaC sodium channels enhance K^+ and H^+ wasting and low plasma levels of both.

8.6
Pseudohypoaldosteronism and Liddle Syndrome

Loss-of-function mutations, primarily produced by deletions, in *SCNN1A*, *SCNN1B*, and *SCNN1G* genes, encoding subunits of the EnaC channel, produce the phenotype of autosomal recessive pseudohypoaldosteronism type 1. This is a severe disease that presents in infancy with life-threatening failure to thrive caused by salt wasting and hypovolemia. They additionally have hyperkalemia and metabolic acidosis. As the name implies, the phenotype appears like that produced by an absence of aldosterone; however, in this case aldosterone (as well as angiotensin and renin) levels are very high. One allele, *SCNN1B G37S*, replaces a conserved glycine residue of the channel with serine, thereby altering gating and reducing the channel open probability. This establishes that channel function *itself* is critical to the phenotype, not some other aspect of the protein. This phenotype powerfully defines the physiology: all of the complex and apparently multifaceted actions of aldosterone can be understood to emanate from increasing the EnaC channel current, since loss of the hormone and loss of the channel appear indistinguishable. The sodium current conducted by the channel is the effector of aldosterone action; *a classical endocrine and hormonal pathway is truly a complex means of ion channel regulation, its diseases truly channelopathies.*

Gain-of-function dominant mutations in *SCNN1B* and *SCNN1G*, altering the C-terminus of the β- or γ-subunits of EnaC channels, cause a pseudohyperaldosterone state called *Liddle syndrome*. The phenotype is hypertension, with low renin, angiotensin and aldosterone levels, suggesting primary excess sodium retention and volume expansion. They also have hypokalemia and metabolic alkalosis, secondary to excessive renal K^+ and H^+ wasting. The mutations mainly result in increased numbers of EnaC channels in the CT; however, some alleles do increase the channel open probability, again showing the sodium current, and not some other aspect of the protein, to be critical. Since the diuretic amiloride blocks the EnaC channel, the drug is useful in treating the disease. Spironolactone, a drug that inhibits aldosterone action, however, has no effect, since the mutant channels are constitutively active and no longer under hormonal control. Not only does amiloride block the excessive sodium resorption, allowing sodium wasting, volume loss, and a normalization of blood pressure, but it also indirectly blocks excessive K^+ and H^+ wasting and corrects those plasma disorders. EnaC does *not* itself transport K^+ or H^+. Blocking the sodium channel in the CT prevents K^+ and H^+ secretion there via effects on the CT transepithelial membrane potential. For this reason, amiloride is a "potassium-sparing" diuretic, in contrast to all of the others to be discussed below.

8.7
Bartter and Gitelman Syndrome

Amiloride, with its ability to block EnaC channels and therefore resorption of only the final 2% of the filtered sodium load, is a relatively weak diuretic. The most potent diuretics, such as furosemide, target the apical Na-K-Cl cotransporter found in the TAL where 25% of the filtered sodium load is resorbed (Fig. 14b). The cotransporter is a solute carrier protein with 12 transmembrane helical spans encoded by *SLC12A1*. It has the ability to couple the electroneutral movement of four ions across the membrane, $1Na^+$, $1K^+$, and $2Cl^-$ (Fig. 6). The carrier's movement is passive in that no ATP energy is directly consumed in transport, but because the flows of all four ions are coupled and must take place in concert for any movement to occur, flux coupling, as was mentioned above, allows the cotransporter to capture the potential energy available in the sodium electrochemical gradient to move K^+ and Cl^- ions along with it as it enters the TAL cell across the apical membrane. This happens despite the fact that K^+ and Cl^- move uphill against their individual electrochemical gradients since it is the net electrochemical potential of the entire complex that dictates the direction of the coupled movements. This is a process called *secondary active transport*, a bioenergetic process coupled to an ion's electrochemical potential, and it is contrasted to *primary active transport*, a process coupled to the chemical potential of ATP hydrolysis and exemplified by the Na/K-ATPase, discussed above.

Loss-of-function alleles of *SLC12A1* were first recognized to cause the severe autosomal recessive disease phenotype of Bartter syndrome (BS), characterized by prenatal polyhydramnios, infantile polyuria, and polydipsia, renal Na^+, K^+, and H^+ wasting with subsequent hypokalemia and hypochloremic metabolic alkalosis, and very high renin, angiotensin, and aldosterone levels. It looked in every way as if these infants were on high doses of furosemide since conception, with the massive renal salt wasting producing the polyhydramnios, polyuria, and volume depletion that the homeostatic aldosterone signaling pathway was mightily struggling to restore. The high aldosterone was forcing a maximal EnaC channel response, but this low-capacity system could not handle the huge load of sodium the dysfunctional TAL had left behind, so sodium continued to be wasted. However, the electrical effect of maximal EnaC channel activation in the CT is present, hence the massive K^+ and H^+ wasting. Just as the EnaC channel has a major impact on the transepithelial electrical potential in the CT, and hence the flux of other ions in that renal segment, the SLC12A1 cotransporter has a major role in creating the TAL transepithelial potential, a very atypical potential in that it is lumen positive, not negative like nearly every other epithelial sheet (Fig. 14b). Physiologically, the lumen positive potential, created by the secondary active transport of Cl^- by the cotransporter, repels cations, particularly the divalent cations Ca^{++} and Mg^{++}, favoring their paracellular resorption and being the major site in the nephron at which they are resorbed. With the loss of SLC12A1 function, the divalent cations are wasted in the urine and the blood levels fall.

Since the SLC12A1 cotransporter carries an electroneutral load of two cations and two anions, it appears puzzling it should be able to create the unusual lumen positive condition. This arises because of two types of critical ion channels in the TAL apical and basolateral membranes (Fig. 14b). Mutations in these channels also produce the BS phenotype, BS type 1 used to specifically designate the disease caused by *SLC12A1* mutations. Since the TAL cells have a high internal K^+, the energetics of cotransport are favored by allowing K^+ to futilely recycle back into the lumen across the apical membrane. This apical potassium channel, the ATP-sensitive ROMK channel encoded by *KCNJ1*, a member of the channel family causing Andersen syndrome (Figs. 5 and 8), was shown to be essential to TAL transepithelial transport since loss-of-function mutations in this gene produced the Bartter syndrome phenotype, BS type 2. The high abundance of ROMK channels in the apical membrane of the TAL contributed half of the explanation for the unusual positive transepithelial potential; abundant K^+ channels gave the apical membrane a hyperpolarized membrane potential nearer the K^+ Nernst potential (Fig. 14b). The last part of the puzzle is the presence of abundant Cl^- channels in the basolateral membrane (Fig. 14b). They depolarize the basolateral membrane potential, and therefore accentuate the difference in potential between the two faces of the TAL cell. They are there to provide a pathway for Cl^- to leave the cell and passively reenter the blood, accompanying Na^+ and effecting net salt resorption in the TAL – the classical "diluting segment" of the tubule where hypotonic luminal fluid is produced. The basolateral Cl^- channel is composed of subunits from the large CLC channel family (Fig. 5) that includes the muscle chloride channel underlying Becker and Thomsen myotonia, discussed above. A crystal structure has recently been obtained for this family of chloride channels, showing them to function as "double-barreled" dimers with each subunit, composed of 18 helical segments and both termini on the cytoplasmic face, having its own ion conduction pathway. The specific subunits found in the kidney are encoded by *CLCNKB* and *CLCNKA*, which are inserted in the membrane accompanied by barttin, their small β-subunit having two transmembrane helices and encoded by *BSND*. Recessive loss-of-function mutations in *CLCNKB* cause classical Bartter syndrome, BS type 3. Bartter syndrome, type 4 is caused by homozygosity for loss-of-function alleles of *BSND*, but this phenotype additionally includes sensorineural deafness. This same set of phenotypes is produced as a *digenic* syndrome with loss-of-function alleles of *both CLCNKB* and *CLCNKA*.

Channelopathies as a cause of deafness were briefly mentioned in the context of recessive cardiac LQT syndromes in Sect. 5.6. Surprisingly, the same channel is expressed selectively in the heart and the inner ear epithelium, where its abundant expression is involved in the production of the potassium-rich endolymph fluid that fills this chamber and participates in the transduction of auditory sensation. Only complete loss of the channel significantly disturbs this secretory process. It is of note that another inner ear *LQT1* homolog, encoded by *KCNQ4*, has dominant alleles that produce only deafness without LQT (*DFNA2*). Just as even a small amount of LQT channel function, such as provided by the dominant LQT alleles, was sufficient for production of the KCl-rich fluid of the inner ear compartment – deafness only occurring with a profound homozygous loss of function – only complete loss of inner ear epithelium chloride channel function causes deafness. The chloride that accompanies potassium in endolymph production passes through the *CLCNKB*- and *CLCNKA*-encoded channel. Without barttin, neither the *CLCNKB*- nor *CLCNKA*-encoded channel can be inserted into either the TAL or inner ear epithelial basolateral membrane. Either channel alone is sufficient to maintain the endolymph production for hearing, but only the *CLCNKB*-encoded channel can support the higher demands of TAL function.

Gitelman syndrome is the result of homozygosity for loss-of-function mutations in *SLC12A3*. It encodes a close relative of the Na-K-Cl cotransporter, called the *thiazide-sensitive Na-Cl cotransporter* to reflect its sensitivity to the thiazide class of diuretics and the K^+ insensitivity of the transport mechanism (Figs. 6 and 14c). This apical cotransporter is found in the DCT of the nephron and is responsible for the resorption of 7% of the filtered sodium load (Fig. 14c). Like Bartter syndrome, the disease presents with renal Na^+ wasting and a secondary wasting of K^+ and H^+ as well as resultant hypokalemia and hypochloremic metabolic alkalosis, as aldosterone and the EnaC channels struggle to correct the volume loss, which ultimately is only minor. Renin, angiotensin, and aldosterone are all elevated; however, the disease is much milder than Bartter syndrome, since the DCT resorbs a much smaller fraction of filtered sodium. Therefore, it classically presents in adolescence, but the presenting picture is *not* one of volume depletion, it is rather a picture predominated by consequences of the chronic plasma electrolyte abnormalities and their impact on excitable tissues. This is commonly manifested as muscle weakness, abdominal pain, and neuromuscular irritability. Patients commonly have hypomagnesemia and accompanying tetany, although the pathogenesis of this abnormality remains to be elucidated.

It should be noted that one would expect hypomorphic alleles in the five genes contributing to Bartter and Gitelman syndrome and the three genes contributing to the EnaC channel to be protective against hypertension. Conversely, hypermorphic alleles in these same eight genes have the potential to contribute to hypertension. Therefore, polymorphic variants of these genes are promising candidates for evaluation in the heritable polygenic trait of blood pressure.

8.8
Hypomagnesemia

Autosomal recessive primary hypomagnesemia has played an important role in beginning to reveal the mechanisms underlying the paracellular pathways of epithelial ion transport, revealing them to be far more complex than simply an unmediated leak between cells (Fig. 14b). The disease presents in early childhood with symptoms similar to Gitelman syndrome, but additionally polyuria and renal stones. It is caused by homozygosity for loss-of-function alleles of *CLDN16*. This gene encodes Paracellin-1, a transmembrane protein with four helical spans found in the TAL and DCT tight junctions where it appears to contribute to a calcium and magnesium selective paracellular path. This gene family appears to encode proteins designated for such paracellular epithelial transport pathways since *CLDN4* and *CLDN15* also influence ion selectivity in other tight junctions.

Autosomal dominant hypomagnesemia appears to arise by a very different mechanism. It was shown in a large extended family to cosegregate with a heterozygous missense mutation in *FYXD2*, the gene encoding the small single membrane-spanning proteolipid γ-subunit of the Na/K-ATPase, a modulatory subunit found associated with the functional $\alpha\beta$ subunits only when the pump is expressed in basolateral membranes of the distal nephron. This subunit alters the transport kinetics of the ion pump, the prime mover setting the gradients and energetics for all coupled transport in these tight epithelia. The pathological *FYXD2 G41R* allele must function as a dominant negative gain-of-function allele, since haploinsufficiency produced by heterozygous deletions spanning the locus does not produce the phenotype. The mutation alters charge in the middle of the transmembrane helix and causes the γ-subunit to fail to traffic beyond the golgi, allowing functional $\alpha\beta$ subunit insertion into the membrane alone. As this is the only pathogenic allele reported to date, the mechanism by which mutations in this locus produce magnesium wasting remains unclear.

Hypomagnesemia with secondary hypocalcemia is caused by homozygosity for loss-of-function mutations in the *TRPM6* gene, over a dozen pathogenic alleles already known. This gene, and other members of its large extended family, are cousins of the genes encoding the pore-containing hexahelical channel subunits of voltage-gated K^+ channels, described in Sect. 3; however, the TRP branch of the family encode an array of selective and nonselective tetrameric cation channels that are coming to be recognized as playing important roles in sensory transduction and epithelial function (Fig. 5). The TRPM6 and TRPM7 channel subunits, each of which contains a C-terminal protein kinase domain, appear to heteromultimerize to form a Mg^{++} channel and pathogenic alleles of *TRPM6* prevent subunit association and channel function. No pathogenic alleles, however, have yet been reported in *TRPM7*.

8.9
Polycystic Kidney Disease

Autosomal dominant polycystic kidney disease (ADPKD) is one of the most common genetic conditions, affecting over 1/800 births and responsible for 10% of chronic dialysis. Mutations – primarily null alleles caused by deletions, nonsense, frameshifts or splicesite alterations – in a relative of the TRP family of genes, PKD2 – encoding polycystin-2 – and in PKD1 – encoding polycystin-1, its homologous heteromultimerizing partner – cause this disease, and produce cystic dilation of kidney tubules, hepatic cysts, and aneurysms in blood vessels. Most of the nephrons appear normal, suggesting a "two-hit" pathogenic model reflecting a germline loss-of-function mutation and probabilistic somatic loss-of-heterozygosity (LOH), similar to that which has proved so powerful in understanding dominant cancers. Coexpression of PKD1 and PKD2 *in vitro* produce functional calcium-permeable nonselective cation channels, but expression of either PKD1 or PKD2 alone does not, and pathogenic alleles prevent heteromultimerization and channel function. The channels are found localized in association with the cilium, a structure of the apical membrane, perhaps transducing mechanical sensory signals it detects. While it is clear that calcium signaling is induced by the cilium and perturbed by the mutant alleles, it is unclear how either alters morphogenesis producing cyst formation. However, recently a clear example of morphogenetic pattern formation, left-right axis asymmetry, was shown to require epithelial PKD2 function.

8.10
Glucosuria

The notion that energy conserved in the sodium gradient across the apical membrane of an epithelial cell can be used in secondary active transport was introduced above in discussing Bartter and Gitelman syndrome. Sodium-coupled transport, however, is used in the gut and nephron in the resorption of many uncharged metabolites, not only ions (Fig. 6). A clear picture of how such energy coupling occurs is provided by two autosomal recessive glucose-wasting disorders. The sodium gradient across the apical membrane arises as a consequence of Na/K pump activity at the basolateral membrane. Secondary active transport of a substance, in this case glucose, occurs "uphill," against its electrochemical gradient by being obligatorily coupled on a carrier protein to the movement of sodium down its electrochemical gradient into the cell, such that the net movement of the coupled system obeys thermodynamics and remains "downhill" (Fig. 14a). A tremendous amount of energy is stored in the sodium gradient. A simple estimation reveals that cellular sodium is about 1/10th that found in the renal tubular fluid and that the membrane potential is ~ -60 mV – the Nernst equation revealing that translates to the equilibrium diffusion potential set up by a 10-fold gradient of a monovalent ion (*see* Sect. 2). Both the 10-fold chemical gradient and the equivalent 10-fold electrical gradient favor movement of Na^+ into the cell, and those potentials are multiplicative. This should allow 1:1 coupling of sodium:glucose to drive cellular glucose to a concentration 100 times that found in tubular fluid. Should one obligatorily couple the movement of two sodium ions to cotransport with one glucose, the multiplicative properties suggest one could achieve the 100-fold enhancement for *each* Na^+ site, and hence a 10 000-fold gradient favoring glucose absorption. These are the properties of SLC5A2, the low affinity 1:1 apical sodium-glucose cotransporter and SLC5A1, the high affinity 2:1 apical sodium-glucose cotransporter, respectively.

Isolated renal glucosuria is caused by homozygous loss-of-function mutations in *SLC5A2*. The high-capacity, low-affinity cotransporter it encodes is glucose-selective and is abundantly expressed in the apical membrane of the initial segment of the proximal tubule of the kidney, the site where the majority of the glucose filtered across the glomerulus is reabsorbed (Fig. 14a). Physiologically, no glucose remains in the urine unless blood levels become so high that they overwhelm the transport capacity of this transporter, as occurs in diabetes mellitus. Therefore, the symptoms of this transport defect are massive glucosuria, and the resultant diuresis of water obligated by this large osmotic load.

Homozygous loss-of-function alleles of *SLC5A1* cause a severe neonatal dehydration syndrome; however, this is because of diarrhea, not the renal loss it produces. The disease caused by these mutations is glucose-galactose malabsorption, and it is characterized by loss of high affinity glucose transport in the kidney and gut. This high-affinity cotransporter is expressed in the apical membrane of the final segment of the proximal tubule where it scavenges the last of the glucose missed by the SLC5A2 cotransporter. Its low capacity, however, is easily overwhelmed, so it cannot compensate for loss of *SLC5A2*. Unlike SLC5A2, it transports galactose in addition to glucose, and is the mechanism that predominates intestinal absorption of these sugars – essential for a newborn dependent on the lactose of milk. Just as sugar left behind in the lumen of the nephron osmotically obligates water loss, so too does this unresorbed intestinal sugar; however, providing the infant a formula of fructose, a sugar that uses a different transporter, is life-saving.

8.11
Nephrogenic Diabetes Insipidus

Ancient physicians recognized that not only could diabetes, characterized by excessive urination and thirst – polyuria and polydipsia – be accompanied by sugary-sweet tasting urine – diabetes mellitus – but could alternatively be accompanied by tasteless urine – diabetes insipidus. Autosomal nephrogenic diabetes insipidus (NDI) results from defects in water channels, called *aquaporins*, and a recent Nobel Prize was awarded for the solution of their nature. Aquaporins reflect a large family of water channels and their function – allowing the flow of water while excluding that of ions, is the mirror image of that of ion channels, which allow the flow of ions without water. Loss-of-function recessive and dominant-negative alleles of *AQP2* result in NDI and in a loss of regulated water channels from the apical membrane of the principal cell of the CT (Fig. 14d), the site of regulated water resorption in the kidney. The peptide hormone alternatively called *vasopressin* or antidiuretic hormone (ADH), and long known to be responsible for the production of concentrated urine is tonically secreted by the pituitary in response to small changes in plasma osmolarity. It was shown to control insertion of the

AQP2 channel into the apical membrane of the principal cell; in the absence of hormone, the aquaporin was sequestered in vesicles beneath the membrane and hormone receptor signaling through cAMP triggered insertion of the channels. The receptor receiving the ADH signal, called the *V2 vasopressin receptor*, is encoded by the X-linked gene *AVPR2* and mutations in this gene were shown to cause the X-linked, more common form of nephrogenic diabetes insipidus. The phenotype in NDI, lethargy progressing to coma, results from effects on excitable tissues of euvolumic hypernatremic hyperosmolarity, as water is wasted by the kidney in excess of salt. Note that both major hormones involved in salt and water balance act independently by directing the insertion of different apical channels into the principal cell of the CT: aldosterone regulates sodium balance and hence plasma volume via the insertion of epithelial sodium channels and ADH regulates water balance and hence plasma osmolarity via the insertion of aquaporin-2.

8.12
Hereditary Kidney Stone

Kidney stone formation, nephrolithiasis, is a common polygenic trait with a lifetime risk of up to 10%. Stones are crystal-like structures that precipitate from urine within the kidney. Their origin is now recognized to be an abnormal urine environment, and recently defects in renal ion channels have been shown to contribute to this phenotype. Above it was briefly mentioned that stone formation was a component of the phenotype in primary hypomagnesemia. Nephrolithiasis is most commonly associated with hypercalciuria and calcium salts are a key component of most stones. Loss-of-function alleles of *CLCN5*, encoding a voltage-gated chloride channel within the family of those causing Bartter syndrome and myotonia (Fig. 5), described above, were recognized to cause a Mendelian X-linked form of hypercalciuria and stone formation, called *Dent disease*, a phenotype that additionally included proteinuria and progressed to renal failure. Additional alleles in the same gene were found to cause two other diseases that had not previously been recognized to be related, X-linked nephrolithiasis and X-linked hypophosphatemic rickets – a phenotype characterized by demineralization of bone. This chloride channel is localized to endosomes within epithelial cells of the proximal tubule, *demonstrating that ion channels play a critical part not only in the function of cellular membranes but also of subcellular compartments*. Defects in the function of the endosomal Cl^- channel impair the process of endocytosis from the apical membrane, secondarily altering the regulation of various apical transporters, but the exact mechanism of pathogenesis remains to be defined. It is of note that homozygous loss-of-function and heterozygous dominant-negative mutations in its close relative *CLCN7*, while causing a distinctly different phenotype, osteopetrosis, ultimately produce the phenotype as a result of a failure in endosomal function. Further, homozygous loss-of-function alleles in *TCIRG1*, encoding a subunit of the endosomal vacuolar ATPase proton pump (Fig. 6), also produce osteopetrosis. Osteopetrosis is a phenotype that includes very dense bone since bone remodeling by the degradative bone osteoclasts fails because endosomes fail to function since they can not acidify. Together, the two osteopetrosis diseases loci suggest that the endosomal Cl^- channels provide a shunt pathway that allows the endosomal H^+-ATPase to pump a proton gradient without stalling because of electrostatic charge accumulation, and suggest a similar role for *CLCN5* in the kidney.

9
Ion Channel Disease of Other Nonexcitable Tissue

In addition to major phenotypes produced by defects directly assignable to abnormalities in the function of heart, muscle, nerve, and epithelia, ion channel defects are already known to produce a wide array of ancillary phenotypes, some of which were mentioned above as additional components to a syndrome, for instance, the skin findings produced by alleles producing deafness in the three gap junction loci, mentioned at the end of Sect. 7.7. Further, new channel phenotypes are coming to be recognized as arising from a failure of internal membranous organelle function, and these organelles play similar functions in most cell types. In these other tissues, channels seem to play pathophysiological roles similar to those played in diseases of excitable tissues or epithelia. For instance, the mirror-image pancreatic β-cell endocrine channelopathies of hyperinsulinism in infancy (HI) and permanent neonatal diabetes (PND) are very similar to neuronal receptor/channel disease, once the special mechanism of glucose sensing by the ATP-sensitive sulfonylurea receptor/channel is taken into account. This receptor/channel contains K^+ channel subunits, encoded by *KCNJ11* (from the family of those involved in Andersen and Bartter syndromes, discussed above) (Figs. 5 and 8), and a regulatory ATP-binding cassette subunit, encoded by *ABCC8*, named *SUR* for the sulfonylurea (SU) drugs that interact with it. Blocking this K^+ channel depolarizes the membrane, activating voltage-activated Ca^{++} channels and a calcium influx that triggers excitation-secretion coupling (as described in Sect. 7.1) with the fusion of insulin-laden vesicles into the β-cell membrane, and hence insulin secretion into the blood. Glucose induces insulin release through its metabolism in the β-cell, the ATP produced serving to block the channel; the SU drugs induce insulin release by directly blocking the channel's conductance. Not surprisingly, HI, which simulates the effect of SU, is caused by homozygosity for loss-of-function mutations in either of the two channel subunits. PND, where insulin can not be released, results from heterozygous dominant gain-of-function missense alleles of *KCNJ11*, which produce constitutively active channels that cannot respond to physiological ATP block induced by glucose, but can respond with insulin secretion when blocked by an SU. Notice how these alleles and the mirror-image phenotypes they produce are reminiscent of the mirror-image "endocrine" EnaC renal channelopathies, discussed in Sect. 8.6. As a second example, osteopetrosis, a defect in bone morphogenesis caused by defects in Cl^- channels and ultimately endosomal function, was briefly mentioned above as an endosomal channelopathy, an initial case of a disease caused by defects in membrane compartments *within* cells. Other membrane compartment channelopathies are beginning to emerge: endoplasmic reticulum channelopathies and MHS, caused by alleles of *RYR1*, discussed above in Sect. 6.4; lysosomal channelopathies and the lysosomal storage disease mucolipidosis type IV, caused by loss-of-function alleles in *MLN1*, encoding a lysosomal TRP-like nonselective cation channel; and mitochondrial channelopathies and severe obesity with diabetes, caused by loss-of-function alleles of *UCP3*, encoding a mitochondrial proton channel that shuts protons across the membrane and thereby uncouples respiration from ATP synthesis, physiologically making mitochondria less efficient, and the disease appearing to result from hyper-efficient mitochondria.

10
Ion Channels as Targets for Drug Therapy in Common Diseases

For the discovery of a pathogenic allele to have full impact in modern medicine, it is necessary to understand its function and to find small molecules able to alter that function – such "rationally designed" novel pharmaceuticals being one of the major hopes for the postgenomic era. While often positionally identified disease genes have been slow to yield either their function or a "rational drug," ion channels are highly amenable and proven targets. For example, discovering ion channel participants in a disease, such as the Mendelian seizure disorder caused by alleles of *KCNQ2* and *KCNQ3*, suggests not only a novel *type* of target for therapy (e.g. suggesting use of acetazolamide – a drug proven effective in a wide range of ion channel diseases), but these channels themselves can further serve as the guides to *finding* entirely novel selective pharmaceuticals. Because the function of channel mechanisms is so well understood, and it is now clear that the *full* range of pathogenic abnormalities in the huge ensemble of channel diseases discussed above is either too much or too little conductance through the channel (not *anything* else), one does not need to guess what has to be changed to remedy the gene defect. It is clear that the therapeutic goal must be to discover a channel closer/blocker or a channel opener/activator, and a number of drugs of both types are already in current use. *In vitro* expression assays of channel function can be used in high-throughput drug screening and in drug optimization. An example of this kind of breakthrough is retigabine, a novel class of seizure medication proven to activate the KCNQ2/KCNQ3 "M current" channel suggested as a seizure drug target by the role of these genes in BFNC. Since the channel mechanisms perturbed in such disorders are likely to be critical participants in producing the relevant membrane potentials leading to disease vulnerability, the drugs targeting them should be useful for treating the disease even in those *not* having an intrinsic defect in the targeted mechanism (i.e. even if a patient has the disease because of a pathologically depolarized critical membrane because of some *other* mechanism, hyperpolarizing the implicated membrane via *any* mechanism should be helpful). Another advantage of pharmacology targeted to channels is the expectation, discussed above, that *a narrow range of tissues express the channel in a physiologically important context, therefore drug specificity is achieved and unwanted side effects are minimized*. Therefore, in this postgenomic world rich in potential therapeutic targets, the potential channels hold for rapid novel drug discovery warrants their prioritization.

Most of the diseases discussed in this chapter are rare channelopathies and despite being promising targets for pharmacological intervention would likely remain "orphan" diseases were it not for the fact that they produce a *phenocopy* of a common disease, and perhaps provide the only molecular lead to a new drug urgently needed by a major market. However, it is likely that many of the loci critical to the development of the common complex polygenic diseases *cannot* be revealed by major-effect Mendelian alleles, such alleles perhaps being early embryonic lethals. An increasingly important approach to the identification of genes underlying complex polygenic disease is to identify "functional candidates" within broad chromosomal regions implicated in the disease and to test disease association (i.e. linkage disequilibrium) with polymorphic markers within these candidates. Ion channel candidates in such diseases could be suggested by the known physiology, pharmacology and pathophysiology, their appropriate or extremely narrow tissue distribution, their membership in a demonstrably pathogenic gene family, contribution to a demonstrably vulnerable component of the action potential, or a salient disease-associated motif. The existence of alleles sharing features common to known pathogenic alleles underlying the monogenic channelopathies, such as dominant-negative inhibition of expressed channel function or incomplete inactivation, would be particularly suggestive.

See also Receptor Biochemistry; Receptor Targets in Drug Discovery.

Bibliography

Books and Reviews

Agre, P., Kozono, D. (2003) Aquaporin water channels: molecular mechanisms for human diseases, *FEBS Lett.* **555**(1), 72–78.

Gargus, J.J. (2003) Unraveling monogenic channelopathies and their implications for complex polygenic disease, *Am. J. Hum. Genet.* **72**(4), 785–803.

Hille, B. (2001) *Ionic Channels of Excitable Membranes*. 2nd edition, Sinauer Associates, Sunderland, Massachusetts.

Jurkat-Rott, K., Lerche, H., Lehmann-Horn, F. (2002) Skeletal muscle channelopathies, *J. Neurol.* **249**, 1493–1502.

Lang, B., Vincent, A. (2003) Autoantibodies to ion channels at the neuromuscular junction, *Autoimmun. Rev.* **2**(2), 94–100.

Lerche, H., Jurkat-Rott, K., Lehmann-Horn, F. (2001) Ion channels and epilepsy, *Am. J. Med. Genet.* **106**, 146–159.

Lifton, R.P., Gharavi, A.G., Geller, D.S. (2001) Molecular mechanisms of human hypertension, *Cell* **104**, 545–556.

MacKinnon, R. (2003) Potassium channels, *FEBS Lett.* **555**(1), 62–65. '

Marban, E. (2002) Cardiac channelopathies, *Nature* **415**, 213–218.

Schultz, S.G. (1998) A century of (epithelial) transport physiology: from vitalism to molecular cloning, *Am. J. Physiol.* **274**(1 Pt 1), C13–C23.

Vander, A.J. (1995) *Renal Physiology*, 5th edition, McGraw-Hill, New York.

Primary Literature

Abbott, G.W., Butler, M.H., Bendahhou, S., Dalakas, M.C., Ptacek, L.J., Goldstein, S.A.N. (2001) MiRP2 forms potassium channels in skeletal muscle with Kv3.4 and is associated with periodic paralysis, *Cell* **104**, 217–231.

Abbott, G.W., Sesti, F., Splawski, I., Buck, M.E., Lehmann, M.H., Timothy, K.W., Keating, M.T., Goldstein, S.A. (1999) MiRP1 forms IKr potassium channels with HERG and is associated with cardiac arrhythmia, *Cell* **97**, 175–187.

Argyropoulos, G., Brown, A.M., Willi, S.M., Zhu, J., He, Y., Reitman, M., Gevao, S.M., Spruill, I., Garvey, W.T. (1998) Effects of mutations in the human uncoupling protein 3 gene on the respiratory quotient and fat oxidation in severe obesity and type 2 diabetes, *J. Clin. Invest.* **102**, 1345–1351.

Barclay, J., Balaguero, N., Mione, M., Ackerman, S.L., Letts, V.A., Brodbeck, J., Canti, C., Meir, A., Page, K.M., Kusumi, K., Perez-Reyes, E., Lander, E.S., Frankel, W.N., Gardiner, R.M., Dolphin, A.C., Rees, M. (2001) Ducky mouse phenotype of epilepsy and ataxia is associated with mutations in the Cacna2d2 gene and decreased calcium channel current in cerebellar Purkinje cells, *J. Neurosci.* **21**, 6095–6104.

Barhanin, J., Lesage, F., Guillemare, E., Fink, M., Lazdunski, M., Romey, G. (1996) K(v)LQT1 and IsK (minK) proteins associate to form the I(Ks) cardiac potassium current, *Nature* **384**, 78–80.

Baulac, S., Huberfeld, G., Gourfinkel-An, I., Mitropoulou, G., Beranger, A., Prud'homme, J.-F., Baulac, M., Brice, A., Bruzzone, R., LeGuern, E. (2001) First genetic evidence of GABA(A) receptor dysfunction in epilepsy: a mutation in the gamma-2-subunit gene, *Nat. Genet.* **28**, 46–48.

Bennett, P.B., Yazawa, K., Makita, N., George, A.L. Jr. (1995) Molecular mechanism for an inherited cardiac arrhythmia, *Nature* **376**, 683–685.

Bernard, C., Anderson, A., Becker, A., Poolos, N.P., Beck, H., Johnston, D. (2004) Acquired dendritic channelopathy in temporal lobe epilepsy, *Science* **305**(5683), 532–535.

Birkenhager, R., Otto, E., Schurmann, M.J., Vollmer, M., Ruf, E.-M., Maier-Lutz, I., Beekmann, F., Fekete, A., Omran, H., Feldmann, D., Milford, D.V., Jeck, N., Konrad, M., Landau, D., Knoers, N.V.A.M., Antignac, C., Sudbrak, R., Kispert, A., Hildebrandt, F. (2001) Mutation of BSND causes Bartter syndrome with sensorineural deafness and kidney failure, *Nat. Genet.* **29**, 310–314.

Browne, D.L., Gancher, S.T., Nutt, J.G., Brunt, E.R., Smith, E.A., Kramer, P., Litt, M. (1994) Episodic ataxia/myokymia syndrome is associated with point mutations in the human potassium channel gene, KCNA1, *Nat. Genet.* **8**, 136–140.

Bulman, D.E., Scoggan, K.A., van Oene, M.D., Nicolle, M.W., Hahn, A.F., Tollar, L.L., Ebers, G.C. (1999) A novel sodium channel mutation in a family with hypokalemic periodic paralysis, *Neurology* **53**, 1932–1936.

Burgess, D.L., Jones, J.M., Meisler, M.H., Noebels, J.L. (1997) Mutation of the Ca(2+) channel beta subunit gene Cchb4 is associated with ataxia and seizures in the lethargic (lh) mouse, *Cell* **88**, 385–392.

Caddick, S.J., Wang, C., Fletcher, C.F., Jenkins, N.A., Copeland, N.G., Hosford, D.A. (1999) Excitatory but not inhibitory synaptic transmission is reduced in lethargic (Cacnb4(lh)) and tottering (Cacna1atg) mouse thalami, *J. Neurophysiol.* **81**, 2066–2074.

Canessa, C.M., Horisberger, J.D., Rossier, B.C. (1993) Epithelial sodium channel related to proteins involved in neurodegeneration, *Nature* **361**(6411), 467–470.

Canessa, C.M., Schild, L., Buell, G., Thorens, B., Gautschi, I., Horisberger, J.D., Rossier, B.C. (1994) Amiloride-sensitive epithelial Na$^+$ channel is made of three homologous subunits, *Nature* **367**(6462), 463–467.

Cascio, M. (2004) Structure and function of the glycine receptor and related nicotinicoid receptors, *J. Biol. Chem.* **279**(19), 19383–19386.

Chandy, K.G., Fantino, E., Wittekindt, O., Kalman, K., Tong, L.L., Ho, T.H., Gutman, G.A., Crocq, M.A., Ganguli, R., Nimgaonkar, V., Morris-Rosendahl, D.J., Gargus, J.J. (1998) Isolation of a novel potassium channel gene hSKCa3 containing a polymorphic CAG repeat: a candidate for schizophrenia and bipolar disorder? *Mol. Psychiatry.* **3**, 32–37.

Chang, S.S., Grunder, S., Hanukoglu, A., Rosler, A., Mathew, P.M., Hanukoglu, I., Schild, L., Lu, Y., Shimkets, R.A., Nelson-Williams, C., Rossier, B.C., Lifton, R.P. (1996) Mutations in subunits of the epithelial sodium channel cause salt wasting with hyperkalaemic acidosis, pseudohypoaldosteronism type 1, *Nat. Genet.* **12**, 248–253.

Charlier, C., Singh, N.A., Ryan, S.G., Lewis, T.B., Reus, B.E., Leach, R.J., Leppert, M. (1998) A pore mutation in a novel KQT-like potassium channel gene in an idiopathic epilepsy family, *Nat. Genet.* **18**, 53–55.

Chen, Q., Kirsch, G.E., Zhang, D., Brugada, R., Brugada, J., Brugada, P., Potenza, D., Moya, A., Borggrefe, M., Breithardt, G., Ortiz-Lopez, R., Wang, Z., Antzelevitch, C., O'Brien, R.E., Schulze-Bahr, E., Keating, M.T., Towbin, J.A., Wang, Q. (1998) Genetic basis and molecular mechanism for idiopathic ventricular fibrillation, *Nature* **392**, 293–295.

Chubanov, V., Waldegger, S., Mederos y Schnitzler, M., Vitzthum, H., Sassen, M.C., Seyberth, H.W., Konrad, M., Gudermann, T. (2004) Free in PMC Disruption of TRPM6/TRPM7 complex formation by a mutation in the TRPM6 gene causes hypomagnesemia with secondary hypocalcemia, *Proc. Natl. Acad. Sci. U.S.A.* **101**(9), 2894–2899.

Claes, L., Del-Favero, J., Ceulemans, B., Lagae, L., Van Broeckhoven, C., De Jonghe, P. (2001) De novo mutations in the sodium-channel gene SCN1A cause severe myoclonic epilepsy of infancy, *Am. J. Hum. Genet.* **68**, 1327–1332.

Cleiren, E., Benichou, O., Van Hul, E., Gram, J., Bollerslav, J., Singer, F.R., Beaverson, K., Aledo, A., Whyte, M.P., Yoneyama, T., deVernejou, M.-C., Van Hul, W. (2001) Albers-Schonberg disease (autosomal dominant osteopetrosis, type II) results from mutations in the ClCN7 chloride channel gene, *Hum. Mol. Genet.* **10**, 2861–2867.

Cummings, C.J., Zoghbi, H.Y. (2000) Trinucleotide repeats: mechanisms and pathophysiology, *Annu. Rev. Genomics Hum. Genet.* **1**, 281–328.

Curran, M.E., Splawski, I., Timothy, K.W., Vincent, G.M., Green, E.D., Keating, M.T. (1995) A molecular basis for cardiac arrhythmia: HERG mutations cause long QT syndrome, *Cell* **80**, 795–803.

Curran, P.F., McIntosh, J.R. (1962) A model system for biological water transport, *Nature* **193**, 347–348.

De Fusco, M., Marconi, R., Silvestri, L., Atorino, L., Rampoldi, L., Morgante, L., Ballabio, A., Aridon, P., Casari, G. (2003) Haploinsufficiency of ATP1A2 encoding the Na^+/K^+ pump alpha-2 subunit associated with familial hemiplegic migraine type 2, *Nat. Genet.* **33**, 192–196.

Deen, P.M.T., Verdijk, M.A.J., Knoers, N.V.A.M., Wieringa, B., Monnens, L.A.H., van Os, C.H., van Oost, B.A. (1994) Requirement of human renal water channel aquaporin-2 for vasopressin-dependent concentration of urine, *Science* **264**, 92–94.

Delmas, P. (2004) Polycystins: from mechanosensation to gene regulation, *Cell* **118**(2), 145–148.

Di sant'agnese, P.A., Darling, R.C., Perera, G.A., Shea, E. (1953) Abnormal electrolyte composition of sweat in cystic fibrosis of the pancreas; clinical significance and relationship to the disease, *Pediatrics* **12**(5), 549–563.

Doyle, D.A., Morais Cabral, J., Pfuetzner, R.A., Kuo, A., Gulbis, J.M., Cohen, S.L., Chait, B.T., MacKinnon, R. (1998) The structure of the potassium channel: molecular basis of K^+ conduction and selectivity, *Science* **280**, 69–77.

Drachman, D.B. (1978) Myasthenia gravis (first of two parts), *N. Engl. J. Med.* **298**(3), 136–142.

Dror, V., Shamir, E., Ghanshani, S., Kimhi, R., Swartz, M., Barak, Y., Weizman, R., Avivi, L., Litmanovitch, T., Fantino, E., Kalman, K., Jones, E.G., Chandy, K.G., Gargus, J.J., Gutman, G.A., Navon, R. (1999) hKCa3/KCNN3 potassium channel gene: association of longer CAG repeats with schizophrenia in Israeli Ashkenazi Jews, expression in human tissues and localization to chromosome 1q21, *Mol. Psychiatry.* **4**, 254–260.

Ducros, A., Denier, C., Joutel, A., Cecillon, M., Lescoat, C., Vahedi, K., Darcel, F., Vicaut, E., Bousser, M.G., Tournier-Lasserve, E. (2001) The clinical spectrum of familial hemiplegic migraine associated with mutations in a neuronal calcium channel, *N. Engl. J. Med.* **345**, 17–24.

Duggal, P., Vesely, M.R., Wattanasirichaigoon, D., Villafane, J., Kaushik, V., Beggs, A.H. (1998) Mutation of the gene for IsK associated with both Jervell and Lange-Nielsen and Romano-Ward forms of long-QT syndrome, *Circulation* **97**, 142–146.

Dunne, M.J., Cosgrove, K.E., Shepherd, R.M., Aynsley-Green, A., Lindley, K.J. (2004) Hyperinsulinism in infancy: from basic science to clinical disease, *Physiol. Rev.* **84**(1), 239–275.

Dutzler, R., Campbell, E.B., Cadene, M., Chait, B.T., MacKinnon, R. (2002) X-ray structure of a ClC chloride channel at 3.0 a reveals the molecular basis of anion selectivity, *Nature* **415**(6869), 287–294.

Dutzler, R., Campbell, E.B., MacKinnon, R. (2003) Gating the selectivity filter in ClC chloride channels, *Science* **300**(5616), 108–112.

Engel, A.G., Ohno, K., Wang, H.-L., Milone, M., Sine, S.M. (1998) Molecular basis of congenital myasthenic syndromes: mutations in the acetylcholine receptor, *The Neuroscientist* **4**, 185–194.

Escayg, A., De Waard, M., Lee, D.D., Bichet, D., Wolf, P., Mayer, T., Johnston, J., Baloh, R., Sander, T., Meisler, M.H. (2000a) Coding and noncoding variation of the human calcium-channel beta(4)-subunit gene CACNB4 in patients with idiopathic generalized epilepsy and episodic ataxia, *Am. J. Hum. Genet.* **66**, 1531–1539.

Escayg, A., MacDonald, B.T., Meisler, M.H., Baulac, S., Huberfeld, G., An-Gourfinkel, I., Brice, A., LeGuern, E., Moulard, B., Chaigne, D., Buresi, C., Malafosse, A. (2000b) Mutations of SCN1A, encoding a neuronal sodium channel, in two families with GEFS + 2, *Nat. Genet.* **24**, 343–345.

Estevez, R., Boettger, T., Stein, V., Birkenhager, R., Otto, E., Hildebrandt, F., Jentsch,

T.J. (2001) Barttin is a Cl- channel beta-subunit crucial for renal Cl- reabsorption and inner ear K$^+$ secretion, *Nature* **414**, 558–561.

Eunson, L.H., Rea, R., Zuberi, S.M., Youroukos, S., Panayiotopoulos, C.P., Liguori, R., Avoni, P., McWilliam, R.C., Stephenson, J.B., Hanna, M.G., Kullmann, D.M., Spauschus, A. (2000) Clinical, genetic, and expression studies of mutations in the potassium channel gene KCNA1 reveal new phenotypic variability, *Ann. Neurol.* **48**, 647–656.

European Polycystic Kidney Disease Consortium. (1994) The polycystic kidney disease 1 gene encodes a 14 kb transcript and lies within a duplicated region on chromosome 16, *Cell* **77**, 881–894.

Fambrough, D.M., Drachman, D.B., Satyamurti, S. (1973) Neuromuscular junction in myasthenia gravis: decreased acetylcholine receptors, *Science* **182**(109), 293–295.

Fletcher, C.F., Lutz, C.M., O'Sullivan, T.N., Shaughnessy, J.D. Jr., Hawkes, R., Frankel, W.N., Copeland, N.G., Jenkins, N.A. (1996) Absence epilepsy in tottering mutant mice is associated with calcium channel defects, *Cell* **87**, 607–617.

Forbush, B. III, Kaplan, J.H., Hoffman, J.F. (1978) Characterization of a new photoaffinity derivative of ouabain: labeling of the large polypeptide and of a proteolipid component of the Na, K-ATPase, *Biochemistry* **17**(17), 3667–3676.

Frattini, A., Orchard, P.J., Sobacchi, C., Giliani, S., Abinun, M., Mattsson, J.P., Keeling, D.J., Andersson, A.-K., Wallbrandt, P., Zecca, L., Notarangelo, L.D., Vezzoni, P., Villa, A. (2000) Defects in TCIRG1 subunit of the vacuolar proton pump are responsible for a subset of human autosomal recessive osteopetrosis, *Nat. Genet.* **25**, 343–346.

Gloyn, A.L., Pearson, E.R., Antcliff, J.F., Proks, P., Bruining, G.J., Slingerland, A.S., Howard, N., Srinivasan, S., Silva, J.M., Molnes, J., Edghill, E.L., Frayling, T.M., Temple, I.K., Mackay, D., Shield, J.P., Sumnik, Z., van Rhijn, A., Wales, J.K., Clark, P., Gorman, S., Aisenberg, J., Ellard, S., Njolstad, P.R., Ashcroft, F.M., Hattersley, A.T. (2004) Activating mutations in the gene encoding the ATP-sensitive potassium-channel subunit Kir6.2 and permanent neonatal diabetes, *N. Engl. J. Med.* **350**(18), 1838–1849.

Gomez, C.M., Maselli, R.A., Vohra, B.P., Navedo, M., Stiles, J.R., Charnet, P., Schott, K., Rojas, L., Keesey, J., Verity, A., Wollmann, R.W., Lasalde-Dominicci, J. (2002) Novel delta subunit mutation in slow-channel syndrome causes severe weakness by novel mechanisms, *Ann. Neurol.* **51**(1), 102–112.

Goodenough, D.A., Paul, D.L. (2003) Beyond the gap: functions of unpaired connexon channels, *Nat. Rev. Mol. Cell Biol.* **4**(4), 285–294.

Grant, A.O., Carboni, M.P., Nepliouava, V., Starmer, C.F., Memmi, M., Napolitano, C., Priori, S. (2002) Long QT syndrome, Brugada syndrome, and conduction system disease are linked to a single sodium channel mutation, *J. Clin. Invest.* **110**, 1201–1209.

Grunder, s., Firsov, D., Chang, S.S., Jaeger, N.F., Gautschi, I., Schild, L., Lifton, R.P., Rossier, B.C. (1997) A mutation causing pseudohypoaldosteronism type 1 identifies a conserved glycine that is involved in the gating of the epithelial sodium channel, *EMBO J.* **16**, 899–907.

Hamill, O.P., Marty, A., Neher, E., Sakmann, B., Sigworth, F.J. (1981) Improved patch-clamp techniques for high-resolution current recording from cells and cell-free membrane patches, *Pflugers Arch.* **391**(2), 85–100.

Hanaoka, K., Qian, F., Boletta, A., Bhunia, A.K., Piontek, K., Tsiokas, L., Sukhatme, V.P., Guggino, W.B., Germino, G.G. (2000) Coassembly of polycystin-1 and -2 produces unique cation-permeable currents, *Nature* **408**, 990–994.

Hansson, J.H., Schild, L., Lu, Y., Wilson, T.A., Gautschi, I., Shimkets, R., Nelson-Williams, C., Rossier, B.C., Lifton, R.P. (1995) A de novo missense mutation of the beta subunit of the epithelial sodium channel causes hypertension and Liddle syndrome, identifying a proline-rich segment critical for regulation of channel activity, *Proc. Natl. Acad. Sci. U.S.A.* **92**, 11495–11499.

Hansson, J.H., Nelson-Williams, C., Suzuki, H., Schild, L., Shimkets, R., Lu, Y., Canessa, C., Iwasaki, T., Rossier, B., Lifton, R.P. (1995) Hypertension caused by a truncated epithelial sodium channel gamma subunit: genetic heterogeneity of Liddle syndrome, *Nat. Genet.* **11**, 76–82.

Hevers, W., Luddens, H. (1998) The diversity of GABAA receptors. Pharmacological and electrophysiological properties of GABAA channel subtypes, *Mol. Neurobiol.* **18**, 35–86.

Hodgkin, A.L., Huxley, A.F. (1952) The components of membrane conductance in the giant axon of Loligo, *J. Physiol.* **116**(4), 473–496.

Ishikawa, K., Tanaka, H., Saito, M., Ohkoshi, N., Fujita, T., Yoshizawa, K., Ikeuchi, T., Watanabe, M., Hayashi, A., Takiyama, Y., Nishizawa, M., Nakano, I., Matsubayashi, K., Miwa, M., Shoji, S., Kanazawa, I., Tsuji, S., Mizusawa, H. (1997) Japanese families with autosomal dominant pure cerebellar ataxia map to chromosome 19p13.1-p13.2 and are strongly associated with mild CAG expansions in the spinocerebellar ataxia type 6 gene in chromosome 19p13.1, *Am. J. Hum. Genet.* **61**, 336–346.

Jentsch, T.J. (2000) Neuronal KCNQ potassium channels: physiology and role in disease, *Nat. Rev. Neurosci.* **1**, 21–30.

Jentsch, T.J. (2002) Chloride channels are different, *Nature* **415**(6869), 276–277.

Jentsch, T.J., Stein, V., Weinreich, F., Zdebik, A.A. (2002) Molecular structure and physiological function of chloride channels, *Physiol. Rev.* **82**, 503–568.

Jiang, Y., Ruta, V., Chen, J., Lee, A., MacKinnon, R. (2003) The principle of gating charge movement in a voltage-dependent K$^+$ channel, *Nature* **423**(6935), 42–48.

Jiang, Y., Lee, A., Chen, J., Cadene, M., Chait, B.T., MacKinnon, R. (2002) The open pore conformation of potassium channels, *Nature* **417**(6888), 523–526.

Jodice, C., Mantuano, E., Veneziano, L., Trettel, F., Sabbadini, G., Calandriello, L., Francia, A., Spadaro, M., Pierelli, F., Salvi, F., Ophoff, R.A., Frants, R.R., Frontali, M. (1997) Episodic ataxia type 2 (EA2) and spinocerebellar ataxia type 6 (SCA6) due to CAG repeat expansion in the CACNA1A gene on chromosome 19p, *Hum. Mol. Genet.* **6**, 1973–1978.

Jouvenceau, A., Eunson, L.H., Spauschus, A., Ramesh, V., Zuberi, S.M., Kullmann, D.M., Hanna, M.G. (2001) Human epilepsy associated with dysfunction of the brain P/Q-type calcium channel, *Lancet* **358**, 801–807.

Jun, K., Piedras-Renteria, E.S., Smith, S.M., Wheeler, D.B., Lee, S.B., Lee, T.G., Chin, H., Adams, M.E., Scheller, R.H., Tsien, R.W., Shin, H.S. (1999) Ablation of P/Q-type Ca(2+) channel currents, altered synaptic transmission, and progressive ataxia in mice lacking the alpha(1A)-subunit, *Proc. Natl. Acad. Sci. U.S.A.* **96**, 15245–15250.

Kaunisto, M.A., Harno, H., Vanmolkot, K.R., Gargus, J.J., Sun, G., Hamalainen, E., Liukkonen, E., Kallela, M., van den Maagdenberg, A.M., Frants, R.R., Farkkila, M., Palotie, A., Wessman, M. (2004) A novel missense ATP1A2 mutation in a Finnish family with familial hemiplegic migraine type 2, *Neurogenetics* **5**(2), 141–146.

Keating, M., Atkinson, D., Dunn, C., Timothy, K., Vincent, G.M., Leppert, M. (1991) Linkage of a cardiac arrhythmia, the long QT syndrome, and the Harvey RAS-1 gene, *Science* **252**, 704–706.

Koefoed-Johnsen, V., Ussing, H.H. (1958) The nature of the frog skin potential, *Acta Physiol. Scand.* **42**(3–4), 298–308.

Kohl, S., Marx, T., Giddings, I., Jagle, H., Jacobson, S.G., Apfelstedt-Sylla, E., Zrenner, E., Sharpe, L.T., Wissinger, B. (1998) Total colourblindness is caused by mutations in the gene encoding the alpha-subunit of the cone photoreceptor cGMP-gated cation channel, *Nat. Genet.* **19**, 257–259.

Konrad, M., Weber, S. (2003) Recent advances in molecular genetics of hereditary magnesium-losing disorders, *J. Am. Soc. Nephrol.* **14**(1), 249–260.

Konrad, M., Schlingmann, K.P., Gudermann, T. (2004) Insights into the molecular nature of magnesium homeostasis, *Am. J. Physiol. Renal Physiol.* **286**(4), F599–F605.

Kornak, U., Kasper, D., Bosl, M.R., Kaiser, E., Schweizer, M., Schulz, A., Friedrich, W., Delling, G., Jentsch, T.J. (2001) Loss of the ClC-7 chloride channel leads to osteopetrosis in mice and man, *Cell* **104**, 205–215.

Kors, E.E., van den Maagdenberg, A.M., Plomp, J.J., Frants, R.R., Ferrari, M.D. (2002) Calcium channel mutations and migraine, *Curr. Opin. Neurol.* **15**, 311–316.

Kubisch, C., Schroeder, B.C., Friedrich, T., Lutjohann, B., El-Amraoui, A., Marlin, S., Petit, C., Jentsch, T.J. (1999) KCNQ4, a novel potassium channel expressed in sensory outer hair cells, is mutated in dominant deafness, *Cell* **96**, 437–446.

Langman, C.B. (2004) The molecular basis of kidney stones, *Curr. Opin. Pediatr.* **16**(2), 188–193.

Leppert, M., Anderson, V.E., Quattlebaum, T., Stauffer, D., O'Connell, P., Nakamura, Y., Lalouel, J.M., White, R. (1989) Benign familial neonatal convulsions linked to genetic markers on chromosome 20, *Nature* **337**, 647–648.

Lerche, C., Scherer, C.R., Seebohm, G., Derst, C., Wei, A.D., Busch, A.E., Steinmeyer, K. (2000) Molecular cloning and functional expression of KCNQ5, a potassium channel subunit that may contribute to neuronal M-current diversity, *J. Biol. Chem.* **275**, 22395–22400.

Letts, V.A., Felix, R., Biddlecome, G.H., Arikkath, J., Mahaffey, C.L., Valenzuela, A., Bartlett, F.S.I.I., Mori, Y., Campbell, K.P., Frankel, W.N. (1998) The mouse stargazer gene encodes a neuronal Ca(2+)-channel gamma subunit, *Nat. Genet.* **19**, 340–347.

Lewis, T.B., Leach, R.J., Ward, K., O'Connell, P., Ryan, S.G. (1993) Genetic heterogeneity in benign familial neonatal convulsions: identification of a new locus on chromosome 8q, *Am. J. Hum. Genet.* **53**, 670–675.

Lin, S.H., Cheng, N.L., Hsu, Y.J., Halperin, M.L. (2004) Intrafamilial phenotype variability in patients with Gitelman syndrome having the same mutations in their thiazide-sensitive sodium/chloride cotransporter, *Am. J. Kidney Dis.* **43**(2), 304–312.

Lipscombe, D., Pan, J.Q., Gray, A.C. (2002) Functional diversity in neuronal voltage-gated calcium channels by alternative splicing of Ca(v)alpha1, *Mol. Neurobiol.* **26**(1), 21–44.

Liu, B.A., Juurlink, D.N. (2004) Drugs and the QT interval – caveat doctor, *N. Engl. J. Med.* **351**(11), 1053–1056.

Lloyd, S.E., Pearce, S.H.S., Fisher, S.E., Steinmeyer, K., Schwappach, B., Schelnman, S.J., Harding, B., Bolino, A., Devoto, M., Goodyer, P., Rigden, S.P.A., Wrong, O., Jentsch, T.J., Craig, I.W., Thakker, R.V. (1996) A common molecular basis for three inherited kidney stone diseases, *Nature* **370**, 445–449.

Lossin, C., Wang, D.W., Rhodes, T.H., Vanoye, C.G., George, A.L. Jr. (2002) Molecular basis of an inherited epilepsy, *Neuron* **34**, 877–884.

Makita, N., Horie, M., Nakamura, T., Ai, T., Sasaki, K., Yokoi, H., Sakurai, M., Sakuma, I., Otani, H., Sawa, H., Kitabatake, A. (2002) Drug-induced long-QT syndrome associated with a subclinical SCN5A mutation, *Circulation* **106**, 1269–1274.

Marubio, L.M., del Mar Arroyo-Jimenez, M., Cordero-Erausquin, M., Lena, C., Le Novere, N., de Kerchove d'Exaerde, A., Huchet, M., Damaj, M.I., Changeux, J.-P. (1999) Reduced antinociception in mice lacking neuronal nicotinic receptor subunits, *Nature* **398**, 805–810.

McClatchey, A.I., McKenna-Yasek, D., Cros, D., Worthen, H.G., Kuncl, R.W., DeSilva, S.M., Cornblath, D.R., Gusella, J.F., Brown, R.H. Jr. (1992) Novel mutations in families with unusual and variable disorders of the skeletal muscle sodium channel, *Nat. Genet.* **2**, 148–152.

McGrath, J., Somlo, S., Makova, S., Tian, X., Brueckner, M. (2003) Two populations of node monocilia initiate left-right asymmetry in the mouse, *Cell* **114**(1), 61–73.

McRory, J.E., Hamid, J., Doering, C.J., Garcia, E., Parker, R., Hamming, K., Chen, L., Hildebrand, M., Beedle, A.M., Feldcamp, L., Zamponi, G.W., Snutch, T.P. (2004) The CACNA1F gene encodes an L-type calcium channel with unique biophysical properties and tissue distribution, *J. Neurosci.* **24**(7), 1707–1718.

Meij, I.C., Koenderink, J.B., van Bokhoven, H., Assink, K.F.H., Groenestege, W.T., de Pont, J.J.H.H.M., Bindels, R.J.M., Monnens, L.A.H., van den Heuvel, L.P.W.J., Knoers, N.V.A.M. (2000) Dominant isolated renal magnesium loss is caused by misrouting of the $Na^+,K(+)$-ATPase gamma-subunit, *Nat. Genet.* **26**, 265–266.

Meisler, M.H., Kearney, J., Ottman, R., Escayg, A. (2001) Identification of epilepsy genes in human and mouse, *Annu. Rev. Genet.* **35**, 567–588.

Meyer-Kleine, C., Steinmeyer, K., Ricker, K., Jentsch, T.J., Koch, M.C. (1995) Spectrum of mutations in the major human skeletal muscle chloride channel gene (CLCN1) leading to myotonia, *Am. J. Hum. Genet.* **57**, 1325–1334.

Miller, M.J., Rauer, H., Tomita, H., Rauer, H., Gargus, J.J., Gutman, G.A., Cahalan, M.D., Chandy, K.G. (2001) Nuclear localization and dominant-negative suppression by a mutant SKCa3 N-terminal channel fragment identified in a patient with schizophrenia, *J. Biol. Chem.* **276**, 27753–27756.

Mochizuki, T., Wu, G., Hayashi, T., Xenophontos, S.L., Veldhuisen, B., Saris, J.J., Reynolds, D.M., Cai, Y., Gabow, P.A., Pierides, A., Kimberling, W.J., Breuning, M.H., Constantinou Deltas, C., Peters, D.J.M., Somlo, S. (1996) PKD2, a gene for polycystic kidney disease that encodes an integral membrane protein, *Science* **272**, 1339–1342.

Mohler, P.J., Splawski, I., Napolitano, C., Bottelli, G., Sharpe, L., Timothy, K., Priori, S.G., Keating, M.T., Bennett, V. (2004) A cardiac arrhythmia syndrome caused by loss of ankyrin-B function, *Proc. Natl. Acad. Sci. U.S.A.* **101**(24), 9137–9142.

Mohler, P.J., Schott, J.J., Gramolini, A.O., Dilly, K.W., Guatimosim, S., duBell, W.H., Song, L.S., Haurogne, K., Kyndt, F., Ali, M.E., Rogers, T.B., Lederer, W.J., Escande, D., Le Marec, H., Bennett, V. (2003) Ankyrin-B mutation causes type 4 long-QT cardiac arrhythmia and sudden cardiac death, *Nature* **421**(6923), 634–639.

Monnier, N., Procaccio, V., Stieglitz, P., Lunardi, J. (1997) Malignant-hyperthermia susceptibility is associated with a mutation of the alpha-1-subunit of the human dihydropyridine-sensitive L-type voltage-dependent calcium channel receptor in skeletal muscle, *Am. J. Hum. Genet.* **60**, 1316–1325.

Morais-Cabral, J.H., Zhou, Y., MacKinnon, R. (2001) Energetic optimization of ion conduction rate by the K^+ selectivity filter, *Nature* **414**, 37–42.

Morimoto, T., Nagao, H., Yoshimatsu, M., Yoshida, K., Matsuda, H. (1993) Pathogenic role of glutamate in hyperthermia-induced seizures, *Epilepsia* **34**, 447–452.

Moslehi, R., Langlois, S., Yam, I., Friedman, J.M. (1998) Linkage of malignant hyperthermia and hyperkalemic periodic paralysis to the adult skeletal muscle sodium channel (SCN4A) gene in a large pedigree, *Am. J. Med. Genet.* **76**, 21–27.

Mulders, S.M., Bichet, D.G., Rijss, J.P.L., Kamsteeg, E.-J., Arthus, M.-F., Lonergan, M., Fujiwara, M., Morgan, K., Leijendekker, R., van der Sluijs, P., van Os, C.H., Deen, P.M.T. (1998) An aquaporin-2 water channel mutant which causes autosomal dominant nephrogenic diabetes insipidus is retained in the Golgi complex, *J. Clin. Invest.* **102**, 57–66.

Nielsen, S., Chou, C.-L., Marples, D., Christensen, E.I., Kishore, B.K., Knepper, M.A. (1995) Vasopressin increases water permeability of kidney collecting duct by inducing translocation of aquaporin-CD water channels to plasma membrane, *Proc. Natl. Acad. Sci. U.S.A.* **92**, 1013–1017.

Noda, M., Takahashi, H., Tanabe, T., Toyosato, M., Kikyotani, S., Furutani, Y., Hirose, T., Takashima, H., Inayama, S., Miyata, T., Numa, S. (1983) Structural homology of Torpedo californica acetylcholine receptor subunits, *Nature* **302**, 528–532.

Ohmori, I., Ouchida, M., Ohtsuka, Y., Oka, E., Shimizu, K. (2002) Significant correlation of the SCN1A mutations and severe myoclonic epilepsy in infancy, *Biochem. Biophys. Res. Commun.* **295**, 17–23.

Ophoff, R.A., Terwindt, G.M., Vergouwe, M.N., van Eijk, R., Oefner, P.J., Hoffman, S.M.G., Lamerdin, J.E., Mohrenweiser, H.W., Bulman, D.E., Ferrari, M., Haan, J., Lindhout, D., van Ommen, G.-J.B., Hofker, M.H., Ferrari, M.D., Frants, R.R. (1996) Familial hemiplegic migraine and episodic ataxia type-2 are caused by mutations in the Ca(2+) channel gene CACNL1A4, *Cell* **87**, 543–552.

Parkinson, N., Brown, S.D. (2002) Focusing on the genetics of hearing: you ain't heard nothin' yet, *Genome Biol.*. **3**(6), COMMENT2006.

Patsalos, P.N. (2000) Antiepileptic drug pharmacogenetics, *Ther. Drug Monit.* **22**(1), 127–130.

Phillips, H.A., Favre, I., Kirkpatrick, M., Zuberi, S.M., Goudie, D., Heron, S.E., Scheffer, I.E., Sutherland, G.R., Berkovic, S.F., Bertrand, D., Mulley, J.C. (2001) CHRNB2 is the second acetylcholine receptor subunit associated with autosomal dominant nocturnal frontal lobe epilepsy, *Am. J. Hum. Genet.* **68**, 225–231.

Piwon, N., Gunther, W., Schwake, M., Bosl, M.R., Jentsch, T.J. (2000) ClC-5 Cl(-)-channel disruption impairs endocytosis in a mouse model for Dent's disease, *Nature* **408**, 369–373.

Preisig-Muller, R., Schlichthorl, G., Goerge, T., Heinen, S., Bruggemann, A., Rajan, S., Derst, C., Veh, R.W., Daut, J. (2002) Heteromerization of Kir2.x potassium channels contributes to the phenotype of Andersen's syndrome, *Proc. Natl. Acad. Sci. U.S.A.* **99**, 7774–7779.

Priori, S.G., Napolitano, C., Schwartz, P.J. (1999) Low penetrance in the long-QT syndrome: clinical impact, *Circulation* **99**, 529–533.

Ptacek, L.J., George, A.L. Jr., Barchi, R.L., Griggs, R.C., Riggs, J.E., Robertson, M., Leppert, M.F. (1992) Mutations in an S4 segment of the adult skeletal muscle sodium channel cause paramyotonia congenita, *Neuron* **8**, 891–897.

Ptacek, L.J., George, A.L. Jr., Griggs, R.C., Tawil, R., Kallen, R.G., Barchi, R.L., Robertson,

M., Leppert, M.F. (1991) Identification of a mutation in the gene causing hyperkalemic periodic paralysis, *Cell* **67**, 1021–1027.

Ptacek, L.J., Tawil, R., Griggs, R.C., Engel, A.G., Layzer, R.B., Kwiecinski, H., McManis, P.G., Santiago, L., Moore, M., Fouad, G., Bradley, P., Leppert, M.F. (1994) Dihydropyridine receptor mutations cause hypokalemic periodic paralysis, *Cell* **77**, 863–868.

Pu, H.X., Scanzano, R., Blostein, R. (2000) Distinct regulatory effects of the Na,K-ATPase gamma subunit, *J. Biol. Chem.* **277**(23), 20270–20276.

Quane, K.A., Healy, J.M., Keating, K.E., Manning, B.M., Couch, F.J., Palmucci, L.M., Doriguzzi, C., Fagerlund, T.H., Berg, K., Ording, H., Bendixen, D., Mortier, W., Linz, U., Muller, C.R., McCarthy, T.V. (1993) Mutations in the ryanodine receptor gene in central core disease and malignant hyperthermia, *Nat. Genet.* **5**, 51–55.

Quinton, P.M., Bijman, J. (1983) Higher bioelectric potentials due to decreased chloride absorption in the sweat glands of patients with cystic fibrosis, *N. Engl. J. Med.* **308**(20), 1185–1189.

Ratjen, F., Doring, G. (2003) Cystic fibrosis, *Lancet* **361**(9358), 681–689.

Raychowdhury, M.K., Gonzalez-Perrett, S., Montalbetti, N., Timpanaro, G.A., Chasan, B., Goldmann, W.H., Stahl, S., Cooney, A., Goldin, E., Cantiello, H.F. (2004) Molecular pathophysiology of mucolipidosis type IV: pH dysregulation of the mucolipin-1 cation channel, *Hum. Mol. Genet.* **13**(6), 617–627.

Rees, M.I., Lewis, T.M., Kwok, J.B.J., Mortier, G.R., Govaert, P., Snell, R.G., Schofield, P.R., Owen, M.J. (2002) Hyperekplexia associated with compound heterozygote mutations in the beta-subunit of the human inhibitory glycine receptor (GLRB), *Hum. Mol. Genet.* **11**, 853–860.

Rho, J.M., Szot, P., Tempel, B.L., Schwartzkroin, P.A. (1999) Developmental seizure susceptibility of kv1.1 potassium channel knockout mice, *Dev. Neurosci.* **21**, 320–327.

Riordan, J.R., Rommens, J.M., Kerem, B., Alon, N., Rozmahel, R., Grzelczak, Z., Zielenski, J., Lok, S., Plavsic, N., Chou, J.L., et al. (1989) Identification of the cystic fibrosis gene: cloning and characterization of complementary DNA, *Science* **245**, 1066–1073.

Rosenthal, W., Seibold, A., Antaramian, A., Lonergan, M., Arthus, M.-F., Hendy, G.N., Birnbaumer, M., Bichet, D.G. (1992) Molecular identification of the gene responsible for congenital nephrogenic diabetes insipidus, *Nature* **359**, 233–235.

Roux, A.F., Pallares-Ruiz, N., Vielle, A., Faugere, V., Templin, C., Leprevost, D., Artieres, F., Lina, G., Molinari, N., Blanchet, P., Mondain, M., Claustres, M. (2004) Molecular epidemiology of DFNB1 deafness in France, *BMC Med. Genet.* **5**(1), 5.

Rowntree, R.K., Harris, A. (2003) The phenotypic consequences of CFTR mutations, *Ann. Hum. Genet.* **67**(Pt 5), 471–485.

Sanders, D.B. (2003) Lambert-eaton myasthenic syndrome: diagnosis and treatment, *Ann. N.Y. Acad. Sci.* **998**, 500–508.

Sanguinetti, M.C., Curran, M.E., Spector, P.S., Keating, M.T. (1996) Spectrum of HERG K^+-channel dysfunction in an inherited cardiac arrhythmia, *Proc. Natl. Acad. Sci. U.S.A.* **93**, 2208–2212.

Sansone, V., Griggs, R.C., Meola, G., Ptacek, L.J., Barohn, R., Iannaccone, S., Bryan, W., Baker, N., Janas, S.J., Scott, W., Ririe, D., Tawil, R. (1997) Andersen's syndrome: a distinct periodic paralysis, *Ann. Neurol.* **42**, 305–312.

Schlingmann, K.P., Konrad, M., Jeck, N., Waldegger, P., Reinalter, S.C., Holder, M., Seyberth, H.W., Waldegger, S. (2004) Salt wasting and deafness resulting from mutations in two chloride channels, *N. Engl. J. Med.* **350**(13), 1314–1319.

Schlingmann, K.P., Weber, S., Peters, M., Niemann Nejsum, L., Vitzthum, H., Klingel, K., Kratz, M., Haddad, E., Ristoff, E., Dinour, D., Syrrou, M., Nielsen, S., Sassen, M., Waldegger, S., Seyberth, H.W., Konrad, M. (2002) Hypomagnesemia with secondary hypocalcemia is caused by mutations in TRPM6, a new member of the TRPM gene family, *Nat. Genet.* **31**(2), 166–170.

Schultz, S.G., Curran, P.F. (1970) Coupled transport of sodium and organic solutes, *Physiol. Rev.* **50**(4), 637–718.

Schwartz, P.J., Priori, S.G., Dumaine, R., Napolitano, C., Antzelevitch, C., Stramba-Badiale, M., Richard, T.A., Berti, M.R., Bloise, R. (2000) A molecular link between the sudden infant death syndrome and the long-QT syndrome, *N. Engl. J. Med.* **343**, 262–267.

Schwartz, P.J., Priori, S.G., Bloise, R., Napolitano, C., Ronchetti, E., Piccinini, A., Goj, C.,

Breithardt, G., Schulze-Bahr, E., Wedekind, H., Nastoli, J. (2001) Molecular diagnosis in a child with sudden infant death syndrome, *Lancet* **358**, 1342–1343.

Schwartz, P.J., Stramba-Badiale, M., Segantini, A., Austoni, P., Bosi, G., Giorgetti, R., Grancini, F., Marni, E.D., Perticone, F., Rosti, D., Salice, P. (1998) Prolongation of the QT interval and the sudden infant death syndrome, *N. Engl. J. Med.* **338**, 1709–1714.

Seeburg, P.H., Single, F., Kuner, T., Higuchi, M., Sprengel, R. (2001) Genetic manipulation of key determinants of ion flow in glutamate receptor channels in the mouse, *Brain Res.* **907**(1–2), 233–243.

Segall, L., Scanzano, R., Kaunisto, M.A., Wessman, M., Palotie, A., Gargus, J.J., Blostein, R. (2004) Kinetic alterations due to a missense mutation in the Na,K-ATPase alpha 2 subunit cause familial hemiplegic migraine type 2, *J. Biol. Chem.* **279**(42), 43692–43696.

Sesti, F., Abbott, G.W., Wei, J., Murray, K.T., Saksena, S., Schwartz, P.J., Priori, S.G., Roden, D.M., George, A.L. Jr., Goldstein, S.A. (2000) A common polymorphism associated with antibiotic-induced cardiac arrhythmia, *Proc. Natl. Acad. Sci. U.S.A.* **97**, 10613–10618.

Shakkottai, V.G., Regaya, I., Wulff, H., Fajloun, Z., Tomita, H., Fathallah, M., Cahalan, M.D., Gargus, J.J., Sabatier, J.M., Chandy, K.G. (2001) Design and characterization of a highly selective peptide inhibitor of the small conductance calcium-activated K^+ channel, SkCa2, *J. Biol. Chem.* **276**(46), 43145–43151.

Shiang, R., Ryan, S.G., Zhu, Y.Z., Hahn, A.F., O'Connell, P., Wasmuth, J.J. (1993) Mutations in the alpha 1 subunit of the inhibitory glycine receptor cause the dominant neurologic disorder, hyperekplexia, *Nat. Genet.* **5**(4), 351–358.

Shieh, C.C., Coghlan, M., Sullivan, J.P., Gopalakrishnan, M. (2000) Potassium channels: molecular defects, diseases, and therapeutic opportunities, *Pharmacol. Rev.* **52**, 557–594.

Shimkets, R.A., Warnock, D.G., Bositis, C.M., Nelson-Williams, C., Hansson, J.H., Schambelan, M., Gill, J.R. Jr., Ulick, S., Milora, R.V., Findling, J.W., Canessa, C.M., Rossier, B.C., Lifton, R.P. (1994) Liddle's syndrome: heritable human hypertension caused by mutations in the beta subunit of the epithelial sodium channel, *Cell* **79**, 407–414.

Simon, D.B., Karet, F.E., Hamdan, J.M., Di Pietro, A., Sanjad, S.A., Lifton, R.P. (1996a) Bartter's syndrome, hypokalemic alkalosis with hypercalciuria, is caused by mutations in the Na-K-2Cl cotransporter NKCC2, *Nat. Genet.* **13**, 183–188.

Simon, D.B., Karet, F.E., Rodriguez-Soriano, J., Hamdan, J.H., DiPietro, A., Trachtman, H., Sanjad, S.A., Lifton, R.P. (1996b) Genetic heterogeneity of Bartter's syndrome revealed by mutations in the K^+ channel, ROMK, *Nat. Genet.* **14**, 152–156.

Simon, D.B., Nelson-Williams, C., Bia, M.J., Ellison, D., Karet, F.E., Molina, A.M., Vaara, I., Iwata, F., Cushner, H.M., Koolen, M., Gainza, F.J., Gitelman, H.J., Lifton, R.P. (1996c) Gitelman's variant of Bartter's syndrome, inherited hypokalaemic alkalosis, is caused by mutations in the thiazide-sensitive Na-Cl cotransporter, *Nat. Genet.* **12**, 24–30.

Simon, D.B., Lu, Y., Choate, K.A., Velazquez, H., Al-Sabban, E., Praga, M., Casari, G., Bettinelli, A., Colussi, G., Rodriguez-Soriano, J., McCredle, D., Milford, D., Sanjad, S., Lifton, R.P. (1999) Paracellin-1, a renal tight junction protein required for paracellular Mg(2+) resorption, *Science* **285**, 103–106.

Simon, D.B., Bindra, R.S., Mansfield, T.A., Nelson-Williams, C., Mendonca, E., Stone, R., Schurman, S., Nayir, A., Alpay, H., Bakkaloglu, A., Rodriguez-Soriano, J., Morales, J.M., Sanjad, S.A., Taylor, C.M., Pilz, D., Brem, A., Trachtman, H., Griswold, W., Richard, G.A., John, E., Lifton, R.P. (1997) Mutations in the chloride channel gene, CLCNKB, cause Bartter's syndrome type III, *Nat. Genet.* **17**, 171–178.

Singh, N.A., Charlier, C., Stauffer, D., DuPont, B.R., Leach, R.J., Melis, R., Ronen, G.M., Bjerre, I., Quattlebaum, T., Murphy, J.V., McHarg, M.L., Gagnon, D., Rosales, T.O., Peiffer, A., Anderson, V.E., Leppert, M. (1998) A novel potassium channel gene, KCNQ2, is mutated in an inherited epilepsy of newborns, *Nat. Genet.* **18**, 25–29.

Smart, S.L., Lopantsev, V., Zhang, C.L., Robbins, C.A., Wang, H., Chiu, S.Y., Schwartzkroin, P.A., Messing, A., Tempel, B.L. (1998) Deletion of the K(V)1.1 potassium channel causes epilepsy in mice, *Neuron* **20**, 809–819.

Splawski, I., Timothy, K.W., Vincent, G.M., Atkinson, D.L., Keating, M.T. (1997a) Molecular basis of the long-QT syndrome associated with deafness, *New. Eng. J. Med.* **336**, 1562–1567.

Splawski, I., Tristani-Firouzi, M., Lehmann, M.H., Sanguinetti, M.C., Keating, M.T. (1997b) Mutations in the hminK gene cause long QT syndrome and suppress IKs function, *Nat. Genet.* **17**, 338–340.

Steinlein, O.K., Mulley, J.C., Propping, P., Wallace, R.H., Phillips, H.A., Sutherland, G.R., Scheffer, I.E., Berkovic, S.F. (1995) A missense mutation in the neuronal nicotinic acetylcholine receptor alpha-4 subunit is associated with autosomal dominant nocturnal frontal lobe epilepsy, *Nat. Genet.* **11**, 201–203.

Stoffel, M., Jan, L.Y. (1998) Epilepsy genes: excitement traced to potassium channels, *Nat. Genet.* **18**, 6–8.

Strom, T.M., Nyakatura, G., Apfelstedt-Sylla, E., Hellebrand, H., Lorenz, B., Weber, B.H.F., Wutz, K., Gutwillinger, N., Ruther, K., Drescher, B., Sauer, C., Zrenner, E., Meitinger, T., Rosenthal, A., Meindl, A. (1998) An L-type calcium-channel gene mutated in incomplete X-linked congenital stationary night blindness, *Nat. Genet.* **19**, 260–263.

Sugawara, T., Mazaki-Miyazaki, E., Fukushima, K., Shimomura, J., Fujiwara, T., Hamano, S., Inoue, Y., Yamakawa, K. (2002) Frequent mutations of SCN1A in severe myoclonic epilepsy in infancy, *Neurology* **58**, 1122–1124.

Sugawara, T., Tsurubuchi, Y., Agarwala, K.L., Ito, M., Fukuma, G., Mazaki-Miyazaki, E., Nagafuji, H., Noda, M., Imoto, K., Wada, K., Mitsudome, A., Kaneko, S., Montal, M., Nagata, K., Hirose, S., Yamakawa, K. (2001) A missense mutation of the Na$^+$ channel alpha II subunit gene Na(v)1.2 in a patient with febrile and afebrile seizures causes channel dysfunction, *Proc. Natl. Acad. Sci. U.S.A.* **98**, 6384–6389.

Sundin, O.H., Yang, J.M., Li, Y., Zhu, D., Hurd, J.N., Mitchell, T.N., Silva, E.D., Maumenee, I.H. (2000) Genetic basis of total colourblindness among the Pingelapese islanders, *Nat. Genet.* **25**, 289–293.

Tanabe, T., Beam, K.G., Powell, J.A., Numa, S. (1988) Restoration of excitation-contraction coupling and slow calcium current in dysgenic muscle by dihydropyridine receptor complementary DNA, *Nature* **336**, 134–139.

Tempel, B.L., Papazian, D.M., Schwarz, T.L., Jan, Y.N., Jan, L.Y. (1987) Sequence of a probable potassium channel component encoded at Shaker locus of Drosophila, *Science* **237**, 770–775.

Thomas, P., Ye, Y., Lightner, E. (1996) Mutation of the pancreatic islet inward rectifier Kir6.2 also leads to familial persistent hyperinsulinemic hypoglycemia of infancy, *Hum. Mol. Genet.* **5**, 1809–1812.

Thomas, P.M., Cote, G.J., Wohllk, N., Haddad, B., Mathew, P.M., Rabl, W., Aguilar-Bryan, L., Gagel, R.F., Bryan, J. (1995) Mutations in the sulfonylurea receptor gene in familial persistent hyperinsulinemic hypoglycemia of infancy, *Science* **268**, 426–429.

Tomita, H., Shakkittai, V.G., Gutman, G.A., Sun, G., Bunney, W.E., Cahalan, M.D., Chandy, K.G., Gargus, J.J. (2003) Novel truncated isoform of SK3 potassium channel is a potent dominant-negative regulator of SK currents: implications in schizophrenia, *Mol. Psychiatry.* **8**(5), 524–535.

Towbin, J.A., Vatta, M. (2001) Molecular biology and the prolonged QT syndromes, *Am. J. Med.* **110**, 385–398.

Tristani-Firouzi, M., Jensen, J.L., Donaldson, M.R., Sansone, V., Meola, G., Hahn, A., Bendahhou, S., Kwiecinski, H., Fidzianska, A., Plaster, N., Fu, Y.H., Ptacek, L.J., Tawil, R. (2002) Functional and clinical characterization of KCNJ2 mutations associated with LQT7 (Andersen syndrome), *J. Clin. Invest.* **110**, 381–388.

Turk, E., Zabel, B., Mundlos, S., Dyer, J., Wright, E.M. (1991) Glucose/galactose malabsorption caused by a defect in the Na(+)/glucose cotransporter, *Nature* **350**, 354–356.

van den Heuvel, L.P., Assink, K., Willemsen, M., Monnens, L. (2002) Autosomal recessive renal glucosuria attributable to a mutation in the sodium glucose cotransporter (SGLT2), *Hum. Genet.* **111**, 544–547.

Van Den Maagdenberg, A.M., Kors, E.E., Brunt, E.R., Van Paesschen, W., Pascual, J., Ravine, D., Keeling, S., Vanmolkot, K.R., Vermeulen, F.L., Terwindt, G.M., Haan, J., Frants, R.R., Ferrari, M.D. (2002) Episodic ataxia type 2 Three novel truncating mutations and one novel missense mutation in the CACNA1A gene, *J. Neurol.* **249**, 1515–1519.

Venter, J.C., Adams, M.D., Myers, E.W., Li, P.W., Mural, R.J., Sutton, G.G., Smith, H.O., et al. (2001) The sequence of the human genome, *Science* **291**, 1304–1351.

Walder, R.Y., Landau, D., Meyer, P., Shalev, H., Tsolia, M., Borochowitz, Z., Boettger, M.B., Beck, G.E., Englehardt, R.K., Carmi, R., Sheffield, V.C. (2002) Mutation of TRPM6 causes familial hypomagnesemia with secondary hypocalcemia, *Nat. Genet.* **31**(2), 171–174.

Wallace, R.H., Marini, C., Petrou, S., Harkin, L.A., Bowser, D.N., Panchal, R.G., Williams, D.A., Sutherland, G.R., Mulley, J.C., Scheffer, I.E., Berkovic, S.F. (2001a) Mutant GABA(A) receptor gamma2-subunit in childhood absence epilepsy and febrile seizures, *Nat. Genet.* **28**, 49–52.

Wallace, R.H., Wang, D.W., Singh, R., Scheffer, I.E., George, A.L. Jr., Phillips, H.A., Saar, K., Reis, A., Johnson, E.W., Sutherland, G.R., Berkovic, S.F., Mulley, J.C. (1998) Febrile seizures and generalized epilepsy associated with a mutation in the Na(+)-channel beta-1 subunit gene SCN1B, *Nat. Genet.* **19**, 366–370.

Wallace, R.H., Scheffer, I.E., Barnett, S., Richards, M., Dibbens, L., Desai, R.R., Lerman-Sagie, T., Lev, D., Mazarib, A., Brand, N., Ben-Zeev, B., Goikhman, I., Singh, R., Kremmidiotis, G., Gardner, A., Sutherland, G.R., George, A.L. Jr., Mulley, J.C., Berkovic, S.F. (2001b) Neuronal sodium-channel alpha-1-subunit mutations in generalized epilepsy with febrile seizures plus, *Am. J. Hum. Genet.* **68**, 859–865.

Wang, H.-S., Pan, Z., Shi, W., Brown, B.S., Wymore, R.S., Cohen, I.S., Dixon, J.E., McKinnon, D. (1998) KCNQ2 and KCNQ3 potassium channel subunits: molecular correlates of the M-channel, *Science* **282**, 1890–1893.

Wang, Q., Shen, J., Splawski, I., Atkinson, D., Li, Z., Robinson, J.L., Moss, A.J., Towbin, J.A., Keating, M.T. (1995) SCN5A mutations associated with an inherited cardiac arrhythmia, long QT syndrome, *Cell* **80**, 805–811.

Wang, Q., Curren, M.E., Splawski, I., Burn, T.C., Millholland, J.M., VanRaay, T.J., Shen, J., Timothy, K.W., Vincent, G.M., de Jager, T., Schwartz, P.J., Towbin, J.A., Moss, A.J., Atkinson, D.L., Landes, G.M., Connors, T.D., Keating, M.T. (1996) Positional cloning of a novel potassium channel gene: KVLQT1 mutations cause cardiac arrhythmias, *Nat. Genet.* **12**, 17–23.

Warmke, J.W., Ganetzky, B. (1994) A family of potassium channel genes related to eag in Drosophila and mammals, *Proc. Natl. Acad. Sci. U.S.A.* **91**, 3438–3442.

Watanabe, H., Nagata, E., Kosakai, A., Nakamura, M., Yokoyama, M., Tanaka, K., Sasai, H. (2000) Disruption of the epilepsy KCNQ2 gene results in neural hyperexcitability, *J. Neurochem.* **75**(1), 28–33.

Watnick, T.J., Torres, V.E., Gandolph, M.A., Qian, F., Onuchic, L.F., Klinger, K.W., Landes, G., Germino, G.G. (1998) Somatic mutation in individual liver cysts supports a two-hit model of cystogenesis in autosomal dominant polycystic kidney disease, *Mol. Cell* **2**, 247–251.

Wilson, P.D. (2004) Polycystic kidney disease, *N. Engl. J. Med.* **350**(2), 151–164.

Wickenden, A.D., Yu, W., Zou, A., Jegla, T., Wagoner, P.K. (2000) Retigabine, a novel anticonvulsant, enhances activation of KCNQ2/Q3 potassium channels, *Mol. Pharmacol.* **58**, 591–600.

Wollnik, B., Schroeder, B.C., Kubisch, C., Esperer, H.D., Wieacker, P., Jentsch, T.J. (1997) Pathophysiological mechanisms of dominant and recessive KVLQT1 K^+ channel mutations found in inherited cardiac arrhythmias, *Hum. Mol. Genet.* **6**, 1943–1949.

Wright, E.M., Turk, E. (2004) The sodium/glucose cotransport family SLC5, *Pflugers Arch.* **447**(5), 510–518.

Yang, P., Kanki, H., Drolet, B., Yang, T., Wei, J., Viswanathan, P.C., Hohnloser, S.H., Shimizu, W., Schwartz, P.J., Stanton, M., Murray, K.T., Norris, K., George, A.L. Jr., Roden, D.M. (2002) Allelic variants in long-QT disease genes in patients with drug-associated torsades de pointes, *Circulation* **105**, 1943–1948.

Yue, Q., Jen, J.C., Nelson, S.F., Baloh, R.W. (1997) Progressive ataxia due to a missense mutation in a calcium-channel gene, *Am. J. Hum. Genet.* **61**, 1078–1087.

Zhuchenko, O., Bailey, J., Bonnen, P., Ashizawa, T., Stockton, D.W., Amos, C., Dobyns, W.B., Subramony, S.H., Zoghbi, H.Y., Lee, C.C. (1997) Autosomal dominant cerebellar ataxia (SCA6) associated with small polyglutamine expansions in the alpha(1A)-voltage-dependent calcium channel, *Nat. Genet.* **15**, 62–69.

Zuberi, S.M., Eunson, L.H., Spauschus, A., De Silva, R., Tolmie, J., Wood, N.W., McWilliam, R.C., Stephenson, J.P., Kullmann, D.M., Hanna, M.G. (1999) A novel mutation in the human voltage-gated potassium channel gene (Kv1.1) associates with episodic ataxia type 1 and sometimes with partial epilepsy, *Brain* **122**, 817–825.

Zwingman, T.A., Neumann, P.E., Noebels, J.L., Herrup, K. (2001) Rocker is a new variant of the voltage-dependent calcium channel gene Cacna1a, *J. Neurosci.* **21**, 1169–1178.

Receptors, Vitamin: *see* Vitamin Receptors

Glossary of Basic Terms

The most basic terms in molecular cell biology are defined below. These, in combination with the key words listed at the head of each article, provide definitions of all essential terms in this Encyclopedia.

Alleles
Alternative forms of a given gene, inherited separately from each parent, differing in nucleotide base sequence and located in a specific position on each homologous chromosome, affecting the functioning of a single product (RNA and/or protein).

Amino Acid
An organic compound containing at least one amino group and one carboxyl group. In the 20 different amino acids that compose proteins, an amino group and carboxyl group are linked to a central carbon atom, the α-carbon, to which a variable side chain is bound (see pages at the back of each volume).

Amplification
The process of replication of specific DNA sequences in disproportionately greater amounts than are present in the parent genetic material, for example, PCR is an *in vitro* amplification technique.

Apoptosis
Regulated process leading to nonpathological animal cell death via a series of well-defined morphological changes; also called *programmed cell death*.

Bacteriophage (phage)
Any virus (containing DNA or RNA) that infects bacterial cells. Some bacteriophages are widely used as cloning vectors.

Base Pair
Association of two complementary nucleotides in a DNA or RNA molecule stabilized by hydrogen bonding between their base components. Adenine pairs with thymine or uracil (A–T; A–U) and guanine pairs with cytosine (G–C) (see pages at the back of each volume).

Bioinformatics
Computational approaches to answer biological questions and enhance the ability of researchers to manipulate, collect, and analyze data more quickly and in new ways. Experts predict that more biologists will do their work *in silico*, using the computer to synthesize, analyze, and interpret the many terabytes of data now being generated.

cDNA (complementary DNA)
A DNA copy of an RNA molecule synthesized from an mRNA template *in vitro* using an enzyme called *reverse transcriptase*; often used as a probe.

Cell Cycle
Ordered sequence of events in which a cell duplicates its chromosomes and divides itself into two. Most eukaryotic cell cycles can be commonly divided into four phases: G_1 (G1) period after mitosis but before DNA synthesis occurs; S-phase when most DNA replication occurs; G_2 (G2) phase period of cell cycle when cells contain twice the G1 complement of DNA; and M-phase when cell division occurs, yielding two daughter cells (mitosis) each with one complete genome.

Cell Differentiation
Progressive restriction of the developmental potential and increasing specialization of function that takes place during the development of the embryo and leads to the formation of specialized cells, tissues, and organs.

Cell Division
Separation of a cell into two daughter cells. In higher eukaryotes, it involves division of the nucleus (mitosis) and of the cytoplasm (cytokinesis); mitosis often is used to refer to both nuclear and cytoplasmic division.

Cell Line
A defined unique population of cells obtained by culture from a primary implant through numerous generations.

Chromatin
The complex of nucleic acids (DNA and RNA) and proteins (histones) comprising eukaryotic chromosomes.

Chromosome
In prokaryotes, the usually circular duplex DNA molecule constituting the genome; in eukaryotes, a threadlike structure consisting of chromatin and carrying genomic information on a DNA double helix molecule. A viral chromosome may be composed of DNA or RNA.

Cloning
Asexual reproduction of cells, organisms, genes, or segments of DNA identical to the original.

Cloning Vector *see* Vector

Codon
Sequence of three nucleotides in DNA or mRNA that specifies a particular amino acid during protein synthesis; also called *triplet*. Of the 64 possible codons, three are stop codons, which do not specify amino acids (see pages at the back of each volume).

Complementary Base Pairing
Nucleic acid sequences on paired polymers with opposing hydrogen-bonded bases adenine (designated A) bonded to thymine (T), guanine (G) to cytosine (C) in DNA and adenine to uracil (U) replacing adenine to thymine in RNA (see pages at the back of each volume).

Complementary DNA see cDNA

Dalton
Unit of molecular mass approximately equal to the mass of a hydrogen atom (1.66×10^{-24} g).

Deoxyribonucleic Acid *see* DNA

Diploid
The number of chromosomes in most cells except the gametes. In humans, the diploid number is 46.

DNA (Deoxyribonucleic Acid)
The molecular basis of the genetic code consisting of a poly-sugar phosphate backbone from which thymine, adenine, guanine, and cytosine bases project. Usually found as two complementary chains (duplex) forming a double helix associated by hydrogen bonds between complementary bases.

DNA Cloning (Gene Cloning)
Recombinant DNA technique in which specific cDNAs or fragments of genomic DNA are inserted into a cloning vector, which then is incorporated into cultured host cells (e.g., *E. coli* cells) and maintained during growth of the host cells.

DNA Library
Collection of cloned DNA molecules consisting of fragments of the entire genome (genomic library) or of DNA copies of all the mRNAs produced by a cell type (cDNA library) inserted into a suitable cloning vector.

DNA Polymerase
Enzymes that catalyze the replication of DNA from the deoxyribonucleotide triphosphates using single- or double-stranded DNA as a template.

DNA Transcription *see* Transcription

E. coli (Escherichia coli)
A colon bacillus, which is the most studied of all forms of life.

Embryonic Stem Cells (ES)
Cultured cells derived from the pluripotent inner cell mass of blastocyst-stage embryos.

Epigenetics
Mechanisms of storing and transmitting cellular information additional to those based on DNA sequences.

Escherichia coli see E. coli

Eukaryotes
Organisms whose cells have their genetic material packed in a membrane-surrounded, structurally discrete nucleus and with well-developed cell organelles. Eukaryotes include all organisms except *archaebacteria* and *eubacteria*.

Expression
The process of making the product of a gene, which is either a specific protein giving rise to a specific trait or RNA forms not translated into proteins (e.g. transfer ribosomal RNAs).

Functional Genomics
A discipline that aims to understand how genes are regulated and what they do, largely through massive parallel studies of gene expression over time and in a variety of tissues.

Gamete
Specialized haploid cell (in animals either a sperm or an egg) produced by meiosis of germ cells; in sexual reproduction, the union of a sperm and an egg initiates the development of a new diploid individual.

Gene Cloning *see* DNA Cloning

Gene
A DNA sequence, located in a particular position on a particular chromosome,

which encodes a specific protein or RNA molecule.

Genomics
Comparative analysis of the complete genomic sequences from different organisms; used to assess evolutionary relations between species and to predict the number and general types of proteins produced by an organism.

Genotype
Entire genetic constitution of an individual cell or organism; also, the alleles at one or more specific loci.

Haploid
The number of chromosomes in a sperm or egg cell, half the diploid number.

Heterozygous
Having two different alleles for a given trait in the homologous chromosomes.

Homologies
Similarities in DNA or protein sequences between individuals of the same species or among different species.

Homologous Chromosomes
Chromosome pairs, each derived from one parent, containing the same linear sequence of genes, and as a consequence, each gene is present in duplicate (e.g., humans have 23 homologous chromosome pairs, but the toad has 11 pairs, the mosquito has three pairs, and so on).

Homozygous
Having two identical alleles for a given trait in the homologous chromosomes.

Hybridization
The formation of a double-stranded polynucleotide molecule when two complementary strands are brought together at moderate temperature. The strands can be DNA or RNA or one of each; a technique for assessing the extent of sequence homology between single strands of nucleic acids.

Ligation
The formation of a phosphodiester bond to join adjacent terminal nucleotides (nicks) to form a longer nucleic acid chain (DNA of RNA); catalyzed by ligase.

Marker
A gene or a restriction enzyme cutting site with a known location on a chromosome and a clear-cut phenotype (expression), or pattern of inheritance, used as a point of reference when mapping a new mutant.

Meiosis
In eukaryotes, a special type of cell division that occurs during maturation of germ cells; comprises two successive nuclear and cellular divisions, with only one round of DNA replication resulting in production of four genetically nonequivalent haploid cells (gametes) from an initial diploid cell.

Messenger RNA *see* **mRNA**

Mitosis
In eukaryotic cells, the process whereby the nucleus is divided, involving condensation of the DNA into visible chromosomes, to produce two genetically equivalent daughter nuclei with the diploid number of chromosomes.

mRNA (messenger RNA)
RNA used to translate information from DNA to ribosome where the information is used to make one or several proteins.

Mutation
The heritable change in the nucleotide sequence of a chromosome.

Nucleotide
The monomer which, when polymerized, forms DNA or RNA. It is composed of a nitrogenous base bonded to a sugar (ribose or deoxyribose), bonded to a phosphate.

Oligonucleotide
A polynucleotide 2 to 20 nucleotide units in length.

Operon
A series of prokaryote genes encoding enzymes of a specific biosynthesis pathway and transcribed into a single RNA molecule.

Organelle
Any membrane-limited structure found in the cytoplasm of eukaryotic cells.

Phage *see* Bacteriophage

Phenotype
The observable characteristics of a cell or organism as distinct from it's genotype.

Plasmid
An extrachromosomal circular DNA molecule found in a variety of bacteria encoding "dispensable functions," such as resistance to antibiotics. Often found in multiple copies per cell and reproduces every time the bacterial cell reproduces. May be used as a cloning vector.

Polymorphism
Difference in DNA sequence among individuals expressed as different forms of a protein in individuals of the same interbreeding population.

Polynucleotide
The polymer formed by condensation of nucleotides.

Probe
A radioactively fluorescent or immunologically labeled oligonucleotide (RNA or DNA) used to detect complementary sequences in a hybridization experiment, for example, identify bacterial colonies that contain cloned genes or detect specific nucleic acids following separation by gel electrophoresis.

Procaryotes (Prokaryotes)
Typically unicellular or filamentous with DNA not located within a nuclear envelope. Prokaryotes include archaebacteria, eubacteria, cyanobacteria, prochlorophytes and mycoplasmas.

Programmed Cell Death *see* Apoptosis

Prokaryotes *see* Procaryotes

Protein
A linear polymer of amino acids linked together in a specific sequence and usually containing more than 50 residues. Proteins form the key structure elements in cells and participate in nearly all cellular activities.

Proteomics
A discipline that promises to determine the identity, function, and structure of each protein in an organelle or cell and to elucidate protein–protein interactions.

Replication
The copying of a DNA molecule duplex yielding two new DNA duplex molecules, each with one strand from the original DNA duplex. Single-stranded DNA

replication results in a single-stranded DNA molecule.

Repressor
A protein that binds to a specific location (operator) on DNA and prevents RNA transcription from a specific gene or operon.

Restriction Fragment Length Polymorphism *see* RFLP

Restriction Mapping
Uses restriction endonuclease enzymes to produce specific cuts (cleavage) in DNA, allowing preparation of a genome map describing the order and distance between cleavage sites.

Reverse Transcription
The synthesis of cDNA from an RNA template as catalyzed by reverse transcriptase.

RFLP (Restriction Fragment Length Polymorphism)
DNA fragment cut by enzymes specific to a base sequence (restriction endonuclease) generating a DNA fragment whose size varies from one individual to another. Used as markers on genome maps and for screening for mutations and genetic diseases.

Ribonucleic Acid *see* RNA

Ribosomes
Small cellular components composed of proteins plus ribosomal RNA that translate the genetic code into synthesis of specific proteins.

RNA (Ribonucleic Acid)
A single-stranded polynucleotide with a phosphate oxyribose backbone and four bases that are identical to those in DNA, with the exception that the base uracil is substituted for thymine.

RNA Interference (RNAi)
Intracellular degradation of RNA that removes foreign RNAs such as those from viruses. These fragments (small, micro, or mini RNA) cleaved from free double-stranded RNA (dsRNA) direct the degradative mechanism to other similar RNA sequences. Used as a technique to silence the expression of targeted genes in a sequence-dependent mode.

RNA Polymerase
The enzyme (peptide) that binds at specific nucleotide sequences, called promoters, in front of genes in DNA, which catalyze transcription of DNA to RNA.

RNA Translation *see* Translation

Stem Cell
A self-renewing cell that divides to give rise to a cell with an identical developmental potential and/or one with a more restricted developmental potential.

Structural Biology
The discovery, analysis and dissemination of three-dimensional structures of protein, DNA, RNA, and other biological macromolecules representing the entire range of structural diversity found in nature.

Transcription (DNA transcription)
Synthesis of an RNA molecule from a DNA template (gene) catalyzed by RNA polymerase.

Transfer RNA *see* tRNA

Translation (RNA translation)
The process on a ribosome by which the sequence of nucleotides in a mRNA

molecule directs the incorporation of amino acids into protein.

tRNA (transfer RNA)
RNA molecules that transport specific amino acids to ribosomes into position in the correct order during protein synthesis.

Vector
A DNA molecule originating from a virus, a plasmid, or a cell of a higher organism into which another DNA fragment can be integrated without loss of the vector's capacity for self-replication. Vectors introduce foreign DNA into host cells where it can be reproduced in large quantities.

Virus
A small parasite consisting of nucleic acid (RNA or DNA) enclosed in a protein coat that can replicate only in a susceptible host cell; widely used in cell biology research.

Wild type
Normal, nonmutant form of a macromolecule, cell or organism.

Zygote
A fertilized egg; a diploid cell resulting from fusion of a male and female gamete.

The Twenty Amino Acids that are Combined to Form Proteins in Living Things

Amino acids with nonpolar side chains

Glycine
Gly
G

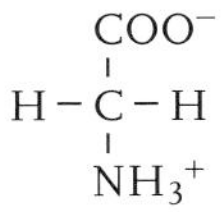

$$\text{H}-\underset{\underset{\text{NH}_3^+}{|}}{\overset{\overset{\text{COO}^-}{|}}{\text{C}}}-\text{H}$$

Alanine
Ala
A

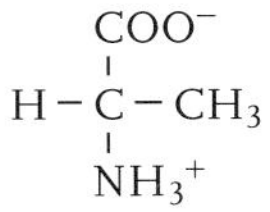

$$\text{H}-\underset{\underset{\text{NH}_3^+}{|}}{\overset{\overset{\text{COO}^-}{|}}{\text{C}}}-\text{CH}_3$$

Valine
Val
V

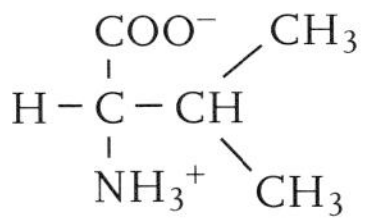

$$\text{H}-\underset{\underset{\text{NH}_3^+}{|}}{\overset{\overset{\text{COO}^-}{|}}{\text{C}}}-\text{CH}\underset{\diagdown\text{CH}_3}{\diagup\text{CH}_3}$$

Leucine
Leu
L

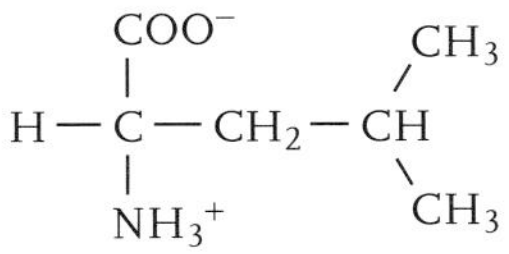

$$\text{H}-\underset{\underset{\text{NH}_3^+}{|}}{\overset{\overset{\text{COO}^-}{|}}{\text{C}}}-\text{CH}_2-\text{CH}\underset{\diagdown\text{CH}_3}{\diagup\text{CH}_3}$$

Isoleucine
Ile
I

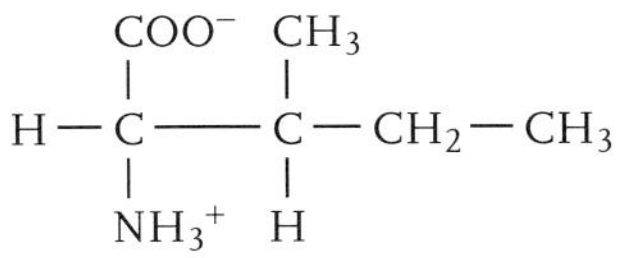

$$\text{H}-\underset{\underset{\text{NH}_3^+}{|}}{\overset{\overset{\text{COO}^-}{|}}{\text{C}}}-\underset{\underset{\text{H}}{|}}{\overset{\overset{\text{CH}_3}{|}}{\text{C}}}-\text{CH}_2-\text{CH}_3$$

Methionine
Met
M

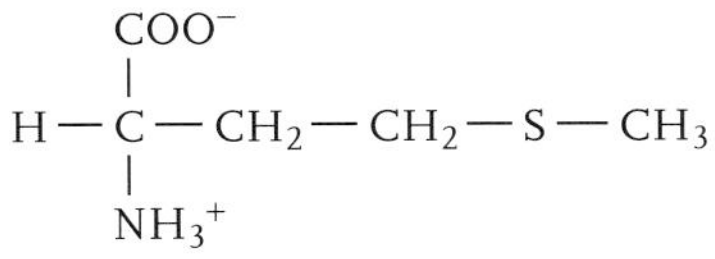

$$\text{H}-\underset{\underset{\text{NH}_3^+}{|}}{\overset{\overset{\text{COO}^-}{|}}{\text{C}}}-\text{CH}_2-\text{CH}_2-\text{S}-\text{CH}_3$$

Proline
Pro
P

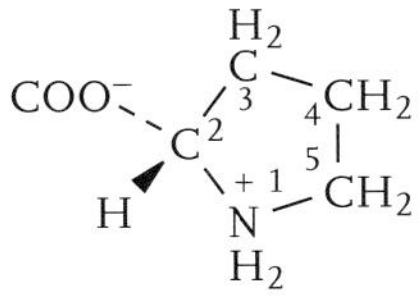

$$\text{COO}^- \diagdown \underset{H \diagup}{C^2} \underset{\underset{H_2}{|}}{\overset{\overset{H_2}{C^3}}{\underset{\displaystyle N^{+1}}{\diagdown}}} \underset{\displaystyle CH_2}{\overset{\displaystyle CH_2}{\underset{5}{|}}}_{4}$$

The Twenty Amino Acids that are Combined to Form Proteins in Living Things

Amino acids with nonpolar side chains (continued)

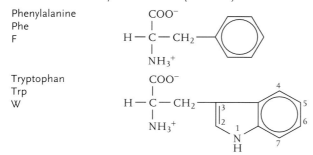

Phenylalanine
Phe
F

Tryptophan
Trp
W

Amino acids with uncharged polar side chains

Serine
Ser
S

Threonine
Thr
T

Asparagine
Asn
N

Glutamine
Gln
Q

Tyrosine
Tyr
Y

Cysteine
Cys
C

Amino acids with charged polar side chains

Lysine
Lys
K

Amino acids with charged polar side chains (continued)

Arginine
Arg
R

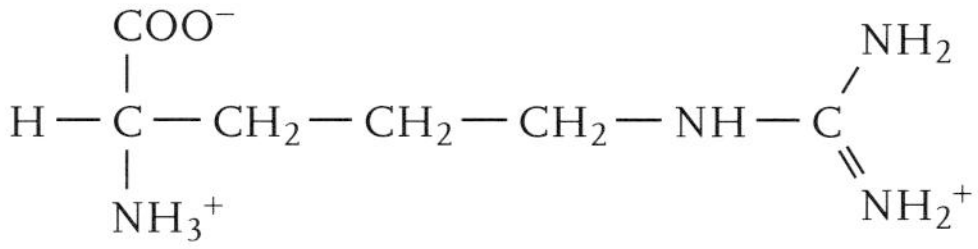

$$\text{H}-\underset{\underset{\text{NH}_3^+}{|}}{\overset{\overset{\text{COO}^-}{|}}{\text{C}}}-\text{CH}_2-\text{CH}_2-\text{CH}_2-\text{NH}-\text{C}\underset{\text{NH}_2^+}{\overset{\text{NH}_2}{\diagup}}$$

Histidine
His
H

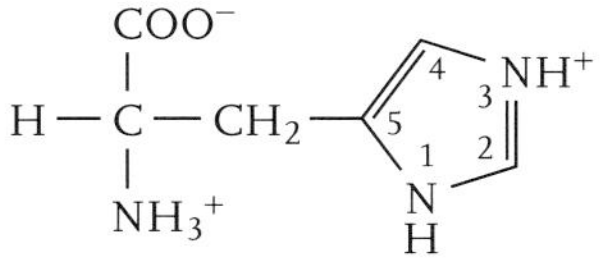

$$\text{H}-\underset{\underset{\text{NH}_3^+}{|}}{\overset{\overset{\text{COO}^-}{|}}{\text{C}}}-\text{CH}_2-\underset{\underset{\text{H}}{\text{N}}}{\overset{4\quad 3}{\underset{1\quad 2}{5}}}\text{NH}^+$$

Aspartic acid
Asp
D

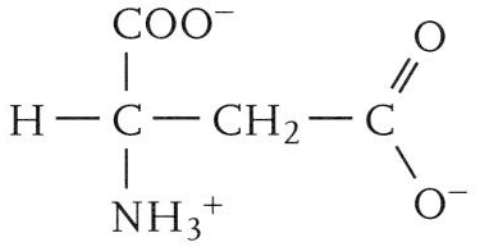

$$\text{H}-\underset{\underset{\text{NH}_3^+}{|}}{\overset{\overset{\text{COO}^-}{|}}{\text{C}}}-\text{CH}_2-\text{C}\underset{\text{O}^-}{\overset{\text{O}}{\diagup\!\!\!\diagup}}$$

Glutamic acid
Glu
E

$$\text{H}-\underset{\underset{\text{NH}_3^+}{|}}{\overset{\overset{\text{COO}^-}{|}}{\text{C}}}-\text{CH}_2-\text{CH}_2-\text{C}\underset{\text{O}^-}{\overset{\text{O}}{\diagup\!\!\!\diagup}}$$

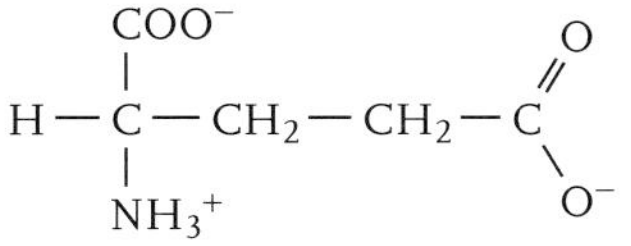

(Figures with kind permission from Voet, D., Voet, J.G., Pratt, C.W. (2001) *Fundamentals of Biochemistry*, Wiley, New York)

The Twenty Amino Acids with Abbreviations and Messenger RNA Code Designations

Amino acid	One letter symbol	Three letter symbol	mRNA code designation
alanine	A	ala	GCU, GCC, GCA, GCG
arginine	R	arg	CGU, CGC, CGA, CGG, AGA, AGG
asparagine	P	asn	AAU, AAC
aspartic acid	D	asp	GAU, GAC
cysteine	C	cys	UGU, UGC
glutamic acid	E	glu	GAA, GAG
glutamine	Q	gln	CAA, CAG
glycine	G	gly	GGU, GGC, GGA, GGG
histidine	H	his	CAU, CAC
isoleucine	I	ile	AUU, AUC, AUA
leucine	L	leu	UUA, UUG, CUU, CUC, CUA, CUG
lysine	K	lys	AAA, AAG
methionine	M	met	AUG
phenylalanine	F	phe	UUU, UUC
proline	P	pro	CCU, CCC, CCA, CCG
serine	S	ser	UCU, UCC, UCA, UCG, AGU, AGC
threonine	T	thr	ACU, ACC, ACA, ACG
tryptophan	W	trp	UGG
tyrosine	Y	tyr	UAU, UAC
valine	V	val	GUU, GUC, GUA, GUG

Complementary Strands of DNA with Base Pairing

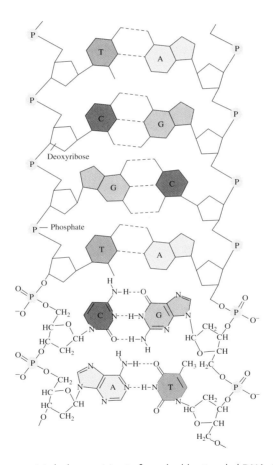

Two nucleotide chains associate by base pairing to form double-stranded DNA. A (Adenine) pairs with T (Thymine), and G (Guanine) pairs with C (Cytosine) by forming specific hydrogen bonds. (Figure with kind permission from Voet, D., Voet, J.G., Pratt, C.W. [2001]: Fundamentals of Biochemistry. Wiley: New York.)